Lecture Notes in Computer Science 6444

Commenced Publication in 1973
Founding and Former Series Editors:
Gerhard Goos, Juris Hartmanis, and Jan van Leeuwen

Kok Wai Wong
B. Sumudu U. Mendis
Abdesselam Bouzerdoum (Eds.)

Neural Information Processing

Models and Applications

17th International Conference, ICONIP 2010
Sydney, Australia, November 22-25, 2010
Proceedings, Part II

Springer

Volume Editors

Kok Wai Wong
Murdoch University
Murdoch, WA, 6150, Australia
E-mail: k.wong@murdoch.edu.au

B. Sumudu U. Mendis
The Australian National University
Canberra, ACT 0200, Australia
E-mail: sumudu.mendis@anu.edu.au

Abdesselam Bouzerdoum
University of Wollongong
Wollongong, NSW 2522, Australia
E-mail: salim@elec.uow.edu.au

Library of Congress Control Number: 2009939833

CR Subject Classification (1998): F.1, I.2, I.4-5, H.3-4, G.3, J.3, C.1.3, C.3

LNCS Sublibrary: SL 1 – Theoretical Computer Science and General Issues

ISSN 0302-9743
ISBN-10 3-642-17533-3 Springer Berlin Heidelberg New York
ISBN-13 978-3-642-17533-6 Springer Berlin Heidelberg New York

springer.com

Printed in Germany

Typesetting: Camera-ready by author, data conversion by Scientific Publishing Services, Chennai, India
Printed on acid-free paper 06/3180

Preface

Welcome to the 17th International Conference on Neural Information Processing (ICONIP 2010) held in Sydney, 22–25 November 2010. In this volume you will find papers presented at this conference. ICONIP is the annual conference of the Asia Pacific Neural Network Assembly (APNNA, http://www.apnna.net). The aim of the Asia Pacific Neural Network Assembly is to promote the interaction of researchers, scientists, and industry professionals who are working in the neural network and related fields in the Asia Pacific region, primarily via the ICONIP conference. This year's theme was hybrid / human centred neural systems.

ICONIP 2010 received 470 excellent submissions. Of these, 146 regular session and 23 special session papers appear in these proceedings by Springer. Many outstanding papers do not appear here due to space limitations. Each paper was assessed by at least three reviewers. The conference will be followed by two associated workshops, the ICONIP International Workshop on Data Mining for Cybersecurity, held in November at the University of Auckland, New Zealand, and the ICONIP International Workshop on Bio-inspired Computing for Intelligent Environments and Logistic Systems, held in March at the Australian National University in Canberra, Australia.

I am very pleased to acknowledge the support of the conference Advisory Board, the APNNA Governing Board and Past Presidents, who gave their advice, assistance and promotion of ICONIP 2010. I gratefully acknowledge the technical sponsorship of the International Neural Network Society (INNS), the Japanese Neural Network Society (JNNS), the European Neural Network Society (ENNS), and the Australian Research Council Network in Human Communication Science (HCSNet).

A special thanks to Kevin Wong, Sumudu Mendis and Sukanya Manna without whom the conference organisation would have been much less smooth.

For the many reviewers who worked hard on giving thorough, tough but fair referee reports, thank you! Finally I would like to thank all the authors of papers, the speakers and panelists, volunteers and audience. With your support ICONIP 2010 will continue the tradition of being an uplifting, educational and enjoyable conference.

October 2010 Tom Gedeon

rogram Committee

ni
l
an
han
vid Cho
cki
ard
s
ng
no
ura
ng
an
ke
o
a
no
o

Ryosuke Hosaka
Chih-Yu Hsu
Xiaolin Hu
Kaizhu Huang
Norhaslinda Kamaruddin
Keisuke Kameyama
Satoru Kato
Jong Kyoung Kim
Kyung-Joong Kim
Min Young Kim
Sungshin Kim
Takanori Koga
Markus Koskela
Ryosuke Kubota
Takio Kurita
James Kwok
Daewon Lee
Hyung-Soo Lee
Nung Kion Lee
Sang-Woong Lee
Soo-Young Lee
Chi Sing Leung
Chunshien Li
Gary C. L. Li
Chee-Peng Lim
Dudy Lim
Heejin Lim
Naiyan Lima
Iuon-Chang Lin
Wilfred Lin
Steve Ling
Qingshan Liu
Weixiang Liu
Zhiyong Liu
Timothy Mann
Sukanya Manna
Nobuo Matsuda
Robert (Bob) McKay
Sadaaki Miyamoto
Takashi Morie
Mitsuteru Nakamura
Wakako Nakamura
Mitsuru Nakazawa
Anto Satriyo Nugroho
Chakarida Nukoolkit

Makoto Ohki
Masaaki Ohkita
Gareth Oliver
Toshiaki Omori
Tetsuya Onoda
Matashige Oyabu
Seiichi Ozawa
Shaoning Pang
Zoltán Petres
Huy Phan
Santitham Prom-on
Shri Rai
Alexander Rast
Napoleon Reyes
Ryo Saegusa
Naoyuki Sato
Stefan Schliebs
Gourab Sen Gupta
Atsushi Shimada
Hayaru Shouno
John Sum
Shiliang Sun
Jaewon Sung
Kenji Suzuki
Masa Takatsuka
Mieko Tanaka
Chan Wai Ting
Heizo Tokutaka
Hiroyuki Torikai
Whye Loon Tung
Eiji Uchino
Hiroshi Wakuya
Liang Wan
Dianhui Wang
Jun Wang
Rongjie Wang
Zhanshan Wang
Yoshikazu Washizawa
Kazuho Watanabe
Bunthit Watanapa
Young-Gul Won
Jiunn-Lin Wu
Liang-chuan Wu
Kui Xiang
Hai Xu

Organiz

Sponsor

Asia Pacific Neural Network Assembly (APNN

Technical Co-sponsors

International Neural Network Society (INNS)
Japanese Neural Network Society (JNNS)
European Neural Network Society (ENNS)
IEEE Computational Intelligence Society (IEEE
ARC Network in Human Communication Scien

Conference Committee

Honorary Chair	Shun-ichi Ar
General Chair	Tamás (Tom)
Technical Program Chairs	Kok Wai (Ke Tom Gedeon, Salim Bouzer
Advisory Board	Irwin King, H Bao-Liang Lu Derong Liu, U Jonathan H. C Jun Wang, Ho Lipo Wang, Si Nikhil R. Pal, Nikola Kasabo Shiro Usui, Jap Soo-Young Le Takeshi Yamak Włodzisław Du
Organising Committee Chairs	Kevin Wong, A B. Sumudu U. M

Technical

Shigeo Abe
Sabri Arik
Hideki Asoh
Hongliang Ba
Sang-Woo Ba
Tao Ban
Andre Barcza
Younès Benn
Michael Bui
Marek Bundz
Chee Seng C
Jonathan Cha
Jonathan H. C
Jianhui Chen
Xinyu Chen
Siu-Yeung D
Sung-Bae Ch
Seungjin Cho
Andrzej Cich
Andrew Cov
Justin Dauw
Zhaohong D
Bin Dong
Kenji Doya
Hiroshi Doz
Ke-Lin Du
Fuqing Dua
Kikuo Fujim
Chun Che F
Andrew Gil
Roland Goe
Eikou Gond
Ping Guo
Shanqing G
Raj Gupta
Amir Hada
Hisashi Ha
Xiong Hao
Pitoyo Hart
Ken Hawic
Hanlin He
Zhaoshui H
Akira Hiro
Kevin Ho
Keiichi Ho

Table of Contents – Part II

Brain Computer Interface

Kernel Methods

Model Generation and Classification

Computational Advance in Bioinformatics

Data Mining for Cybersecurity

Self-organizing Maps and Their Applications

Machine Learning Applications to Image Analysis

Computational Neuroscience and Cognitive Science

Data and Text Processing

Adaptive Algorithms

Bio-inspired Algorithms

Hierarchical Methods

Utilizing Fuzzy-SVM and a Subject Database to Reduce the Calibration Time of P300-Based BCI

Sercan Taha Ahi[1], Natsue Yoshimura[2], Hiroyuki Kambara[2], and Yasuharu Koike[3]

[1] Department of Computational Intelligence and System Science
[2] Precision and Intelligence Laboratory
[3] Solution Science Research Laboratory
Tokyo Institute of Technology,
4259-R2-15, Nagatsuta, Midori-ku, Yokohama 226-8503 Japan
taha.ahi@hi.pi.titech.ac.jp, yoshimura@cns.pi.titech.ac.jp,
hkambara@hi.pi.titech.ac.jp, koike@pi.titech.ac.jp

Abstract. Current Brain-Computer Interfaces (BCI) suffer the requirement of a subject-specific calibration process due to variations in EEG responses across different subjects. Additionally, the duration of the calibration process should be long enough to sufficiently sample high dimensional feature spaces. In this study, we proposed a method based on Fuzzy Support Vector Machines (Fuzzy-SVM) to address both issues for P300-based BCI. We conducted P300 speller experiments on 18 subjects, and formed a subject-database using a leave-one-out approach. By computing weight values for the data samples obtained from each subject, and by incorporating those values into the Fuzzy-SVM algorithm, we achieved to obtain an average accuracy of 80% with only 4 training letters. Conventional subject-specific calibration approach, on the other hand, needed 12 training letters to provide the same performance.

Keywords: Brain-Computer Interfaces, P300, EEG, Subject-Database, Fuzzy Support Vector Machines.

1 Introduction

Linear soft margin support vector machines (C-SVM) assume that each data sample in the training set is of equal importance, while constructing the model parameters and hence the decision boundary. However, in some pattern recognition problems, the samples do not represent assigned class labels equally well. Especially in Brain-Computer Interfaces (BCI), the intersubject variance in Electroencephalogram (EEG) responses is so significant that incorporating samples from a pool of subjects does not help to sample feature space of the test subject, and it may even deteriorate the system performance [1]. This is the main reason why subject-specific calibration has been assumed to be indispensable for current BCI.

Fuzzy support vector machines (Fuzzy-SVM) [2] provide this problem with a solution by assigning each input data sample to a different degree of penalty

K.W. Wong et al. (Eds.): ICONIP 2010, Part II, LNCS 6444, pp. 1–8, 2010.

when they lie on the wrong side of the boundary. Different penalty degrees represent different quantities of importance, or weight, and therefore, each sample contributes differently to the construction of the decision boundary. In this study, we employed P300 recordings obtained from 18 different people. Using a leave-one-out approach, we formed a database of 17 subjects, assigned different weight values to the samples acquired from different subjects and validated the performance of Fuzzy-SVM. As long as there is a sufficient amount of training data, P300-based BCI operate with acceptable accuracy [3], [4]. Our goal in this study was to obtain around 80% of system accuracy with minimum amount of subject-specific calibration data.

2 Linear Soft Margin Support Vector Machines (C-SVM)

Given a set of training data samples and corresponding class labels $\mathbf{D} = \{(\mathbf{x}_n, y_n) | \mathbf{x}_n \in \mathbf{R}^p, y_n \in \{-1, +1\}\}_{n=1}^N$, C-SVM attempt to find a hyperplane that separates the data samples of two different classes with maximal margin, while allowing errors during separation [5]. A user defined constant C controls the tradeoff between the maximization of the margin and number of errors on the training set. The task can be modeled by the following optimization problem.

$$\begin{aligned} \min_{(\mathbf{w},b,\boldsymbol{\xi})} \quad & \frac{1}{2}\|\mathbf{w}\|^2 + C\sum_{n=1}^{N} \xi_n \\ \text{subject to} \quad & \forall_{n=1}^{N}\ y_n(\mathbf{w}^T\mathbf{x}_n + b) \geq 1 - \xi_n\ , \\ & \forall_{n=1}^{N}\ \xi_n \geq 0 \end{aligned} \tag{1}$$

where $\mathbf{w} = [w_1, ..., w_p]^T$ and b are the model parameters, $\{\xi_n\}_{n=1}^N \geq 0$ are the *slack variables.* To solve (1), we first construct the Lagrangian L_P by introducing the multipliers $\{\alpha_n\}_{n=1}^N \geq 0$ and $\{r_n\}_{n=1}^N \geq 0$ with the corresponding set of KKT conditions.

$$\begin{aligned} \min_{(\mathbf{w},b,\boldsymbol{\xi},\boldsymbol{\alpha},\mathbf{r})} \quad & \frac{1}{2}\|\mathbf{w}\|^2 + C\sum_{n=1}^{N} \xi_n - \sum_{n=1}^{N} \alpha_n[y_n(\mathbf{w}^T\mathbf{x}_n + b) - 1 + \xi_n] - \sum_{n=1}^{N} r_n\xi_n \\ \text{subject to} \quad & \alpha_n \geq 0 \\ & y_n(\mathbf{w}^T\mathbf{x}_n + b) - 1 + \xi_n \geq 0 \\ & \alpha_n[y_n(\mathbf{w}^T\mathbf{x}_n + b) - 1 + \xi_n] = 0 \\ & r_n \geq 0 \\ & \xi_n \geq 0 \\ & r_n\xi_n = 0 \end{aligned} \tag{2}$$

Setting the partial derivatives of L_P with respect to $\mathbf{w}$, b, and $\boldsymbol{\xi}$ to zero, and substituting obtained equations into (2), we form the Lagrangian L_D, which only depends on the unknown multipliers $\boldsymbol{\alpha}$. L_D represents a quadratic optimization problem and can be solved by different approaches such as chunking or sequential minimal optimization (SMO) [6]. Once the optimization problem in (1) is solved and the model parameters are computed, the label y^* of a new feature vector $\mathbf{x}^*$ is assigned to $sign(\mathbf{w}^T\mathbf{x}^* + b)$.

3 Fuzzy Support Vector Machines (Fuzzy-SVM)

Fuzzy-SVM are an extension to the C-SVM that assign weight values $\{m_n\}_{n=1}^N \in (0, 1]$ to the slack variables of the training data samples [2]. As these weight values are also multiplied by the C-parameter, they adjust the weight of each data sample during computation of the separating hyperplane. A small m_n decreases the cost for the misclassification of the corresponding $\mathbf{x}_n$, hence lowers the importance of the data sample.

$$\begin{aligned} \min_{(\mathbf{w},b,\boldsymbol{\xi})} \quad & \frac{1}{2}\|\mathbf{w}\|^2 + C\sum_{n=1}^{N} m_n\xi_n \\ \text{subject to} \quad & \forall_{n=1}^{N}\; y_n(\mathbf{w}^T\mathbf{x}_n + b) \geq 1 - \xi_n \\ & \forall_{n=1}^{N}\; \xi_n \geq 0 \\ & \forall_{n=1}^{N}\; 1 \geq m_n > 0 \end{aligned} \tag{3}$$

Comparing (1) with (3), we see that the complexity of the problems are almost identical. The only difference between Fuzzy-SVM and C-SVM is the boundary of Lagrangian multipliers $\{\alpha_n\}$. With the introduction of m_n, each multiplier is bounded with $0 \leq \alpha_n \leq Cm_n$. During our evaluations, we used LIBSVM implementation of Fuzzy-SVM [7].

4 P300 Spellers

P300 spellers are brain-computer interfaces that enable subjects to spell text through the so-called oddball paradigm [8]. In the oddball paradigm, subjects are randomly presented with two types of stimuli, one of which is frequent. The infrequent, or oddball, stimulus generates a positive peak in the EEG recordings at around 300 ms after the stimulus onset assuming that the subject reacted to the stimulus by noticing or silently counting it. P300 spellers present subjects a matrix of letters (Fig. 1) and initiate random flashes on the rows and columns at a frequency of about 2-5 Hz. Flashes on the target letter generate a particular response in the EEG recordings, and the system attempts to identify those responses using a linear classifier. A complete cycle of flashes in which all of the 5 rows and 6 columns flash once are called as one *trial*.

From a pattern recognition point of view, the task of identifying the target row or column is a binary classification problem, where the classifier outputs are added up for T number of trials. Feature vectors are derived from the EEG responses of the subject to each row- or column-flash. During calibration, i.e. adjustment of the model parameters, feature vectors that correspond to a target row or target column flash are labeled as $+1$. Others are labeled as -1. Given

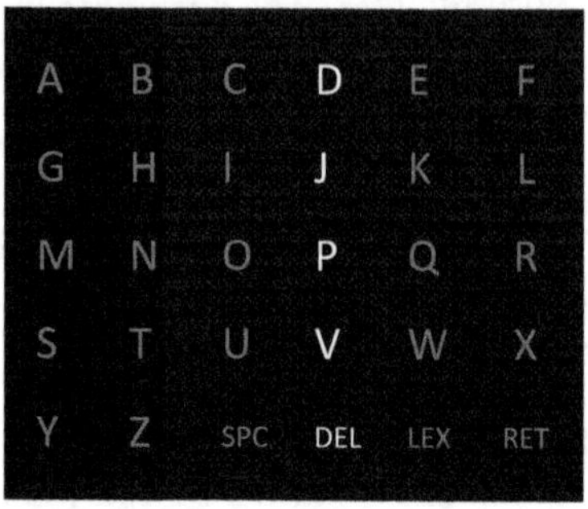

Fig. 1. Conventional P300 speller interface, where the fourth column col_4 is flashing

the speller interface in Fig. 1, and the probabilistic classifier outputs $\{\Phi^t_{row}, \Psi^t_{col}\}$ $t = 1, ..., T$, the target row and column are simply identified by (4a) and (4b).

$$\text{row}_{(y_n=+1)} = \arg\max_{row\in[1,5]} \sum_{t=1}^{T} \Phi^t_{row} \tag{4a}$$

$$\text{col}_{(y_n=+1)} = \arg\max_{col\in[1,6]} \sum_{t=1}^{T} \Psi^t_{col} \tag{4b}$$

5 Collected Data

To evaluate the efficacy of the subject-database and Fuzzy-SVM, we conducted P300 copy-spelling experiments with 18 able-bodied voluntary subjects (mean = 29.0, std = 4.1, range 23-38 years old). We acquired the EEG data through a 64-channel Biosemi ActiveTwo system at 256 Hz but employed only the data recorded from the Fz, Cz, Pz, Oz, C3, C4, P3, P4, PO7, and PO8 channels according to the international 10-20 system [9].

Targets, namely English letters from-A-to-Z, were placed in a 5x6 matrix (Fig. 1). The matrix also included four special characters, but those characters were not assigned as targets. The task for the subject was to attend to the specified letter, while the rows and columns were flashing in random order. The interval between two flashes was 175 ms, and the flash duration was 100 ms. 5 trials were performed for each target letter, and after each 5 trial block, a 6 s break was given to start with the next target letter. Using the 5x6 matrix, subjects copy-spelled the following 15 words: *SAME*, *PIPE*, *HINT*, *CHAT*, *KEEN*, *RARE*, *USSR*, *BARE*, *UNIX*, *FARE*, *VIII*, *CARE*, *ECHO*, *MARE*, *EMMA*, which corresponds to 60 target-letter/subject. The raw EEG data acquired through the 10 selected channels were initially passed through a 1-18 Hz bandpass filter, and downsampled by a factor of 8. Each trial consisted of 11 flashes, and as a result, yielded 11 epochs. Epoch data were extracted using a time window of 600 ms following stimulus onsets, corresponding $\lceil 256/8*0.6 \rceil = 20$ sample points per channel. Extracted epoch data were normalized to the [-1,+1] interval, and channel data are concatenated to obtain the 200 (20 sample-points/channel x 10 channels) dimensional feature vectors $\{\mathbf{x}_n\}$.

6 Database

We collected data from 18 subjects and formed the database using a leave-one-out approach. At each run, one of the 18 subjects was selected as the test subject, and the corresponding data set was denoted as $\mathbf{D}_0$. The remaining sets were denoted as $\{\mathbf{D}_k\}$ $k = 1, ..., 17$. Individual data samples corresponding to $letter_1$-to-$letter_{20}$ were grouped as the training set $\{\mathbf{D}_k^{train}\}$ $k = 0, ..., 17$, and the data samples corresponding to $letter_{21}$-to-$letter_{60}$ were grouped as the test set $\{\mathbf{D}_k^{test}\}$ $k = 0, ..., 17$. The database was constructed using the training sets of the 17 subjects $\mathbf{D}_{database} = \{\mathbf{D}_1^{train}, ..., \mathbf{D}_{17}^{train}\}$, and $\{\mathbf{D}_k^{test}\}$ $k = 1, ..., 17$ were discarded for a more realistic evaluation scenario. $\mathbf{D}_{database}$ included 18700 feature vectors (11 flashes/trial x 5 trials/target-letter x 20 target-letters/subject x 17 subjects), where $|\mathbf{D}_{database} : y_n = +1| = 3400$ and $|\mathbf{D}_{database} : y_n = -1| = 15300$.

7 Estimation of the *C*-Parameter and the Weight Values

During evaluations, we set the C-parameter of the support vector machines to 1 and computed the weight values $\{m_n\}$ $n = 1, ..., N$ accordingly. Computation of the weight values was a key step before solving the optimization problem in (3). Initially, we assumed that the weight values are the same for the data samples obtained from the same subject, and the value for the test subject is always equal to the ratio of $|\mathbf{D}_k^{train}|/|\mathbf{D}_0^{train}|$, $k \in [1, 17]$ (note that all $|\mathbf{D}_k^{train}|$ are the same). Since there were 17 different subjects whose training sets were available in the database, we needed to compute only 17 different weight values. These values are calculated in two steps. At first, we trained 17 C-SVM using $\{\mathbf{D}_k^{train}\}$ $k = 1, ..., 17$ and tested each C-SVM on $\mathbf{D}_0^{train}$ leading to individual accuracy scores of $\{c_k\}$. At the second step, we employed a mapping function $g(x) : \mathbf{R} \mapsto \mathbf{R}$ to obtain $\{m_k\}$ from $\{c_k\}$. We used a mapping function in the form of Fig. 2(b). We set c_{min} and m_{min} to 0.05, and c_{max} to 0.95. Keeping the endpoints constant, we also tried logarithmically and exponentially increasing functions, but found that monotonically increasing functions with similar endpoints produce almost identical performances, so we omitted those results for brevity.

The main properties of the mapping function $g(x)$ are twofold. (1) However low $\{c_k\}$ become, corresponding weight values should be nonzero, because zero weight values make the matching data samples in the database invisible to the classifier. (2) As $\{c_k\}$ increase, the weight values should also increase, meaning that samples of the similar subjects should count more during computation of the SVM model parameters. The upper limit for the increase (m_{max}) can be determined by cross-validation. During our experiments, we tried $\{0.05, 0.10, ..., 1.00\}$ and observed that a value of 0.20 generates satisfactory results.

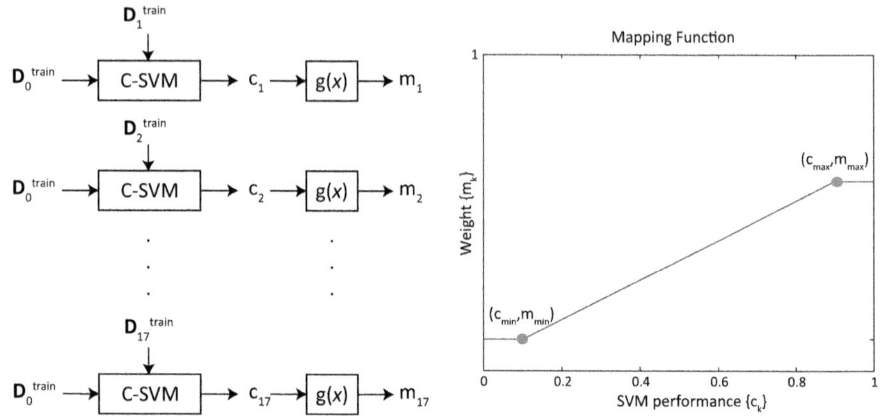

(a) (b)

Fig. 2. (a) Computation of the weight values is illustrated. 17 different C-SVM were trained using $\{\mathbf{D}_k^{train}\}$ $k = 1, ..., 17$ and tested on $\mathbf{D}_0^{train}$. Corresponding accuracy values $\{c_k\}$ $k = 1, ..., 17$ were later mapped to the weight values by the function $g(x)$. (b) Form of the proposed mapping function $g(x)$, $g : \mathbf{R} \mapsto \mathbf{R}$.

8 Results

During evaluations, we compared four different calibration approaches as summarized in Table 1. In the first approach, we trained a single C-SVM using only the test subject's training set $\mathbf{D}_0^{train}$ and tested the trained C-SVM on the subject's test set $\mathbf{D}_0^{test}$ (conventional approach). In the second approach, we combined $\mathbf{D}_0^{train}$ with the training sets of the 17 other subjects $\{\mathbf{D}_1^{train}, ..., \mathbf{D}_{17}^{train}\}$ to train a single C-SVM. While computing the separating hyperplane, all data samples were assumed to be of equal weight. In the third approach, we did not use any data from the test subject's training set, and therefore, this case represented zero calibration. We trained a single C-SVM with $\mathbf{D}_{database}$, and tested the classifier with $\mathbf{D}_0^{test}$. In the fourth approach, we combined $\mathbf{D}_0^{train}$ with $\{\mathbf{D}_1^{train}, ..., \mathbf{D}_{17}^{train}\}$, but different than the second approach, we trained a Fuzzy-SVM by multiplying the data samples with the computed weight values (proposed approach). As shown in Fig. 3, the proposed approach provided a 2%~20% increase in the average accuracy.

We also measured the training times of the classifiers and presented the results at Table 2. We observed that the improved performance of the proposed approach comes with a computational cost of about 10 minutes of training time, compared to about one second for the conventional subject-specific calibration (*No database*) approach. Note that these measurements do not include the time needed to collect the corresponding training data.

Table 1. Evaluation details

Approach	Training Set ($\mathbf{D}^{train}$)	Test Set	Classifier	C-const
No database	$\mathbf{D}_0^{train}$	$\mathbf{D}_0^{test}$	C-SVM	1
All subjects	$\mathbf{D}_0^{train} \bigcup \{\mathbf{D}_1^{train}, ..., \mathbf{D}_{17}^{train}\}$	$\mathbf{D}_0^{test}$	C-SVM	1
Database only	$\{\mathbf{D}_1^{train}, ..., \mathbf{D}_{17}^{train}\}$	$\mathbf{D}_0^{test}$	C-SVM	1
Weighted subjects	$\mathbf{D}_0^{train} \bigcup \{m_1\mathbf{D}_1^{train}, ..., m_{17}\mathbf{D}_{17}^{train}\}$	$\mathbf{D}_0^{test}$	Fuzzy-SVM	1

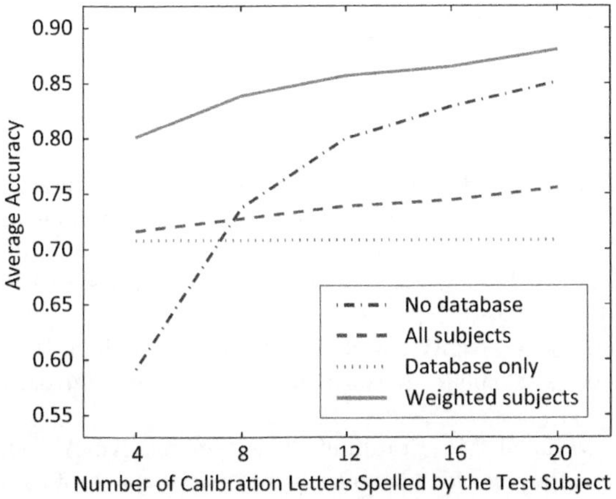

Fig. 3. Average accuracy values obtained with four different training approaches. Note that *Database only* corresponds to zero calibration, and therefore the system performance does not depend on the amount of the subject-specific calibration data.

Table 2. Number of feature vectors in the corresponding training set and the average training time (in seconds) of the machine learning algorithm. The measurements were conducted on a computer with Intel Core2 Duo 2.66 GHz processor and 2GB of RAM.

Approach	# of Calibration Letters Spelled by the Test Subject				
	4	**8**	**12**	**16**	**20**
No database	220 0.42s	440 0.47s	660 0.80s	880 1.25s	1100 1.80s
All subjects	220+18700 786s	440+18700 807s	660+18700 824s	880+18700 845s	1100+18700 864s
Database only	18700 765s	18700 765s	18700 765s	18700 765s	18700 765s
Weighted subjects	220+18700 645s	440+18700 656s	660+18700 669s	880+18700 682s	1100+18700 694s

9 Conclusion

BCI systems provide improved system performance with increasing amounts of subject-specific calibration data. Therefore, current systems keep the calibration time long enough to acquire satisfactory amounts of training data. Prolonged calibration processes represent a major inconvenience for the end-users. In this study, we showed that incorporation of a pool of subjects' training data into the calibration process along with the fuzzy support vector machines alleviates this inconvenience. By assigning subjects that show similar responses with the user to larger weights, we managed to reach, for example, the average accuracy of 80% with only 4 training letters.

References

1. Lu, S., Guan, C., Zhang, H.: Unsupervised brain computer interface based on inter-subject information and online adaptation. IEEE Transactions on Neural Systems and Rehabilitation Engineering 17, 135–145 (2009)
2. Lin, C.-F., Wang, S.-D.: Fuzzy support vector machines. IEEE Transactions on Neural Networks 13, 464–471 (2002)
3. Krusienski, D.J., Sellers, E.W., McFarland, D.J., Vaughan, T.M., Wolpaw, J.R.: Toward enhanced p300 speller performance. Journal of Neuroscience Methods 167, 15–21 (2008)
4. Hoffmann, U., Vesin, J.-M., Ebrahimi, T., Diserens, K.: An efficient p300-based brain-computer interface for disabled subjects. Journal of Neuroscience Methods 167, 115–125 (2008)
5. Cortes, C., Vapnik, V.: Support-vector networks. Machine Learning 20, 273–297 (1995)
6. Platt, J.: Using analytic qp and sparseness to speed training of cupport vector machines. In: Advances in Neural Information Processing Systems 11, pp. 557–563. MIT Press, Cambridge (1999)
7. Chang, C.-C., Lin, C.-J.: LIBSVM: A Library for Support Vector Machines (2001), Software available at `http://www.csie.ntu.edu.tw/~cjlin/libsvm`
8. Donchin, E., Spencer, K.M., Wijesinghe, R.: The mental prosthesis: Assessing the speed of a p300-based braincomputer interface. IEEE Transactions on Rehabilitation Engineering 8, 174–179 (2000)
9. Finke, A., Lenhardt, A., Ritter, H.: The mindgame: A p300-based brain-computer interface game. Neural Networks 22, 1329–1333 (2009)

Importance Weighted Extreme Energy Ratio for EEG Classification

Wenting Tu and Shiliang Sun

Department of Computer Science and Technology, East China Normal University
500 Dongchuan Road, Shanghai 200241, P.R. China
w.tingtu@gmail.com, slsun@cs.ecnu.edu.cn

Abstract. Spatial filtering is important for EEG signal processing since raw scalp EEG potentials have a poor spatial resolution due to the volume conduction effect. Extreme energy ratio (EER) is a recently proposed feature extractor which exhibits good performance. However, the performance of EER will be degraded by some factors such as outliers and the time-variances between the training and test sessions. Unfortunately, these limitations are common in the practical brain-computer interface (BCI) applications. This paper proposes a new feature extraction method called importance-weighted EER (IWEER) by defining two kinds of weight termed *intra-trial importance* and *inter-trial importance*. These weights are defined with the density ratio theory and assigned to the data points and trials respectively to improve the estimation of covariance matrices. The spatial filters learned by the IWEER are both robust to the outliers and adaptive to the test samples. Compared to the previous EER, experimental results on nine subjects demonstrate the better classification ability of the IWEER method.

Keywords: Brain-computer interface (BCI), Feature extraction, Extreme energy ratio (EER), Density ratio.

1 Introduction

A brain-computer interface (BCI) is a system for controlling a device, e.g, a computer, a wheelchair or a neuroprothesis by human intentions. BCI technology relies on the ability of individuals to voluntarily and reliably produce changes in their electroencephalogram (EEG) signal activities. Classification of electroencephalogram (EEG) signals is an important problem in the development of EEG-based brain computer interfaces (BCIs) and spatial filtering can improve classification performance considerably. Extreme energy ratio (EER) [1] can efficiently calculate spatial filters for brain signal classification. It is theoretically equivalent and computationally superior to a highly successful algorithm called common spatial patterns (CSP) [2].

However, the effectiveness of EER method may be disturbed by some limitations. First, it is sensitive to outliers. If the few training samples that are measured within the 'calibration' time are contaminated by such artifacts, a suboptimal feature extractor or classifier can be the consequence [3]. Thus the feature extractors and classifiers should be robust to the outliers, e.g., by reducing or eliminating their negative influences. Moreover, the data still prove to be inherently non-stationary after outlier elimination [4]. The difference of the spatial distribution of the activation patterns between

K.W. Wong et al. (Eds.): ICONIP 2010, Part II, LNCS 6444, pp. 9–16, 2010.

calibration and feedback sessions will also strongly degrades the performance of EER features. Thus adaptive learning is necessary to boost up the performance of existing classifiers and feature extractors. Due to the time-varying characteristic of EEG signals during different sessions, it is reasonable to utilize samples in the test session to increase the classification ability on test samples. One way to realize this goal is to combine the brain signals recorded recently in the test session and samples which were labeled in the training session to enhance the classification ability. However, up to now, most work addressing this problem are based on labeled feedback samples. This assumption is hard to achieve in practical BCI systems since it needs disturbing users and may upset them. As a result, it is worthwhile to study how to make feature extractors or classifiers adaptive to the test samples with no labels.

In this paper, we define two kinds of weight: *intra-trial importance* and *inter-trial importance* by considering the distribution of samples in test session and obtain an improved EER algorithm called importance weighted EER (IWEER). We expect these two kinds of weight will be small on the samples that are outliers or strongly dissimilar with the distribution of test samples (we call the later one as the "misleading" sample since it may degrade the performance of feature extraction or classification on the test session). Thus, the negative influences of outliers and "misleading" samples to the estimation of covariance matrices can be reduced. As a result, the spatial filters obtained by IWEER can not only be robust to the outliers but also adaptive to the test samples.

The rest of this paper is organized as follows. Section 2 describes previous work on the EER algorithm. Subsequently, in Section 3, we introduce the proposed IWEER method and its computational details. The experimental results on data sets from nine subjects are presented in Section 4. Finally, the conclusion and promising work for IWEER are showed in Section 5.

2 EER Algorithm: A Brief Review

EER tries to discover source signals whose average energy of two conditions are most different. In other words, it learns the spatial filters maximizing the variance of band-pass filtered EEG signals under one condition while minimizing it for the other condition. Though having the same motivation as CSP, it simplifies the CSP algorithm to a Rayleigh quotient. EER has been proven to be theoretically equivalent and computationally superior to the CSP method in [1].

Assume only one latent signal source from each class is to be recovered. For an EEG sample X, the spatially filtered signal with a spatial filter denoted by $\phi_{(N\times 1)}$ will be $\phi^\top X$. The signal energy after filtering can be represented by the sample variance as $(\phi^\top X)(\phi^\top X)^\top \propto \phi^\top C\phi$, where C is the normalized covariance of the EEG sample X and can be written as:

$$C = \frac{1}{T-1}\frac{XX^\top}{tr(XX^\top)}. \tag{1}$$

Without loss of generality, we can ignore the multiplicative factor $1/(T-1)$ in the following calculation of covariances. As a result, in order to maximize the difference of energy features under two conditions, EER finds a spatial filter which maximizes or minimizes their ratio. Thus, the discriminative EER criterion is defined as follows:

$$\max / \min \frac{\phi^\top C_A \phi}{\phi^\top C_B \phi}, \tag{2}$$

where C_A and C_B are the covariances of specific classes that can be computed as the average of all single covariances belonging to one class:

$$C_A = \frac{1}{T_A} \sum_{p}^{T_A} \frac{X_{A_p} X_{A_p}^\top}{tr(X_{A_p} X_{A_p}^\top)}, \quad C_B = \frac{1}{T_B} \sum_{q}^{T_B} \frac{X_{B_q} X_{B_q}^\top}{tr(X_{B_q} X_{B_q}^\top)}. \tag{3}$$

By optimizing (2), we can obtain two optimal spatial filters ϕ^*_{max} and ϕ^*_{min} which maximize and minimize the objective function in (2). It turns out that ϕ^*_{max} and ϕ^*_{min} are two eigenvectors corresponding to the maximal and minimal eigenvalues of the matrix $(C_B^{-1} C_A)$, respectively.

For classification, when we wish to extract m sources, EER will seek $2m$ spatial filters. Half of them maximize the objective function (2) while the other half minimize it. Thus, ϕ^*_{max} consists of m generalized eigenvectors of the matrix pairs (C_A, C_B) which correspond to the m maximal eigenvalues: $\phi^*_{max} \triangleq [\phi_1, \cdots, \phi_m]$. Similar, the m entries of ϕ^*_{min} are m generalized eigenvectors of matrix pair (C_A, C_B) whose eigenvalues are minimal. For a new EEG sample, it can be filtered by $2m$ spatial filters coming from two filter banks ϕ^*_{max} and ϕ^*_{min}. Thus, the energy feature vector consists of the $2m$ energy values.

However, if the EEG training set is noisy or different strongly from the test set, the covariance matrices may be poor or non-representative estimates of the mental states involved, and thus the spatial filters learned by EER will be poor.

3 Importance Weighted EER

In this subsection we introduce two kinds of weight which can define the "importance" of each data point and each trial respectively. A data point or trial is less important when it is strongly different from the data distribution in test session since it may be the outlier or misleading sample. By integrating the weights into the estimation of covariance matrices, we obtain IWEER method which can be robust to the outliers and adaptive to the test samples.

Density ratio estimation is a recently proposed algorithm in various statistical data processing tasks such as non-stationarity adaptation, outlier detection, feature selection, and independent component analysis [5]. Based on the density ratio theory, two kinds of weight called *intra-trial importance* and *inter-trial importance* are defined and integrated into the EER method. They are assigned to data points and trials respectively.

Let $X_{A_p} = \{x_1^{A_p}, \cdots, x_T^{A_p}\}$, $(p = 1, \cdots, T_A)$ and $X_{B_q} = \{x_1^{B_q}, \cdots, x_T^{B_q}\}$, $(q = 1, \cdots, T_B)$ respectively define samples from two different classes A and B, where T_A and T_B are their corresponding sample sizes. Analogously, $X_{U_t} = \{x_1^{U_t}, \cdots, x_T^{U_t}\}$, $(t = 1, \cdots, T_C)$ is used to denote the samples form test set with T_C being the corresponding sample number. Suppose we have a training snapshot set $\{X_A\} \cup \{X_B\} = \{x_1, \cdots x_{l_1}\}$ with $l_1 = T(T_A + T_B)$ and a test snapshot set $\{X_U\} = \{x_1, \cdots x_{l_2}\}$ with $l_2 = T \times T_C$.

We define *intra-trial importance* which are different for each data point in a trial. The *intra-trial importance* of data point $x_i^{A_p}$ is formulated as:

$$w_i^{A_p} = \frac{p^{te}(x_i^{A_p})}{p^{tr}(x_i^{A_p})}, \tag{4}$$

where $p^{te}(\cdot)$ is the test density estimated by the test snapshot set $\{X_U\}$, and $p^{tr}(\cdot)$ is the training density estimated by the set $\{X_A\} \cup \{X_B\}$. The calculation of the *intra-trial importance* depends on the density ratio estimation. A naive approach to estimate the density ratio would be to first estimate the training and test densities separately from the training and test input samples, and then estimate the ratio of the estimated densities. However, density estimation is known to be a difficult problem, especially for high-dimensional situations. Therefore, this approach is not feasible and alternatives should be used. Here we adopt a method called Kullback–Leibler importance estimation procedure (KLIEP) [6] to directly estimate the density ratio. KLIEP directly estimates the density ratio as follows:

$$\hat{w}(x) = \frac{\hat{p}_{te}(x)}{\hat{p}_{tr}(x)} = \sum_{l=1}^{b} \alpha_l \text{exp}\left(-\frac{||x - c_l||^2}{2\sigma^2}\right), \tag{5}$$

where $\{\alpha_l\}_{l=1}^{b}$ are coefficients to be learned ($\alpha_l \geq 0$ for $l = 1, 2, \cdots, b$), $\{c_l\}_{l=1}^{b}$ are chosen randomly from $\{x_j^{te}\}_{j=1}^{n_{te}}$, and the number of parameters is set to $b = min(100, n_{te})$ in the experiments. The kernel width σ can be optimized by cross-validation.

By the aforementioned model, the test input density can be formulated as:

$$\hat{p}_{te}(x) = \hat{w}(x) p_{tr}(x). \tag{6}$$

With this expression, $\{\alpha_l\}_{l=1}^{b}$ are then determined by minimizing the Kullback–Leibler divergence between $p_{te}(x)$ to $\hat{p}_{te}(x)$:

$$\begin{aligned} KL(p_{te}(x), \hat{p}_{te}(x)) &= \int_D p_{te}(x) \text{log} \frac{p_{te}(x)}{\hat{w}(x) p_{tr}(x)} dx \\ &= \int_D p_{te}(x) \text{log} \frac{p_{te}(x)}{p_{tr}(x)} dx - \int_D p_{te}(x) \text{log} \hat{w}(x) dx, \end{aligned} \tag{7}$$

where D is the domain of x. Note that the first term of (16) is a constant with regard to $\{\alpha_l\}_{l=1}^{b}$ and minimizing the second term is equivalent to maximizing its negative form. As a result, the optimization criterion of KLIEP is given as follows:

$$\max \sum_{j=1}^{n_{te}} \log \left[\sum_{l=1}^{b} \alpha_l \text{exp}\left(-\frac{||x_j^{te} - c_l||^2}{2\sigma^2}\right)\right], \tag{8}$$

subject to

$$\sum_{i=1}^{n_{tr}} \sum_{l=1}^{b} \alpha_l \text{exp}\left(-\frac{||x_i^{tr} - c_l||^2}{2\sigma^2}\right) = n_{tr}, \text{ and } \alpha_1, \alpha_2, \cdots, \alpha_b \geq 0. \tag{9}$$

After obtaining the *intra-trial importance* for each data point, we calculate the normalized covariance of one EEG sample as:

$$\tilde{C}^{A_p} = \frac{1}{T-1}\sum_{i=1}^{T} w_i^{A_p} x_i^{A_p} (x_i^{A_p})^\top, \quad p = 1, \cdots, T_A. \tag{10}$$

Analogously, we can get $\tilde{C}^{B_q}$ for $q = 1, \cdots, T_B$.

On top of the *intra-trial improtance*, we also assign a kind of weight to each trial to define its total importance, which is called *inter-trial importance* . Here we just use the sum of all weights of the data points in a trial to represent the importance of the corresponding trial:

$$I^{A_p} = \sum_{i=1}^{T} w_i^{A_p}, \quad p = 1, \cdots, T_A. \tag{11}$$

Analogously, we can obtain I^{B_q} for $q = 1, \cdots, T_B$.

Similarly, the improved estimations of covariances for specific classes can be computed as

$$\tilde{C}_A = \frac{1}{T_A}\sum_{p=1}^{T_A} I^{A_p}\tilde{C}^{A_p}, \tag{12}$$

$$\tilde{C}_B = \frac{1}{T_B}\sum_{q=1}^{T_B} I^{B_q}\tilde{C}^{B_q}, \tag{13}$$

Then the criterion of IWEER can be defined as

$$\max / \min \frac{\phi^T \tilde{C}_A \phi}{\phi^T \tilde{C}_B \phi}. \tag{14}$$

Consequently, we can learn the spatial filters by eigen-decomposing of matrix pair $(\tilde{C}_A, \tilde{C}_B)$.

Thus we assign less importance value to the training samples that are largely different from test samples. They may be outliers by measurement artifacts and non-standard noise sources, or misleading samples generalized by the strong differences in the training and test session. Due to the lighter weights, they will be deemphasized in the estimation of the covariance matrices. As a result, our IWEER can be robust to the outliers and adaptive to the test data distribution.

4 Experiment

4.1 Data Description and Experimental Setup

The EEG data used in this study were made available by Dr. Allen Osman of University of Pennsylvania during the NIPS 2001 BCI workshop [7]. There were a total of nine subjects denoted $S_1, S_2, \cdots, S_9$, respectively. For each subject, the task was to imagine moving his or her left or right index finger in response to a highly predictable visual cue.

EEG signals were recorded with 59 electrodes mounted according to the international 10C10 system. A total of 180 trials were recorded for each subject. Ninety trials with half labeled left and the other half right were used for training, and the other 90 trials were for testing. Each trial lasted six seconds with two important cues. The preparation cue appeared at 3.75 s indicating which hand movement should be imagined, and the execution cue appeared at 5.0 s indicating it was time to carry out the assigned response.

Signals from 15 electrodes over the sensorimotor area are used in this paper, and for each trial the time window from 4.0 s to 6.0 s is retained for analysis. Other preprocessing operations include common average reference, 8–30 Hz bandpass filtering, and signal normalization to eliminate the energy variation of different recording instants [9].

All the epochs were filtered with pass band 8-30Hz and spatially by common average reference [8]. Firstly, we employed the EER and IWEER criterions respectively to calculate the projection directions with parameters σ determined by 10-fold crossvalid on the training set. Then corresponding energies (variances) are extracted as features for later classification. Fisher linear discriminant classification method was used to classify EEG signals to be tested.

4.2 Results and Performance Analysis

The classification results for nine subjects with the feature extraction methods EED and CSP are shown in Fig. 1. From this figure, we see that for almost all subjects IWEER outperforms EER consistently, which demonstrates the better classification ability of

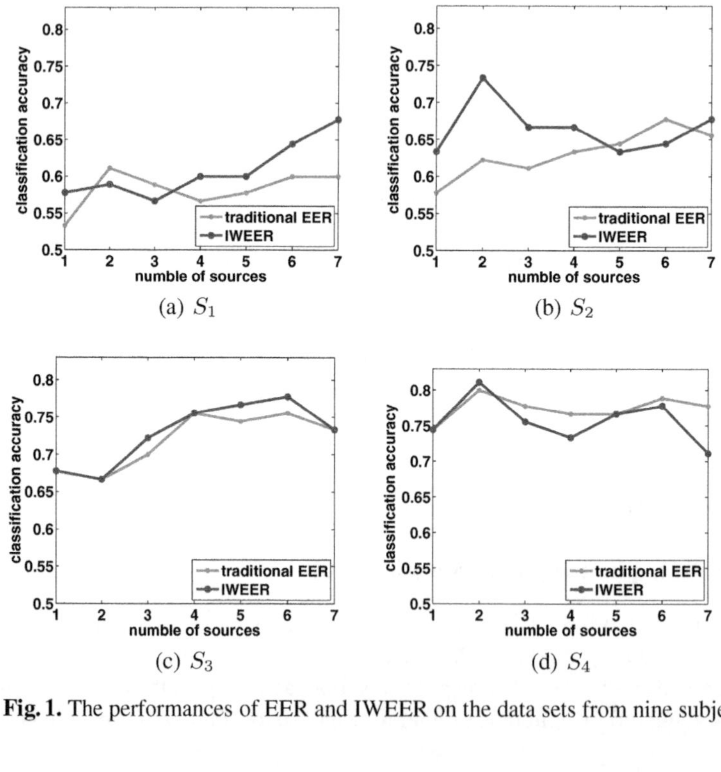

Fig. 1. The performances of EER and IWEER on the data sets from nine subjects

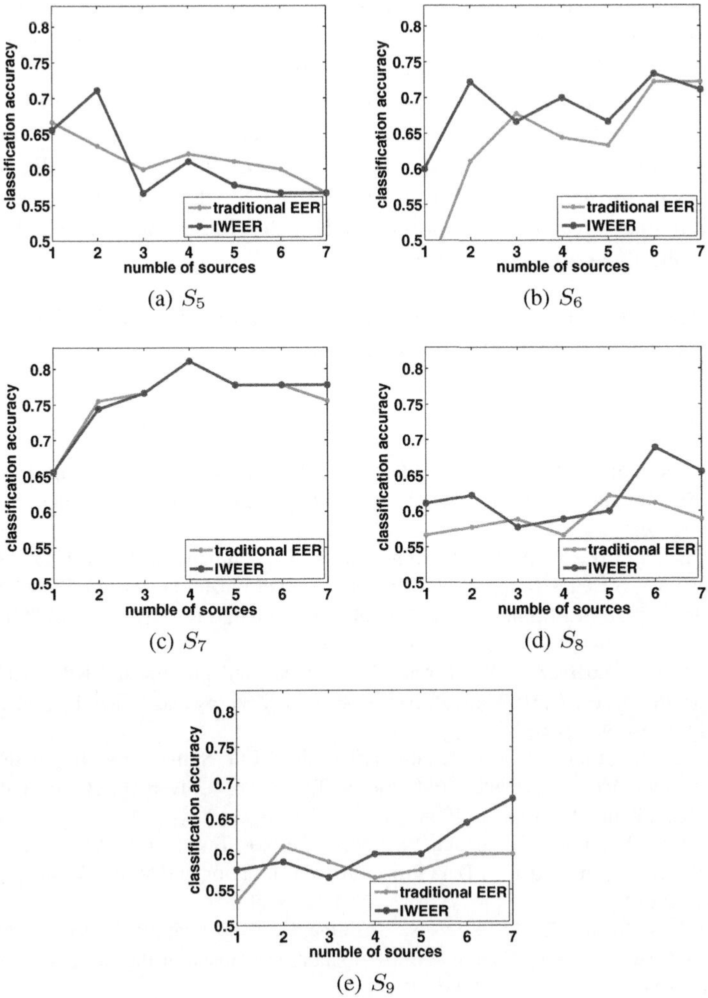

(a) S_5 (b) S_6 (c) S_7 (d) S_8 (e) S_9

Fig. 1. *(continued)*

spatial filters obtained by IWEER. Note on S_4 and S_9, the previous EER can obtain good results ($\geq 80\%$) and the improvements of performances are relatively trivial, compared with the large improvements on other data sets.

5 Conclusion and Future Work

In this paper, we have proposed a new feature extractor which can utilize the distribution knowledge of the samples in the test session to improve the classification ability of the spatial filters. Specially, we define two kinds of weight called *intra-trial importance* and *inter-trial importance* and assign them to the data points and trials. Thus the covariance matrixes estimated by these weighted samples can be improved and the obtained spatial filters are both robust to outliers and adaptive into the test samples. By integrating the weights into the EER method, we proposed IWEER. The experimental results on the data sets from nine subjects demonstrate the better performance of the IWEER.

Though the current research of IWEER is based on off-line analysis, it can be easily extended to the on-line scenario [10], which is meaningful for practical BCI applications. Moreover, the samples weighting strategy of IWEER is promising for selecting samples from other sessions and subjects to reduce the calibration time, which has been proposed explicitly as a task, e.g., by Schalk et al [11].

Acknowledgments. This work is supported by the National Natural Science Foundation of China under Projects 60703005 and 61075005.

References

1. Sun, S.: The Extreme Energy Ratio Criterion for EEG Feature Extraction. In: Kůrková, V., Neruda, R., Koutník, J. (eds.) ICANN 2008, Part II. LNCS, vol. 5164, pp. 919–928. Springer, Heidelberg (2008)
2. Hill, N.J., Lal, T.N., Schröder, M., Hinterberger, T., Widman, G., Elger, C.E., Schölkopf, B., Birbaumer, N.: Classifying Event-Related Desynchronization in EEG, ECoG and MEG Signal. In: Franke, K., Müller, K.-R., Nickolay, B., Schäfer, R. (eds.) DAGM 2006. LNCS, vol. 4174, pp. 404–413. Springer, Heidelberg (2006)
3. Müller, K.-R., Anderson, C.W., Birch, G.E.: Linear and non-linear Methods for Brain-Computer interfaces. IEEE Transactions on Neural Systems and Rehabilitation Engineering 11(2), 165–169 (2003)
4. Krauledat, M.: Analysis of Nonstationarities in EEG Signals for Improving Brain-Computer interface Performance. PhD thesis, Technische Universitfit Berlin, Fakultfit IV –Elektrotechnik und Informatik (2008)
5. Sugiyama, M., Kanamori, T., Suzuki, T., Hido, S., Sese, J., Takeuchi, I., Wang, L.: A Density-Ratio Framework for Statistical Data Processing. Information and Media Technologies 4(4), 962–987 (2009)
6. Sugiyama, M., Suzuki, T., Nakajima, S., Kashima, H., von Bünau, P., Kawanabe, M.: Direct importance Estimation for Covariate Shift Adaptation. Annals of the Institute of Statistical Mathematics 60(4), 699–746 (2008)
7. Sajda, P., Gerson, A., Mller, K.R., Blankertz, B., Parra, L.: A Data Analysis Competition to Evaluate Machine Learning Algorithms for Use in Brain-Computer interfaces. IEEE Transactions on Neural Systems and Rehabilitation Engineering 11(2), 184–185 (2003)
8. Nunez, P.L., Srinivasan, R., Westdorp, A.F., Wijesinghe, D.M., Tucker, R.B., Cadusch, P.J.: EEG coherency I: Statistics, Reference Electrode, volume conduction, Laplacians, cortical imaging, and interpretation at multiple scale. Electroenceph. Clinical Neurophysiology 103, 499–515 (1997)
9. Müller-Gerking, J., Pfurtscheller, G., Flyvbjerg, H.: Designing Optimal Spatial Filters for Single-trial EEG Classification in a Movement Task. Clinical Neurophysiology 110(5), 787–798 (1999)
10. Millán, J.R.: On the Need for On-Line Learning in Brain-Computer interfaces. In: IEEE International Conference on Neural Networks - Conference Proceedings, vol. 4, pp. 2877–2882 (2004)
11. Schalk, G., Blankertz, B., Chiappa, S., et al.: BCI competition III (2004–2005), `http://ida.first.fraunhofer.de/projects/bci/competitioniii/`

Toward Automated Electrode Selection in the Electronic Depth Control Strategy for Multi-unit Recordings

Gert Van Dijck[1], Ahmad Jezzini[2], Stanislav Herwik[3], Sebastian Kisban[3], Karsten Seidl[3], Oliver Paul[3], Patrick Ruther[3], Francesca Ugolotti Serventi[2], Leonardo Fogassi[2,4], Marc M. Van Hulle[1], and Maria Alessandra Umiltà[2,4]

[1] Computational Neuroscience Research Group, Laboratorium voor Neuro- en Psychofysiologie, Katholieke Universiteit Leuven, Herestraat 49, 3000 Leuven, Belgium
[2] Department of Neuroscience, University of Parma, Via Volturno 39, 43100 Parma, Italy
[3] Microsystem Materials Laboratory, Department of Microsystems Engineering (IMTEK), University of Freiburg, Georges-Koehler-Allee 103, 79110 Freiburg, Germany
[4] Italian Institute of Technology, Section of Parma, Parma, Italy
gert.vandijck@med.kuleuven.be, leonardo.fogassi@unipr.it, marc.vanhulle@med.kuleuven.be, mariaalessandra.umilta@unipr.it

Abstract. Multi-electrode arrays contain an increasing number of electrodes. The manual selection of good quality signals among hundreds of electrodes becomes impracticable for experimental neuroscientists. This increases the need for an automated selection of electrodes containing good quality signals. To motivate the automated selection, three experimenters were asked to assign quality scores, taking one of four possible values, to recordings containing action potentials obtained from the monkey primary somatosensory cortex and the superior parietal lobule. Krippendorff's alpha-reliability was then used to verify whether the scores, given by different experimenters, were in agreement. A Gaussian process classifier was used to automate the prediction of the signal quality using the scores of the different experimenters. Prediction accuracies of the Gaussian process classifier are about 80% when the quality scores of different experimenters are combined, through a median vote, to train the Gaussian process classifier. It was found that predictions based also on firing rate features are in closer agreement with the experimenters' assignments than those based on the signal-to-noise ratio alone.

Keywords: Continuous wavelet transform, Electronic depth control, Gaussian process classifier, Inter-rater reliability, Multi-unit recordings, Spike detection.

1 Introduction

Multi-electrode arrays (MEAs) are able to monitor the simultaneous spiking activity of many neurons [1] and are, therefore, in a position to provide valuable insights into how multiple neurons process information and how different brain areas interact. MEAs typically consist of tens of electrodes with inter-electrode distances ranging from 100 to 500 μm [2]. Recent advances in CMOS-based microprobes [3], support

K.W. Wong et al. (Eds.): ICONIP 2010, Part II, LNCS 6444, pp. 17–25, 2010.

in-vivo recordings up to 500 electrodes with inter-electrode distances as small as 40.7 μm [3], hence, at a spatial range of an individual neural cell's soma diameter. Signal quality can degrade over time due to apoptosis, tissue drift, relaxation, inflammation and reactive gliosis, among other reasons. Hence, there is a need for (re)selecting or (re)positioning the electrodes. Essentially 2 different systems can be used for obtaining and maintaining a good signal quality among different electrodes: movable microelectrodes [4-7] and the electronic depth control system [3]. The movable microelectrode systems mechanically position single electrodes independently, whereas the electronic depth control system switches electronically between different microelectrodes on a single shaft (up to 500 electrodes [3]) once the array is implanted. In the electronic depth control system, 8 electrodes, out of the total number of electrodes, can be read simultaneously from a single shaft. This is due to technological limitations in the switching electronics and the lithography. So far, the electrode selection was performed manually by the experimenter or semi-automatically. In the latter case, electrodes are sorted according to the signal-to-noise ratio (SNR) [3]. To fully automate the electrode selection, it is of interest to know whether experimenters can agree on the quality of a signal. This is studied in section 3 using Krippendorff's alpha inter-rater reliability. So far, the SNR was used as the quality metric in both the movable microelectrode systems [4,6-7] and the electronic depth control [3]. It is studied in section 4 whether signal features other than the SNR may be needed to reflect the experimenter's assignments of a quality score to signals containing action potentials. This is performed by studying the accuracy of a Gaussian process classifier (GPC) [8] as a function of the SNR feature and firing rate features.

2 Experiments

The neural activity used in this study was collected from two series of recording sessions performed using a new generation microprobes [9], semi-chronically implanted in the cortex of an awake macaque monkey.

The silicon-based probes applied in this study comprise four slender, 8-mm-long probe shafts arranged as a comb, as shown in Figure 1(a). Each probe shaft has a width and thickness of 140 μm and 100 μm, respectively, and carries nine circular Pt electrodes with a diameter of 35 μm (cf. Figure 1(a)). Out of these 36 electrodes, the first eight electrodes of each shaft, i.e. 32 electrodes, as counted from the electrode tip are accessible for the recording experiments. The inter-electrode pitch and the distance between the probe shafts were set to 250 μm and 550 μm, respectively. The two-dimensional electrode array as shown in Figure 1(a) is fabricated using microsystem technology detailed elsewhere [9]. Figure 1(b) shows the silicon-based probe assembled to a U-shaped polyimide cable comprising the interconnection part for a zero insertion force (ZIF) connector. The probe insertion into the brain tissue is performed using the insertion device in Figure 1(b). During insertion, the probe and its cable are fixed on the insertion device using vacuum suction. They are released for operating of the probe after retraction of the insertion device. The two semi-chronic recording sessions were performed using the same neural device. In the first series of recording sessions, the neural device was implanted in the primary somatosensory cortex (SI) and it was kept in the cortex for 8 days.

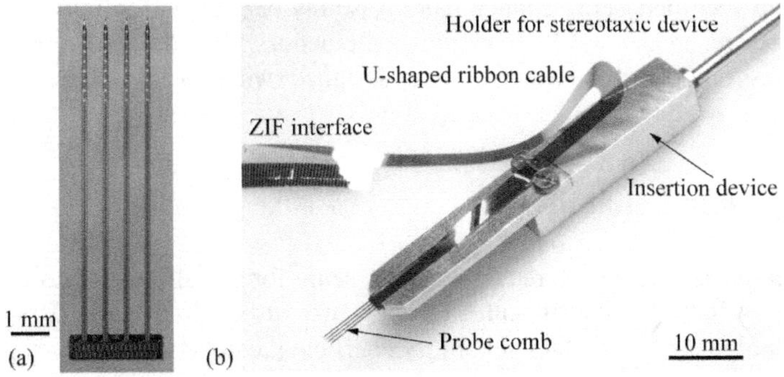

Fig. 1. (a) Silicon-based probe comb with four slender shafts with planar electrodes and a connector pads for cable assembly and (b) probe comb assembled to a U-shaped ribbon cable temporarily mounted onto an insertion device using vacuum for insertion

The second implantation was located in the superior parietal lobule (SPL) and the neural device was kept in the cortex for 10 days. The acquisition time of each trial lasted 4 seconds that corresponded to 2 seconds before and 2 seconds after the somatosensory stimulation. In order to assess the quality of the recorded signal from the implanted neural device, the neuronal activity of each electrode was carefully investigated on-line by the experimenter. In addition to the on-line recorded signal assessment, three experimenters performed an off-line quality signal assessment.

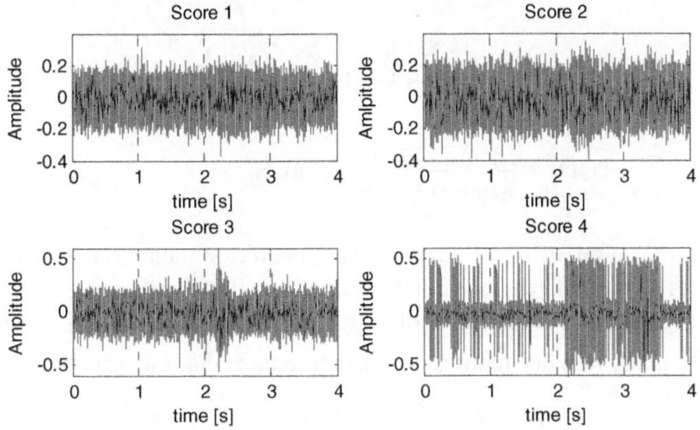

Fig. 2. Example signals of the four different scores. All 3 raters agreed independently on the score of each of the 4 signals. The stimulus was presented starting from 2 seconds.

This assessment was done separately and independently by each experimenter in order to avoid any possible bias in the evaluation of the neural signal quality. The assessment was done by a visual inspection of the neuronal discharge in single trials, on the basis of the personal experience of the experimenter. The assignment was based on a four levels scale, from 1 to 4. Each number corresponded to a different quality of the recorded neuronal activity, see Figure 2. Score 1 was assigned when no activity was recorded, score 2 when there was only background activity, score 3 when many different spikes were recorded simultaneously but the isolation of a single neuron was not possible. Score 4 was assigned when well isolated spikes were recorded.

3 Inter-rater Reliability

The degree to which raters agree on a quality score for signals, can be computed by an inter-rater reliability coefficient. Hereto, we used Krippendorff's agreement coefficient alpha (α) [10]. This reliability coefficient can be interpreted as: the degree to which independent observers, here using quality scores on a scale of 1 to 4, respond identically to each individual signal. Properties of α that are important in our case are the ability to deal with more than 2 raters, the use of confidence intervals, the correction for small sample sizes and the possibility to penalize more heavily larger disagreements between raters. To illustrate the latter: a 1-4 confusion between raters, i.e. one rater gives the lowest score 1 and another gives the highest score 4, is not as forgivable as 1-2 confusion or a 2-3 confusion between raters.

The alpha reliability can be computed as:

$$\alpha = 1 - \frac{D_o}{D_e}, \tag{1}$$

where D_o is the observed disagreement and D_e is the expected disagreement. D_o and D_e are computed as:

$$D_o = \frac{1}{n} \sum_c \sum_k o_{ck} \delta_{ck}^2, \tag{2}$$

$$D_e = \frac{1}{n(n-1)} \sum_c \sum_k n_c \cdot n_k \delta_{ck}^2. \tag{3}$$

Here, n is the total number of values that have been assigned over all raters, i.e. n is equal to #raters x #signals (here 3 raters x 256 signals). Further, o_{ck} is the observed number of c-k pairs, i.e. the number of times one rater gave quality 'c' and another quality 'k'. δ_{ck}^2 is the penalty factor for a disagreement that one rater rated a signal as quality 'c' and another as quality 'k' ($\delta_{ck}^2 = 0$ for c = k), n_c and n_k are the number of times score 'c' and score 'k' have been assigned. For details on the computation of the observed coincidences o_{ck} we refer to chapter 11 of [10].

The α value can range between -1 and 1, with a -1 indicating systematic disagreement between raters and 1 indicating systematic agreement. A small value of α around 0 indicates that raters do not agree more or disagree less than expected by chance level. In Table 1, we report the reliability scores for different choices of the penalty function δ_{ck}^2.

Table 1. Krippendorff's alpha reliability scores for different metric values. 1000 bootstraps were taken to obtain the 95% confidence interval. α_{123} is the reliability coefficient computed between all 3 raters.

Metric δ^2_{ck}	α_{123} coefficient	95% confidence interval
Nominal: $\delta^2_{ck} = 1, c \neq k$; $\delta^2_{ck} = 0, c = k$	0.6458	[0.5989,0.6908]
Ordinal: see [10]	0.8882	[0.8691,0.9048]
Interval: $\delta^2_{ck} = (c-k)^2$	0.9006	[0.8863,0.9144]
Ratio: $\delta^2_{ck} = \left(\frac{c-k}{c+k}\right)^2$	0.8658	[0.8435,0.8859]

The first column of Table 1 shows that the nominal metric punishes every mismatch between the scores of raters in an equally severe way, while in the other metrics the penalty depends on the degree of mismatch. The ordinal metric is likely the most suitable metric here, because it considers the values of the raters as a rank. It is observed that the nominal metric leads to the smallest reliability value, while the other metrics are about the same. This suggests that raters disagree sometimes on the exact quality score, but their assignments will seldom differ by more than 1 value. Larger confusions almost never happen. Indeed, in only 6 out of the 256 signals, a difference in quality larger than 1 was observed between any 2 observers. Under the ordinal (0.8882), interval (0.9006) and ratio (0.8658) metric, the assignments can be considered as reliable.

4 Prediction of Quality Scores

Our goal is now to predict the quality scores assigned by the different raters, since these predictions can then be used in the electronic depth control. Hereto, we train and test a Gaussian process classifier using features extracted from the neural signals. The features are based on the spikes which are detected with a continuous wavelet transform. We consider the following features: (1) the signal-to-noise ratio (SNR) which has been considered so far as the standard in electrode positioning [4,6-7], and selection [3], (2) the maximal firing rate in 20 ms bins around stimulus presentation and (3) the average firing rate in the same interval around stimulus presentation.

First, spikes are detected off-line in the neural signals using a continuous wavelet transform [11]. We used the Daubechies 2 wavelet, detecting spikes that are about 0.5 ms wide. We tried 2 detection thresholds: one equal to 0 and a more conservative one that only detects the larger spikes, see Figure 3. A higher detection threshold will in general lead to a lower probability of detection (PD) and to a lower probability of false alarms (PFA). For an interpretation of the detection thresholds the reader is referred to [11].

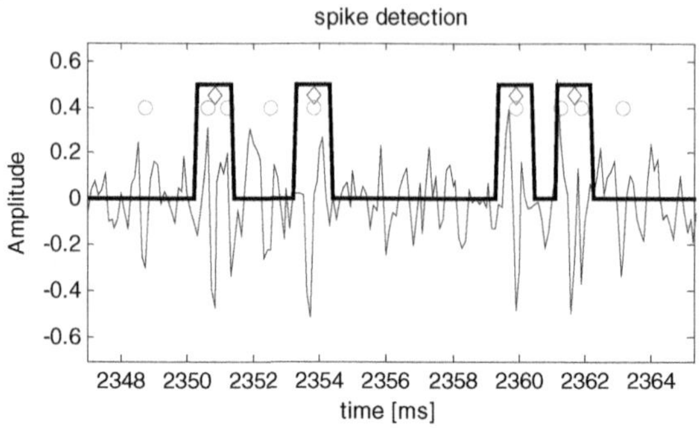

Fig. 3. The circles locate the spikes detected with a continuous wavelet transform (Daubechies 2 mother wavelet) using a threshold equal to 0, the diamonds are spike locations with a more conservative threshold equal to 0.2. The boxes are centered on the detected spikes (spikes detected at the diamonds) taking 0.5 ms before and 0.5 ms after the spike location. The average root-mean-square (RMS) of the signal within the boxes (spikes) is divided by the standard deviation outside the boxes (which are the noise segments) in the computation of the signal-to-noise ratio.

The SNR is computed as the average root-mean-square (RMS) of the spikes, N in total, divided by the median absolute deviation (MAD) $\hat{\sigma}_{noise}$ of the noise segments (see also Figure 3):

$$SNR_{dB} = 20 \cdot \log_{10} \frac{\frac{1}{N}\sum_{n=1}^{N} RMS(spike_n(t))}{\hat{\sigma}_{noise}} . \quad (4)$$

The median absolute deviation was used as a robust estimator for the noise [11]. The SNR quantifies how large spikes are compared to the background noise. However, it has to be noted that only a few, but large spikes can result in a high SNR. Experimenters often are not only interested in obtaining large spikes, but also in how responsive the cell is towards a stimulus. Therefore, also firing rate features were computed. For the analyses performed in the present study, we considered 1 second before and one second after the somatosensory stimulation, in order to be sure to include the whole neuronal response, since the discharge onset varied in relation to the different properties of the neurons recorded from different cortical layers and cortical areas. We segmented this interval into bins of 20 ms, the firing rate of the bin with the highest firing rate was considered as a feature. Furthermore, the average firing rate over the same interval was computed as the third feature.

We used a Gaussian process classifier (GPC) [8] to predict the quality scores. The variational Bayesian approach [8], using a radial basis function (RBF) kernel, was used. The kernel hyperparameters as well as the other parameters of the GPC are found by maximizing the variational lower bound on the marginal likelihood [8]. The advantage of using this GPC compared to some support vector machines is that no extra cross-validation cycles are required for tuning the kernel hyperparameters, they are inferred elegantly in a Bayesian way from the data [8]. We used the leave-one-out method for validation. Before training and testing the GPC, the SNR and firing rate features were standard normalized. The classification test accuracies for each rater separately, are shown in Figure 4. The SNR and the bin with maximal firing rate were used to train and test the GPC for the results shown in Figure 4.

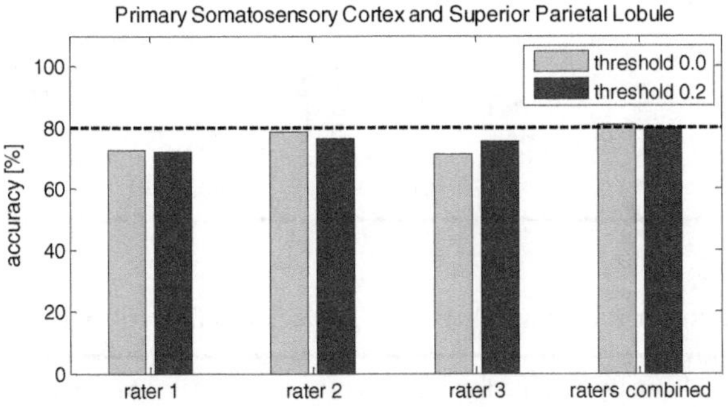

Fig. 4. Classification accuracies for the prediction of the quality scores using the variational Bayesian GPC and a leave-one-out validation. The last pair of bars, labeled 'raters combined', was obtained by taking the median vote of all 3 raters for each signal and training and testing the GPC on this median vote. For this median vote the accuracy is about 80%, which is higher than the accuracies of each rater separately.

To study the effect of different feature combinations, we tested (1) the SNR only, (2) the SNR and Max. Fr (bin with maximal firing rate), (3) the SNR and Avg. Fr (average firing rate) and (4) the SNR, Max. Fr and Avg. Fr. The results are shown in Figure 5.

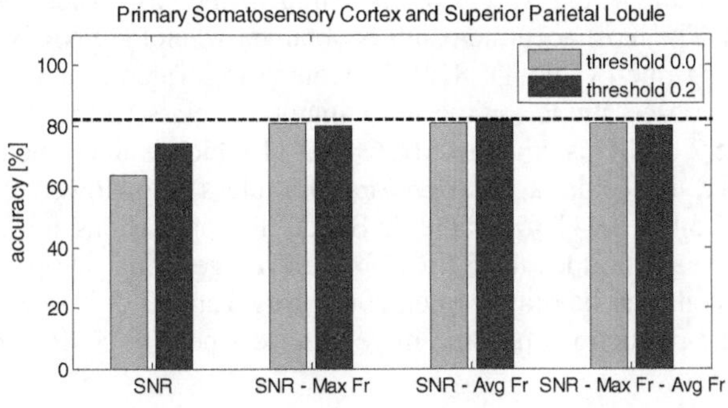

Fig. 5. Accuracies of different parameter combinations in the variational Bayesian GPC. The median vote of all 3 raters was used to provide the quality scores to the GPC. Clearly, the accuracies are higher when the SNR is combined with the bin containing the maximal firing rate (the SNR-Max Fr pair of bars) or with the average firing rate (the SNR-Avg Fr pair of bars) compared to the SNR alone (the SNR pair of bars).

Table 2. Confusion matrix SNR

		Manually assigned scores			
	74.22 %	score 1	score 2	score 3	score 4
Predicted scores	score 1	**107**	17	1	0
	score 2	2	**4**	4	1
	score 3	2	20	**28**	4
	score 4	7	6	2	**51**

Table 3. Confusion matrix SNR – Max Fr

		Manually assigned scores			
	80.08 %	score 1	score 2	score 3	score 4
Predicted scores	score 1	**111**	19	1	0
	score 2	6	**24**	11	2
	score 3	0	4	**19**	3
	score 4	1	0	4	**51**

Table 4. Confusion matrix SNR – Avg Fr

		Manually assigned scores			
	82.03 %	score 1	score 2	score 3	score 4
Predicted scores	score 1	**112**	19	1	2
	score 2	6	**25**	9	1
	score 3	0	3	**22**	2
	score 4	0	0	3	**51**

Table 5. Conf. matrix SNR – Max Fr – Avg Fr

		Manually assigned scores			
	79.69 %	score 1	score 2	score 3	score 4
Predicted scores	score 1	**110**	19	1	0
	score 2	6	**24**	11	2
	score 3	0	4	**19**	3
	score 4	2	0	4	**51**

Comparing previous tables, we observe that the predictions of the quality scores become more accurate when a firing rate parameter is used in combination with the SNR (Table 3, Table 4 and Table 5), compared to the case when the SNR is used alone (Table 2). The most accurate result is obtained when the SNR is combined with the average firing rate (Table 4): 82.03%. Comparing Table 2 and Table 4, we can make following observations for the most important class (score 4). The *recall* for class 4 in Table 2 (SNR) is equal to 51/56 = 0.911 which is the same as for the other tables. However, in Table 2 the *precision* for class 4 is lower, 51/66 = 0.773, compared to 51/54 = 0.944 for Table 4. Hence, the predictions for class 4 become much more precise if besides the SNR also the average firing rate parameter is used. Similar conclusions can be drawn when comparing Table 3 or Table 5 with Table 2: in both cases the predictions become more precise when the SNR is combined with firing rate features.

5 Conclusion

Quality score assignments by experimental neuroscientists seldom differ more than 1 quality value, on a scale of 4 possible values. This was reflected in the high (inter-rater) Krippendorff's alpha reliability of about 0.88 using the ordinal metric. In previous research, the signal-to-noise ratio (SNR) was used as the quality metric in movable microelectrode systems [4,6-7]. In this research a variational Bayesian Gaussian process classifier was used to predict the quality scores, leading to an accuracy of 82% if the SNR and the average firing rate are used together. These results suggest that the selection of the electrodes that capture signals of the highest quality can be performed by using the predictions of the quality scores.

Acknowledgments. We thank Richard Csercsa, Peter Pazmany Catholic University, Budapest, for taking part in assigning quality scores to the neural signals. This research was performed within the framework of the Information Society Technologies (IST) Integrated Project NeuroProbes of the 6th Framework Program (FP6) of the European Commission (Project number IST-027017). Gert Van Dijck and Marc M. Van Hulle are sponsored by the CREA financing program (CREA/07/027) of the K.U. Leuven and the Belgian Fund for Scientific Research – Flanders (G.0588.09).

References

1. Brown, E.N., Kass, R.E., Mitra, P.P.: Multiple Neural Spike Train Data Analysis: State-of-the-Art and Future Challenges. Nat. Neurosci. 7, 456–461 (2004)
2. Pine, J.: A History of MEA Development. In: Taketani, M., Baudry, M. (eds.) Advances in Network Electrophysiology: Using Multi-Electrode Arrays, pp. 3–23. Springer-Verlag New York Inc., New York (2006)
3. Seidl, K., Torfs, T., De Mazière, P.A., Van Dijck, G., Csercsa, R., Dombovari, B., et al.: Control and Data Acquisition Software for High-density CMOS-based Microprobe Arrays Implementing Electronic Depth Control. Biomed. Tech. 55, 183–191 (2010)
4. Sato, T., Suzukia, T., Mabuchi, K.: A New Multi-electrode Array Design for Chronic Neural Recording, with Independent and Automatic Hydraulic Positioning. J. Neurosci. Methods 160, 45–51 (2007)
5. Fee, M.S., Leonardo, A.: Miniature Motorized Microdrive and Commutator System for Chronic Neural Recording in Small Animals. J. Neurosci. Methods 112, 83–94 (2001)
6. Cham, J.G., Branchaud, E.A., Nenadic, Z., Greger, B., Andersen, R.A., Burdick, J.W.: Semi-chronic Motorized Microdrive and Control Algorithm for Autonomously Isolating and Maintaining Optimal Extracellular Action Potentials. J. Neurophysiol. 93, 570–579 (2005)
7. Jackson, N., Sridharan, A., Anand, S., Baker, M., Okandan, M., Muthuswamy, J.: Long-term Neural Recordings using MEMS based Movable Microelectrodes in the Brain. Front. Neuroeng. 3(10), 1–10 (2010)
8. Girolami, M., Rogers, S.: Variational Bayesian Multinomial Probit Regression with Gaussian Process Priors. Neural Comput. 18, 1790–1817 (2006)
9. Herwik, S., Kisban, S., Aarts, A., Seidl, K., Girardeau, G., Benchenane, K., et al.: Fabrication Technology for Silicon Based Microprobe Arrays used in Acute and Subchronic Neural Recording. J. Micromech. Microeng. 19, 074008 (2009)
10. Krippendorff, K.: Content Analysis: an Introduction to its Methodology. Sage Publications Inc., California (2004)
11. Nenadic, Z., Burdick, J.W.: Spike Detection Using the Continuous Wavelet Transform. IEEE Trans. Biomed. Eng. 52, 74–87 (2005)

Tensor Based Simultaneous Feature Extraction and Sample Weighting for EEG Classification

Yoshikazu Washizawa[1], Hiroshi Higashi[2,1], Tomasz Rutkowski[1], Toshihisa Tanaka[2,1], and Andrzej Cichocki[1]

[1] RIKEN Brain Science Institute, Japan
[2] Department of Electrical and Electronic Engineering, Tokyo University of Agriculture and Technology, Japan

Abstract. In this paper we propose a Multi-linear Principal Component Analysis (MPCA) which is a new feature extraction and sample weighting method for classification of EEG signals using tensor decomposition. The method has been successfully applied to Motor-Imagery Brain Computer Interface (MI-BCI) paradigm. The performance of the proposed approach has been compared with standard Common Spatial Pattern (CSP) as well with a combination of PCA and CSP methods. We have achieved an average accuracy improvement of two classes classification in a range from 4 to 14 percents.

Keywords: Feature extraction, classification, tensor decomposition, multi-linear PCA.

1 Introduction

Many contemporary signal processing applications are characterized by multiway structures of the recorded datasets. Biomedical datasets belong to the most common examples of such multiway structures which for example could be indexed by *subjects*, *trials*, *sensor channels*, *time* or *frequency bins*, etc. It is often a case that datasets are reconstructed in form of a vector or a matrix to apply next vector/matrix based signal processing methods. Feature extraction and selection are key factors in classification and model reduction problems. Usually, sampled data are represented by vectors or matrices and model reduction is performed by PCA. However, in many applications, original sampling data such as multi-modal brain data sets can be naturally represented by tensors (multi-way arrays) [1]-[4]. Tensors provide natural and convenient representations of such multidimensional data sets allowing us to preform feature extraction and classification.

Multilinear PCA (MPCA) and Higher Order SVD (HOSVD) for tensor data have been proposed (see, for example [2,3]). A pioneering work is the so-called Higher order singular value decomposition (HOSVD) [3] which is a decomposition method which performs decomposition of a tensor into a dense core-tensor and unitary matrices. The HOSVD is a tensor decomposition technique rather than feature extraction under the approximation model. Biomedical data sets

K.W. Wong et al. (Eds.): ICONIP 2010, Part II, LNCS 6444, pp. 26–33, 2010.

very often contain noisy periods which should be classified as outliers, for example, in the EEG experiments, subjects often don't obey the instructions by not focusing their attention on the presented stimuli. For the accurate machine learning results, it is thus necessary to automatically detect and reject such outliers. For this purpose we extend in this paper the multilinear principal component analysis (MPCA) [2] to extract features and to perform simultaneously weighting of data samples.

The following notation and mathematical operations for tensors (multi-way data) are used in the paper (note that tensor has multiple indexes - usually more than three):

- **Tensor** - denoted by a calligraphic large letter e.g., $\mathcal{A}$ or $\mathcal{A}(i_1, i_2, \ldots, i_N)$ ($i_1 = 1, 2, \ldots, I_1, \ldots, i_N = 1, 2, \ldots, I_N$), where N is the number of modes. Supposed that each index of a mode is positive integer, and the maximum number of the index is called the dimension of the mode $(I_1, I_2, \ldots, I_N)$.
- **Fiber** - an I_j-dimensional vector obtained by fixing all modes except the jth mode, is called the fiber of the jth mode,
- **Unfolding matrices** - (I_j) by $(\prod_{k \neq j} I_k)$ matrix laying all possible fibers of the jth mode is called the unfolding matrix of the jth mode. We denote the unfolding matrix of the jth mode by $\boldsymbol{A}_{(j)}$. The inverse operation of the unfolding is called the folding. We denote the unfold and the fold operator of the jth mode by $\text{Unfold}_j(\cdot)$ and $\text{Fold}_j(\cdot)$ respectively ($\boldsymbol{A}_{(j)} = \text{Unfold}_j(\mathcal{A})$, $\mathcal{A} = \text{Fold}_j(\boldsymbol{A}_{(j)})$).
- **Tensor-matrix multiplication** - suppose $\boldsymbol{a}$ be a fiber vector of a tensor $\mathcal{A}$. Given m by I_j matrix $\boldsymbol{B}$, the jth multiplication $\mathcal{A} \times_j \boldsymbol{B}$ is done by replacing all possible fibers $\boldsymbol{a}$ by $\boldsymbol{B}\boldsymbol{a}$. The dimension of the jth mode of $(\mathcal{A} \times_j \boldsymbol{B})$ is m. Note that $\mathcal{A} \times_j \boldsymbol{B} = \text{Fold}_j(\boldsymbol{B}\boldsymbol{A}_{(j)})$.
- We denote $\mathcal{A} \times_1 \boldsymbol{B}_1 \times_2 \boldsymbol{B}_2 \cdots \times_N \boldsymbol{B}_N = \prod_{i=1}^{N} \mathcal{A} \times_i \boldsymbol{B}_i$.
- **Frobenius norm** of a tensor $\mathcal{A}$ is defined by

$$\|\mathcal{A}\|_F^2 = \sum_{i_1=1}^{I_i} \sum_{i_2=1}^{I_2} \cdots \sum_{i_N=1}^{I_N} [\mathcal{A}(i_1, i_2, \ldots, i_N)]^2. \tag{1}$$

2 Multilinear Weighted PCA

2.1 Matrix Based PCA and Weighted PCA

We explain basic concept for the standard PCA and the weighted PCA

Let an L-dimensional vector $\boldsymbol{x}_k \in \mathbb{R}^L$ be the k-th sample, and a matrix $\boldsymbol{X} = [\boldsymbol{x}_1, \boldsymbol{x}_2, \ldots \boldsymbol{x}_M]$ represents all available samples, where M is the number of samples.

The standard PCA can be performed by the following optimization problem (applying the least mean squared error (LMSE) criterion)

$$\underset{\boldsymbol{P}}{\text{minimize}}\ f_0(\boldsymbol{P}) = \sum_{i=1}^{M} \|\boldsymbol{x}_i - \boldsymbol{P}\boldsymbol{x}_i\|^2 = \|\boldsymbol{X} - \boldsymbol{P}\boldsymbol{X}\|_F^2, \quad \text{subject to } \text{rank}(\boldsymbol{P}) \leq r,$$

where r is a parameter that specifies the dimension of the subspace. The optimum solution $\boldsymbol{P}$ is the projection matrix onto principal subspace of the all samples, hence there exists an orthogonal transform matrix $\boldsymbol{U} \in \mathbb{R}^{L \times r}$ such that $\boldsymbol{P} = \boldsymbol{U}\boldsymbol{U}^\top$ and $\boldsymbol{U}^\top \boldsymbol{U} = \boldsymbol{I}$, where $\boldsymbol{I}$ is the identity matrix. Note that $\boldsymbol{U}$ has an unitary ambiguity.

The standard PCA is also often formulated by using maximum variance criterion. Then the problem cab be reduced smaller dimensional space. However, the optimization problem has an unitary constraint and some ambiguity mentioned above.

In general for standard PCA, all samples are handled evenly. However, in real applications there are some outliers or bad data that do not represent features well. In order to extract proper features from samples, we should down-rate such samples. In such cases the cost function for weighted PCA can be expressed,

$$f_1(\boldsymbol{P}, \boldsymbol{D}) = \sum_{i=1}^{M} d_i^2 \|\boldsymbol{x}_i - \boldsymbol{P}\boldsymbol{x}_i\|^2 = \|\boldsymbol{X}\boldsymbol{D} - \boldsymbol{P}\boldsymbol{X}\boldsymbol{D}\|_F^2, \tag{2}$$

where d_i is the weight for the i-th sample, $\boldsymbol{D} = \mathrm{diag}(d_1, \ldots, d_M)$. The case of $\boldsymbol{D} = \boldsymbol{I}$ is reduced to the standard PCA.

If we do not have *a priori* information about quality of each sample, we have to estimate optimal weighting diagonal matrix $\boldsymbol{D}$ with positive entries d_i . The samples that do not have proper feature contribute large cost in the cost function $f_0(\boldsymbol{P})$, and we should downrate such samples. Hence, we introduce following optimization problem,

$$\begin{aligned} &\underset{\boldsymbol{P},\boldsymbol{D}}{\text{minimize}} \quad f_1(\boldsymbol{P}, \boldsymbol{D}) = \|\boldsymbol{X}\boldsymbol{D} - \boldsymbol{P}\boldsymbol{X}\boldsymbol{D}\|_F^2 \\ &\text{subject to} \quad \mathrm{rank}(\boldsymbol{P}) \le r, \ \sum_{k=1}^{M} {d_k}^2 = M, \ 0 \le l \le d_k \le u, \ \text{for } k = 1, \ldots, M \end{aligned} \tag{3}$$

where l and u are lower and upper bound of the weight respectively.

2.2 Multilinear PCA for Tensor Samples

In [2], data tensor is separated to trials or samples, i.e., $\mathcal{X}_1, \ldots, \mathcal{X}_M$ and M is the number of samples. However, we consider such index is a one of modes because integrated form enables us to extract feature and weight samples simultaneously.

MPCA [2] was originally defined by using maximum variance criterion, that is equivalent to LMSE problem.

$$\underset{\boldsymbol{P}_1,\boldsymbol{P}_2,\ldots,\boldsymbol{P}_N}{\text{minimize}} \quad \|\mathcal{X} - \prod_{i=1}^{N} \mathcal{X} \times_i \boldsymbol{P}_i\|_F^2, \quad \text{subject to} \quad \mathrm{rank}(\boldsymbol{P}_i) \le r_i, \quad \text{for all } i, \tag{4}$$

where the mode that corresponds to the index of samples or trials is fixed to $\boldsymbol{P}_i = \boldsymbol{I}$.

2.3 Details of the Proposed Method

In our approach, we combine standard MPCA with the idea of weighting sample (3). Let $A \subset \{1,\ldots,N\}$ be a set of modes that its dimension is reduced by PCA, e.g., channel or frequency, and $B \subset \{1,\ldots,N\}$ be a set of modes that its dimension is weighted, e.g., trial or subject. Here, $A \cap B = \phi$. The optimization problem including (4) and (3) can be written as

$$\begin{aligned}\underset{\boldsymbol{P}_a,\boldsymbol{D}_b}{\text{minimize}}\ \ & f_2 = \left\| \prod_{b\in B}(\mathcal{X}\times_b \boldsymbol{D}_b) - \prod_{b\in B}\{\prod_{a\in A}(\mathcal{X}\times_a \boldsymbol{P}_a)\times_b \boldsymbol{D}_b\}\right\|_F^2 \\ \text{subject to}\ \ & \text{rank}(\boldsymbol{P}_a)\le r_a,\ \forall a\in A,\quad \boldsymbol{D}_b = \text{diag}(d_1^b, d_2^b,\ldots,d_{L_b}^b), \\ & \sum_{j=1}^{L_b}(d_j^b)^2 = L_b,\ 0\le l_b\le d_j^b\le u_b,\ j=1,\ldots,L_b,\ \forall b\in B,\end{aligned} \tag{5}$$

where, l_b and u_b are the lower and upper bound of the weights d_j^b in the bth-mode, respectively. The cost function is an extension of the problem (3). If $\mathcal{X}$ has only two modes and $|A| = |B| = 1$, then this problem is equivalent to the problem (3). Note that in (3), the cost function can be simplified to $\|(\boldsymbol{I}-\boldsymbol{P})\boldsymbol{X}\boldsymbol{D}\|_F^2$. However, for higher-order tensors, we cannot simplify the cost function.

Since there is no way to solve (5) directly, we exploit alternating least square (ALS) strategy in a similar way to MPCA. In each step, for each mode $i = 1,2,\ldots,N$, we solve the problem (5) for the matrix of the mode i ($\boldsymbol{P}_i$ or $\boldsymbol{D}_i$), while fixing the matrices of the other modes $j \neq i$. We consider two sub-problems that are the cases $i \in A$ and $i \in B$.

i) The case of $i \in A$. We obtain optimal $\boldsymbol{P}_i$ while fixing all the other matrices $\boldsymbol{D}_j$ and $\boldsymbol{P}_j$, $j \neq i$. Then the sub-problem is

$$\underset{\boldsymbol{P}_i}{\text{minimize}}\ \ f_3(\boldsymbol{P}_i) = \|\boldsymbol{F}_i - \boldsymbol{P}_i\boldsymbol{Z}_i\|_F^2,\quad \text{subject to}\ \ \text{rank}(\boldsymbol{P}_i) = r_i, \tag{6}$$

where $\boldsymbol{F}_i = \text{Unfold}_i(\prod_{j\in B}\mathcal{X}\times_j \boldsymbol{D}_j)$. and $\boldsymbol{Z}_i = \text{Unfold}_i(\prod_{k\in B}\{\prod_{j\in A,j\neq i}(\mathcal{X}\times_j \boldsymbol{P}_j)\times_k \boldsymbol{D}_k\})$. Let $\boldsymbol{R}_{FZ} = \boldsymbol{F}_i\boldsymbol{Z}_i^\top$ and $\boldsymbol{R}_Z = \boldsymbol{Z}_i\boldsymbol{Z}_i^\top$. Then $f_3(\boldsymbol{P}_i)$ can be transformed to $f_3(\boldsymbol{P}_i) = \|\boldsymbol{F}_i - \boldsymbol{P}_i\boldsymbol{Z}_i\|_F^2 = \|\boldsymbol{P}_i\boldsymbol{R}_Z^{1/2} - \boldsymbol{R}_{FZ}\boldsymbol{R}_Z^{-\top/2}\|_F^2 + \|\boldsymbol{F}_i\|_F^2$, where $\boldsymbol{R}_Z^{1/2}$ is one of matrices that satisfies $\boldsymbol{R}_Z = \boldsymbol{R}_Z^{1/2}\boldsymbol{R}_Z^{\top/2}$. From the Schmidt approximation theorem (also called the Eckart-Young theorem), $f_3(\boldsymbol{P}_i)$ is minimized when $\boldsymbol{P}_i\boldsymbol{R}_Z^{1/2} = \sum_{i=1}^{r_i}\boldsymbol{u}_j\boldsymbol{u}_j^\top\boldsymbol{R}_{FZ}\boldsymbol{R}_Z^{-\top/2}$ where $\boldsymbol{u}_j$ is the eigenvector corresponding to the jth largest eigenvalue of $\boldsymbol{R}_{FZ}\boldsymbol{R}_Z^{-1}\boldsymbol{R}_{FZ}^\top$. Hence $f_3(\boldsymbol{P}_i)$ is minimized by

$$\boldsymbol{P}_i = \sum_{j=1}^{r_i}\boldsymbol{u}_j\boldsymbol{u}_j^\top\boldsymbol{R}_{FZ}\boldsymbol{R}_Z^{-1}. \tag{7}$$

If all $\boldsymbol{D}_i = \boldsymbol{I}$ (in the case of MPCA), all column vectors of $\boldsymbol{Z}$ are projections of corresponding column vectors of $\boldsymbol{F}$. Hence, $\boldsymbol{P}_i$ becomes also projection matrix, and there exists a partial unitary matrix $\boldsymbol{V}$ that satisfies $\boldsymbol{P}_i = \boldsymbol{V}\boldsymbol{V}^\top$ and $\boldsymbol{V}^\top\boldsymbol{V} = \boldsymbol{I}$ ($\boldsymbol{V}$ is not unique). In the case of $\boldsymbol{D}_i \neq \boldsymbol{I}$, $\boldsymbol{P}_i$ in eq. (7) is not a projector any more, therefore, we should consider LMSE criterion but not maximum variance criterion under the unitary constraint. In the right hand side of eq. (7), the first factor (matrix $\boldsymbol{U}$) is the partial unitary that pulls back a vector from $\mathbb{R}^{r_i}$ to $\mathbb{R}^{L_i}$. The essential feature extraction is done in the second factor represented by $\boldsymbol{U}^\top \boldsymbol{R}_{FZ}\boldsymbol{R}_Z^{-1}$. We utilize this factor as the feature extractor in a similar way as was used $\boldsymbol{V}^\top$ in the standard MPCA.

ii) The case of $i \in B$. In the case of $i \in B$, the problem is reduced to following optimization sub-problem,

$$\begin{aligned} &\underset{\boldsymbol{D}_i}{\text{minimize}} \;\; f_4(\boldsymbol{D}_i) = \|\boldsymbol{D}_i(\boldsymbol{F}_i - \boldsymbol{Z}_i)\|_F^2 = \|\boldsymbol{D}_i\boldsymbol{G}_i\|_F^2, \\ &\text{subject to} \;\; \sum_{j=1}^{L_i} {d_j^i}^2 = L_i,\; 0 \le l_i \le d_j^i \le u_i, \;\; \text{for } j = 1, \dots, L_i, \end{aligned} \tag{8}$$

where $\boldsymbol{F}_i = \text{Unfold}_i(\prod_{j\in B, j\neq i} \mathcal{X} \times_j \boldsymbol{D}_j)$, $\boldsymbol{Z}_i = \text{Unfold}_i(\prod_{k\in B,k\neq i}\{\prod_{j\in A}(\mathcal{X} \times_j \boldsymbol{Q}_j) \times_k \boldsymbol{D}_k\})$, $\boldsymbol{G}_i = \boldsymbol{F}_i - \boldsymbol{Z}_i$, $\boldsymbol{D}_i = \text{diag}(d_1^i, d_2^i, \dots, d_{L_i}^i)$.

Let $\boldsymbol{g}_j$ be the jth row vector of $\boldsymbol{G}_i$ ($\boldsymbol{G}_i = [\boldsymbol{g}_1, \boldsymbol{g}_2, \dots, \boldsymbol{g}_{L_i}]^\top$) and $\boldsymbol{h} = [\|\boldsymbol{g}_1\|^2, \|\boldsymbol{g}_2\|^2, \dots, \|\boldsymbol{g}_{L_i}\|^2]^\top$. $f_4(\boldsymbol{D}_i)$ can be rewritten to $\|\boldsymbol{D}_i\boldsymbol{G}_i\|_F^2 = \sum_{j=1}^{L_i} {d_j^i}^2 \|\boldsymbol{g}_i\|^2 = \boldsymbol{d}^\top\boldsymbol{h}$, where $\boldsymbol{d} = [{d_1^i}^2, {d_2^i}^2, \dots, {d_{L_i}^i}^2]^\top$. Let $\boldsymbol{1}_{L_i}$ be the L_i-dimension vector whose elements are one ($\boldsymbol{1}_{L_i} = [1, 1, \dots, 1]^\top$), $\boldsymbol{C} = [\boldsymbol{1}_{L_i}| - \boldsymbol{1}_{L_i}|\boldsymbol{I}_{L_i}| - \boldsymbol{I}_{L_i}]^\top$, and $\boldsymbol{r} = [L_i, -L_i|u_i \cdot \boldsymbol{1}_{L_i}^\top| - l_i \cdot \boldsymbol{1}_{L_i}^\top]^\top$. From these definitions, the minimization problem (8) can be formulated as the standard linear programming (LP) problem,

$$\underset{\boldsymbol{d}}{\text{minimize}} \;\; \boldsymbol{d}^\top\boldsymbol{h}, \quad \text{subject to } \boldsymbol{C}\boldsymbol{d} \le \boldsymbol{r}. \tag{9}$$

The problem can be solved by existing algorithms such as simplex method.

The ALS procedure monotonically decreases the cost function of (5), and leads to a local minimum solution. Initialization for $\boldsymbol{P}_i$ and $\boldsymbol{D}_i$ is also important problem. For $\boldsymbol{P}_i$, initialization by identity matrix ($\boldsymbol{P}_i = \boldsymbol{I}$) or simple PCA of the unfolding matrix $\boldsymbol{X}_{(i)}$ are feasible. For $\boldsymbol{D}_i$, initialization by identity matrix ($\boldsymbol{D}_i = \boldsymbol{I}$) is feasible.

The pseudo-code of the final procedure is presented Algorithm 1.

3 Experiment

We used EEG data containing two classes: right hand and right foot motor-imageries. They were provided by Fraunhofer FIRST (Intelligent Data Analysis Group) and Campus Benjamin Franklin of the Charité - University Medicine Berlin (Department of Neurology, Neurophysics Group) [6]. The EEG signals were recorded from five subjects. 118 EEG channels are measured at positions of the extended international 10/20-system ($N_{ch} = 118$). Signals were band-pass

Algorithm 1. Tensor based feature extraction/sample weighting

Input: $\mathcal{X}$
Parameter: r_a , l_b and u_b ($a \in A$, $b \in B$)
Output: $\boldsymbol{P}_a$, $\boldsymbol{D}_b$ ($a \in A$, $b \in B$)

Initialization: $\boldsymbol{P}_a$ ($\forall a \in A$): identity matrix or simple PCA, $\boldsymbol{D}_b$ ($\forall b \in B$): identity matrix, $k = 0$: no. of iteration.
calculate initial cost L_0 from the cost function f_2.
repeat
  $k \leftarrow k + 1$
  for all $a \in A$ **do**
    Update $\boldsymbol{P}_a$ by eq. (7).
  end for
  for all $b \in B$ **do**
    Update $\boldsymbol{D}_b$ by solving linear programming (9).
  end for
  Calculate cost L_k from the cost function f_2.
until $(L_{k-1} - L_k)$ is sufficiently small.

filtered between 0.05–200 Hz and then digitized at 1000 Hz with 16 bit (0.1 μV). During each experiment, the subject was given visual cues that indicated for 3.5 seconds which of the three motor imagery should be performed: left hand, right hand, and right foot. The resting interval between two trials was randomized from 1.75–2.25 seconds. Only EEG trials for right hand and right foot were provided. Each class of EEG signals consists of 140 trials ($N_{tr} = 2 \times 140 = 280$). We used pre-processed EEG data that were down-sampled to 100 Hz, and we applied a band-pass filter between 5-45 Hz.

The data is described in a three-mode tensor. Three modes represent time, channel, and trial. The data tensor $\mathcal{X}$ is in $\mathbb{R}^{T \times N_{\mathrm{ch}} \times N_{\mathrm{tr}}}$. In order to remove phase component, we applied the Fourier transform for mode 1, and took absolute values. The transformed data is denoted by $\hat{\mathcal{X}} \in \mathbb{R}^{T/2 \times N_{\mathrm{ch}} \times N_{\mathrm{tr}}}$. Then we applied our feature extraction and sample weighting method. The sets are $A = \{1, 2\}$ and $B = \{3\}$, i.e., features with respect to frequency and channels are extracted and samples with respect to trials are weighted simultaneously. Our method outputs three matrices $\boldsymbol{P}_1$, $\boldsymbol{P}_2$ and $\boldsymbol{D}_3$. As we mentioned, we utilize only a part, $\boldsymbol{U}^\top \boldsymbol{R}_{FZ} \boldsymbol{R}_Z^{-1}$ in eq. (7). We denote the parts of $\boldsymbol{P}_1$ and $\boldsymbol{P}_2$ by $\boldsymbol{U}_1 \in \mathbb{R}^{r_1 \times T/2}$ and $\boldsymbol{U}_2 \in \mathbb{R}^{r_2 \times N_{\mathrm{ch}}}$ respectively. We used data of both two classes for this procedure.

As a reference, the common spatial pattern (CSP) method is usually used since it is the basic tool for classification of EEG signals for M-I-BCI paradigms [5]. Since CSP is the technique for time-domain signals we use the data $\mathcal{X}$ instead of $\hat{\mathcal{X}}$. Each row vector of $\boldsymbol{U}_1$ represents a frequency filter. Therefore, we applied these frequency filters for all fibers of mode 1. Note that since this frequency filter is FIR filter, it is relatively easy to apply to real time streaming signals. We denoted the filtered signal of the ith row of $\boldsymbol{U}_1$ by $\tilde{\mathcal{X}}_i$ ($i = 1, \ldots, r_1$). Then, we applied the feature extraction for channels ($\tilde{\mathcal{X}}_i \times_2 \boldsymbol{U}_2$). Finally, by fixing mode 3

Table 1. Classification accuracies [%] and standard deviation of BCI benchmark over five-fold CV

	Accuracies and standard deviations			Optimal parameters	
Subj.	Proposed	CSP	PCA + CSP	r_1 r_2 u/l	r (PCA+CSP)
aa	72.14 ± 6.75	52.14 ± 2.93	61.07 ± 5.98	7 20 1.1	20
al	86.79 ± 5.14	73.93 ± 4.11	80.71 ± 4.62	1 25 2.5	20
av	59.64 ± 4.66	52.86 ± 4.82	57.86 ± 2.04	5 20 3.5	25
aw	65.71 ± 4.26	59.64 ± 2.70	61.79 ± 9.99	2 50 4.0	20
ay	79.64 ± 6.00	55.36 ± 5.05	78.93 ± 3.66	2 20 1.5	20
Total	72.79 ± 11.04	58.79 ± 8.97	68.07 ± 11.27		

(trial) to n, we obtained a matrix, $\boldsymbol{X}_{(i,n)} \in \mathbb{R}^{T\times r_2}$ $(n = 1, \ldots, N_{\rm tr}, i = 1, \ldots, r_1)$. By rearranging matrices, we have $\boldsymbol{X}_n = [\boldsymbol{X}_{(1,n)}|\boldsymbol{X}_{(2,n)}|\cdots|\boldsymbol{X}_{(r_1,n)}] \in \mathbb{R}^{T\times(r_1 r_2)}$.

We denote the two-classes labeled training sets by $\boldsymbol{X}^1_{n_1}$ and $\boldsymbol{X}^2_{n_2}$, and corresponding weight by $d^{(1)}_{n_1}$ and $d^{(2)}_{n_2}$ that are corresponding diagonal elements of the weight matrix $\boldsymbol{D}_3$ $(n_1 = 1, \ldots, N_1, n_2 = 1, \ldots, N_2)$.

The standard CSP utilizes variance-covariance matrix of the data. To make fair comparison, we exploited the weight matrix $\boldsymbol{D}_3$ to extend the CSP to weighted CSP using weighted variance-covariance matrix,

$$\boldsymbol{\Sigma}_c = \sum_{n=1}^{N_c} (d^{(c)}_n)^2 (\boldsymbol{X}^c_n)^\top \boldsymbol{X}^c_n, \quad c = 1, 2, \tag{10}$$

$\boldsymbol{\Sigma}_c \in \mathbb{R}^{(r_1 r_2)\times(r_1 r_2)}$. The weight vectors of CSP maximizes the Fisher's criterion, $(\boldsymbol{w}_c^\top \boldsymbol{\Sigma}_c \boldsymbol{w}_c)/(\boldsymbol{w}_c^\top(\boldsymbol{\Sigma}_1 + \boldsymbol{\Sigma}_2)\boldsymbol{w}_c)$ for $c = 1, 2$. Testing data $\boldsymbol{X} \in \mathbb{R}^{T\times(r_1 r_2)}$ is classified the class, such that $\|\boldsymbol{X}\boldsymbol{w}_c\|$ is larger.

We conducted five-fold cross validation (CV). The data set was divided into five subsets, and four subsets are used for training (both feature extraction/sample weighting and CSP), and remaining one is used for testing. The optimal parameter is chosen for each subject (the parameters are the same among CV). We compared three methods; our method, standard CSP, and PCA+CSP. For our method, we gave the one parameter u/l for upper and lower bound of the linear programming. Table 1 shows the classification accuracies, standard deviations, and the optimal parameters.

The proposed method outperforms standard CSP and CSP with PCA for all subjects in the dataset. In case of $u/l = 1$, all training samples are handled evenly. For subjects 'aa' and 'ay', smaller values of u/l resulted in better performance. This means that EEG signals obtained from those subjects have fewer outliers or noisy samples. In contrast, data of subjects 'av' and 'aw' have more outliers.

We coded our algorithm in GNU Octave compiled with Intel Math Kernel Library. GLPK was used for linear programming. For one CV, our algorithm takes 50 sec. to obtain $\boldsymbol{U}_1$, $\boldsymbol{U}_2$ and $\boldsymbol{D}_3$ (four iterations), 17 sec. to obtain $\boldsymbol{X}_n$ on a PC that has Intel Core i7 2.8GHZ CPU.

As we discussed in the above sections, the feature extractor of the mode 1, $\boldsymbol{P}_1$, works as a FIR filter. Even though the extractor does not only result with positive values, we consider the lager magnitude elements of $\boldsymbol{P}_1$ to represent important feature for classification since the effect of the negative sign vanishes in eq. (10). We believe that still a detailed study and alternative methods such as non-negative matrix factorization (NMF) should be discussed in future research.

4 Conclusions

We proposed the extension of the Multilinear PCA method for the classification of EEG signals in application to the motor-imagery-paradigm. The main contribution of the technique is to incorporate the weighting of tensor samples and to convert the problem to the standard linear programming. The included in the paper simulation results are very promising since we have achieved a considerable improvement of classification performance of the benchmark BCI EEG data with motor-imagery-paradigm (right hand and foot movements) for five subjects. We expect to achieve further improvements by further application of time-frequency EEG preprocessing techniques as well 4D or 5D Tucker models with various constraints.

References

1. Cichocki, A., Zdunek, R., Phan, A.H., Amari, S.: Nonnegative Matrix Factorization and Tensor Factorizations. John Wiley and Sons, Chichester (2009)
2. Lu, H., Plataniotis, K.N., Venetsanopoulos, A.N.: MPCA: Multilinear principal component analysis of tensor objects. IEEE Trans. Neural Network 19(1), 18–39 (2008)
3. Lathauwer, L.D., Moor, B.D., Vandewalle, J.: A multilinear singular value decomposition. SIAM Journal of Matrix Analysis and Application 21(4), 1253–1278 (2001)
4. Phan, A.H., Cichocki, A.: Tensor Decompositions for Feature Extraction and Classification of High Dimensional Datasets. IEICE NOLTA E93-N(10) (October 2010)
5. Ramoser, H., Mueller-Gerking, J., Pfurtscheller, G.: Optimal spatial filtering of single trial EEG during imagined hand movement. IEEE Trans. Rehabilitation Engineering 8(4), 441–446 (2000)
6. Dornhege, G., Blankertz, B., Curio, G., Müller, K.-R.: Boosting bit rates in noninvasive EEG single-trial classifications by feature combination and multiclass paradigms. IEEE Trans. Biomedical Engineering 51(6), 993–1002 (2004)

A Tongue-Machine Interface: Detection of Tongue Positions by Glossokinetic Potentials

Yunjun Nam[1], Qibin Zhao[2], Andrzej Cichocki[2], and Seungjin Choi[1,3,4]

[1] School of Interdisciplinary Bioscience and Bioengineering
Pohang University of Science and Technology
San 31 Hyoja-dong, Nam-gu, Pohang 790-784, Korea
druid@postech.ac.kr
[2] Lab for Advanced Brain Signal Processing
Brain Science Institute, RIKEN
2-1 Hirosawa, Wako-shi, Saitama, 351-0198, Japan
{qbzhao,a.cichocki}@brain.riken.jp
[3] Department of Computer Science
[4] Division of IT Convergence Engineering
Pohang University of Science and Technology
San 31 Hyoja-dong, Nam-gu, Pohang 790-784, Korea
seungjin@postech.ac.kr

Abstract. Artifacts are electrical activities that are detected along the scalp by an electroencephalography (EEG) but that originate from non-cerebral origin, which often need to be eliminated before further processing of EEG signals. Glossokinetic potentials are artifacts related to tongue movements. In this paper we use these glossokinetic artifacts (instead of eliminating them) to automatically detect and classify tongue positions, which is important in developing a tongue-machine interface. We observe that with a specific selection of a few electrode positions, glossokinetic potentials show contralateral patterns, so that the magnitude of potentials is linearly proportional to the tongue positions flicking at the left to the right inside of cheek. We design a simple linear model based on principal component analysis (PCA) to translate glossokinetic potentials into tongue positions. Experiments on cursor control confirm the validity of our method for tongue position detection using glossokinetic potentials.

1 Introduction

Various assistive technologies have been developed for persons with limb motor disabilities, in order to help them to carry out daily tasks. Assistive devices are controlled by detecting brain waves, muscle activities, eye positions, or tongue motions. Tongue has recently been recognized as a promising man-machine interface, due to several reasons [1,2]. A tongue is directly connected to a brain by

K.W. Wong et al. (Eds.): ICONIP 2010, Part II, LNCS 6444, pp. 34–41, 2010.

cranial nerves and the distance from a brain is relatively short. A tongue generally escapes severe damage in spinal cord injuries and the last to be affected in most neuromuscular degenerative disorders, such as amyotrophic lateral sclerosis (ALS). Furthermore a tongue consists of special muscles suitable for complex vocalization and ingestion tasks, so it can move very quickly and accurately and does not fatigue easily. Furthermore, tongue movements can be hidden by a mouth cavity, and it brings a cosmetic advantage.

There are several tongue-operated assistive devices that utilize above advantages, such as the Tongue Drive [2], inductive tongue computer interface (ITCI) [3], Think-A-Move [4], Tongue-mouse [5], TongueTouchKeypad (TTK) [6], and Tonguepoint [7]. In contrast to most of existing devices, we solely use glossokinetic artifacts recorded on a specific selection of electrodes by an EEG, in order to detect and classify tongue positions which are allowed to flick at the left to the right inside of cheek.

2 Glossokinetic Potentials

A glossokinetic artifact (GKA) [8,9] is a slow wave response caused by tongue movements, which can be considered as of the major EEG artifact. As the cause of the glossokinetic potential, the potential difference between the base and the tip of the tongue was regarded as main reason for the phenomenon. The tip of the tongue has a negative electrical charge with respect to the root, and if the tongue touches a palate or a cheek, significant discharge that is detectable at scalp, is generated [9]. However, Vanhatalo et al. [10] suggested that rearrangement of conduction pathway could be another origin of glossokinetic potential. If a tongue touches a cheek or a palate, a new current pathway between a ground and a reference is created, and then this conductive change could be observed as the altered potential of EEG channels. They also reported that the this kind of glossokinetic artifact can be removed by insulating surface of the tongue [10].

During our extensive experiments for searching new tongue-related mental tasks, we have found out that the glossokinetic potential (GKP) shows contralateral patterns (we will call them as contralateral glossokinetic potentials), when ground and reference electrodes are located along a longitudinal line of the head, as illustrated in Fig. 1 (a). In other words, we observed that when the tongue touches the cheek, DC levels of EEG signals recorded from two earlobes move to opposite directions and its magnitude is linearly proportional to the angle of the tongue as shown in Fig. 2.

It can be explained by resistance change caused by contact between a cheek and a tongue (see Fig. 1 (b)). When tongue touches cheek, a new current pathway is created and this pathway act as a variable resistance that can increase or decrease DC levels of EEG signals. We analyze this difference to find contact position of cheek and tongue, which also corresponds the direction of the tongue.

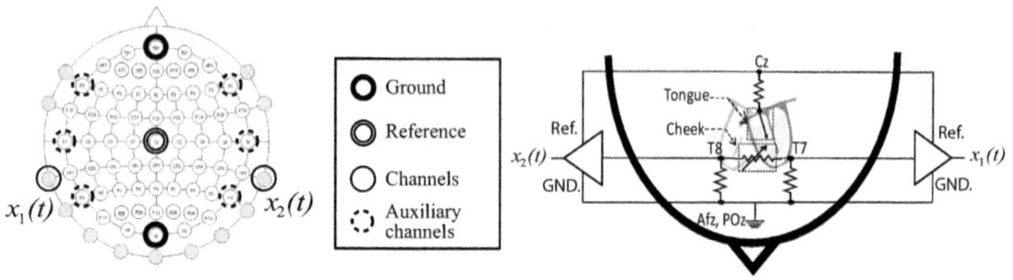

(a) Channel location (b) Circuit model for GKP

Fig. 1. (Left) Channel location for EEG recording. Signals are recorded from temporal regions (F7/8, T7/8, P7/8 and left/ right ear lobes) for measuring contralateral glossokinetic potential. Signals from left and right ear lobes are mainly used for the analysis. However if the subject's performance is not good enough, auxiliary channels have been also used for improving the performance. **(Right) Circuit model for GKP.** The reference electrode was mounted on top of the head (Cz) and two coupled ground electrodes were mounted on the forehead (AFz) and back of the head (POz). EEG signals from left earlobe (or T7) and right earlobe (or T8) were also recorded. If the tongue's position is changed while maintaining contact with a cheek, the conductive condition between the ground and the reference will be changed, and these changes can be interpreted as a variable resistance between electrodes for generating distinguishable signals to detect the tongue's position.

3 Method

3.1 Contralateral Glossokinetic Potentials

To detect the tongue's direction, we used potential difference between different EEG channels recorded from left and right side of the head. EEG signals were recorded using g.Tec system with 5 Ag/AgCl electrodes. The reference electrode was mounted on top of the head (Cz) and two coupled ground electrodes were mounted on the forehead (AFz) and back of the head (POz). Then the signals from left earlobe and right earlobe were recorded. Instead of ear lobes, other channels on temporal region such as T7 or T8 can be used, but pattern differences are not as clear as using earlobes. We will denote signal from left earlobe as $\boldsymbol{x}_1(t)$ and signal from right earlobe as $\boldsymbol{x}_2(t)$.

For validating the relationship between EEG signals and tongue's direction, we recorded EEG signals during real tongue movement from 2 healthy subjects. The subjects are requested to move their tongue to the direction where the cue is located while maintaining contact between tongue and cheek. At the beginning of the each trial, the cue was appeared from right side, and move to left side along the semicircle line. Then the cue was returned back to right side by same path as shown in Fig. 2. Traveling from one side to other side took 6 seconds.

When tongue is moving from right side to left side, signal of $\boldsymbol{x}_1(t)$ is decreased while $\boldsymbol{x}_2(t)$ is increased. In opposite, if the tongue is moving to right side $\boldsymbol{x}_1(t)$ is increased while $\boldsymbol{x}_2(t)$ is decreased. These changes are almost linearly proportional to the direction of tongue. Small fluctuations between each trials (on 17~20, 32~35 and so on) are caused by EOG (eye blinking).

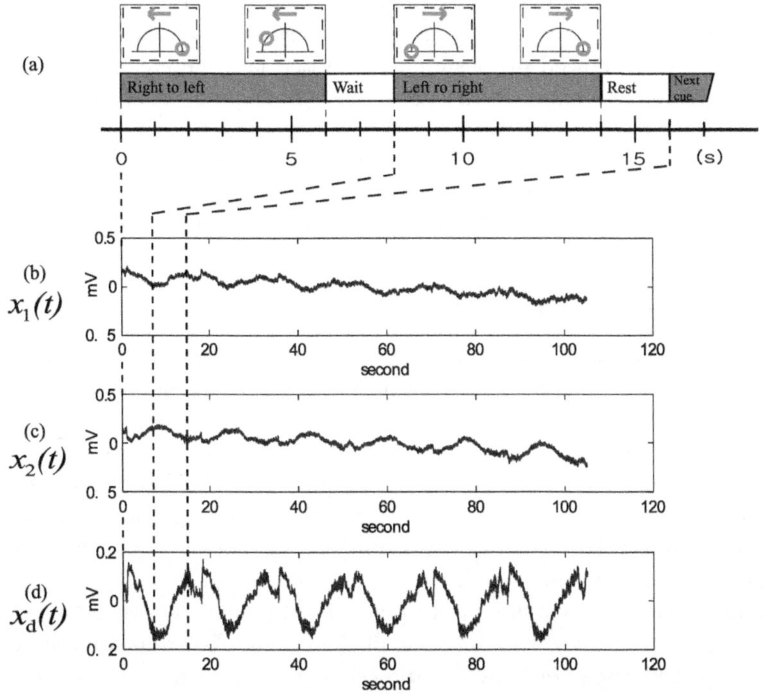

Fig. 2. (a) Visual cue, (b) $x_1(t)$: Signal recorded on left side channel (Left earlobe), (c) $x_2(t)$: Signal recorded on right side channel (Right earlobe), (d) Difference between $x_1(t)$ and $x_2(t)$, $x_d(t) = x_1(t) - x_2(t)$. EEG signals during repetitive left-right side tongue movements. The subjects are requested to move their tongues to the direction where the cue is located while maintaining contact between tongue and cheek. At the beginning of each trial, the cue was appeared from right side and move to left side along the semicircle line. Then the cue was returned back to right side by same path. Traveling from one side to other side took 6 seconds.

From Fig. 2 (b), (c), we can see that the potential of each signal is gradually decreasing. These patterns, can be considered as drift, is general characteristic of electric sensors. The magnitude of EEG signal are evaluated by relative potential difference between the channel, the reference and the ground. However, conductive properties of human body are unstable and such instabilities are observed by drift, which is unwanted and unwarranted signal changes over time.

Although this artifact, the relation between tongue's direction and difference between $x_1(t)$ and $x_2(t)$ was significantly maintained, during the experiments. We developed the model that can translate this potential difference to the direction of the tongue for the novel human computer interface.

3.2 Tip for Experiments

The proposed paradigm exploits resistance change caused by contact between two tissues: a tongue and a cheek, so property of tissue's surface are important factor. At first, the humidity inside mouth is quite important. If inside mouth is dry, the pattern change was insignificant. Otherwise inside mouth is filled with water, the signals showed more significant potential difference (but we did not use water during the experiment). Even though there is no water, the subject could wet their tongue with their saliva for improving performance.

Moreover, we observed that contralateral patterns for GKP are similar to electrooculogram (EOG) for horizontal eye-ball movements. So, we request to the subjects not to move eyes for horizontal direction during the experiments.

3.3 Detection of Tongue Positions

Electrical conditions of body and skin could be varied over time, so amount of difference between channels also could be changed. To build robust interface system that can follow up this change, we applied short training procedure to the interface. During the training phase, we recorded 3 kinds of signal for different tongue's position: left (L), front (F) and right (R). The length of each signal was 4 seconds, so 12 seconds were required for recording entire training data as shown in Fig. 3. By ignoring transition state between tasks, only the last 3 seconds of signals are considered.

As mentioned in previous section, potential difference between two channels is linearly proportional to the direction of the tongue. To translate the potential difference to the angle for tongue's direction, we used simple linear model. At first, each signal was low-pass filtered with cut-off frequency 4 Hz by using butterworth filter (green line on Fig. 3 (a), (b)) to avoid unstable fluctuation. From the filtered signals, the features corresponding to potential difference were obtained. Simple subtraction could be an acceptable choice, but for maximizing discriminability, we applied principle component analysis (PCA)-based linear filter. By using the PCA, we can find the vector $\widehat{\boldsymbol{w}}$ that can maximize feature's temporal differences, as follows,

$$\widehat{\boldsymbol{w}} = \arg\max_{\boldsymbol{w}} \boldsymbol{w}^\top \boldsymbol{X}\boldsymbol{X}^\top \boldsymbol{w}, \tag{1}$$

where $\boldsymbol{X} = [\boldsymbol{x}(1), \ldots, \boldsymbol{x}(N)]$ and $\boldsymbol{x} = [\boldsymbol{x}_1^\top, \boldsymbol{x}_2^\top]^\top$ and N is the time length of EEG signals. $\boldsymbol{X}$ is EEG signals recorded during the left, front and right cues of the training procedure. From the vector $\widehat{\boldsymbol{w}}$, we can calculate projected signal $\boldsymbol{z}(t) = \widehat{\boldsymbol{w}}^\top X$, as it illustrated on Fig. 3 (c). We can see that obtained feature is increased whenever tongue is moved to next position on right side. If the vector $\boldsymbol{z}(t)$ is segmented into $\boldsymbol{z}_L(t)$, $\boldsymbol{z}_F(t)$, $\boldsymbol{z}_R(t)$, then we can obtain their mean values μ_L, μ_F, μ_R.

The next step was translating obtained feature $\boldsymbol{z}(t)$ with range from μ_L to μ_R, to corresponding angle $\boldsymbol{\theta}(t)$ with range from $-\pi/2$ to $\pi/2$, as represented on Fig. 3. (d). The value of $\boldsymbol{\theta}(t)$ means the relative angle of tongue's direction, when the angles for positions L, F and R is fixed to $-\pi/2$, 0 and $\pi/2$. Since relations between potential difference and corresponding angle are linearly proportional,

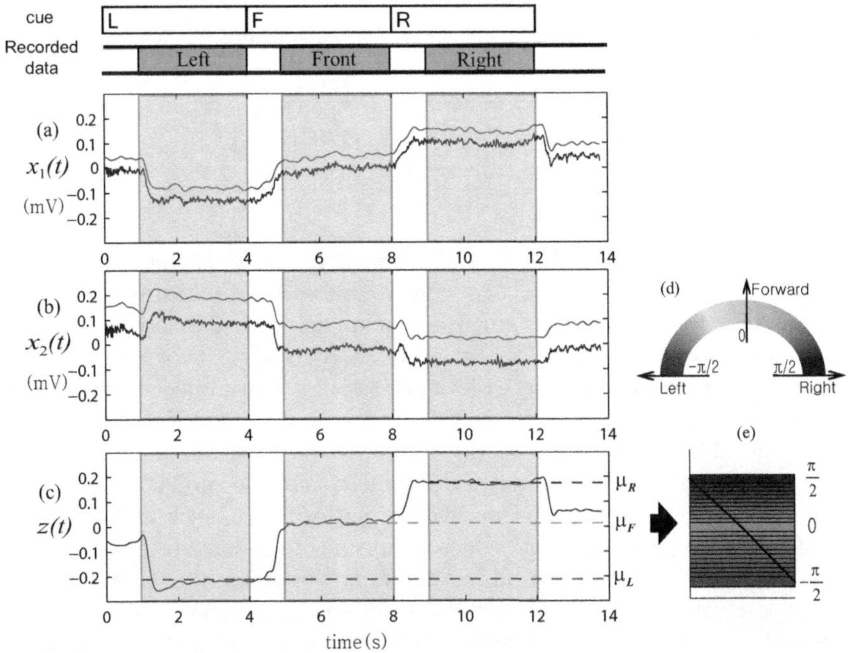

Fig. 3. Training procedure. (a) Signal recorded on left side channel. (b) Signal recorded on right side channel. (c) Obtained feature by PCA for predicting tongue's direction. (d) A color wheel for representing predicted angles. (e) Linear model for translating potential difference to angle. Measured EEG signals (blue line on (a), (b)) are band-pass filtered (green line on (a), (b)). Then feature values (c) are obtained by projecting the signals to the linear base from PCA. The mean values of each time segment for left, front and right (μ_L, μ_F, μ_R) was calculated. From these values, the linear model (e) that translates the feature value of (c) to the tongue's angle of (d) was generated. Newly obtained feature value on (c) was translated to the angle on (e) with the same height.

if the new feature value is given by z_t, its corresponding angle θ_t can be calculated by

$$\theta_t = \begin{cases} -\frac{\pi}{2} \cdot \frac{z_t - \mu_F}{\mu_L - \mu_F}, & (\mu_L \leq z_t \leq \mu_F) \\ \frac{\pi}{2} \cdot \frac{z_t - \mu_F}{\mu_R - \mu_F}, & (\mu_F \leq z_t < \mu_R) \\ \text{Error}, & \text{etc.} \end{cases} \tag{2}$$

4 Experiments

The following experiments were designed to evaluate the accuracy of the developed interface. Set of random angles were given to the subjects and the subjects moved their tongue to match direction of the interface to the given cue as close as possible. In Fig. 4, the given cue is plotted by blue line and the predicted angel from the signal is plotted by green line. For each session, subjects repeated

Table 1. Performance for the cue following experiment. The mean of absolute error was measured for each trial.

	1	2	3	4	5	6
Subject 1	20.4°	26.1°	13.4°	14.2°	14.3°	17.3°
Subject 2	38.7°	48.8°	39.7°	43.1°	42.2°	42.1°

training and testing procedure. As mentioned in the implementation section, training procedure took about 12 seconds and testing procedure takes almost 40 second for testing 15 trials for different angles with 2.5 seconds of length. To measure error of the interface, mean of absolute error between the cue and the predicted result was measured for each session. The subject 1 was a trained subject who has experiences for this interface, and the subject 2 was a naïve subject with 3 hours of the instruction and practices procedure. As shown in Table 1, the subject 2 showed worse performances than the subject 1. The subject 1 reported a few tips that can improve the performance, such as "Increase the area of contacting surfaces." or "Touch by upside of the tongue.". However these tips were not beneficial for the subject 2. We are assuming that individual physiological conditions are related with these differences and personal adaptation is required to guarantee the performance. Furthermore, during the experiment for the subject 2, the severe drift was observed and the subject reported that the direction of the interface is slowly biasing to the single direction. To eliminate this interference, we reset the baselines of signals whenever trials are begun. We are still working to find better experimental conditions and more sophisticated model to solve above problems.

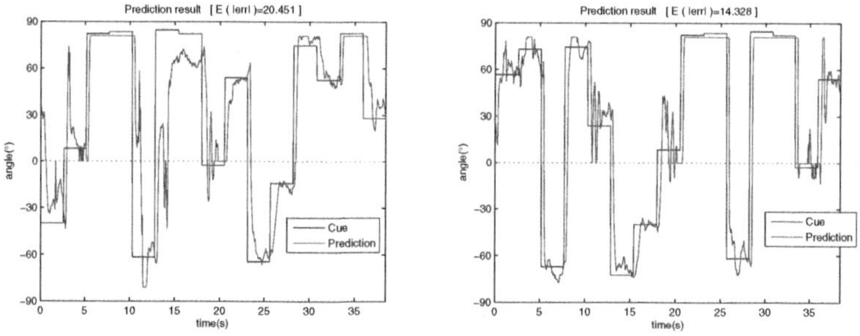

(a) Subject 1, session 1 (b) Subject 1, session 5

Fig. 4. Results for cue following experiments

5 Conclusions

We have proposed a novel interface paradigm based on glossokinetic potential. Until now, the glossokinetic artifact was a vague concept that includes all artifacts evoked by tongue-related tasks such as speaking, touching, licking or even swallowing. In this paper, we specified the contralateral glossokinetic potential that means potential change triggered by contact between tongue and cheek. We also showed that the patterns of GKP is linearly proportional to the tongue's direction with specific arrangement of EEG channels. This patterns can be used for designing new interface that can input the angle by the tongue. We hope that the interface has unique advantages over other tongue-based or EEG-based interfaces. Unlike known tongue-based interfaces, the system is able to obtain signals from outside of the mouth in completely noninvasive way, so it causes less discomfort. The interface enables analog control like stirring wheel, because the system uses subtle potential variations modulated by the tongue's position. Our future research topic is implementation of practical interface that can control external devices by tongue movements while optimizing above advantages. We already implemented this idea to control an electric wheelchair, but due to space limit, this is out of the scope of this paper.

Acknowledgments. This work was supported by National Research Foundation (NRF) of Korea (No. 2010-0018828 and 2010-0018829) and WCU Program (Project No. R31-2008-000-10100-0).

References

1. Krishnamurthy, G., Ghovanloo, M.: Tongue drive: A tongue operated magnetic sensor based wireless assistive technology for people with severe disabilities. In: Proceedings of the IEEE International Symposium on Circuits and Systems (ISCAS), Island of Kos, Greece, pp. 5551–5554 (2006)
2. Huo, X., Ghovanloo, M.: Using unconstrained tongue motion as an alternative control mechanism for wheeled mobility. IEEE Transactions on Biomedical Engineering 56(6), 1719–1726 (2009)
3. Struijk, L.N.S.A.: An inductive tongue computer interface for control of computers and assistive devices. IEEE Transactions on Biomedical Engineering 53(12), 2594–2597 (2006)
4. ThinkAMove: (Introduction to think-a-move's technology), `http://www.think-a-move.com/pdfs/Intro_to_TAM_Technology.pdf` (accessed 6/14/10)
5. Nutt, W., Arlanch, C., Nigg, S., Staufert, G.: Tongue-mouse for quadriplegics. Journal of Micromechanics and Microengineering 8, 155–157 (1998)
6. TongueTouchKeypad (Tonguetouch keypad^{TM}), `http://newabilities.com`
7. Salem, C., Zhai, S.: An isometric tongue pointing device. In: Proceedings of the ACM Conference on Human Factors in Computing Systems (CHI), pp. 538–359 (1997)
8. Klass, D., Bickford, R.G.: Glossokinetic potentials appearing in the electroencephalogram. EEG and Clinical Neurophysilology 12 (1960)
9. Fisch, B.J.: Fisch and Spehlmann's EEG Primer, 3rd edn. Elsevier, Amsterdam (1999)
10. Vanhatalo, S., Dewaraja, A., Holmes, M.D., Miller, J.W.: Topography and elimination of slow EEG responses related to tongue movements. NeuroImage 20, 1419–1423 (2003)

Practical Surface EMG Pattern Classification by Using a Selective Desensitization Neural Network

Hiroshi Kawata[1], Fumihide Tanaka[1], Atsuo Suemitsu[2], and Masahiko Morita[1]

[1] Graduate School of Systems and Information Engineering, University of Tsukuba
1-1-1 Tennodai, Tsukuba, Ibaraki, Japan
[2] School of Information Science, Japan Advanced Institute of Science and Technology
1-1 Asahidai, Nomi, Ishikawa, Japan
kawata@bcl.esys.tsukuba.ac.jp, fumihide@iit.tsukuba.ac.jp,
sue@jaist.ac.jp, mor@bcl.esys.tsukuba.ac.jp

Abstract. Real-time pattern classification of electromyogram (EMG) signals is significant and useful for developing prosthetic limbs. However, the existing approaches are not practical enough because of several limitations in their usage, such as the large amount of data required to train the classifier. Here, we introduce a method employing a selective desensitization neural network (SDNN) to solve this problem. The proposed approach can train the EMG classifier to perform various hand movements by using a few data samples, which provides a highly practical method for real-time EMG pattern classification.

Keywords: EMG Pattern Classification, Selective Desensitization Neural Network, Prosthetic Limb, Hand Movement Classification.

1 Introduction

Hands play an important role in our lives. The classification of hand movements by using surface electromyogram (EMG) signals is an important research issue in the development of prosthetic limbs. Although there is an extensive history of research in this field, the real-time robust implementation of this methodology is still practically very difficult [1,2]. First, because each hand movement is associated with multiple muscles, the surface EMG signal obtained from a sensor is the superposition of all the signals obtained from the related muscle activity; hence, complicating the correspondence relationship between movements and signals. Second, surface EMG signals are not reproducible, because there is a large difference between individuals, and even within a person the signals tend to fluctuate on every trial. As a result, in order for the existing approaches to work, the following conditions have been assumed:

- Collect sufficient data samples from the subject.
- Choose the number of sensors carefully in order to avoid redundancy, which often causes harmful effects while learning the data.

K.W. Wong et al. (Eds.): ICONIP 2010, Part II, LNCS 6444, pp. 42–49, 2010.

- Choose the positions of the sensors carefully.
- The subject needs to be trained in advance, so that he/she can deliver stable EMG signals.
- Preprocess the obtained signals carefully, so that the data-learning algorithm can produce a satisfactory EMG classifier.
- Extract suitable features from the data samples (for the same reason as above).

All these requirements made the real-time EMG pattern classification practically difficult to implement.

On the other hand, a selective desensitization neural network (SDNN) [3] performs significantly better in approximating a wide range of functions by using few training data samples. Therefore, in this paper, we will exploit the SDNN for the classification of the surface EMG pattern. In particular, we will apply this method to the problem of hand-movement classification, wherein real-time performance is crucial, particularly for prosthetic limbs.

2 Research Background

2.1 Electromyogram

Muscle contraction is triggered by the excitement of muscle fibers, which is invoked by a signal from the alpha motor neurons in the spinal cord. The electrical potential difference measured through the muscle contraction is called a myogenic potential, and its time-series signal is called an EMG. Since an EMG occurs 30–100 ms before the muscle contraction, it is considered theoretically possible to estimate the occurrence of the corresponding bodily movement from the EMG signals before the actual movement (muscle contraction) occurs.

For measuring the EMG signals, two types of electrodes can be used: needle electrodes and surface electrodes. The needle electrodes target specific muscle fibers and measure EMG signals with precision. However, they are accompanied with a physical pain to the subject, because the needle has to be inserted into the subject's skin. On the other hand, in the case of surface electrodes, there is little pain, as there is no needle insertion involved to measure the EMG signals. Instead, the electrical potential measured by the surface electrodes is a summation of the local electrical potentials, which makes the exact estimation of the corresponding bodily movement more difficult than that in the case of using needle electrodes.

In this study, we will use surface electrodes, considering the advantage and to try overcoming the disadvantage described above by introducing the SDNN.

2.2 Selective Desensitization Neural Network

The SDNN [3] is known to have overcome the several limitations of the multilayer perceptron, and to ably approximate a wide range of functions by using few training data samples. Here, we will be illustrating an example of approximating a function $y = f(x)$ by employing the SDNN, given a continuous-valued input vector $\boldsymbol{x} = (x_1, \ldots, x_m)$, where $m \geq 2$.

The input layer of the SDNN consists of m neuronal groups $(G_1^1, \ldots, G_m^1)$. Each group is composed of n neurons and represents an input variable x_μ, i.e., the input variable is represented in a distributed manner by the activity patterns of the neurons. Then, the middle layer of the SDNN consists of $m(m-1)$ neuronal groups $G_{\mu,\nu}^2$ $(\mu, \nu = 1, \ldots, m; \mu \neq \nu)$. The neurons in $G_{\mu,\nu}^2$ are connected with both the neurons in G_μ^1 and G_ν^1 $(\mu \neq \nu)$, in the input layer. This realizes a procedure called desensitization, which neutralizes the output of the neuron regardless of its input and inner potential. For example, if a neuron is configured to output either 1 or -1 with equal probabilities as its default output, it will output 0 in the case that the neuron is desensitized. Finally, the output layer of the SDNN consists of n' neurons, each of which is connected with all the neurons in the middle layer. The output of the i-th neuron in the output layer is calculated by

$$y_i = g\left(\sum_{\mu,\nu(\neq\mu)} \sum_{j=1}^{n} \omega_{ij}^{\mu,\nu} x_j^{\mu,\nu} - h_i \right), \tag{1}$$

where h_i is a threshold, $\omega_{ij}^{\mu,\nu}$ is a synaptic weight from the j-th neuron of $G_{\mu,\nu}^2$ in the middle layer, and $g(u)$ is the activation function, where $g(u) = 1$ for $u > 0$ and 0 for $u \leq 0$.

Learning of this network is performed using a target vector $\boldsymbol{p} = (p_1, \ldots, p_{n'})$. The threshold and the synaptic weights between the middle layer and the output layer are specifically updated by

$$\omega_{i,j}^{\mu,\nu} \leftarrow \omega_{i,j}^{\mu,\nu} + c(p_i - y_i)x_j^{\mu,\nu}, \tag{2}$$
$$h_i \leftarrow h_i - c(p_i - y_i), \tag{3}$$

where c is a learning coefficient.

3 Methods

3.1 Signal Measurement

Personal-EMG (Oisaka Electronic Device Ltd. [4]) equipment is used to measure the surface EMG signals. This can measure the integral of the EMG signal (IEMG) and the original EMG at the same time. In this study, the EMG and IEMG signals are sampled at 3 kHz by using a 12-bit A/D converter, and for the classification of hand movements, an IEMG signal is used, which is low-pass filtered with a cut-off frequency of 4.8 Hz.

Regarding the myoelectric sensors, we use 10 pairs of wet-electrodes, which are pasted around the subject's right arm (Fig. 1). The sensors target the following six muscles: flexor carpi radialis, flexor digitorum profundus, flexor carpi ulnaris, extensor digitorum, flexor carpi long radialis, brachioradialis, and biceps brachii [5]. However, the sensors do not need to be positioned accurately.

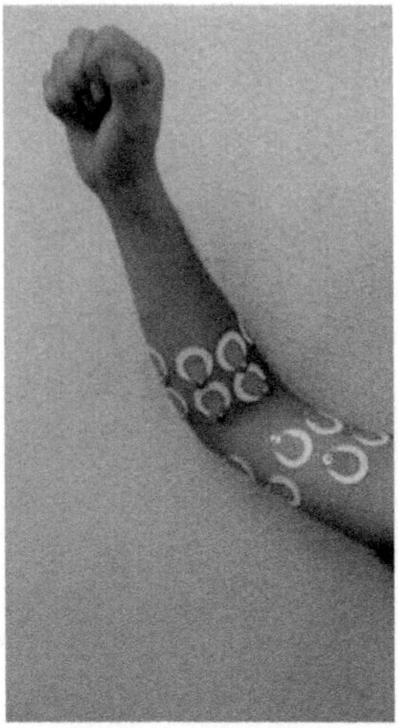

Fig. 1. Placement of wet electrodes

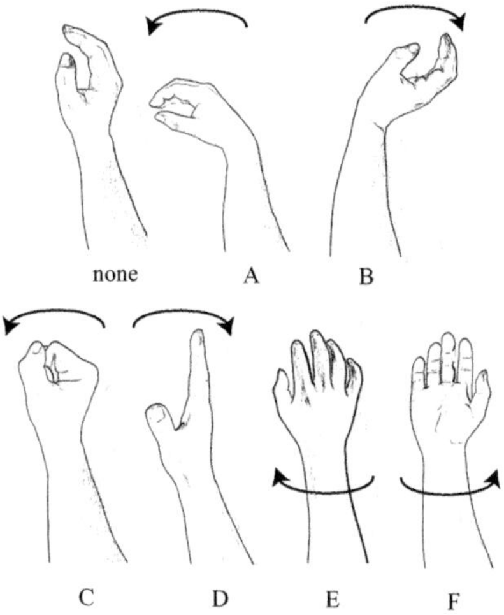

Fig. 2. Seven categories to be classified from the EMG signals [6]

3.2 Target Hand Movements

In this study, six hand movements (wrist flexion, wrist extension, grasping, opening up, wrist supination, and wrist pronation) and no-movement conditions are targeted for the classification. In the following sections, we denote the no-movement condition by "basic position (none)," wrist flexion by "movement-A," wrist extension by "movement-B," grasping by "movement-C," opening up by "movement-D," wrist supination by "movement-E," and wrist pronation by "movement-F" (Fig. 2).

3.3 Preprocessing of IEMG Signals

Preprocessing is performed to handle the IEMG signals with the SDNN (Fig. 3). First, each IEMG signal is normalized by the maximum value at each channel, and the normalized IEMG signals are then normalized again by the maximum value at each time step. Next, each IEMG channel is connected to a neuronal group in the input layer of the SDNN, and each neuronal group is composed of multiple neurons, as described in the previous section. In consequence, we code the value of the IEMG signal so that only 50% of the neurons can be in a continual excited state, and the pattern of excitement can depict the continuous change in the IEMG value consecutively (Fig. 4).

3.4 Learning of the SDNN

The internal structure of the SDNN is shown in Fig. 5. In this study, the input layer of the SDNN is composed of 360 neurons: 300 neurons for 10 IEMG channels, 30 neurons for the total value of all the IEMG signals, and 30 neurons

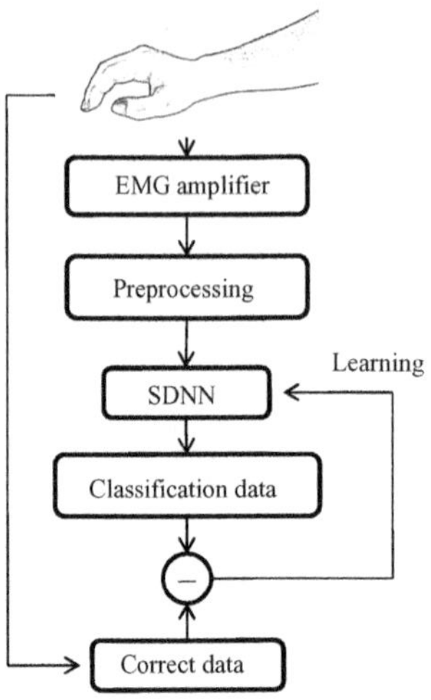

Fig. 3. Training process of the proposed system

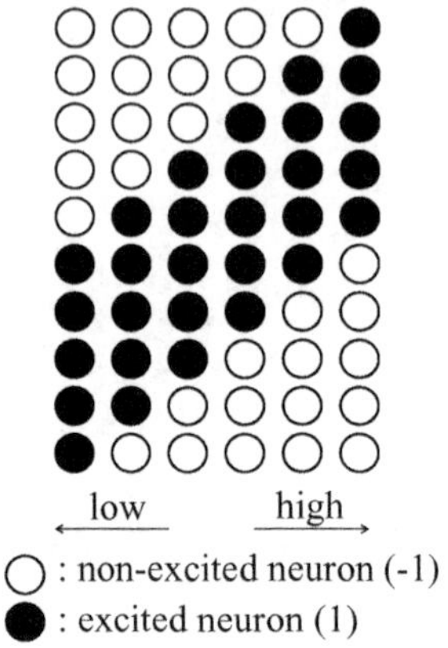

Fig. 4. Distributed coding of IEMG signals

for the difference of the total IEMG value from a step in the past. There are two middle layers composed of a total of 6600 neurons: in the first layer, half of the neurons are desensitized by the corresponding neurons in the input layer, except for the neurons representing the total value of the IEMG signals; and in the second layer, the desensitization procedure is repeated by the neurons representing the total value of the IEMG signals. The output layer is composed of six neurons, each of which corresponds to the classifier of each movement.

In the learning cycle, we train the SDNN by supplying the preprocessed input signals greater than the noise threshold and the target patterns representing the corresponding movement. The synaptic weights from the middle layer to the output layer and the thresholds in the output layer are specifically modified according to Eqs. (2) and (3). Here, the training is repeated 10 times and the learning coefficient c is set to 0.1.

3.5 Evaluation of Classification

In order to evaluate the classification ability of the proposed system after learning, we define a classification rate for each movement as follows. First, a test data sample is fed into the system and movement detection is performed in every frame. Second, if any movement has been detected more than six times, the test data sample is classified into the movement detected most frequently; otherwise, it is classified into "none". Third, we judge whether the classification is correct or not. For example, the classification is regarded as correct if the classified movement is the same as that corresponding to the test data sample. Finally, we apply this procedure to all the test data samples corresponding to the same movement, and then calculate the classification rate as a percentage of the number of correct classifications to the total number of the test data samples.

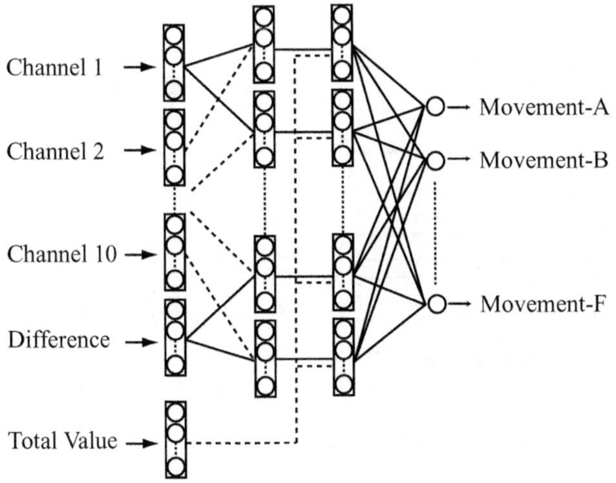

Fig. 5. Structure of SDNN

4 Experiment

To collect the IEMG data, five male subjects are asked to execute one movement for 2 s and repeat all the six movements (Fig. 2), three times in the same order in a session. The session is repeated nine times, which provides the total number of data samples.

After the measurement, a cross validation is performed to calculate the final classification rate: (1) pick up one session data (which contains three data samples for every movement) to train the SDNN whose classification rates are calculated by using the other eight session data as test data, (2) repeat it by changing the training data samples for all combinations, and (3) compute the total average as the final classification rate.

Figure 6 shows an example of the IEMG signals obtained from one subject when the subject performs the six movements. Each line corresponds to the signal from a channel, and each shaded box represents the movement that is labeled. From this figure, it can be seen that the signals are very unstable and fluctuate at every trial.

Figure 7 plots the final classification rate for each movement of each subject. The average classification rates over the six movement categories are (s1) 86.73%, (s2) 100.00%, (s3) 92.44%, (s4) 97.15%, and (s5) 100.00%. The total average classification rate over the five subjects and the six movements is 95.26%.

Figure 8 shows an example of the total value of the IEMG signals together with the outputs (classified movements) of six neurons in the output layer of the SDNN. The shaded regions represent the movements classified by the SDNN.

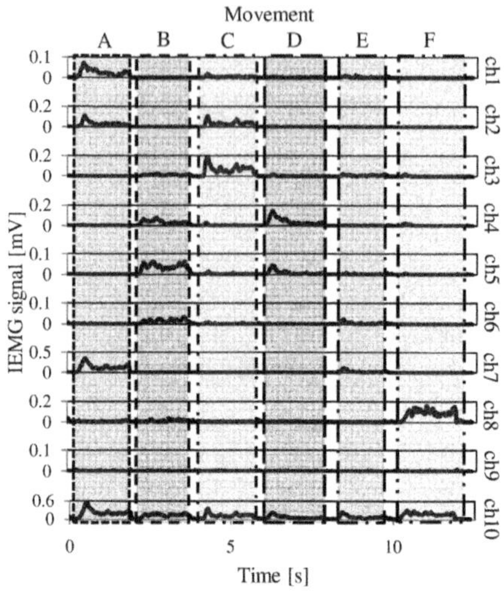

Fig. 6. Example of IEMG signals from a subject

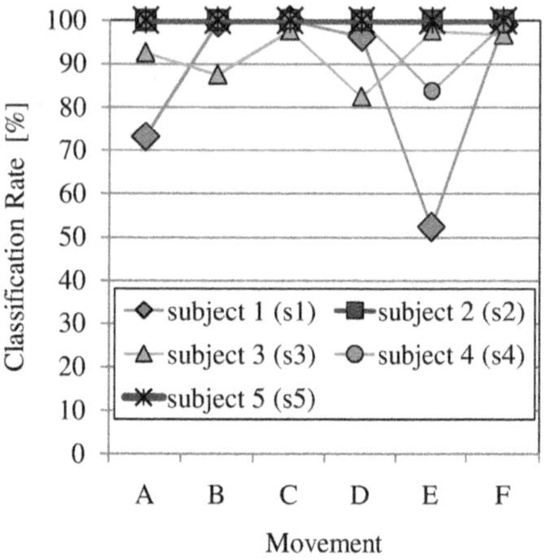

Fig. 7. Final classification rate for movements A–F for five subjects

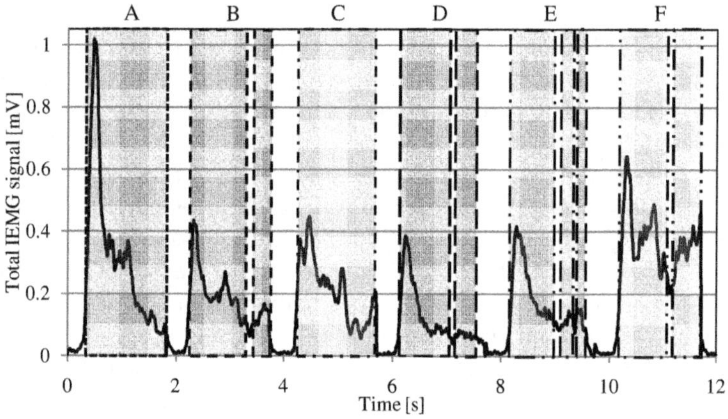

Fig. 8. Example of real-time classification of six movements

Each classification is computed together with an increase in the total IEMG value, implying that the real-time classification is achieved (see the video at [7]).

5 Conclusion

By introducing the SDNN into the pattern classifier, the real-time pattern classification of multiple hand movements was presented. The experimental results from the five human subjects showed that only three training data samples for each movement are sufficient for the proposed system to output a classification accuracy of >95% (average) for the six targeted hand movements. This approach is considered to be more practical than the existing methods for the following reasons:

- It does not require large number of training data samples to obtain a good classifier.
- It does not require the user to position sensors on optimal locations.
- It does not require complicated preprocessing of the signal data.
- It does not require the subject to be trained or to be given detailed instructions in advance.

Future work includes more detailed analyses on both the number of training data samples and sensors. Furthermore, because the SDNN exhibits high performance in approximating a wide range of functions, it is considered to be able not only to classify the categories of movements but also to estimate the speed/force of each movement.

Acknowledgments

This study was supported by a Grant-in-Aid for Scientific Research on Priority Areas (21013007) and a Grant-in-Aid for Young Scientists (B) (21700576) from the Ministry of Education, Culture, Sports, Science and Technology of Japan (MEXT), and a Grant-in-Aid for Scientific Research (B) (22300079) from the Japan Society for the Promotion of Science (JSPS).

References

1. Tsuji, T., Fukuda, O., Bu, N.: New Developments in Biological Signal Analysis (in Japanese). Journal of the Japan Society of Applied Electromagnetics 13(3), 201–207 (2005)
2. Yokoi, H., Chiba, R.: Now and Future of Cyborg Technology (in Japanese). Journal of the Society of Instrument and Control Engineers 47(4), 351–358 (2008)
3. Morita, M., Murata, K., Morokami, S., Suemitsu, A.: Information Integration Ability of Layered Neural Networks with the Selective Desensitization Method (in Japanese). The IEICE Transactions on Information and Systems J87-D-II(12), 2242–2252 (2004)
4. Oisaka Electronic Device Ltd., `http://www.oisaka.co.jp/P-EMG.html`
5. Fujita, T.: Guide Anthropotomy (in Japanese), pp. 88–92. Nankodo (2003)
6. Yoshikawa, M., Mikawa, M., Tanaka, K.: Real-Time Hand Motion Classification and Joint Angle Estimation Using EMG Signals (in Japanese). The IEICE Transactions on Information and Systems J92-D(1), 93–103 (2009)
7. Real-Time Classification of Multiple Hand Movements, `http://volga.esys.tsukuba.ac.jp/~kawata/demovideo/demovideo.wmv`

Reliability-Based Automatic Repeat reQuest with Error Potential-Based Error Correction for Improving P300 Speller Performance

Hiromu Takahashi, Tomohiro Yoshikawa, and Takeshi Furuhashi

Nagoya University, Furo-cho, Chikusa-ku, Nagoya 464-8603, Japan

Abstract. The P300 speller allows users to select letters just by thoughts. However, due to the low signal-to-noise ratio of the P300 response, signal averaging is often performed, which improves the spelling accuracy but degrades the spelling speed. The authors have proposed *reliability-based automatic repeat request* (RB-ARQ) to ease this problem. RB-ARQ could be enhanced when it is combined with the error correction based on the error-related potentials. This paper presents how to combine both methods and how to optimize parameters to maximize the performance of the P300 speller. The result shows that the performance was improved by 40 percent on average.

Keywords: Brain-computer interface, P300 speller, automatic repeat request, error-related potential.

1 Introduction

Brain-computer interfaces (BCIs) are promising technologies to restore control and communication to severely paralyzed people, and appealing to healthy people as well [10]. The P300 speller is one of the BCI applications, which allows users to select letters just by thoughts [5]. However, due to the low signal-to-noise ratio of the P300 response, signal averaging is often performed, which improves the spelling accuracy but degrades the spelling speed. The authors have proposed *reliability-based automatic repeat request* (RB-ARQ) to ease this problem [9], i.e., one can spell letters faster while the accuracy remains at a high level. Meanwhile, it was reported that the error-related potentials (ErrPs) elicited when a user notices some error could be used for error correction in BCIs [6,3]. This suggests that RB-ARQ could be enhanced when combined with the ErrP-based error correction (hereafter referred to as error correction).

In RB-ARQ, there is a trade-off between the accuracy and the speed, and they are controllable by changing the threshold for the repeat requests; thus there exists such a threshold that balances them and maximizes *Utility* [2], which is a performance measure of the P300 speller calculated using the accuracy and the speed. On the other hand, it was reported that higher accuracy of the fed back results led to larger amplitude of ErrPs [6], which would result in a greater discriminability of ErrPs and accordingly an even higher spelling accuracy in the

K.W. Wong et al. (Eds.): ICONIP 2010, Part II, LNCS 6444, pp. 50–57, 2010.

P300 speller. Thus, such a threshold that maximizes the Utility when RB-ARQ is solely applied does not always lead to the maximum Utility when combined with the error correction. Meanwhile, Dal Seno et al. [3] conducted an on-line P300 speller experiment with the error correction. However, the relation between the Utility and the performance measures for ErrPs detection, i.e., the true-positive (TP) and true-negative (TN) rates, was not taken into account; thus, the TP and TN rates were not necessarily tuned to maximize the Utility. Therefore, the present study has three purposes: firstly to investigate the relation between the spelling accuracy and the discriminability of ErrPs, secondly to show how to tune the TP and TN rates to maximize the Utility, and lastly to evaluate the combination of RB-ARQ with the error correction.

2 Methods

2.1 P300 Speller

This paper utilizes the P300 speller in the BCI2000 [8] with an interface as in Fig. 1(a), which allows spelling a letter per trial. A trial proceeds as follows: subjects are given 3 s to turn their gaze on the target letter, and then the stimulus presentation, i.e., successive and random intensifications of each row and column with a stimulus duration of 100 ms and an inter-stimulus interval of 75 ms, follows. Subjects are instructed to count how many times the target flashed. Only the EEG corresponding to the target should contain the P300 response. Each sequence consists of 12 flashes, i.e., 6 rows and 6 columns, and 5 sequences are repeated per letter. A result is presented for 1 s immediately after the stimuli finishes as in Fig. 1(b), which enables seeing the result without gaze shift whichever the target is. However, only in the first two experiments involving subject A and B, a fixation point appeared at the center of the monitor after the stimuli and it was replaced by the result 1 s later. Note that the white square in the bottom left appears synchronously with the result presentation so that a photo-sensor attached to the monitor can detect the exact timing of the result presentation, which is delayed by roughly 30 ms due to a display lag. In addition, a typical number of sequences per letter is 15; however it is set to be 5 to obtain sufficient ErrP samples.

2.2 Proposed Method

Automatic repeat request (ARQ) is an error control scheme, in which the receiver asks the sender for re-transmission on error detection, and reliability-based automatic repeat request (RB-ARQ) [9] is a variant of ARQ, which employs a reliability of a classification result, i.e., the maximum posterior probability, as the repeat criterion. Specifically, the stimulus presentation continues until the maximum posterior probability λ_i, calculable after ith sequence, is larger than an arbitrary threshold λ. A large threshold λ leads to a better accuracy p and more sequences per letter, i.e., a longer trial duration d (see details in [9]). As a result, RB-ARQ makes BCIs faster and more accurate than the standard averaging.

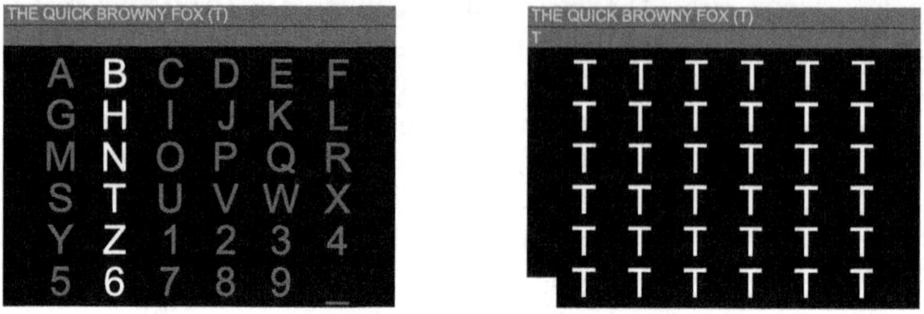

(a) Stimulus presentation (b) Result presentation

Fig. 1. Illustration of the user interface used in this experiment

ErrP-based error correction rejects the selected letter on detection of ErrPs, consequently the accuracy p can be improved. This paper proposes to combine RB-ARQ with the error correction, in which the stimulus presentation continues until the stopping criterion is satisfied, and the selected letter is rejected on detection of ErrPs.

2.3 Performance Evaluation

The discriminability [4] defined in (1) is utilized to measure how separable the ErrP and the non-ErrP are,

$$d' = \frac{|\mu_0 - \mu_1|}{\sigma} , \tag{1}$$

where μ_0, μ_1, and σ respectively denote the mean discriminant score [7] of non-ErrP, that of ErrP, and the standard deviation of them, in this paper.

The Utility [2] defined in (2) is used to evaluate the performance of the speller because it is said to be a more practical measure than the popular information transfer rate (ITR) [10],

$$U = \frac{(2p-1)}{d} \log_2(N-1) , \tag{2}$$

where N is the number of selectable letters, but $U = 0$ if $p \leq 0.5$. RB-ARQ can maximize U by adjusting the threshold λ, on which p and d depend. By contrast, when the error correction is applied, the Utility becomes

$$U' = \frac{pc-(1-p)(1-e)}{d} \log_2(N-1) , \tag{3}$$

where e and c respectively denote the TP and TN rate, i.e., the probability of correct classification of EEGs after erroneous result presentation and that after correct result presentation, respectively; but $U' = 0$ if $pc \leq (1-p)(1-e)$. Apparently U' is a function of p, e, and c.

2.4 Utility Maximization

Now let us consider how to maximize U'. To this end, the receiver operating characteristic (ROC) curve [4] is utilized, because p is thought to depend on e

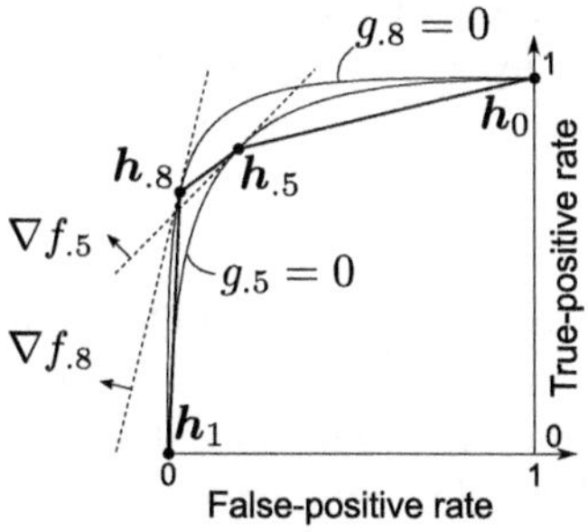

Fig. 2. ROC curves (thin lines), optimum e and c (filled circles) for Utility, and linear interpolation of optimum points (thick line)

and c as described just bellow, but their mathematical relation is unknown. The ROC curve is a plot of e versus $1 - c$, as the decision threshold for ErrPs varies. It passes points of $(0, 0)$ and $(1, 1)$, and the greater d' is, the more top left corner it approaches; thus, e and c depend not only on the decision threshold but also on d', and on p assuming that d' is proportional to p.

Certain p and d' are obtained given a threshold λ in RB-ARQ, then let $g_p(e, c) = 0$ be the resultant ROC curve and $f_p(e, c)$ be the Utility function (3) given p. Consequently, the maximization of U' given λ is expressed as follows:

$$\max_{(e,c)} f_p(e, c) \quad \text{subject to} \quad g_p(e, c) = 0 \ . \tag{4}$$

The solution of (4), denoted by $\boldsymbol{h}_p$, yields a local maxima under a certain λ; thus, the global maxima can be found by tuning λ. Figure 2 illustrates how to find $\boldsymbol{h}_{.5}$ and $\boldsymbol{h}_{.8}$ from a geometrical viewpoint, which tells that e and c are not always equal in order to maximize the Utility. It is worth noting that $\boldsymbol{h}_p$ converges to $(0, 0)$ and $(1, 1)$ when p approaches 1 and 0, respectively, though ROC curves in both cases are not plottable. Suppose $h(e, c) = 0$ denotes the curve that $\boldsymbol{h}_p$ moves along, then the polygonal line $\tilde{h}(e, c) = 0$ connecting $\{\boldsymbol{h}_0, \boldsymbol{h}_{.5}, \boldsymbol{h}_{.8}, \boldsymbol{h}_1\}$ is a underestimation of $h(e, c) = 0$. Therefore, in this paper, an arbitrary $\boldsymbol{h}_p$ is approximately obtained by an interpolation using $\tilde{h}(e, c) = 0$.

3 Experiment

3.1 Data Collection

Five male volunteer subjects: A, B, C, D, and E in their early 20's participated in this experiment. Their EEGs were recorded from Fz, C3, Cz, C4, and Pz referenced to the linked-ears with the sampling rate of 1000 Hz using a Polymate AP216 (DIGITEX LAB. CO., LTD, Tokyo, Japan). A session consisted of spelling predetermined twenty target letters, or twenty trials, and each subject performed eleven sessions. The results were selected independently from his or her EEGs and presented except for the first session. Namely they were simply the same as the target with a given accuracy: $p = 0.8$ in a randomly chosen half and $p = 0.5$ in the rest, to investigate the relation between the spelling accuracy and the discriminability of ErrPs, which the users were not informed of.

3.2 Data Processing

In the off-line waveform analysis, their EEGs were down-sampled to 100 Hz and filtered with a pass-band of 1 Hz to 10 Hz since both the P300 and the ErrPs are relatively slow potential changes. The EEGs after the result presentation were extracted with a time-window of 500 ms, and a threshold rejection of $\pm 50\ \mu V$ was applied, not to include those contaminated by artifacts. In the off-line classification analysis, the rejection was not applied, and the data were further down-sampled to 20 Hz, and a time-window from 200 ms to 500 ms was used. The linear discriminant analysis (LDA) was employed, and the classifier was trained separately for each subject. The ten-fold cross validation [7] has been performed to estimate the accuracy of ErrPs, i.e., e and c, and this procedure was repeated with different decision thresholds to draw a ROC curve for each subject. The P300 data were similarly collected except that a time-window of 650 ms was used instead. The datasets of the first two sessions were used to train a classifier, which was different for each subject. The rest nine were used for evaluation: successive three sessions were assumed to be a single session so that the results could be comparable to the conventional studies. Hence, the number of sequences was 15 and that of spelled letters was 60.

4 Results and Discussions

Figure 3 illustrates the grand average error-minus-correct difference waveforms at Cz after the result presentation. This figure shows that the waveforms had three peaks at roughly 200 ms, 290 ms, and 410 ms, respectively; and the 410-ms component was observably larger at $p = 0.8$ than at $p = 0.5$. Taking the display lag of around 30 ms into account, the characteristics of these peaks are consistent with [6]. Additionally, Fig. 4 shows the average discriminabilities calculated by (1), telling that the discriminability was larger at $p = 0.8$; however, they were not statistically significantly different at the significance level of 5% (the p-value was 0.076), according to the paired t-test. Nonetheless, considering the proportionality of the amplitude of ErrPs to the accuracy, shown by earlier studies including [6], and the number of samples, we could conclude that the discriminability d' is also proportional to the accuracy p.

The spelling accuracies on the basis of the 15 sequences per trial for the five subjects were 95.0%, 98.3%, 83.3%, 91.7%, and 91.7%, respectively; but remember that these accuracies did not affect the result presentation. These accuracies can compare with the result of the data set II of the BCI competition III, where several competitors achieved the accuracy of over 90 percents [1]; thus, the attained accuracies are satisfactory enough to proceed to a further analysis.

Figures 5(a) and 5(b) show the obtained ROC curves for subject B and C, whose spelling accuracies were the best and the worst, respectively. Basic

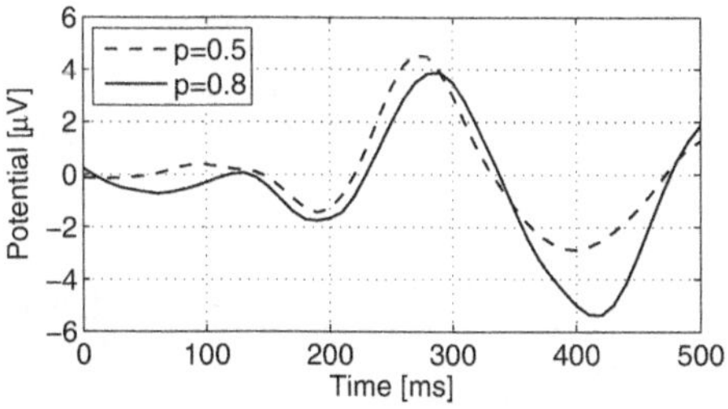

Fig. 3. Grand average waveforms of ErrPs

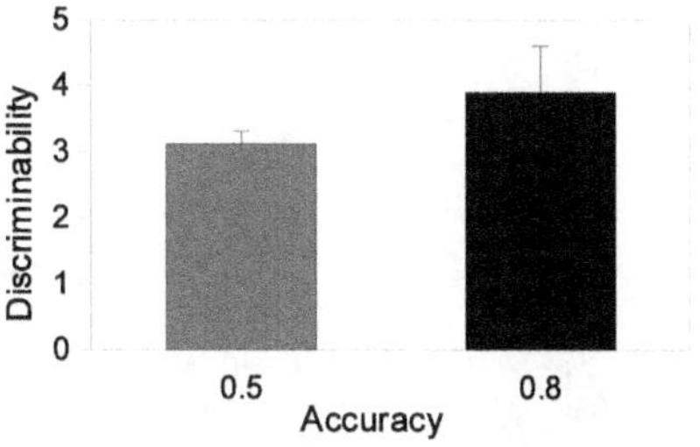

Fig. 4. Discriminability of ErrPs

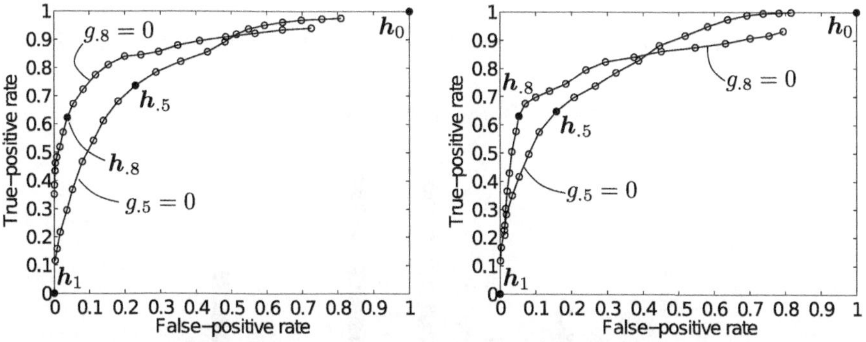

(a) Subject B (b) Subject C

Fig. 5. ROC curves and optimum points $\boldsymbol{h}_p$ $(p = 0, 0.5, 0.8, 1)$

characteristics of these figures, e.g., $g_{.8} = 0$ approaches the $(0, 1)$ point closer than $g_{.5} = 0$, are in agreement with those of the expected figure as in Fig. 2. Figures 6(a) and 6(b) show the performance curves of subject B and C, respectively. Note that when RB-ARQ is applied, the number of sequences varies from trial to trial, and the average number depends on the threshold λ; thus, such a λ_i $(i = 1, 2, \ldots, 15)$ was set that the average became i. Also note that when the error correction was applied, the Utility was calculated by using (3), where approximate e and c obtained from the interpolation were used. Figures 6(a) and 6(b) tell that the Utility reached the maximum at a certain average number of sequences, i.e., a certain threshold λ_i in the case of RB-ARQ, and that the proposed combination yielded the best Utility in both cases. In addition, when the Utility was maximized in the case of the combination, the accuracies p were 78.3% and 71.7% for subject B and C, respectively, and the average over all subjects was $80.0 \pm 6.0\%$; thus, TP rate e and TF rate c for the maximum Utility were tolerant of the approximation error caused by the interpolation. Moreover, Fig. 7 shows the performance gain of each method from Averaging: the gain was defined as the rate of increase of the maximum Utility obtained by each method to that obtained by Averaging, assuming that in practice the system operator would select the best performance in each method. This figure

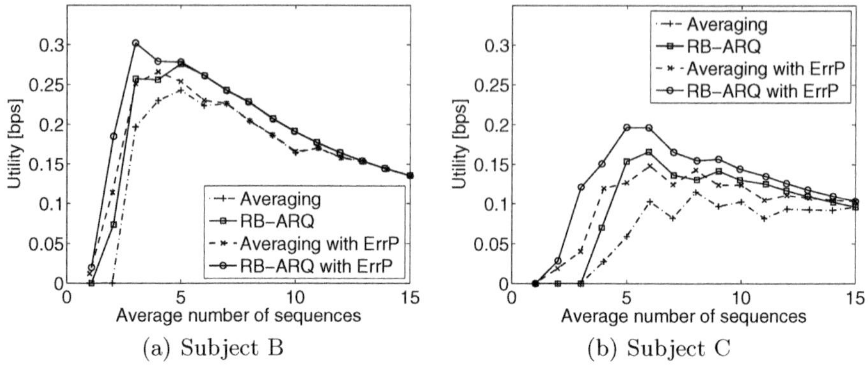

Fig. 6. Performance curves of the Utility versus the number of sequences

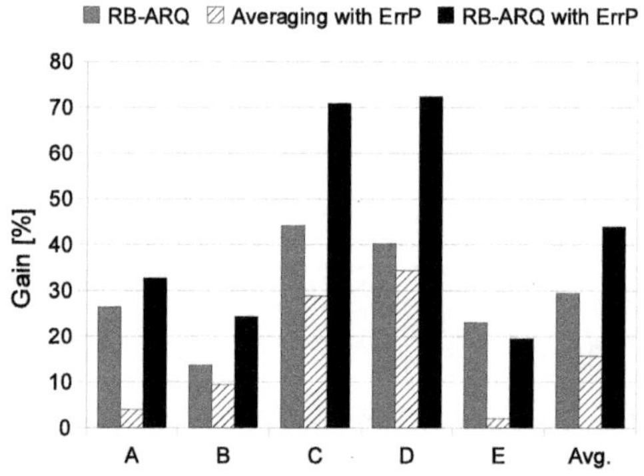

Fig. 7. Performance gain from Averaging

tells that RB-ARQ improved the Utility by 30% on average and the combination further improved it by up to over 40% on average.

However, these results are conditional on the ROC curves estimated by the cross-validations and on the classifier for the P300 trained using the datasets recorded in the same day, also on the fact that the successive three sessions were assumed to be a single session; thus, further analyses and experiments are necessary before drawing a final conclusion.

5 Conclusion

This paper proposed to combine RB-ARQ with the ErrP-based error correction, and showed how to tune the true-positive and true-negative rates of ErrP detection to maximize the performance of the P300 speller. To evaluate the proposed method, this paper conducted the P300 speller experiment where five subjects participated. The results showed that the discriminability of ErrPs was proportional to the spelling accuracy, and that the proposed combination method yielded the best performance, which was greater than the standard averaging by 40% on average. However, these results are conditional; therefore, further analyses and experiments are necessary.

Acknowledgements

This work was supported by the Grant-in-Aid for JSPS Fellows No. 22-8417 and for Scientific Research (C) No. 22500200 from the Ministry of Education, Culture, Sports, Science and Technology (MEXT), Japan.

References

1. Blankertz, B., Müller, K.R., Krusienski, D.J., et al.: The BCI competition III: validating alternative approaches to actual BCI problems. IEEE Trans. Neural Syst. Rehab. Eng. 14(2), 153–159 (2006)
2. Dal Seno, B., Matteucci, M., Mainardi, L.: The utility metric: A novel method to assess the overall performance of discrete brain-computer interfaces. IEEE Trans. Neural Syst. Rehab. Eng. 18(1), 20–28 (2009)
3. Dal Seno, B., Matteucci, M., Mainardi, L.: On-line detection of p300 and error potentials in a BCI speller. Computational Intelligence and Neuroscience 2010, Article ID 307254, 5 pages (2010)
4. Duda, R.O., Hart, P.E., Stork, D.G.: Pattern Classification. New Technology Communications, 2nd edn. (2001)
5. Farwell, L.A., Donchin, E.: Talking off the top of your head: toward a mental prosthesis utilizing event-related brain potentials. Electroencephalogr. Clin. Neurophysiol. 70(6), 510–523 (1988)
6. Ferrez, P.W., Millan, J.d.R.: Error-related EEG potentials generated during simulated brain-computer interaction. IEEE Trans. Biomed. Eng. 55(3), 923–929 (2008)
7. Hastie, T., Tibshirani, R., Friedman, J.: The Elements of Statistical Learning: Data Mining, Inference, and Prediction, 2nd edn. Springer, Heidelberg (2009)
8. Schalk, G., McFarland, D., Hinterberger, T., et al.: BCI2000: a general-purpose brain-computer interface (BCI) system. IEEE Trans. Biomed. Eng. 51(6), 1034–1043 (2004)
9. Takahashi, H., Yoshikawa, T., Furuhashi, T.: Application of support vector machines to reliability-based automatic repeat request for brain-computer interfaces. In: Proc. 31st Annual Int. Conf. IEEE EMBS, pp. 6457–6460 (2009)
10. Wolpaw, J.R., Birbaumer, N., Heetderks, W.J., et al.: Brain-computer interface technology: a review of the first international meeting. IEEE Trans. Rehab. Eng. 8(2), 164–173 (2000)

An Augmented-Reality Based Brain-Computer Interface for Robot Control

Alexander Lenhardt and Helge Ritter

Bielefeld University, CITEC,
Universitaetsstr. 2123, 33615 Bielefeld, Germany
{alenhard,helge}@techfak.uni-bielefeld.de
http://www.cit-ec.de

Abstract. In this study we demonstrate how the combination of Augmented-Reality (AR) techniques and an asynchronous P300-based Brain-Computer Interface (BCI) can be used to control a robotic actuator by thought. We show results of an experimental study which required the users to move several objects placed on a desk by concentrating on a specific object. Competitive communication speed of up to 5.9 correct symbols per minute at a 100% accuracy level could be achieved for one subject using an asynchronous paradigm which enables the user to start communicating a command at an arbitrary time and thus mitigating the drawbacks of the standard cue based P300 protcols.

Keywords: P300 BCI asynchronous speller augmented-reality robot.

1 Introduction

Recent years have brought up a tremendous growth of contributions to the field of brain-computer interfaces (BCI). In general, BCI allow to infer mental commands of the user by measuring their brain potentials with *electroencephalography (EEG)*. Such a system enables a direct communication between a computer and the user without using any motor function as it is required for speaking, eye-tracking or using a mouse or keyboard. Nowadays the application of BCIs, once intended for people with severe motor disabilities , spans the full range of medical to entertainment scenarios.

Two popular strategies are employed in modern BCIs which control artificial actuators. As reviewed by [7], control strategies can be distinguished by *Process Control* and *Goal Selection* (see figure 1). Thereby, process control strategies aim at direct control of the available motors or muscles which result in an action while goal selection approaches focus on recognizing the intent of a user. The intent serves as input for an execution unit which translates it to a sequence of motor actions. Obviously, the goal selection approach does not require the user to achieve full control over the output device via brain signals. A rather simple *goal selection* scheme would be sufficient which requires only a one dimensional selection of a target symbol corresponding to the desired action while the process control approach would require the user to be able to control at least 3 DOF

K.W. Wong et al. (Eds.): ICONIP 2010, Part II, LNCS 6444, pp. 58–65, 2010.

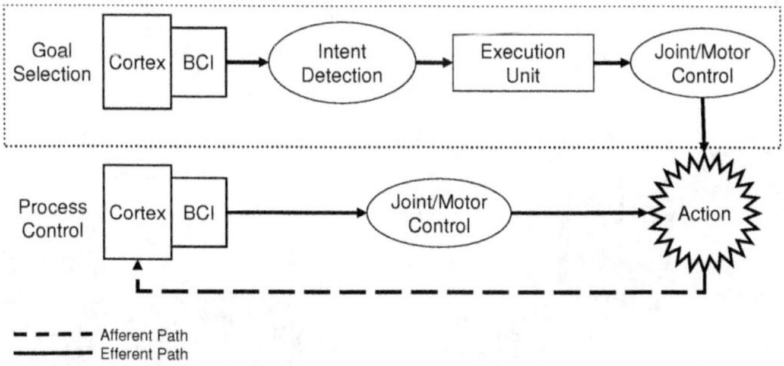

Fig. 1. Depicted are the two main approaches when using BCIs to control a robotic device. Process control establishes a closed feedback-loop between action and the brain and offer full control over the robot. Goal selection approaches are easy to use since actual motor control is delegated to autonomous subsystems.

to navigate freely in space. With goal selection approaches, actual control is delegated to system subcomponents (e.g. robot controller) which do not require control via higher cognitive functions. This reflects the normal output pathways of the brain and its underlying systems for motor control. Thus, it can be stated that goal selection approaches are currently the most natural way to control robotic output devices by brain signals. A popular type of goal-selection BCIs is the P300-Speller paradigm [2] which presents visual stimuli in rapid succession and random order. The subjects are required to attend to a specific stimulus and mentally count whenever the focused stimulus appears. The appearance rate of this *relevant stimulus* is low compared to the other *background stimuli*. Whenever the relevant stimulus appears, a positive deflection in the EEG occurring at 300ms after stimulus onset can be observed. This component, called P300, is utilized by P300 BCIs to predict the user's selection. A drawback of most goal-selection systems (primarily P300-based BCIs) is the assumption that the user is trying to communicate with the BCI at all times. While this assumption makes it possible to easily implement the selection method via visual or auditive cues that signal the start of a selection round (i.e. trial), it is an unreasonable constraint for a practical application. Even though P300-based BCI require external stimulation, and thus are reliant on cues to some degree, it is possible to omit the trial-based nature of these paradigms. A system which is able to continuously present stimuli and detect whether a user is currently trying to communicate with the system is considered an *asynchronous BCI.* Operating in an asynchronous mode is vital for any practical application involving control of robotic actuators since it allows to intervene or start communicating with the system at any time.

Recent research has brought up new methods for stimulus dependent BCIs to achieve this kind of behavior. In [6] the control of a wheelchair is realized by using buttons on a screen as selection targets for a P300 BCI. Each button is associated with a preprogrammed path the wheelchair should take. Their system features

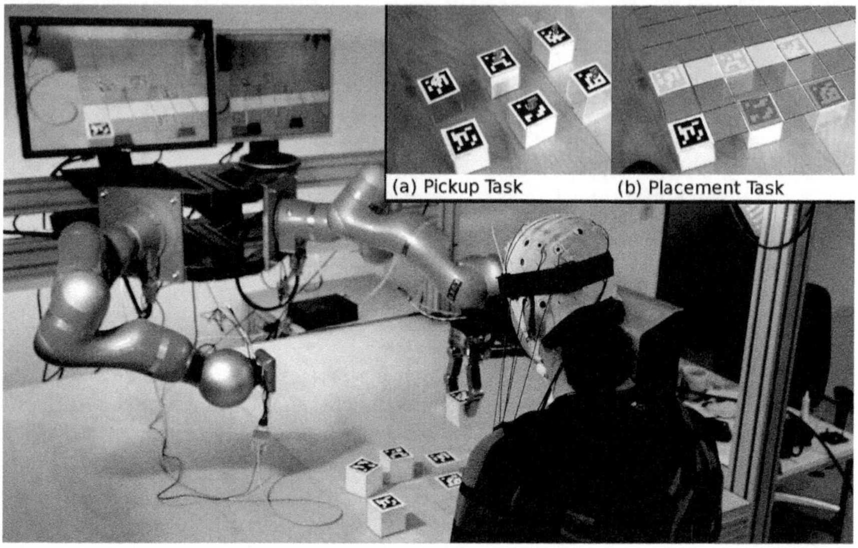

Fig. 2. Scene setup with robot arm and marker objects. (a) Scene as viewed through the HMD during the *pickup* task. (b) Scene during the *placement* task.

an asynchronous protocol that allows the subject to issue commands at any time. Similar work related to the detection of *no-control* states has been presented by [8]. Instead of exploiting the distribution of class scores, they proposed a method which computes a probabilistic model of EEG data during no-control periods.

In this paper we present an asynchronous P300 BCI designed to grasp and place arbitrary objects using a robot arm with an attached gripper. A method to dynamically adapt the stimulus content which represents the physical scene in front of the subject using augmented-reality techniques will be described. Further, an extended version of the asynchronous protocol as presented by [6] will be evaluated. A preliminary study with 4 healthy subjects was conducted to test the usability of the system.

2 Augmented-Reality BCI Design

The BCI setup consists of a Kuka robot arm with an attached Schunk SDH-1 gripper, a stereoscopic video see-through head-mounted display (HMD) with two integrated firewire cameras and a g.tec gUSBAmp 16 channel EEG amplifier. Five plastic cubes with attached markers on the upper side were placed on a desk. When the scene is observed through the HMD, 3D models of numbers are augmented on top of these cubes as seen in figure 2. The markers are special 2D bar code images which code a unique number. These bar codes are recognized by the vision component of the BCI system. Interaction with the scene is a two-step process consisting of an *object pickup* and and *object placement* task.

2.1 Experimental Protocol

The selection of an object is achieved by flashing up all numbers one-by-one in random order (Fig. 2 (a)) while the user mentally counts whenever the desired object flashes. At the beginning of each flash a short EEG time window (epoch) of 700ms is extracted and passed to the classification method. An object is selected when the classification method reports sufficient confidence in the current prediction. On a successful classification, the 3D coordinates and orientation of the object are extracted using the available methods of ARToolkit [4]. Since the extracted coordinates are relative to the camera position but are required to conform with the robots coordinate frame, a special reference marker with known coordinates in the robot frame is used to calculate the objects position relative to the reference marker. This step is necessary since the camera position (i.e. the head of the user) is not tracked. The extracted coordinates are sent to the robot backend with the instruction to grasp the object. Placing an object works in a similar way. After an object has been grasped, the reference marker serves as the origin for a semi-transparent 8×8 chessboard model as shown in Fig. 2 (b). Each cell of the board corresponds to a spatial location on the desk. Selecting a location is done in the same way as in the well known P300-Speller paradigm [2] by flashing up rows and columns in random order. The intersection of the row and column with the highest classification score will correspond to the selected grid cell.

3 Methods

The classification of epochs containing P300 event-related potentials requires an initial training step in which EEG data of the user is collected. During the training phase, the system starts in object selection mode and marks a random cube by highlighting its associated stimulus in green for 3 seconds. The user is advised to mentally count whenever this stimulus is flashed. The BCI then starts to flash the stimuli in random order in a 200ms interval. After all stimuli have been flashed 5 times, the BCI switches to grid mode and repeats the procedure analogously. The flashing of all rows and columns are flashed in random order once is repeated 5 times resulting in 80 stimulations per trial in this mode.

3.1 Data Acquisition

EEG data from the position *10-20 locations* [3] O1, O2, Pz, P3, P4, Cz, C3, C4, FCz, Fz, F3 and F4 were collected at 256Hz sampling rate using a 16-channel gUSBAmp EEG amplifier with reference and ground electrodes attached to the left and right mastoids. The derivation method was set to common reference which measures the potential differences between the active and the reference electrode. During the training phase, data of 20 stimulus presentation rounds (trials) for each task (*pickup/placement*) were collected. All data were highpass filtered to at 1Hz and subsequently downsampled by a factor of 16. No effort has been made to remove eye blink and muscle artifacts from the training set.

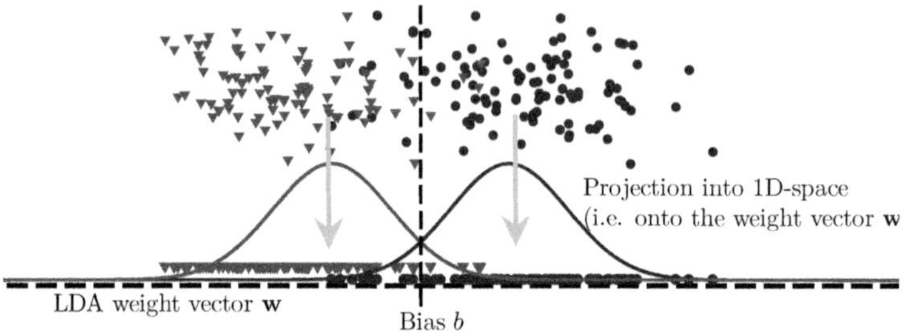

Fig. 3. Classification score distribution for P300 and non-P300 class

3.2 P300 Classification

Using the data obtained during the training phase, a two-class dataset containing both P300 (P^+) and non-P300 (P^-) time windows of 700ms (epochs) was extracted. The set was balanced to contain an equal number of observation for both classes. Each epoch was concatenated channel wise to form a single epoch vector $\mathbf{x}$. A *Fisher Linear Discriminant* classifier with regularized covariance matrices was trained on these data by computing

$$\mathbf{w} = \mathbf{S}_W^{-1}(\mu_+ - \mu_-) \tag{1}$$

with $\mathbf{S}_w$ being the regularized *within scatter matrix* which is defined as the sum of sample covariance matrices of both classes $\mathbf{S}_+ + \mathbf{S}_-$. The mean vectors of the P300 class are represented by μ_+ while the non-P300 means are represented by μ_-. A classification score is obtained by projecting an epoch vector $\mathbf{x}$ onto $\mathbf{w}$ resulting in a scalar value. Positive values correspond to a P^+ class assignment while negative values are associated with the P^- class.

3.3 Asynchronous Control for P300 BCIs

To mitigate the drawbacks of the trial based nature of classical P300 BCIs, we developed a novel flexible method similar to [6] extending our previous work [5]. Classical P300 BCIs use a fixed number of stimulus presentation round to acquire multiple EEG segments for each stimulus (epochs) which are subsequently averaged to improve the signal-to-noise ratio for improved classification accuracy. Using this method, a cue (marking the start of a new trial) is required that instructs the user to start focusing on the symbol she wants to select. In contrast to the classical approach of using dedicated time intervals during which data is collected for a classification, we are continuously presenting stimuli and collecting EEG epochs for each stimulus. While the first approach can be considered as a *batch method* as it needs to acquire multiple EEG epochs and further assumes that the subject is focused on the BCI, our new method is an *online method* that successively adds EEG epochs and dynamically decides when to output a classification. In that sense, our new method has two major advantages over the classical approach. Given a function P that measures the confidence of a classification result based on n epochs, a trial can be ended whenever P exceeds a certain threshold. As shown in [5] this can significantly reduce the number of stimulus presentations. As a second advantage, our method does not rely on the assumption that the subject is trying to communicate with the BCI at all time and thus can be considered as an asynchronous BCI. We will show that both problems of detecting *no-control/intentional-control* states of the subject and detection of the optimal number of subtrials are in fact closely related problems which can be solved by computing a single metric based distribution of classification scores.

Depicted in figure 3 is a simplified sketch where the points represent the EEG data epochs which are being projected onto the weight vector $\mathbf{w}$. These scores will be interpreted as features and the gaussian PDF properties μ and σ^2 for both classes in the feature space are estimated from the training set. In its simplest form, the method works by iteratively adding scores for each corresponding target i to an observation vector $\mathbf{x}_i$. A subsequent two-sided z-test is conducted with the $H0$ hypothesis that the observation mean $\overline{\mathbf{x}_i}$ is equal to the P300 class mean μ_+. A sequence of classification scores is considered *reliable* whenever the z-test can not reject H0 at a given significance level. Using this approach, the problem of finding the optimal number of stimulus presentations is solved since with increasing subtrials $\overline{\mathbf{x}}_i$ will converge towards μ_+ if i is the attended symbol and towards μ_- if i does not correspond to an attended symbol. At the same time, the second problem of detecting *no-control* states is solved. When the subject is not focusing on the BCI, it is unlikely that mean scores will converge towards μ_+ and thus all targets will be classified as P^-.

More formally, the method utilizes the standardized z-statistic

$$z(\mathbf{x}, \mu, \sigma^2) = \frac{\overline{\mathbf{x}} - \mu}{\sigma^2/\sqrt{n}} \tag{2}$$

to estimate the observation mean value's deviation from the true mean μ. A decision function that determines the end of a trial can be formulated as

$$\mathcal{D}(\mathbf{x}_i, \alpha, \mu, \sigma^2) = P(z(\mathbf{x}_i, \mu, \sigma^2) \leq z_{p=\alpha}|H_0) < \alpha \tag{3}$$

with z_p being the quantile function which can be derived by inversion of the standard normal cumulative density function. This formulation allows to define a *confidence level* α which can be used to tune the BCI for higher speed or higher accuracy. With increasing α, the acceptance interval around the P^+ mean is getting smaller which in turn mean that fewer scores will be accepted for the P^+ class assignment. With very high confidence levels it is even possible that no classification occurs at all. Thus, the general aim is to minimize false positives[1] while keeping the acceptance rate at a reasonable level.

---

[1] i.e. Labeling a sequence of classification score for one symbol as P300 epoch when it is in fact belonging to the non-P300 class.

4 Experimental Results

An experiment was conducted with 4 mixed male and female subjects with the aim to evaluate the overall usability of the system and the feasibility of the asynchronous protocol. The task for the subject was to move objects placed in front of them to a different location on the table. Both, the object and the target location were chosen by the subject. Whenever the robot picked up or placed an object, the subject was asked to report on the correctness of the robots action. The experiment ended when the subject successfully moved 10 objects to different locations which required 20 selection commands (i.e. 10 for *pickup* and 10 for the *placement* task). The task performance was measured by calculating the communication rate in correct symbols per minute, as well as the overall accuracy. To evaluate the feasibility of the asynchronous protocol, the subject's focus had to be distracted from the BCI stimulus presentation. For this reason, they were asked to fill out a short questionnaire and answer to question of the experiment supervisor after the experiment ended. During that time the BCI was still running and ready to receive commands. Further, to evaluate how long the system takes to recognize that the subject is now actively communicating with the system, one more object relocation had to be carried out. Table 1 summarizes the number of wrongly conducted actions per minute during questionnaire period as well as the time it took the system to recognize a voluntary selection command of the user (i.e. *time to active (TTA)*).

Table 1. BCI performance achieved in the study. The measures accuracy (Acc.), correct symbols per minute (Sym./min), actions per minute during the *no-control* period and *time to active* (TTA) are shown.

	Pickup Task		**Placement Task**		**No-Control Task**	
Subject	Acc.	Sym./min	Acc.	Sym./min	Act./min	TTA
S1	80%	3.3 (2:20)	70%	1.1 (6:30)	0.4	12s
S2	90%	2.3 (2:50)	70%	1.4 (5:00)	0.6	18s
S3	100%	5.9 (1:40)	90%	1.7 (5:50)	0.2	8s
S4	80%	2.3 (3:40)	60%	0.7 (8:20)	0.8	10s
Mean	87.5%	3.45	72.5%	1.23	0.5	12

5 Discussion

For the *pickup* task, promising accuracies of 80% up to 100% across all subjects could be achieved. The communication speed is roughly between 2 and 3 symbols/min except for subject 3 who performed exceptionally well compared to the other users. These values however dropped significantly during the *placement* task. While the performance loss in terms of communcation speed was expected since this task contains 64 stimuli in contrast to the 5 stimuli of the *pickup* task, the accuracy loss is remarkable. We explain this loss of accuracy with the fact that all subjects noted on the questionnaire that it is difficult to keep focused on a specific grid cell due to their equal appearance. All of them remarked that during the training and online phase their attention slipped to adjacent cells from time to time. One subject also noted that the cell he was focusing seemed to visually fade when he was focusing it for a sustained time. Contrary, none of the users had problems to stay focused during the *pickup* task which used unique 3D models of numbers as stimuli. We assume that this issue could be related to the rather poor performance of subject 4 during the placement task. The evaluation of the asynchronous protocol shows an average misclassification rate of 0.5 symbols per minute. For this experiment, the α value for the decision function (Eq. 3) was set to 0.2 which seemed to be a reasonable tradeoff for speed and accuracy. On average, our method achieves a false positive rate (FPR) of 0.5 symbols per minute which is comparable to the work of [8] with an FPR of 0.71. Similar to Zhang et al. the accuracy of the system can be optimized at the cost of communication speed. The overall performance of the system is sufficient for practical use (e.g. for motion impaired people). The current limitation of marker based pose estimation can also be replaced by a more sophisticated method using natural features [1]. Practical tasks could consist of picking up a variety of objects like telephones, books or using a TV remote control. In the near future, this method could also be extended to aid in *hands-busy tasks* for healthy subjects as mentioned in [9].

References

[1] Collet, A., Berenson, D., Srinivasa, S.S., Ferguson, D.: Object recognition and full pose registration from a single image for robotic manipulation. In: IEEE International Conference on Robotics and Automation (ICRA), Kobe, Japan (2009)

[2] Farwell, L.A., Donchin, E.: Talking off the top of your head: toward a mental prosthesis utilizing event-related brain potentials. Electroencephalogr. Clin. Neurophysiol. 70(6), 510–523 (1988)

[3] Jasper, H.H.: Report on the commitee on methods of clinical examination in electroencephalography. Electroenceph. Clin. Neurophysiol. 10(370) (1958)

[4] Kato, H., Billinghurst, M.: Marker tracking and hmd calibration for a video-based augmented reality conferencing system. In: Iwar, p. 85. IEEE Computer Society, Los Alamitos (1999)

[5] Lenhardt, A., Kaper, M.: HJ Ritter. An adaptive p300-based online brain-computer interface. IEEE Transactions on Neural Systems and Rehabilitation Engineering: A Publication of The IEEE Engineering in Medicine and Biology Society 16(2), 121 (2008)

[6] Rebsamen, B., Teo, C.L., Zeng, Q., Ang Jr., M.H., Burdet, E., Guan, C., Zhang, H., Laugier, C.: Controlling a wheelchair indoors using thought. IEEE Intelligent Systems, 18–24 (2007)

[7] Wolpaw, J.R.: Brain–computer interfaces as new brain output pathways. The Journal of Physiology 579(3), 613 (2007)

[8] Zhang, H., Guan, C., Wang, C.: Asynchronous p300-based brain-computer interfaces: A computational approach with statistical models. IEEE Transactions on Biomedical Engineering 55(6), 1754–1763 (2008)

[9] Zhu, D., Gedeon, T., Taylor, K.: Keyboard before head tracking depresses user success in remote camera control. In: Gross, T., Gulliksen, J., Kotzé, P., Oestreicher, L., Palanque, P., Prates, R.O., Winckler, M. (eds.) INTERACT 2009. LNCS, vol. 5727, pp. 319–331. Springer, Heidelberg (2009)

Brain Computer Interfaces: A Recurrent Neural Network Approach

Gareth Oliver and Tom Gedeon

Australian National University
{gareth.oliver,tom.gedeon}@anu.edu.au
http://www.cs.anu.edu.au

Abstract. This paper explores the use of recurrent neural networks in the field of Brain Computer Interfaces(BCI). In particular it looks at a recurrent neural network, an echostate network and a CasPer neural network and attempts to use them to classify data from BCI competition IIIs dataset IVa. In addition it proposes a new method, EchoCasPer, which uses the CasPer training scheme in a recurrent neural network. The results showed that temporal information existed within the BCI data to be made use of, but further pre-processing and parameter exploration was needed to reach competitive classification rates.

Keywords: BCI, RNN, CasPer, Echostate Network.

1 Introduction

Brain Computer Interfaces (BCI) is attempts to use the brain as a control device for a computer. Long term, BCIs have potential commercial application due to their nature as a new mode of control for computers. Given sufficient reliability they would allow the control of electronic devices with just a thought! More immediately, Brain Computer Interfaces can be used to improve the quality of life of patients suffering from advanced Amyotrophic Lateral Sclerosis (ALS). Patients with advanced ALS suffer from complete paralysis while still receiving information from the environment and having active brains. A BCI would allow such people to communicate, as well as potentially control entire devices.

This paper investigates various classifiers that take advantage of the time series nature of the EEG data, and the temporal information that should be contained within it. In particular it will investigate various recurrent neural networks, which have typically been successful in exploiting temporal information.

2 Classifiers

Four different classification methods were used. A Recurrent Neural Network, trained using the Backpropogation Through Time(BPTT) initially proposed by [1]. An Echostate Network, based on the work of [2], [3]. Thirdly a CasPer Neural

K.W. Wong et al. (Eds.): ICONIP 2010, Part II, LNCS 6444, pp. 66–73, 2010.

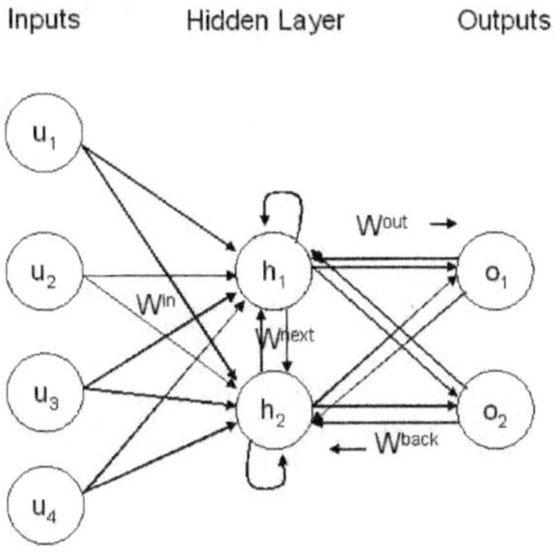

Fig. 1. Topography of a RNN with 4 Input neurons, 2 Hidden neurons and a single Output neuron

Network, following the design proposed in [4],[5]. Finally a new, hybrid extension of the CasPer neural network, EchoCasPer Neural Network, was designed to extend the principles behind the training of the CasPer neural network to a network whose structure could take advantage of the temporal information inherent in time series data.

2.1 Recurrent Neural Network

The RNN has three sets of neurons, Input (I), Hidden (H) and a single Output neuron (Fig. 1 shows the topography). The following weight matrices define the connections between these layers.

- Input to Hidden neurons: a $I \times H$ matrix **W**in
- Hidden to Hidden neurons: a $H \times H$ matrix **W**next
- Input and Hidden to Output neurons: a $(I + H) \times 1$ matrix **W**out
- Output neurons to Hidden neurons: a $1 \times H$ matrix **W**back

Using these weight matrices, the activation functions were defined as follows:

$$h(t+1) = tanh(u(t+1) \times \mathbf{W}in + h(t) \times \mathbf{W}next + o(t) \times \mathbf{W}back) \quad (1)$$

$$o(t+1) = [u(t+1), h(t+1)] \times \mathbf{W}out \quad (2)$$

To train the RNN the activation function is used to calculate the value of each hidden neuron and each output neuron at all times t. The error of the output neurons γ and hidden neurons δ are calculated for the last time step by:

$$\gamma(t) = class(t) - o(t) \quad (3)$$

$$\delta(t) = (1 - h(t)^2) \times \gamma(t) \times \mathbf{W}out \quad (4)$$

The errors of each previous time step are then calculated by:

$$\gamma(t) = class(t) - o(t) + \sum h(t+1) \times \mathbf{W}back \tag{5}$$

$$\delta(t) = (1 - h(t)^2) \times \gamma(t) \times \mathbf{W}out + \sum h(t+1) \times \mathbf{W}back \tag{6}$$

Finally the weights are updated by these error values using:

$$\mathbf{W}in_{i,j} + = l \sum_{t=0}^{T} \delta_i(t) \times I_j(t) \tag{7}$$

$$\mathbf{W}out_i + = l \sum_{t=0}^{T} \gamma(t) \times h_i(t) \tag{8}$$

$$\mathbf{W}next_i + = l \sum_{t=0}^{T-1} o(t) \times \delta_i(t+1) \tag{9}$$

$$\mathbf{W}back_{i,j} + = l \sum_{t=0}^{T-1} \delta_j(t+1) \times h_i(t) \tag{10}$$

Where l is the learning rate. To classify using a trained RNN, the error between each individual class is calculated by

$$error(class) = \sum_{t=0}^{T} (class - o(t))^2 \tag{11}$$

and the smallest is selected.

2.2 Echostate Network

The Echostate Network topography is the same as that of the RNN, see Fig. 1. As such the activation function and weights can be defined by equations 1 and 2. To ensure the echostate property[6], after initialising the weight matrices,

$$\mathbf{W}next = \frac{\mathbf{W}next}{\lambda_{max}} \tag{12}$$

$$\mathbf{W}next = \mathbf{W}next \times \alpha \tag{13}$$

where λ_{max} is the maximum absolute eigenvalue of Wnext and α is a value between 0 and 1 which governs the speed of attenuation within the hidden neurons.

To train the Echostate Network, BPTT is again used to calculate the error of each output neuron and each element of the hidden layer as per equations 3 to 6. The weights are then updated as follows:

$$\mathbf{W}in_{i,j} + = l \sum_{t=0}^{T} \delta_i(t) \times I_j(t) \tag{14}$$

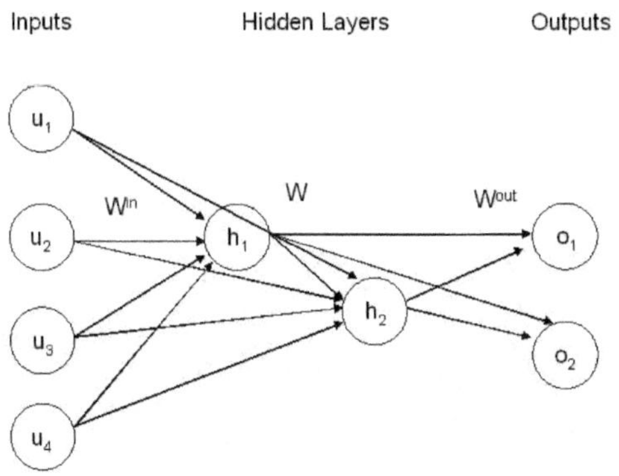

Fig. 2. Topography of a CasPer Neural Network with 4 Input neurons, 2 Hidden neurons and a single Output neuron

$$\mathbf{W}out_i + = l \sum_{t=0}^{T} \gamma(t) \times h_i(t) \tag{15}$$

Notice that the weights connecting the hidden layer remain fixed. Once again classification is done by calculating the error for each individual class by equation 11.

2.3 CasPer Neural Network

A full description of the CasPer Neural Network training method can be seen in [4]. The topography of a CasPer Neural Network is shown in in Fig. 2. The hypobolic tangent function was used as the activation function for neurons within the CasPer Neural Network. The classification is done by calculating the error for each individual class and selecting the minimum by:

$$error(class) = (class - o)^2 \tag{16}$$

2.4 EchoCasPer Neural Network

The topography of the EchoCasPer Neural Network can be seen in Fig. 3. It consists of Input(I), Hidden(H), Casper(C) Neurons and an Output Neuron, (Fig. 3 shows an example topography). The connections are defined by the weight matrices:

- Input layer to the Reservoir: $I \times R$ matrix **W**in
- Reservoir to itself: $R \times R$ matrix **W**next
- To i^{th} neuron of the CasPer layer: $(I + R + i) \times 1$ matrix **W**i
- To the Output neuron: $(I + R + C) \times O$ matrix **W**out

The Wnext matrix has the Echostate property enforced on it as in the Echostate Network section. The activation functions are defined for each time t by:

$$r(t+1) = tanh((r)(t) \times \mathbf{W}next + (u)(t+1) \times \mathbf{W}in) \tag{17}$$

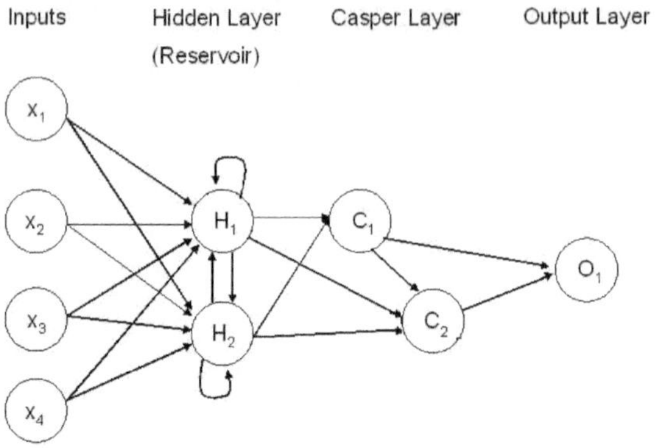

Fig. 3. Topography of a EchoCasPer Neural Network with 4 Input neurons, 2 Hidden neurons, 2 CasPer layer Neuron and a single Output neuron

$$c_i(t+1) = tanh([(u)(t+1), r(t+1), c_1(t+1)...c_{i-1}(t+1)] \times \mathbf{W}i) \quad (18)$$

$$o(t+1) = ([(u)(t+1), r(t+1), c(t+1)] \times \mathbf{W}out) \quad (19)$$

The training takes place in a series of iterations equal to C. At the beginning of each iteration the weight matrices need to be updated for the new neuron. **W**i needs to be initialised, and a new column needs to be added to **W**out. Additionally, matrices of learning rates, L1, L2 and L3 need to be assigned to each weight going to and from the neurons in the CasPer layer. The weights in the reservoir are not adaptable. The learning rates should be defined so that $L1 > L2 >> L3$.

All weights going to into the newly added neuron should be given a learning rate of L1 so that they adapt the fastest. Weights going from the newly added neuron to the output layer should be given a learning rate of L2. All other weights going to and from neurons in the CasPer or output layer should be given a learning rate of L3 so that the weights are mostly fixed.

Once the network is set up for the next set of training, it is trained to within a level of convergence using a backpropagation through time algorithm (BPTT). This is done by first running the EchoCasPer network for a training data for n time iterations, collecting the activations of the neurons at each time period. The errors for the CasPer layer(δ), Output Layer (γ) neurons are given by:

$$\gamma(t) = class - o(t) \quad (20)$$

$$\delta_i(t) = (1 - c_i(t)) \times \gamma(t) \times \mathbf{W}out_{I+R+i} + \sum_{m=i}^{C} \delta_m(t) \times \mathbf{W}m_i \quad (21)$$

Using the calculated errors the weights can be updated for each time step using the appropriate learning rates by:

$$\mathbf{W}out{+} = \sum_{t=0}^{T} gradout[u(t), h(t), c(t)] \times \gamma(t) - sign(\mathbf{W}out)\mathbf{W}out^2 e^{-0.01it} \quad (22)$$

$$\mathbf{W}i{+} = \sum_{t=0}^{T} grad_i[u(t), h(t), c_1(t)..c_i(t)] \times \delta_i(t) - sign(\mathbf{W})\mathbf{W}^2 e^{-0.01it} \quad (23)$$

Here the matrices grad and gradout contain the learning rates for the appropriate weight. The final term in each of the above expression is a regularisation term, used to prevent overfitting. Sign returns the sign of the passed number, so that the regularisation is performed in the correct direction. The decay term is a parameter that controls the amount of regularisation applied and it is the number of iterations of training have passed since the last neuron was added.

Once the weights have been updated, the error is calculated and convergence is checked. The convergence can be checked with:

$$0.01 < \frac{Elast - Ecurr}{Elast} \times (15 + pnodes) \quad (24)$$

Where p is an input parameter, increasing the steps trained before convergence and nodes is the number of neurons currently inserted in the hidden layer of the network. By having the convergence dependent on the number of neurons in the network the neural network will train longer. After convergence is reached the next neuron is added to the CasPer layer. Classification can be done once again by equation 11.

3 Pre-processing

The data for each classifier was pre-processed using two pre-processing techniques. Firstly the channels across the motor cortex were selected from the total number of channels, using expert knowledge to determine the correct ones to remove, reducing the number of channels to 43. Secondly a FIR filter was constructed to reduce frequencies outside of the beta (11 to 25Hz) rhythm as this is the primary band in which responses to motor tasks occur.

4 Experiment

The dataset used in these tests comes from the the Brain Computer Interface III, which is a competition that was held in association with Brain-Computer Interface Technology: Third International Meeting of june 2005 [7]. The specific dataset used was that of dataset IVa. This was provided by Fraunhofer FIRST, Intelligent Data Analysis Group. The dataset is a two class classification problem, with varying amount of training data and induced noise.

The data was recorded by a 118 electrode EEG. Only the right foot and right hand tasks were given as the training and test data for this classification

Table 1. BCI Compeon III Dataset IVa

Subject	Test	Training	Induced Noise
aa	168	112	yes
al	224	56	no
av	84	196	no
aw	56	224	yes
ay	28	252	no

problem. Each of the 5 subjects was broken into a separate test and training datasets, with characteristics described in the below table.

After initial testings, the following parameters for the topography of each network were used. The RNN had 7 Hidden Layer Neurons and a learning rate of 0.01. The Echostate Network also had 7 Hidden Layer Neurons and a learning rate of 0.01 with an alpha value of 0.5. The CasPer Neural Network had 5 Hidden Layer Neurons and values of 0.001, 0.0001 and 0.00001 for the learning rates L1, L2 and L3 respectively. The EchoCasPer Network had 7 Hidden Layer and 5 CasPer Layer Neurons, values of 0.1, 0.01 and 0.001 for L1, L2 and L3 respectively and finally an alpha value of Alpha.

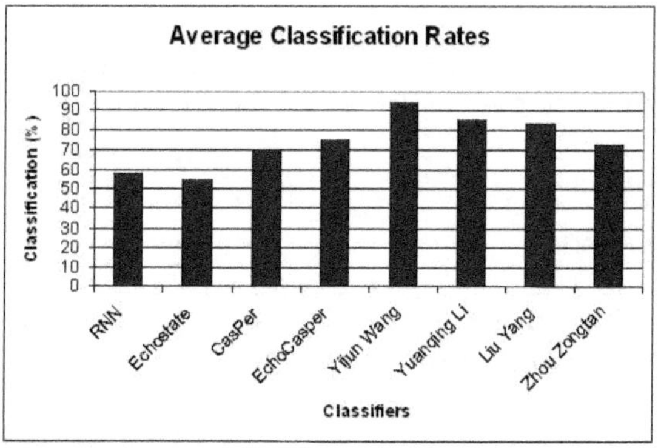

Fig. 4. Average classification rates over all subjects

Fig. 4 shows the average results of each classifier, in addition to four of the results of the best classifiers in the BCI competition. More details of their algorithms can be found in [7],[8],[9]. It can be clearly seen that, of the designed classifiers, the EchoCasPer Network performed the best, followed by the CasPer Network. All of these results fell short of the literature classifiers.

5 Conclusion

Both the RNN and the echostate network performed relatively poorly. The CasPer neural network was found to be significantly more successful than these two but fell short of the best literature classifiers. In addition, the large number of inputs from the time series data resulted in extremely long training times, which also limited the amount of fine tuning of the multiple parameters could take place. The EchoCasPer network was designed as a way to allow the iterative method of dynamically creating the topography of a neural network used by the CasPer algorithm be used in a network that maintains a state. The resulting network made use of a reservoir from the echostate network to store the state while allowing a neural network to be built above it that preserved the CasPer properties. The results on the dataset were close to the literature results, and so shows promise if parameters and preprocessing methods are further fine-tuned. Additionally the classification rate was a 5% improvement over the CasPer algorithm. This implied that the EchoCasPer network was able to make use of some of the temporal information in the resevoir so further investigation is warrented.

References

1. Werbos, P.J.: Backpropagation through time: What it does and how to do it. Proceedings of IEEE 78, 1550–1560 (1990)
2. Ozturk, M.C., Xu, D., Principe, J.C.: Analysis and design of echostate networks. Neural Computation 19, 111–138 (2006)
3. Mezzano, T.: Echo State Networks application on maze problems. PhD thesis, Katholieke Universiteit Leuven (2007)
4. Treadgold, N.K., Gedeon, T.D.: A cascade network employing progressive rprop. In: International Work Conference on Artificial and Natural Neural Networks, pp. 733–742 (1997b)
5. Treadgold, N.K., Gedeon, T.D.: Extending and Benchmarking the CasPer Proceedings of the 10th Australian Joint Conference on Artificial Intelligence: Advanced Topics in Artificial Intelligence (1997a)
6. Jaeger, H.: Short term memory in echostate networks. Technical report (2002)
7. Blankertz, B., Muller, K.R., Krusienski, D., Schalk, G., Wolpaw, J.R., Schlogl, A., Pfurtscheller, G., Millan, J.R., Schroder, M., Birbaumer, N.: The bci competition iii: Validating alternative approachs to actual bci problems. IEEE Transactions on Neural Systems and Rehabilitation Engineering 14, 153–159 (2006)
8. Wang, Y., Zhang, Z., Li, Y., Gao, X., Gao, S., Yang, F.: Bci competition 2003data set iv: An algorithm based on cssd and fda for classifying single-trial eeg. IEEE Transactions on Biomedical Engineering 51, 1081–1086 (2004)
9. Zhu, X., Guan, C., Wu, J., Cheng, Y., Wang, Y.: Expectation maximization method for eeg-based continuous cursor control. EURASIP Journal on Advances in Signal Processing (2007)

Research on Relationship between Saccade-Related EEG Signals and Selection of Electrode Position by Independent Component Analysis

Arao Funase[1,2], Motoaki Mouri[1,2], Andrzej Cichocki[2], and Ichi Takumi[1]

[1] Graduate School of Engineering, Nagoya Institute of Technology, Gokiso-cho, Showa-ku, Nagoya, 466-8555, Japan
[2] Brain Science Institute, RIKEN, 2-1, Hirosawa, Wako, 351-0198, Japan

Abstract. Our goal is to develop a novel BCI based on an eye movements system employing EEG signals on-line. Most of the analysis on EEG signals has been performed using ensemble averaging approaches. However, in signal processing methods for BCI, raw EEG signals are analyzed.

In order to process raw EEG signals, we used independent component analysis(ICA).

Previous paper presented extraction rate of saccade-related EEG signals by five ICA algorithms and eight window size.

However, three ICA algorithms, the FastICA, the NG-FICA and the JADE algorithms, are based on 4th order statistic and AMUSE algorithm has an improved algorithm named the SOBI. Therefore, we must re-select ICA algorithms.

In this paper, Firstly, we add new algorithms; the SOBI and the MILCA. Using the Fast ICA, the JADE, the AMUSE, the SOBI, and the MILCA. The SOBI is an improved algorithm based on the AMUSE and uses at least two covariance matrices at different time steps. The MILCA use the independency based on mutual information. We extract saccade-related EEG signals and check extracting rates.

Secondly, we check relationship between window sizes of EEG signals to be analyzed and extracting rates.

Thirdly, we researched on relationship between Saccade-related EEG signals and selection of electrode position by ICA. In order to develop the BCI, it is important to use a few electrode. In previous studies, we analyzed EEG signals using by 19 electrodes. In this study, we checked various combination of electrode.

1 Introduction

Brain-computer interfaces (BCIs) have been researched as a novel human interface for a few decades. The capabilities of BCIs allow them to be used in situations unsuitable for the conventional interfaces. BCIs are used to connect a user and a computer via an electroencephalogram (EEG).

K.W. Wong et al. (Eds.): ICONIP 2010, Part II, LNCS 6444, pp. 74–81, 2010.

EEG related to saccadic eye movements have been studied by our group toward developing a BCI eye-tracking system [1]. In previous research, saccade-related EEG signals were analyzed using the ensemble averaging method. Ensemble averaging is not suitable for analyzing raw EEG data because the method needs many repetitive trials.

Recording EEG data repetitively is a critical problem to develop BCIs. It is essential to overcome this problem in order to realize practical use of BCIs for single trial EEG data.

Recently, the independent component analysis (ICA) method has been introduced in the field of bio-signal processing as a promising technique for separating independent sources. The ICA method can process raw EEG data and find features related to various one's activity. Therefore, the ICA algorithm overcomes the problems associated with ensemble averaging, and the ICA analyzes the waveforms of the EEG data.

There are many algorithms to compute independent components [2]. In previous studies[3], we used the FastICA, the NG-FICA, the AMUSE, the JADE to analyze saccade-related EEG signals. However, we must re-select an ICA algorithm since three ICA algorithms: the FastICA, the NG-FICA and the JADE algorithms are based on the 4th order statistic and the AMUSE algorithm has an improved algorithm named the SOBI[8].

In this research, we add new algorithms: the SOBI and the MILCA[9]. The SOBI is an improved algorithm based on the AMUSE and uses the independency based on two covariance matrices at different time steps. The MILCA uses the independency based on mutual information. Using the Fast ICA, the JADE, the AMUSE, the SOBI, and the MILCA, we extract saccade-related EEG signals and check extracting rates.

Secondly, we focus on window sizes of EEG signals to be analyzed. In order to analyze EEG signals in on-line system, we must choose an appropriate window size to extract continuous EEG signals. In this paper, we separate window sizes into two groups: the windows excluding EEG signals after eye movements and the windows include EEG signals after eye movements.

Thirdly, we researched on relationship between Saccade-related EEG signals and selection of electrode position by ICA. In order to develop the BCI, it is important to use a few electrode. In previous studies, we analyzed EEG signals using by 19 electrodes. In this study, we checked various combination of electrode.

2 Independent Component Analysis (ICA)

The ICA method is based on the following principles. Assuming that the original (or source) signals have been linearly mixed, and that these mixed signals are available, ICA recognizes in a blind manner a linear combination of the mixed signals, and recovers the original source signals, possibly re-scaled and randomly arranged in the outputs.

The $\mathbf{s} = [s_1, s_2, \cdots, s_n]^{\mathbf{T}}$ means n independent signals from mutual EEG sources in the brain, for example. The mixed signals $\mathbf{x}$ are thus given by $\mathbf{x} = \mathbf{A}\mathbf{s}$, where $\mathbf{A}$ is an $n \times n$ invertible matrix. $\mathbf{A}$ is the matrix for mixing independent signals. In the ICA method, only $\mathbf{x}$ is observed. The value for $\mathbf{s}$ is calculated by $\mathbf{s} = \mathbf{W}\mathbf{x}$ ($\mathbf{W} = \mathbf{A}^{-1}$). However, it is impossible to calculate $\mathbf{A}^{-1}$ algebraically because information for $\mathbf{A}$ and $\mathbf{s}$ are not already known. Therefore, in the ICA algorithm, $\mathbf{W}$ is estimated non-algebraically. The assumption of the ICA algorithm is that $\mathbf{s}$ is mutually independent. In order to calculate $\mathbf{W}$, different cost functions are used in the literature, usually involving a non-linearity that shapes the probability destiny function of the source signals.

3 Experimental Settings

There were two tasks in this study (See Fig.1). The first task was to record the EEG signals during a saccade to a visual target when a subject moves its eyes to a visual stimulus that is on his right or left side. The second task was to record the EEG signals as a control condition when a subject dose not perform a saccade even though a stimulus has been displayed. Each experiment was comprised of 50 trials in total: 25 on the right side and 25 on the left side.

The experiments were performed in an electromagnetically shielded dark room to reduce the effect of electromagnetic noise and any visual stimuli in the environment. The visual targets were three LEDs placed in a line before the subject. One was located 30 [cm] away from the nasion of the subject. The other two LEDs were placed to the right and left of the center LED, each separated by 25 degrees from the nasion. They were illuminated randomly to prevent the subjects from trying to guess which direction the next stimulus would be coming form next.

The subject were five 25-26 year-old male and subjects all have normal vision.

The EEG signals were recorded through 19 electrodes (Ag-AgCl), which were placed on the subject's head in accord with the international 10-20 electrode position system (see Fig.2). The Electrooculogram (EOG) signals were simultaneously recorded through two pairs of electrodes (Ag-AgCl) attached to the top-bottom side and right-left side of the right eye.

All data were sampled at 1000 [Hz], and stored on a hard disk for off-line data processing after post-amplification. The raw EEG data was filtered by a high-pass filter (cut-off 0.53 [Hz]) and a low-pass filter (cut-off 120 [Hz]). The EOG data was recorded through a high-pass filter (cut-off 0.1 [Hz]) and a low-pass filter (cut-off 15 [Hz]).

Recorded EEG signals were calculated by five ICA algorithms: FastICA, AMUSE, JADE, SOBI, MILCA. In order to calculate independent components, we must decide the window length. In this paper, there were 8 size windows.

- Group A
  - Window A: -999[ms] to 1000[ms]
  - Window B: -499[ms] to 500[ms]
  - Window C: -349[ms] to 350[ms]
- Group B
  - Window D: -999[ms] to 0[ms]

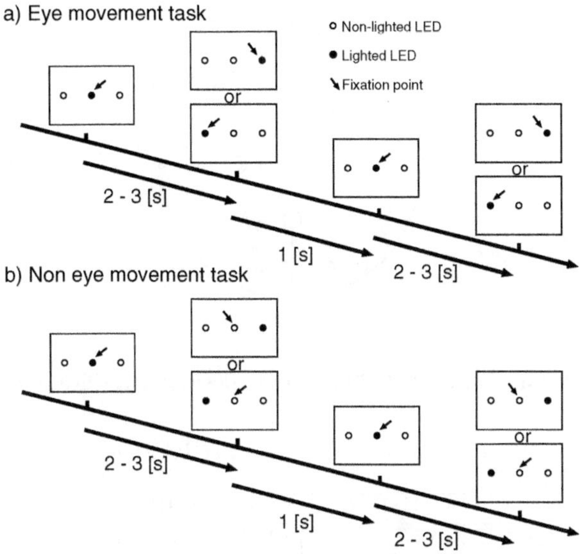

Fig. 1. Experimental tasks

- Window E: -499[ms] to 0[ms]
- Window F: -349[ms] to 0[ms]
- Window G: -249[ms] to 0[ms]
- Window H: -99[ms] to 0[ms]

0[ms] indicates the starting point of saccade. After eye movements, EEG signals include big noises caused by EOG signals. In order to observe influence of noises caused by EOG signals, we separated window size into two groups: Window A to C including EEG signals after saccade and window D to H excluding EEG signals after saccade.

In using five algorithms and eight windows, we calculated saccade-related independent components.

4 Experimental Results

First, we define two words: an extracting rate and saccade-related IC. The extraction rate is defined by the following ratio:

(the number of trials in which saccade-related IC are extracted)
/ (The total number of trials).

We make assumption that a saccade-related IC has a positive peak from -50 [ms] $\sim$ -1 [ms]. The peak-amplitude n is larger than 3; $n = \frac{\bar{x}-\mu}{s}$; where $\bar{x}$ is mean of

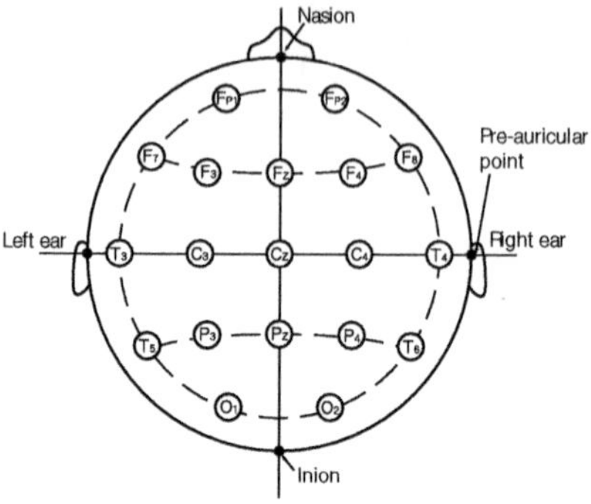

Fig. 2. International 10-20 electrode position classification system

Table 1. Extracted rate by five ICA algorithms

	AMUSE	FICA	JADE	SOBI	MILCA
A	14%	98%	100%	70%	50%
B	18%	82%	94%	76%	46%
C	30%	94%	96%	80%	62%
D	30%	98%	98%	66%	50%
E	24%	94%	96%	70%	46%

Table 2. Extracted rate by six window size

category	Window size	FastICA	JADE
A	-999 ~ 1000 [ms]	37.2%	38%
	-499 ~ 500 [ms]	29.6%	27.2%
	-349 ~ 350 [ms]	22.4%	26.4%
B	-999 ~ 0 [ms]	90%	93.6%
	-499 ~ 0 [ms]	93.2%	96.4%
	-349 ~ 0 [ms]	99.4%	99.2%
	-249 ~ 0 [ms]	93.2%	93.6%
	-99 ~ 0 [ms]	99.4%	99.2%

EEG potential during 1000 [ms] before saccade, μ is maximum amplitude, and s is standard deviation during 1000 [ms] before saccade.

Table 1 represents the rate for extracting saccade-related ICs from the raw EEG data by each algorithm in the case of window E. From these results, the FastICA and JADE got good performance in extracting saccade-related independent components. However, the results of the AMUSE and SOBI and MILCA algorithm were not good. From these results, in order to extract saccade-related EEG signals, it is not suitable to use independency of 2nd order statistics and the mutual information.

Next, we focus on extracting rate in each windows (see Table 2). From Table 2, extracting rates in group 1 were lower than those in group 2. Therefore, we should not use EEG signals after saccade. because the signals in group 1 include EOG noise. In the case of group 2, the results of small window size is better. From these result, we can get good results in the case of short window size excluding signals after saccade. Fig 3 is extracted signals by FastICA and window D,E,F. Each extracted signals are not the same although input signals are the same. However, each signals denote the same tendency.

5 Selection of Electrode

Next, we researched on relationship between Saccade-related EEG signals and selection of electrode position by ICA. In order to develop the BCI, it is important to use a few electrode. In previous studies, we analyzed EEG signals using by 19 electrodes.

In results by ensemble averaging, Fig.4 is a series of a topographical map of the EEG at three different time period; 45, 25, 5[ms] advance to the saccade in the visual experiment above. The upper series indicate the maps in the right directional saccade, and the lower, the left. The gray scale in the figure indicates the amplitude. (The darker shows the higher amplitude.) Besides, each map is viewed from the top. (The upper of the diagram is a frontal lobe, and the lower, an occipital lobe.) This figure shows clearly that the sharp drop is observed at the right occipital cerebrum. On the contrary, the EEG at the left occipital cerebrum decreases in the left saccade.

From this results, electrodes in occipital lobe are more important than another electrodes. Therefore, we select 4 sets of the electrode position.

- All electrodes (19 electrodes)
- O1, O2, P3, P4, Pz, Cz (6 electrodes)
- P3, P4, Pz (3 electrodes)
- O1, O2 (2 electrodes)

Table 3 shows relationship between the extracting rates and the electrode positions. From this results, results using only 3 electrodes have good extracting rates

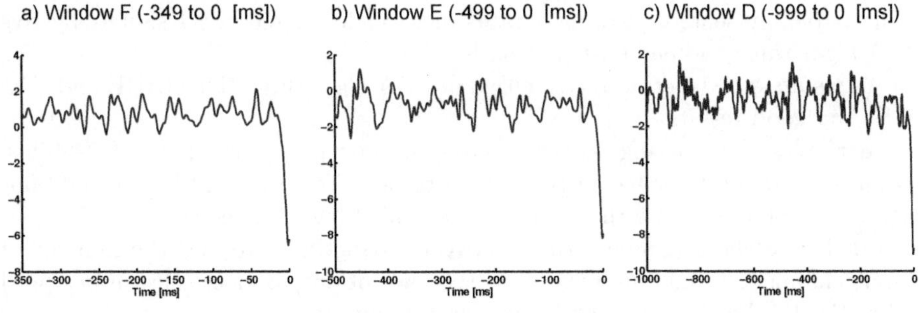

Fig. 3. Extracted signals for FastICA by each window size

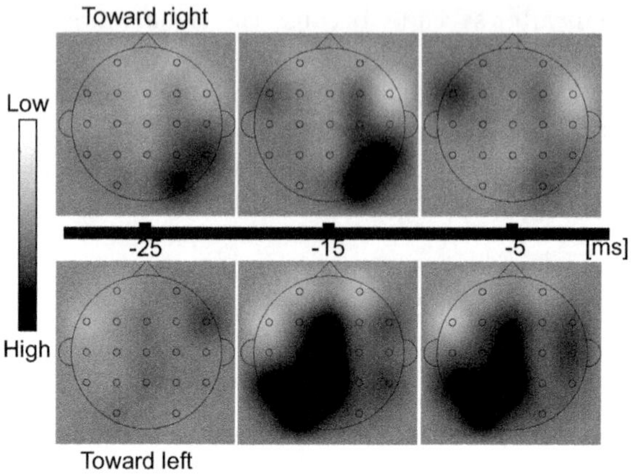

Fig. 4. Topography of EEG electric potential

Table 3. Relationship between extracting rate and electrode position

Electrode position	Number of electrodes	Extracting rate by FastICA	Extracting rate by JADE
All	19	99.4%	99.2%
O1, O2, P3, P4, Pz, Cz	6	100%	100%
P3, P4, Pz	3	92%	92%
O1, O2	2	80%	80%

and results using only 3 electrodes have better extracting rates than results using all electrodes. Therefore, important electrodes for extracting saccade-related EEG signals stay on only occipital lobe.

6 Conclusion

This paper presented extraction rates of saccade-related EEG signals by five ICA algorithms and eight window sizes.

As results of extracting rate focused on ICA algorithms, The JADE and Fast ICA had good results.

As results of extracting rates focused on window sizes, the window H (-99[ms] $\sim$ 0[ms]) had good results. In the case of the window A,B, and C, we could not get good results because these windows included big EOG noise.

Finaly, we checked relationship between extracting rate and the number of input channels. From this results, in the case of only 3 electrodes in occipital lobe (P3, P4, Pz), we can obtain best extracting rate.

In this paper, we checked only 4 sets of electrodes. In the future, we must check all combination of electrodes.

References

1. Funase, A., Yagi, T., Kuno, Y., Uchikawa, Y.: A study on electro-encephalo-gram (EEG) in eye movement. Studies in Applied Electromagnetics and Mechanics 18, 709–712 (2000)
2. Cichocki, A., Amari, S.: Adaptive blind signal and image processing. Wiley, Chichester (2002)
3. Funase, A., Hashimoto, T., Yagi, T., Barros, A.K., Cichocki, A., Takumi, I.: Research for estimating direction of saccadic eye movements by single trial processing. In: Proc. of 29th Annual International Conference of the IEEE EMBS, pp. 4723–4726 (2007)
4. Hyvärinen, A., Oja, E.: A fast fixed-point algorithm for independent component analysis. Neural Computation (9), 1483–1492 (1997)
5. Choi, S., Cichocki, A., Amari, S.: Flexible independent component analysis. Journal of VLSI Signal Processing 26(1), 25–38 (2000)
6. Tong, L., Soon, V., et al.: Indeterminacy and indentifiability of blind indentification. IEEE Trans. CAS 38, 499–509 (1991)
7. Cardoso, J.-F., Souloumiac, A.: Blind beam-forming for non Gaussian signals. IEE Proceedings-F 140, 362–370 (1993)
8. Belouchrani, A., Abed-Meraim, K., Cardoso, J.F., Moulines, E.: Second-order blind separation of temporally correlated sources. In: Proc. Int. Conf. on Digital Sig. Proc., (Cyprus), pp. 346–351 (1993)
9. Stogbauer, H., Kraskov, A., Astakhov, S.A., Grassberger, P.: Least Dependent Component Analysis Based on Mutual Information. Phys. Rev. E 70(6), 066123 (2004)
10. Barros, A.K., Vigàrio, R., Jousmäki, V., Ohnishi, N.: Extraction of event-related signal form multi-channel bioelectrical measurements. IEEE Transaction on Biomedical Engineering 47(5), 61–65 (2001)

Application of SVM-Based Filter Using LMS Learning Algorithm for Image Denoising

Tzu-Chao Lin, Chien-Ting Yeh, and Mu-Kun Liu

Department of Computer Science and Information Engineering
WuFeng University, Chiayi, Taiwan 62153, R.O.C.
tclin@wfu.edu.tw

Abstract. In this paper, a novel adaptive filter based on support vector machines (SVMs) that preserves image details and effectively suppresses impulsive noise is proposed. The filter employs an SVM impulse detector to judge whether an input pixel is noisy. If a noisy pixel is detected, a median filter is triggered to replace it. Otherwise, it stays unchanged. To improve the quality of the restored image, an adaptive LUM filter based on scalar quantization (SQ) is activated. The optimal weights of the adaptive LUM filter are obtained using the least mean square (LMS) learning algorithm. Experimental results demonstrate that the proposed scheme outperforms other decision-based median filters in terms of noise suppression and detail preservation.

Keywords: support vector machine, least mean square, impulsive noise, image restoration.

1 Introduction

Impulsive noise is a common type of noise that often corrupts digital images during acquisition or transmission processes over open communication channels. Denoising is an essential step before image segmentation, edge detection and object recognition for image processing [1]. Image restoration is concerned with not only how to efficiently remove impulsive noise but also how to preserve image details. The median filter is a well-known nonlinear filter. However, while suppressing impulsive noise, the median filter sometimes removes fine details.

In recent years, variants of the median filter such as weight median (WM) filters, fuzzy-rule-based filters, and decision-based filters have been developed in an attempt to improve the median filter [2-6]. Satisfactory results have been achieved using these filters. Nevertheless, many WM filters tend to mistakenly alter noise-free pixels; the generalization capability of fuzzy-rule-based filters is poor; and the parameters of decision-based filters inflexibly depend on a pre-assumed noise density level. The latest advancement is the adaptive two-pass median (ATM) filter based on support vector machines (SVMs) by Lin and Yu, called the support vector classifier (SVC) based filter [3]. The SVC-based filter can be also regarded as a decision-based filter. It first utilizes SVMs to classify the signal as either noise-corrupted or noise-free and then applies the noise-free reduction filter to remove the corrupted pixels. However,

K.W. Wong et al. (Eds.): ICONIP 2010, Part II, LNCS 6444, pp. 82–90, 2010.

to improve the filtering performance, the SVC-based filter also needs proper threshold values for a pre-assumed noise density level in second pass filtering.

In this paper, we propose a novel adaptive SVC-based (ASVC) filter based on SVMs and a least mean square (LMS) learning algorithm to overcome the drawbacks of previous methods. The proposed ASVC filter consists of an SVM impulse detector and LUM (low-upper-middle) smoothers [10]. The proposed adaptive LUM filter uses an adjustable weight to best balance the tradeoff between impulse noise suppression and image detail preservation. The scalar quantizer (SQ) method and a learning approach based on the LMS algorithm are employed to obtain the optimal weight for each block independently [9]. With this filtering framework, the proposed ASVC filter can perform significantly better than other median-based filters in terms of noise suppression and detail preservation.

The rest of this paper is organized as follows. In Section 2, the concept of SVMs is reviewed. In Section 3, the design of the proposed ASVC filter is presented in detail. Section 4 presents the results of some extensive experiments. Finally, the conclusion is given in Section 5.

2 Support Vector Machines

Support vector machines (SVMs) have been recently proposed as a kind of feedforward network for pattern recognition because of their high generalization capability without priori knowledge [7]. In the present study, the SVM technique is employed to classify a signal as either noise-corrupted or noise-free.

Let $\{(x_i, y_i), i = 1, \cdots, N\}$ be a training data set of N data points, where $x_i \in R^n$ and $y_i \in \{-1, 1\}$. The training goal of the SVM is to find an optimal hyperplane that maximally separates the training data set into two classes. For the linearly inseparable case, the optimal classification hyperplane is found by solving the following quadratic programming (QP) problem [8]:

$$\begin{aligned} &\min\ J(W, \xi) = \frac{1}{2}\|W\|^2 + C\sum_{i=1}^{N}\xi_i \\ &s.t.\ \ y_i[W \cdot x_i + b] \geq 1 - \xi_i, \\ &\xi_i \geq 0, \quad i = 1, 2, \cdots, N, \end{aligned} \tag{1}$$

where W is a weight vector, ξ_i is a non-negative variable, b is a bias of the hyperplane, and C is a predefined positive constant. The optimization problem (1) can be rewritten as:

$$\begin{aligned} &\max\ M(\alpha) = -\frac{1}{2}\sum_{i,\,j=1}^{N}\alpha_i\alpha_j y_i y_j \langle x_i \cdot x_j \rangle + \sum_{i=1}^{N}\alpha_i \\ &s.t.\ \ \sum_{i=1}^{N}\alpha_i y_i = 0, \\ &\alpha_i \in [0, C], i = 1, 2, \cdots, N. \end{aligned} \tag{2}$$

By solving (2), we can get the optimal hyperplane:

$$\sum_{i=1}^{N} \alpha_i y_i \langle x \cdot x_i \rangle + b = 0. \tag{3}$$

Therefore, the linearly inseparable discrimination function is in the form:

$$f(x) = \mathrm{sgn}\,[\sum_{i=1}^{N} \alpha_i y_i \langle x_i \cdot x \rangle + b], \tag{4}$$

where $\mathrm{sgn}[\cdot]$ stands for the bipolar sign function and non-negative variables α_i are the Lagrange multipliers.

For the nonlinearly inseparable case, the original data is projected onto a certain high dimensional Euclidean space H by a nonlinear map $\Phi : R^n \to H$. The kernel function $K(x_i, x_j)$ is introduced such that it is not necessary to get to the bottom of $\Phi(\cdot)$. Hence, the optimal decision can be found by solving the following dual problem:

$$\begin{aligned} &\max\ Q(\alpha) = \sum_{i=1}^{N} \alpha_i - \frac{1}{2} \sum_{i=1}^{N} \sum_{j=1}^{N} \alpha_i \alpha_j y_i y_j K(x_i, x_j) \\ &s.t.\ \sum_{i=1}^{N} \alpha_i y_i = 0, \\ &0 \le \alpha_i \le C, \quad i = 1, 2, \cdots, N. \end{aligned} \tag{5}$$

By solving (5), we can get the optimal hyperplane with the maximal margin:

$$\sum_{i=1}^{N} \alpha_i y_i K(x, x_i) + b = 0. \tag{6}$$

Therefore, the nonlinearly inseparable discrimination function that separates the training data into two classes is in the form:

$$f(x) = \mathrm{sgn}\,[\sum_{i=1}^{N} \alpha_i y_i K(x, x_i) + b]. \tag{7}$$

3 Design of the ASVC Filter

Let $I = \{k = (k_1, k_2) \mid 1 \le k_1 \le H, 1 \le k_2 \le W\}$ denote the pixel coordinates of the image, where H and W are the image height and width, respectively. Let $x(k)$ represent the input pixel value of the image at location $k \in I$. The filter window $w\{k\}$ is defined in terms of the image coordinates symmetrically surrounding the current pixel $x(k)$. The window size $S = 2n+1$ (n is a non-negative integer) can be given by

$$w\{k\} = \{x_f(k) : f = 1, 2, \cdots, n, n+1, \cdots, S\}, \tag{8}$$

where the input pixel $x(k) = x_{n+1}(k)$ is the central pixel. Nonlinear LUM smoothers have been shown to be equivalent to center-weighted medians (CWM) [10]. LUM smoothers are defined as follows:

$$y(k) = MED\{x_{(c)}(k), x(k), x_{(S-c+1)}(k)\}, \tag{9}$$

where MED denotes the median operation, $1 \le c \le (S+1)/2$, $x(k)$ is the central sample from the filter window $w\{k\}$ and $x_{(1)}(k) \le x_{(2)}(k) \le \cdots \le x_{(S)}(k)$ is the rank-ordered set of $w\{k\}$.

3.1 Structure of the ASVC Filter

The framework of the proposed ASVC filter is illustrated in Fig. 1. At first, the SVM impulse detector is used to efficiently determine whether the median filter or the identity filter will be on. The first median filter can remove most of the noise, but smaller impulse noise might remain. To alleviate this problem, an adaptive LUM filter is used to remove noisy pixels that are missed, false alarms, or over-correction errors.

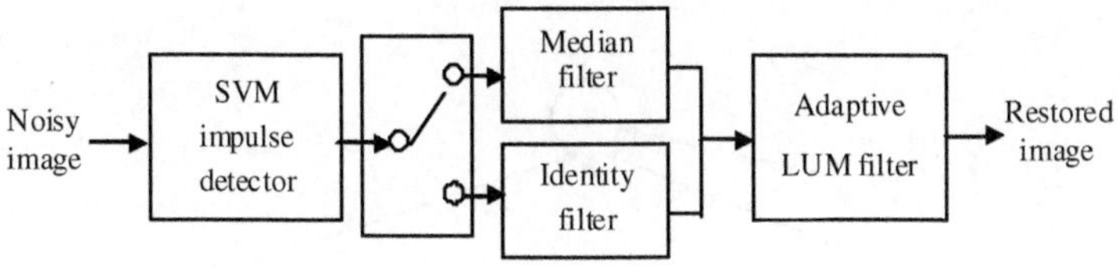

Fig. 1. Structure of ASVC filter

3.2 SVM Impulse Detector

3.2.1 Feature Extraction

Before noise filtering begins, the local features of the filter window $w\{k\}$ must be extracted to identify noisy pixels [3]. We take into account the local features in the filter window, such as prominent signals and the possible presence of details and edges. The following three variables can be defined to generate the feature vector $f\{k\}$ as the input data of the SVM impulse detector.

Definition 1: The variable $c(k)$ denotes the absolute difference between the input $x(k)$ and the median value of $w\{k\}$ as follows [1]:

$$c(k) = |x(k) - MED(w\{k\})|.$$

Definition 2: $c^{w_0}(k) = MED\{x_1(k), \cdots, x_n(k), w_0 \Diamond x_{n+1}(k), \cdots, x_S(k)\}$.

Here, w_0 denotes the non-negative integer weight, and $w_0 \Diamond x_{n+1}(k)$ means that there are w_0 copies of input pixel $x(k) = x_{n+1}(k)$.

Definition 3: $l(k) = \left| x(k) - c^3(k) \right|$.

Definition 4: $e(k) = \left| x(k) - c^5(k) \right|$.

Note that $c(k)$, $l(k)$ and $e(k)$ are a measure for detecting the probability whether the input $x(k)$ is noisy. In the present study, the feature vectors are given by:

$$f\{k\} = \{c(k), l(k), e(k)\} . \quad (10)$$

The feature vectors $f\{k\}$ serve as the input data set to the SVM impulse detector.

3.2.2 Training the SVM Classifier

The optimal separating hyperplane can be obtained through a training process by using a set of supervised class labels for the training corrupted image. The input in the training process is the set of unsupervised features $f\{k\}$. Figure 2 shows the feedforward network architecture of the SVM impulse detector that identifies noise-free pixels and noise-corrupted pixels [3].

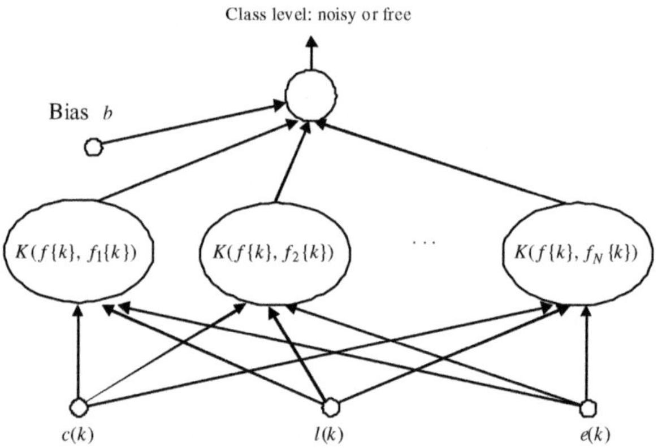

Fig. 2. Feedforward network architecture of support vector machines

3.3 Adaptive Weight of the LUM Filter

The adaptive weight c can help the LUM filter perform various degrees of noise suppression and image detail preservation. To decide the adaptive weight c, the weight controller shown in Fig. 3 is proposed in this work [9]. Note that $d(\cdot)$ shown in Fig. 3, which is defined as a function of the feature vector, is a classifier used to determine the partitioning, and $\beta_i(k), i \in \{1, 2, \cdots, M\}$ serves as the adaptive weight c for the LUM filter.

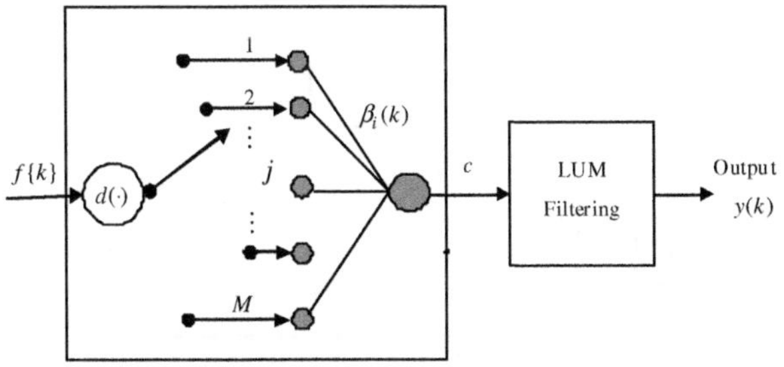

Fig. 3. Structure of the adaptive LUM filter

3.3.1 Partitioning of the Feature Space

The feature vector space $\Re^3$ is where the feature vector exists:

$$f\{k\} = \{c(k), l(k), e(k)\} \in \Re^3. \tag{11}$$

The weight controller, shown in Fig. 3, decides that $\Re^3$ is partitioned into M mutually exclusive blocks Ω_i, $i = 1, 2, \cdots, M$. Then, each weight $\beta_i(k)$ is associated with the i block in the partition given by:

$$\Omega_i = \{f\{k\} \in \Re^3 : d(f\{k\}) = i\}, \quad i = 1, 2, \cdots, M, \tag{12}$$

where the classifier $d(\cdot)$ is now defined as a function of the feature vector $f\{k\}$. As a result, the M blocks satisfy

$$\Re^3 = \bigcup_{i=1}^{M} \Omega_i \text{ and } \Omega_i \cap \Omega_j = \phi, \text{ for } i \neq j. \tag{13}$$

Each input $x(k)$ corresponding to its $f\{k\}$ is only classified into one of the M blocks by the classifier $d(\cdot)$. Due to the low computational complexity of the partitioning indices, the classifier $d(\cdot)$ can be designed using simple scalar quantization (SQ) [1], [9]. Each scalar component of $f\{k\}$ can be classified independently using SQ, which involves an encoder mapping process and a decoder mapping process [9]. The encoder mapping process includes receiving the input value $f_j\{k\}$ and providing an output codeword, which is determined using the interval in which the value falls. The decoder mapping process transforms the codeword into a representative value z. In the present study, the encoder mapping process divides the range [0, 255] into five intervals such that each scalar component $f_j\{k\}$ belongs to one of the five intervals, as shown in Fig. 4 [9]. Other quantization intervals are also possible. The three-dimensional array P can be used to perform the partitioning as: $P[z_1][z_2][z_3]$, $1 \le z_1 \le 5$, $1 \le z_2 \le 5$, $1 \le z_3 \le 5$.

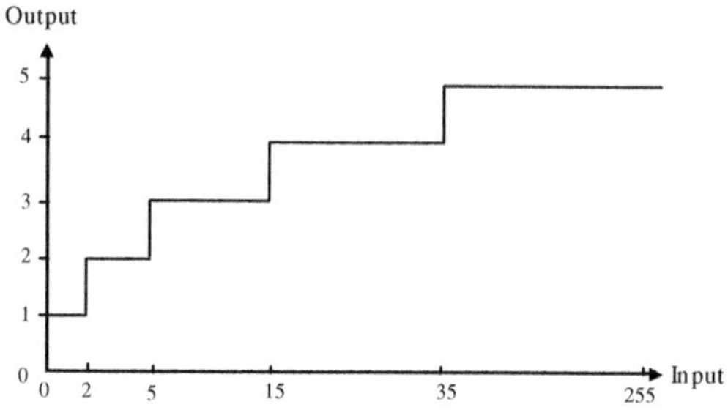

Fig. 4. Quantizer input-output map for $c(k)$, $l(k)$, and $e(k)$ scalar feature vectors

3.3.2 Setting Weights Using LMS Algorithm

Designing the optimal weight $\beta_i(k)$, $i = 1, 2, \cdots, M$ for the adaptive LUM filter requires minimizing the mean square error (MSE). The value of $\beta_i(k)$ can be obtained independently by executing the LMS algorithm, which is capable of minimizing the error function with respect to the block Ω_i [7]. The weights $\beta_i(k)$ corresponding to their block Ω_i can be adjusted in an iterative fashion along with the error surface toward the optimal solution. For each input $x(k)$ associated with block Ω_i, the value of $\beta_i(k)$ is updated iteratively in a gradient way:

$$\beta_i^{(t+1)}(k) = \begin{cases} \beta_i^{(t)}(k) - \eta_i \,|\, e(k) \,\|\, x(k) - o(k) \,|, & \beta_i^{(t+1)} \geq 0 \\ 0, & \beta_i^{(t+1)} < 0 \end{cases}. \tag{14}$$

Here, the error $e(k)$ is the difference between the desired output $o(k)$ and the physical output $y(k)$. η_i denotes the learning rate (with constant), $\beta_i^{(0)}(k)$ represents the initial weight, and $\beta_i^{(t)}(k)$ is the weight after the t-th iteration.

3.4 Adaptive LUM Filtering

At first, the median filter is activated for noisy pixels. The SVM impulse detector might make mistakes, so undetection and misdetection problems might occur with the ASVC filter. Undetected noisy pixels may remain in the restored image because the SVM impulse detector did not detect them as noisy and misdetected pixels may be mistakenly modified although they are noise-free. Therefore, we incorporated adaptive LUM filtering in the ASVC filter to reduce the number of undetected and misdetected pixels. Since the ASVC filter adaptively selects an optimized weight to carry out the filtering operation for each input pixel $x(k)$ corresponding to its $\beta_i(k)$ of block Ω_i, better noise attenuation can be achieved without degrading the quality of fine details.

4 Experimental Results

The optimal separating hyperplane and optimal weight $\beta_i(k), i = 1, 2, \cdots, M$ were obtained using a training image 'Couple' corrupted by 20% impulsive noise in the training process. Several experiments were conducted to compare the proposed ASVC filter with the standard median (MED) filter, the tri-state median (TSM) filter [2], the fuzzy median (FM) filter [4], the partition fuzzy median (PFM) filter [1], the fast peer group filter (FPGF) [6], and the adaptive two-pass median (ATM) filter [3] in terms of noise removal capability (measured in PSNR). 3×3 filter windows were used in all the experiments. Table 1 compares the PSNR results of removing both the fixed-valued and random-valued impulsive noise at 20%. As the table shows, the proposed ASVC filter performs significantly better than the other schemes. Figure 5 shows the restoration result comparison for the image 'Lake' corrupted by 20% random-valued impulsive noise. The ASVC filter produces a better subjective visual quality restored image by offering more noise suppression and detail preservation.

Table 1. Comparative restoration results in PSNR (dB) for 20% impulsive noise. (a) fix-valued impulse, (b) random-valued impulse.

Filter	Cameraman		F16		Boat		Lake		Lena	
	(a)	(b)	(a)	(b)	(a)	(b)	(a)	(b)	(a)	(b)
MED	33.76	31.51	29.62	30.93	29.20	30.14	27.19	27.84	30.18	31.72
TSM	34.71	31.62	32.76	31.56	31.16	32.29	29.73	30.22	31.84	34.03
FM	34.82	32.15	31.45	30.79	30.86	32.11	28.61	29.76	31.32	33.40
PFM	36.08	34.04	32.92	32.03	33.34	32.43	31.13	30.11	35.52	33.88
FPGF	34.11	29.65	29.66	28.52	29.94	29.46	27.58	28.15	31.06	30.11
ATM	36.64	33.34	32.99	32.00	33.05	32.42	30.62	29.96	35.77	34.26
ASVC	36.81	34.15	33.58	32.20	33.06	32.52	31.15	30.26	35.94	34.45

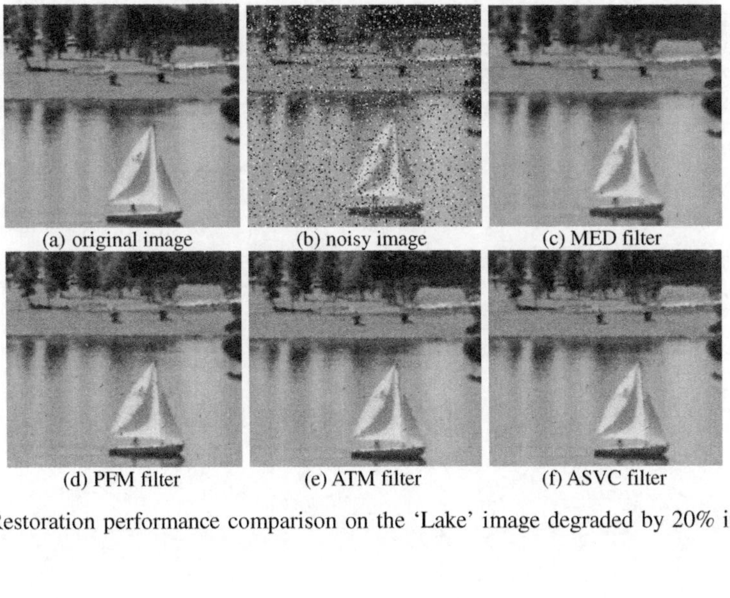

(a) original image (b) noisy image (c) MED filter

(d) PFM filter (e) ATM filter (f) ASVC filter

Fig. 5. Restoration performance comparison on the 'Lake' image degraded by 20% impulsive noise

5 Conclusion

In this paper, a novel decision-based median filter based on SVMs was developed to preserve more image details while effectively suppressing impulsive noise. A new impulse detector design based on SVMs is a part of the proposed ASVC filter; it is responsible for judging whether the input pixel is noisy. In addition, an adaptive LUM filter was proposed to efficiently improve the detection error of the SVM impulse detector. The excellent generalization capability of SVMs and the optimal weight of each block allow the mean square error of the filter output to be minimized. The experimental results demonstrate that the proposed ASVC filter is superior to a number of well-accepted decision-based median filters in the literature.

References

1. Lin, T.-C., Yu, P.-T.: Partition fuzzy median filter based on fuzzy rules for image restoration. Fuzzy Sets and Systems 147, 75–97 (2004)
2. Chen, T., Ma, K.K., Chen, L.H.: Tri-state median filter for image denoising. IEEE Trans. Image Processing 8, 1834–1838 (1999)
3. Lin, T.-C., Yu, P.-T.: Adaptive two-pass median filter based on support vector machines for image restoration. Neural Computation 16, 333–354 (2004)
4. Arakawa, K.: Median filters based on fuzzy rules and its application to image restoration. Fuzzy Sets and Systems 77, 3–13 (1996)
5. Lin, T.-C.: Partition belief median filter based on Dempster-Shafer theory in image processing. Pattern Recognition 41, 139–151 (2008)
6. Smolka, B., Chydzinski, A.: Fast detection and impulsive noise removal in color images. Real-Time Imaging 11, 389–402 (2005)
7. Haykin, S.: Neural Networks A Comprehensive Foundation, 2nd edn. Prentice Hall, Englewood Cliffs (2005)
8. Lin, K.M., Lin, C.J.: A study on reduced support vector machines. IEEE Trans. Neural Networks 14, 1449–1459 (2003)
9. Lin, T.-C.: A new adaptive center weighted median filter for suppressing noise in images. Information Sciences 177, 1073–1087 (2007)
10. Lukac, R.: Binary LUM smoothing. IEEE Signal Processing Letters 19, 400–403 (2002)

Tuning N-gram String Kernel SVMs via Meta Learning

Nuwan Gunasekara, Shaoning Pang, and Nikola Kasabov

KEDRI, AUT University, Private Bag 92006, Auckland 1020, New Zealand
{spang,nkasabov}@aut.ac.nz

Abstract. Even though Support Vector Machines (SVMs) are capable of identifying patterns in high dimensional kernel spaces, their performance is determined by two main factors: SVM cost parameter and kernel parameters. This paper identifies a mechanism to extract meta features from string datasets, and derives a *n-gram* string kernel SVM optimization method. In the method, a *meta model* is trained over computed string meta-features for each dataset from a string dataset pool, learning algorithm parameters, and accuracy information to predict the optimal parameter combination for a given string classification task. In the experiments, the *n-gram* SVM were optimized using the proposed algorithm over four string datasets: spam, Reuters-21578, Network Application Detection and e-News Categorization. The experiment results revealed that the proposed algorithm was able to produce parameter combinations which yield good string classification accuracies for *n-gram* SVM on all string datasets.

Keywords: Meta learning, n-gram String Kernels, SVM, Text Categorization, SVM Optimization.

1 Introduction

Support Vector Machines (SVMs) are a set of supervised learning techniques, using statistical learning principles, kernel mapping, and optimization techniques for classification and regression, SVM in its simplest form learns a separating hyperplane which maximizes the distance between the hyperplane and its closest point by solving a convex quadratic optimization problem. In practice, the effectiveness of SVM is heavily dependent upon three main factors: kernel selection, SVM cost parameter, and kernel parameters.

Apparently, as diverse sets of kernel functions are available, identifying the most suitable one for a given pattern recognition task is quite challenging, where the researcher spends considerable time on kernel identification. On the other hand, for a suitable kernel function, the performance is again influenced by two factors: SVM cost parameter and kernel parameters. Also, for string kernel parameters like substring length in *n-gram* kernel and subsequence size in fixed length subsequence kernel affects the performance of SVM. Thus, SVM kernel parameter optimization is a challenging difficulty for pattern recognition, due to the higher dimensionality of the parameter space.

K.W. Wong et al. (Eds.): ICONIP 2010, Part II, LNCS 6444, pp. 91–98, 2010.

While various optimization techniques are being used in SVM kernel parameter optimization, namely trial and error method, grid optimization, leave-one out cross validation, generalization error estimation using gradient descent and evolutionary algorithms [1]. It is noticeable that all the above SVM optimization methods are merely for numeric kernels, but not sustainable for string kernels. For string kernel SVM optimization, researchers are confronted with two main obstacles. Firstly, there is little or no literature pertaining to string kernel SVM optimization. Secondly, string characteristics influence string kernel SVM optimization, while it is often difficult to learn string characteristics from data.

The paper explains a novel string kernel SVM optimization method by adopting string characteristics extraction, meta model learning, and performance prediction regression. Consider the widely accepted success of *n-gram* SVM for string classification, the paper addresses only *n-gram* kernel as a starting work of our string kernel SVM optimization research.

2 N-gram String Kernel

For *n-gram* kernel, a string s is defined from alphabet Σ of $|\Sigma|$ symbols, and is presented in a feature space F, where each dimension is a string [2]. Also, Σ^* represents the set of all strings and Σ^n represents the string set of length n. Furthermore, ss' represents the concatenation of strings s and s'. Now, the *substrings*: u, v_1, v_2 of string s, are defined such that:

$$s = v_1 u v_2,$$

where, if $v_1 = \varepsilon$ (ε is the empty string of 0 length) then, u is called to be the *prefix* of s and if $v_2 = \varepsilon$, then u is called to be the *suffix* of s. Now, a feature map Φ is defined in feature space F, with below embedding,

$$\Phi_u^n(s) = |\{(v_1, v_2) : s = v_1 u v_2\}|, u \in \Sigma^n.$$

The associated kernel is defined as:

$$K_n(s,t) = \langle \Phi^n(s), \Phi^n(t) \rangle = \sum_{u \in \Sigma^n} \Phi_u^n(s) \Phi_u^n(t),$$

and the computational complexity of *n-gram* kernel is written as $O(n|s|t|)$ [2].

3 The Proposed Meta Learning Approach to N-gram SVM Optimization

Motivated by [3], this paper defines a set of string meta-features. With a pool of string datasets, a meta model is built on extracted string meta-features, string kernel SVM parameter combinations and accuracy information. The obtained meta model is used to predict the string classification accuracy of a given *n-gram* SVM parameter combination for a string data classification task.

```
input  :
         L_TR =Training String Dataset Pool
         L_TS =Testing String Dataset Pool
         C =Parameter Combination Pool for Training (c ∈ C)
         C' =Parameter Combination Pool for Testing (c' ∈ C')
         LA =SVM with n-gram String Kernel
output: Parameter combination ĉ_l which yields the best accuracy for string
         dataset D_lTS

for l ← 1 to l' do
    Pick D_lTR from L_TR
    for p ← 1 to p' do
        Compute f'_{p,D_lTR}
    end
    repeat
        Pick a parameter combination c from C
        Do 10-fold cross validation on D_lTR, using LA with parameter
        combination c which yields Y_{D_lTR,c} accuracy
    until no more parameter combinations in C;
end

Build a regression model (meta model) using f'_{p,D_lTR}, c, and Y_{D_lTR,c}
for l ← 1 to l' do
    Pick D_lTS from L_TS
    for p ← 1 to p' do
        Compute f'_{p,D_lTS}
    end
    repeat
        Pick a parameter combination c' from C'
        Predict accuracy Y_{D_lTS,c'} for LA with parameter combination c' using
        build meta model
        if Y_{D_lTS,c'} is maximum then
            ĉ_l = c'
        end
    until no more parameter combinations in C;
end
```

Algorithm 1. The proposed meta learning algorithm for *n-gram* SVM optimization

3.1 String Meta-features

In order to use the meta-features discussed in Lam et al. [3] for string classification, the string dataset needs to be presented as terms and term frequencies. We accomplish this in a string dataset by using splitting characters: "+':(){}[]. ,-\" to split a string into set of terms or synonymously *tokens*. This approach is referred as 'tokenization' in the literature [2]. Each token is associated with a frequency, which is how many times it occurs in the dataset. This token-frequency information are used to compute the string meta-features explained in this paper. Note that tokenization here is used only for string meta feature calculation, but not for n-gram SVM classification.

Assume a string dataset which has n number of instances, the seven string meta-features are:

1. **AvgInstanceLen:** The average instance length of the dataset. The instance length refers to number of tokens in an instance. The average is taken across all the instances. If i^{th} instance has N_i tokens, then the average instance length for that dataset is $\frac{\sum_{i=1}^{n} N_i}{n}$.
2. **AvgTokenVal:** The average token weight of an instance across a string dataset. Initially, the token weight is calculated for each token and the average is computed for single instance. Then, the average token weights for each instance are summed and the average is computed for all the instances. If there are m unique tokens in i^{th} instance, the average token weight for a string dataset is written as:

$$\textit{Average token weight of the string dataset} = \frac{\sum_{i=1}^{n}\sum_{j=1}^{m} TW(j,i)}{mn}, \quad (1)$$

where $TW(j,i)$ is the token weight of jth token in i^{th} instance. According to [4]s interpretation of *term weight*, the $TW(j,i)$ can be written as:

$$TW(j,i) = TF(j,i) \times IDF(j), \quad (2)$$

where $IDF(j)$ is the inverse document frequency of jth token, and $TF(j,i)$ is the frequency of jth token in instance i. Furthermore, according to [4], the $IDF(j)$ is computed as:

$$IDF(j) = \log\frac{n}{TF(j)} + 1, \quad (3)$$

where $TF(j)$ is the frequency of the jth token in the dataset. Now, considering (2) and (3), equation (1) is rewritten as:

$$\textit{Average token weight of the string dataset} = \frac{\sum_{i=1}^{n}\sum_{j=1}^{m} TF(j)\left(\log\frac{n}{TF(j)} + 1\right)}{mn} \quad (4)$$

3. **AvgMaxTokenVal:** The average maximum token weight of an instance across a string dataset. Maximum token weights of an instance are summed and the average is taken across all instances.
4. **AvgMinTokenVal:** The average minimum token weight of an instance across a given string dataset. Minimum token weights of an instance are summed and the average is taken across all instances.
5. **AvgTokenThre:** The average number of tokens above a token weight threshold for a given string dataset. The token weight threshold is set globally. The number of tokens where the token weight is above the threshold are summed and the average is taken across all instances.

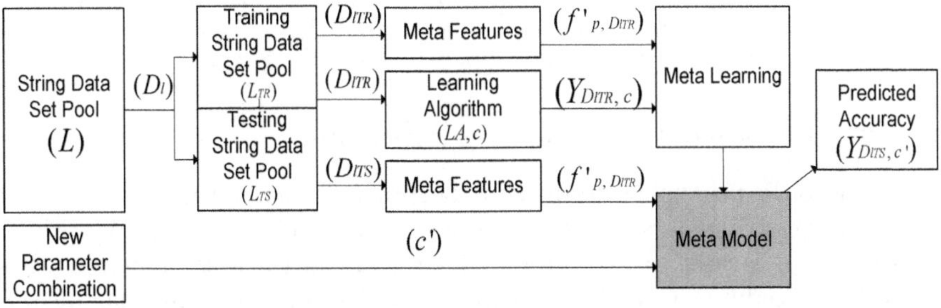

Fig. 1. The procedure to employ meta learning for string classification

6. **AvgTopInfoGain:** The average information gain of the top r tokens in the string dataset. The information gain of each individual token is computed for each instance and raked. Then, the average is taken across top r terms with highest information gain.
7. **NumInfoGainThres:** The average number of tokens in an instance where the information gain value exceeds a globally specified threshold.

3.2 Meta Learning on String Categorization

Consider a string dataset D, which is represented as a vector in *token-frequency space* Ω, where each dimension in Ω is associated with one *token*. Now, the dataset D is represented via function ω in this new Ω token-frequency space:

$$\omega(D) = (TF(t_1, D), TF(t_2, D), ..., TF(t_N, D)) \in \Omega,$$

where $TF(t_j, D)$ is the token frequency of jth token in the string dataset D, and N is the number of unique tokens in the dataset. Now one can derive a function $f_p : \Omega \rightarrow \mathbb{R}$:

$$f_p(D) = f'_{p,D}$$

where $f'_{p,D}$ represents the value for the pth string-meta feature for D. For the string dataset D, there are p' finite meta-features, where all string meta-features $f_p(D)$ $(p = 1, 2, 3, ..., p')$ are well defined.

Using the above discussed string meta-features, Figure 1 explains the principle of meta learning for string classification. Assume there is a string dataset pool L with l' datasets, where, each string dataset D_l $(D_l \in L, l = 1, ..., l')$ is again subdivided into unique D_{lTR} (training) and D_{lTS} (testing) datasets, which creates training (L_{TR}) and testing (L_{TS}) dataset pools. The string meta-feature $f'_{p,D_{lTR}}$ is computed for dataset D_{lTR}. Also, for D_{lTR}, the machine learning algorithm LA with parameter combination c, generates $Y_{D_{lTR},c}$ classification accuracy. These computed string meta-features ($f'_{p,D_{lTR}}$), parameter combinations (c) and accuracy information ($Y_{D_{lTR},c}$) generate a meta model via regression, which is able to predict the classification accuracy for a new string dataset, given the computed string meta-features and the parameter combination. Hence, for a new string dataset D_{lTS}, the meta model predicts the accuracy $Y_{D_{lTS},c'}$ for parameter combination c' by computing string-meta features $f'_{p,D_{lTS}}$.

3.3 The Proposed Meta Learning Algorithm

According to the principle introduced in Section 3.2, the built meta model is able to predict the string classification accuracy for a machine learning algorithm on a novel string dataset, using computed string meta-features. This section explains the procedure to use this principle (meta learning for string classification) to optimize string kernel SVMs, which is shown in Algorithm 1. Algorithm 1 explains the procedure to use meta learning to optimize *n-gram* string kernel with SVM.

4 Experiments

The proposed algorithm was experimented for *n-gram* SVM optimization. The algorithm was trained using training string dataset pool L_{TR}, and was tested on testing string dataset pool L_{TS}. In the experiments, SVM cost parameter (c) was selected as $2^0, 2^1, ..., 2^{16}$ and the substring length in *n-gram* string kernel was selected as $1, 2, ..., 8$. The string meta-feature, $AvgMinTokenVal$ was not considered in the training stage, as it was having the value 0 for all datasets. Also, the global threshold for the $AvgTokenThr$ was set to 2 in all the experiments. Support Vector Regression (SVR) was used to build the meta model. In the training stage, the parameters which yield lowest cross validation RMSE for SVR, were considered in regression (in building the meta model). 10 fold- cross validation was done for the top 10 predicted parameter combinations, on each string dataset. The performance evaluation was done considering Root Mean Squared Error (RMSE) for the top 10 predicted parameters on each dataset.

In the experiments, the *n-gram* string kernel was implemented using shogun octave interface [5]. The string meta-feature computation program was coded using C++ language. All the experiments were run on a PC having Intel Core2 Duo 3GHz processor and 2.96 Gb RAM.

4.1 Datasets

Four string datasets were used in the string dataset pool $L = \{$*Spam*, *Reuters-21578*, *Network Application Detection*, *e-News Categorization*$\}$. In a strict meta learning concept, datasets for training and testing should be from different sources. For a simple meta learning experiment, we construct training dataset pool (L_{TR}) and testing dataset pool (L_{TS}) by partitioning respectively *Spam*, *Reuters-21578*, *Network Application Detection*, *e-News Categorization* into two parts, and taking one part for training and the other part for testing.

The spam dataset consists of 696 ham messages and 384 spam messages from [6]. The Reuters dataset used in the experiments has the exact split to Lodhi et al. [7]. The network application detection dataset consists of network traffic

Table 1. Experimental Results for *N-gram* SVM Optimization on each string dataset)

(a) Spam Data

cost	substring length	rank	predicted%	actual%
16384	8	1	99.3074	98.33333
4096	8	2	99.2982	98.33333
32768	7	3	99.2683	98.33333
16384	7	4	99.2652	98.33333
4096	7	5	99.2628	98.33333
4096	6	6	99.2524	98.33333
4096	5	7	99.2066	98.33333
2048	8	8	99.1935	98.33333
32768	6	9	99.1862	98.33333
4096	2	10	99.1803	97.81251
root mean squared error				0.971167898

(b) Reuters-21578

cost	substring length	rank	predicted%	actual%
4096	6	1	92.6961	95.36587
2048	6	2	92.6842	95.36587
8192	6	3	92.6771	95.36587
1024	6	4	92.6636	95.36587
512	6	5	92.6474	95.36587
16384	6	6	92.6405	95.36587
256	6	7	92.6376	95.36587
128	6	8	92.6322	95.36587
64	6	9	92.6294	95.36587
32	6	10	92.6279	95.36587
root mean squared error				2.712372587

(c) Network Application Detection

cost	substring length	rank	predicted%	actual%
16384	2	1	99.6656	98.22917
2048	2	2	99.5895	98.22917
1024	2	3	99.5776	98.22917
32768	2	4	99.5685	98.22917
4096	2	5	99.5634	98.22917
512	2	6	99.5607	98.22917
256	2	7	99.5489	98.22917
128	2	8	99.542	98.22917
64	2	9	99.5384	98.22917
32	2	10	99.5365	97.81252
root mean squared error				1.386738588

(d) e-News Categorization

cost	substring length	rank	predicted%	actual%
4096	5	1	90.8189	74.35295
4096	6	2	90.6225	75.05881
4096	4	3	90.5267	73.76471
8192	5	4	90.484	73.88235
32768	5	5	90.3336	73.64706
16384	5	6	90.3317	73.64706
8192	6	7	90.3088	75.17647
4096	3	8	90.2765	74.70588
8192	4	9	90.275	73.41176
32768	4	10	90.1778	73.17646
root mean squared error				16.34502652

Table 2. Root Mean Squared Error (RMSE) for String Kernel SVM Optimization on each String Dataset (for top 10 predicted parameter combinations)

String Kernel	Dataset	RMSE	Avg RMSE
N-gram	Spam	0.971168	5.353826
	Reuters-21578	2.712373	
	Network Application Detection	1.386739	
	e-News Categorization	16.345027	

data produced by network applications. The e-News Categorization dataset is collected from four electronic newspapers: New Zealand Herald, The Australian, The Independent and The Times, on five news topics: business, education, entertainments, sport and travel.

5 Results on *N-gram* SVM Optimization

In the experiment, the algorithm attempts to find optimized parameters (substring length and SVM cost) for *n-gram* SVM. The algorithm was trained on string dataset pool L_{TR} and tested on testing string dataset pool L_{TS}. The SVR parameters: $\gamma = 0.95$ and *SVR Cost=500* were used in regression. 10-fold cross validation was done for the top 10 predicted parameter combinations on each string dataset. Table 1 summarizes the experiment results for the top 10 predicted combinations on each dataset. As seen in Table 1, the proposed algorithm produces optimized parameters, which yield good string classification accuracies for *n-gram* SVM, on all four string datasets. The algorithm has a very low RMSE for top 10 predicted on spam, Reuters-21578 and network application detection datasets (see Table 1a- 1c). Even though in Table 2, the proposed algorithm is seen with a quite high RMSE on the e-News categorization dataset, the top 10 predicted parameter combinations yield still good string classification accuracies on the e-News dataset (see Table 1d).

6 Conclusions and Future Work

Addressing *n-gram* string kernel SVM optimization, this paper proposes a novel meta learning algorithm. In this study, we define *string meta-features* for extracting meta knowledge from any string dataset, apply obtained string meta features for meta learning on string classification, and finally predict optimum string kernel SVM parameters for *n-gram* string kernel SVM on a string dataset. The experimental results show that the proposed algorithm produces parameter combinations which yield good string classification accuracies on most of the datasets. In practice, the proposed algorithm is applicable to all types of string kernel optimization, even though some string kernel SVMs (e.g. *edit-distance* SVM) may not produce the same good classification result as the *n-gram* string SVM. One limitation of the presented research is that, depending upon the SVR parameters, the proposed algorithm sometime yields accuracies above 100% on a given string dataset. This can be resolved by setting upper and lower bounds for the SVR regression. We will leave this as our future work along with developing new string kernel optimization techniques.

References

1. Zhang, X.L., Chen, X., He, Z.: An ACO-based algorithm for parameter optimization of support vector machines. Expert Systems with Applications (9), 6618–6628 (2010)
2. Shawe-Taylor, J., Cristianini, N.: Kernel methods for pattern analysis. Cambridge University Press, Cambridge (2004)
3. Lam, W., Lai, K.: A meta-learning approach for text categorization. In: Proceedings of the 24th Annual International ACM SIGIR Conference on Research and Development in Information Retrieval, pp. 303–309. ACM, New York (2001)
4. Hersh, W.: Information retrieval: A health and biomedical perspective. Springer, New York (2008)
5. Sonnenburg, S., Raetsch, G., Schaefer, C., Schoelkopf, B.: Large scale multiple kernel learning. The Journal of Machine Learning Research 7, 1531–1565 (2006)
6. Spam assassin public mail corpus (2002), `http://spamassassin.apache.org/publiccorpus/` (Retrieved December 23, 2009)
7. Lodhi, H., Saunders, C., Shawe-Taylor, J., Cristianini, N., Watkins, C.: Text classification using string kernels. The Journal of Machine Learning Research 2, 419–444 (2002)

Bilinear Formulated Multiple Kernel Learning for Multi-class Classification Problem

Takumi Kobayashi and Nobuyuki Otsu

National Institute of Advanced Industrial Science and Technology,
1-1-1 Umezono, Tsukuba, Japan

Abstract. In this paper, we propose a method of multiple kernel learning (MKL) to inherently deal with multi-class classification problems. The performances of kernel-based classification methods depend on the employed kernel functions, and it is difficult to predefine the optimal kernel. In the framework of MKL, multiple types of kernel functions are linearly integrated with optimizing the weights for the kernels. However, the multi-class problems are rarely incorporated in the formulation and the optimization is time-consuming. We formulate the multi-class MKL in a bilinear form and propose a scheme for computationally efficient optimization. The scheme makes the method favorably applicable to large-scaled samples in the real-world problems. In the experiments on multi-class classification using several datasets, the proposed method exhibits the favorable performance and low computation time compared to the previous methods.

Keywords: Kernel methods, multiple kernel learning, multi-class classification, bilinear form.

1 Introduction

The kernel-based methods have attracted keen attentions, exhibiting the state-of-the-art performances, such as in support vector machines (SVM) [10] and kernel multivariate analyses [8]. These methods are applied in various real-world tasks, e.g., in the fields of computer vision and signal processing. In the kernel-based methods, the input vectors are implicitly embedded in a high dimensional space (called kernel feature space) via kernel functions which efficiently compute inner products of those vectors in the kernel feature space. Thus, the performance of the kernel-based methods depends on how to construct the kernel functions.

In recent years, Lanckriet et al. [5] proposed the method to integrate different kernel functions with optimizing the weights for the kernels, which is called multiple kernel learning (MKL). By combining multiple types of kernels, the heterogeneous information, which is complementary to each other, can be effectively incorporated, possibly improving the performance. The composite kernel is successfully applied to, for example, object recognition [11].

In MKL, the weights for combining the kernels are obtained via the optimization processes based on a certain criterion, mainly for classification. Since the criterion can be defined in different formula, various methods for MKL have been

K.W. Wong et al. (Eds.): ICONIP 2010, Part II, LNCS 6444, pp. 99–107, 2010.

proposed by treating different optimization problems in different approaches; e.g., semi-definite programming [5] and semi-infinite linear program [9,13]. Most of the methods, however, are intended for classifying binary classes, while real-world problems contain multi classes in general. In addition, for application to practical problems, the optimization process should be computationally efficient.

In this paper, we propose a MKL method for multi-class classification problems. Without decomposing the multi-class problem into several binary class problems, the proposed method inherently deals with it based on the formulation of Crammer & Singer [2] which first proposed multi-class SVM using a single kernel. The contributions of this paper are as follows:

- We extend the formulation of multi-class classification in [2] to cope with multiple kernel functions, and formulate multi-class MKL (MC-MKL) in a bilinear form. In the formulation, the optimal weights for kernel functions are obtained in respective classes.
- We propose a scheme to effectively optimize the bilinear formulated problem, which makes the method applicable to large-scaled samples.
- In the experiments on various datasets, we demonstrate the effectiveness of the proposed method, compared to the existing MKL methods [7,13].

While Zien & Ong [13] proposed the method of MC-MKL based on a similar formulation, we employ a different criterion for the margin of the multi-class classifiers and propose a more efficient optimization scheme.

2 Bilinear Formulation for MC-MKL

To consider multiple kernels, we introduce multiple types of features $\boldsymbol{x}^{(r)}$ ($r \in \{1, .., R\}$, where R is the number of feature types). The inner products of those features can be replaced with respective types of kernels via kernel tricks: $\boldsymbol{x}_i^{(r)\prime}\boldsymbol{x}_j^{(r)} \to k_r(\boldsymbol{x}_i^{(r)}, \boldsymbol{x}_j^{(r)})$. Crammer & Singer [2] have proposed a formulation for multi-class SVM, considering only a single type of feature $\boldsymbol{x}$. We extend the formulation to incorporate the multiple types of features (kernels). We additionally introduce the weights $\boldsymbol{v}$ for feature types as well as the weights $\boldsymbol{w}$ within features similarly in MKL methods [13]. These two kinds of weights are mathematically integrated into the following bilinear form to constitute multi-class classification [2]:

$$\overset{*}{c} = \arg\max_{c \in \{1,..,C\}} \left\{ \sum_{r=1}^{R} v_c^{(r)} \boldsymbol{w}_c^{(r)\prime} \boldsymbol{x}^{(r)} = \boldsymbol{w}_c' \boldsymbol{X} \boldsymbol{v}_c = \langle \boldsymbol{X}, \boldsymbol{w}_c \boldsymbol{v}_c' \rangle_F \right\}, \tag{1}$$

where C is the number of classes, $\langle \, , \rangle_F$ indicates Frobenius inner product, $v_c^{(r)}$ is a weight for the r-th type of feature, $\boldsymbol{w}_c^{(r)}$ is a classifier vector for the r-th type of feature vector in class c, and these variables are concatenated into long vectors, respectively;

$$\boldsymbol{v}_c \triangleq \left[v_c^{(1)}, \cdots, v_c^{(R)}\right]', \; \boldsymbol{X} \triangleq \begin{bmatrix} \boldsymbol{x}^{(1)} & 0 & 0 \\ 0 & \ddots & 0 \\ 0 & 0 & \boldsymbol{x}^{(R)} \end{bmatrix}, \; \boldsymbol{w}_c \triangleq \begin{bmatrix} \boldsymbol{w}_c^{(1)} \\ \vdots \\ \boldsymbol{w}_c^{(R)} \end{bmatrix}. \tag{2}$$

Then, we can consider the margin of the above-defined bilinear classifiers (projections) by using the Frobenius norm $||\boldsymbol{w}_c\boldsymbol{v}_c'||_F$, and define the following optimization problem based on a large margin criterion:

$$\begin{aligned} \text{OP:} \quad & \min_{\{\boldsymbol{w}_c,\boldsymbol{v}_c\},\{\xi_i\}} \sum_{c=1}^{C} ||\boldsymbol{w}_c\boldsymbol{v}_c'||_F + \kappa \sum_{i=1}^{N} \xi_i \\ \text{s.t.} \quad & \forall i, c \quad \langle \boldsymbol{X}_i, \boldsymbol{w}_{y_i}\boldsymbol{v}_{y_i}' \rangle_F - \langle \boldsymbol{X}_i, \boldsymbol{w}_c\boldsymbol{v}_c' \rangle_F + \delta_{y_i,c} \geq 1 - \xi_i, \end{aligned} \tag{3}$$

where N is the number of samples, y_i indicates the class label of the i-th sample; $y_i \in \{1, .., C\}$, $\delta_{y_i,c}$ equals to 1 if $c = y_i$ and 0 otherwise, ξ_i is a slack variable for soft margin, and κ is a parameter to control the trade-off between the margin and the training errors. Note that we minimize the Frobenius norm, not squared one in SVM. Since this problem is difficult to directly optimize, we employ the upper bound of the Frobenius norm:

$$||\boldsymbol{w}_c\boldsymbol{v}_c'||_F = ||\boldsymbol{w}_c||||\boldsymbol{v}_c|| \leq \frac{1}{2}(||\boldsymbol{w}_c||^2 + ||\boldsymbol{v}_c||^2). \tag{4}$$

Therefore, the optimization problem OP is modified to

$$\begin{aligned} \text{P':} \quad & \min_{\{\boldsymbol{w}_c,\boldsymbol{v}_c\},\{\xi_i\}} \frac{1}{2}\left(\sum_{c=1}^{C} ||\boldsymbol{w}_c||^2 + ||\boldsymbol{v}_c||^2\right) + \kappa \sum_{i=1}^{N} \xi_i \\ \text{s.t.} \quad & \forall i, c \quad \boldsymbol{w}_{y_i}'\boldsymbol{X}_i\boldsymbol{v}_{y_i} - \boldsymbol{w}_c'\boldsymbol{X}_i\boldsymbol{v}_c + \delta_{y_i,c} \geq 1 - \xi_i. \end{aligned} \tag{5}$$

The weights $\boldsymbol{w}$ and $\boldsymbol{v}$ separately emerge as standard squared norm, which facilitates the optimization. It can be shown that the OP and P' have the identical optimum solution by using rescaling technique described in Sec. 3.3.

In the problem P', if the optimal weights $\overset{*}{\boldsymbol{v}}$ are obtained, the optimal classifier vectors are represented as $\overset{*}{\boldsymbol{w}}_c = \sum_{i=1}^{N} \overset{*}{\tau}_{ic}\boldsymbol{X}_i\overset{*}{\boldsymbol{v}}_c$, where $\overset{*}{\tau}_{ic}$ are the optimal dual variables [2]. Thus, the multi-class bilinear classifier in Eq.(1) results in

$$\begin{aligned} \overset{*}{\boldsymbol{w}}{}_c'\boldsymbol{X}\overset{*}{\boldsymbol{v}}_c &= \sum_{i=1}^{N} \overset{*}{\tau}_{ic}\overset{*}{\boldsymbol{v}}{}_c'\boldsymbol{X}_i'\boldsymbol{X}\overset{*}{\boldsymbol{v}}_c = \sum_{i=1}^{N} \overset{*}{\tau}_{ic} \sum_{r=1}^{R} \overset{*}{v}{}_c^{(r)2} \boldsymbol{x}_i^{(r)'}\boldsymbol{x}^{(r)} \\ &\rightarrow \sum_{i=1}^{N} \overset{*}{\tau}_{ic} \sum_{r=1}^{R} \overset{*}{v}{}_c^{(r)2} k_r(\boldsymbol{x}_i^{(r)}, \boldsymbol{x}^{(r)}), \end{aligned} \tag{6}$$

where $k_r(\boldsymbol{x}_i^{(r)}, \boldsymbol{x}^{(r)})$ is a kernel function on behalf of the inner-product of the r-th type of features, $\boldsymbol{x}_i^{(r)'}\boldsymbol{x}^{(r)}$, in kernel tricks. Note that the kernel functions can be differently defined for respective feature types. The squared weights $v_c^{(r)2}$ play a role to weight the kernel functions as in MKL, and produce the composite kernel function specialized to class c. In this case, we can introduce alternative nonnegative variables $d_c^{(r)} = v_c^{(r)2} \geq 0$ without loss of generality. The variables $\boldsymbol{d}$ are the weights for kernel functions, and therefore the above bilinear formulation is applicable to MC-MKL. The primal problem P' is reformulated to

$$\begin{aligned}\text{P:} \min_{\{\boldsymbol{w}_c,\boldsymbol{d}_c\},\{\xi_i\}} & \frac{1}{2}\left(\sum_{c=1}^{C}||\boldsymbol{w}_c||^2 + \mathbf{1}'\boldsymbol{d}_c\right) + \kappa\sum_{i=1}^{N}\xi_i \quad (7)\\ \text{s.t. } \forall i, c\ & \boldsymbol{w}'_{y_i}\boldsymbol{X}_i\boldsymbol{d}_{y_i}^{\frac{1}{2}} - \boldsymbol{w}'_c\boldsymbol{X}_i\boldsymbol{d}_c^{\frac{1}{2}} + \delta_{y_i,c} \geq 1-\xi_i,\ \ \boldsymbol{d}_c \geq 0,\end{aligned}$$

where $\boldsymbol{d}_c = [d_c^{(1)}, .., d_c^{(R)}]'$, and $\boldsymbol{d}_c^{\frac{1}{2}}$ is a component-wise square root of the vector $\boldsymbol{d}_c$. In the problem P, non-negativity constraint is additionally introduced to the problem P' (or OP). The bilinear classifier is finally obtained by

$$\overset{*}{\boldsymbol{w}}{}_c'\boldsymbol{X}\overset{*}{\boldsymbol{d}}{}_c^{\frac{1}{2}} = \sum_{i=1}^{N}\overset{*}{\tau}_{ic}\sum_{r=1}^{R}\overset{*}{d}{}_c^{(r)}k_r(\boldsymbol{x}_i^{(r)}, \boldsymbol{x}^{(r)}). \quad (8)$$

We describe the scheme to efficiently optimize P in the following section.

3 Optimization Methods

The primal problem P in Eq.(7) has the following dual form, similarly to [11]:

$$\max_{\{\boldsymbol{\tau}_i\}} \sum_{i=1}^{N}\boldsymbol{e}'_{y_i}\boldsymbol{\tau}_i, \text{ s.t. } \forall i\ \ \boldsymbol{\tau}_i \leq \kappa\boldsymbol{e}_{y_i},\ \mathbf{1}'\boldsymbol{\tau}_i = 0,\ \forall r, c\ \ \frac{1}{2}\sum_{i,j}k_r(\boldsymbol{x}_i^{(r)}, \boldsymbol{x}_j^{(r)})\tau_{ic}\tau_{jc} \leq \kappa.$$

where $\boldsymbol{\tau}_i$ is the i-th C-dimensional dual variable, $\boldsymbol{e}_{y_i}$ is a C-dimensional vector in which only the y_i-th element is 1 and the others are 0, and $\mathbf{1}$ is a C-dimensional vector of which all elements are 1. This is a convex problem having the global optimum. However, it is actually solved by second order cone programming, which requires exhaustive computational cost, and it is not applicable to large-scaled samples. Therefore, we take an alternative scheme to optimize the primal problem P in a manner similar to [7,11]. The scheme is based on the iterative optimization for $\boldsymbol{w}$ and $\boldsymbol{d}$, with applying projected gradient descent.

3.1 Optimization with Respect to $\boldsymbol{w}$

At the t-th iteration with fixing the variable $\boldsymbol{d}$ to $\boldsymbol{d}^{[t]}$, in a manner similar to [2], the problem P results in the dual form:

$$\begin{aligned}&\max_{\{\boldsymbol{\tau}_i\}}\ -\frac{1}{2}\sum_{i,j=1}^{N}\sum_{c=1}^{C}(\boldsymbol{v}_c^{[t]\prime}\boldsymbol{X}'_i\boldsymbol{X}_j\boldsymbol{v}_c^{[t]})\tau_{ic}\tau_{jc} + \sum_{i=1}^{N}\boldsymbol{e}'_{y_i}\boldsymbol{\tau}_i\\ \Leftrightarrow\ \text{D}_w\text{:}\ &\max_{\{\boldsymbol{\tau}_i\}}\ -\frac{1}{2}\sum_{i,j=1}^{N}\boldsymbol{\tau}'_i\boldsymbol{\Lambda}_{ij}\boldsymbol{\tau}_j + \sum_{i=1}^{N}\boldsymbol{e}'_{y_i}\boldsymbol{\tau}_i,\quad \text{s.t. } \forall i\ \ \boldsymbol{\tau}_i \leq \kappa\boldsymbol{e}_{y_i},\ \ \mathbf{1}'\boldsymbol{\tau}_i = 0,\ (9)\end{aligned}$$

where $\boldsymbol{\Lambda}_{ij}$ is a C-dimensional diagonal matrix, $\{\boldsymbol{\Lambda}_{ij}\}_{cc} = \sum_{r=1}^{R} d_c^{(r)[t]}k_r(\boldsymbol{x}_i^{(r)}, \boldsymbol{x}_j^{(r)})$. In this dual problem, the constants derived from $\boldsymbol{d}^{[t]}$ are omitted. This is optimized by iteratively solving the decomposed small subproblem [2,3], as follows.

Algorithm 1. Optimization for subproblem SubD_w

Require: Reindex $\tilde{b}_c = \frac{b_c}{\lambda_c}$ and λ_c such that $\tilde{b}_c$ are sorted in decreasing order

Initialize $c = 2,\ \ \zeta_{num} = \lambda_1^2\tilde{b}_1 - \kappa,\ \ \zeta_{den} = \lambda_1^2.$

while $c \leq C,\ \ \zeta = \frac{\zeta_{num}}{\zeta_{den}} \leq \tilde{b}_c$ **do**

$\zeta_{num} \leftarrow \zeta_{num} + \lambda_c^2\tilde{b}_c,\ \ \zeta_{den} \leftarrow \zeta_{den} + \lambda_c^2,\quad c \leftarrow c+1.$

end while

Output: $\tilde{\tau}_c = \min(b_c, \zeta\lambda_c), \qquad \therefore \tau_c = \min\{\kappa\delta_{y,c}, \lambda_c^2(\zeta - \beta_c)\}.$

The dual problem D_w is decomposed into N small subproblems; the i-th subproblem focuses on the dual variable $\boldsymbol{\tau}_i$ associated with the i-th sample, while fixing the others $\boldsymbol{\tau}_j$ $(j \neq i)$:

$$\text{SubD}_w\text{: } \max_{\boldsymbol{\tau}_i}\ -\frac{1}{2}\boldsymbol{\tau}_i'\boldsymbol{\Lambda}_{ii}\boldsymbol{\tau}_i - \boldsymbol{\beta}'\boldsymbol{\tau}_i - \gamma, \quad \text{s.t. } \boldsymbol{\tau}_i \leq \kappa\boldsymbol{e}_{y_i},\ \ \boldsymbol{1}'\boldsymbol{\tau}_i = 0, \tag{10}$$

where

$$\boldsymbol{\beta} = \sum_{j\neq i}\boldsymbol{\Lambda}_{ij}\boldsymbol{\tau}_j - \boldsymbol{e}_{y_i}, \quad \gamma = \frac{1}{2}\sum_{j\neq i, k\neq i}\boldsymbol{\tau}_j\boldsymbol{\Lambda}_{jk}\boldsymbol{\tau}_k - \sum_{j\neq i}\boldsymbol{e}_{y_j}'\boldsymbol{\tau}_j.$$

For optimization in D_w, the process to solve SubD_w works in rounds for all i and the dual variables $\boldsymbol{\tau}_i$ are updated until convergence. The subproblems are rather more complex than those in [2] since they include not scalar value $\boldsymbol{x}_i'\boldsymbol{x}_j$ but the diagonal matrix $\boldsymbol{\Lambda}_{ij}$ derived from multiple features (kernels). However, it is noteworthy that they are solved at a quite low computational cost, as follows.

Optimization for Subproblem SubD_w

In the following, we omit the index i for simplicity. By ignoring the constant, the subproblem SubD_w in Eq.(10) is reformulated to

$$\min_{\tilde{\boldsymbol{\tau}}}\ \frac{1}{2}||\tilde{\boldsymbol{\tau}}||^2, \quad \text{s.t. } \tilde{\boldsymbol{\tau}} \leq \kappa\lambda_y^{-1}\boldsymbol{e}_y + \boldsymbol{\Lambda}^{-\frac{1}{2}}\boldsymbol{\beta}, \quad \boldsymbol{\lambda}'\tilde{\boldsymbol{\tau}} = \boldsymbol{\lambda}'\boldsymbol{\Lambda}^{-\frac{1}{2}}\boldsymbol{\beta},$$

where $\tilde{\boldsymbol{\tau}} = \boldsymbol{\Lambda}^{\frac{1}{2}}\boldsymbol{\tau} + \boldsymbol{\Lambda}^{-\frac{1}{2}}\boldsymbol{\beta}$, $\boldsymbol{\lambda}$ is a C-dimensional vector composed of diagonal elements of $\boldsymbol{\Lambda}^{-\frac{1}{2}}$, and λ_y is the y-th element of the vector $\boldsymbol{\lambda}$. By using $\boldsymbol{b} = \kappa\lambda_y^{-1}\boldsymbol{e}_y + \boldsymbol{\Lambda}^{-\frac{1}{2}}\boldsymbol{\beta}$, the constraints are rewritten as

$$\text{s.t. } \tilde{\boldsymbol{\tau}} \leq \boldsymbol{b}, \quad \boldsymbol{\lambda}'\tilde{\boldsymbol{\tau}} = \boldsymbol{\lambda}'\boldsymbol{b} - \kappa. \tag{11}$$

The Lagrangian for this problem is

$$L = \frac{1}{2}||\tilde{\boldsymbol{\tau}}||^2 - \boldsymbol{\alpha}'(\boldsymbol{b} - \tilde{\boldsymbol{\tau}}) - \zeta(\boldsymbol{\lambda}'\tilde{\boldsymbol{\tau}} - \boldsymbol{\lambda}'\boldsymbol{b} + \kappa), \tag{12}$$

where $\boldsymbol{\alpha} \geq 0$, ζ are Lagrangian multipliers. When the subproblem is optimized, the followings hold:

$$\frac{\partial L}{\partial\tilde{\boldsymbol{\tau}}} = \tilde{\boldsymbol{\tau}} + \boldsymbol{\alpha} - \zeta\boldsymbol{\lambda} = 0, \quad \text{KKT: } \forall c\ \ \alpha_c(b_c - \tilde{\tau}_c) = 0. \tag{13}$$

Therefore, we obtain

$$\alpha_c = 0 \Rightarrow \tilde{\tau}_c = \zeta\lambda_c,\ \ \zeta \leq \frac{b_c}{\lambda_c}, \qquad \alpha_c > 0 \Rightarrow \tilde{\tau}_c = b_c,\ \zeta > \frac{b_c}{\lambda_c}. \tag{14}$$

By using the above, the second constraint in Eq.(11) results in

$$\boldsymbol{\lambda}'\tilde{\boldsymbol{\tau}} = \zeta \sum_{c|\alpha_c=0} \lambda_c^2 + \sum_{c|\alpha_c>0} \lambda_c b_c = \sum_{c=1}^{C} \lambda_c b_c - \kappa, \quad \therefore \zeta = \frac{\sum_{c|\alpha_c=0} \lambda_c b_c - \kappa}{\sum_{c|\alpha_c=0} \lambda_c^2}. \tag{15}$$

Thus, for solving the subproblem, we only seek ζ satisfying Eq.(14) and (15), and the simple algorithm is constructed in Algorithm 1.

The optimization of D_w is the core and most exhaustive process for the whole optimization in P. Therefore, the effective algorithm (Algorithm 1) to solve the subproblem SubD_w makes the whole optimization process computationally efficient.

3.2 Optimization with Respect to $\boldsymbol{d}$

Then, the optimization of P is performed with respect to $\boldsymbol{d}$. In this study, we simply employ projected gradient descent approach, although the other method such as in [12] would be applicable. In this approach, the objective cost function is minimized by a line search [6] along the projected gradient under the constraints $\boldsymbol{d} \geq 0$. Based on the principle of strong duality, the primal P is represented by using D_w in Eq.(9) with the optimal dual variables $\boldsymbol{\tau}^{[t]}$ as

$$\min_{\{\boldsymbol{d}_c\}} \left\{ \left(\sum_{c=1}^{C} \frac{1}{2}\mathbf{1}'\boldsymbol{d}_c - \boldsymbol{\theta}_c'\boldsymbol{d}_c \right) + \sum_{i=1}^{N} \boldsymbol{e}_{y_i}'\boldsymbol{\tau}_i^{[t]} = W(\boldsymbol{d}) \right\}, \qquad \text{s.t. } \forall c\ \ \boldsymbol{d}_c \geq 0,$$

where $\boldsymbol{\theta}_c$ is a R-dimensional vector of $\theta_c^{(r)} = \frac{1}{2}\sum_{i,j} \tau_{ic}^{(r)[t]}\tau_{jc}^{(r)[t]} k_r(\boldsymbol{x}_i^{(r)}, \boldsymbol{x}_j^{(r)})$. In this case, W is differentiable with respect to $\boldsymbol{d}$ (ref. [7]), and thus the gradients are obtained as $\nabla W = \frac{1}{2}\mathbf{1} - \boldsymbol{\theta}_c$. Thereby, the optimization in P is performed by using projected gradient descent, $\boldsymbol{d}^{[t+1]} = \boldsymbol{d}^{[t]} - \epsilon\nabla W$. We apply a line search [6] to greedily seek the parameter ϵ such that W, i.e., the objective cost function in P, is minimized while ensuring $\boldsymbol{d} \geq 0$. Note that, in this greedy search, the cost function is evaluated several times via optimization of D_w.

3.3 Rescaling

After optimization of D_w with the fixed $\boldsymbol{d}$, the cost function is further decreased by simply rescaling the variables of $\boldsymbol{\tau}$ and $\boldsymbol{d}$, so as to reach the lower bound in Eq.(4). The rescaling, $\hat{\tau}_{ic} = s_c\tau_{ic}$, $\hat{\boldsymbol{d}}_c = \frac{1}{s_c}\boldsymbol{d}_c$, does not affect the bilinear projection in Eq.(8) and thus the constraints in P are kept:

$$\hat{\boldsymbol{w}}_c'\boldsymbol{X}\hat{\boldsymbol{d}}_c^{\frac{1}{2}} = \sum_{i=1}^{N} s_c\tau_{ic} \sum_{r=1}^{R} \frac{d_c^{(r)}}{s_c} k_r(\boldsymbol{x}, \boldsymbol{x}_i) = \boldsymbol{w}_c'\boldsymbol{X}\boldsymbol{d}_c^{\frac{1}{2}}, \tag{16}$$

while the first term in the cost function is transformed to

$$\frac{1}{2}\sum_{c=1}^{C}||\hat{\boldsymbol{w}}_c||^2+\mathbf{1}'\hat{\boldsymbol{d}}_c=\frac{1}{2}\sum_{c=1}^{C}s_c||\boldsymbol{w}_c||^2+\frac{1}{s_c}\mathbf{1}'\boldsymbol{d}_c. \tag{17}$$

The optimal rescaling that minimizes the above is analytically obtained as $\overset{*}{s}_c = \sqrt{\mathbf{1}'\boldsymbol{d}_c}/||\boldsymbol{w}_c||$, and Eq.(17) equals to the lower bound (the Frobenius norm):

$$\frac{1}{2}\sum_{c=1}^{C}||\hat{\boldsymbol{w}}_c||^2+\mathbf{1}'\hat{\boldsymbol{d}}_c=\sum_{c=1}^{C}\sqrt{\mathbf{1}'\boldsymbol{d}_c}||\boldsymbol{w}_c||=\sum_{c=1}^{C}||\boldsymbol{w}_c\boldsymbol{d}_c^{\frac{1}{2}'}||_F. \tag{18}$$

Although the rescaled $\hat{\boldsymbol{\tau}}$ is not necessarily the solution of the problem D_w with the rescaled $\hat{\boldsymbol{d}}$, in the greedy optimization for $\boldsymbol{d}$, the gradients using $\hat{\boldsymbol{\tau}}$ are employed as the approximation for $\nabla W(\hat{\boldsymbol{d}})$. This rescaling contributes to fast convergence.

4 Experimental Result

We show the classification performances and computation time of the proposed methods in comparison with the other MKL methods [7,13] on various datasets. We employed the 1-vs-all version of [7] to cope with multi-class problems. The proposed method is implemented by using MATLAB with C-mex on Xeon 3GHz PC. For the methods of [7,13], we used the MATLAB codes provided by the authors and combined them with libsvm [1] and MOSEK optimization toolbox in order to speed up those methods as much as possible. In this experiment, the parameter values in the all methods are set as follows: κ is determined from $\kappa \in \{0.5, 1, 10\}$ based on 3-fold cross validation and the maximum number of iterations is set to 40 iterations for fair comparison of computation time. These methods almost converge on various datasets within 40 iterations.

First, we used four benchmark datasets: *waveform* from UCI Machine Learning Repository, *satimage* and *segment* from the STATLOG project, and *USPS* [4]. The multiple RBF kernels with 10 σ's (uniformly selected on the logarithmic scale over $[10^{-1}, 10^2]$) were employed. We drew 1000 random training samples and classified the remained samples. The trial is repeated 10 times and the average performance is reported. Fig. 1(a,b) shows the classification results (error rates) and computation time on those datasets. While the performances of the proposed method are competitive to the others, the computation time is much more reduced; especially, more than 20 times faster than the method of [13].

Next, we applied the proposed method to the other practical classification problems in cell biology. The task is to predict the sub-cellular localizations of proteins, and in this case it results in multi-class classification problems. We employed a total of 69 kernels of which details are described in [13]. MKL would

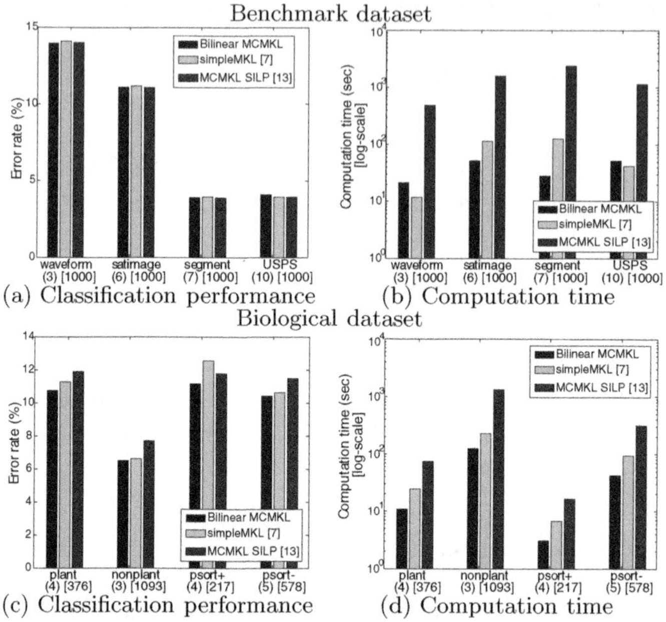

Fig. 1. The classification performances (error rates) and computation time on the four benchmark datasets. The number of classes is indicated in parentheses and that of training samples is in brackets. The left bar shows the result of the proposed methods.

be effectively applied to these substantial types of kernel. In this experiment, we used four biological datasets [13]: *plant*, *nonplant*, *psort+*, and *psort-*. We randomly split the dataset into 40% for training and 60% for testing. The trial is repeated 10 times and the average performance is reported. The results are shown in Fig. 1(c,d), demonstrating that the proposed method is quite effective; the proposed method is superior and faster to the methods of [7,13]. The experimental result shows that the proposed method effectively and efficiently combines a lot of heterogeneous kernel functions.

5 Conclusion

We have proposed a multiple kernel learning (MKL) method to deal with multi-class problems. In the proposed method, the multi-class classification using multiple kernels is formulated in the bilinear form, and the computationally efficient optimization scheme is proposed in order to be applicable to large-scaled samples. In the experiments on the benchmarks and the biological datasets, the proposed method exhibited the favorable performances and computation time compared to the previous methods of MKL.

References

1. Chang, C.C., Lin, C.J.: LIBSVM: a library for support vector machines (2001), Software available at `http://www.csie.ntu.edu.tw/~cjlin/libsvm`
2. Crammer, K., Singer, Y.: On the algorithmic implementation of multiclass kernel-based vector machines. Journal of Machine Learning Research 2, 265–292 (2001)
3. Fan, R.E., Chang, K.W., Hsieh, C.J., Wang, X.R., Lin, C.J.: Liblinear: A library for large linear classification. Journal of Machine Learning Research 9(6), 1871–1874 (2008)
4. Hull, J.: A database for handwritten text recognition research. IEEE Trans. Pattern Analysis and Machine Intelligence 16(5), 550–554 (1994)
5. Lanckriet, G., Cristianini, N., Bartlett, P., Ghaoui, L.E., Jordan, M.I.: Learning the kernel matrix with semidefinite programming. Journal of Machine Learning Research 5, 27–72 (2004)
6. Nocedal, J., Wright, S. (eds.): Numerical optimization. Springer, New York (1999)
7. Rakotomamonjy, A., Bach, F.R., Canu, S., Grandvalet, Y.: Simplemkl. Journal of Machine Learning Research 9, 2491–2521 (2008)
8. Schölkopf, B., Smola, A.: Learning with Kernels. MIT Press, Cambridge (2001)
9. Sonnenburg, S., Rätsch, G., Schäfer, C., Schölkopf, B.: Large scale multiple kernel learning. Journal of Machine Learning Research 7, 1531–1565 (2006)
10. Vapnik, V.: Statistical Learning Theory. Wiley, Chichester (1998)
11. Varma, M., Ray, D.: Learning the discriminative power-invariance trade-off. In: Proceedings of IEEE 11th International Conference on Computer Vision (2007)
12. Xu, Z., Jin, R., King, I., Lyu, M.R.: An extended level method for efficient multiple kernel learning. In: Advances in Neural Information Processing Systems, vol. 21, pp. 1825–1832 (2008)
13. Zien, A., Ong, C.S.: Multiclass multiple kernel learning. In: Proceedings of the 24th International Conference on Machine Learning (2007)

Feature Extraction Using Support Vector Machines

Yasuyuki Tajiri, Ryosuke Yabuwaki, Takuya Kitamura, and Shigeo Abe

Graduate School of Engineering
Kobe University
Rokkodai, Nada, Kobe, Japan

Abstract. We discuss feature extraction by support vector machines (SVMs). Because the coefficient vector of the hyperplane is orthogonal to the hyperplane, the vector works as a projection vector. To obtain more projection vectors that are orthogonal to the already obtained projection vectors, we train the SVM in the complementary space of the space spanned by the already obtained projection vectors. This is done by modifying the kernel function. We demonstrate the validity of this method using two-class benchmark data sets.

Keywords: Feature extraction, kernel discriminant analysis, pattern recognition, support vector machines.

1 Introduction

Feature selection and feature extraction are important in realizing high generalization ability of a classifier. Especially, feature extraction extracts features from the original features by linear or nonlinear transformation. The most well-used feature extraction methods are kernel principal component analysis (KPCA) [1,2] and kernel discriminant analysis (KDA) [3]. To obtain more than one projection vector for two-class linear discriminant analysis (LDA), nonparametric discriminant analysis (NDA) [4] is proposed in which the projection vectors in the complementary space of the space spanned by the already obtained projection vector are determined. But for KDA, to obtain a projection vector for two-class problems, we need to calculate the inverse of the matrix whose dimension is the number of training data. Therefore, for large size problems, calculation time will also be large.

In this paper, we propose feature extraction by support vector machines (SVMs) [5,6]. In training an SVM, we determine the separating hyperplane so that the margin is maximized. The coefficient vector of the hyperplane is orthogonal to the hyperplane and because the training data of different classes projected onto the coefficient vector are maximally separated, we can interpret the SVM as a feature extractor. In addition, to obtain more than one orthogonal projection vector, we iteratively determine the projection vectors in the complementary space of the space spanned by the already obtained projection vectors

K.W. Wong et al. (Eds.): ICONIP 2010, Part II, LNCS 6444, pp. 108–115, 2010.

borrowing the idea from [4]. This is done by training the SVM with the modified kernel function.

This paper is organized as follows. In Section 2, we explain KDA and its problems. In Section 3, we explain classification by an SVM and the proposed method using the SVM. In Section 4, we experimentally compare the proposed method with KDA using two-class problems. And Section 5 concludes the work.

2 Kernel Discriminant Analysis

In this section we briefly describe kernel discriminant analysis (KDA). Let the m-dimensional training data for Class 1 be $\{\mathbf{x}_1^1, \ldots, \mathbf{x}_{M_1}^1\}$, and the data for Class 2 be $\{\mathbf{x}_1^2, \ldots, \mathbf{x}_{M_2}^2\}$. We obtain the coefficient vector $\mathbf{w}$ that maximizes the difference of class centers and minimizes total scatter in the feature space mapped by $\boldsymbol{\phi}(\mathbf{x})$.

Let the difference of squares of the centers of mapped data be d^2. Then d^2 is given by

$$\begin{aligned} d^2 &= (\mathbf{w}^\top(\mathbf{c}_1 - \mathbf{c}_2))^2 \\ &= \mathbf{w}^\top(\mathbf{c}_1 - \mathbf{c}_2)(\mathbf{c}_1 - \mathbf{c}_2)^\top \mathbf{w}, \end{aligned} \tag{1}$$

where $\mathbf{c}_i$ is the center of the class i data:

$$\mathbf{c}_i = \frac{1}{M_i}\sum_{j=1}^{M_i} \boldsymbol{\phi}(\mathbf{x}_j^i) = (\boldsymbol{\phi}(\mathbf{x}_1^i), \ldots, \boldsymbol{\phi}(\mathbf{x}_{M_i}^i)) \begin{bmatrix} \frac{1}{M_i} \\ \vdots \\ \frac{1}{M_i} \end{bmatrix} \quad \text{for } i = 1, 2. \tag{2}$$

And we define the between-class scatter matrix Q_B as follows:

$$Q_B = (\mathbf{c}_1 - \mathbf{c}_2)(\mathbf{c}_1 - \mathbf{c}_2)^\top. \tag{3}$$

The variance of mapped data s_i^2 is defined by

$$s_i^2 = \mathbf{w}^T Q_T \mathbf{w} \quad \text{for } i = 1,\ 2, \tag{4}$$

where Q_T is the total scatter matrix:

$$\begin{aligned} Q_T &= \frac{1}{M}\sum_{j=1}^{M} (\boldsymbol{\phi}(\mathbf{x}_j) - \mathbf{c})(\boldsymbol{\phi}(\mathbf{x}_j) - \mathbf{c})^\top \\ &= \frac{1}{M}\left[\boldsymbol{\phi}(\mathbf{x}_1), \ldots, \boldsymbol{\phi}(\mathbf{x}_M)\right](I_M - \mathbf{1}_M) \begin{bmatrix} \boldsymbol{\phi}(\mathbf{x}_1)^\top \\ \vdots \\ \boldsymbol{\phi}(\mathbf{x}_{M_i})^\top \end{bmatrix}. \end{aligned} \tag{5}$$

Here, $\mathbf{c}$ is the center of the training data, I_M is the $M \times M$ unit matrix, and $\mathbf{1}_M$ is the matrix with all elements being $1/M$. We consider maximizing the following function:

$$J(\mathbf{w}) = \frac{\mathbf{w}^\top Q_B \mathbf{w}}{\mathbf{w}^\top Q_T \mathbf{w}}, \tag{6}$$

where $\mathbf{w}$ is defined as follows:

$$\mathbf{w} = (\boldsymbol{\phi}(\mathbf{x}_1), \ldots, \boldsymbol{\phi}(\mathbf{x}_M))\boldsymbol{\alpha}. \tag{7}$$

Here $\boldsymbol{\alpha} = (\alpha_1, \ldots, \alpha_M)^\top$. Substituting (7) into (6), we obtain

$$J(\boldsymbol{\alpha}) = \frac{\boldsymbol{\alpha}^T K_B \boldsymbol{\alpha}}{\boldsymbol{\alpha}^T K_T \boldsymbol{\alpha}}, \tag{8}$$

where K_B and K_T are

$$K_B = (\mathbf{k}_{B_1} - \mathbf{k}_{B_2})(\mathbf{k}_{B_1} - \mathbf{k}_{B_2})^\top, \tag{9}$$

$$\mathbf{k}_{B_i} = \frac{1}{M_i} \begin{pmatrix} \sum_{j=1}^{M_i} K(\mathbf{x}_1, \mathbf{x}_j^i) \\ \cdots \\ \sum_{j=1}^{M_i} K(\mathbf{x}_M, \mathbf{x}_j^i) \end{pmatrix} \quad \text{for} \quad i = 1, 2, \tag{10}$$

$$K_T = \frac{1}{M} \begin{pmatrix} K(\mathbf{x}_1, \mathbf{x}_1), \cdots, K(\mathbf{x}_1, \mathbf{x}_M) \\ \cdots \\ K(\mathbf{x}_M, \mathbf{x}_1) \cdots, K(\mathbf{x}_M, \mathbf{x}_M) \end{pmatrix} (I_M - \mathbf{1}_M) \times \begin{pmatrix} K(\mathbf{x}_1, \mathbf{x}_1), \cdots, K(\mathbf{x}_1, \mathbf{x}_M) \\ \cdots \\ K(\mathbf{x}_M, \mathbf{x}_1) \cdots, K(\mathbf{x}_M, \mathbf{x}_M) \end{pmatrix}. \tag{11}$$

Here, $K(\mathbf{x}, \mathbf{x}') = \boldsymbol{\phi}^\top(\mathbf{x})\, \boldsymbol{\phi}(\mathbf{x}')$ is a kernel. If K_T is positive semi-definite, the solution that maximizes (6) is given by

$$\boldsymbol{\alpha} = (K_T + \varepsilon I_M)^{-1}(\mathbf{k}_{B_1} - \mathbf{k}_{B_2}), \tag{12}$$

where ε is a small positive value. Using (7) and (12), we can obtain the projection vector, but because K_T is an $M \times M$ matrix, the calculation time is of the cubic order of the number of training data.

3 Support Vector Machines

3.1 Classification by Support Vector Machines

In this section, we discuss classification by the SVM. For a two-class problem, we consider the following decision function:

$$D(\mathbf{x}) = \mathbf{w}^\top \boldsymbol{\phi}(\mathbf{x}) + b = 0, \tag{13}$$

where $\mathbf{w}$ is the coefficient vector and b is the bias term. We use the following L2 SVM:

$$\text{maximize} \quad Q(\boldsymbol{\alpha}) = \sum_{i=1}^{M} \alpha_i - \frac{1}{2} \sum_{i,j=1}^{M} \alpha_i \alpha_j y_i y_j \left(K(\mathbf{x}_i, \mathbf{x}_j) + \frac{\delta_{ij}}{C} \right) \tag{14}$$

$$\text{subject to} \quad \sum_{i=1}^{M} y_i \alpha_i = 0 \quad \alpha_i \geq 0 \quad \text{for} \ \ i = 1, \ldots, M, \tag{15}$$

where $(\mathbf{x}_i, y_i)$ $(i = 1, \ldots, M)$ are M training input-output pairs, with $y_i = 1$ if $\mathbf{x}_i$ belongs to Class 1, and $y_i = -1$ if Class 2, C is the margin parameter that determines the tradeoff between the maximization of the margin and minimization of the classification error, and α_i are Lagrange multipliers. Solving (15) for α_s yields

$$\alpha_s = - \sum_{i \neq s, i \in S} y_s y_i \alpha_i, \tag{16}$$

where S is the set of support vector indices. Substituting (16) into (14), we obtain

$$\text{maxmize} \quad Q(\boldsymbol{\alpha}_S) = \mathbf{c}_S^\top \boldsymbol{\alpha}'_S - \frac{1}{2} \boldsymbol{\alpha}'^\top_S K_S \boldsymbol{\alpha}'_S \tag{17}$$

$$\text{subject to} \quad \boldsymbol{\alpha}_S \geq \mathbf{0}, \tag{18}$$

where $\boldsymbol{\alpha}'_S$ and $\boldsymbol{c}_S$ are $(|S| - 1)$ dimensional vectors, K_S is a $(|S| - 1) \times (|S| - 1)$ positive definite matrix, and

$$\alpha'_{S_i} = \alpha_i \quad \text{for} \quad i \neq s, \ i \in S, \tag{19}$$

$$c_{S_i} = 1 - y_s y_i \quad \text{for} \quad i \neq s, \ i \in S, \tag{20}$$

$$K_{S_{ij}} = y_i y_j (K(\mathbf{x}_i, \mathbf{x}_j) - K(\mathbf{x}_i, \mathbf{x}_s) - K(\mathbf{x}_s, \mathbf{x}_j) + K(\mathbf{x}_s, \mathbf{x}_s) + \frac{1 + \delta_{ij}}{C}) \quad \text{for} \quad i, j \neq s, \ i, j \in S. \tag{21}$$

If S is given we can obtain the solution solving (17) for $\boldsymbol{\alpha}'_S$. Here, we estimate S by active set training [7]. Starting from the initial set of S, we calculate $\boldsymbol{\alpha}'_S$ by

$$\boldsymbol{\alpha}'_S = K_S^{-1} \mathbf{c}_s. \tag{22}$$

We delete subscripts with non-positive α_i from S and add subscripts associated with violating data to S and iterate the above procedure until the same set of support vectors is obtained for the consecutive iterations.

The coefficient vector $\mathbf{w}$ and bias term b are obtained by

$$\mathbf{w} = \sum_{i \in S} y_i \alpha_i \boldsymbol{\phi}(\mathbf{x}_i),$$

$$b = y_i - \sum_{j \in S} \alpha_j y_j \left(K(\mathbf{x}_i, \mathbf{x}_j) + \frac{\delta_{ij}}{C} \right) \quad \text{for} \quad i \in S. \tag{23}$$

3.2 Feature Extraction by SVM

Usually, the decision function obtained by training the SVM is used to classify data. Instead of considering the SVM as a classifier, we can interpret it as a feature extractor. Two classes are separated by the optimal separating hyperplane and the coefficient vector $\mathbf{w}$ is orthogonal to the optimal hyperplane. Therefore, the training data of different classes projected onto $\mathbf{w}$ are maximally separated.

Because the coefficient vector in the optimal separating hyperplane works as a feature extractor, like KDA, by $\mathbf{w}^\top \boldsymbol{\phi}(\mathbf{x}) + b$, we obtain one feature. To define more than one projection vector, we consider obtaining the projection vector that is orthogonal to the previously obtained projection vectors. This is done by determining the projection vectors in the complementary space of the space spanned by the already obtained projection vectors.

Namely, we obtain the first projection vector $\mathbf{w}_{(1)}$ training the SVM. Then we obtain the second projection vector $\mathbf{w}_{(2)}$ that is orthogonal to $\mathbf{w}_{(1)}$ in the feature space. Likewise, we obtain the kth projection vector $\mathbf{w}_{(k)}$ that satisfies

$$\mathbf{w}_{(k)}^\top \mathbf{w}_{(i)} = 0 \quad \text{for} \quad i = 1, \ldots, k-1. \tag{24}$$

Now assuming that projection vectors $\mathbf{w}_{(i)} (i = 1, \ldots, k-1)$ are obtained, consider determining $\mathbf{w}_{(k)}$. We obtain this vector from the complementary space of the space spanned by the already obtained projection vectors. Let $\mathbf{z}_{(k)}$ be the vector of $\boldsymbol{\phi}(\mathbf{x})$ in the complementary space of the space spanned by $\mathbf{w}_{(i)} (i = 1, \ldots, k-1)$ as follows:

$$\mathbf{z}_{(k)} = \boldsymbol{\phi}(\mathbf{x}) - \sum_{i=1}^{k-1} \mathbf{w}_{(i)}^\top \frac{\boldsymbol{\phi}(\mathbf{x})}{||\mathbf{w}_{(i)}||^2} \mathbf{w}_{(i)}, \tag{25}$$

where $\mathbf{z}_{(1)} = \boldsymbol{\phi}(\mathbf{x})$, and $\mathbf{w}_{(i)}$ is obtained by

$$\mathbf{w}_{(i)} = \sum_{j \in S_{(i)}} y_j \alpha_{j(i)} \boldsymbol{\phi}(\mathbf{x}_j). \tag{26}$$

Here, $S_{(i)}$ and $\alpha_{j\,(i)}$ are the set of support vector indices and support vectors for $\mathbf{w}_{(i)}$. In determining $\mathbf{w}_{(k)}$, we only need to replace $K(\mathbf{x}, \mathbf{x}')$ in (14) with $\mathbf{z}_{(k)}^\top\, \mathbf{z}'_{(k)}$. Here, we define $K_{(k)}(\mathbf{x}, \mathbf{x}')$ by

$$\begin{aligned} K_{(k)}(\mathbf{x}, \mathbf{x}') &= \mathbf{z}_{(k)}^\top \mathbf{z}'_{(k)} \\ &= K(\mathbf{x}, \mathbf{x}') - \sum_{i=1}^{k-1} \mathbf{w}_{(i)}^\top \phi(\mathbf{x}) \mathbf{w}_{(i)}^\top \phi(\mathbf{x}') ||\mathbf{w}_{(i)}||^{-2}, \end{aligned} \tag{27}$$

where $K_{(1)}(\mathbf{x}, \mathbf{x}') = K(\mathbf{x}, \mathbf{x}')$. Using (27), we can iteratively obtain the new projection vector.

The advantage of our method is that we can calculate the projection vectors by modifying the kernel calculation of SVM software by (27).

In LDA or KDA, to determine the projection vector we solve (12) by the $M \times M$ matrix inversion, which is of the order of M^3. However, to obtain the projection vector in SVM, at each iteration of active set training we only need to obtain an $|S| \times |S|$ inverse matrix, which is of the order of $|S|^3$. Usually $|S|$ is much smaller than M. Therefore, feature extraction by the SVM is expected to be much faster.

Table 1. Two-class bench mark data sets

Data	Inputs	Train	Test	Sets
Banana	2	400	4900	100
B. cancer	9	200	77	100
Diabetes	8	468	300	100
German	20	700	300	100
Heart	13	170	100	100
Image	18	1300	1010	20
Ringnorm	20	400	7000	100
F. solar	9	666	400	100
Splice	60	1000	2175	20
Thyroid	5	140	75	100
Titanic	3	150	2051	100
Twonorm	20	400	7000	100
Waveform	21	400	4600	100

Table 2. Parameter values for two-class problems

	KDA		SVM		
Data	γ	C (Rec)	γ	C	C (Rec)
Banana	10	500	10	100	0.1
B. cancer	0.1	0.5	15	0.5	50000
Diabetes	0.1	1	0.1	5000	0.5
German	0.1	0.5	10	5	50000
Heart	0.1	0.5	3	1	50000
Image	15	5	15	100000	10
Ringnorm	0.1	100	15	10	5
F. solar	0.1	0.1	1	1000	3
Splice	15	50	10	5	100
Thyroid	5	50	15	500	10
Titanic	1	0.5	0.5	10	0.1
Twonorm	0.1	0.5	10	0.5	1
Waveform	0.1	0.5	15	1	50000

4 Computer Experiment

We evaluated the proposed method as a feature extractor for the linear SVM (SVM with linear kernels) used as a classifier and compare the proposed method with KDA.

We used two-class benchmark data sets [8,9] shown in Table 1. In the table, "Inputs," "Train," and "Test" are the numbers of input variables, training data, and test data, respectively, and "Sets" is the number of training and test data set pairs. We used RBF kernels: $K(\mathbf{x}, \mathbf{x}') = \exp(-\gamma||\mathbf{x} - \mathbf{x}'||^2)$, where γ is a positive parameter and set 10^{-12} to ε in (12).

To compare the proposed method with the conventional method, we need to determine the values of the kernel parameter γ, and the margin parameter C. For KDA + linear SVM, we determined the values of γ for KDA and C for the linear SVM by fivefold cross-validation using the first five data sets. And for the proposed method, we determined the values of γ and C for the feature extractor, by training the SVM with RBF kernels and then we extracted ten features in maximum setting the values of γ and C determined by cross-validation. In classification, for the ten extracted features, we determined the value of C for the linear SVM by cross-validation and used the C value for one to ten features. The value of γ was selected from {0.1, 0.5, 1, 1.5, 3, 5, 10, 15} and the value of C was selected from {0.1, 0.5, 1, 5, 10, 50, 100, 500, 10^3, 5 10^3, 10^4, 5 10^4, 10^5}.

Table 2 shows the parameter values determined by the above procedure. In the table, the "γ" columns in KDA and SVM list the γ values determined for KDA and SVM and the "C (Rec)" columns list the C values for the linear SVM. The "C" column in SVM lists the C values for the SVM feature extractor.

Using the parameter values listed in Table 2, we extracted features, performed classification by the linear SVM, and calculated the average recognition rates and

Table 3. Recognition performance for test data sets

	KDA	SVM (1)	SVM (C)		SVM (T)	
Data	Test	Test	Test	Feature	Test	Feature
Banana	88.74±0.56	**89.27±0.51**	89.21±0.47	5	89.25±0.47	10
B. cancer	**74.30**±4.51	73.04±4.86	72.04±5.06	8	73.04±4.86	1
Diabetes	**76.87**±1.69	75.79±1.60	76.79±1.62*	10	76.91±1.63	2
German	75.71±2.10	**76.62±2.18**	75.04±2.18	10	76.69±2.18	1
Heart	83.24±3.14	**84.47±3.08**	82.69±3.08	8	84.36±3.08	1
Image	95.48±0.60	**96.12±0.69**	96.26±0.58*	3	96.34±0.59	7
Ringnorm	**98.01**±0.28	97.67±0.34	97.92±0.28*	2	98.03±0.26	9
F. solar	**66.35**±1.62	66.02±1.76	66.12±1.56*	3	66.15±1.61	2
Splice	87.60±0.55	**89.08±0.74**	89.29±0.73*	10	89.29±0.73	10
Thyroid	93.71±2.81	**95.86±2.20**	95.97±2.39*	8	95.98±2.24	7
Titanic	**77.52**±1.12	77.44±0.61	77.41±0.57	10	77.44±0.61	5
Twonorm	97.29±0.23	**97.57±0.27**	97.57±0.27*	1	97.57±0.27	1
Waveform	**90.12**±0.47	89.98±0.46	88.56±0.37	5	90.07±0.37	1

Table 4. Feature extraction time (s)

Data	KDA	SVM (1)	SVM (5)	SVM (10)
Banana	0.61	0.18	0.97	2.14
B. cancer	0.04	0.01	0.05	0.10
Diabetes	0.62	0.11	0.54	1.12
German	2.21	0.33	1.66	3.89
Heart	0.03	0.01	0.05	0.09
Image	24.18	0.31	1.58	3.29
Ringnorm	0.82	0.20	1.19	3.00
F. solar	1.91	0.29	1.47	3.02
Splice	10.77	1.12	7.27	15.48
Thyroid	0.02	0.01	0.03	0.05
Titanic	0.06	0.06	0.29	0.63
Twonorm	0.88	0.12	0.69	1.58
Waveform	0.66	0.22	1.31	3.15

the standard deviations. Table 3 shows the average recognition rates and their standard deviations of the test data sets for KDA and the proposed method. In the table, the "SVM (1)" column shows the recognition rates when one feature was extracted by SVM and the C value for the linear SVM was determined using one feature by cross-validation. In KDA and SVM (1), the better recognition rate is shown in boldface. The "SVM (C)" column lists the recognition rates and the number of features selected by cross-validation. This means that we selected the features that realized the maximum recognition rate for the validation data set among one to ten features. The asterisk denotes that the recognition rate of SVM (C) is higher than or equal to that of SVM (1). The "SVM (T)" column lists the maximum recognition rates among one to ten features and the associated number of features. Thus, the recognition rate of SVM (T) is better than or equal to that of SVM (C).

The recognition rates of SVM (1) are comparable with those of KDA. For SVM (T) more than one feature were better than one feature for eight data sets.

But the recognition rates of SVM (C) were better than or equal to those of SVM (1) for seven data sets and for other six data sets worse. This meant that the cross-validation did not work well for these data sets.

Table 4 shows the time for extracting one feature for KDA and one, five, and ten features for the proposed method. Compared to KDA by the proposed method with one feature we could reduce feature extraction time for all the data sets. Especially, large data sets such as German, Image, and Splice data sets, the proposed method was much faster than KDA. Feature extracting time increases linearly as five and ten features were extracted.

5 Conclusions

In this paper, we proposed feature extraction using SVMs. We regard the coefficient vector orthogonal to the separating hyperplane as a projection vector, and obtain a new projection vector in the complementary space of the space spanned by the already obtained projection vectors.

According to the computer experiment, the recognition rates of many data sets were almost equal to the conventional method. But in some cases we could not optimize the number of feature by cross-validation. The proposed method with one feature could reduce calculation cost for almost all data sets. Especially, we could reduce much calculation cost for large data sets.

References

1. Schölkopf, B., Mika, S., Burges, C.J.C., Knirsch, P., Möller, K.R., Smola, A.J.: Input space vs. feature space in kernel-based methods. IEEE Transactions on Neural Networks 10(5), 1000–1017 (1999)
2. Schölkopf, B., Smola, A.J., Möller, K.R.: Nonlinear component analysis as a kernel eigenvalue problem. Neural Computation 10, 1299–1319 (1998)
3. Mike, S., Rötsch, G., Weston, J., Schölkopf, B., Möller, K.R.: Fisher discriminant analysis with kernels. In: Neural Networks for Signal Processing IX, pp. 41–48 (1999)
4. Raducanu, B., Vitrià, J.: Online nonparametric discriminant analysis for incremental subspace learning and recognition. Pattern Analysis & Applications 11, 259–268 (2008)
5. Vapnik, V.N.: The Nature of Statistical Learning Theory. Springer, New York (1995)
6. Abe, S.: Support Vector Machines for Pattern Classification (Advances in Pattern Recognition), 2nd edn. Springer, London (2010)
7. Yabuwaki, R., Abe, S.: Convergence improvement of active set support vector training. In: Proc. International Joint Conference on Neural Networks, pp. 1426–1430 (2010)
8. Rätsch, G., Onda, T., Müller, K.R.: Soft margins for AdaBoost. Machine Learning 42(3), 287–320 (2001)
9. `http://ida.first.fraunhofer.de/projects/bench/benchmarks.htm`

Class Information Adapted Kernel for Support Vector Machine

Tasadduq Imam and Kevin Tickle

CINS, CQUniversity, Rockhampton, QLD 4702, Australia

Abstract. This article presents a support vector machine (SVM) learning approach that adapts class information within the kernel computation. Experiments on fifteen publicly available datasets are conducted and the impact of proposed approach for varied settings are observed. It is noted that the new approach generally improves minority class prediction, depicting it as a well-suited scheme for imbalanced data. However, a SVM based customization is also developed that significantly improves prediction performance in terms of different measures. Overall, the proposed method holds promise with potential for future extensions.

Keywords: SVM, Class Informed Kernel, RBF, Sensitivity, Imbalanced data.

1 Introduction

Support Vector Machine (SVM) [1,2] has positioned itself as a state-of-the-art pattern classification technique in many contemporary research areas including brain informatics (e.g. [3,4]). Given a set of inputs with known class labels (i.e., supervised learning), SVM maps the input space to a high-dimensional feature space such that the training data become linearly separable. The outcome of training SVM is a decision hyperplane that maximizes margin from the class boundaries and, thereby, produces a classifier with high generalization capacity. The explicit mapping from input space to feature space is unknown and is controlled by a function, termed as kernel function, that computes the dot product between the mapped input vectors in the feature space (and dot product is the only processing step, in the unknown feature space, required for SVM training and prediction). While several kernel functions have been proposed and employed in literature, radial basis kernel are often used due to robust performance. These kernel functions can also be viewed as measuring similarity between the feature vectors [5]. The optimization process involved in the SVM training [6,7] also considers the similarity between feature vectors in its underlying philosophy. However, the similarity (i.e., kernel) is computed based on the input vectors' attributes only and class information of the corresponding vectors are not involved. Viewing kernel value as a similarity measure, this article presents a kernel that takes into consideration the class information of the corresponding vectors. The aim is to conceptualize the impact of including class information during kernel

K.W. Wong et al. (Eds.): ICONIP 2010, Part II, LNCS 6444, pp. 116–123, 2010.

computation on the classifier's performance. The study reveals that the proposed approach improves the performance of SVM in terms of different measures.

The rest of this article is organized as follows. In Section 2, we present a brief survey on Support Vector Machine classification technique. Section 3 then details our proposed learning approach. A set of experiments, outlining the different characteristics of the proposed approach, and corresponding discussions on outcomes are then highlighted in Section 4. Lastly Section 5 provides a summary of the findings and indicates future potential research.

2 Support Vector Machine

Support Vector Machine is a robust classifier that derives the maximal margin decision hyperplane during training and use it to discriminate test data to one of the two classes (i.e., SVM is a binary classifier working with two class labels only) during prediction [8]. Let the training dataset comprises of tuple $(\mathbf{x}_i, y_i)$ $for\, i = 1...N$ where, N is the total number of data, $\mathbf{x}_i$ the i-th attribute vector and y_i (where $y_i \epsilon \{-1, 1\}$) the corresponding class label. Then SVM training can be expressed as the following optimization problem (dual form):

$$min_{\boldsymbol{\alpha}} J_D(\boldsymbol{\alpha}) = \frac{1}{2} \sum \alpha_i \alpha_j y_i y_j K(\mathbf{x}_i.\mathbf{x}_j) - \sum \alpha_i \tag{1}$$

$subject\ to, \sum \alpha_i y_i = 0;\ 0 \leq \alpha_i \leq C \ \ for\, \forall \alpha_i;$

The function $K : \Re x \Re \rightarrow \Re$ is known as the kernel function, that computes the dot product between the data vectors in high dimensional feature space. Several kernel functions have been proposed in literature. Two of these most commonly used kernels are:

- Linear: $(\mathbf{x}_1.\mathbf{x}_2)$
- RBF: $(e^{-\gamma \|\mathbf{x}_1 - \mathbf{x}_2\|^2}$ for $\gamma > 0)$.

The parameter C in the optimization problem (Eq. 1) is a user-defined penalty assignment on training errors. Together, the parameters to kernel (ex., γ for RBF) and C are referred to as the hyper-parameters. SVM training, basically, computes a weight (α_i) associated to each of the data points. In the final solution, data points with $\alpha_i > 0$ are the only important points for classification and are termed as support vectors. SVM training also computes an intercept b for the decision hyperplane. The prediction on a test data $\mathbf{x}$ is given by: $sign\,(\sum_{i=1}^{nsv} \alpha_i^* y_i K(\mathbf{x}_i.\mathbf{x})\ + b)$, where nsv is the number of support vectors (i.e., data points with non-zero α).

3 Class Informed Kernel

Noting that dot product relates to cosine similarity, the kernel values (that denotes dot product between vectors in feature space), can be viewed as indicating similarity between the data vectors. But, the calculation of this similarity is based on the attribute vectors only and does not take into account the respective class labels. Although literature exists that have attempted different kernel modifications to improve prediction performance (e.g. [9]), a kernel that adapts class information in the computation and also addresses the issue that arises during prediction from the use of such kernels (explained in a subsequent paragraph), to the best of our knowledge, is still lacking. Viewing the kernel values as similarity measures, a class informed kernel is as proposed below:

$$K((\mathbf{x}_1, y_1)\,,\,(\mathbf{x}_2, y_2)) = e^{-\gamma(\|\mathbf{x}_1-\mathbf{x}_2\|^2\,+\,(y_1-y_2)^2}\ for\ \gamma > 0; \tag{2}$$

Here, $\mathbf{x}_1, \mathbf{x}_2$ are the input vectors' attributes and $y_1, y_2 \epsilon \{-1\,,\,+1\}$ are the respective class labels.

The kernel expression in Eq. 2 is similar to the expression of RBF kernel. The difference is the additional term $(y_1 - y_2)^2$. Assuming that class labels are either $+1$ or -1, the value of this additional term results in 0 when both $\mathbf{x}_1, \mathbf{x}_2$ belong to the same class and 4 when $\mathbf{x}_1, \mathbf{x}_2$ belong to the different class. Thus based on the class labels of the vectors for which kernel is computed, an additional weight is added to the expression of RBF. Further, it is to be noted that, the term $\| \mathbf{x}_1 - \mathbf{x}_2 \|^2$ denotes the distance between the two vectors in input space. For same class input vectors and the proposed kernel K, this distance remains the same as that for RBF kernel. However, for different class input vectors, the addition of the positive weight in effect increases the distance between the vectors (i.e., artificially increases pairwise margin and thus reduces overlap between the different class data).

An issue with use of this class informed kernel lies in the application of it during prediction. While class labels of the support vectors, derived from training SVM, are known, that of the prediction vectors are unknown. To address this issue, we employ a learning framework outlined in Fig. 1. During training, a SVM model is learnt using the proposed class informed kernel. In addition, a second classifier is trained on the training data. This second classifier (termed as *support classifier*) is used to provide an estimate of class label during test phase. The estimated class label coupled with the prediction input vector is then given as input to the trained SVM model and the outcome from the model is the final prediction.

4 Experiment Setups and Results

4.1 Datasets and Software

We perform experiments on five publicly available datasets [10]: diabetes, glass, iris, liver and vehicle. For running SVM, we employ the LibSVM classification technique as implemented in R (through the package kernlab) [11]. Other than the diabetes and liver datasets, the rest of the datasets have originated from multi-class domain. Since SVM is primarily a binary classification technique, we convert the multi-class datasets to binary by considering one of the class as positive class and the rest as negative class. Doing this conversion for each of the

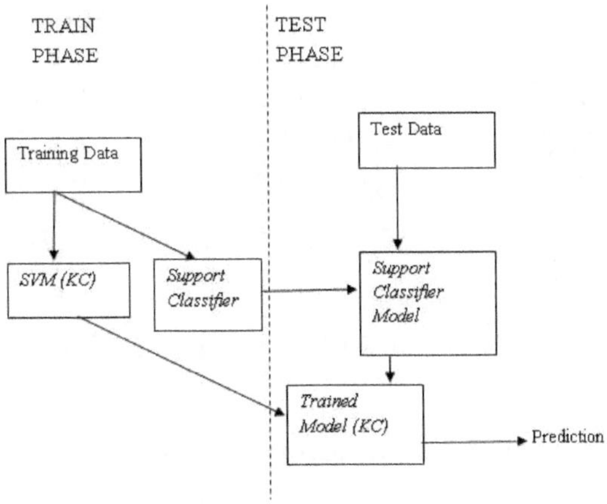

Fig. 1. Framework for learning with class-informed kernel

multiclass data results in total the 15 datasets outlined in Table 1. Table 1 also indicates the total number of data, number of positive and negative class samples and the percentage of the class representation in the datasets. It is noteworthy that some of the datasets are imbalanced (i.e., skewed) in terms of representation of the classes. Imbalanced dataset often arises in many practical applications and it is well known that many classifiers make more prediction errors on minority class samples than that belonging to majority class [12]. Accuracy (ratio of the total number of correctly classified data and the total number of data) is not an appropriate performance measure when datasets is imbalanced. Sensitivity (accuracy for the positive class data) and gmean (geometric mean of the accuracy for positive class data and the accuracy of negative class data) are often employed as performance metric for imbalanced datasets [13, 14]. In our experiments, we focus on all three of these prediction performance measures.

Table 1. Datasets used in the experiments

Datasets	Total Data	# Positive	# Negative	% (+)	% (-)
diabetes	768	268	500	34.90	65.10
glass_1	214	70	144	32.71	67.29
glass_2	214	76	138	35.51	64.49
glass_3	214	17	197	7.94	92.06
glass_5	214	13	201	6.07	93.93
glass_6	214	9	205	4.21	95.79
glass_7	214	29	185	13.55	86.45
iris_1	150	50	100	33.33	66.67
iris_2	150	50	100	33.33	66.67
iris_3	150	50	100	33.33	66.67
liver	345	145	200	42.03	57.97
vehicle_1	846	212	634	25.06	74.94
vehicle_2	846	217	629	25.65	74.35
vehicle_3	846	218	628	25.77	74.23
vehicle_4	846	199	647	23.52	76.48

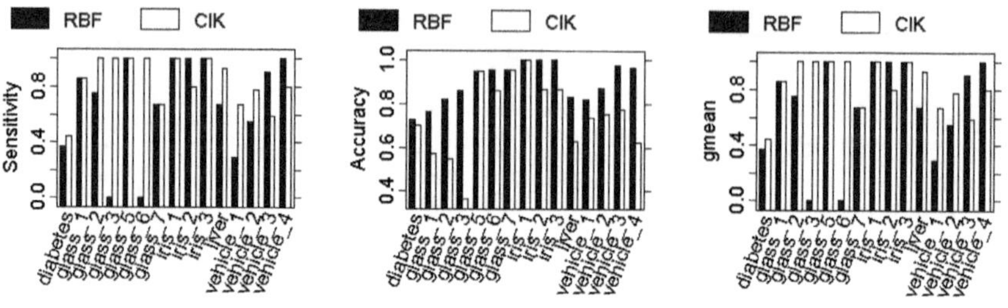

Fig. 2. Performance measures for RBF kernel and CIK (Class informed kernel)

4.2 Experiment with Class Informed Kernel

For our experiments, we randomly split each of the datasets into a train and a test data file. 90% of the total data is used for training and the rest for prediction. Stratified sampling is used to preserve the ratio of positive and negative class data in the train and test files. We analyse the impact of class informed kernel by comparing its performance against a SVM trained on the dataset using RBF kernel. Focus is made on RBF kernel due to its wide popularity and well known robust performance, and also due to the similarity of class informed kernel to the RBF kernel. For RBF kernel, a 10-fold cross validation technique is employed to determine the best parameters (γ and C) for SVM training on the training data file, and the trained model is employed to note prediction performance on test data file. For class informed kernel (CIK), γ and C are set to the best parameter values identified for the RBF kernel. For this initial experiment, we use Naive Bayes classifier (due to its simplicity and high training speed) as the support classifier. Fig. 2 denotes comparison of RBF and CIK. We note that while the CIK (with Naive Bayes as support classifier) based learning does not perform well against the RBF learning in terms of accuracy, in terms of sensitivity the CIK emerges as a clear winner (performs better than or comparable to RBF for 12 datasets out of 15). In terms of gmean, however, there is no clear winner (CIK performs better than or comparable to RBF for 8 datasets out of 15). Thus, we observe that CIK has a positive impact on prediction performance, especially when the dataset is imbalanced (i.e., CIK results in higher prediction of minority class).

4.3 Varied Support Classifier

In the previous experiment, we have used Naive Bayes as the support classifier. In this section, we present the impact of other different support classifier on prediction outcomes. In particular, we experiment with recursive partitioning and regression trees [15] and a single-hidden-layer neural network. For each of these different support classifiers, prediction performance of CIK is compared against that for RBF. Fig. 3 illustrates the results. We note that the performance for CIK noticeably varies depending on the support classifier. For recursive partitioning and regression trees (RT), CIK consistently performs better or comparable to

RBF for all the datasets in terms of sensitivity. In terms of accuracy and gmean, however, CIK (with RT as support classifier) performs worse than RBF. For single-hidden-layer neural network (NN), CIK performs better than or comparable to RBF for 10 of the 15 datasets in terms of sensitivity. In terms of accuracy, CIK with NN as support performs better than CIK with Naive Bayes as support. However, compared to RBF, CIK with NN performs slightly worse than RBF in terms of accuracy and gmean (CIK with NN is comparable or better than RBF in terms of both accuracy and gmean for 6 of the 15 datasets). Overall, CIK with NN performs better than CIK with other support classifiers focused on so far. In the next subsection, we present CIK with another support classifier that significantly depicts performance improvement over RBF.

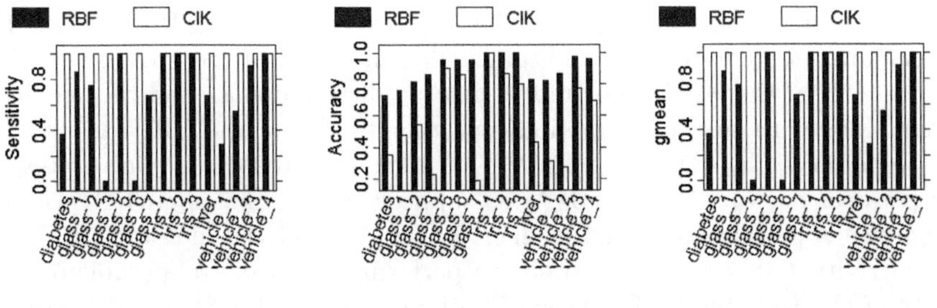

(a) RT

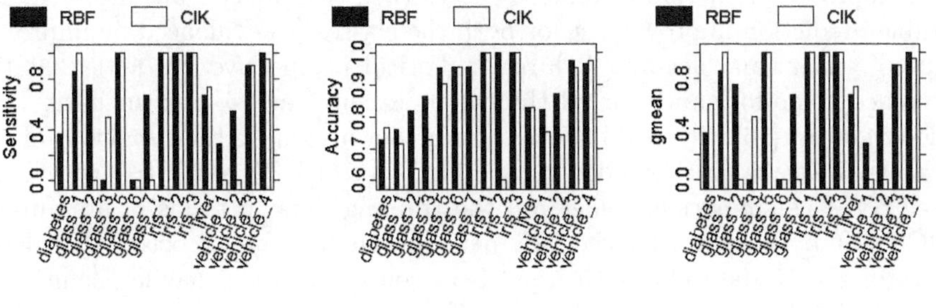

(b) NN

Fig. 3. Performance measures for RBF kernel and CIK (Class informed kernel) with (a) recursive regression and partition tree (RT) and (b) single-hidden-layer neural network (NN) as support classifier

4.4 SVM as Support Classifier

In the previous sub-sections we have experimented with different support classifiers and noted varied effects on prediction performance. More specifically, CIK has generally performed better than RBF in terms of sensitivity, but depicted variations in terms of the other two measures. In this section, we present results for RBF kernel based SVM being used as the support classifier. Thus, for a

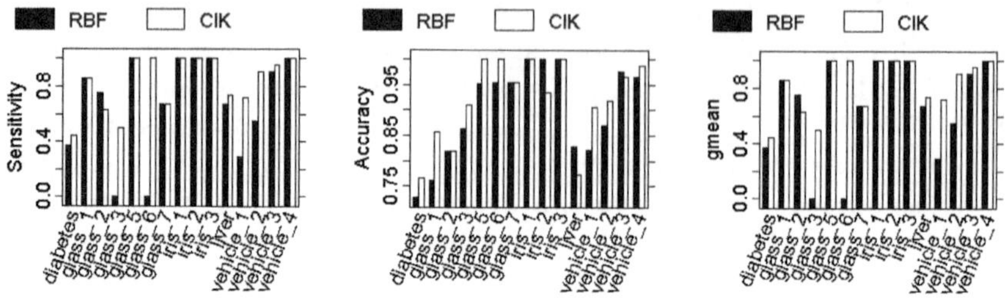

Fig. 4. Performance measures for RBF kernel and CIK (Class informed kernel) with SVM as support classifier

given train data file, SVM is first trained on the input using the class informed kernel formulation of Eq. 2, and another SVM is trained using RBF kernel. The hyper-parameters are kept at the same values for both of these trainings. During prediction, the RBF based model first predicts the class and the predicted labels along with respective attribute vectors are fed to the CIK based model for prediction. The outcome from CIK is the final prediction. Fig. 4 presents the performance of CIK with SVM as support against that for RBF. We note a significant performance improvement in terms of all the measures. In terms of sensitivity, CIK (with SVM support) performs better or comparable to RBF for 14 of the 15 datasets. In terms of both accuracy and gmean, CIK performs better or comparable to RBF for 12 of the 15 datasets. Thus, not only CIK with SVM improves prediction of minority class (i.e., sensitivity), but also achieves notable prediction improvement for both the classes (as evidenced by improved value of gmean and accuracy). From statistical perspective, we note that the difference in performance for all the three measures are significant using two-tailed sign test [16] with $p < 0.05$. To get further insight on the behaviour of the classifier, we have also recorded the performance of RBF and CIK (with RBF based SVM as support classifier) on the training data in terms of area under ROC (AUC). We note that the CIK performs comparable or better than RBF for 14 of the 15 datasets (with iris_3 being only exception, having slight drop in AUC). We have also noted the ratio of training data incorrectly classified by both RBF and CIK (*II*), incorrectly classified by RBF but correctly classified by CIK (*IC*), correctly classified by RBF but incorrectly classified by CIK (*CI*) and correctly classified by both RBF and CIK (*CC*). We observe that for majority of the datasets, *IC* is greater than *CI*. These findings imply that CIK gains better separability between the class representatives than RBF and which, in turn, provides an explanation of its better prediction performance on the test set in terms of the different measures.

5 Conclusion

This article has presented a new learning approach along with a kernel formulation for SVM incorporating class information. An integral part of this approach is the training of a second (support) classifier and results have been presented for varied support classification schemes. Overall, the proposed kernel based learning (CIK) improves prediction performance in terms of sensitivity and thereby is well suited for imbalanced data classification. Experiments are also conducted using SVM as support classifier and statistically significant prediction performance improvement is noted. The proposed kernel is based on RBF kernel formulation. Future research possibilities lie in the extension of the formulation in terms of other kernels and varied added weights.

Acknowledgments. The authors would like to thank the anonymous reviewers for their insightful recommendations.

References

1. Vapnik, V.: The Nature of Statistical Learning Theory. Springer, Heidelberg (2000)
2. Smola, A.: Advances in Large Margin Classifiers. MIT Press, Cambridge (2000)
3. Xiang, J., Chen, J., Zhou, H., Qin, Y., Li, K., Zhong, N.: Using SVM to Predict High-Level Cognition from fMRI Data: A Case Study of 4* 4 Sudoku Solving. Brain Informatics, 171–181 (2009)
4. Yang, J., Zhong, N., Liang, P., Wang, J., Yao, Y., Lu, S.: Brain activation detection by neighborhood one-class SVM. Cognitive Systems Research 11(1), 16–24 (2010)
5. Xu, J., Li, H., Zhong, C.: Relevance ranking using kernels. Technical Report MSR-TR-2009-80, Microsoft Research Technical Report (2009)
6. Chang, C.C., Lin, C.J.: LIBSVM: a library for support vector machines (2001), Software available at `http://www.csie.ntu.edu.tw/~cjlin/libsvm`
7. Fan, R., Chen, P., Lin, C.: Working set selection using second order information for training support vector machines. The Journal of Machine Learning Research 6, 1918 (2005)
8. Cristianini, N., Shawe-Taylor, J.: An introduction to support Vector Machines: and other kernel-based learning methods. Cambridge Univ. Pr., Cambridge (2000)
9. Wu, G., Chang, E.: KBA: kernel boundary alignment considering imbalanced data distribution. IEEE Transactions on Knowledge and Data Engineering, 786–795 (2005)
10. Libsvm data: Classification, regression, and multi-label (2010)
11. Karatzoglou, A., Smola, A., Hornik, K., Zeileis, A.: Kernlab – an S4 package for kernel methods in R. Journal of Statistical Software 11(9), 1–20 (2004)
12. Chawla, N., Japkowicz, N., Kotcz, A.: Editorial: special issue on learning from imbalanced data sets. ACM SIGKDD Explorations Newsletter 6(1), 1–6 (2004)
13. Kubat, M., Holte, R., Matwin, S.: Machine Learning for the Detection of Oil Spills in Satellite Radar Images. Machine Learning 30(2), 195–215 (1998)
14. Tan, P., Steinbach, M., Kumar, V.: Introduction to data mining. Pearson Addison Wesley, Boston (2006)
15. Breiman, L.: Classification and regression trees. Chapman & Hall/CRC, Boca Raton (1984)
16. Demšar, J.: Statistical comparisons of classifiers over multiple data sets. The Journal of Machine Learning Research 7, 1–30 (2006)

Gaze Pattern and Reading Comprehension

Tan Vo, B. Sumudu U. Mendis, and Tom Gedeon

School of Computer Science, The Australian National University,
Acton, Canberra, ACT 0200, Australia
{tan.vo,sumudu.mendis,tom.gedeon}@anu.edu.au

Abstract. Does the way a person read influence the way they understand information or is it the other way around? In regard to reading of English text, just how much we can learn from a person's gaze pattern? It is known that while reading, we inadvertently form rational connections between pieces of information we pick up from the text. That reflects in certain disruptions in the norms of reading paradigm and that gives us clues to our interest level in reading activities.

In this paper, we validate the above statement and then propose a novel method of detecting the level of engagement in reading based on a person's gaze-pattern. We organised some experiments in reading tasks of over thirty participants and the experimental outputs are classified with Artificial Neural Networks with an approximately 80 percent accuracy. The design of this approach is simple and computationally feasible enough to be applied in a real-life system.

"Your eyes are the windows to your soul"

Keywords: Eye-gaze Pattern, Artificial Neural Network, SVM, Grid-based Clustering.

1 Introduction

Everyone is taught to read (in English) the same way: Read a line form left to right and then drop down to the next line once the end of the current line is reached. As a beginner, we followed this simple rule very closely but as we get more adept in reading English text, it is no longer the case. What we have found is that people develop their own personal behaviours when reading, that they probably do not notice they even do. One aim of this research is to characterise these behaviours and to identify and abstract significant model that can show how engaged a person is in the reading activity.

We conducted an experiment where we capture test participants gaze activities while they perform reading tasks. We analysed a set of features of those data: reading time, fixations time, differences in X and Y coordinates, etc.... to identify the key factors to indelicate user engagement level in reading. With the positive result obtained from that, we would like to introduce a simple but affective approach to measure a person's interest in the materials he is reading.

We use Artificial Neural Network (ANN) method to validate the effectiveness of the proposed approach. By combing our solution with a reasonably simple

K.W. Wong et al. (Eds.): ICONIP 2010, Part II, LNCS 6444, pp. 124–131, 2010.

ANN, we could introduce a very achievable real life system. This solution is also flexible in combining with other classifying techniques. We further strengthen our claim by achieving a comparably accurate results with Support Vector Machine (SVM).

2 Proposed Method

2.1 Background

Comprehending the meaning of words in sentences and paragraphs is a great (unnoticed) strain on a persons cognitive process. In order to comprehend text a person needs to be able to read quickly because a person can, generally, only keep seven pieces of information (± 2) in their short-term memory [6]. Any additional information is quickly lost and cannot be recalled. This general rule stands for many different kinds of information from the very simple (letters, or words) to the very complex (entire sentences or a word and 20 Literature Survey all its associated contexts). This allows a fluent reader to be able to "chunk" related information together so that they can get more words into their short term memories.

The above phenomenon results in certain disruptions in reading patterns. We believe that these stochastic behaviours are the keys to effectively quantify the reading engagement load level of a person. Previously in 2009, a study was organised in our research group to investigate using eye-tracking to analyse reading behaviour. Even still in the preliminary stages, it showed the potential in using machine learning approached to classify eye gaze patterns. The purpose of the research behind this paper is to consolidate the previous studies result and to propose a feasible model for classifying gaze-pattern with machine learning methods such as Neural Networks.

The method we are proposing here is for detecting user engagement in reading and is based on the aforementioned gaze features. We also introduce three design principles to make it a lightweight yet effective method for this purpose.

2.2 Effective Reduction of Data Resolution

The gaze-tracking equipment that we use provides us the gaze points in term of a series of X and Y coordinates. These coordinates identify the locations of the gaze points on the screen and have been used to calculate the horizontal and vertical movements. In previous experiments[2][3], fixation points have been produced by filtering those gaze points, resulting in a more interpretable form for later data processing.

By observation, we found that most of the movements of fixations, i.e. saccades are just small and subtle position changes caused by the fact that the eyes do not actually focus on one place. Those saccades are considered irrelevant for this purpose and we can afford to omit them in the pattern recognition stage, hence further reduce the sample size.

We proposed a simple but effective method by dividing the screen into smaller cells using a m-by-n grid. This effectively replaces change in positions of any two fixations with the difference in position of the cells that contain them. We refer to this as **cell movements**. In the cases when the fixation movements are contained within a cell, we consider it a no change in cell position. The benefit of this is it will be less computationally demanding to perform any processing/analysis because the number of data points have been greatly reduced. We can also adjust the resolution of the grid (m and n) for finer or coarser filtering.

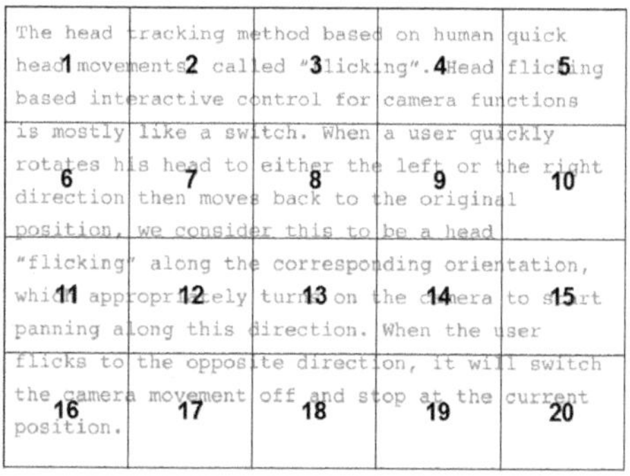

Fig. 1. A paragraph is divided into cells by a 4-by-5 grid. Each identified by a cell number

We examined the data sample from the experiment with and without using this data reducing method. Both yield comparable results except for computational speed.

2.3 Focus on Back Tracking and Forward Tracking

The most significant disruption in reading flow are the skipping forward and back-tracking activities found in the gaze. As participants try to “link” information up, they shift their eyes' focus back-and-forth to achieve a better understanding of the information.

Back tracking and forward tracking are two activities that we would like to qualify as the main factor to detect engagement in reading tasks. To quantify if a gaze movement is a back/forward tracking patterns, we consider if it belongs to the two “extreme” of cell movement groups. If we established a normal distribution of the distances of movement, the “extreme” groups are the one that did not fall within the 68-th confidence interval (a margin of one standard error). Figure 2 below demonstrates this idea:

The figure depicts the distribution of all cell movement distances of a person reading of one paragraph. It shows that if the distance of a cell movement is less than -5 ($\mu - 1\sigma$), that saccade is considered a backtracking. On the other hand, a forward tracking saccade is one that has the distance greater than 5($\mu + 1\sigma$). These thresholds are expected to be different on a case by case basis.

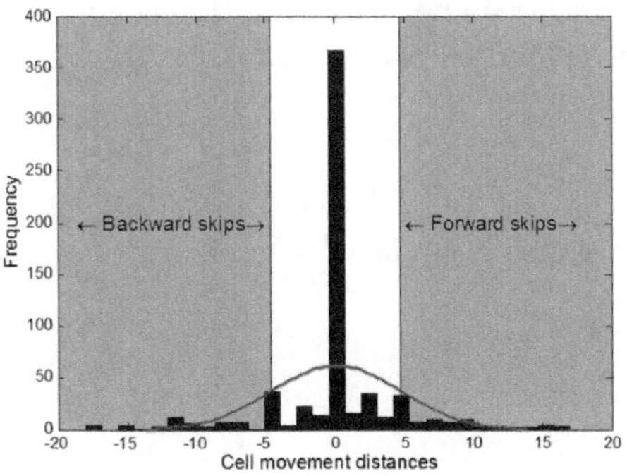

Fig. 2. A distribution of cell movement distances throughout reading activity of a paragraph

2.4 Levenberg-Marquardt Based Neural Network

Previously, an experiment carried out by Zhu et al.[2] to evaluate the performance of Levenberg-Marquardt optimization, combined with fuzzy signature, in classifying gaze patterns. What they found is that this optimization algorithm performs well with the gaze-pattern classification problem and on par with SVM in the two classes test.

In this paper, we evaluate the performance of Levenberg-Marquardt optimization as the training function in a Neural Network to classify eye gaze. The neural network we construct is a two-layer, feed-forward back-propagation that has one single output node. Hence the output value regarding to a pattern T is described as[4],[5],[7]:

$$y_1^T = g_O(b_1 + \sum_j W_{1j} \cdot g_H(b_j + \sum_k w_{jk} \cdot x_k^T)), \tag{1}$$

- b_1, b_j: the bias
- w_{1j} is the weight of the jth hidden neuron to the single output neuron
- w_{jk} is the weight of kth input neuron to the jth hidden neuron
- x_k^T the kth element of the input pattern T
- g_O transfer function on the output layer - linear transfer function
- g_H transfer function on the hidden layers - sigmoid transfer function

We evaluate the training performance of the network with this error function (mean square error):

$$E = \frac{1}{N} \sum_{k=1}^{N} (y_E - y_P)^2, \tag{2}$$

where y_E is the vector of predict outcomes and y_P represents the vector of predicted outcome.

The back-propagation training algorithm, being Levenberg-Marquardt optimization, will be represented by the formula[5]:

$$\delta w = (J^T J + I \cdot \mu)^{-1} J^T e \tag{3}$$

where J is the Jacobian matrix of the error function calculated in equation(2), μ is the learning rate which is updated after iteration. *diag* being the diagonal of $J^T J$.

3 Experiment

3.1 Background

We have selected 35 participants for this experiment. The experiment involves the participant reading some paragraphs from a computer screen while the computer gathers their eyes'(gaze) movements with gaze-tracking equipment.

In total there were ten paragraphs for the participants to read. Seven of the paragraphs were taken from the paper "Keyboard before Head Tracking Depresses User Success in Remote Camera Control" by Zhu et al.[1]. The remaining three paragraphs were extracts from various sources (miscellaneous paragraphs). Five of the paragraphs from the paper were chosen for the amount of useful information that was contained within. The other two paragraphs from the paper and the three miscellaneous paragraphs describing that paper were chosen because of their generality and lack of specific technical information, the paragraphs being introductory in nature. That is, care was taken to make sure that this fact was not obvious.

3.2 Setup

Within the 35 volunteered participants, we divided them into two groups. Group A were people that had been informed that they would have to answer questions about the paragraphs they read toward the end of the experiment session. Group B, however, were allowed to read as if they were just reading any piece of text - and that they would not be questioned at the end. Group A contained 13 people while group B had 22 participants.

The paragraphs were presented to participants in different orders to prevent any specific paragraph ordering from affecting the results. This was an experiment design choice to help show which participants could look at the bigger picture even when the information is out of sequence and scattered.

The screen used was a 19 inch LCD and was set to a resolution of 1280 by 1024. All the paragraphs are displayed in full-screen. To stabilise the head position, we use a chin rest, which positions the participant faces about 72 cm away from the centre of the screen.

3.3 Data Collection and Preparation

The gaze points are collected at about 1/60th of a second rate, and we produce fixation points from them. We "group" gaze points into clusters with size of 15 pixels radius[3] to define fixations. The fixation length (ms) is worked out by the number of gaze points within each cluster.

Below are a visualisation of gaze data being projected onto their correspondent paragraphs. The solid circles represents the fixation points. The shade of the circles indicates the fixation length - with the darker one indicate a longer fixation point. The lines that connect the fixations points represent the saccades.

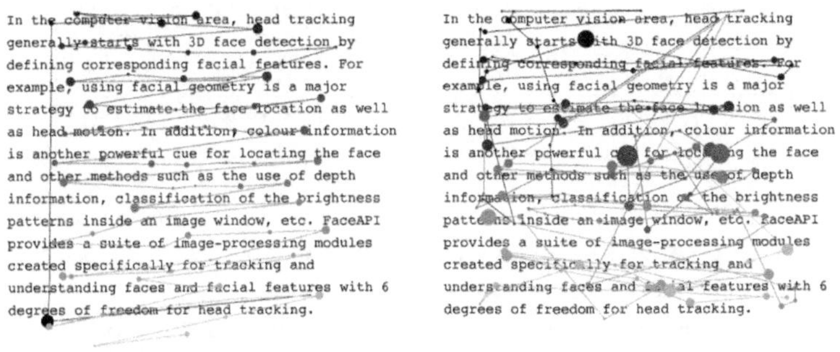

(a) (b)

Fig. 3. An example of 2 read patterns of the same paragraph by two different participant

We further filter out the data using the aforementioned grid-based method with 4-by-5 grid. Based on that we calculate numbers of back-tracking and forward-tracking of each paragraph. For the classification task, we use these three following feature to be evaluated with ANN:

- Average fixation length for each paragraph and each participant
- Back-tracking count for each paragraph and each participant
- Forward-tracking count for each paragraph and each participant

4 Evaluation and Comparison

4.1 Neural Network Results

A two-layer neural network with one output neuron is use for classifying data. The transfer function of the output layer is a linear transfer function while the hidden layer is equipped with a tangent sigmoid transfer function. The hidden layer comprise of 5 neurons. We designed this to be a binary classification problem where the target values were 1 for *relevant paragraph* and 0 for *irrelevant paragraph*. The neural network was back propagation trained with Levenberg-Marquardt optimization. The LM parameters are configured with an initial μ value of 0.001 with the increase and decrease factors as 10 and 0.1 respectively. The network performance is evaluated by Mean of Square Error method.

We performed 9-folds cross-validation and obtained the average of classification accuracies from every fold. Due to the relatively small sample size available

(35 participants - with 10 paragraphs per participant), 9-fold cross-validation is preferred to the conventional 10-fold method. For each fold, the training set is divided as followed: 60% for data training, 20% was used to generalise the network and prevent over-fitting and the last 20% was used as the test data.

Table 1. ANN Results for Eye-gaze Feature Pattern Recognition

Experiment	Classification Error Rate	Sensitivity	Specificity
Group A	0.2586	0.7241	0.7586
Group B	0.1717	0.7879	0.8687
Both Groups	0.1975	0.7898	0.8153

With this ANN setup, We were able to achieve about 80% classification accuracy as seen in Table 1. This is encouraging because we only provided three gaze parameters as classification categories.

As we can see the classification performance achieved with Group A data is slightly lower than with group B data. The hypothesis behind that is that with group A, where participants were expected to answer questions about the paragraph, that lead to a more "careful" reading behaviour for all paragraphs. That would results to a less disruptive forward, backward movements in the reading patterns. But nevertheless, the classification results in all cases are positive considered the small number of classification features.

4.2 Support Vector Machine Comparison

Support Vector Machines (SVM) are well-established method for this type of classification problems[4].

We constructed a conventional Support Vector Machine with a linear kernel. We used the same dataset as we did with ANN. The labels we chose are "1" for *relevant paragraph* and "0" for *irrelevant paragraph* . To have a fair comparisons, we also cross-validated the results using 9-folds and obtains the average Classification Error Rate (CER) after 9 iteration.

Table 2. SVM Results for Eye-gaze Feature Pattern Recognition

Experiment	SVM Classification Error Rate(CER)	ANN CER	SVM Sensitivity	SVM Specificity
Group A	0.2538	0.2586	0.7077	0.7846
Group B	0.1773	0.1717	0.8091	0.8364
Both Groups	0.2086	0.1975	0.7600	0.8229

Table 2 compares the results we got by using SVM technique with the previous results by ANN. We found that ANN and SVM performance in term of accuracy almost exactly match each other. Both methods (SVM and ANN) result in a very high accuracy rate and with further optimisations on both, we believe we can attain even more positive results.

5 Conclusion

In this paper, we demonstrated the effectiveness of ANN in recognising gaze-patterns. The findings are encouraging because ANN combined with our proposed method for data preprocessing has resulted in a low computational model that achieves highly accurate results: An ANN being trained with only three classification categories is able to achieve 80% accuracy is very encouraging. It has also consolidate the outcome of our previous experiment[3] as well as the use of Levenberg-Marquardt optimization as the training algorithm for this types of problem[2].

References

1. Zhu, D., Gedeon, T., Taylor, K.: Keyboard before Head Tracking Depresses User Success in Remote Camera Control. In: Gross, T., Gulliksen, J., Kotzé, P., Oestreicher, L., Palanque, P., Prates, R.O., Winckler, M. (eds.) INTERACT 2009. LNCS, vol. 5727, pp. 319–331. Springer, Heidelberg (2009)
2. Zhu, D., Mendis, B.S., Gedeon, T., Asthana, A., Goecke, R.: A Hybrid Fuzzy Approach for Human Eye Gaze Pattern Recognition. In: Köppen, M., Kasabov, N., Coghill, G. (eds.) ICONIP 2008. LNCS, vol. 5507, pp. 655–662. Springer, Heidelberg (2009)
3. Fahey, D.: A Preliminary Investigation into using Eye-Tracking to Analyse a Persons Reading Behaviour. B.S. thesis. The School of Computer Science Australian National University (2009)
4. Byvatov, E., Fechner, U., Sadowski, J., Schneider, G.: Comparison of Support Vector Machine and Artificial Neural Network Systems for Drug/Nondrug Classification. Journal of Chemical Information and Computer Sciences, 1882–1889 (2003)
5. Hagan, M.T., Menhaj, M.B.: Training feedforward networks with the Marquardt algorithm. IEEE Transactions on Neural Networks 5(6), 989–993 (1994), doi:10.1109/72.329697
6. Miller, G.A.: The Magical Number Seven, Plus or Minus Two: Some Limits on Our Capacity for Processing Information. In: The Psychology of Communication: Seven Essays. Penguin Books, Inc. (1970)
7. Mendis, B.S.U., Gedeon, T.D., Koczy, L.T.: Learning Generalized Weighted Relevance Aggregation Operators Using Levenberg-Marquardt Method. In: International Conference on Hybrid Intelligent Systems (2006)

A Theoretical Framework for Multi-sphere Support Vector Data Description

Trung Le, Dat Tran, Wanli Ma, and Dharmendra Sharma

Faculty of Information Sciences and Engineering
University of Canberra, ACT 2601, Australia
{trung.le,dat.tran,wanli.ma,dharmendra.sharma}@canberra.edu.au

Abstract. In support vector data description (SVDD) a spherically shaped boundary around a normal data set is used to separate this set from abnormal data. The volume of this data description is minimized to reduce the chance of accepting abnormal data. However the SVDD does not guarantee that the single spherically shaped boundary can best describe the normal data set if there are some distinctive data distributions in this set. A better description is the use of multiple spheres, however there is currently no investigation available. In this paper, we propose a theoretical framework to multi-sphere SVDD in which an optimisation problem and an iterative algorithm are proposed to determine model parameters for multi-sphere SVDD to provide a better data description to the normal data set. We prove that the classification error will be reduced after each iteration in this learning process. Experimental results on 28 well-known data sets show that the proposed multi-sphere SVDD provides lower classification error rate comparing with the standard single-sphere SVDD.

Keywords: Support vector data description, spherically shaped boundary, imbalanced data, one-class classification, novelty detection.

1 Introduction

In One-Class Classification, also called Novelty Detection (ND), the problem of data description is to make a description of a normal data set and to detect new sample that resembles this data set [9]. Data description can be used for outlier detection to detect abnormal samples from a data set. Data description is also used for a classification problem where one class is well sampled while other classes are severely undersampled. In real-world applications, collecting the normal data is cheap and easy while the abnormal data is expensive and is not available in several situations [14]. For instance, in case of machine fault detection, the normal data under the normal operation is easy to obtain while in faulty situation the machine is required to devastate completely. Therefore one-class classification is more difficult than conventional two-class classification because the decision boundary of one-class classification is mainly constructed

K.W. Wong et al. (Eds.): ICONIP 2010, Part II, LNCS 6444, pp. 132–142, 2010.

from samples of only the normal class and hence it is hard to decide how strict decision boundary should be. ND is widely applied to many application domains such as network intrusion, currency validation, user verification in computer systems, medical diagnosis [3], and machine fault detection [16].

There are two main approaches to solving the data description problem which are density estimation approach [1][2][12] and kernel based approach [13][14][20]. In density estimation approach, the task of data description is solved by estimating a probability density of a data set [11]. This approach requires a large number of training samples for estimation, in practice the training data is not sufficient and hence does not represent the complete density distribution. The estimation will mainly focus on modeling the high density areas and can result in a bad data description [14]. Kernel-based approach aims at determining the boundaries of the training set rather than at estimating the probability density. The training data is mapped from the input space into a higher dimensional feature space via a kernel function. Support Vector Machine (SVM) is one of the well-known kernel-based methods which constructs an optimal hyperplane between two classes by focusing on the training samples close to the edge of the class descriptors [17]. These training samples are called support vectors. In One-Class Support Vector Machine (OCSVM), a hyperplane is determined to separate the normal data such that the margin between the hyperplane and outliers is maximized [13]. Support Vector Data Description (SVDD) is a new SVM learning method for one-class classification [14]. A hyperspherically shaped boundary around the normal data set is constructed to separate this set from abnormal data. The volume of this data description is minimized to reduce the chance of accepting abnormal data. SVDD has been proven as one of the best methods for one-class classification problems [19].

Some extensions to SVDD have been proposed to improve the margins of the hyperspherically shaped boundary. The first extension is Small Sphere and Large Margin (SSLM) [20] which proposes to surround the normal data in this optimal hypersphere such that the margin—distance from outliers to the hypersphere, is maximized. This SSLM approach is helpful for parameter selection and provides very good detection results on a number of real data sets. We have recently proposed a further extension to SSLM which is called Small Sphere and Two Large Margins (SS2LM) [7]. This SS2LM aims at maximising the margin between the surface of the hypersphere and abnormal data and the margin between that surface and the normal data while the volume of this data description is being minimised.

Other extensions to SVDD regarding data distribution have also been proposed. The first extension is to apply SVDD to multi-class classification problems [5]. Several class-specific hyperspheres that each encloses all data samples from one class but excludes all data samples from other classes. The second extension is for one-class classification which proposes to use a number of hyperspheres to describe the normal data set [19]. Normal data samples may have some distinctive distributions so they will locate in different regions in the feature space and hence if the single hypersphere in SVDD is used to enclose all normal data, it will also enclose abnormal data samples resulting a high false positive error rate. However this work was not presented in detail, the proposed method is heuristic and there is no theoretical investigation provided to show that the multi-sphere approach can provide a better data description.

We propose in this paper a new theoretical framework to the multi-sphere SVDD. A set of hyperspheres is proposed to describe the normal data set assuming that normal data samples have distinctive data distributions. We formulate the optimisation problem for multi-sphere SVDD and prove how SVDD parameters are obtained through solving this problem. An iterative algorithm is also proposed for building data descriptors, and we also prove that the classification error will be reduced after each iteration. Experimental results on 28 well-known data sets show that the proposed multi-sphere SVDD provides lower classification error rates comparing with the standard single-sphere SVDD.

2 Support Vector Data Description (SVDD)

Let $X = \{x_1, x_2, \ldots, x_n\}$ be the normal data set. SVDD [14] aims at determining an optimal hypersphere including all normal data points in this data set X while abnormal data points are not included. The optimisation problem is as follows

$$\min_{R,c,\xi} \left(R^2 + C \sum_{i=1}^{n} \xi_i \right) \tag{1}$$

subject to

$$\begin{aligned} &||\phi(x_i) - c||^2 \leq R^2 + \xi_i \qquad i = 1, \ldots, n \\ &\xi_i \geq 0, \quad i = 1, \ldots, n \end{aligned} \tag{2}$$

where R is radius of the hypersphere, C is a constant, $\xi = [\xi_i]_{i=1,\ldots,n}$ is vector of slack variables, $\phi(.)$ is the nonlinear function related to the symmetric, positive definite kernel function $K(x_1, x_2) = \phi(x_1)^T \phi(x_2)$, and c is centre of the hypersphere.

For classifying an unknown data point x, the following decision function is used: $f(x) = sign(R^2 - ||\phi(x) - c||^2)$. The unknown data point x is normal if $f(x) = +1$ or abnormal if $f(x) = -1$.

3 One-Class Support Vector Machine (OCSVM)

In OCSVM [13], a hyperplane is determined to separate all normal data and at the same time maximise the margin between the normal data and the hyperplane. OCSVM can be modelled as follows

$$\min_{w,\rho} \left(\frac{1}{2}||w||^2 - \rho + \frac{1}{\nu n} \sum_{i=1}^{n} \xi_i \right) \tag{3}$$

subject to

$$w^T\phi(x_i) \geq \rho - \xi_i \qquad i = 1, \ldots, n$$
$$\xi_i \geq 0, \quad i = 1, \ldots, n \tag{4}$$

where w is the normal vector of the hyperplane, ρ is the margin and ν is a positive constant.

The decision function is $f(x) = sign(w^T\phi(x) - \rho)$. The unknown data point x is normal if $f(x) = +1$ or abnormal if $f(x) = -1$.

4 Multi-sphere SVDD

4.1 Problem Formulation

Consider a set of m hyperspheres $S_j(c_j, R_j)$with center c_j and radius R_j, $j = 1, \ldots, m$. This hypershere set is a good data description of the normal data set $X = \{x_1, x_2, \ldots, x_n\}$ if each of the hyperspheres describes a distribution in this data set and the sum of all radii $\sum_{j=1}^{m} R_j^2$ should be minimised.

Let matrix $U = [u_{ij}]_{n \times m}$, $i = 1, \ldots, n$, $j = 1, \ldots, m$ where u_{ij} is the hard membership representing the belonging of data point x_i to hypersphere S_j, $u_{ij} = 0$ if x_i is not in S_j and $u_{ij} = 1$ if x_i is in S_j. The optimisation problem of multi-sphere SVDD can be formulated as follows

$$\min_{R,c,\xi} \left(\sum_{j=1}^{m} R_j^2 + C \sum_{i=1}^{n} \xi_i \right) \tag{5}$$

subject to

$$\sum_{j=1}^{m} u_{ij}||\phi(x_i) - c_j||^2 \leq \sum_{j=1}^{m} u_{ij}R_j^2 + \xi_i \qquad i = 1, \ldots, n$$
$$\xi_i \geq 0, \quad i = 1, \ldots, n \tag{6}$$

where $R = [R_j]_{j=1,\ldots,m}$ is vector of radii, C is a constant, $\xi = [\xi_i]_{i=1,\ldots,n}$ is vector of slack variables, $\phi(.)$ is the nonlinear function related to the symmetric, positive definite kernel function $K(x_1, x_2) = \phi(x_1)^T\phi(x_2)$, and $c = [c_j]_{j=1,\ldots,m}$ is vector of centres.

Minimising the function in (5) over variables R, c and ξ subject to (6) will determine radii and centres of hyperspheres and slack variables if the matrix U is given. On the other hand, the matrix U will be determined if radii and centres of hyperspheres are given. Therefore an iterative algorithm will be applied to find the complete solution. The algorithm consists of two alternative steps: 1) Calculate radii and centres of hyperspheres and slack variables, and 2) Calculate membership U.

We present in the next sections the iterative algorithm and prove that the classification error in the current iteration will be smaller than that in the previous iteration.

For classifying a data point x, the following decision function is used

$$f(x) = sign\Big(\max_j \big\{R_j^2 - ||\phi(x) - c_j||^2\big\}\Big) \tag{7}$$

The unknown data point x is normal if $f(x) = +1$ or abnormal if $f(x) = -1$.

4.2 Calculating Radii, Centres and Slack Variables

The Lagrange function for the optimisation problem in (5) subject to (6) is as follows

$$L(R, c, \xi, \alpha, \beta) = \sum_{j=1}^{m} R_j^2 + C\sum_{i=1}^{n}\xi_i + \sum_{i=1}^{n}\alpha_i\Big(||\phi(x_i) - c_{s(i)}||^2 - R_{s(i)}^2 - \xi_i\Big) - \sum_{i=1}^{n}\beta_i\xi_i \tag{8}$$

where $s(i)$ is index of the hypersphere to which data point x_i belong and satisfies $u_{is(i)} = 1$ and $u_{ij} = 0\ \forall j \neq s(i)$.

Setting derivatives of $L(R, c, \xi, \alpha, \beta)$with respect to primal variables to 0, we obtain

$$\frac{\partial L}{\partial R_j} = 0 \quad \Rightarrow \sum_{i \in s^{-1}(j)} \alpha_i = 1 \tag{9}$$

$$\frac{\partial L}{\partial c_j} = 0 \quad \Rightarrow c_j = \sum_{i \in s^{-1}(j)} \alpha_i\phi(x_i) \tag{10}$$

$$\frac{\partial L}{\partial \xi_j} = 0 \quad \Rightarrow \alpha_i + \beta_i = C, \quad i = 1, \ldots, n \tag{11}$$

$$\alpha_i \geq 0, \quad ||\phi(x_i) - c_{s(i)}||^2 - R_{s(i)}^2 - \xi_i \geq 0,$$
$$\alpha_i\Big(||\phi(x_i) - c_{s(i)}||^2 - R_{s(i)}^2 - \xi_i\Big) = 0 \tag{12}$$

$$\beta_i \geq 0, \quad \xi_i \geq 0, \quad \beta_i\xi_i = 0 \tag{13}$$

To get the dual form, we substitute (9)-(13) to the Lagrange function in (8) and obtain the following:

$$\begin{aligned} L &= \sum_{i=1}^{n}\alpha_i||\phi(x_i) - c_{s(i)}||^2 \\ &= \sum_{i=1}^{n}\alpha_i K(x_i, x_i) - 2\sum_{i=1}^{n}\alpha_i\phi(x_i)c_{s(i)} + \sum_{i=1}^{n}\alpha_i||c_{s(i)}||^2 \\ &= \sum_{i=1}^{n}\alpha_i K(x_i, x_j) - \sum_{j=1}^{m}\sum_{i \in s^{-1}(j)}\alpha_i\phi(x_i)c_j + \sum_{j=1}^{m}\sum_{i \in s^{-1}(j)}\alpha_i||c_j||^2 \end{aligned}$$

$$
\begin{aligned}
&= \sum_{i=1}^{n} \alpha_i K(x_i, x_j) - 2\sum_{j=1}^{m}\Big(c_j \sum_{i \in s^{-1}(j)} \alpha_i \phi(x_i)\Big) + \sum_{j=1}^{m}\Big(||c_j||^2 \sum_{i \in s^{-1}(j)} \alpha_i\Big) \\
&= \sum_{i=1}^{n} \alpha_i K(x_i, x_j) - \sum_{j=1}^{m} ||c_j||^2 \\
&= \sum_{j=1}^{m}\Big(\sum_{i \in s^{-1}(j)} \alpha_i K(x_i, x_i) - ||c_j||^2\Big) \\
&= \sum_{j=1}^{m}\Big(\sum_{i \in s^{-1}(j)} \alpha_i K(x_i, x_i) - ||\sum_{i \in s^{-1}(j)} \alpha_i \phi(x_i)||^2\Big) \\
&= \sum_{j=1}^{m}\Big(\sum_{i \in s^{-1}(j)} \alpha_i K(x_i, x_i) - \sum_{i,i' \in s^{-1}(j)} \alpha_i \alpha_{i'} K(x_i, x_{i'})\Big) \qquad (14)
\end{aligned}
$$

The result in (14) shows that the optimisation problem in (5) is equivalent to m individual optimisation problems as follows

$$
\min\Big(\sum_{i \in s^{-1}(j)} \alpha_i K(x_i, x_i) - \sum_{i,i' \in s^{-1}(j)} \alpha_i \alpha_{i'} K(x_i, x_{i'})\Big) \quad j = 1, \ldots, m \qquad (15)
$$

subject to

$$
\sum_{i \in s^{-1}(j)} \alpha_i = 1 \quad \text{and} \quad 0 \leq \alpha_i \leq C \quad j = 1, \ldots, m \qquad (16)
$$

After solving all of these individual optimization problems, we can calculate the updating radii $R = [R_j]$ and centres $c = [c_j]$, $j = 1, \ldots, m$ using the equations in SVDD.

The following theorem is used to consider the relation of slack variables to data points classified.

Theorem 1. *Assume that (R, c, ξ) is a solution of the optimisation problem (5) and x_i is classified to hypersphere $S_k(c_k, R_k)$. If x_i is missclassified then $\xi_i = ||\phi(x_i) - c_k||^2 - R_k^2$. If x_i is correctly classified then $\xi_i = 0$ when $x_i \in S_k$ and $\xi_i = ||\phi(x_i) - c_k||^2 - R_k^2$ when $x_i \notin S_k$.*

Proof. From (12) and $\xi_i \geq 0$, we have $\xi_i = \max\Big\{0, ||\phi(x_i) - c_k||^2 - R_k^2\Big\}$. The data point x_i is misclassified if it is not in any of the hypersheres, it follows that $||\phi(x_i) - c_j||^2 > R_j^2, \forall j$. So $\xi_i = ||\phi(x_i) - c_k||^2 - R_k^2$ with some k. If x_i is correctly classified then the proof is obtained using (7).

The following empirical error can be defined

$$
error(i) = \begin{cases} \min_j \Big\{||\phi(x_i) - c_j||^2 - R_j^2\Big\} & \text{if } x_i \text{ is misclassified} \\ 0 \text{ otherwise} \end{cases} \qquad (17)
$$

Referring to Theorem 1, it is easy to prove that $\sum_{i=1}^{n} \xi_i$ is an upper bound of $\sum_{i=1}^{n} error(i)$.

4.3 Calculating Membership U

We use radii and centres of hyperspheres to update the membership matrix U. The following algorithm is proposed:

For $i = 1$ to n do
  If x_i is misclassified then
    Let $j_0 = \arg\min_j \left\{||\phi(x_i) - c_j||^2 - R_j^2\right\}$
    Set $u_{ij_0} = 1$ and $u_{ij} = 0$ if $j \neq j_0$
  End if
  Else
    Denote $J = \{j : x_i \in S(c_j, R_j)\}$
    Let $j_0 = \arg\min_{j \in J} \left\{||\phi(x_i) - c_j||^2\right\}$
    Set $u_{ij_0} = 1$ and $u_{ij} = 0$ if $j \neq j_0$
  End Else
End For

4.4 Iterative Learning Process

The proposed iterative learning process for multi-sphere SVDD will run two alternative steps until a convergence is reached as follows

Initialise U by clustering the normal data set in the input space
Repeat the following
  Calculate R, c and ξ using U
  Update U using R and c
Until a convergence is reached

We can prove that the classification error in the current iteration will be smaller than that in the previous iteration through the following key theorem.

Theorem 2. *Let $(R,\ c,\ \xi,\ U)$ and $(\overline{R},\ \overline{c},\ \overline{\xi},\ \overline{U})$ be solutions at the previous iteration and current iteration, respectively. The following inequality holds*

$$\sum_{j=1}^{m} \overline{R}_j^2 + C\sum_{i=1}^{n} \overline{\xi}_i \leq \sum_{j=1}^{m} R_j^2 + C\sum_{i=1}^{n} \xi_i \tag{18}$$

Proof. We prove that (R, c, ξ, U) is a feasible solution at current iteration.

Case 1: x_i is misclassified.

$$\sum_{j=1}^{n} u_{ij}\left(||\phi(x_i) - c_j||^2 - R_j^2\right) - \sum_{j=1}^{n} \overline{u}_{ij}\left(||\phi(x_i) - c_j||^2 - R_j^2\right) =$$
$$u_{is(i)}\left(||\phi(x_i) - c_{s(i)}||^2 - R_{s(i)}^2\right) - \min_j \left(||\phi(x_i) - c_j||^2 - R_j^2\right) \geq 0 \tag{19}$$

Hence

$$\sum_{j=1}^{n} \overline{u}_{ij}\left(||\phi(x_i) - c_j||^2 - R_j^2\right) \le \sum_{j=1}^{n} u_{ij}\left(||\phi(x_i) - c_j||^2 - R_j^2\right) \le \xi_i \tag{20}$$

(20) is reasonable due to (R, c, ξ, U) is solution at the previous step.

Case 2: x_i is correctly classified.
Denote $J = \{j : x_i \in S(c_j, R_j)\}$ and $j_0 = \arg\min_{j \in J}\left\{||\phi(x_i) - c_j||^2\right\}$ then

$$\sum_{j=1}^{n} \overline{u}_{ij}\left(||\phi(x_i) - c_j||^2 - R_j^2\right) = ||\phi(x_i) - c_{j_0}||^2 - R_{j_0}^2 \le 0 \le \xi_i \tag{21}$$

From (20) - (21), we can conclude that (R, c, ξ, U) is a feasible solution at current iteration. In addition, $(\overline{R}, \overline{c}, \overline{\xi}, \overline{U})$ is optimal solution at current iteration. That results in our conclusion.

The validity of Theorem 2 is key to resolve the rationale of our algorithm. The reason is that $\sum_{j=1}^{m} R_j^2$ stands for general error whereas $\sum_{i=1}^{n} \xi_i$ is a quite tight upper bound of empirical error according to (17). Thus the inequality (18) shows that the error in current iteration is less than that in previous iteration.

5 Experimental Results

We performed our experiments on 28 well-known data sets related to machine fault detection and bioinformatics. These data sets were originally balanced data sets and some of them contain several classes. For each data set, we picked up a class at a time and divided the data set of this class into two equal subsets. One subset was used as training set and the other one with data sets of other classes were used for testing. We repeated dividing a data set ten times and calculated the average classification rates. We also compared our multi-sphere SVDD method with SVDD and OCSVM. The classification rate *acc* is measured as [6]

$$acc = \sqrt{acc^{+}acc^{-}} \tag{22}$$

where acc^{+} and acc^{-} are the classification accuracy on normal and abnormal data, respectively.

The popular RBF kernel function $K(x, x') = e^{-\gamma||x-x'||^2}$ was used in our experiments. The parameter γ was searched in $\{2^k : k = 2l + 1, l = -8, -7, \ldots, 2\}$. For SVDD and multi-sphere SVDD, the trade-off parameter C was searched over the grid $\{2^k : k = 2l + 1, l = -8, -7, \ldots, 2\}$. For OCSVM, the parameter ν was searched in $\{0.1k : k = 1, \ldots, 9\}$. For multi-sphere SVDD, the number of

Table 1. Number of data points in 28 data sets. #normal: number of normal data points, #abnormal: number of abnormal data points and d: dimension.

Data set	#normal	#abnormal	d
Arrhythmia	237	183	278
Astroparticle	2000	1089	4
Australian	383	307	14
Breast Cancer	444	239	10
Bioinformatics	221	117	20
Biomed	67	127	5
Colon cancer	40	22	2000
DelfPump	1124	376	64
Diabetes	500	268	8
DNA	464	485	180
Duke	44	23	7129
Fourclass	307	555	2
Glass	70	76	9
Heart	303	164	13
Hepatitis	123	32	19
Ionosphere	255	126	34
Letter	594	567	16
Liver	200	145	6
Protein	4065	13701	357
Sonar	97	111	67
Spectf	254	95	44
Splice	517	483	60
SvmGuide1	2000	1089	4
SvmGuide3	296	947	22
Thyroid	3679	93	21
USPS	1194	6097	256
Vehicle	212	217	18
Wine	59	71	13

hyperspheres was changed from 1 to 10 and 50 iterations were applied to each training.

Table 2 presents classification results for OCSVM, SVDD and multi-sphere SVDD (MS-SVDD). Those experimental results over 28 data sets show that MS-SVDD always performs better than SVDD. The reason is that SVDD is regarded as a special case of MS-SVDD when the number of hyperspheres is 1. MS-SVDD provides the highest accuracies for data sets except for Colon cancer and Biomed data sets. For some cases, MS-SVDD obtains the same result as SVDD. This could be explained as only one distribution for those data sets. Our new model seems to attain the major improvement for the larger data sets. It is quite obvious since the large data sets could have different distributions and can be described by different hyperspheres.

Table 2. Classification results (in %) on 28 data sets for OCSVM, SVDD and Multi-sphere SVDD (MS-SVDD)

Data set	OCSVM	SVDD	MS-SVDD
Arrhythmia	63.16	70.13	70.13
Astroparticle	89.66	90.41	93.23
Australian	77.19	80.00	81.80
Breast Cancer	95.25	98.64	98.64
Bioinformatics	68.34	68.10	72.00
Biomed	74.98	63.83	74.76
Colon cancer	69.08	67.42	67.42
DelfPump	63.20	70.65	75.27
Diabetes	68.83	72.30	78.72
DNA	76.08	73.70	83.01
Duke cancer	62.55	65.94	65.94
FourClass	93.26	98.48	98.76
Glass	80.60	79.21	79.21
Heart	73.40	77.60	79.45
Hepatitis	76.82	80.17	81.90
Ionosphere	90.90	88.73	92.12
Letter	91.42	95.86	98.03
Liver	73.80	62.45	74.12
Protein	63.65	70.68	72.11
Sonar	65.97	72.91	72.91
Spectf	77.10	70.71	77.36
Splice	64.43	70.51	70.51
SVMGuide1	89.56	87.92	93.05
SvmGuide3	63.14	70.63	70.63
Thyroid	87.88	87.63	91.44
USPS	92.85	92.83	96.23
Vehicle	64.50	70.38	75.04
Wine	88.30	98.31	98.31

6 Conclusion

We have proposed a new theoretical framework to multi-sphere support vector data description. A data set is described by a set of hyperspheres. This is an incremental learning process and we can prove theoretically that the error rate obtained in current iteration is less than that in previous iteration. We have made comparison of our proposed method with support vector data description and one-class support vector machine. Experimental results have shown that our proposed method provided better performance than those two methods over 28 well-known data sets.

References

1. Bishop, C.M.: Novelty detection and neural network validation. In: IEE Proceedings of Vision, Image and Signal Processing, pp. 217–222 (1994)
2. Barnett, V., Lewis, T.: Outliers in statistical data, 3rd edn. Wiley, Chichester (1978)

3. Campbell, C., Bennet, K.P.: A linear programming approach to novelty detection. In: Advances in Neural Information Processing Systems, vol. 14 (2001)
4. Chang, C.-C., Lin, C.-J.: LIBSVM: A Library for Support Vector Machines, `http://www.csie.ntu.edu.tw/~cjlinlibsvm`
5. Hao, P.Y., Liu, Y.H.: A New Multi-class Support Vector Machine with Multi-sphere in the Feature Space. In: Okuno, H.G., Ali, M. (eds.) IEA/AIE 2007. LNCS (LNAI), vol. 4570, pp. 756–765. Springer, Heidelberg (2007)
6. Kubat, M., Matwin, S.: Addressing the curse of imbalanced training set: One-sided selection. In: Proc. 14th International Conference on Machine Learning, pp. 179–186 (1997)
7. Le, T., Tran, D., Ma, W., Sharma, D.: An Optimal Sphere and Two Large Margins Approach for Novelty Detection. In: Proc. IEEE World Congress on Computational Intelligence (WCCI) (accepted 2010)
8. Lin, Y., Lee, Y., Wahba, G.: Support vector machine for classification in nonstandard situations. Machine Learning 15, 1115–1148 (2002)
9. Moya, M.M., Koch, M.W., Hostetler, L.D.: One-class classifier networks for target recognition applications. In: Proceedings of World Congress on Neural Networks, pp. 797–801 (1991)
10. Mu, T., Nandi, A.K.: Multiclass Classification Based on Extended Support Vector Data Description. IEEE Transactions on Systems, Man, And Cybernetics Part B: Cybernetics 39(5), 1206–1217 (2009)
11. Parra, L., Deco, G., Miesbach, S.: Statistical independence and novelty detection with information preserving nonlinear maps. Neural Computation 8, 260–269 (1996)
12. Roberts, S., Tarassenko, L.: A Probabilistic Resource Allocation Network for Novelty Detection. Neural Computation 6, 270–284 (1994)
13. Schlkopf, B., Smola, A.J.: Learning with kernels. The MIT Press, Cambridge (2002)
14. Tax, D.M.J., Duin, R.P.W.: Support vector data description. Machine Learning 54, 45–56 (2004)
15. Tax, D.M.J.: Datasets (2009), `http://ict.ewi.tudelft.nl/~davidt/occ/index.html`
16. Towel, G.G.: Local expert autoassociator for anomaly detection. In: Proc. 17th International Conference on Machine Learning, pp. 1023–1030. Morgan Kaufmann Publishers Inc., San Francisco (2000)
17. Vapnik, V.: The nature of statistical learning theory. Springer, Heidelberg (1995)
18. Vert, J., Vert, J.P.: Consistency and convergence rates of one class svm and related algorithm. Journal of Machine Learning Research 7, 817–854 (2006)
19. Xiao, Y., Liu, B., Cao, L., Wu, X., Zhang, C., Hao, Z., Yang, F., Cao, J.: Multi-sphere Support Vector Data Description for Outliers Detection on Multi-Distribution Data. In: Proc. IEEE International Conference on Data Mining Workshops, pp. 82–88 (2009)
20. Yu, M., Ye, J.: A Small Sphere and Large Margin Approach for Novelty Detection Using Training Data with Outliers. IEEE Transaction on Pattern Analysis and Machine Intelligence 31, 2088–2092 (2009)

Fast Implementation of String-Kernel-Based Support Vector Classifiers by GPU Computing

Yongquan Shi[1], Tao Ban[2,*], Shanqing Guo[1], Qiuliang Xu[1], and Youki Kadobayashi[2]

[1] Shandong University Jinan 250101, Shandong, China
[2] Information Security Research Center, National Institute of Information and Communications Technology Tokyo 184-8795, Japan

Abstract. Text categorization is widely used in applications such as spam filtering, identification of document genre, authorship attribution, and automated essay grading. The rapid growth in the amount of text data gives rise to the urgent need for fast text classification algorithms. In this paper, we propose a GPU based SVM solver for large scale text datasets. Using Platt's Sequential Minimal Optimization algorithm, we achieve a speedup of 5–40 times over LibSVM running on a high-end traditional processor. Prediction time based on the paralleled string kernel computing scheme shows 5–90 times faster performance than the CPU based implementation.

1 Introduction

Standard learning systems such as neural networks and decision trees operate on input data that are represented as feature vectors. There are many cases, however, where the input data cannot readily be represented as explicit feature vectors, e.g., bio-sequences, images, and text documents. An effective alternative to explicit feature extraction is provided by kernel methods (KM) [1]. The most well-known KM is the Support Vector Machine (SVM) [2][3], which implements the maximum margin principle by means of a convex optimization algorithm in the dual form. Other classic learning algorithms such as Perceptron, Principal Component Analysis (PCA), and Fisher Discriminant Analysis (FDA) can also presented in their dual forms [4][5][6] .

Recently, many string kernels which can incorporate domain-specific knowledge have been studied. String kernel based Support vector machines have shown competitive performance in tasks such as text classification [7] and protein homology detection [8]. A common feature of kernel based methods is their dependency on the kernel matrix computation: training an algorithm in the dual form usually invokes quadratic computation cost in terms of the number of kernel evaluations. For full-matrix based methods such as PCA and FDA, an $N \times N$ kernel matrix need to be computed and stored, and for sparse methods such as SVM and Perceptron, a $V \times N$ sub-matrix are needed, where V is usually in linear relation to N . Popular string kernels such as edit distance kernel and gap-weighted subsequences kernel are computed by dynamic programming and has a computational complexity of $O(|x||y|)$. The computational expense hinders the application of string-kernel based methods on large scale applications. Moreover, many text-related applications require real-time

K.W. Wong et al. (Eds.): ICONIP 2010, Part II, LNCS 6444, pp. 143–151, 2010.

system response, which also calls for an efficient implementation, computationally as well as economically, of string kernel computing algorithms.

The programmable Graphic Processor Unit (GPU) has already been used to implement many algorithms including computational geometry, image processing, as well as computer graphics [10][11]. Specifically, with regards to high-performance computing for pattern recognition, some previous works [12][13] show that general purpose numeric kernel functions including Linear kernel, Radial Basis Function kernel, Polynomial kernel, and Hyperbolic kernel can be paralleled easily with a high speedup compared with CPU based implementations.

Aiming at effectively treating with large scale text-based applications, this paper presents an implementation of string-kernel-based SVM on the CUDA C programming interface. Following the previous works in [12], we modify the Sequential Minimization Optimization algorithm [14] for nonlinear L1 soft-margin SVM algorithm and adopt a justifiable kernel sub-matrix caching method. Emphasis is given on how to parallelize and compute a group of string kernels simultaneously with highly parallelized GPU computing threads. Explored string kernels include p -spectrum kernel, gap-weighted subsequence kernel, and edit distance kernel. We verify the operation and evaluate the performance of the proposed integrated system using Reuters-21578 [15] and SpamAssassin Public Corpus [16]. Our gpuSKSVM achieves speedups of 5-90x over LibSVM running on a high-end traditional processor.

The organization of the paper is as follows. Section 2 describes the SVM and string kernels briefly. Section 3 gives an overview of the architectural and programming features of the GPU. Section 4 presents the details of implementation of the parallel string kernels on the GPU. Section 5 describes the experimental results. Section 6 concludes the paper.

2 Support Vector Machines and String Kernels

2.1 Support Vector Machines

We now focus on the standard two-class soft-margin SVM problem (C-SVM), which classifies a given data point $x \in R^n$ by assigning a label $y \in \{-1, 1\}$. The main task in training SVM is to solve the following quadratic optimization problem with respect to α:

$$\begin{aligned} &\min_{\alpha} && \frac{1}{2}\alpha^T Q\alpha - 1^T\alpha \\ &\text{Subject to} && y^T\alpha = 0 \\ & && 0 \le \alpha_i \le C, i = 1,\dots,m \end{aligned}$$

where Q is the $N \times N$ positive semi-definite kernel matrix, $Q_{ij} = y_i y_j K(x_i, x_j)$; and $K(x_i, x_j) = (x_i)^T (x_j)$ is the kernel function; and 1 is a vector of all ones.

SVM predict the class label of a new data point x to be classified by the following decision function:

$$f(x) = sign(\sum_{i=1}^{m} y_i \alpha_i K(x_i, x) + b)$$

where b is a bias constant.

2.2 String Kernels

Regular kernels for SVM work mainly on numerical data, which is unsuitable for text-based applications. To extend SVM for text data analysis, we implemented the following string kernels algorithms in our experiments.

2.2.1 Gap-Weighted Subsequence Kernel

Gap Weighted Subsequence (GWS) kernel is to compare strings by way of the common subsequences they share – the more common subsequences and the less gaps (the degree of contiguity between the subsequences), the more the contribute to the inter-string similarity. A decay factor λ is used to weight the presence of a gap in a string. We weight the occurrence of the subsequence u with the exponentially decaying weight $\lambda^{l(i)}$. The feature space for the gap-weighted subsequences kernel of length p is indexed by I = $\sum$p, with the mapping given by

$$\varphi_u^p(s) = \sum_{i:u=s(i)} \lambda^{l(i)}, u \in \Sigma^p$$

where i is an index sequence identifying the occurrence of a subsequence $u = s(i)$ in string s and $l(i)$ is the length of s. The corresponding kernel is defined as

$$k_p(s,t) = \langle \varphi^p(s), \varphi^p(t) \rangle = \sum_{u \in \Sigma^p} \varphi_u^p(s) \varphi_u^p(t)$$

We consider computing an intermediate dynamic programming table DP_p. The complexity of the computation required to compute the table DP_p is clearly $O(p\,|s||t|)$ making the overall complexity of $k_p\ (s, t)$ equal to $O(p\,|s||t|)$ [1].

2.2.2 Edit Distance Kernel

Edit distance (or Levenshtein Distance, abbreviated as LD hereinafter) is a measure of the similarity between two strings. More specifically, LD between two strings is the number of insertions, deletions, and substitutions required to transforms a source string s into a target string t. Clearly, the greater the LD, the more different the strings are. For s and t with respective length of n and m, the calculation of LD is a recursive procedure. First set $d(i,0)$ to i for $i = 0, \ldots, n$, and $d(0, j)$ for $j = 0, \ldots, m$,. Then, for other pairs i, j we have

$$d(i, j) = \min(d(i-1, j) + 1, d(i, j-1) + 1, d(i-1, j-1) + r(s(i), t(j)))$$

where for characters a and b, $r(a,b) = 0$ if $a = b$ and $r(a,b) = 1$, otherwise.

Similar to the above GWS kernel, the computational complexity of LD is $O(|s|\,|t|)$.

2.2.3 P-spectrum Kernel

p -spectrum transforms strings into high dimensional feature vectors where each feature corresponds to a contiguous substring. Comparing the p -spectra of two strings can give important information about their similarity in applications where contiguity plays an important role. The p -spectrum kernel is defined as the inner product of their p-spectra.

We adopt an idea for p-spectrum kernel from the open source software Shogun [18], which is a large scale machine learning toolbox. This idea for p-spectrum kernel has a time complexity of $O(|s|+|t|)$ after the pretreatment on the input string data.

3 Graphics Processors

Now GPUs are mainly used to accelerate specific parts of an application, being attached to a host CPU that performs most of the control-dominant computation. In this study, the algorithm is implemented on GeForce 9800GT GPU, which is widely available on the market. Related parameters are listed in Table 1.

Table 1. NVIDIA GeForce 9800GT Parameters

Number of multiprocessors	14
Multiprocessor width	8
Multiprocessor share memory size	16KB
Number of stream processors	112
Clock rate	1.57 GHZ
Memory capacity	1024 MB
Memory bandwidth	57.6 GB/S
Compute capacity	1.1

4 Kernels on GPU

We use the SMO, which is first proposed by Platt [14], with the improved first-order variable selection heuristic proposed by Keerthi [9]. In one iteration step of SMO we need to calculate kernel values between a single string and all the training strings. For large data sets, the amount of kernel calculations is very large. Obviously we can use the inherent features of GPU to simplify these calculations, which is the idea used in our study. Although parallelization of a specific kernel algorithm may result in better speedup, we are more interested in developing a generic methodology that can be applicable to string kernels with different functional characteristics. This is especially useful where parallelization of the kernel algorithm is knotty, e.g., LD and subsequence kernels which require dynamic programming.

The efficient implementation relies on one key factor of GPU programming – the trade off between memory access and the number of computing threads running in parallel. The shared memory of a Streaming Multiprocessor (SM), is much faster than local and global memory. However, there is a inherent limit on the volume of shared memory. Unlike the numerical input data, string data take up much more space. Employing the shared memory monotonously will result in very few active threads in an SM. To express thousands of threads making use of the hardware capabilities effectively, we have to employ GPU global memory for input string data and intermediate data. In general, one thread is created for computing $k(x_i,x_j)$, where x_i and x_j stand for two input strings. If some input operand resides in off-chip memory, the latency is much higher. As a result, reducing the number of global memory access and accessing global memory efficiently are two factors for speeding up string kernel.

4.1 p-spectrum Kernel

We adopt the idea for p -spectrum kernel from Shogun. This idea has a time complexity of O (|s| +|t|) after some pretreatment on the input string data.

The pretreatment consists of four steps. The first step is to extract p -mers from input strings. All threads of one block load current input string s into the shared memory concurrently. After S is loaded into shared memory, all the threads of the block extract p -mers of s independently. The extracted p -mers are stored into the shared memory. The second step is to sort the p -mers for sting s . It is probable that one input string contains the same p -mers, so the third step is removing repeated p -mers from the sorted p -mer list of input string s . When detecting a unique p -mer in the sorted p -mer list, occurrence number of the p -mer is attached to the p -mer value. During the repeatability statistics the self kernel $k(s,s)$ of s can be computed. The last step is storing the new sorted p -mer lists for the subsequent kernel evaluation.

In the SMO iteration when a kernel function call is launched, the kernel value $k(x_s, x_i)$ where $i \in [1, N]$ are to be evaluated. Here N stands for the total number of source strings and x_s stands for the common source string between the N kernel value evaluations. We load x_s into the GPU's per SM shared memory. This is the key to the performance improvement, since accessing the shared memory is orders of magnitude faster than accessing the global memory. In order to make use of GPU resources as far as possible every thread is responsible for single kernel value, which means the other source string of each thread is distinct. Because of the limit on the shared memory, x_i for each thread is loaded from global memory during the kernel value computation. Now the p -spectrum kernel value can be achieved easily by traversal of the two p -mer lists. However, the traversal pace of each thread is inconsistent in nature, which destroys the coalesced memory access. This thread branch problem may cause one-third decline in performance, which is a very serious bottleneck.

In consideration of the nature characteristics of desynchronization, it is almost impossible to solve thread branch and memory access latency problem together. We mainly focus on the memory access latency due to its great impact on efficiency. At the beginning all threads in a warp access global memory in coalesced way. Each thread has an independent index variable indicating the current shifting position in shared p -mer list. After one thread gets its global p -mer we let the index of this thread moves forward as far as possible, which means examining the shared p -mer continuously until the next shared p -mer is equal or bigger than the loaded global p -mer element. Thus, each global memory access can be carried out in coalesced way. In view of the above improvements the p -spectrum kernel evaluation on GPU can be much faster than that on CPU.

4.2 LD Kernel and Gap-Weighted Subsequence Kernel

LD kernel and gap-weighted Subsequence Kernel are both dynamic programming applications. A dynamic programming application solves an optimization problem by storing and reusing the results of its sub-problem solutions.

The computational complexity of LD is $O(|s||t|)$. For two strings s and t, the shared string s among all the threads is loaded into shared memory by threads belonging to one block and one $(n+1)(m+1)$ matrix D is required to evaluate the LD $L(s,t)$ between s and t. The potential pairs of sequences are organized in this 2D matrix. The algorithm fills the matrix from top left to bottom right, step-by-step. The value of each data element depends on the values of its northwest-, north- and west-adjacent elements. Matrix row is filled in turn, so only the previous line and the current line of the matrix D is useful. Using this feature one space of length n is enough for storing matrix D, which saves a lot of memory space. Even though we have this shortcut, the intermediate space of each thread is still too much to be saved in the shared memory. In this realization we encounter no thread branch problem. Without loss of generality we compute $D(i, j)$ using $D(i-1, j-1)$, $D(i, j-1)$, and $D(i-1, j)$ from a previous step. Next we compute $D(i, j+1)$, which in turn depends on $D(i-1, j)$, $D(i, j)$, and $D(i-1, j+1)$. Now we find that there is no need to load $D(i-1, j)$ and $D(i, j)$ used by $D(i, j+1)$ from global memory for computing $D(i, j+1)$ after computing $D(i, j)$. This finding halves the memory access operation. The GPU implement of LD kernel is fully compliance with CUDA's parallel constraint. It stands to reason that GPU LD may achieve great speed up.

The GWS kernel is a little bit more complicated than the LD kernel. It involves two intermediate matrices. All the skills employed to compute the LD kernel are applied to compute the GWS kernel. One difference is that there is a simple thread branch in GWS kernel, which affects the parallelism. Due to more time-consuming global memory operations than the LD kernel, speedup on GWS kernel algorithm are not expected to be as significant as the LD kernel.

5 Experiment Results

The current gpuSKSVM system is tested on GeForce 9800 GT. Our gpuSKSVM's performance is compared with LibSVM, which also employs SMO. Experiments are done on three text datasets. Detailed information on these datasets are listed in Table 2. LibSVM was run on an Intel Core 2 Duo 2.33 GHz processor and given a cache size of 800M, which is almost the same as the memory volume limit of our gpuSKSVM. File I/O time was not included for all the solvers. Computation time of gpuSKSVM includes data transfer between GPU and CPU memory.

Table 2. Text datasets for benchmarking

DATASET	# POINTS	# LENGTH
Reuters-21578 Earn vs Acq	6295	100
Reuters-21578 Earn vs Rest	10753	100
SpamAssassin Public Corpus	6047	100

Table 3 presents training performance results for the two solvers. We can see that the GPU based implementation in all cases achieves a speedup from 5-90 times compared with the CPU based implementation. Although for p-spectrum kernel and

Table 3. Comparison of GPU and LibSVM on computation time and accuracy

			Training Time(SEC)			Test Time(SEC)			Accuracy	
		Parameters	**CPU**	**GPU**	**Speedup**	**CPU**	**GPU**	**Speedup**	**CPU**	**GPU**
Spam	P-spec.	C=10, L=100,p=4	12.30	2.48	4.96	2.2	0.07	31.43	95.2	95.2
	LD	C=10, L=100,g=0.005	354.66	18.70	18.97	66.42	1.78	37.31	93.7	91.7
	GWS	C=10, L=100,p=5	5748	149.45	38.46	670	44.56	15.04	97.5	95.9
E.V.A	P-spec.	C=10, L=100,p=4	12.60	1.45	8.69	4.80	0.13	36.92	98.06	98.06
	LD	C=10, L=100,g=0.005	317.99	19.87	16.00	100.41	5.89	17.05	96.39	97.50
	GWS	C=10, L=100,p=5	3035	234.56	12.94	805	8.94	90.10	90.22	92.00
E.V.R.	P-spec.	C=10, L=100,p=4	31.70	3.15	10.06	11.50	0.34	33.82	98.87	98.87
	LD	C=10, L=100,g=0.005	761.8	59.15	12.88	215.18	11.97	17.98	94.63	94.82
	GWS	C=10, L=100,p=5	7838	639.61	12.25	2012	359	5.59	98.74	98.8

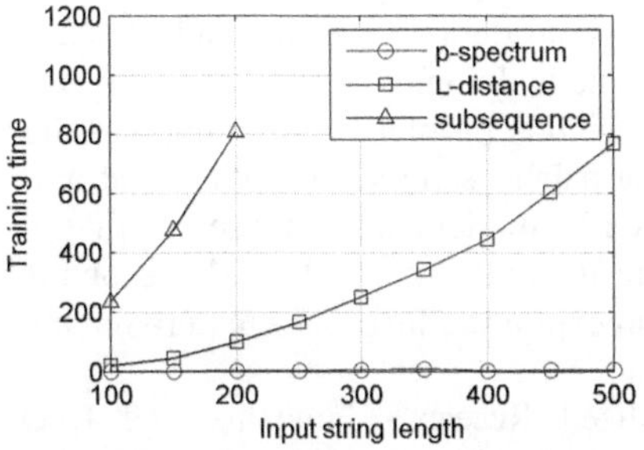

(a)

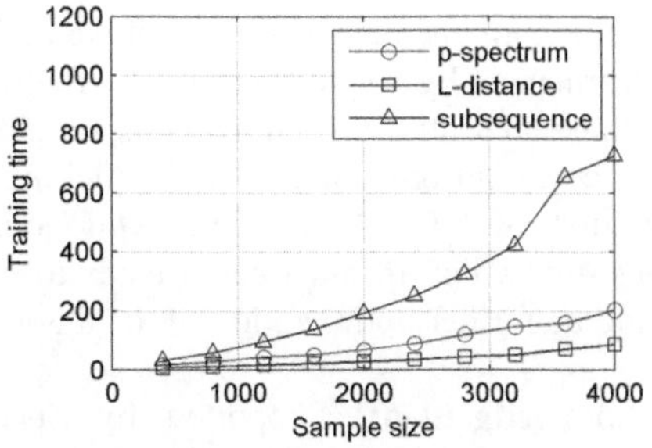

(b)

Fig. 1. Computation times with different sample and input string length

subsequence kernel, GPU based implementation can only use single-precision float, further detailed inspection on the results shows that there is little difference with respect to the number of iterations until converge. The number of support vector obtained by the two implementations are almost the same. This indicates that the computation cost for prediction are comparable -- recall that the computation cost of prediction heavily depend on kernel computation between support vectors. So the speedup on the prediction can also be deemed as speed improvement on the kernel function evaluations. For p -spectrum kernel, we achieve a speedup on kernel computation with factor of around 30; for GWS kernel, the factor is from 5 to 90, depend on the nature of the problem; for the LD kernel, the CPU algorithm happens to not converge for dataset with sample size over 4,000, and we are inspecting the reason of this problem, and the numerical result will be reported.

In the second group of experiments, we examine how computation time of gpuSKSVM changes with the sample size and input string length. Generally the prediction accuracy increases as the input strings increase in length, because of the increment of presented information. However, as we know, the computation complexity of the kernels also increases along with the string length. In Figure 1(a) we show the influence of the substring length on the computation time. Theoretically the p-spectrum kernel increase in linear proportion to, and the LD kernel and subsequence kernel increase in quadratic to the length of the input strings, the curves in Figure 1(a) clearly verifies these properties. Moreover, the figure also indicates that the complexity of this learning algorithm does not significantly increase with the input string length -- the number of invoked iteration before convergence remain almost the same when the string length varies. Generally as shown in Figure 1(b), the training time should always increase as the sample size increases. This is more about the nature of the SMO algorithm rather than a feature of its GPU implementation. However, by Figure 1(b), we wan to show that, GPU programming based implementation will speedup the training time even for datasets with considerably small sample size.

6 Conclusion and Future Work

In this paper we presented a high-speed SVM learning algorithm based on CUDA. For the implemented string kernel we achieve considerable speedup and expect better performance by exploiting more effective methodologies. First, we have not yet implemented all the optimizations possible for this problem. For example, LibSVM uses a second order heuristic for picking the pair of samples to optimize at a single iteration of QP optimization, while our GPU implementation uses a first order heuristic, which in most cases leads to more iterations than LibSVM. Second, there could be further optimization of the algorithm exploiting fast speed texture memory.

Acknowledgement. Supported by Specialized Research Fund for the Doctoral Program of Higher Education (Grant No. 20090131120009), Natural Science Foundation of Shandong Province for Youths (Grant No. Q2008G01), Outstanding Young Scientists Foundation Grant of Shandong Province(Grant No. BS2009DX018), Independent Innovation Foundation of Shandong University(Grant No. 2009TS031), The Key Science-Technology Project of Shandong Province of Shandong(Grant No. 2010GGX10117).

References

[1] Shawe-Taylor, J., Cristianini, N.: Kernel Methods for Pattern Analysis. Cambridge University Press, Cambridge (2004)
[2] Vapnik, V.: The Nature of Statistical Learning Theory. Springer, Heidelberg (1995)
[3] Cristianini, N., Shawe-Taylor, J.: An introduction to Support Vector Machines. Cambridge University Press, Cambridge (2000)
[4] Oei, C., Friedland, G., Janin, A.: Parallel Training of a Multi-Layer Perceptron on a GPU. ICSI Technical Report (2009)
[5] Andrecut, M.: Parallel GPU Implementation of Iterative PCA Algorithms. Journal of Computational Biology 16(11), 1593–1599 (2009)
[6] Hall, J.D., Hart, J.C.: GPU Acceleration of Iterative Clustering (2004)
[7] Lodhi, H., Saunders, C., Shawe-Taylor, J., Cristianini, N., Watkins, C.: Text classification using string kernels. Journal of Machine Learning Research (2), 419–444 (2002)

[8] Saigo, H., Vert, J., Ueda, N., Akutsu, T.: Protein homology detection using string alignment kernels. Oxford University Press, Oxford (2004)

[9] Keerthi, S.S., Shevade, S.K., Bhattacharyya, C., Murthy, K.R.K.: Improvements to Platt's SMO Algorithm for SVM Classifier Design. Neural Comput. 13, 637–649 (2001)

[10] Moravanszky, A.: Linear algebra on the GPU. In: Engel, W.F. (ed.) Shader X 2. Wordware Publishing, Texas (2003)

[11] Manocha, D.: Interactive geometric & scientific computations using graphics hardware. In: SIGGRAPH, Tutorial Course#11 (2003)

[12] Catanzaro, B., Sundaram, N., Keutzer, K.: Fast support vector machine training and classification on graphics processors. In: ICML 2008: Proceedings of the 25th International Conference on Machine Learning, pp. 104–111. ACM, New York (2008)

[13] Carpenter, A.: CUSVM: A cuda implementation of support vector classification and regression (2009)

[14] Platt, J.C.: Fast training of support vector machines using sequential minimal optimization. In: Advances in Kernel Methods: Support Vector Learning, pp. 185–208. MIT Press, Cambridge (1999)

[15] http://www.cs.umb.edu/~smimarog/textmining/datasets/index.html

[16] http://spamassassin.apache.org/publiccorpus/

Classification of Imbalanced Data by Combining the Complementary Neural Network and SMOTE Algorithm

Piyasak Jeatrakul, Kok Wai Wong, and Chun Che Fung

School of Information Technology
Murdoch University
South Street, Murdoch, Western Australia 6150
{p.jeatrakul,k.wong,l.fung}@murdoch.edu.au

Abstract. In classification, when the distribution of the training data among classes is uneven, the learning algorithm is generally dominated by the feature of the majority classes. The features in the minority classes are normally difficult to be fully recognized. In this paper, a method is proposed to enhance the classification accuracy for the minority classes. The proposed method combines Synthetic Minority Over-sampling Technique (SMOTE) and Complementary Neural Network (CMTNN) to handle the problem of classifying imbalanced data. In order to demonstrate that the proposed technique can assist classification of imbalanced data, several classification algorithms have been used. They are Artificial Neural Network (ANN), k-Nearest Neighbor (k-NN) and Support Vector Machine (SVM). The benchmark data sets with various ratios between the minority class and the majority class are obtained from the University of California Irvine (UCI) machine learning repository. The results show that the proposed combination techniques can improve the performance for the class imbalance problem.

Keywords: Class imbalanced problem, artificial neural network, complementary neural network, classification, misclassification analysis.

1 Introduction

In recent years, many research groups have found that an imbalanced data set could be one of the obstacles for many Machine Learning (ML) algorithms [1], [2], [3], [4]. In the learning process of the ML algorithms, if the ratio of minority classes and majority classes is significantly different, ML tends to be dominated by the majority classes and the features of the minority classes are recognize slightly. As a result, the classification accuracy of the minority classes may be low when compared to the classification accuracy of the majority classes. Some researchers have examined this problem under the balancing of the bias and variance problems [5].

According to Gu et al. [4], there are two main approaches to deal with imbalanced data sets: data-level approach and algorithm approach. While the data-level approach aims to re-balance the class distribution before a classifier is trained, the algorithm level approach aims to strengthen the existing classifier by adjusting algorithms to recognize the smaller classes. There are three categories of data-level approach. These

K.W. Wong et al. (Eds.): ICONIP 2010, Part II, LNCS 6444, pp. 152–159, 2010.

are the under-sampling technique, the over-sampling technique and the combined technique. For the under-sampling techniques, many algorithms have been proposed, for example Random under-sampling [1], Tomek links [6], Wilson's Edited Nearest Neighbor Rule (ENN) [7], and Heuristic Pattern Reduction (HPR) [8]. There are also several techniques applied for over-sampling methods such as Random over-sampling [1], and Synthetic Minority Over-sampling Technique (SMOTE) [3].

In order to evaluate the classification performance of an imbalanced data set, the conventional classification accuracy cannot be used for this purpose because the minority class has minor impact on the accuracy when compared to the majority class [4]. Therefore, alternative measures have to be applied. The Geometric mean (G-mean) and the area under the Receiver Operating Characteristic (ROC) curve have been applied to evaluate the classification performance for imbalanced data set [4]. They are good indicators for the class imbalance problem because they attempt to maximize and balance the performance of ML between the minority class and the majority class. G-mean and the area under ROC curve (AUC) are also independent of the imbalanced distribution [9].

In the reported literature, most research dealt with this problem with an aim to increase the classification performance of imbalanced data. They focused on examining the feasibility of re-distribution techniques for handling imbalanced data [1], [2], [3], [9]. Furthermore, several cases in the literature have presented that the combination of under-sampling and over-sampling techniques generally provides better results than a single technique [1]. By considering in a similar direction, this paper takes an approach by proposing alternative re-distribution techniques to enhance the classification performance. A combined technique based on both sampling techniques is also proposed.

In this paper, in order to re-balance the class distribution, the combined approaches of two techniques, Complementary Neural Network (CMTNN) and Synthetic Minority Over-Sampling Technique (SMOTE), are proposed. While CMTNN is applied as an under-sampling technique, SMOTE is used as an over-sampling technique. CMTNN is used because of its special feature of predicting not only the "truth" classified data but also the "false" data. SMOTE is applied because it can create new instances rather than replicate the existing instances. SMOTE is also the successful over-sampling technique applied commonly to the class imbalanced problem in the literature [1], [4].

2 The Proposed Techniques

In this section, the concepts of CMTNN and SMOTE are described. The proposed combined techniques will then be presented.

2.1 Complementary Neural Network (CMTNN)

CMTNN [10] is a technique using a pair of complementary feedforward backpropagation neural networks called Truth Neural Network (Truth NN) and Falsity Neural Network (Falsity NN) as shown in Fig 1. While the Truth NN is a neural network that is trained to predict the degree of the truth memberships, the

Falsity NN is trained to predict the degree of false memberships. Although the architecture and input of Falsity NN are the same as the Truth NN, Falsity NN uses the complement outputs of the Truth NN to train the network. In the testing phase, the test set is applied to both networks to predict the degree of truth and false membership values. For each input pattern, the prediction of false membership value is expected to be the complement of the truth membership value. Instead of using only the truth membership to classify the data, which is normally done by most convention neural network, the predicted results of Truth NN and Falsity NN are compared in order to provide the classification outcomes [11].

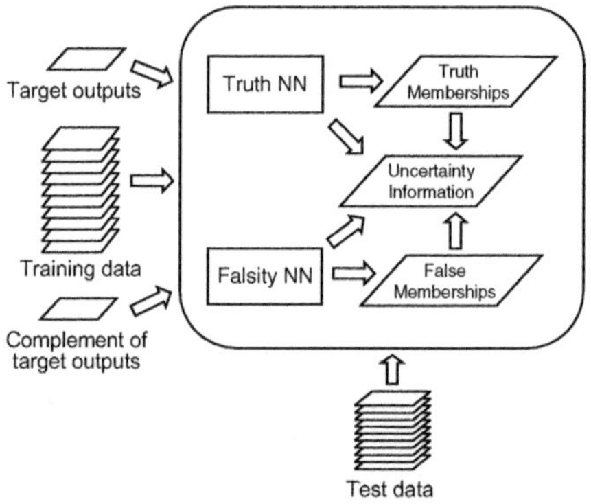

Fig. 1. Complementary neural network [11]

In order to apply CMTNN for under-sampling problem, Truth NN and Falsity NN are employed to detect and remove misclassification patterns from a training set. There are basically two ways to perform under-sampling [12].

Under-Sampling Technique I

a. The Truth and Falsity NNs are trained by truth and false membership values.

b. The prediction outputs *(Y)* on the training data *(T)* of both NNs are compared with the actual outputs *(O)*.

c. The misclassification patterns of Truth NN and Falsity NN $(M_{Truth}, M_{Falsity})$ are also detected if the prediction outputs and actual outputs are different.

For Truth NN : If $Y_{Truth\ i} \neq O_{Truth\ i}$ then $M_{Truth} \leftarrow M_{Truth}\ U\ \{T_i\}$ (2)

For Falsity NN : If $Y_{Falsity\ i} \neq O_{Falsity\ i}$ then $M_{Falsity} \leftarrow M_{Falsity}\ U\ \{T_i\}$ (3)

d. In the last step, the under-sampling for the new training set (T_c) is performed by eliminating the misclassification patterns detected by both the Truth NN (M_{Truth}) and Falsity NN $(M_{Falsity})$. $T_c \leftarrow T - (M_{Truth} \cap M_{Falsity})$ (4)

Under-Sampling Technique II

a. Repeat the step *a.* to *b.* of under-sampling technique I.

b. The under-sampling for the new training set (T_c) is performed by eliminating all misclassification patterns detected by the Truth NN (M_{Truth}) and Falsity NN $(M_{Falsity})$ respectively. $T_c \leftarrow T - (M_{Truth} \cup M_{Falsity})$ (5)

2.2 Synthetic Minority Over-Sampling Technique (SMOTE)

SMOTE [3] is an over-sampling technique. This technique increases a number of new minority class instances by interpolation method. The minority class instances that lie together are identified before they are employed to form new minority class instances. This technique is able to generate synthetic instances rather than replicate minority class instances; therefore, it can avoid the over-fitting problem. The algorithm is described in Fig. 2.

```
O is the original data set
P is the set of positive instances (minority class instances)
For each instance x in P
    Find the k-nearest neighbors (minority class instances) to x in P
    Obtain y by randomizing one from k instances
    difference = x – y
    gap = random number between 0 and 1
    n = x + difference * gap
    Add n to O
End for
```

Fig. 2. The Synthetic Minority Oversampling Technique (SMOTE) [3]

2.3 The Proposed Combined Techniques

In order to obtain the advantages of using the combination between under-sampling and over-sampling techniques as presented in the literature [1] and [3], in this paper, CMTNN is applied as under-sampling while SMOTE is used for over-sampling. They are combined in order to better handle the imbalanced data problem. Four techniques can be derived by the combination as follows.

1. Under-sampling only the majority class using the CMTNN under-sampling technique I and then over-sampling the minority class using SMOTE technique
2. Under-sampling only the majority class using the CMTNN under-sampling technique II and then over-sampling the minority class using SMOTE technique
3. Over-sampling the minority class using SMOTE technique before under-sampling both classes using the CMTNN under-sampling technique I
4. Over-sampling the minority class using SMOTE technique before under-sampling both classes using the CMTNN under-sampling technique II

For all the proposed techniques mentioned above, the ratio between the minority and majority class instances after implementing SMOTE algorithm is 1:1.

3 Experiments and Results

Four data sets from the UCI machine learning repository [13] are used in the experiment. The data sets for binary classification problems include Pima Indians Diabetes data, German credit data, Haberman's Survival data, and SPECT heart data. These data sets are selected because they are imbalanced data sets with various ratios between the minority class and the majority class. The characteristics of these four data sets are shown in Table 1.

Table 1. Characteristics of data sets used in the experiment

Name of data set	No. of instances	No. of attributes	Minority class (%)	Majority class (%)
Pima Indians Diabetes data	768	8	34.90	65.10
German Credit data	1000	20	30.00	70.00
Haberman's Survival data	306	3	26.47	73.53
SPECT Heart data	267	22	20.60	79.40

For the purpose of establishing the classification model and testing it, each data set is first split into 80% training set and 20% test set. Furthermore, the cross validation method is used in order to reduce inconsistent results. Each data set will be randomly split ten times to form different training and test data sets. For the purpose of this study, the results of the ten experiments of each data set will be averaged.

In the experiment, after the training sets are applied by the proposed combined techniques, three different learning algorithms, which are ANN, SVM (kernel function = Radial Basis Function (RBF)), and k-NN (k=5) are used for the classification. The classification performance is then evaluated by G-mean and AUC. Furthermore, in order to compare the performance of the proposed techniques to others, the over-sampling technique, SMOTE, will be compared as the base technique. The other two under-sampling approaches, Tomek links [6] and ENN [7], are also used for this purpose. These comparison techniques are selected because they have been applied widely to the class imbalance problem [1], [9].

Table 2. The results of G-Mean and AUC for each data set classified by ANN

	Pima Indian Diabetes data		German Credit data		Haberman's Survival data		SPECT Heart data	
Techniques	GM	AUC	GM	AUC	GM	AUC	GM	AUC
Original Data	70.12	0.8276	63.92	0.7723	33.11	0.5885	64.05	0.7590
a. ENN	72.64	0.8298	70.74	0.7794	50.45	0.6305	71.80	0.7895
b. Tomek links	73.11	0.8288	70.48	0.7793	51.88	0.6323	72.88	0.8178
c. SMOTE	74.30	0.8281	71.48	0.7777	58.60	0.6345	73.59	0.8241
d. Technique I (Majority) + SMOTE	75.55	0.8332	72.03	0.7855	60.00	0.6452	73.86	0.8374
e. Technique II (Majority) + SMOTE	74.53	0.8300	73.32	0.7873	62.78	0.6770	74.32	0.8273
f. SMOTE + Technique I	75.00	0.8285	71.52	0.7844	61.41	0.6653	73.00	0.8264
g. SMOTE + Technique II	74.96	0.8300	72.07	0.7860	58.59	0.6248	74.04	0.8373
Best technique	d	d	e	e	e	e	e	d
Second best	f	e & g	g	g	f	f	g	g

The experimental results in Table 2, 3 and 4 show that four proposed techniques combined CMTNN and SMOTE generally performs better than other techniques, in terms of G-mean and AUC in each learning algorithm (ANN, SVM, and k-NN). They improve the performance significantly when comparing to the results of original data sets. The proposed techniques f (SMOTE + CMTNN technique I) can improve G-mean up to 45.41% on Haberman's Survival data classified by SVM. Moreover, technique g (SMOTE + CMTNN technique II) generally present the better technique in the experiments.

The results of the ANN classifier in Table 2 show that the combined technique d (CMTNN technique I (Majority) + SMOTE) and technique e (CMTNN technique II (Majority) + SMOTE) present the best results of G-mean and AUC. Technique g also presents the second best performance in most cases. The proposed combined techniques (technique d e f g) show the improvement significantly when comparing to the results of G-mean on original test sets from 5.43% to 29.67%. In addition, when the results of technique d. and e. are compared to the base technique (SMOTE), the results of G-mean show the improvement from 0.73% to 4.73%.

In Table 3, SVM is employed as a classifier. The results show that technique g (SMOTE + CMTNN technique II) presents the best performance on two test sets. The significant improvement by technique g is up to 13.19% on German Credit data when compared to the base technique, SMOTE. ENN and Tomek links technique also perform well on some test sets. This is because they can broaden the margin between two classes by eliminating instances near the separating hyperplane [1].

Table 3. The results of G-Mean and AUC for each data set classified by SVM

	Pima Indian Diabetes data		German Credit data		Haberman's Survival data		SPECT Heart data	
Techniques	GM	AUC	GM	AUC	GM	AUC	GM	AUC
Original Data	67.81	0.8294	56.78	0.7660	19.13	0.6520	71.81	0.7249
a. ENN	73.04	0.8281	70.01	0.7842	53.16	0.7105	77.15	0.7717
b. Tomek links	72.83	0.8231	70.73	0.7846	49.61	0.6982	76.72	0.7681
c. SMOTE	74.32	0.8247	58.03	0.7381	58.33	0.6336	71.59	0.7253
d. Technique I (Majority) + SMOTE	74.75	0.8144	60.03	0.7573	61.16	0.6505	73.08	0.7349
e. Technique II (Majority) + SMOTE	74.89	0.8177	66.84	0.7626	60.92	0.6732	74.80	0.7503
f. SMOTE + Technique I	74.11	0.8262	67.87	0.7805	64.54	0.6843	74.39	0.7466
g. SMOTE + Technique II	75.57	0.8306	71.22	0.7902	58.32	0.6204	75.33	0.7555
Best technique	g	g	g	g	f	a	a	a
Second best	e	Origin	b	b	d	b	b	b

In Table 4, k-NN (k=5) is used as a classifier. Technique g (SMOTE + CMTNN technique II) show the best and the second best performance in every test set. While Technique f (SMOTE + CMTNN technique I) show the best outcome in two test sets, ENN perform well only on SPECT Heart data.

In order to explain why the proposed combined techniques outperform other techniques, the characteristics of the both techniques need to be discussed. On one hand, SMOTE technique gains the benefits of avoiding the over-fitting problem of the minority class by interpolating new minority class instances rather than duplicating the existing instances [1]. On the other hand, the misclassification analysis using CMTNN can enhance the quality of the training data by removing possible misclassification patterns from data sets.

Table 4. The results of G-Mean and AUC for each data set classified by k-NN (k=5)

	Pima Indian Diabetes data		German Credit data		Haberman's Survival data		SPECT Heart data	
Techniques	GM	AUC	GM	AUC	GM	AUC	GM	AUC
Original Data	65.27	0.7665	59.35	0.7483	40.11	0.5741	68.00	0.8121
a. ENN	71.15	0.7817	64.40	0.7566	46.47	0.5915	77.56	0.8369
b. Tomek links	72.06	0.7865	67.42	0.7625	47.57	0.5918	74.10	0.8148
c. SMOTE	71.78	0.7742	68.69	0.7518	55.82	0.5836	74.20	0.8005
d. Technique I (Majority) + SMOTE	72.11	0.7938	69.32	0.7572	56.28	0.5927	74.64	0.8264
e. Technique II (Majority) + SMOTE	73.17	0.7956	69.94	0.7686	57.50	0.6050	74.53	0.8030
f. SMOTE + Technique I	73.95	0.8104	72.35	0.7785	56.39	0.6226	74.13	0.8121
g. SMOTE + Technique II	73.42	0.8058	71.21	0.7719	59.30	0.6302	75.30	0.8179
Best technique	f	f	f	f	g	g	a	a
Second best	g	g	g	g	e	f	g	d

For generalization, when the proposed techniques are compared, technique g (SMOTE + CMTNN technique II) constantly presents the best or the second best in most cases among different classification algorithms. This is because when the training data is applied by SMOTE technique, it can create larger and less specific decision boundaries for the minority class [3]. Consequently, when the data is applied by CMTNN as under-sampling, the training data is eliminated all possible misclassification patterns detected by both the Truth NN and Falsity NN. Moreover, when a number of instances removed from the training sets are compared, it is found that misclassification instances eliminated by technique g are greater than other combined techniques. The lesser noise the training set retains the better performance the learning algorithm performs.

However, in some cases, for example Haberman's Survival data, technique g cannot gain the better results than other techniques. This is because it removes lots of instances from the training set. While technique g removes misclassification instances between 14% and 24% in other data sets, it eliminates instances up to 55% in Haberman's Survival data. As a consequence, a number of remaining instances of this data is not enough for the learning algorithms (ANN and SVM) to generalize the correct results. Therefore, in summary, although the combined technique g consistently presented better results in this paper, the number of instances removed by technique g is also a major constraint which is able to affect the classification performance on the class imbalanced problem.

4 Conclusions

This paper presents the proposed combined techniques to re-distribute the data in classes to solve the class imbalance problem. They are the integration of under-sampling techniques using Complementary Neural Network (CMTNN) and the over-sampling technique using Synthetic Minority Over-sampling Technique (SMOTE). The experiment employs three types of machine learning algorithms for classifying the test sets including ANN, SVM, and k-NN. The results of classification are evaluated and compared in terms of performance using the widely accepted measures for the class imbalance problem, which are G-mean and AUC. The results obtained from the experiment indicated that the proposed combined technique by SMOTE and CMTNN generally performs better than other techniques in most test cases.

References

[1] Batista, G.E.A.P.A., Prati, R.C., Monard, M.C.: A study of the behavior of several methods for balancing machine learning training data. SIGKDD Explorations Newsletter 6, 20–29 (2004)

[2] Laurikkala, J.: Improving identification of difficult small classes by balancing class distribution. In: Quaglini, S., Barahona, P., Andreassen, S. (eds.) AIME 2001. LNCS (LNAI), vol. 2101, p. 63. Springer, Heidelberg (2001)

[3] Chawla, N.V., Bowyer, K.W., Hall, L.O., Kegelmeyer, W.P.: SMOTE: Synthetic minority over-sampling technique. Journal of Artificial Intelligence Research 16, 321–357 (2002)

[4] Gu, Q., Cai, Z., Zhu, L., Huang, B.: Data mining on imbalanced data sets. In: International Conference on Advanced Computer Theory and Engineering, ICACTE 2008, pp. 1020–1024 (2008)

[5] Gedeon, T.D., Wong, P.M., Harris, D.: Balancing bias and variance: Network topology and pattern set reduction techniques. In: Sandoval, F., Mira, J. (eds.) IWANN 1995. LNCS, vol. 930, pp. 551–558. Springer, Heidelberg (1995)

[6] Tomek, I.: Two Modifications of CNN. IEEE Transactions on Systems, Man and Cybernetics 6, 769–772 (1976)

[7] Wilson, D.L.: Asymptotic properties of nearest neighbor rules using edited Data. IEEE Transactions on Systems, Man and Cybernetics 2, 408–421 (1972)

[8] Gedeon, T.D., Bowden, T.G.: Heuristic pattern reduction. In: International Joint Conference on Neural Networks, Beijing, vol. 2, pp. 449–453 (1992)

[9] Barandela, R., Sanchez, J.S., Garcia, V., Rangel, E.: Strategies for learning in class imbalance problems. Pattern Recognition 36, 849–851 (2003)

[10] Kraipeerapun, P., Fung, C.C., Nakkrasae, S.: Porosity prediction using bagging of complementary neural networks. In: Yu, W., He, H., Zhang, N. (eds.) ISNN 2009. LNCS, vol. 5551, pp. 175–184. Springer, Heidelberg (2009)

[11] Kraipeerapun, P., Fung, C.C.: Binary classification using ensemble neural networks and interval neutrosophic sets. Neurocomput. 72, 2845–2856 (2009)

[12] Jeatrakul, P., Wong, K.W., Fung, C.C.: Data cleaning for classification using misclassification analysis. Journal of Advanced Computational Intelligence and Intelligent Informatics 14(3), 297–302 (2010)

[13] Asuncion, A., Newman, D.J.: UCI Machine Learning Repository, University of California, Irvine, School of Information and Computer Sciences (2007)

Generalization Error of Faulty MLPs with Weight Decay Regularizer

Chi Sing Leung[1], John Sum[2], and Shue Kwan Mak[1]

[1] Department of Electronic Engineering, City University of Hong Kong, Hong Kong
[2] Institute of Technology Management, National Chung Hsing University, Taiwan

Abstract. Weight decay is a simple regularization method to improve the generalization ability of multilayered perceptrons (MLPs). Besides, the weight decay method can also improve the fault tolerance of MLPs. However, most existing generalization error results of using the weight decay method focus on fault-free MLPs only. For faulty MLPs, using a test set to study the generalization ability is not practice because there are huge number of possible faulty networks for a trained network. This paper develops a prediction error formula for predicting the performance of faulty MLPs. Our prediction error results allows us to select an appropriate model for MLPs under open node fault situation.

1 Introduction

One concern in the multilayer perceptron (MLP) training is how well a trained MLP perform on an unseen test set [1]. Therefore, many methods, such as subset selection and regularization, were proposed for improving the generalization ability. One simple method is the weight decay technique [2–4]. However, the performance of weight decay depends on whether an appropriate decay parameter is chosen.

One approach to select the appropriate decay parameter is to use a test set. We train a number of networks with different decay parameters and then select the best trained network based on the test set. However, in many real situations, data are very scarce. Also, the process to investigate the performance of the trained networks based on the test set is very time consuming. Another approach is mean prediction error (MPE)[1]. We train a number of networks with different decay parameters and then select the best trained network based on a so-called prediction error formula which is a function of training error and trained weights. Although there are a lot of theoretical results related to the generalization ability, many of them focus on faulty-free networks only.

In the implementation of a neural network, network faults take place unavoidably [5–8]. One of important fault models is open node fault, where some hidden nodes are disconnected to the output layer. Several algorithms for this fault model have been investigated [6–10]. Unfortunately, most of them focused on the training error. With a test set, they can be modified to handle the generalization error but the modifications may not b practice because of huge number of possible faulty networks.

K.W. Wong et al. (Eds.): ICONIP 2010, Part II, LNCS 6444, pp. 160–167, 2010.

From [9], the weight decay method is able to handle the open node fault of MLPs. To optimize the generalization ability under the open node fault situation, we can use different weight decay parameters to train a number of MLPs. For each trained MLP, we generate a huge number of faulty MLPs. Afterwards, we use the test set to study the performance of those faulty networks to study the generalization ability. Clearly, this test set method is very time consuming.

This paper investigates the generalization ability of MLPs under the open node fault situation. We develop a MPE formula for MLPs to estimate the performance of MLPs. Based on this formula, we can perform the model selection (appropriate weight decay parameter) for MLPs with open node fault. The background knowledge about the MLP model, the weight decay technique, and fault model will be presented in Section 2. Section 3 presents the MPE formula. Section 4 presents the simulation result. Section 5 concludes our results.

2 MLP and Fault Model

The training dataset is denoted as $\mathcal{D}_t = \{(\boldsymbol{x}_j, y_j) : \boldsymbol{x}_j \in \Re^K, y_j \in \Re\}$, where $\boldsymbol{x}_j$ and y_j are the training input and desired output, respectively, and $j = 1, \cdots, N$. The dataset $\mathcal{D}_t$ is generated by a stochastic system, given by $y_j = f(\boldsymbol{x}_j) + \epsilon_j$, where $f(\cdot)$ is the unknown system mapping, and ϵ_j's, being the measurement noise, are identical independent zero-mean Gaussian variables with variance S_ϵ. We denote $\mathcal{D}_f = \{(\boldsymbol{x}'_{j'}, y'_{j'}), j' = 1, \cdots, N'\}$ as the test set.

In the MLP model, the unknown mapping $f(\cdot)$ is realized by

$$f(\boldsymbol{x}) = f(\boldsymbol{x}, \boldsymbol{w}^o, b^o, \boldsymbol{w}^I, \boldsymbol{b}^I) = \boldsymbol{w}^{oT} \boldsymbol{\phi}(\boldsymbol{w}^I \boldsymbol{x} + \boldsymbol{b}^I) + b^o, \quad (1)$$

where $\mathbf{w}^o = (w_1^o, \cdots, w_M^o)^T$ is the output weight vector, b^o is the output basis, $\boldsymbol{\phi} = (\phi_1, \cdots, \phi_M)^T$ is the hidden output vector, $\boldsymbol{w}^I$ is the input weight matrix, and $\boldsymbol{b}^I$ is the input bias vector. This paper uses the hyperbolic tangent as the activation function $\phi(\cdot)$. We can denote $\mathbf{w}$ as the collection of all the parameters. The network output $f(\boldsymbol{x}, \boldsymbol{w}^o, b^o, \boldsymbol{w}^I, \boldsymbol{b}^I)$ can then be written as $f(\boldsymbol{x}, \mathbf{w})$. The training set error of a given weight vector is

$$\mathcal{E}_t(\mathbf{w}) = \frac{1}{N} \sum_{j=1}^{N} (y_j - \boldsymbol{w}^{oT} \boldsymbol{\phi}(\boldsymbol{w}^I \boldsymbol{x}_j + \boldsymbol{b}^I) - b^o)^2 = \frac{1}{N} \sum_{j=1}^{N} (y_j - f(\boldsymbol{x}_j, \mathbf{w}))^2. \quad (2)$$

In weight decay, the training objective is

$$\mathcal{L}(\mathbf{w}) = \frac{1}{N} \sum_{j=1}^{N} (y_j - f(\boldsymbol{x}_j, \mathbf{w}))^2 + \lambda \mathbf{w}^T \mathbf{w}. \quad (3)$$

When open node fault appears in a hidden node, it is equivalent to set the output of the faulty node to zero. Hence, we can use a weight multiplicative model to describe the open node fault situation, given by

$$\tilde{w}_{i,\beta}^o = \beta_i w_i^o, \ \forall \ i = 1, \cdots, M \quad (4)$$

where the fault factor β_i describes whether the i-th hidden node operates properly. If $\beta_i = 0$, the i-th node is out of work. If $\beta_i = 1$, the i-th node operates properly. The training set error of an implementation $\tilde{\mathbf{w}}_\beta$ is given by

$$\mathcal{E}_{t,\beta}(\mathbf{w}) = \frac{1}{N}\sum_{j=1}^{N}(y_j - \sum_{i=1}^{M}\beta_i w_i^o \phi_{j,i} - b^o)^2 \tag{5}$$

where $\phi_{j,i}$ is the i-th element of $\boldsymbol{\phi}(\boldsymbol{w}^I\boldsymbol{x}_j + \boldsymbol{b}^I)$.

3 Mean Prediction Error of Fault MLPs

In this section, we will first estimate the train set error of a faulty MLP and then test set error of a faulty MLP. We assume that the faulty factors β_j's are identical independent binary random variables with $\text{Prob}(\beta_j = 0) = p$ and $\text{Prob}(\beta_j = 1) = 1 - p$. The training set error over all faulty vectors becomes

$$\overline{\mathcal{E}}_{t,\beta}(\mathbf{w}) = \frac{1}{N}\sum_{j=1}^{N}\left((y_i - b^o)^2 - 2(y_i - b^o)\left\langle\sum_{i=1}^{M}\beta_i\, w_i^o\, \phi_{j,i}\right\rangle_\beta + \left\langle(\sum_{i=1}^{M}\beta_j\, w_j^o\, \phi_{j,i})^2\right\rangle_\beta\right], \tag{6}$$

where $< \cdot >$ is the expectation operator. Since $\langle\beta_i\rangle = \langle{\beta_i}^2\rangle = 1-p$ and $\langle\beta_i\beta_{i'}\rangle = (1-p)^2$ for $i \neq i'$, we have

$$\left\langle\sum_{i=1}^{M}\beta_i w_i^o \phi_{j,i}\right\rangle_\beta = (1-p)\sum_{i=1}^{M} w_i^o\, \phi_{j,i} \tag{7}$$

$$\frac{1}{N}\sum_{j=1}^{N}\left\langle(\sum_{i=1}^{M}\beta_i\, w_i^o\, \phi_{j,i})^2\right\rangle_\beta = \boldsymbol{w}^{oT}\left((1-p)\boldsymbol{G} + (1-p)^2\,(\boldsymbol{\Phi} - \boldsymbol{G})\right)\boldsymbol{w}^o\,. \tag{8}$$

where $\boldsymbol{\Phi} = \frac{1}{N}\sum_{j=1}^{N}\boldsymbol{\phi}_j\,\boldsymbol{\phi}_j^T$, $\boldsymbol{\phi}_j = \boldsymbol{\phi}(\boldsymbol{w}^I\boldsymbol{x}_j + \boldsymbol{b}^I)$, and $\boldsymbol{G} = \mathbf{diag}(\boldsymbol{\Phi})$. The expectation on $\mathcal{E}_{t,\beta}(\mathbf{w})$ over all possible fault patterns is then given by

$$\overline{\mathcal{E}}_{t,\beta}(\mathbf{w}) = \frac{p}{N}\sum_{j=1}^{N}(y_j - b^o)^2 + \frac{1-p}{N}\sum_{j=1}^{N}(y_j - f(\boldsymbol{x}_j, \mathbf{w}))^2 + (p-p^2)\boldsymbol{w}^{oT}\left\{\boldsymbol{G} - \boldsymbol{\Phi}\right\}\boldsymbol{w}^o\,. \tag{9}$$

Similarly, we have the test set error over all possible fault patterns, given by

$$\overline{\mathcal{E}}_{f,\beta}(\mathbf{w}) = \frac{p}{N'}\sum_{j'=1}^{N}(y_{j'} - b^o)^2 + \frac{1-p}{N'}\sum_{j'=1}^{N'}(y_{j'} - f(\boldsymbol{x}_{j'}, \mathbf{w}))^2 + (p-p^2)\boldsymbol{w}^{oT}\left\{\boldsymbol{G}' - \boldsymbol{\Phi}'\right\}\boldsymbol{w}^o \tag{10}$$

where $\boldsymbol{\Phi}' = \frac{1}{N'}\sum_{j'=1}^{N'}\boldsymbol{\phi}_j'\,{\boldsymbol{\phi}_j'}^T$, $\boldsymbol{\phi}_{j'}' = \boldsymbol{\phi}(\boldsymbol{w}^I\boldsymbol{x}_{j'} + \boldsymbol{b}^I)$, and $\boldsymbol{G}' = \mathbf{diag}(\boldsymbol{\Phi}')$. Let $\mathbf{w}_*$ be the true weight vector. The datasets are then generated by

$$y_j = f(\boldsymbol{x}_j, \mathbf{w}_*) + \epsilon_j \quad \text{and } y_{j'} = f(\boldsymbol{x}_{j'}, \mathbf{w}_*) + \epsilon_{j'}\,. \tag{11}$$

In weight decay, our training objective is $\mathcal{L}(\mathbf{w}) = \frac{1}{N}\sum_{j=1}^{N}(y_j - f(\boldsymbol{x}, \mathbf{w}))^2 + \lambda \mathbf{w}^T\mathbf{w}$. So, there is a derivation $\Delta_{\mathbf{w}}$ between the trained vector $\mathbf{w}$ from the true weight vector $\mathbf{w}_*$. After training, we have

$$\frac{\partial \mathcal{L}(\mathbf{w})}{\partial \mathbf{w}} = \mathbf{0}\,. \tag{12}$$

Considering the first order approximation [11] on (12), we have

$$\frac{\partial \mathcal{L}(\mathbf{w}_*)}{\partial \mathbf{w}} + \frac{\partial^2 \mathcal{L}(\mathbf{w}_*)}{\partial \mathbf{w} \partial \mathbf{w}} \Delta_{\mathbf{w}} = \mathbf{0}. \tag{13}$$

Hence,

$$\Delta_{\mathbf{w}} = (\mathbf{H} + \lambda \mathbf{I})^{-1} \left(\frac{1}{N} \sum_{j=1}^{N} \epsilon_j \boldsymbol{\varphi}_j + \lambda \mathbf{w}_* \right) \tag{14}$$

where $\mathbf{I}$ is an identity matrix, $\boldsymbol{\varphi}_j = \frac{\partial f(\boldsymbol{x}_j, \mathbf{w}_*)}{\partial \mathbf{w}}$, and $\mathbf{H} = \frac{1}{N}\sum_{j=1}^{N} \boldsymbol{\varphi}_j \boldsymbol{\varphi}_j^T$. Considering the first order approximation on $f(\boldsymbol{x}_j, \mathbf{w})$ around $\mathbf{w}_*$, we have

$$f(\boldsymbol{x}_j, \mathbf{w}) = f(\boldsymbol{x}_j, \mathbf{w}_*) + \boldsymbol{\varphi}_j^T (\mathbf{w} - \mathbf{w}_*) = f(\boldsymbol{x}_j, \mathbf{w}_*) + \boldsymbol{\varphi}_j^T \Delta_{\mathbf{w}}. \tag{15}$$

From (11) and (15), we have

$$y_j - f(\boldsymbol{x}_j, \mathbf{w}) = \epsilon_j - \boldsymbol{\varphi}_j^T \Delta_{\mathbf{w}}\,. \tag{16}$$

Applying (16) into (9), we have

$$\overline{\mathcal{E}}_{t,\beta}(\mathbf{w}) = \frac{p}{N} \sum_{j=1}^{N} (y_j - b^o)^2 + \frac{1-p}{N} \sum_{j=1}^{N} (\epsilon_j - \boldsymbol{\varphi}_j^T \Delta_{\mathbf{w}})^2 + (p - p^2) \boldsymbol{w}^{oT} \left\{ \boldsymbol{G} - \boldsymbol{\Phi} \right\} \boldsymbol{w}^o\,. \tag{17}$$

Applying (14) into (17) and taking average over ϵ_j, we can rewrite the training set error of faulty MLP, i.e., (17), as

$$\begin{aligned}
\overline{\mathcal{E}}_{t,\beta}(\mathbf{w}) = & \frac{p}{N} \sum_{j=1}^{N} (y_j - b^o)^2 + (1-p) S_\epsilon - \frac{2(1-p) S_\epsilon}{N} \mathbf{Tr}(\mathbf{H}(\mathbf{H} + \lambda \mathbf{I})^{-1}) \\
& + \frac{(1-p) S_\epsilon}{N} \mathbf{Tr}(\mathbf{H}(\mathbf{H} + \lambda \mathbf{I})^{-1} \mathbf{H} (\mathbf{H} + \lambda \mathbf{I})^{-1}) \\
& + (1-p) \lambda^2 \mathbf{w}_*^T (\mathbf{H} + \lambda \mathbf{I})^{-1} \mathbf{H} (\mathbf{H} + \lambda \mathbf{I})^{-1} \mathbf{w}_* \\
& + (p - p^2) \boldsymbol{w}^{oT} \left\{ \boldsymbol{G} - \boldsymbol{\Phi} \right\} \boldsymbol{w}^o
\end{aligned} \tag{18}$$

where $\mathbf{Tr}(\cdot)$ is the trace operator. Using the similar technique, we can rewrite the test set error of faulty MLPs, i.e., (10), as

$$\begin{aligned}
\overline{\mathcal{E}}_{f,\beta}(\mathbf{w}) = & \frac{p}{N} \sum_{j'=1}^{N'} (y_{j'} - b^o)^2 + (1-p) S_\epsilon \\
& + \frac{(1-p) S_\epsilon}{N'} \mathbf{Tr}(\mathbf{H}'(\mathbf{H}' + \lambda \mathbf{I})^{-1} \mathbf{H}' (\mathbf{H}' + \lambda \mathbf{I})^{-1}) \\
& + (1-p) \lambda^2 \mathbf{w}_*^T (\mathbf{H}' + \lambda \mathbf{I})^{-1} \mathbf{H}' (\mathbf{H}' + \lambda \mathbf{I})^{-1} \mathbf{w}_* \\
& + (p - p^2) \boldsymbol{w}^{oT} \left\{ \boldsymbol{G} - \boldsymbol{\Phi} \right\} \boldsymbol{w}^o
\end{aligned} \tag{19}$$

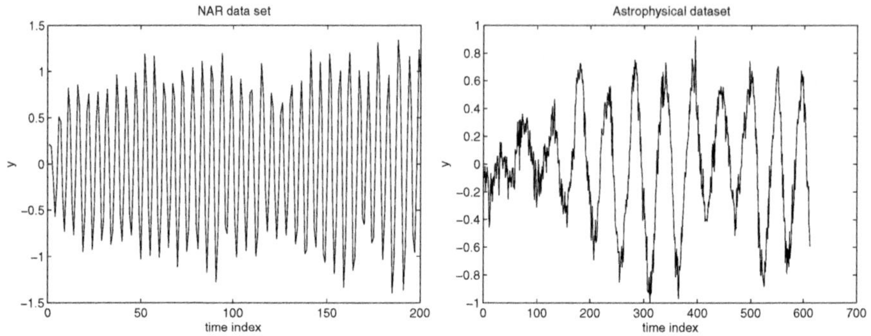

Fig. 1. Datasets. (a) The NAR dataset. (b) The astrophysical data.

$\mathbf{H}' = \frac{1}{N'}\sum_{j'=1}^{N'} \boldsymbol{\varphi}'_j \boldsymbol{\varphi}'^{T}_j$, and $\boldsymbol{\varphi}'_j = \frac{\partial f(\boldsymbol{x}_{j'}, \mathbf{w}_*)}{\partial \mathbf{w}}$. For large N and N', $\mathbf{H}' = \mathbf{H}$, $\boldsymbol{G}' = \boldsymbol{G}$, and $\boldsymbol{\Phi}' = \boldsymbol{\Phi}$. Compared (19) with (18), the test set error of faulty MLPs can be approximated as

$$\begin{aligned}\overline{\mathcal{E}}_{f,\beta}(\mathbf{w}) &= \frac{p}{N}\sum_{j=1}^{N}(y_j - b^o)^2 + \frac{1-p}{N}\sum_{j'=1}^{N'}(y_j - f(\boldsymbol{x}_j, \mathbf{w}))^2 \\ &\quad + \frac{2(1-p)S_\epsilon}{N}\mathbf{Tr}(\mathbf{H}(\mathbf{H}+\lambda\mathbf{I})^{-1}) + (p-p^2)\boldsymbol{w}^{oT}\left\{\boldsymbol{G}' - \boldsymbol{\Phi}'\right\}\boldsymbol{w}^o \qquad (20)\\ &= \frac{p}{N}\sum_{j=1}^{N}(y_j - b^o)^2 + (1-p)\mathcal{E}_t(\mathbf{w}) + \frac{2(1-p)S_\epsilon}{N}\mathbf{Tr}(\mathbf{H}(\mathbf{H}+\lambda\mathbf{I})^{-1}) \\ &\quad + (p-p^2)\boldsymbol{w}^{oT}\left\{\boldsymbol{G} - \boldsymbol{\Phi}\right\}\boldsymbol{w}^o . \qquad (21)\end{aligned}$$

In the above, the term $\mathcal{E}_t(\mathbf{w})$ is the training set error of the trained fault-free MLP. The matrices $\mathbf{H}$, $\boldsymbol{G}$, and $\boldsymbol{\Phi}$ can be estimated from the training set. The fault rate p is assumed to be known. The only unknown is the variance of the measurement noise S_ϵ. However, it can be obtained from the Fedorov's method [12], given by $S_e \approx \frac{N}{N-M}\mathcal{E}_t(\mathbf{w})$ or Moody's method [1], given by $S_e \approx \frac{N}{N-M_{eff}}\mathcal{E}_t(\mathbf{w})$, where $M_{eff} = \mathbf{Tr}(\mathbf{H}(\mathbf{H}+\lambda\mathbf{I})^{-1})$. From (21), we can directly estimate the generalization error of a faulty MLP under the open node fault situation based on the training set error of a fault-free MLP, the trained weights, and the training set.

4 Simulations

To verify our result, we consider two datasets. The first dataset, shown in Figure 1(a), is generated from a nonlinear autoregressive time (NAR) series [13], given by

$$\begin{aligned}y(i) &= \left(0.8 - 0.5\exp(-y^2(i-1))\right) y(i-1) - \left(0.3 + 0.9\exp(-y^2(i-1))\right) y(i-2) \\ &\quad + 0.1\sin(\pi y(i-1)) + \epsilon(i), \qquad (22)\end{aligned}$$

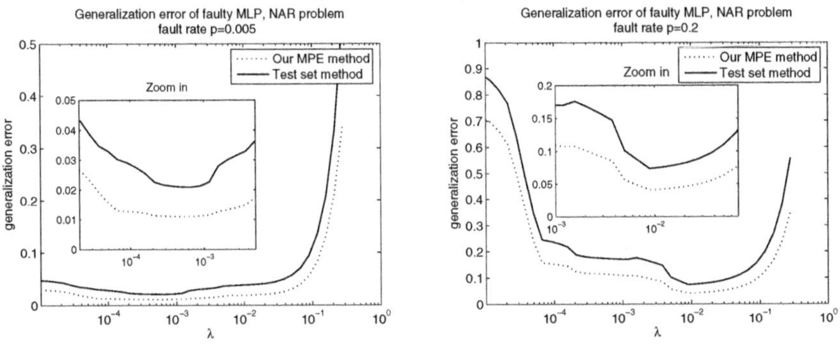

(a) fault rate $p = 0.005$.

(a) fault rate $p = 0.2$.

Fig. 2. Generalization error of faulty MLPs for the NAR problem. We can observe that the MPE formula can locate an appropriate weight decay parameter λ to minimize the test set error of faulty MLPs.

where $\epsilon(i)$ is a mean zero Gaussian random variable that drives the series. Its variance is equal to 0.01. Two hundred samples were generated given $y(0) = y(-1) = 0$. The first 50 data points were used for training. The other 150 samples were used for testing. Our MPL model is used to predict $y(i)$ based on the past observations, $\{y(i-1), y(i-2)\}$. The MLP model has two input nodes, 12 hidden nodes, and one output node.

The second dataset, shown in Figure 1(b), is the astrophysical data. It is the time variation of the intensity of the white dwarf star PG1159-035 during March 1989 [14]. The data samples are noisy and nonlinear in nature. Part one of this dataset is selected. There are 618 data samples. Our task is to train MLPs to predict the current value $y(i)$ based on six past values $\{y(i-1), \cdots, y(i-6)\}$. The MLP model has six input nodes, 16 hidden nodes, and one output node. There are 612 input-output pairs. The first 300 pairs are the training data and the remaining pairs are the test data.

In weight decay, the turning parameter is λ. We illustrate how our MPE result can help us to select an appropriate value of λ for minimizing the test error of faulty MLPs. We training MLPs under different λ values. For each λ value, we try 20 MLPs with different initial conditions. After training, we calculate the MPE value based on our formula. To verify our estimation, we also measure the test error of faulty MLPs based on the test sets. For each fault rate and each trained MLP, we randomly generate 10,000 faulty networks.

The simulation results are presented in Figures 2–3. From the figure, although there are differences between the true test error and MPE value of faulty MLPs, the general shape of the two curves are quite similar. Also, a large range of λ produces the good MPE values/generalization error.

Our MPE results can help us to select an appropriate value of λ for minimizing the true test error of faulty networks. For example, for the NAR problem with $p = 0.005$, the optimal value of λ based on the test set method is around 0.0005. With our MPE formula, the good λ value to minimize the MPE values is also

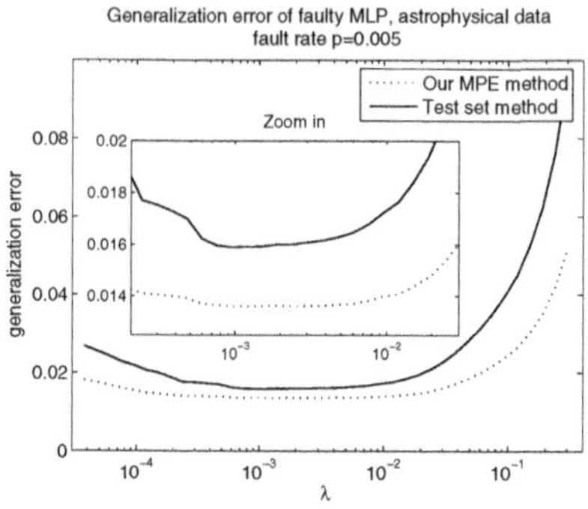

(a) fault rate $p = 0.005$.

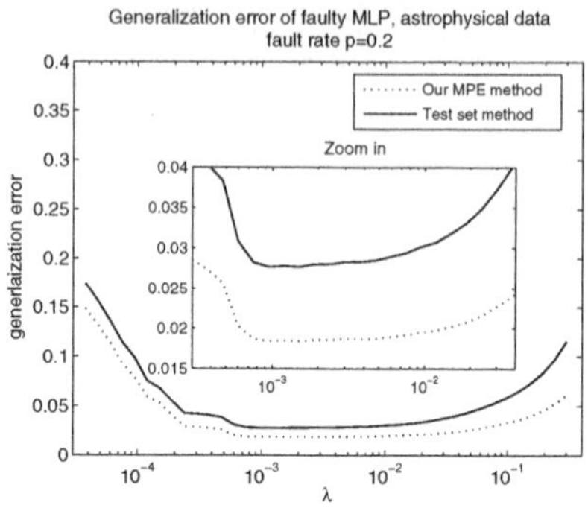

(a) fault rate $p = 0.2$.

Fig. 3. Generalization error of faulty MLPs for the astrophysical data. We can observe that the MPE formula can locate an appropriate weight decay parameter λ to minimize the test set error of faulty MLPs.

around 0.0005. For the NAR problem with $p = 0.2$, the optimal value of λ based on the test set method is around 0.009. With our MPE formula, the good λ value to minimize the MPE values is also around 0.009.

5 Conclusion

This paper perform an error analysis on faulty MLPs. A MPE formula with weight decay for open node fault were derived. Simulation results show that our MPE results can help us to select an appropriate value of decay parameter for minimizing the test error of faulty networks. Since the generalization error plays an important role in MLPs, we will further explore theoretical issues of the MPE result for other fault modes, such as multiplicative weight noise and open weight fault.

Acknowledgment

The work described in this paper was supported by a grant from City University of Hong Kong (Project No. 7002588).

References

1. Moody, J.E.: Note on generalization, regularization, and architecture selection in nonlinear learning systems. In: Proc. First IEEE-SP Workshop on Neural Networks for Signal Processing, pp. 1–10 (September 1991)
2. Krogh, A., Hertz, J.A.: A simple weight decay can improve generalization. In: Advances in Neural Information Processing Systems 4,[Neural Information Processing Systems Conference], pp. 950–957. Morgan Kaufmann, San Francisco (1992)

3. Leung, C.S., Young, G.H., Sum, J., Kan, W.K.: On the regularization of forgetting recursive least square. IEEE Trans. Neural Netw. 10(6), 1482–1486 (1999)
4. Leung, C.S., Tsoi, A., Chan, L.: Two regularizers for recursive least square algorithms in feedforward multilayered neural networks. IEEE Trans. Neural Netw. 12(6), 1314–1332 (2001)
5. Leung, C.S., Sum, J.: 'A fault-tolerant regularizer for RBF networks. IEEE Trans. Neural Netw. 19(3), 493–507 (2008)
6. Zhou, Z.H., Chen, S.F.: Evolving fault-tolerant neural networks. Neural Computing and Applications 11(3-4), 156–160 (2003)
7. Phatak, D.S., Koren, I.: Complete and partial fault tolerance of feedforward neural nets. IEEE Trans. Neural Netw. 6(2), 446–456 (1995)
8. Chandra, P., Singh, Y.: Fault tolerance of feedforward artificial neural networks – a framework of study. In: Proceedings of the International Joint Conference on Neural Networks 2003, Portland, OR, vol. 1, pp. 489–494 (July 2003)
9. Bernier, J.L., Ortega, J., Ros, E., Rojas, I., Prieto, A.: A quantitative study of fault tolerance, noise immunity, and generalization ability of MLPs. Neural Comput. 12(12), 2941–2964 (2000)
10. Ahmadi, A., Fakhraie, S.M., Lucas, C.: Behavioral fault model for neural networks. In: Proceedings of the 2009 International Conference on Computer Engineering and Technology, pp. 71–75. IEEE Computer Society, Washington (January 2009)
11. Larsen, J.: Design of neural network filters. Ph.D. dissertation, Technical University of Denmark, Denmark (July 1993)
12. Fedorov, V.V.: Theory of optimal experiments. Academic Press, London (1972)
13. Chen, S.: Local regularization assisted orthogonal least squares regression. Neurocomputing, 559–585 (2006)
14. Singh, S.: Noise impact on time-series forecasting using an intelligent pattern matching technique. Pattern Recognition 32, 1389–1398 (1999)

The Effect of Bottlenecks on Generalisation in Backpropagation Neural Networks

Xu Zang

School of Computer Science
Australian National University
Acton, ACT 0200 Australia
u4651788@anu.edu.au

Abstract. Many modifications have been proposed to improve back-propagation's convergence time and generalisation capabilities. Typical techniques involve pruning of hidden neurons, adding noise to hidden neurons which do not learn, and reducing dataset size. In this paper, we wanted to compare these modifications' performance in many situations, perhaps for which they were not designed. Seven famous UCI datasets were used. These datasets are different in dimension, size and number of outliers. After experiments, we find some modifications have excellent effect of decreasing network's convergence time and improving generalisation capability while some modifications perform much the same as unmodified back-propagation. We also seek to find a combine of modifications which outperforms any single selected modification.

Keywords: bottleneck, backpropagation, neural network, pruning, noise.

1 Introduction

One difficulty of carrying out back-propagation networks is deciding the number of hidden neurons. There are two methods to solve this problem. One is starting the network with extra hidden neurons and excising weights which have similar or complementary functions with others. This method is called 'pruning' [1] the other is using less hidden neurons and gradually adding more hidden neurons to the network [2]. The neural networks' learning can be viewed as 'curve fitting' [3] and good generalisation capabilities come from a smoother data fitting curve. In other words, networks with less hidden neurons generalise better. However, some research points out that by using extra units, thus adding dimensions to the error surface, the network's training can avoid local minima [4]. So one of our experimental goals is to observe this dilemma.

It is claimed [5] claims that adding noise to the training can dramatically improve a network's ability to recognize noisy data as well as clean data. In another paper [6] it is concluded that with noise in the training, the error converges faster.

K.W. Wong et al. (Eds.): ICONIP 2010, Part II, LNCS 6444, pp. 168–176, 2010.

All the modifications mentioned above are concerned with the inside of the neural network black box, while some research puts forward algorithms for reducing the input patterns [7-10]. This kind of modification assumes that outliers exist in real datasets and the training period can be shortened by discarding these outliers.

2 Algorithm Descriptions

2.1 Sensitivity

The basic idea of 'sensitivity' is excising weights without which the network is affected least (Karnin, 1990). The sensitivity is evaluated as:

$$S_{ij} = -\sum_{0}^{N-1} [\Delta w_{ij}(n)]^2 \frac{w_{ij}^f}{w_{ij}^f - w_{ij}^i}$$

where w_{ij}^f is the value of weight when the training stops, w_{ij}^i is the initial value of weight, and n is the number of running epochs.

2.2 Distinctiveness

The distinctiveness idea prunes neurons which have similar or complementary effect corresponding to other neurons [1, 12-14]. For each hidden neuron, construct a vector containing the output activations for all input patterns during a selected epoch. Then calculate the angle between each vector. All of the angles are normalised to 0 to 180 degrees. If the angle is less than 15 degrees, the relative hidden neurons are considered as similar and one of those could be removed with its connected weights added to the other one. On the other hand, if the angle is larger than 165 degrees, these two neurons are considered to have complementary functions, and both could be removed.

2.3 Badness

The badness factor 'indicates badness of each hidden unit. It is the sum of back propagated error component over all patterns for each hidden unit' [2].

The badness factor is evaluated as

$$BAD_i^{k-1} = \sum_p (e_i^{k-1,p})^2 = \sum_p (\sum_j w_{ij}^{k-1\,k} \delta_j^k)^2$$

where

$$\delta_j^k = (t_j - o_j^k) f'(net_j^k) \text{ (output layer)}$$

$$\delta_j^{k-1} = (\sum_l w_{jl}^{k-1\,k} \delta_j^k) f'(net_j^{k-1}) \text{ (hidden layer)}$$

net_j^k is the input to j^th neuron, t_j is the target output, and o_j^k is the desired output.

The neuron whose badness factor is largest is considered the worst neuron and the weights connected to it will be set to small random values.

2.4 Adding Noise

In this paper, we add a kind of noise [6] which is evaluated as:

$$\text{noise}_{i,j}(m) = \beta_{i,j}\left(x_{i,j}(m) - 0.5\right)$$

and

$$x_{i,j}(m+1) = \alpha x_0(m)\left(1 - x_0(m)\right).$$

At each epoch of the training, each weight is changed with noise added. In our experiment, α is set to be 4.0 and β is fixed as 0.01.

2.5 Heuristic Pattern Reduction

If we had before training some prior knowledge of what percentage of outliers exists in the dataset, then we could discover and eliminate them relatively readily. This situation rarely occurs.

The heuristic pattern reduction algorithm [7-8] eliminates outliers by assuming that the number of outliers are significantly less than normal patterns, and that a trained network can provide information as to the approximate likelihood that a pattern is an outlier. That is, that the error on a pattern is roughly correlated with its likelihood of being an outlier. Thus, we just remove half of the data in the approximate middle of the training, with the patterns sorted by the pattern error we discard every second pattern.

2.6 Bimodal Distribution Removal

The Bimodal Distribution Removal algorithm [9] also does not need any prior knowledge about the input dataset, and uses the network itself to recognise the outliers. The algorithm processes as follows:

(1) Begin training with the whole training set.
(2) Wait until the normalised variance of errors over the training set v_{ts} is below 0.1.
(3) Calculate the mean error $\bar{\delta}_{ts}$ over the training set.
(4) Take from the training set those patterns error greater than error $\bar{\delta}_{ts}$.
(5) Calculate the mean $\bar{\delta}_{ss}$, and standard deviation σ_{ss} of this subset.
(6) Permanently remove all patterns from the training set with $\text{error} \geq \bar{\delta}_{ss} + \sigma ss$ where $0 \leq \alpha \leq 1$
(7) Repeat steps 2-6 every 50 epochs, until normalised variance of errors over the training set $v_{ts} \leq 0.01$."

3 Experiments

Table 1. The 7 UCI datasets used with their attribute information

Dataset	Number of instances	Number of attributes	Number of classes
Breast Cancer	569	32	2
Ecoli	336	8	8
Ionosphere	351	34	2
Iris	150	4	3
Sonar	208	60	2
Survival	306	3	2
Wine	178	13	3

3.1 Method

In the experiments, we used all algorithms mentioned above as well as normal back-propagation to train networks using all of the datasets. The training was halted at 3000 epochs, well past the best generalisation on all data sets.

The evaluation criteria are final errors of the network and the best prediction accuracy on test data. Each training used ten-fold cross validation. Due to lack of space, we show an example of each of pruning and adding noise, and show substantial experimental results for reducing data sets.

3.2 Results

We report all the results here, with the baseline normal back-propagation (BP) results line being repeated in the tables for consistent ease of comparison.

An example of pruning of the neurons or connections is shown below.

Table 2. The average accuracy on test data by normal BP and badness

	breast	ecoli	ionosphere	iris	sonar	survival	wine
BP	0.5165	0.5806	0.8528	0.9667	0.7827	0.0290	0.3438
noise	0.4797	0.5461	0.8389	0.9600	0.7594	0.0065	0.3319

The experiment results of badness algorithm show that it does not improve the generalisation on any of these seven datasets. We can see that 'badness' is bad.

An example of adding noise is shown below.

Table 3. The average accuracy on test data by normal BP and adding noise

	breast	ecoli	ionosphere	iris	sonar	survival	wine
BP	0.5165	0.5806	0.8528	0.9667	0.7827	0.0290	0.3438
noise	0.5202	0.6247	0.8667	0.9600	0.7885	0.0355	0.3722

Adding noise has a slight improvement of the net's generalisation capability, normally 1% to 5%.

A set of examples of reducing datasets is shown below.

Table 4. The average accuracy on test data by normal BP, BDR and HPR

	breast	ecoli	ionosphere	iris	sonar	survival	wine
BP	0.5165	0.5806	0.8528	0.9667	0.7827	0.0290	0.3438
heuristic	0.5956	0.7090	0.8380	0.9600	0.7218	0.7110	0.5847
bimodal	0.6874	0.6520	0.7287	0.7133	0.6378	0.7323	0.6354

Bimodal distribution removal (BDR) and heuristic pattern reduction (HPR) generally outperform normal BP in the experiment. In some datasets, the accuracy on test data is markedly improved by 22%. Furthermore, the survival dataset which seems hard to be recognised by normal BP, is well recognised by BDR and HPR.

We can detect some pattern in the results, in that the better the result using BP, the less beneficial effect either technique has, with BDR reducing the correctness in these cases. We can predict that this is due to the nature and amount of noise in the dataset. While BP can cope with noise, the less noise in the dataset the better it will perform, and will be negatively impacted by the elimination of hard-to-learn training patterns.

Since the HPR method will only remove half of these, any negative effect is reduced. Thus, if BP performs well, we should use only HPR, while if BP performs badly, we should try both HPR and BDR.

The training process on the Survival dataset by nomal BP, BDR and HDR are shown in Figures 1, 2, 3.

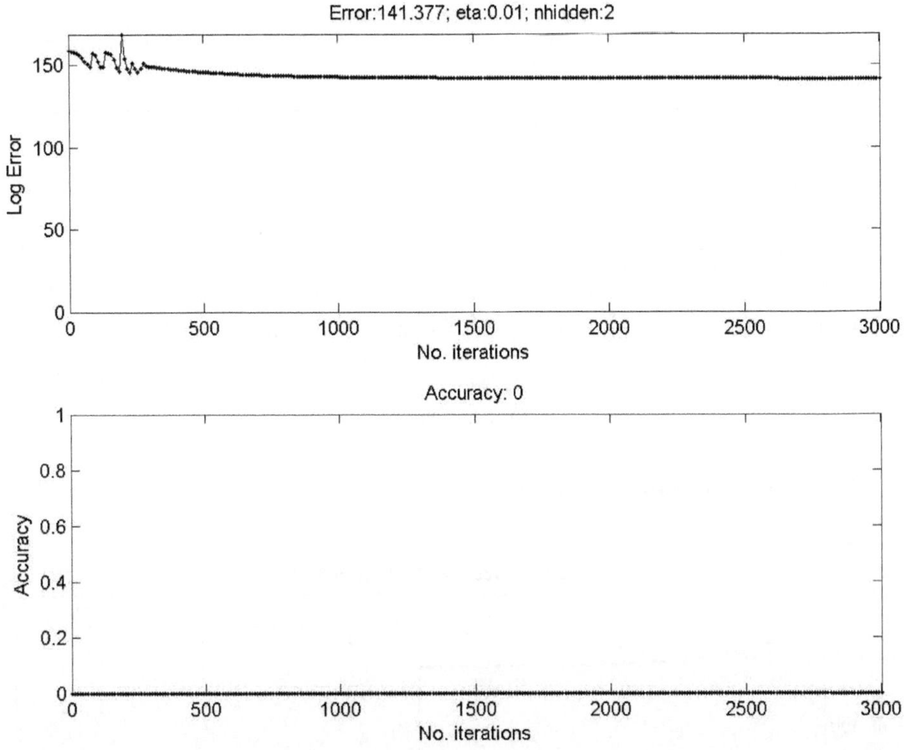

Fig. 1. Survival dataset training by normal BP

We can see that the training error does not decrease and the test set accuracy stays at zero throughout. This run is clearly stuck in a very bad local mininimum.

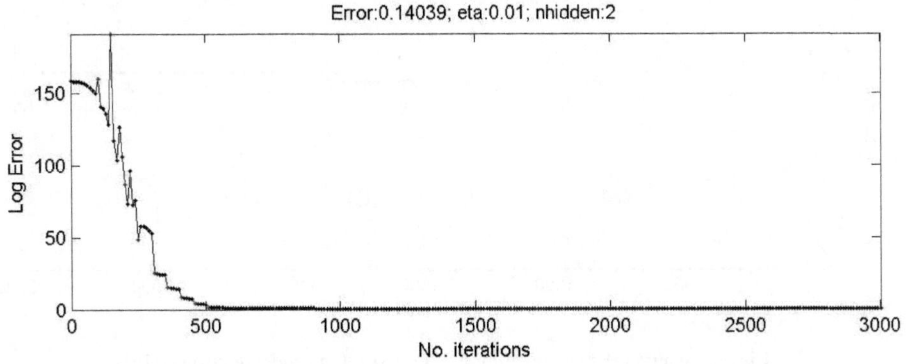

Fig. 2. Survival dataset training by Bimodal Distribution Removal (BDR)

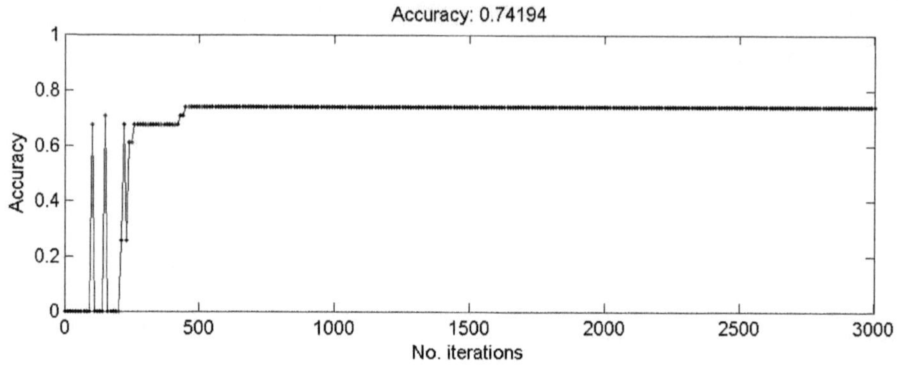

Fig. 2. (*continued*)

The network is clearly now learning well.

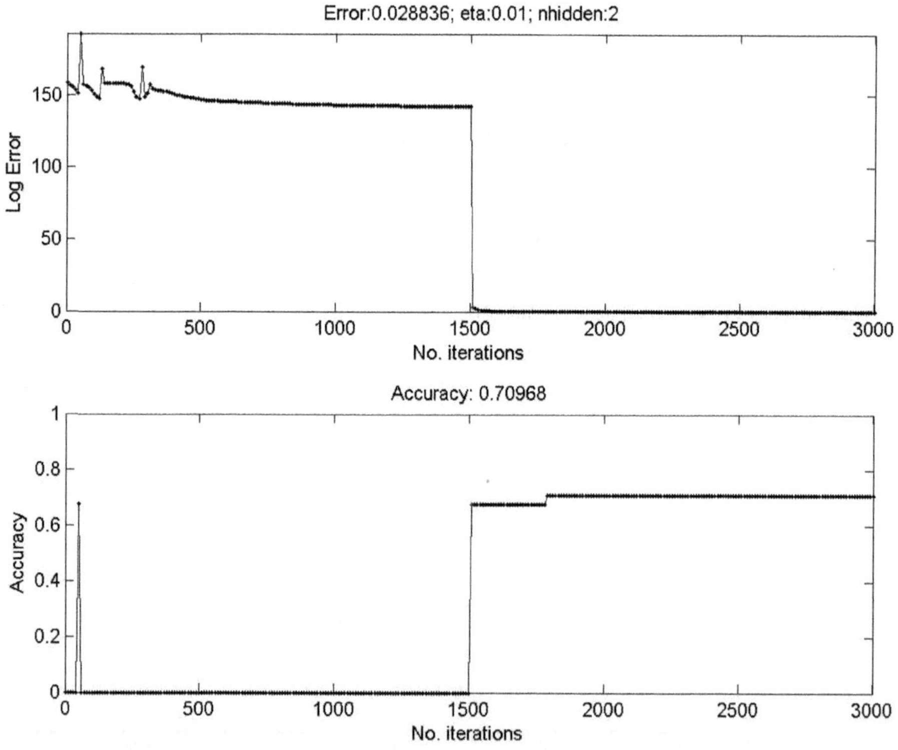

Fig. 3. Survival dataset training by Heuristic Pattern Reduction (HPR)

At 1,500 epochs, half the training patterns are eliminated and the network learns within a very small number of epochs. Clearly, the relatively slow and otherwise insufficient learning up to this point is suitable to sort the training patterns meaningfully and so the technique can eliminate half the noisy points.

Eliminating half the noisy points appears to work since there are enough normal patterns that the underlying function can still be learnt but with only half as much distraction from confounding noise / outliers and allows us to balance the bias and variance dilemma [15] in neural network training.

One interesting observation of these two algorithms is that they discard a great percentage of the original dataset, and clearly not all of them are outliers. In the example we have shown, HPR eliminates exactly 50% of the patterns, producing datasets which improve training. For the BDR technique, for some datasets, the remaining training data are composed by parts of classes which represented by the original data; however, the network is still able to recognise the classes which are removed since that information is still encoded within its trained weights.

Table 5. The average % of the dataset removed by BDR and accuracy

	breast	ecoli	ionosphere	iris	sonar	survival	wine
BP accur.	0.5165	0.5806	0.8528	0.9667	0.7827	0.0290	0.3438
removed	**0.1481**	**0.9412**	**0.9630**	**0.6074**	**0.8254**	**0.2724**	**0.7222**
BDR accur.	0.6874	0.6520	0.7287	0.7133	0.6378	0.6897	0.6354

For the survival dataset we can see that only 27% is removed, to substantially improve accuracy. On that dataset the HPR technique was better, removing 50%. We can not otherwise conclude that the amount removed is correlated with the results beyond our previous observations linking success to the amount of noise in the data.

4 Conclusions

Many authors have proposed various modifications to the backpropagation algorithm, and usually compare it against one or two other modifications at most, and sometimes only on their own datasets. To our knowledge, this is the first thorough comparison of a reasonable number of such modifications, on a reasonable number of standard datasets.Our results presented here showed brief examples of pruning and adding noise, which notwithstanding their prominence in the literature are not the lowest hanging fruit. We showed a substantial set of results on reducing data sets and have derived heuristics for when the two main techniques can be expected to have a significant beneficial effect on neural network learning and generalization.

References

1. Gedeon, T.D., Harris, D.: Network Reduction Techniques. In: International Conference on Neural Networks Methodologies and Applications, vol. 1, pp. 119–126. AMSE, San Diego (1991)
2. Hagiwara, M.: Novel back propagation algorithm for reduction of hidden units and acceleration of convergence using artificial selection. In: International Joint Conference on Neural Networks, pp. 625–630. IEEE Press, San Diego (1990)
3. Wieland, A., Leighton, R.: Geometric analysis of neural network capabilities. In: ICNN 1987, vol. III, pp. 385–392 (1987)
4. Rumelhart, D.E., Hinton, G.E., Williams, R.J.: Learning representations by back-propagating errors. Nature 323, 533–536 (1986)
5. Sietsma, J., Dow, R.F.: Creating Artificial Neural Networks That Generalize. Neural Networks 4, 67–79 (1991)
6. Azamimi, A., Uwate, Y., Nishio, Y.: Good Learning Performance of Backpropagation Algorithm with Chaotic Noise Features. In: SSJW 2008, Sanuki, Japan, pp. 36–38 (2008)
7. Gedeon, T.D., Bowden, T.G.: Heuristic Pattern Reduction. In: IJCNN 1992, Beijing, vol. 2, pp. 449–453 (1992)
8. Gedeon, T.D., Bowden, T.G.: Heuristic Pattern Reduction II. In: 3rd ICYCS, Beijing, vol. 3, pp. 43–45 (1993)
9. Slade, P., Gedeon, T.D.: Bimodal Distribution Removal. In: Mira, J., Cabestany, J., Prieto, A.G. (eds.) IWANN 1993. LNCS, vol. 686, pp. 249–254. Springer, Heidelberg (1993)
10. Gedeon, T.D., Slade, P.: Reducing Training Set Size to Improve Learning. In: 2nd European Congress on Intelligent Techniques and Soft Computing (EUFIT 1994), Aachen, pp. 1232–1236 (1994)
11. Karnin, E.D.: A simple procedure for pruning back-propagation trained neural networks. IEEE Transactions on Neural Networks 1, 239–242 (1990)
12. Gedeon, T.D., Harris, D.: Creating Robust Networks. In: International Joint Conference on Neural Networks, Singapore, pp. 2553–2557 (1991)
13. Gedeon, T.D., Harris, D.: Progressive Image Compression. In: International Joint Conference on Neural Networks, Baltimore, vol. 4, pp. 403–407 (1992)
14. Gedeon, T.D., Harris, D.: Hidden Units in a Plateau. In: 1st International Conference on Intelligent Systems, Singapore, pp. 391–395 (1992)
15. Gedeon, T.D., Wong, P.M., Harris, D.: Balancing Bias and Variance: Network Topology and Pattern Set Reduction Techniques. In: Sandoval, F., Mira, J. (eds.) IWANN 1995. LNCS, vol. 930, pp. 551–558. Springer, Heidelberg (1995)

Lagrange Programming Neural Networks for Compressive Sampling

Ping-Man Lam[1], Chi Sing Leung[1], John Sum[2], and A.G. Constantinides[3]

[1] Department of Electronic Engineering, City University of Hong Kong, Hong Kong
[2] Institute of Technology Management, National Chung Hsing Univesity, Taiwan
[3] Imperial College, UK

Abstract. Compressive sampling is a sampling technique for sparse signals. The advantage of compressive sampling is that signals are compactly represented by a few number of measured values. This paper adopts an analog neural network technique, Lagrange programming neural networks (LPNNs), to recover data in compressive sampling. We propose the LPNN dynamics to handle three sceneries, including the standard recovery of sparse signal, the recovery of non-sparse signal, and the noisy measurement values, in compressive sampling. Simulation examples demonstrate that our approach effectively recovers the signals from the measured values for both noise free and noisy environment.

1 Introduction

Compressive sampling is a novel sampling paradigm in data acquisition for sparse signals [1,2]. It suggests that many signals can be concisely represented by a few measured values when a proper basis is chosen. The basic principle is that a signal can be represented by a number of measured values, which is much smaller than the length of the signal. To obtain those measured values, signals are measured by a set of pseudo random functions. Afterwards, we can recover the original signal by these measured values. Traditional approaches to recover signals from compressive sampling are based on Newton's method.

Analog computational circuits [3] have long been used to linearly approximate the nonlinear response of a system in the neural networks community. The advantages of using analog neural circuits as specific purposes include the inherent parallelism of the weight sum operations [4] and the compact size[5]. Hopefield [6] investigated an analog neural circuit for solving quadratic optimization problems. In [7], a canonical nonlinear programming circuit was proposed to solve nonlinear programming problems with inequality constraints. Apart from optimization, neural circuits can also be used for searching the maximum of a set of numbers [8]. In [9], the Lagrange programming neural network (LPNN) model was proposed to solve general nonlinear constrained optimization problems based on the well-known Lagrange multiplier method. A LPNN consists of two types of neurons: variable and Lagrangian neurons. The variable neurons seeks for a state with the minimum cost in a system while the Lagrange neurons

K.W. Wong et al. (Eds.): ICONIP 2010, Part II, LNCS 6444, pp. 177–184, 2010.

are trying to constrain the system state of the system such that the system state falls into the feasible region.

This paper adopts the LPNN model to recover data in compressive sampling. We propose the LPNN dynamics to handle three sceneries, including the standard recovery of sparse signal, the recovery of non-sparse signal, and the noisy measurement values, in compressive sampling. We formulates the signal recovering process in compressive sampling as a set of differentiate equations which are derived from a Lagrange function. From the dynamics, we find out that the operation of the LPNN model for compressive sampling can be considered as a special form of bidirectional associative memories. Another interesting thing is that the connection matrix of the LPNN model for compressive sampling is random matrix.

This paper is organized as follows. In Section 2, the background of compressive sampling and the LPNN model are reviewed. Section 3 introduces our neural model for recovering compressive sampled signals. In Section 4, we use two examples to verify our LPNN model for compressing sampling. Section 5 concludes our results.

2 Compressive Sampling and LPNN

In compressive sampling, a sparse[1] signal $\boldsymbol{x} \in \Re^n$ is measured by a set of m random valued vectors, $\{\boldsymbol{\phi}_1, \cdots, \boldsymbol{\phi}_m : \boldsymbol{\phi}_j \in \Re^n\}$, where $m < n$. The m measured values are given by

$$y_1 = \langle \boldsymbol{x}, \boldsymbol{\phi}_1 \rangle, \quad \cdots, y_m = \langle \boldsymbol{x}, \boldsymbol{\phi}_m \rangle, \tag{1}$$

where $\langle \cdot, \cdot \rangle$ is the inner product operator. The compressive sampling process can be written in the matrix form, given by $\boldsymbol{y} = \boldsymbol{\Phi x}$. Practically, $\boldsymbol{\phi}_i$'s are pseudo random valued vectors, such as noiselets [10]. One of advantages of using noiselets is that the matrix $\boldsymbol{\Phi}$ can be decomposed as a multiscale manner. Hence, the computation of $\boldsymbol{\Phi x}$ can be performed in an efficient manner. If the sparse signal $\boldsymbol{x}$ has high sparsity feature in the original domain, it can be recovered from the measured signal $\boldsymbol{y}$ by solving the following constrained optimization problem, given by

$$\min \ \|\hat{\boldsymbol{x}}\|_{l_1} \ \text{subject to} \ \boldsymbol{y} = \boldsymbol{\Phi}\hat{\boldsymbol{x}}\,, . \tag{2}$$

From [1,2], if $m < C_o \log n$, then with probability very close to one the signal can be recovered, where C_o is constant depending on $\boldsymbol{\Phi}$. If the signal $\boldsymbol{x}$ is not sparse in the original domain, we can transform it into a basis $\boldsymbol{\Psi}$, such as wavelet basis. Then, the recovery of the signal can be formulated:

$$\min \ \|\boldsymbol{\Psi}\hat{\boldsymbol{x}}\|_{l_1} \ \text{subject to} \ \boldsymbol{y} = \boldsymbol{\Phi}\hat{\boldsymbol{x}}\,. \tag{3}$$

In (3), the objective is to search an n-component signal $\hat{\boldsymbol{x}}$ with a sparse transformed signal $\boldsymbol{\Psi x}$. The constraints $\boldsymbol{y} = \boldsymbol{\Phi}\hat{\boldsymbol{x}}$ limit the searching signals.

[1] A signal is sparse if a few components are large in magnitude and other components are nearly zero.

A LPNN aims at minimizing the following constrained optimization problem:

$$\text{Minimize } f(\hat{\boldsymbol{x}}) \text{ subject to } \boldsymbol{h}(\hat{\boldsymbol{x}}) = \mathbf{0}, \tag{4}$$

where $\hat{\boldsymbol{x}} \in \Re^n$ is the state of system, $f : \Re^n \to \Re$ is the objective function, and $\boldsymbol{h} : R^n \to R^m$ describes the m equality constraints, where $m < n$. The LPNN model uses the Lagrange multiplier approach to obtain the optimized solution of (4). The Lagrange objective function is given by

$$\mathcal{L}(\hat{\boldsymbol{x}}, \lambda) = f(\hat{\boldsymbol{x}}) + \boldsymbol{\lambda}^T \boldsymbol{h}(\hat{\boldsymbol{x}}), \tag{5}$$

where $\boldsymbol{\lambda} = [\lambda_1, \cdots, \lambda_m]^T$ is the Lagrange multiplier vector. To realize the optimization, a LPNN consists of two types of neurons: variable and Lagrange neurons. Intuitively, the variable neurons are seeking for a minimum point of (4) and the Lagrange neurons are trying to constrain the state of the system. The transient behavior of those neurons is given by

$$\frac{d\hat{\boldsymbol{x}}}{dt} = -\nabla_x L(\hat{\boldsymbol{x}}, \boldsymbol{\lambda}) \text{ and } \frac{d\boldsymbol{\lambda}}{dt} = \nabla_\lambda L(\hat{\boldsymbol{x}}, \boldsymbol{\lambda}). \tag{6}$$

With (6), the network will be settled down at an stable state [9] when the gradient vectors $\nabla_{\hat{\boldsymbol{x}}} \boldsymbol{h}(\hat{\boldsymbol{x}}) = \{\nabla_{\hat{\boldsymbol{x}}} h_1(\hat{\boldsymbol{x}}), \cdots, \nabla_{\hat{\boldsymbol{x}}} h_m(\hat{\boldsymbol{x}})\}$ of $\boldsymbol{h}(\hat{\boldsymbol{x}})$ are linear independent.

3 LPNNs for Compressive Sampling

3.1 Sparse Signal

To recover a sparse signal from the measured vector $\boldsymbol{y}$ by LPNNs, one may suggest that we can define the objective function of the LPNN model as

$$\mathcal{L}(\hat{\boldsymbol{x}}, \boldsymbol{\lambda}) = \|\hat{\boldsymbol{x}}\|_{l_1} + \boldsymbol{\lambda}^T (\boldsymbol{y} - \boldsymbol{\Phi}\hat{\boldsymbol{x}}). \tag{7}$$

However, the norm–1 measure, involving absolute operator $|x|$, is not differentiable at $x = 0$. Hence, we need an approximation on the absolute operator. In this paper, we use the following approximation:

$$|x| \approx \alpha(x) = \frac{\log \cosh ax}{a} \tag{8}$$

to approximate the absolute operator, where $a > 1$. For a large a, the approximation is quite accurate.

With the approximation, the Lagrange objective function for recovering the signal is

$$\mathcal{L}(\hat{\boldsymbol{x}}, \boldsymbol{\lambda}) = f(\hat{\boldsymbol{x}}) + \boldsymbol{\lambda}^T (\boldsymbol{y} - \boldsymbol{\Phi}\hat{\boldsymbol{x}}). \tag{9}$$

where $f(\hat{\boldsymbol{x}}) = \alpha(\hat{x}_1) + \cdots + \alpha(\hat{x}_n) = \frac{\log \cosh a\hat{x}_1}{a} + \cdots + \frac{\log \cosh a\hat{x}_n}{a}$. The dynamics of the variable and Lagrange neurons are given by

$$\frac{d\hat{\boldsymbol{x}}}{dt} = -\tanh(a\hat{\boldsymbol{x}}) + \boldsymbol{\Phi}^T \boldsymbol{\lambda}, \text{ and } \frac{d\boldsymbol{\lambda}}{dt} = \boldsymbol{y} - \boldsymbol{\Phi}\hat{\boldsymbol{x}}. \tag{10}$$

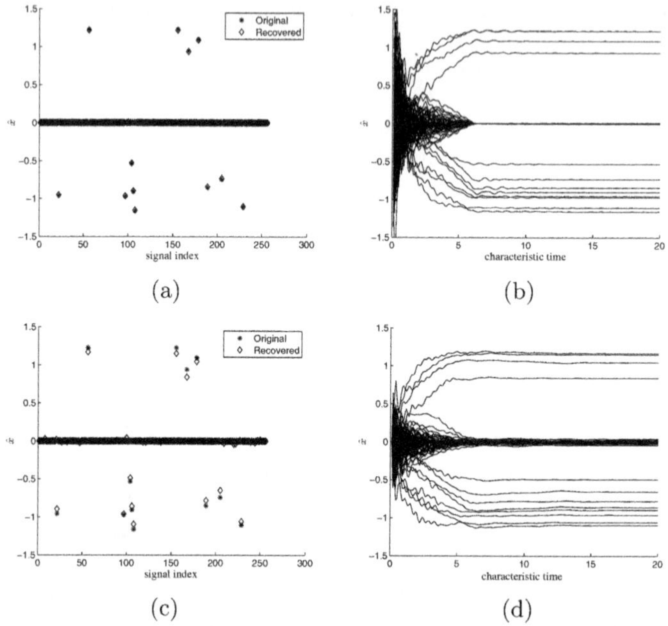

Fig. 1. The 1D sparse artificial signal recovery. (a) The sparse signal and its recovery without measurement noise. (b) The dynamics of recovered signal without measure noise.(c) The original sparse signal and its recovery with measurement noise. (d) The dynamics of recovered signal with measurement noise.

In the above, $\boldsymbol{\Phi}$ is the interconnection matrix between the variable neurons and Lagrange neurons. This kind of dynamics can be considered as a special form of bi-directional associative memories (BAMs) [11,12]. Besides, practically, the interconnection matrix $\boldsymbol{\Phi}$ can be a random $\{+1, -1\}$ matrix. This suggests that the implementation of the analog neural circuit could be very simple because we have no need to precisely construct the interconnection matrix.

3.2 Non-sparse Signal

Our neural circuit can also be used for recovering a signal $\boldsymbol{x}$ with low sparsity. Let $\boldsymbol{x} = x_1, \cdots, x_n$ be an 1D signal in the spatial domain. One of the simplest methods for handling this case is to consider the spatial domain gradient, given by $\nabla_s \hat{\boldsymbol{x}} = [\hat{x}_1 - \hat{x}_0, \hat{x}_2 - \hat{x}_1, \cdots, \hat{x}_n - \hat{x}_{n-1}]^T$. The Lagrange objective function for recovering the signal becomes

$$\mathcal{L}(\hat{\boldsymbol{x}}, \boldsymbol{\lambda}) = f(\nabla_s \hat{\boldsymbol{x}}) + \boldsymbol{\lambda}^T(\boldsymbol{y} - \boldsymbol{\Phi}\hat{\boldsymbol{x}}) . \tag{11}$$

The dynamics of neurons are given by

$$\frac{d\hat{x}_i}{dt} = -\tanh[a(\hat{x}_i - \hat{x}_{i-1})] + \tanh[a(\hat{x}_{i+1} - \hat{x}_i)] + [\boldsymbol{\Phi}^T \boldsymbol{\lambda}]_i , \tag{12}$$

$$\frac{d\boldsymbol{\lambda}}{dt} = \boldsymbol{y} - \boldsymbol{\Phi}\hat{\boldsymbol{x}} , \tag{13}$$

where $[\cdot]_i$ is the i-th component of a vector.

3.3 Measurement Noise

When there is some measurement noise in $\boldsymbol{y}$, the compressive sampling process can be represented as

$$\boldsymbol{y} = \boldsymbol{\Phi x} + \boldsymbol{\xi}\,, \tag{14}$$

where $\boldsymbol{\xi}_j$'s are independently identical random noise with zero mean and variance σ^2. In this case, the estimated signal $\hat{\boldsymbol{x}}$ can be recovered based on the optimization problem:

$$\min f(\hat{\boldsymbol{x}}) \text{ subject to } \boldsymbol{y} - \boldsymbol{\Phi}\hat{\boldsymbol{x}} = \hat{\boldsymbol{\xi}}_j \ \forall j = 1, \cdots, m \ \text{ and } \ \hat{\boldsymbol{\xi}}^T \hat{\boldsymbol{\xi}} \leq m\sigma^2. \tag{15}$$

Again, the Lagrange objective function (15) is

$$\mathcal{L}(\hat{\boldsymbol{x}}, \hat{\boldsymbol{\xi}}, \boldsymbol{\lambda}, \mu) = f(\hat{\boldsymbol{x}}) + \boldsymbol{\lambda}^T(\boldsymbol{y} - \boldsymbol{\Phi}\hat{\boldsymbol{x}} - \hat{\boldsymbol{\xi}}) + \mu(\hat{\boldsymbol{\xi}}^T \hat{\boldsymbol{\xi}} - m\sigma^2)\,, \tag{16}$$

where $\hat{\boldsymbol{x}}$ is the state vector of the network, $\hat{\boldsymbol{\xi}}$ is the estimation of the noise vector, $\boldsymbol{\lambda}$ is the Lagrange multiplier vector, and μ is another Lagrange multiplier. With (16), the dynamics of the network are

$$\begin{array}{ll} \frac{d\hat{\boldsymbol{x}}}{dt} = -\tanh(a\hat{\boldsymbol{x}}) + (\boldsymbol{\Phi}^T\boldsymbol{\lambda}), & \frac{d\hat{\boldsymbol{\xi}}}{dt} = -2\mu\hat{\boldsymbol{\xi}} + \boldsymbol{\lambda} \\ \frac{d\boldsymbol{\lambda}}{dt} = (\boldsymbol{y} - \boldsymbol{\Phi}\hat{\boldsymbol{x}}) - \hat{\boldsymbol{\xi}}, & \frac{d\mu}{dt} = \hat{\boldsymbol{\xi}}^T \hat{\boldsymbol{\xi}} - m\sigma^2 \end{array}. \tag{17}$$

4 Simulation Results

4.1 Artificial Data

The first example is an artificial 1D signal which has 256 values. 244 values are of zero value while 12 values are wof non-zero value. These non-zero values are located randomly distributed. Figure 1 shows the recovery of the 1D sparse artificial signal from 60 measured values with no measurement noise. From the figure, the recovered signal has the nearly the same to the original signal as shown in Figure 1(a). Moreover, the recovered signal has converged after 10 characteristic times, as shown in Figure 1(b). Figures 1(c)-(d) show the recovery of the 1D sparse artificial signal from 60 noisy measured values. The noise added to the measured values is independently identical random noise with zero mean and variance $\sigma^2 = 0.04$. In this experiment, the values of recovered signal are also close to the original signal as shown in Figure 1(c), and the recovered signal is converged after 10 characteristic times, as shown in Figure 1(d).

4.2 Non Sparse Signal

We use a large scale problem, 2D image, to verify our LPNN approach. Its resolution is $256 \times 256 = 65536$ pixels. Its values are non-sparse in nature. We generalize the 1D method presented in Section 3.2 to the 2D case. There are $n = 65536$ variable neurons and we vary the number m of measured values.

Fig. 2. Image recovery by our LPNN approach. (a) Performance curve. (b) Original image. (c) Recovered image using DCT with 20000 measured values. The PSNR is equal to 29.0dB. (d) Recovered image using our LPNN approach with 20000 measured values. The PSNR is equal to 31.6dB.

Our proposed approach is compared with an 2D DCT-based approach. In the DCT-based approach, the image is transformed to the DCT domain. The selection order of DCT coefficients is based on the zig-zag ordering. In our approach, the first 1000 measured values are the low-frequency DCT coefficients and the rest of measured values are obtained from the compressive sampling process. Figure 2 shows the result. The result verifies that our LPNN approach is successful in recovering the signal for this large scale problem. Figure 2 (a) shows the peak signal-to-noise rate (PSNR). When the number of measured values are less than 10000, both approaches have nearly the same PSNR value. When the number of measured values used increases, our LPNN approach outperforms the

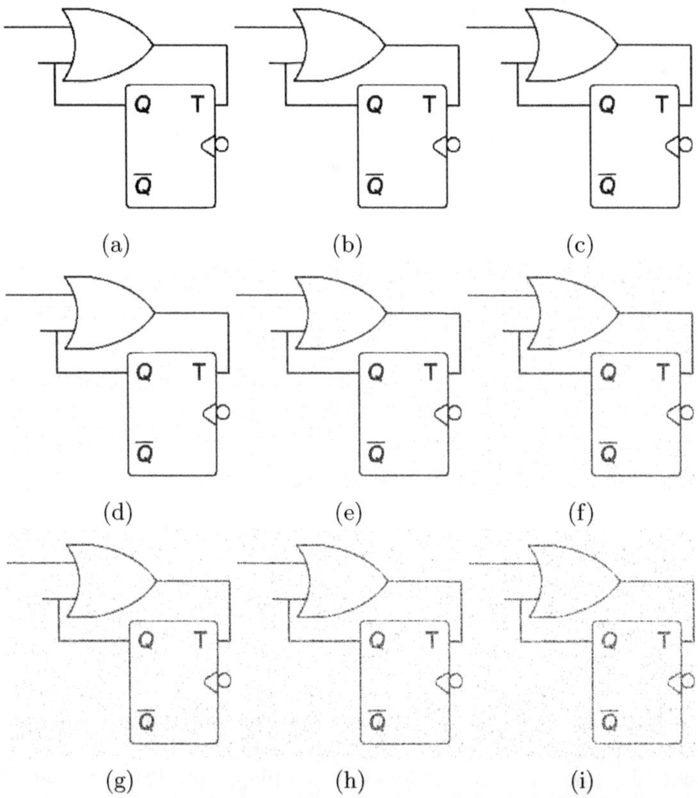

Fig. 3. Image recovery by our LPNN approach for noisy measurement values. (a) Original image. (b) Recovery with no measurement noise. (c)–(i) Recovery with measurement noise. (c) $\sigma^2 = 0.0001$. (d) $\sigma^2 = 0.0004$. (e) $\sigma^2 = 0.0009$. (f) $\sigma^2 = 0.0016$. (g) $\sigma^2 = 0.0025$. (h) $\sigma^2 = 0.0036$. (i) $\sigma^2 = 0.0049$.

DCT-based approach. For example, when the number of measured values used is equal to 20000, the DCT-based approach has 29.0dB and our approach has 31.6dB. Figures 2(b)-(d) shows the original image, the recovered images from the DCT-based approach and our approach. Ringing artifacts are observed for the DCT-based approach as shown in Figure 2(c). Besides, in the DCT-based approach, there is a large visual distortion around the arm of the camera stand. On the other hand, in our LPNN approach, there are no ringing artifact and no distortion around the arm of the camera stand.

4.3 Measurement Noise

We use a 2D image with resolution equal to $256 \times 256 = 65536$ pixels to verify our LPNN model for recovering signal from noisy measurements. The image, shown in Figure 3, is sparse. There are $n = 65536$ variable neurons. $15,000$ measured values are used for recovering. After the measurement, we find that the average power of the measured values is equal to 0.0492. We add Gaussian measurement noise to the measured values. The noise variances are equal to $\{0, 0.0001, 0.0004, 0.0009, 0.0016, 0025, 0.0036, 0.0049\}$. We use the Section 3.3's method to recover the image. Figure 3 shows the recovered images. From the figure, the LPNN method can recover the image when there are noise in the measured values. As expected, the recovered signal is degraded when the variance of the noise increases.

5 Conclusion

In this paper, we formulate the LPNN model to handle the signal recovery in compressive sampling. We propose the LPNN dynamics to handle three sceneries, including the standard recovery of sparse signal, the recovery of non-sparse signal, and the noisy measurement values. For the standard recovery of sparse signal, the dynamics of our LPNN model can be considered as a special case of bidirectional memories. Simulation results verify that our approach can be applied for recovering 1D and 2D data in compressive sampling for both noise free and noisy environments.

Acknowledgment. The work described in this paper was supported by a grant from City University of Hong Kong (Project No. 7002588).

References

1. Candès, E.J., Wakin, M.B.: An introduction to compressive sampling. In: IEEE Signal Processing, vol. 25, pp. 21–30 (March 2008)
2. Romberg, J.: Imaging via compressive sampling. In: IEEE Signal Processing, vol. 25, pp. 14–20 (March 2008)
3. Dennis, J.B.: Mathematical Programming and Electrical Networks. Wiley, New York (1959)
4. Graf, H., Jackel, L.: Analog electronic neural network circuits. IEEE Circuits Devices Mgag. 5(4), 44–49 (1989)
5. Mead, C.: Analog VLSI and Neural Systems. Wesley, London (1989)
6. Hopfield, J.J.: Neural networks and physical systems with emergent collective computational abilities. Proc. of the National Academy of Sciences 79, 2554–2558 (1982)
7. Chua, L., Lin, G.: Nonlinear programming without computation. IEEE Trans. on Circuits Syst. 31, 182–188 (1984)
8. Sum, J., Leung, C.-S., Tam, P., Young, G., Kan, W., Chan, L.-W.: Analysis for a class of winner-take-all model. IEEE Trans. Neural Networks 10(1), 64–71 (1999)
9. Zhang, S., Constantinidies, A.G.: Lagrange programming neural networks. IEEE Trans. on Circuits and Systems II 39, 441–452 (1992)
10. Coifman, R., Geshwind, F., Meyer, Y.: Noiselets. Appl. Comp. Harmon. Anal. 10, 27–44 (2001)
11. Kosko, B.: Bidirectional associative memories. IEEE Trans. Syst. Man Cybern. 18(1), 49–60 (1988)
12. Leung, C.-S., Chan, L.-W., Lai, E.: Stability and statistical properties of second-order bidirectional associative memory. IEEE Trans. Neural Networks 8(2), 267–277 (1997)

Input and Output Mapping Sensitive Auto-Associative Multilayer Perceptron for Computer Interface System Based on Image Processing of Laser Pointer Spot

Chanwoong Jung[1], Sang-Woo Ban[2], Sungmoon Jeong[3], and Minho Lee[1,3]

[1] Department of Sensor and Display Engineering,
Kyungpook National University, 1370 Sankyuk-Dong, Puk-Gu, Taegu 702-701 Korea
[2] Department of Information and Communication Engineering, Dongguk University
707 Seokjang-Dong, Gyeongju, Gyeongbuk 780-714, Korea
[3] School of Electronics Engineering,
Kyungpook National University, 1370 Sankyuk-Dong, Puk-Gu, Taegu 702-701 Korea
cwjung@ee.knu.ac.kr, swban@dongguk.ac.kr,
jeongsm@ee.knu.ac.kr, mholee@knu.ac.kr

Abstract. In this paper, we propose a new auto-associative multilayer perceptron (AAMLP) that properly enhances the sensitivity of input and output (I/O) mapping by applying a high pass filter characteristic to the conventional error back propagation learning algorithm, through which small variation of input feature is successfully indicated. The proposed model aims to sensitively discriminate a data of one cluster with small different characteristics against another different cluster's data. Objective function for the proposed neural network is modified by additionally considering an input and output sensitivity, in which the weight update rules are induced in the manner of minimizing the objective function by a gradient descent method. The proposed model is applied for a real application system to localize laser spots in a beam projected image, which can be utilized as a new computer interface system for dynamic interaction with audiences in presentation or meeting environment. Complexity of laser spot localization is very wide, therefore it is very simple in some cases, but it becomes very tough when the laser spot area has very slightly different characteristic compared with the corresponding area in a beam projected image. The proposed neural network model shows better performance by increasing the input-output mapping sensitivity than the conventional AAMLP.

Keywords: Auto-associative multilayer perceptron, input and output mapping sensitivity, laser pointer detection, computer interface using a laser pointer.

1 Introduction

A lot of researchers are currently focusing on new computer interfacing methods [1-5]. In a beam projection for presentation and meeting, it has been tried to develop easy interface methods based on gesture and/or speech recognition. Those studies have the same goal to overcome the limitation of interaction using conventional computer keyboard and mouse. When a presenter tries to manipulate his/her presentation system, he or she must closely stand to the control devices such as keyboard and mouse. Also, the presentation slides are

K.W. Wong et al. (Eds.): ICONIP 2010, Part II, LNCS 6444, pp. 185–192, 2010.

mostly controlled by a presenter only. It hinders active discussion based on controlling the presentation slides by people who attend the meeting. In this paper, we propose a new laser pointer recognition based computer interface system to help a beam project based meeting, which is a kind of digital convergence system for a laser pointer and a computer mouse as well as joysticks for a game.

In early stage of laser pointer interface system development [1-4], image processing based approaches have been generally considered. Kirstein, C. et al propose a laser pointer system based on a camera tracked system, in which mouse move up and down events are controlled based on X-Window system of the Linux. This system takes too long time to localizing a laser pointer [1]. Olsen Jr. et al describes an inexpensive laser pointer interaction system working on XWeb system for laser pointer interaction. This system is uncomfortable because the user had to preset a specific mode, therefore it is not suitable for getting a fast response [2]. Alternatively, Lapointe et al reported an architecture of laser pointer interaction system with fast detection of a laser pointer. But they didn't mention about the performance of their system [3]. Another research is to apply a hardware platform for a laser pointer system. This simple idea has a separated function of a camera-tracked laser pointer with a gyro sensor. It is natural that a new physical switch on a laser pointer has been considered seriously considering the necessity of a switch to turn on/off cursor echo. However a physical switch is not directly related to image processing improvement, in particular, the improvement of laser pointer detection reliability. Also, this system still has several disadvantages; if brightness of environment is too strong, the system will not operate very well [4]. Moreover, all of image processing based laser pointer interaction systems have common problems, which are easily affected by illumination noise and white balance of a camera as well as image sensor noises. Moreover, it is almost impossible to detect a laser spot if a spot area in a background pointed by a specific color of a laser pointer (red or green) has similar color with that of a laser pointer.

In order to overcome those limitations of conventional laser pointer detection system based on an image processing, we newly propose a new auto-associative multilayer perceptron (AAMLP) neural network with an input and output (I/O) mapping sensitive error back propagation (EBP) learning algorithm. The proposed neural network learns the background patch without a laser spot of an input camera image to resemble to the same patch of an image frame showing on a computer monitor screen. In test mode, the trained neural network generates an output values according to an input patch of a camera image, and checks whether the produced outputs are similar to the same patch values that is used for training of AAMLP. If a patch contains a laser pointer spot, the input will be different from that of training phase, then the produced output will generate different value from the input patch values. However, when a background patch has similar color with a laser pointer, the variation of the input patch value is not too big, and resultantly the neural network output by the conventional error back propagation is not much different from the case that the background does not contain a laser pointer spot. Thus, we develop a new auto-associative multilayer perceptron model with a new learning algorithm that can generate a big difference of output value according to a small variation of input pattern, in which an I/O mapping sensitivity term is additionally considered in an objective function generally defined as a sum of squares of output errors.

Section 2 and Section 3 describe the proposed new neural network model and its application for a laser pointer based computer interface system, respectively. Section 4 presents experimental results and discussion. Conclusion and future works follow in Section 5.

2 Proposed Input-Output Mapping Sensitive AAMLP(I/OS-AAMLP)

Many two-class clustering and recognition problems such as face detection and laser spot localization, which is considered as an application problem have tremendous within-class variability. Such a two-class recognition problem might be one of the partially-exposed environments problems which are well known problems where training data from on-class is very little or non-existent.

An Auto-associative neural network has been used successfully in many such partially-exposed problems [6-7]. An auto-associative neural network is basically a neural network whose input and target vectors are the same as shown in Fig. 1.

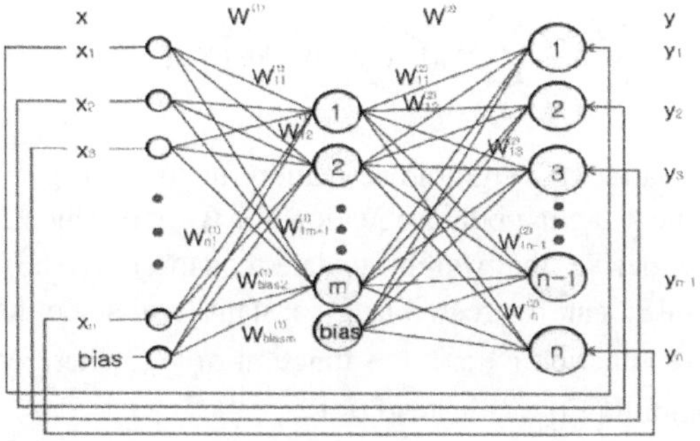

Fig. 1. Architecture of an auto-associative multilayer perceptron

For the AAMLP with an error back propagation learning algorithm, the objective function, E_p^s, is generally defined as a sum of squares of output errors as shown in Eq. (1) [6]. The weight update learning algorithm for the AAMLP is derived in the course of minimizing the objective function. Eq (2) presents an output equation of the AAMLP.

$$E_p^s = \frac{1}{2}\sum_i (t_i^s - y_i^s)^2 \tag{1}$$

where t_i^s is a target value of an output node *i* and, *s* and *p* represent a sample index of a specific pattern class and a pattern class index, respectively.

$$y_i^s = \sum_j w_{ij} h_j \tag{2}$$

where h_j is an output of a hidden node *j*, of which the equation is described in Eq.(4).

In our newly proposed AAMLP with a new learning algorithm that reflects an input/output sensitivity term in the objective function as $\widetilde{E}_p^s$ in Eq. (3). In Eq. (3), the denominator reflects an input/output sensitivity, which is obtained from Eq. (5) derived from Eq. (4) of an input/output sensitivity equation. In order to make a positive definite cost function, we consider only $f'(\sum w_{jk}x_k)$ term in Eq. (5), which plays a role for getting the better sensitivity of I/O mapping.

$$\widetilde{E}_p^s = \frac{E_p^s}{f'(\sum_k w_{jk}x_k)} \tag{3}$$

where w_{jk} is a weight between j^{th} hidden node and k^{th} input node and x_k is an input value of the kth input node.

$$\frac{\partial y_i}{\partial x_k} = \frac{\partial y_i}{\partial h_j}\frac{\partial h_j}{\partial x_k}, \left(y_i = \sum_j w_{ij}h_j, h_j = f\left(\sum_k w_{jk}^s x_k^s\right)\right) \tag{4}$$

$$\frac{\partial y_i}{\partial x_k} = \sum_j w_{ij} f'\left(\sum w_{jk}x_k\right) w_{jk} \tag{5}$$

where $f'(\cdot)$ is a derivative of a sigmoid activation function $f(\cdot)$.

Eq. (6) shows the weight update equation for w_{ij}, in which $\Delta\omega_{ij}$ is obtained by Eq. (7). When we derive the weight update equation as shown in Eq. (7), the denominator term of $\widetilde{E}_p^s$ can be treated as a constant term since the denominator term of the new objective function $\widetilde{E}_p^s$ is not a function of w_{ij}. Therefore, E_p^s is considered for derivation equation in Eq. (7) instead of $\widetilde{E}_p^s$.

$$w_{ij}(n+1) = w_{ij}(n) + \sum_{s=1}^{M} \Delta w_{ij}^s \tag{6}$$

$$\begin{aligned} \Delta w_{ij}^s &= -\eta \frac{\partial E_p^s}{\partial w_{ij}} \\ &= \eta(t_i^s - y_i^s) f'(\sum w_{ij}h_j^s) h_j^s \\ &\triangleq \delta_i^s h_j^s \end{aligned} \tag{7}$$

where $f'(\cdot)$ is a derivative of a sigmoid activation function $f(\cdot)$ and η is a learning rate.

Eq. (8) is the weight update equation for w_{jk}, in which Δw_{jk} is derived by Eq. (9).

$$w_{jk}(n+1) = w_{jk}(n) + \Delta w_{jk} \tag{8}$$

$$\Delta w_{jk} = \frac{\partial \widetilde{E}_p^s}{\partial w_{jk}} = \frac{\partial}{\partial w_{jk}}\left(\frac{E_p^s}{f'(\sum_k w_{jk} x_k)}\right)$$

$$= \frac{\left(\frac{\partial}{\partial w_{jk}} E_p^s\right) \cdot f'(\sum_k w_{jk} x_k) - E_p^s \cdot \left(\frac{\partial}{\partial w_{jk}} f'(\sum_k w_{jk} x_k)\right)}{(f'(\sum_k w_{jk} x_k))^2} \tag{9}$$

$$= \left(\eta_1 (t_i^s - y_i^s) w_{ij} - \frac{\eta_2 \frac{1}{2} \sum_i (t_i^s - y_i^s)^2 \left(1 - 2f\left(\sum_k w_{jk} x_k\right)\right)}{f\left(\sum_k w_{jk} x_k\right)\left(1 - f\left(\sum_k w_{jk} x_k\right)\right) + \varepsilon} \right) x_k$$

where η_1 and η_2 are learning rates which are experimentally defined, and η_1 should be greater than η_2 to increase the I/O sensitive effect. ε is a constant slack value to avoid being divided by zero in Eq. (9).

3 Laser Pointer Interfacing System Using I/OS-AAMLP

Figs. 2 (a) and (b) shows a new computer interfacing environment and the overall processes of the proposed computer interface system using I/OS-AAMLP, respectively. The proposed system mainly consists of three parts. First part is to localize a beam projected area (ROI) from an input beam projected screen image captured using a camera. Second one is to localize a laser spot in a localized beam projected area, which is the most important function for the proposed laser pointer based computer interfacing system. Third one is to recognize moving patterns of laser spots obtained from input image stream in order to generate proper interaction commends for interfacing.

For localizing a beam projected area from an input camera image, an iterative lighting condition adjustment method and a geometric based rectangular area detection algorithm using edge features are applied. In addition, a warping algorithm is used for coordinate transformation from a real coordinate obtained from localized beam projected area of a camera input image to a computer frame coordinate [8]. Next, a laser spot area in each localized beam projected area is detected using the proposed I/OS-AAMLP with a new input-output mapping sensitive error back propagation learning algorithm, which enhances performance for localizing even a difficult laser spot area such as having low sensitivity against a background. The three channel color intensities (red, green, blue) are considered as input features for training each area. Instead of directly using the three color features as input of the I/OS-AAMLP, we considered a principal component analysis (PCA) [9] for dimension reduction of input features for enhancing computation time as well as for obtaining better features for training each area of a beam projected image. Finally, the proposed interface system can generate five different interfacing commands (scroll up/down/left/right and double click) based on recognizing the laser pointer moving patterns using the coordinates of the localized laser spots obtained from an input image stream, which is conducted by a conventional multilayer perceptron [10].

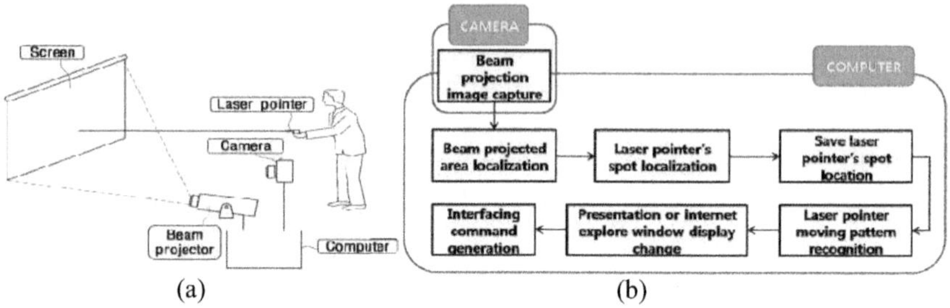

Fig. 2. A new computer interfacing environment (a) and the process outline of the laser pointer based computer interface system using the proposed I/OS-AAMLP (b)

4 Experiments

We verified the performance of the proposed I/OS-AAMLP using the accuracy of laser spots detection in a laser pointer based computer interfacing system. Each patch size in Fig. 3 for localizing a laser spot is 5 by 5 pixels. Therefore a color feature vector with 75 values (3 color channels for 5 by 5 pixels) is considered as a feature for recognizing a laser spot. Each extracted feature vector is transformed by projecting on a number of selected eigenvectors obtained from PCA. The PCA was applied to reduce the computation time of the proposed model with small dimensions of the input data, and also to extract more plausible robust features against noise. The number of eigenvectors is experimentally decided by comparison of performance according to the different cases with varying number of principal components. According to performance comparison result, 25 principal components are considered for dimension reduction of 75-dimension feature vectors. Therefore, the reduced 25-dimension vectors are used as input of the I/OS-AAMLP for deciding whether a laser spot area or not during a test mode. 100 sample patches without a laser spot are used for training the I/OS-AAMLP. For testing the I/OS-AAMLP, 10, 44, 54, 37, 10, 30 21, 40, 26 and 30 samples are considered for both each 10 without a laser spot data sets and each 10 with a laser spot data sets. Figure 3 shows a training process of the I/OS-AAMLP, which has 26 input nodes including one constant bias node, 16 hidden nodes including one constant bias node and 25 output nodes as the same number of input nodes since the I/OS-AAMLP model follows the structure of the conventional AAMLP.

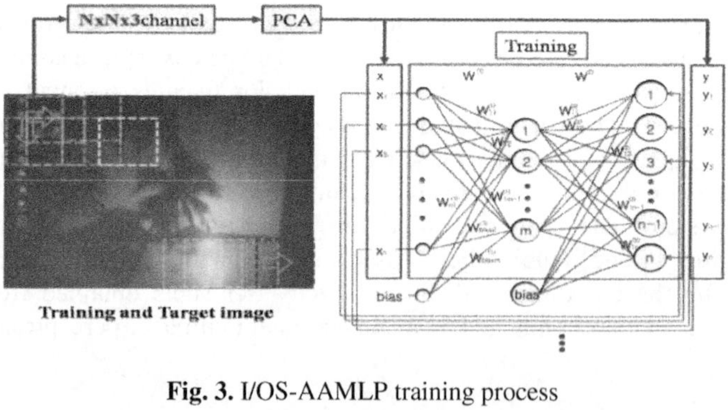

Fig. 3. I/OS-AAMLP training process

Table 1 compares experimental results between a conventional AAMLP and the proposed I/OS-AAMLP. The same test data samples are used for the conventional AAMLP and the I/OS-AAMLP. In Table 1, the relative error ratio is calculated by Eq. (10), in which an error means the difference between input and corresponding output of the neural network and the relative error ratio reflects sensitivity about input and output mapping of the neural network.

$$\text{relative error ratio} = \frac{\text{average error for all test images with a laser spot}}{\text{average error for all test images without a laser spot}} \tag{10}$$

The larger value of the relative error ratio means that an output of the neural network more sensitively responds about an input, which means that a laser spot with small variance can be more properly indicated by the neural network model. As shown in Table 1, the proposed I/OS-AAMLP shows better localization performance than the conventional AAMLP even in the cases having very low discrimination property between a laser spot area and a corresponding background area. Moreover, the proposed I/OS-AAMLP shows robust performance in terms of a threshold of error for deciding whether a trained non-laser spot area or a laser spot area. The conventional AAMLP shows more sensitive performance according to change of a threshold value. In the experiments, learning rates η_1 and η_2 for I/OS-AAMLP are 0.005 and 0.000001, respectively. As shown in Table 1, for both the conventional AAMLP and the proposed I/OS-AAMLP, better performance was obtained from the method considering PCA for dimension reduction and feature enhancement. Moreover, the proposed I/OS-AAMLP applying PCA shows the best performance among four different methods, which are due to the input-output sensitivity enhancement by the I/OS-AAMLP as well as feature enhancement by PCA. In addition, the I/OS-AAMLP can take far less computation time when it applies PCA for reducing dimension of input data than the model without applying PCA.

Table 1. Performance comparison of the proposed I/OS-AAMLP with the conventional AAMLP for laser spot localization

Data set No.	Relative error ratio				Laser spot detection accuracy performance			
	AAMLP (without PCA)	AAMLP (with PCA)	Proposed I/OS-AAMLP (without PCA)	Proposed I/OS-AAMLP (with PCA)	AAMLP (without PCA)	AAMLP (with PCA)	Proposed I/OS-AAMLP (without PCA)	Proposed I/OS-AAMLP (with PCA)
1	1.8785	3.4600	2.0348	3.7339	100%	100%	100%	100%
2	1.9222	1.5451	2.1785	1.6044	79.55%	79.55%	81.82%	81.82%
3	1.8606	2.8947	2.0886	2.9007	94.44%	100%	96.30%	100%
4	1.3843	3.9265	1.6179	2.9469	72.97%	100%	94.59%	100%
5	1.6751	2.2431	1.5679	2.5015	100%	50%	70%	100%
6	1.5649	7.8424	1.6108	9.4989	73.33%	100%	63.33%	100%
7	2.0005	1.5076	1.8297	1.5184	100%	85.71%	100%	90.48%
8	2.0693	14.0511	2.2017	15.2526	82.5%	100%	92.5%	100%
9	1.0545	1.7205	1.2826	1.9362	80.77%	100%	92.31%	100%
10	1.7333	7.4001	1.8360	7.8044	90%	100%	93.33%	100%
	Total				87.36%	91.53%	88.42%	97.23%

5 Conclusion

We propose a new AAMLP with increased input-output mapping sensitivity. The proposed laser pointer based interface system with a novel I/O sensitive EBP learning algorithm works well under illumination change and complex background images with similar color of laser pointer. The proposed algorithm can not only improve the laser pointer detection but also can be used for an inspection system like TFT-LCD manufacturing. Although experimental data is not sufficient in this work, our approach guides a new concept to solve difficult problems of the laser pointer detection which has illumination change, complex background, distortion of image sensor and other nonlinear environment variation.

Acknowledgments. This research was financially supported by the Ministry of Education, Science Technology(MEST) and Korea Institute for Advancement of Technology(KIAT) through the Human Resource Training Project for Regional Innovation.

References

1. Kirstein, C., Müller, H.: Interaction with a Projection Screen Using a Camera-tracked Laser Pointer. In: Int. Conf. on Multimedia Modeling, pp. 191–192. IEEE Computer Society Press, Los Alamitos (1998)
2. Olsen Jr., D.R., Nielsen, T.: Laser pointer interaction. In: SIGCHI, pp. 17–22 (2001)
3. Lapointe, J.-F., Godin, G.: On-Screen Laser Spot Detection for Large Display Interaction. In: IEEE Int. Work. on Haptic Audio Environments and their Applications, pp. 72–76 (2005)
4. Lim, G.W., Sharifi, F., Kwon, D.: Fast and Reliable Camera-tracked Laser Pointer System Designed for Audience. In: 5th Int. Conf. on Ubiquitous Robots and Ambient Intelligence, pp. 529–534 (2008)
5. Kim, N., Lee, S., Lee, J., Lee, B.: Laser Pointer Interaction System Based on Image Processing. Journal of Korea Multimedia Society 11(3), 373–385 (2008)
6. Baek, J., Cho, S.: Time jump in:long rising pattern detection in KOSPI200 future suing an auto-associative neural network. In: 8th International Conference on Neural Information Processing, pp. 160–165 (2001)
7. Ban, S., Lee, M., Yang, H.: A face detection using biologically motivated bottom-up saliency map model and top-down perception model. Neurocomputing 56, 475–480 (2004)
8. Bradski, G.R., Kaehler, A.: Learning OpenCV: Computer Vision with the OpenCV Library, pp. 236–237. O'Reilly Media, Inc., Sebastopol (2008)
9. Jolliffe, I.T.: Principal Component Analysis, 2nd edn. Series: Springer Series in Statistics, XXIX, 487, p. 28 illus. Springer, NY (2002)
10. Haykin, S.: Neural Networks: a comprehensive foundation, 2nd edn., pp. 156–252. Prentice Hall International, Inc., New Jersey (1998)

Improving Recurrent Neural Network Performance Using Transfer Entropy

Oliver Obst[1,2], Joschka Boedecker[3,4], and Minoru Asada[3,4]

[1] CSIRO ICT Centre, Adaptive Systems, P.O. Box 76, Epping, NSW 1710, Australia
[2] School of Information Technologies, The University of Sydney, NSW 2006, Australia
[3] Department of Adaptive Machine Systems, Osaka University, Suita, Osaka, Japan
[4] JST ERATO Asada Synergistic Intelligence Project, Suita, Osaka, Japan
Oliver.Obst@csiro.au
joschka.boedecker@ams.eng.osaka-u.ac.jp
asada@ams.eng.osaka-u.ac.jp

Abstract. Reservoir computing approaches have been successfully applied to a variety of tasks. An inherent problem of these approaches, is, however, their variation in performance due to fixed random initialisation of the reservoir. Self-organised approaches like intrinsic plasticity have been applied to improve reservoir quality, but do not take the task of the system into account. We present an approach to improve the hidden layer of recurrent neural networks, guided by the learning goal of the system. Our reservoir adaptation optimises the information transfer at each individual unit, dependent on properties of the information transfer between input and output of the system. Using synthetic data, we show that this reservoir adaptation improves the performance of offline echo state learning and Recursive Least Squares Online Learning.

Keywords: Machine learning, recurrent neural network, information theory, reservoir computing, guided self-organisation.

1 Introduction

Reservoir Computing (RC) is a recent paradigm in the field of recurrent neural networks (for a recent overview, see [1]). RC computing approaches have been employed as mathematical models for generic neural microcircuits, to investigate and explain computations in neocortical columns (see e.g. [2]). A key element of reservoir computing approaches is the randomly constructed, fixed hidden layer – typically, only connections to output units are trained. Despite their impressive performance for some tasks (e.g. [3]), their fixed random connectivity can lead to significant variation in performance [4]. To address this issue, approaches like Intrinsic Plasticity (IP) [5, 6] can help to improve randomly constructed reservoirs. IP is based on the idea to maximise available information at each internal unit in a self-organised way by changing the behaviour of individual units. This is contrast to, for example, Hebbian learning [7], which strengthens connections between two units if their firing patterns are temporally correlated.

K.W. Wong et al. (Eds.): ICONIP 2010, Part II, LNCS 6444, pp. 193–200, 2010.

Both adaptation of individual units as well as adaptation of connections are phenomena that occur in biological units.

IP learning has been used as an approach to optimise reservoir encoding specific to the input of the network [6]. It is, however, *only* dependent on the input data, and does not take the desired output of the system into account, i.e., it is not guaranteed to lead to optimised performance with respect to the learning task of the network [4]. Ideally, we would like to retain the principle of a self-organised approach to optimise reservoirs, but to guide self-organisation [8] based on the overall learning goal.

The approach presented in this paper for the first time leads to a method that optimises the information transfer at each individual unit, dependent on properties of the information transfer between input and output of the system. The optimisation is achieved by tuning self-recurrent connections, i.e., the means to achieve this optimisation can be viewed as a compromise between Hebbian and IP learning. Using synthetic data, we show that this reservoir adaptation improves the performance of offline echo state learning, and is also suitable for online learning approaches like backpropagation-decorrelation learning [9] or recursive least squares (RLS, see e.g. [10]).

2 Echo State Networks

ESN provide a specific architecture and a training procedure that aims to solve the problem of slow convergence [11, 3] of earlier recurrent neural network training algorithms. ESN are normally used with a discrete-time model, i.e. the network dynamics are defined for discrete time-steps t, and they consist of inputs, a recurrently connected hidden layer (also called *reservoir*) and an output layer (see Fig. 1).

We denote the activations of units in the individual layers at time t by $\mathbf{u}_t$, $\mathbf{x}_t$, and $\mathbf{o}_t$ for the inputs, the hidden layer and the output layer, respectively. The matrices $\mathbf{w}^{\text{in}}$, $\mathbf{W}$, $\mathbf{w}^{\text{out}}$ specify the respective synaptic connection weights. Using $\mathrm{f}(x) = \tanh x$ as output nonlinearity for all hidden layer units, the network dynamics is defined as:

$$\mathbf{x}_t = \mathbf{f}(\mathbf{W}\mathbf{x}_{t-1} + \mathbf{w}^{\text{in}}\mathbf{u}_t) \tag{1}$$

$$\mathbf{o}_t = \mathbf{w}^{\text{out}}\mathbf{x}_t \tag{2}$$

The main differences of ESN to traditional recurrent network approaches are the setup of the connection weights and the training procedure. To construct an ESN, units in the input layer and the hidden layer are connected randomly. Only connections between the hidden layer and the output units are trained, usually with a supervised offline learning approach using linear regression. Here, the output weights $\mathbf{w}^{\text{out}}$ are calculated using the collection of desired output states $\mathbf{D}$, and the pseudoinverse of a matrix $\mathbf{S}$ collecting the states of the system over a number of steps as $\mathbf{w}^{\text{out}} = \mathbf{S}^{\dagger}\mathbf{D}$ (see [11] for details). An online learning procedure, like RLS, adapts the output weights while training input is fed into

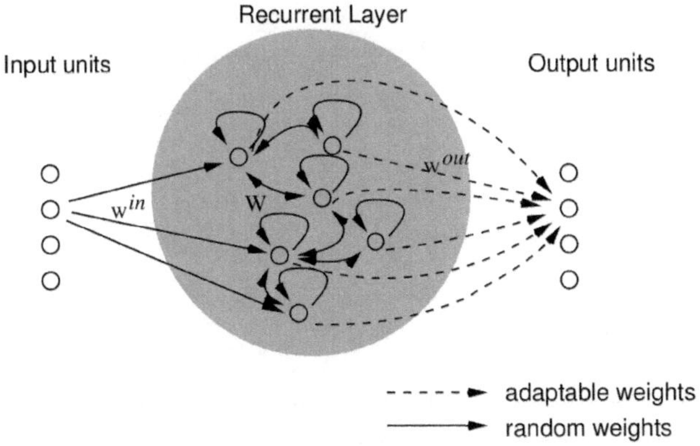

Fig. 1. In echo state networks, only output weights (represented by dashed lines) are trained, all other connections are setup randomly and remain fixed. The recurrent layer is also called a *reservoir*, analogously to a liquid, which has fading memory properties in response to perturbations (like e.g. ripples caused by a rock thrown into a pond).

the network, i.e., no states need to be collected. The RLS update rule can be described with the following set of equations:

$$\alpha_t = \mathbf{d}_t - \mathbf{w}^{\text{out}}_{t-1} \cdot \mathbf{x}_t \ , \tag{3}$$

$$\mathbf{g}_t = \mathbf{p}_{t-1} \cdot \mathbf{x}_t / (\lambda + {\mathbf{x}_t}^T \cdot \mathbf{p}_{t-1} \cdot \mathbf{x}_t) \ , \tag{4}$$

$$\mathbf{p}_t = (\mathbf{p}_{t-1} - \mathbf{g}_t \cdot {\mathbf{x}_t}^T \cdot \mathbf{p}_{t-1})/\lambda \ , \tag{5}$$

$$\mathbf{w}^{\text{out}}_t = \mathbf{w}^{\text{out}}_t + (\alpha_t \cdot \mathbf{g}^T_t) \ , \tag{6}$$

where α_t represents the *a priori error* vector between desired output $\mathbf{d}_t$ and current input, $\mathbf{p}_t$ the inverse of the autocorrelation, and λ is close to 1 and is an exponential forgetting factor. RLS has been applied to ESN learning in [12].

Even though the reservoir weights are randomly initialised and remain fixed, these connections cannot be completely random; they are typically designed to have the *echo state property*. The definition of the echo state property has been outlined in [11] and is summarised in the following section.

2.1 The Echo State Property

The Echo State Property is reflected in the following definition. In simple terms, the system has echo state property if different initial states converge to each other for all inputs. Consider a time-discrete recursive function:

$$\mathbf{x}_{t+1} = \mathbf{F}(\mathbf{x}_t, \mathbf{u}_{t+1}) \tag{7}$$

that is defined at least on a compact sub-area of the vector-space $\mathbf{x} \in R^n$, with n the number of internal units. The $\mathbf{x}_t$ are to be interpreted as internal states and $\mathbf{u}_t$ is some external input sequence, i.e. the stimulus.

Definition 1. *Assume an infinite stimulus sequence:* $\bar{\mathbf{u}}^\infty = \mathbf{u}_0, \mathbf{u}_1, \ldots$ *and two random initial internal states of the system* $\mathbf{x}_0$ *and* $\mathbf{y}_0$*. From both initial states* $\mathbf{x}_0$ *and* $\mathbf{y}_0$ *the sequences* $\bar{\mathbf{x}}^\infty = \mathbf{x}_0, \mathbf{x}_1, \ldots$ *and* $\bar{\mathbf{y}}^\infty = \mathbf{y}_0, \mathbf{y}_1, \ldots$ *can be derived from the update equation Eq. (7) for* $\mathbf{x}_{t+1}$ *and* $\mathbf{y}_{t+1}$*. The system* $F(\cdot)$ *will have the echo state property if, independently of the set* $\mathbf{u}_t$*, for any* $(\mathbf{x}_0, \mathbf{y}_0)$ *and all real values* $\epsilon > 0$*, there exists a* $\delta(\epsilon)$ *for which* $d(\mathbf{x}_t, \mathbf{y}_t) \leq \epsilon$ *for all* $t \geq \delta(\epsilon)$*, where* d *is a square Euclidean metric.*

3 Transfer Entropy

To improve the reservoir based on the learning goal, we are interested in detecting the characteristics of the information transfer between input and desired output of the system. Transfer Entropy [13] is an information-theoretic measure for the information provided by a source about the next state of the destination which was not already contained in its own history. It is similar to mutual information [see e.g. 2], but asymmetric (i.e. directed), and takes the dynamics of information transfer into account. The transfer entropy from a source node Y to a destination node X is the mutual information between previous l states of the source $y_n^{(l)}$ and the next state of the destination x_{n+1},

$$T_{Y \to X} = \lim_{k,l \to \infty} \sum_{\mathbf{u}_n} p(x_{n+1}, x^{(k)}, y_n^{(l)}) \log_2 \frac{p(x_{n+1} | x_n^{(k)}, y_n^{(l)})}{p(x_{n+1} | x_n^{(k)})}. \tag{8}$$

where $\mathbf{u}_n$ is the state transition tuple $(x_{n+1}, x^{(k)}, y_n^{(l)})$.

For our purposes, $T_{Y \to X}(k, l)$ represents finite k, l approximation.

4 Reservoir Dynamics

In the following, we consider the case where we have an one-dimensional input vector $\mathbf{u}$. The learning goal for our system is a one step-ahead prediction of an one-dimensional output vector $\mathbf{v}$. Departing from the usual reservoir dynamics described above, we use

$$\mathbf{x}(k+1) = \mathrm{diag}(\mathbf{a})\mathbf{W}\mathbf{y}(k) + (\mathbf{I} - \mathrm{diag}(\mathbf{a}))\mathbf{y}(k) + \mathbf{w}^{\mathrm{in}}\mathbf{u}(k) \tag{9}$$

$$\mathbf{y}(k+1) = \mathbf{f}(\mathbf{x}(k+1)) \ , \tag{10}$$

where x_i, $i = 1, ..., N$ are the neural activations, $\mathbf{W}$ is the $N \times N$ reservoir weight matrix, $\mathbf{w}^{\mathrm{in}}$ the input weight, $\mathbf{a} = (\alpha_1, ..., \alpha_N)^T$ a vector of local decay factors, $\mathbf{I}$ is the identity matrix, and k the discrete time step. In this work, we use $\mathbf{f}(\mathbf{x}) = \tanh(\mathbf{x})$.

The α_i represent a decay factor, or coupling of a unit's previous state with the current state; they are computed as:

$$\alpha_i = \frac{2}{1 + m_i} \ , \tag{11}$$

where m_i represents the memory length of unit i ($m_i \in \{1, 2, 3, ...\}$). All memory lengths are initialised to $m_i = 1$, so that $\alpha_i = 1$, i.e. the reservoir has the usual update rule. Increasing individual m_i during an adaptation will increase the influence of a unit's past states on its current state.

5 Adaptation of Information Transfer

Adaption of the reservoir to the learning goal introduces two extra steps to the learning procedure. In a first step, we determine the required history size l to maximise the information transfer from input $\mathbf{u}$ to output $\mathbf{v}$, i.e. a first idea may be to look for a value

$$l_{\max} = \arg\max_{l} \ \mathrm{T}_{\mathbf{u}\rightarrow\mathbf{v}}(1, l) \ .$$

Using increasingly larger history sizes may, however, always increase the transfer entropy (by possibly smaller and smaller values). To optimise the information transfer, we will instead be looking for the smallest value $\hat{l}$ that does not increase the transfer entropy $\mathrm{T}_{\mathbf{u}\rightarrow\mathbf{v}}(1, \hat{l}-1)$ by more than a threshold ϵ, i.e.

$$\mathrm{T}_{\mathbf{u}\rightarrow\mathbf{v}}(1, \hat{l}+1) \leq \mathrm{T}_{\mathbf{u}\rightarrow\mathbf{v}}(1, \hat{l}) + \epsilon \quad \textbf{and} \tag{12}$$

$$\mathrm{T}_{\mathbf{u}\rightarrow\mathbf{v}}(1, l) > \mathrm{T}_{\mathbf{u}\rightarrow\mathbf{v}}(1, l-1) + \epsilon \quad \textbf{for all } l < \hat{l} \ . \tag{13}$$

From this first step, we learn the contribution of the size of the input history to the desired output (the learning goal of the system): some input-output pairs may require a larger memory of the input history to be informative about the next output state, other outputs may be more dynamic, and be dependent on the current input state only.

We take this information into the second step, which consists of a pre-training of the reservoir. Here, the local couplings of the reservoir units are adapted so that the transfer entropy from the input of each unit to its respective output is optimised *for the particular input history length* $\hat{l}$. The idea behind this step is to locally adjust the memory at each unit to approximate the required memory for the global task of the system. Pre-training is done in epochs of length ℓ over the training data. Over each epoch θ, we compute, for each unit i, the transfer entropy from activations $x_i^{(\ell)}$ to output $y_i^{(\ell)}$:

$$\mathrm{te}_i^{\theta} = \mathrm{T}_{x_i^{(\ell)}\rightarrow y_i^{(\ell)}}(1, \hat{l}). \tag{14}$$

If the information transfer during the current epoch θ exceeds the information transfer during the past epoch by a threshold (i.e., $\mathrm{te}_i^{\theta} > \mathrm{te}_i^{\theta-1} + \epsilon$), the local memory length m_i is increased by one. Likewise, if $\mathrm{te}_i^{\theta} < \mathrm{te}_i^{\theta-1} - \epsilon$, the local memory length is decreased by one, down to a minimum of 1.

After each epoch, all m_i and α_i are adapted according to this rule, and used to compute activations over the next epoch. Once the training data is exhausted, pre-training of the reservoir is finished and the α_i are fixed. For the subsequent training we compute the output weights by linear regression with data as used in the pre-training. In additional experiments, we use RLS online learning, where adaptation and training of output weights were run in the same loop.

6 Experimental Results

We tested our method using a one-step ahead prediction of unidirectionally coupled maps, and a one-step ahead prediction of the Mackey-Glass time series.

6.1 Prediction of Autoregressive Coupled Processes

As first experiments we studied our approach using a one-step ahead prediction of two unidirectionally coupled autoregressive processes:

$$u_{t+1} = 0.7\, u_t + 0.7 \cos(0.3t) + n_t^x(0, \sigma^2) \text{ and} \tag{15}$$

$$v_{t+1} = 0.7\, v_t + e\, u_{t-\omega+1} + n_t^y(0, \sigma^2) \ , \tag{16}$$

where the parameter $e \in [0, 1]$ regulates the coupling strength, $\omega \in \{0, 1, 2, ...\}$ an order parameter, and $n_t^x(0, \sigma^2)$ and $n_t^y(0, \sigma^2)$ are independent Gaussian random processes with zero mean and standard deviation $\sigma = 0.4$. For each trial, we generated time series $\mathbf{u}$ and $\mathbf{v}$ (random initial conditions; time series divided into 10000 values for training and 1200 values for testing; the first 200 values of both training and testing were used to prime the reservoir), where the task of our system was a one-step ahead prediction of $\mathbf{v}$ using $\mathbf{u}$. The reservoir was initialised using a random, sparse recurrent weight matrix ($|\lambda| = 0.95$), with 40 internal units. Figure 2 (a) displays the mean square errors of the prediction over the test data for different coupling strengths and fixed $\omega = 0$ for both echo state learning with and without adaptation of information transfer in the reservoir. All values are averaged over 50 trials; for each individual trial the same reservoir and time series have been used once with and without adaptation. The prediction using the reservoir adaptation is better over almost the entire range of e, with the improvement becoming more significant as the influence of the input time series becomes larger. Figure 2 (b) is a plot of the mean square error for different ω using a fixed coupling of $e = 0.75$. In all but one cases the reservoir adaptation improves results.

6.2 Prediction of Mackey-Glass Time Series

A further experiment was prediction of the widely used Mackey-Glass time series (see e.g. [11, 14, 6]) with parameter τ set to 17. The first task using this time series was again a one-step ahead prediction using a reservoir size of 40 units. For this task, the transfer entropy between input and output time series is maximised already for smaller values of ℓ compared to our first experiment (ℓ was typically around 2 for Mackey-Glass one-step ahead prediction), i.e., the information used from the previous state to predict the next state is already quite high. The

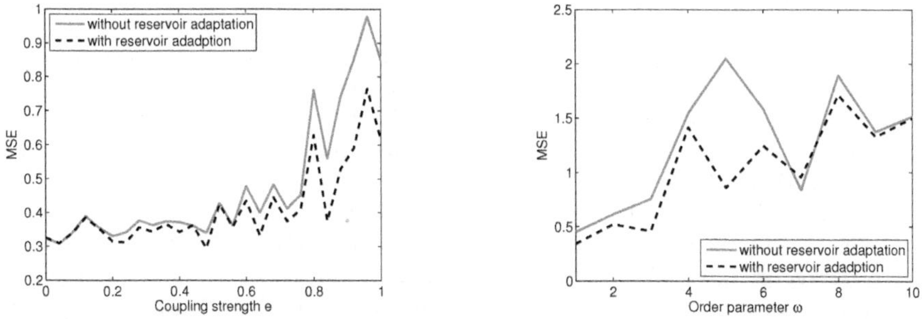

Fig. 2. (a) Left: mean square errors of the prediction over the test data for different coupling strengths and fixed $\omega = 0$. (b) Right: mean square error for different ω using a fixed coupling of $e = 0.75$. Reported results are averages over 50 runs.

reservoir adaptation lead to an average improvement of the MSE (averaged over 50 runs) from $0.4530 \cdot 10^{-6}$ to $0.0751 \cdot 10^{-6}$. Individually, in 48 of the 50 runs, the same reservoir performed better with adaptation than without adaptation.

Instead of offline learning, we also used RLS in the same loop with our reservoir adaptation. To less consider data from earlier stages of the adaptation, we used a forgetting factor $\lambda = 0.995$. Again, the adaptation improved performance, from $9.1 \cdot 10^{-6}$ to $7.2 \cdot 10^{-6}$; a fine-tuning of λ may further improve the results.

7 Conclusions

We presented an information-theoretic approach to reservoir optimisation. Our approach uses a local adaptation of a units internal state, based on properties of the information transfer between input and desired output of the system. The approach has shown to improve performance in conjunction with offline echo-state regression, as well as with RLS online learning. In our experiments we have used only a small number of internal units – our goal was to show the capability of our approach compared to standard echo state learning. In first additional experiments (not reported here), we have shown that for a larger number of units our adaptation leads to an even larger improvement compared to echo state learning without adaptation. A further investigation of statistical properties of coding in the reservoir obtained by our adaptation may provide useful insights. Moreover, other information-theoretic measures such as the active information storage [15] may be useful to further improve the local adaptation rule.

Acknowledgments. The Authors thank the Australian Commonwealth Scientific and Research Organization's (CSIRO) Advanced Scientific Computing group for access to high performance computing resources used for simulation and analysis.

References

[1] Lukosevicius, M., Jaeger, H.: Reservoir computing approaches to recurrent neural network training. Computer Science Review 3(3), 127–149 (2009)
[2] Maass, W., Natschläger, T., Markram, H.: Real-time computing without stable states: A new framework for neural computation based on perturbations. Neural Computation 14(11), 2531–2560 (2002)
[3] Jaeger, H., Haas, H.: Harnessing Nonlinearity: Predicting Chaotic Systems and Saving Energy in Wireless Communication. Science 304(5667), 78–80 (2004)
[4] Boedecker, J., Obst, O., Mayer, N.M., Asada, M.: Initialization and self-organized optimization of recurrent neural network connectivity. HFSP Journal 3(5), 340–349 (2009)
[5] Triesch, J.: A gradient rule for the plasticity of a neuron's intrinsic excitability. In: Duch, W., Kacprzyk, J., Oja, E., Zadrożny, S. (eds.) ICANN 2005. LNCS, vol. 3696, pp. 65–70. Springer, Heidelberg (2005)
[6] Steil, J.J.: Online reservoir adaptation by intrinsic plasticity for backpropagation-decorrelation and echo state learning. Neural Networks 20(3), 353–364 (2007)
[7] Hebb, D.O.: The organization of behavior: a neuropsychological theory. Lawrence Erlbaum Associates, Mahwah (1949)
[8] Prokopenko, M.: Guided self-organization. HFSP Journal 3(5), 287–289 (2009)
[9] Steil, J.J.: Backpropagation-decorrelation: Recurrent learning with O(N) complexity. In: Proceedings of the International Joint Conference on Neural Networks (IJCNN), vol. 1, pp. 843–848 (2004)
[10] Hayes, M.H.: Chapter 9.4 Recursive Least Squares. In: Statistical Digital Signal Processing and Modeling. Wiley, Chichester (1996)
[11] Jaeger, H.: The "echo state" approach to analysing and training recurrent neural networks. Technical Report 148, GMD – German National Research Institute for Computer Science (2001)
[12] Jaeger, H.: Adaptive nonlineaer systems identification with echo state networks. In: Advances in Neural Information Processing Systems, pp. 609–615 (2003)
[13] Schreiber, T.: Measuring information transfer. Physical Review Letters 85(2), 461–464 (2000)
[14] Hajnal, M., Lőrincz, A.: Critical echo state networks. In: Kollias, S.D., Stafylopatis, A., Duch, W., Oja, E. (eds.) ICANN 2006. LNCS, vol. 4131, pp. 658–667. Springer, Heidelberg (2006)
[15] Lizier, J.T., Prokopenko, M., Zomaya, A.Y.: Detecting non-trivial computation in complex dynamics. In: Almeida e Costa, F., Rocha, L.M., Costa, E., Harvey, I., Coutinho, A. (eds.) ECAL 2007. LNCS (LNAI), vol. 4648, pp. 895–904. Springer, Heidelberg (2007)

Design of Artificial Neural Networks Using Differential Evolution Algorithm

Beatriz A. Garro[1], Humberto Sossa[1], and Roberto A. Vázquez[2]

[1] Centro de Investigación en Computación – IPN
Av. Juan de Dios Batiz, esquina con Miguel de Othon de Mendizábal
Ciudad de México, 07738, México
[2] Escuela de Ingeniería – Universidad La Salle
Benjamín Franklin 47 Col. Condesa CP 06140 México, D.F
bgarrol@ipn.mx, hsossa@cic.ipn.mx, ravem@ipn.mx

Abstract. The design of an Artificial Neural Network (ANN) is a difficult task for it depends on the human experience. Moreover it needs a process of testing and error to select which kind of a transfer function and which algorithm should be used to adjusting the synaptic weights in order to solve a specific problem. In the last years, bio-inspired algorithms have shown their power in different non-linear optimization problems. Due to their efficiency and adaptability, in this paper we explore a new methodology to automatically design an ANN based on the Differential Evolution (DE) algorithm. The proposed method is capable to find the topology, the synaptic weights and the transfer functions to solve a given pattern classification problems.

1 Introduction

Differential Evolution (DE) is an algorithm based on the classical steps of the Evolutionary Computation. However, it does not use binary encoding as an ordinary genetic algorithm; DE uses real number vectors, [1]. It does not use a probability density function to self-adapt its parameters as an Evolution Strategy [2]. Instead, DE performs mutation based on the distribution of the solutions in the current population. In this way, search directions and possible step-sizes depend on the location of the individuals selected to calculate the mutation values [3].

A feed-forward artificial neural network (ANN) is a powerful tool widely used in the field of pattern recognition and time series analysis. However, despite their power in some practical problems, ANNs cannot reach an optimum performance in several non-linear problems. This fact is caused because the parameters, used during learning phase such as learning rate, momentums, among others, do not allow compute the best set of synaptic weights.

Several works that use evolutionary strategies for training ANNs have been reported in the literature. Refer for example to [4], [5] and [6]. In general, the authors present modified PSO algorithms as an alternative for training an ANN [7], [8] and [9]; however most of the research is focused only in the evolution of the synaptic weights and sometimes in the optimum selection of the number of neurons in hidden layers [10] and [11].

K.W. Wong et al. (Eds.): ICONIP 2010, Part II, LNCS 6444, pp. 201–208, 2010.

DE algorithm has been less used in this kind of work. For example, in [12] and [13] the authors proposed a modified DE algorithm for the training of a multilayer neural network. In [14] three neural networks' architectures with different training techniques are evaluated and trained with a DE algorithm applied to a forecasting weather problem.

In this paper we explore a new methodology to automatically design an ANN based on a DE algorithm. This research includes not only the problem of finding the optimal set of synaptic weights of an ANN but also its topology and the transfer functions for each neuron. In other words, given a set of inputs and desired patterns, the proposed method will be capable to find the best topology, the number of neurons, the transfer function for each neuron and the synaptic weights in order to design an ANN that can be used to solve a given problem. Finally, the proposed method is tested with several non-linear problems and compared against PSO and back propagation algorithms.

2 Basics on Feed-Forwards Neural Networks

A neural network is a massively parallel-distributed processor made up from simple processing units. Each value of an input pattern $\mathbf{A} \in \mathbb{R}^N$ is associated with its weight value $\mathbf{W} \in \mathbb{R}^N$, which is normally between 0 and 1. The output of the neurons will be then performed as,

$$y = f\left(\sum_{i=1}^{N} a_i w_i + \theta\right) \tag{1}$$

where $f(x)$ is the transfer function which generates the output from the neuron.

Basically, learning is a process by synaptic weights $\mathbf{W}$ and bias levels θ of a neural network are adapted through a continuing process based on a labeled set of training data made up of p input-output samples:

$$\mathbf{T}^{\xi} = \left\{\left(\mathbf{x}^{\xi} \in \mathbb{R}^N, \mathbf{d}^{\xi} \in \mathbb{R}^M\right)\right\} \forall \xi = 1,\ldots,p \tag{2}$$

where $\mathbf{x}$ is the input pattern and $\mathbf{d}$ the desired response.

Given the training sample $\mathbf{T}^{\xi}$, the requirement is to compute the free parameters of the neural network so that the actual output $\mathbf{y}^{\xi}$ of the neural network due to $\mathbf{x}^{\xi}$ is close enough to $\mathbf{d}^{\xi}$ for all ξ in a statistical sense. In this sense, we might use the mean-square error given in eq. 3 as the objective function to be minimized:

$$e = \frac{1}{p \cdot M} \sum_{\xi=1}^{p} \sum_{i=1}^{M} \left(d_i^{\xi} - y_i^{\xi}\right)^2 \tag{3}$$

One of the most commonly used supervised ANN model is feed-forward network that uses backpropagation (BP) learning algorithm [15-16] to minimize the objective function given by eq. 3.

3 Basics on Differential Evolution

In 1995 an adaptive and efficient scheme emerged: Differential Evolution (DE) algorithm, proposed by Kenneth Price and Rainer Storn [17]. Due to its exploration capacity over a search space of a given problem, the DE algorithm avoids staying in a local optimum. It has few parameters and it converges to the optimum faster than others evolutionary techniques (the solution's representation is given by vectors of real numbers). All these characteristics convert the DE in an excellent algorithm for optimization of a complex, non-differential and non-continuous problems. [18].

The algorithm consists in randomly choosing a target vector and a base vector, in addition, two different members of the population must be randomly chosen. In order to realize the mutation is necessary to do a subtraction between these last two vectors. And then, the result is multiplied by a constant parameter denoted by F. Immediately after, the result of this operation and the base vector, chosen at the beginning of the algorithm, are summed. This new vector is called the mutated vector. At once, it is realized the crossover operation which involves a comparison (variable by variable) between the mutated vector and the target vector, creating another vector called trial vector. The comparison consists of a simple rule: If a random number is either equal or higher than a crossover rate CR it is preserved the variable of the trial vector, otherwise is preserved the variable of the target vector. Finally the last step is the selection of the vector that has to generate the new population: the trial vector or the target vector. Only the vector with the best fitness is selected. The pseudo code of "DE/rand/1/bin" is shown in the next algorithm, adapted from [3].

```
1. Randomly select two vectors from the current generation.
2. Use these two vectors to compute a difference vector.
3. Multiply the difference vector by weighting factor F.
4. Form the new trial vector by adding the weighted difference vec-
   tor to a third vector randomly selected from the current popula-
   tion.
```

4 Design of an ANN Using DE

In this section it is described how given a set of patterns $\mathbf{T}$, the synaptic weights, the architecture and the transfer function of each neuron of an ANN can be automatically adjusted by means of a DE algorithm.

In order to achieve this goal, we codify this information as follows: each vector will be represented by a matrix $\mathbf{x} \in \mathbb{R}^{(MNN+2)\times MNN}$ composed by three parts: topology, synaptic weights and transfer function.

The topology of the ANN is codified based on the binary square matrix representation of a graph $\mathbf{x}$ where each component x_{ij} represents the connections between neuron i and neuron j when $x_{ij} = 1$. However, instead of evolving this binary information we decided to codify this information into its decimal base value and then evolve it. For example suppose that next binary code "01101" represents the connections of a i-th neuron to 5 neurons where only neurons 2, 3, and 5 are connected to i. This binary code is transformed into its decimal base value resulting in "13", which will be the number that we will evolve instead of the binary value. This scheme is much faster. Instead of evolving a string of MNN bits, we evolve only a decimal base number. The synapses weights of the ANN are codified based on the square matrix representation of a graph $\mathbf{x}$ where each component x_{ij} represents the synaptic weight between neuron i and neuron j. Finally, the transfer function for each neuron will be represented by an integer from 0 to 5 representing one of the 6 transfer functions used in this research: logsig, tansig, sin, gaussina, linear and hard limit. These functions were selected because they are the most popular and useful transfer functions in several kinds of problems. Figure 1 shows the individual representation of the ANN.

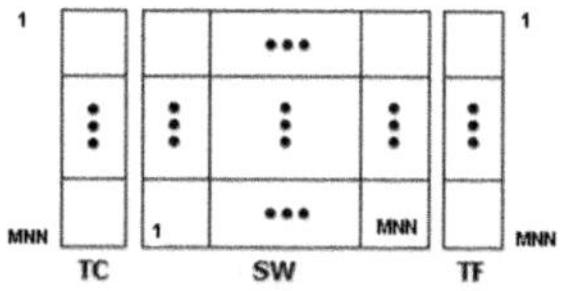

Fig. 1. Individual representation of an ANN. Note that TC represents the topology of the network, SW represents the synaptic weights and TF represents TF of each neuron.

Another alternative to codify the topology could be based on the synaptic weights: values less than a given threshold means no connection. However, determining the best threshold could bring other kind of problems; that is why in this paper we will not focus in this scheme.

The fitness function which measures the performance of an individual is given by eq. 3 where the output y_i of the ANN is computed as follows (note that the restriction $j < i$ is used to avoid the generation of cycles in the ANN):

1) For the first $nfin$ neurons, the output $o_i = a_i$.
2) For each neuron n_i, $i = nfin, \ldots, MNN$.
   a) Get connections by using individual $\mathbf{x}_{1,i}$.
   b) For each neuron $j < i$ connected to n_i, $o_i = f(o)$ where f is a transfer function given by individual $\mathbf{x}_{nfout,i}$ and o is compute using eq. 1.
3) Finally, $y_i = o_i$, $i = MNN - nfouy, \ldots, MNN$.

5 Experimental Results

In order to evaluate the accuracy of the proposed method, several experiments were performed using 4 data sets [19]: XOR, iris plant, wine, and breast cancer.

All data sets were partitioned into two sets: a training set and a testing set. For the iris plant data set, the first 120 examples were used for the training set, and the remaining 30 examples for the testing set. For the wine set, the first 90 examples were used for training set, and the remaining 89 examples for the testing set. For the breast cancer, the first 600 examples were used for training set, and the remaining 83 for testing set. For the XOR problem, the first 4 examples of the 2-D version were used for training, and noisy versions of these examples were used for the testing set.

The input features of all data set were rescaled in a range between $[0,1]$. The outputs were encoded encoded by the 1-to- c representations for c classes.

The parameters for the DE algorithm were set as follows: population size NP was set to 50, the number of generations was set to 2000, the population was initialized in the range $[-5,5]$, $CR = 0.9$, and $F = rand[0.3, 0.7]$.

Ten expertiments over each data set using the same parameters were performed. It is important to notice that the topology, the transfer functions for each neuron and the set of synaptic weights for each ANN were automatically determined by the DE algorithm. Transfer functions were labeled as *logsig* (LS), *tansig* (HT), *sin* (SN), *radbas* (GS), *pureline* (LN) and *hard limit* (HL).

Fig. 2 shows 2 ANNs generated with the proposed method for the XOR problem. Something important to metion about this set of experimets is that most of the generated ANNs had the same topology and transfer function. Fig. 3 shows 2 of the 10 ANNs designed by the basic DE algorithm. As you can appreciate, the ANNs generated with DE algorithm use different transfer functions and the connections between neurons are completely different. Fig. 4 shows 2 ANNs generated with the proposed method for the wine data set. Finally, 2 of the 10 ANNs obtained by the basic DE algorithm for the breast cancer data set are shown in Fig. 5.

It is worth mentioning that topologies of the ANNs designed with the proposed method are different to the traditional feed-forward; the topologies obtained present lateral connections and connections between input and output layers. Moreover, in Fig. 5(a) we can observe that one of the features from the input pattern (neuron 7) does not contribute to the output of the ANN. This effect can be seen as a reduction of the dimensionality.

Table 1 shows the average MSE found by four algorithms during the design of the ANN (topololy, synaptic weights and transfer functions): Basic PSO, second generation of PSO (SGPSO), a modified PSO (MPSO) algorithm [11] and the basic DE algorithm. We can observe that the proposed methodology provides better results using the DE algorithm under the same conditions during training phases as equal as during testing phase. Compared against BP algorithm for an ANN composed by 3 layers, learning rate was of 0.1 for the same data sets and 2000 epochs. The proposed method provided comparable results. However, the advantage of our metodology is that we do not need to apriori select the architecture, transfer function and learning algorithm of the ANN; althoght the proposed method seems to require more epochs (2000x50) than BP (2000), it automatically designs the ANN by itself.

Table 2 shows the classification error achieved by the proposed methodology using DE and MPSO which was the PSO algorithm that provides the best results.

Based on the previous results we can consider DE algorithms as an alternative to automatically design an ANN using only a labeled set of training data. Furthermore, if we modify the individual coding scheme, the proposed methodology could generated other kinds of ANN as feed-forward networks (without lateral conections), radial basis NN and even recurrent NN.

Table 1. Comparision of different methods in terms of the minimum square error (MSE)

DB	Basic PSO		2GPSO		MPSO		BP		DE/rand/1/bin	
	Tr. Er.	Te. Er.	Tr. Er.	Te. Er.	Tr. Er.	Te. Er.	Tr. Er.	Te. Er.	Tr. Er.	Te. Er.
XOR	0	0	0	0	0	0	0.0097	0.011	6.1E-9	0.0120
Iris	0.202	0.237	0.182	0.092	0.057	0.173	0.0075	0.128	0.028	0.0146
Wine	0.233	0.305	0.276	0.058	0.076	0.295	0.0001	0.016	0.072	0.0748
Breast Cancer	0.032	0.018	0.325	0.013	0.035	0.316	0.0085	0.013	0.021	0.0068

Table 2. Average accuracy of the proposed method in terms of the classification error (CER)

DB	MPSO		DE/rand/1/bin	
	Tr. Er.	Te. Er.	Tr. Er.	Te. Er.
XOR	0.025	0.025	0	0.025
Iris	0.043	0.0165	0.0425	0.0033
Wine	0.201	0.268	0.0811	0.0764
Breast Cancer	0.032	0.014	0.0225	0.0036

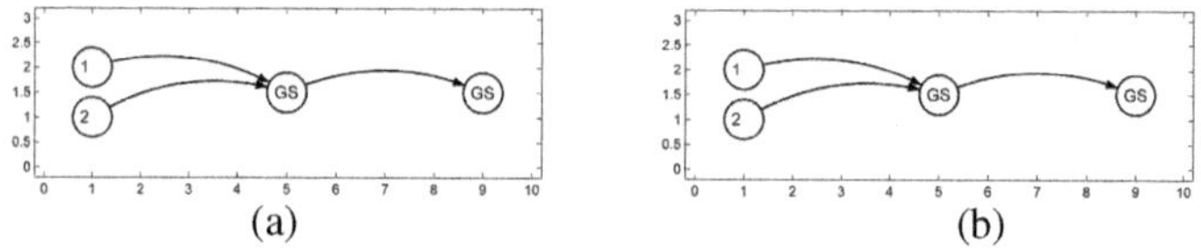

Fig. 2. Two topologies obtained with their transfer function for each neuron using DE/rand/1/bin algorithm for the XOR problem

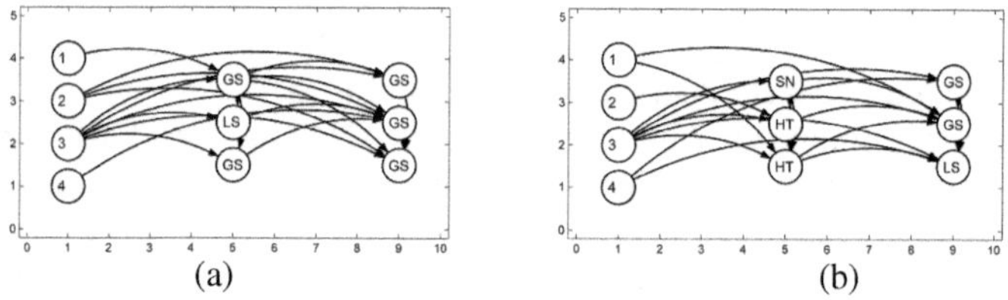

Fig. 3. Two topologies obtained with their transfer function for each neuron using DE/rand/1/bin algorithm for the Iris database

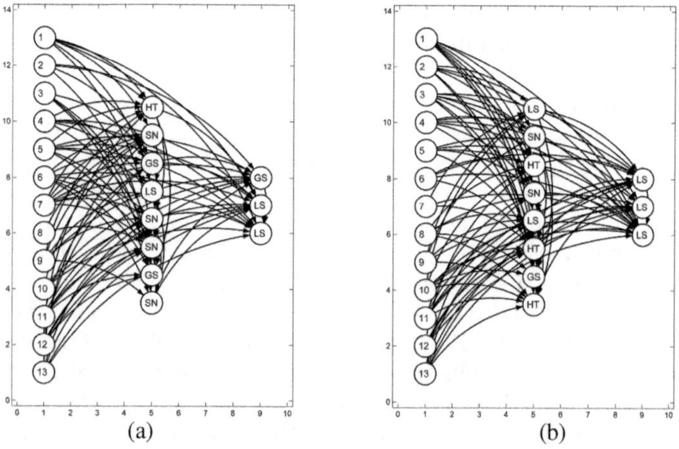

Fig. 4. Two topologies obtained with their transfer function for each neuron using DE/rand/1/bin algorithm for the Wine database

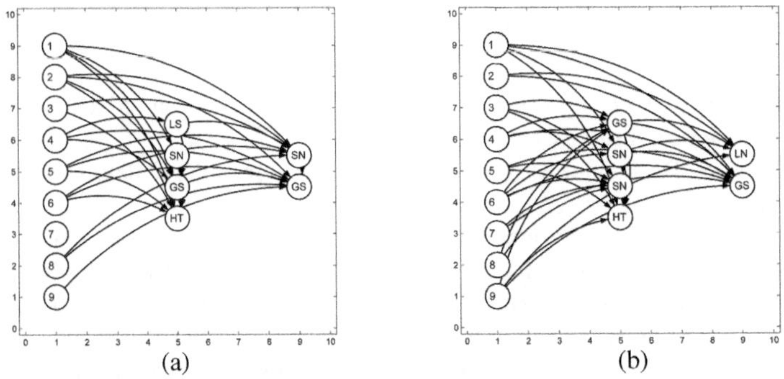

Fig. 5. Two topologies obtained with their transfer function for each neuron using DE/rand/1/bin algorithm for the Cancer database

6 Conclusions

In this paper a methodology to automatically design an ANN was proposed. This new alternative is based on bio-inspired algorithms. Particularly in this paper, the Differential Evolution (DE) algorithm was addopted. The design of a neural network can be seen as an optimization problem, which consists on finding the best values of the synapses, the best topology and the best transfer functions for each neuron which minimize an error function. For that reason, DE algorithm is suitable to automaticly design the ANN bases on the optimization of an error function (minimization of an objective function).

The accuracy of the proposal was tested using several non-linear problems and the results show a clear advantage over traditional schemes which involve the selection of a learning algorithm, a topology and the transfer functions. Compared against BP our proposed method provides comparable results. Moreover, the results achieved using the proposed methodology combined with DE were better compared to those obtained with PSO based algorithms. As was observed in the above figures, the proposed method generates different topologies with different transfer functions and in some cases is posible to reduce the dimentionality of the input patterns.

Nowadays we are studying other kind of error measures such as fuzzy classification error [20] in order to compare different algorithms and prove the robustness of the methodology. On the other hand, we are designing a new objective function based on these error measures. Moreover, we are implementing other bioinspired techniques such as bee colony optimization to automatically desing an ANN.

Acknowledgements. Authors thank SIP-IPN under grant 20100468 and COFAA for the economical support. They also thank the European Union, the European Commission and CONACYT for the economical support. This paper has been prepared by economical support of the European Commission under grant FONCICYT 93829. The content of this paper is an exclusive responsibility of the CIC-IPN and it cannot be considered that it reflects the position of the European Union. We thank also the reviewers for their comments for the improvement of this paper.

References

[1] Goldberg, D.E.: Genetic Algorithms in Search, Optimizationand Machine Learning. Addison-Wesley Publishing Co., Reading (1989)
[2] Schwefel, H.-P.: Evolution and Optimization Seeking. John Wiley & Sons, Chichester (1995)
[3] Mezura-Montes, E., Velazquez-Reyes, J., Coello Coello, C.A.: A Comparative Study of Differential Evolution Variants for Global Optimization. In: GECCO (2006)
[4] Tejen, S., et al.: A Hybrid Artificial Neural Networks and Particle Swarm Optimization for Function Approximation. In: ICIC, vol. 4, pp. 2363–2374 (2008)
[5] Chau, K.W.: Application of a PSO-based neural network in analysis of outcomes of construction claims. Automation in Construction 16, 642–646 (2007)
[6] Chatterjee, A., et al.: A Particle Swarm Optimized Fuzzy-Neural Network for Voice Controlled Robot Systems. IEEE Trans. on Ind. Electronics 52, 1478–1489 (2005)
[7] Wang, Z., et al.: Particle Swarm Optimization and Neural Network Application for QSAR. In: Proceedings 18th Parallel and Distributed Processing Symposium (2004)
[8] Zhao, L., Yang, Y.: PSO-Based Single Multiplicative Neuron Model for Time Series Prediction. Expert Systems with Applications 36, 2805–2812 (2009)
[9] Da, Y., Ge, X.R.: An improved PSO–based ANN with simulated annealing technique. Neurocomput. Lett. 63, 527–533 (2005)
[10] Yu, J., Xiand, L., Wang, S.: An Improved Particle Swarm Optimization for Evolving Feedforward Artificial Neural Networks. Neural Processing Letters 26, 217–231 (2007)
[11] Garro, B.A., Sossa, H., Vazquez, R.A.: Design of Artificial Neural Networks using a Modified Particle Swarm Optimization Algorithm. In: Proc. IEEE Int. Joint Conf. Neural Networks, pp. 938–945 (2009)
[12] Ilonen, J., Kamarainen, J.-K., Lampinen, J.: Differential Evolution Training Algorithm for Feed-Forward Neural Networks. Neural Processing Letters 17(1), 93–105 (2003)
[13] Guiying, N., Yongquan, Z.: A Modified Differential Evolution Algorithm for Optimization Neural Network. In: Proceedings of ISKE (2007)
[14] Abdul –Kader, H.M.: Neural Networks Training Based on Differential Evolution Algorithm Compared with Other Architectures for Weather Forecasting34. IJCSNS 9(3) (March 2009)
[15] Anderson, J.A.: Introduction to Neural Networks. MIT Press, Cambridge (1995)
[16] Werbos, P.J.: Backpropagation through time: What it does and how to do it. Proc. IEEE 78, 1550–1560 (1990)
[17] Storn, R., Price, K.: Differential Evolution - a Simple and Efficient Adaptive Scheme for Global Optimization over Continuous Spaces, TR-95-012, ICSI (1995)
[18] Price, K.V., Storn, R.M., Lampinen, J.A.: Differential evolution: a practical approach to global optimization. Springer, Heidelberg (2005)
[19] Murphy, P.M., Aha, D.W.: UCI repository of machine learning databases. Dept. Inf. Comput. Sci., Univ. California, Irvine, CA (1994)
[20] Mendis, B.S.U., Gedeon, T.D.: A comparison: Fuzzy Signatures and Choquet Integral. IEEE Congress on Computational Intelligence, 1464–1471 (2008)

ESNs with One Dimensional Topography

N. Michael Mayer, Matthew Browne, and Horng Jason Wu

Nat'l Chung Cheng University,
168 University Road,
Min-Hsiung, Chia-Yi,
Taiwan
nmmayer@gmail.com

Abstract. In this paper the standard Echo State approach is combined with a topography, i.e. it is assigned with a position which implies certain constraints of the mutual connectivity between these neurons. The overall design of the network allows certain neurons to process new information earlier than others. As a consequence the connectivity of the trained output layer can be analyzed; conclusions can be drawn regarding which reservoir depth is sufficient to process the given task. In particular we look at connection strengths of different locations of the reservoir as a function of the test error which can be influenced by using ridge regression.

Keywords: Reservoir computing, topographic maps, recurrent neural networks.

1 Introduction

Echo state networks (ESNs) have been an interesting subject for investigations in the field of recurrent neural networks (RNN) in recent years [3]. In this type of RNN the connectivitiy in the lower layers usually are chosen in a random fashion, learning is done in a single layer similar to some types of preceptrons and support vector machines [4]. It is no surprise that although in principle the network is functional for almost all kinds of connectivity, certain schemes (i.e. ortho-normal matrices in the recurrent layer) perform significantly better then other, in particular for a given input statistics (see [1] for a general investigation of different non-topological connectivity restrictions).

One possible version of such a scheme is to assign locations on a to the neurons and restrict the connectivity in such a way that only nearby located neurons can connect with each other. This concept has been successfully applied in various approaches (e.g. [5]). In the following a similar concept is discussed in the framework of reservoir computers. The intention is have a look in which way the topography affects the functionality and how it can be used for the analysis.

Deng and Zhang [2] investigate small world connectivity in ESNs which can be seen as a very special type of topography. Naturally, ESNs with certain topographies occur in distributed ESNs [7]. In this work a much simpler example

K.W. Wong et al. (Eds.): ICONIP 2010, Part II, LNCS 6444, pp. 209–216, 2010.

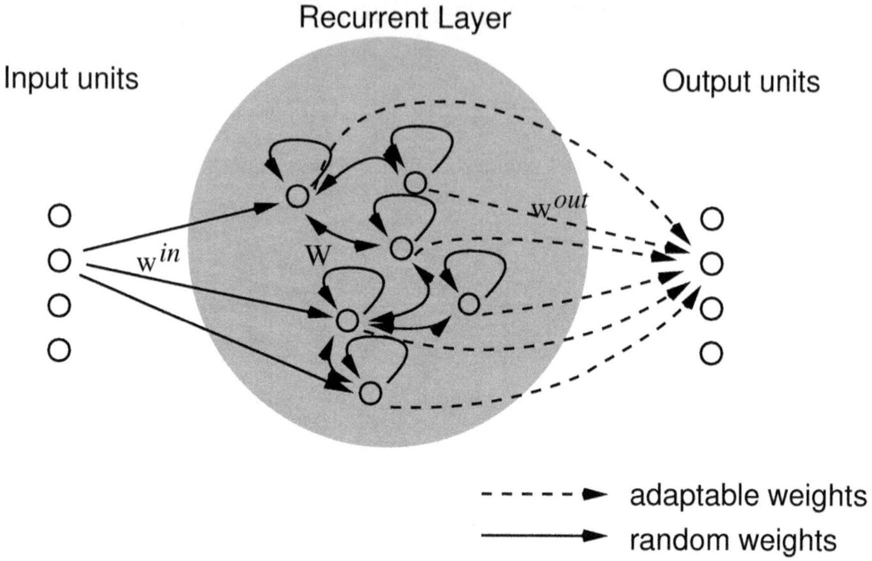

Fig. 1. ESN networks: Principle setup

of a topology is used: A one dimensional chain, where the input is fed into the reservoir in at the beginning of the chain, and propogates temporally to the end. Here it is of particular interest to check at which point the highest weights to the output layer result from the learning algorithm. It is also interesting in which way the weights relate to the available information.

In the next section the general design of ESNs is described, then the special restrictions for the one dimensional connectivity is introduced. A detailed description of the simulation details follows. Finally results a reported and discussed.

2 Model

2.1 Echo State Networks

Echo State Networks (ESN) are an approach to address the problem of slow convergence in recurrent neural network learning. ESN consist of three layers (see Fig. 1): a) an input layer, where the stimulus is presented to the network; b) a randomly connected recurrent hidden layer; and c) the output layer. Connections in the output layer are trained to reproduce the training signal. The network dynamics is defined for discrete time-steps t, with the following equations:

$$\mathbf{x}_{\mathsf{lin},t+1} = \mathbf{W}\mathbf{x}_t + \mathbf{w}^{in}\mathbf{u}_t \quad (1)$$

$$\mathbf{x}_{t+1} = \tanh\left(\mathbf{x}_{\mathsf{lin},t+1}\right) \quad (2)$$

$$\mathbf{o}_t = \mathbf{w}^{out}\mathbf{x}_t \quad (3)$$

where the vectors $\mathbf{u}_t, \mathbf{x}_t, \mathbf{o}_t$ are the input and the neurons of the hidden layer and output layer respectively, and $\mathbf{w}^{in}$, $\mathbf{W}$, $\mathbf{w}^{out}$ are the matrices of the respective synaptic weight factors.

Connections in the hidden layer are random but the system needs to fulfil the so-called echo state condition. Jaeger [4] gives a definition; a slightly more compact form of the echo state condition is repeated here:

Consider a time-discrete recursive function $\mathbf{x}_{t+1} = F(\mathbf{x}_t, \mathbf{u}_t)$ that is defined at least on a compact sub-area of the vector-space $\mathbf{x} \in R^n$. and where $\mathbf{x}_t$ are to be interpreted as internal states and $\mathbf{u}_t$ is some external input sequence, i.e. the stimulus.

The definition of the echo-state condition is the following: Assume an infinite stimulus sequence: $\bar{\mathbf{u}}^\infty = \mathbf{u}_0, \mathbf{u}_1, \ldots$ and two random initial internal states of the system $\mathbf{x}_0$ and $\mathbf{y}_0$. To both initial states $\mathbf{x}_0$ and $\mathbf{y}_0$ the sequences $\bar{\mathbf{x}}^\infty = \mathbf{x}_0, \mathbf{x}_1, \ldots$ and $\bar{\mathbf{y}}^\infty = \mathbf{y}_0, \mathbf{y}_1, \ldots$ can be assigned.

$$\mathbf{x}_{t+1} = F(\mathbf{x}_t, \mathbf{u}_t) \tag{4}$$

$$\mathbf{y}_{t+1} = F(\mathbf{y}_t, \mathbf{u}_t) \tag{5}$$

Then the system $F(.)$ fulfils the echo-state condition if independent from the set $\mathbf{u}_t$ and for any ($\mathbf{x}_0$,$\mathbf{y}_0$) and all real values $\epsilon > 0$ there exists a $\delta(\epsilon)$ for which

$$d(\mathbf{x}_t, \mathbf{y}_t) \leq \epsilon \tag{6}$$

for all $t \geq \delta$. The ESN is designed to fulfil the echo state condition.

2.2 Design of the Connectivity Matrix

The connectivity of the hidden layer is restricted to

$$\mathbf{W}_{ij,raw} = \frac{\exp(-(i-j)^2)}{2\sigma^2} A_{ij}, \tag{7}$$

where $A_{ij} = R_{ij} - R_{ji}$ is an anti-symmetric matrix constructed from the square random matrix from white noise that is equally distributed in the range $[-0.5, 0.5)$. The resulting matrix was normalized after measuring its largest singular value $s_{max} : \mathbf{W}_{ij} = \mathbf{W}_{ij,raw}/s_{max}(\mathbf{W}_{ij,raw})$.

The input matrix is defined as

$$\mathbf{w}_i^{in} = \frac{exp(-i^2)}{2\sigma^2} r_i, \tag{8}$$

where r_i is equally distributed noise in the same way as the values of R_{ij}. An example for the resulting matrix is illustrated in Fig. 2. It is important to note that a signal that is presented to the network takes some time to propagate from the lower indices to the higher indices. In this way, strong connectivity to higher indexed neurons indicates higher relevance or older data.

Solving the equation

$$\mathbf{w}^{out} M_{lin} = V_{lin} \tag{9}$$

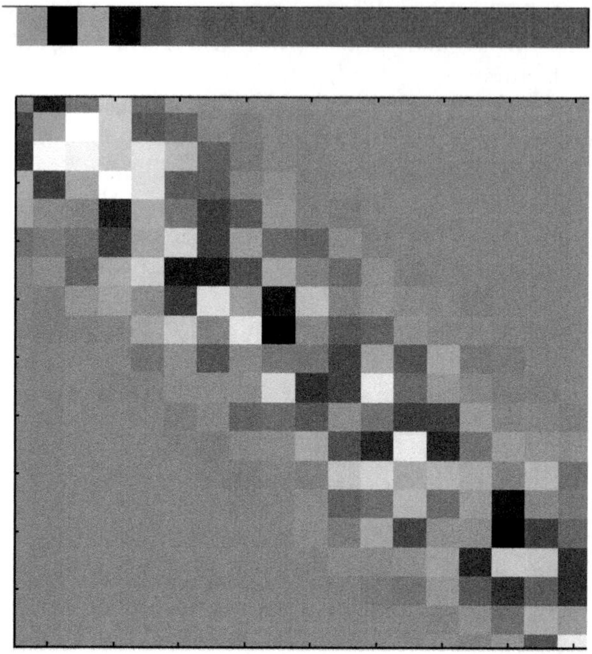

Fig. 2. Input matrix $\mathbf{w}^{in}$ (top) and recurrent matrix $\mathbf{W}$ (bottom) as gray valued plots. If a scalar valued input signal is presented to the network it propagates through the ESN, where the neurons with the higher indices receive the information later then the neurons with the lower indices.

yields the trained output matrix $\mathbf{w}^{out}$. In dependence on the given task values of the matrix $\mathbf{w}^{out}$ can have high values (i.e. the optimal solution of the mean square error problem). In order to limit the weights, ridge regression is used. ridge regression is realized by adding a square matrix $m = \lambda I$, where I is the identity matrix and λ indicates the strength of the force that keeps the absolute values of $\mathbf{w}^{out}$ small. The concatenated matrix $M_{lin,ridge} = [M_{lin}, m]$ and $V_{lin,ridge} = [V_{lin}, 0..0]$. The method is well known (see [8]). Instead Eq. 9 the system uses

$$\mathbf{w}^{out} M_{lin,ridge} = V_{lin,ridge}. \tag{10}$$

Small values of λ tolerate larger values in $\mathbf{w}^{out}$, whereas large values enforce smaller values.

2.3 Measuring the Information Transfer in the Reservoir

For estimating the information between parts of the reservoir a probabilistic approach [4,6] is used. Instead of training the output $\omega_t \in \Omega$ directly, we model a probability that a specific event has occurred with regard to the output. In other words: the aim is to train the network (or that particular part of it) is

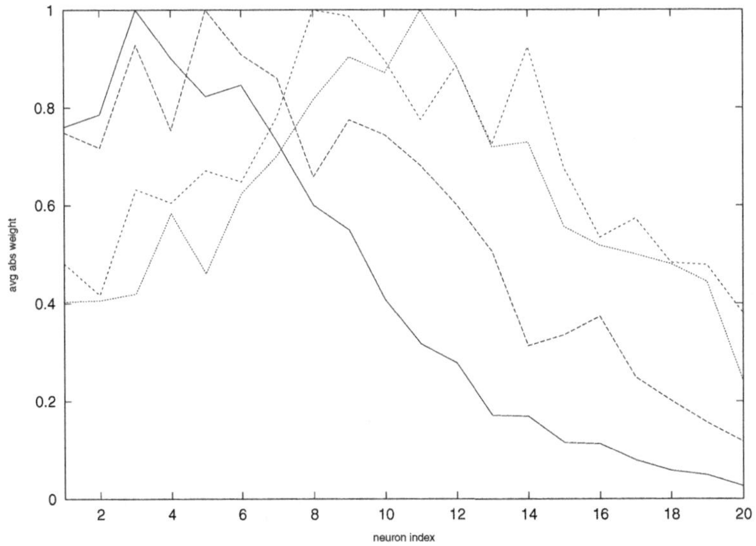

Fig. 3. Average connectivity of neurons at different positions. The graph shows results from different values λ of the ridge regression constraint. The values of λ are 0.1 (full line), 0.01 (long dashes), 0.001 (short dashes), 0.0 (normal linear regression, dotted). The taught delay is 3 in this case.

trained in that way that each of the output units represents the probability of an event. The simplest way to do this is to teach the output o_r of the network to reproduce the probability that the as a range of the – statistical – output variable $\Omega_r \subset \Omega$ that is of interest for the given task. The task of network is to find $p(\Omega_r|\mathbf{x}_t(\bar{\mathbf{u}}^\infty))$ in the following written short $p(\Omega_r)$. We define the teaching signal d_r as:

$$\begin{aligned} if\ (\omega_t \in \Omega_r)\ d_r &= 1 \\ else\ d_r &= 0 \end{aligned}$$

The mean square error (MSE) is the

$$E_{mse} =< (d_{r,t} - o_r)^2 >= p(\Omega_r)(1 - o_r)^2 + (1 - p(\Omega_r))o_r^2 \tag{11}$$

The derivative $\partial E_{mse}/\partial o_r$ set equal to zero yields the point at which E_{mse} is minimal:

$$o_r - p(\Omega_r) = 0 \tag{12}$$

Thus, the MSE is reached when $o_r = p(\Omega_r)$; we can assume

$$o_r \rightarrow p(\Omega_r), \tag{13}$$

for sufficiently long learning sequences. Since – with the common restrictions of reservoir computing– the full information of the input history is encoded

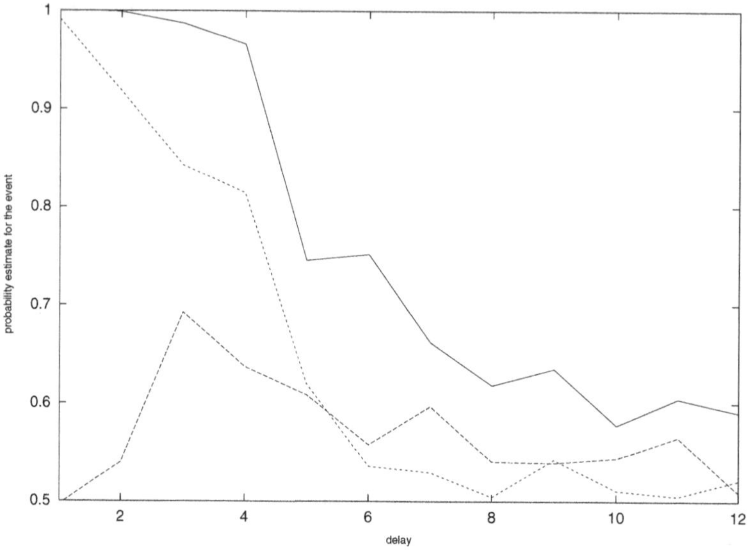

Fig. 4. The figure shows the reconstruction (probability) for different delays. The results were learned from different parts of the network. First, all neurons of the hidden layer were used (full line), second the 15 neurons with the highest indices were used (long dashes), third the 5 neurons with lowest indices were used (short dashes).

in the activity state of the reservoir. Thus, –without additional efforts in the hidden layer– more information about statistical variables can be retrieved from additional output units: Because the optimal solution (absolute minimum of the MSE) can be derived, the network is going to find the true probability **as far as it is detectable by linear regression from the current state of the reservoir**. Usually, the quality of the network performance and the learning progress can be checked by measuring E_{mse}, where values close to zero represent a good network performance. It should be noted that for the learning rule outlined above the theoretical limit is above zero. Under the assumption that the $p(\Omega_r)$ is the true probability we get:

$$E_{min}(\Omega_r) = min_{o_r}(E_{mse}) = p(\Omega_r)(1 - p(\Omega_r)). \tag{14}$$

However, since in fact the true value $p(\Omega_r)$ is unknown, it is not a good idea to use $E_{mse} - E_{min}$ as a measure. However, it can be used to find out if the output node is deterministic (i.e. the output node takes either 0 or 1). In this case the minimal error is in fact 0 again.

Instead one could go the following way in that we can get a set of outputs that covers a complete range of the random variable in the way that for a range $r \in R$:

$$\cup_{r \in R} \Omega_r = \Omega, \tag{15}$$

$$\Omega_i \cap \Omega_j = \emptyset, \tag{16}$$

for all $i \neq j$. Obviously, we have $\sum_{r \in R} p(\Omega_r) = 1$. We can test the constraint in the network. We test the quality of the network output by testing measuring $\sum_r o_r$ which should be close to one if the network has adapted sufficiently. For this particular task, instead of continuous values random sequences of 1 and 0 are used.

3 Results

A first test task (which has been already used by [4] is to let the network reconstruct the sequence of input values after a delay (in this case 4 steps. The simulation was conducted as the initial states of the neurons in the hidden layer were random numbers in the range $[-0.5, 0.5)$. Throughout the simulation the input of the network were random numbers in the range of $[-0.5, 0.5)$. The network of 30 neurons was then run in a transient phase of 500 cycles which was followed by a training phase of 280 cycles. During the training the states $\mathbf{x}_t$ of the recurrent layer by using them as line vectors of a composed matrix M_{lin}. Also a vector V is composed from the output that is trained output. The σ value of the neighbourhood with had a value of 2.8.

Comparing ridge regression and unconstrained linear regression, one can see significant differences between the areas on the reservoir with respect to strength of connectivity (see Fig. 3). Whereas non-constraint regression tends to have its largest output weights towards the higher indices, the examples of ridge regression have larger connections to the lower indices with maximum around of the taught delay. Preliminary test show that simulations with large constraint factors λ tend to have its maximum values for the connectivity around the value of the taught delay.

The second simulation was done with an array of 80 neurons, the neighbourhood width was σ chosen to be 15.5 , a small noise was added. Again the network should reproduce the input sequence with various delays. For practical reasons the range of the out put unit was not chosen between one and zero but between -1 and 1, since this is a linear transformation arguments from section 2.3 still apply. Figure 4 shows results from the simulations for different delays and different subsets of the hidden layer. The simulation showed that the 5 neurons with lowest indices suffice to produce almost the same performance as the whole network. The 15 neurons with the highest indices show a bad performance for small delays since the information could not reach them in time. For a delay of around 3 a maximum performance was reached. For higher delays the performance decayed again.

4 Discussion

ESNs are interesting feature extractors because they project instaneous input data into a potentially infite space of temporally persistent, random, and nonlinear 'echo' feature reservoir. Applying topographies to reservoirs can provide new insights into the relation between the input statistics and the learned task.

Because ESN neurons in this scheme have a spatial or temporal location, their relative contribution to solving an estimation problem can be interpreted. Constraining the network of recurrent interconnections may result in more a efficient feature extraction. Although the estimation task in this case study was certainly trivial, it is intended to represent a first step to attempting more complex spatial and temporal topologies, than can then be applied to non-trivial estimation tasks. Applicable real-world applications include distributed sensor networks and short-term memory models of agent behaviour.

Acknowledgements

The work has been supported by the National Science Council grant number 98-2218-E-194-003 -MY2.

References

1. Boedecker, J., Obst, O., Mayer, N.M., Asada, M.: Initialization and self-organized optimization of recurrent neural network connectivity. Human Frontier Science Program (HFSP) Journal 3, 340–349 (2009)
2. Deng, Z., Zhang, Y.: Collective behavior of a small-world recurrent neural system with scal-free distribution. IEEE Transactions on Neural Networks 18, 1364–1375 (2007)
3. Hammer, B., Schrauwen, B., Steil, J.J.: Recent advances in ecient learning of recurrent networks. In: Proceedings of the European Symposium of Artificial Neural Networks, ESANN (2009)
4. Jaeger, H.: The 'echo state' approach to analysing and training recurrent neural networks. In: GMD Report 148, GMD German National Research Insitute for Computer Science (2001)
5. Kohonen, T.: Self-Organizing Maps, 3rd edn. Springer Series in Information Sciences, vol. 30. Springer, Heidelberg (2001)
6. Michael Mayer, N., Obst, O., Chang, Y.-C.: Time series causality inference using echo state networks. In: LVA/ICA Conference Proceedings (to appear, 2010)
7. Obst, O.: Distributed fault detection using a recurrent neural network. In: Proceedings of the International Conference on Information Processing in Sensor Networks (IPSN 2009). IEEE Computer Society, Los Alamitos (2009)
8. Tarantola, A.: Inverse Problem Theory. Society for Industrial and Applied Mathematics (2004) (available online as a free PDF version), ISBN 0-89871-572-5

iGAPK: Improved GAPK Algorithm for Regulatory DNA Motif Discovery

Dianhui Wang[1,*] and Xi Li[1,2]

[1] Department of Computer Science and Computer Engineering, La Trobe University, Melbourne, VIC, 3086, Australia
dh.wang@latrobe.edu.au
[2] Department of Primary Industries, Bioscience Research Division, Victorian AgriBiosciences Centre, Bundoora, VIC, 3083, Australia

Abstract. Computational DNA motif discovery is one of the major research areas in bioinformatics, which helps to understanding the mechanism of gene regulation. Recently, we have developed a GA-based motif discovery algorithm, named as GAPK, which addresses the use of some identified transcription factor binding sites extracted from orthologs for algorithm development. With our GAPK framework, technical improvements on background filtering, evolutionary computation or model refinement will contribute to achieving better performances. This paper aims to improve the GAPK framework by introducing a new fitness function, termed as relative model mismatch score (RMMS), which characterizes the conservation and rareness properties of DNA motifs simultaneously. Other technical contributions include a rule-based system for filtering background data and a "most one-in-out" (MOIO) strategy for motif model refinement. Comparative studies are carried out using eight benchmark datasets with original GAPK and two GA-based motif discovery algorithms, GAME and GALF-P. The results show that our improved GAPK method favorably outperforms others on the testing datasets.

1 Introduction

Transcription factor binding sites (TFBSs or DNA motifs) are short and subtle genomic segments that are usually found in promoter regions of genes. The interaction between TFBSs and a specified group of proteins (e.g. transcription factors) determines the transcriptional activity and dominates the gene expression level. The study of DNA motifs refers to identify TFBS locations from a set of co-regulated genes using experimental or computational approaches. As a major complement to the traditional wet-lab identification methods, computational algorithms have shown the good potential on the problem solving in terms of time and cost [1]. Literature studies usually cluster the searching algorithms into statistical approaches such as AlignACE [2] and machine learning methods such as MEME [3]. According to the performance assessment from Hu

* Corresponding author.

K.W. Wong et al. (Eds.): ICONIP 2010, Part II, LNCS 6444, pp. 217–225, 2010.

et.al. [1] and Tompa et.al. [4], developing advanced algorithms remains as a challenge to computational biologists.

In the domain of computational motif discovery, the concept of prior knowledge (PK) has been investigated from different perspectives and successfully applied to improve the system performances. Based on the ways that the PK is interpreted and used, they can be classified into model-driven and data-driven. The model-driven approaches concentrate on developing appropriate motif models which interact with prior knowledge. To the best of our knowledge, the term of prior knowledge in motif discovery was first introduced to the work of MEME in [5]. Authors utilized the importance of motif palindrome: the inverse complement of a DNA sequence is the same as its original sequence. When given the prior knowledge of motif palindrome, the nucleotide appearances of the corresponding position in the motif model are considered as consistent. Also, the usage of Dirichlet mixture with the background letter distributions as prior could help to estimate the probabilities of the possible pattern occurrences in the expected motif positions. Results showed the chosen of appropriate prior knowledge into MEME could improve prediction performance. In [6], the existing poor motif models were considered as PK to iteratively optimize the model quality by combing the ChIP sequences with a novel genetic algorithm. On the other hand, the data-driven approaches specialize in extracting and parsing the known biological features as prior knowledge to favor the prediction. By undertaking the support from comparative genomic studies and the availability of massive genomic sequences, the importance of motif evolutionary conservation has draw more attention than usual. In [7], the conserved orthologous blocks in the promoter regions of multiple species were treated as PK. The searching of motifs then focused on those conserved blocks. According to the theory that transcription factor binding sites usually have specific distances against the specific biological landmarks (such as transcription start site), the authors considered the localization information of possible true binding sites as prior knowledge to improve the motif prediction accuracy [8].

Genetic Algorithm (GA) has been employed to resolve motif discovery problem with some favorable results [9], [10]. Wei and Jensen [9] proposed a GA-based motif discovery framework GAME with some unique features, especially a Bayesian-based posterior distribution was developed as the fitness function. Another GA approach named as GALF-P was presented in [10] that employed a local filtering operation during the searching stage and an adaptive model post-processing. In our previous work reported in [11], we developed a GA-based motif discovery framework GAPK, which highlights the use of prior knowledge on binding sites with the purpose of search space reduction and population initialization. Results have indicated GAPK can achieve a better prediction performance, comparing with GAME and GALF-P.

This paper aims to further improve the framework of GAPK with some fundamental contributions. A relative model mismatch score (RMMS), which measures the motif model by considering the help of background information, is developed as the fitness function. Using the identified binding sites extracted from orthologs, we develop a PK based filtering mechanism to improve the computing environment in terms of background kmers reduction. Also, a novel model refinement scheme is proposed for optimizing the obtained motif models. To assess the performance of the improved GAPK (iGAPK), we focus on making comparisons among the original GAPK, GAME and GALF-P. Eight benchmarked datasets are employed in the simulations. Results show that our proposed iGAPK outperforms both GAME and GALF-P, and performs slightly better than the original GAPK. Taking the number of system parameters into consideration, iGAPK improves the original GAPK with better robustness.

2 Relative Model Mismatch Score and α-Ratio

Due to the significance of Position Frequency Matrix (PFM) in computational motif discovery, we employ PFM as the motif model representation in this study. Suppose we have a collection of subsequences denoted as S with length k, each subsequence is considered as a k-mer, $B_1B_2\cdots B_k$, where $B_i \in \Sigma = \{A, C, G, T\}, i = 1, 2, \cdots, k$. Then, the collection is denoted by $\{K_p : p = 1, 2, \cdots\}$. To construct the PFM model of S, each k-mer is first encoded to a binary matrix [12], that is, e(k-mer) $= [a_{ij}]_{4\times k}$, $a_{ij} = 1$ if $B_j = V_i$, otherwise $a_{ij} = 0$, where $(V_1, V_2, V_3, V_4) = (A, C, G, T)$. Then, the motif PFM model of S can be given by:

$$M = \frac{1}{|S|} \sum_{K_p \in S} e(K_p), \tag{1}$$

where $|S|$ represents the cardinality of the set S.

2.1 Relative Model Mismatch Score

Wang and Lee recently proposed a Model Mismatch Score (MMS) as a quality metric quantifying the conservation property of the motif model [12].

$$\text{MMS} = \frac{1}{|S|} \sum_{K_p \in S} d(K_p, M). \tag{2}$$

where $d(\cdot, \cdot)$ defined in [12] is the generalized Hamming distance function that measures the mismatch between a k-mer $\in S$ and the PFM M of S.

The MMS score represents the conservation property of cognate binding sites due to the evolutionary selection [12]. To reflect the rareness of true binding sites with respect to the background k-mers, a relative model mismatch score (RMMS) is developed recently in [13]. Experiments show that this quantitive metric for measuring the quality of motif models works well, and its discrimination power on motif models is at least comparable to some widely used metrics such as the well-known information content (IC) [14]. The following gives the RMMS expression:

$$\text{RMMS} = \frac{1}{|S|} \sum_{K_p \in S} R(K_p, M_s), \tag{3}$$

where

$$R(K_p, M_s) = \frac{d(K_p, M_s)}{d(K_p, M_b)}. \tag{4}$$

Here,$R(\cdot,\cdot)$ denotes the relative mismatch score named as RMS of a k-mer, where M_s is the PFM of the motif model S and M_b is the PFM of the background model B. In this study, we construct M_b by using the whole input sequences. A motif model with a small RMMS stands for a high degree of conservation associated with the model rareness.

2.2 α-Ratio

By given the dataset D and the motif model S, we present a ratio based function α-ratio to model a given k-mer K from D under a mixture of two components, which are the relative mismatch score (RMS) and 3rd-order Markov background model [15]:

$$\alpha(K) = \frac{log(p_0(K))}{R(K, M_s)}, \tag{5}$$

where $p_0(K)$ is an estimated background probability evaluated by the 3rd-order Markov model for a given K.

Intuitively, a given k-mer with a smaller α-ratio to the motif model indicates it has a greater possibility to be a true binding site than others.

3 Improved GAPK Algorithm

The improved GAPK algorithm iGAPK is composed of three components: search space reduction, evolutionary computations for identifying candidate motifs and a model refinement for finalizing the outputs. Details are described in the following subsections.

3.1 Search Space Reduction

Our filtering mechanism aims at narrowing down the entire search space and increasing the signal-to-noise ratio by discarding noisy k-mers from the background with the usage of PK.

Given the PK model M_{pk} that is made up of a set of true TFBSs associated with the considered transcription factor, we first calculate the α-ratio of each k-mer in the dataset D against M_{pk}. In here, a normalization process is applied:

$$\alpha_n(K) = \frac{\alpha_{\max} - \alpha(K)}{\alpha_{\max} - \alpha_{\min}}, \tag{6}$$

where K is a k-mer in D, α(K) is the α-ratio of K against M_{pk}, $\alpha_{\max}$ and $\alpha_{\min}$ are the maximum and the minimum α-ratio from D, respectively.

The normalized α-ratio is regarded as the filtering score. For the k-mers in the PK model M_{pk}, the maximum filtering score among them can be obtained by:

$$\lambda = \max_{K_p \in M_{pk}} \alpha_n(K_p). \tag{7}$$

To decide whether a given k-mer from D should be eliminated, the filtering rule is proposed as following:

$$\text{If } \alpha_n(K) > \lambda, \text{Then, } K \text{ is discarded, where } K \in D. \tag{8}$$

The reduced dataset composed of the remaining k-mers after filtering, denoted by R_D. Evolutionary computations will be only applied for the set R_D rather than the original set D.

3.2 Evolutionary Computations

Suppose that there are n sequences in the considered dataset and each sequence consists of at least one binding sites. Let a chromosome be represented as a vector $(vc_1, vc_2, \cdots, vc_n)$, where vc_i is a k-mer extracted from the i-th input sequence. Each chromosome indeed is regarded as a candidate of motif model, which is composed of n kmers. By minimizing a defined fitness function, we can obtain some potential motif models. The final motif model will be determined by refinement step followed by a model ranking process. Therefore, the cardinality of the final motif model can be larger than the number of sequence n. In our iGAPK, the fitness function is taken as the relative model mismatch score $RMMS$.

The population is initialized with assistance of the prior knowledge. The standard roulette-wheel method is employed to choose chromosomes as parents for reproduction. Two genetic operators are used in iGAPK to reproduce children chromosomes, which are *Crossover* and *Replacement*. The reproduction will not be terminated until the size of the population is doubled. A combination of winners-take-all selection and tournament selection are applied to maintain the population size for each generation. Detailed explanations of the reproduction process can be found in [11].

3.3 Model Refinement

In iGAPK, we proposed a model post-processing after the GA process [11]. To enhance the prediction accuracy and system reliability, we further improve the post-processing stage by proposing a novel model refinement approach which consists of three components: Merging, Adding and Removing. The details are given below.

A two-step process on model merging is applied. Firstly, an alternative pattern for each candidate is generated by composing of the dominant nucleotide (highest frequency) from each column in its PFM model. The merging starts with populating new models by grouping candidates with the same alternative pattern. Secondly, these merged models are ranked according to their RMMS scores. Then, the model with lowest RMMS score is selected as a starting point to compare the IC values with the rest of models. All models with a limited IC difference, i.e., ($|IC_a - IC_b| \leq \delta$), are merged together. This process is repeated recursively until there is no model left for merging. In iGAPK, the default value for the δ is set as 0.01. During the merging process, a k-mer might accidentally appear more than one time in the merged model. It would cause negative effect in the model quality. Therefore, we only keep one and remove the redundancies.

The Proposed MOIO Scheme for Model Refinement

BEGIN
  Input: Model $T =: M_T$, Reduced Dataset R_D.
  Set $\Delta IC_T^* =: 0.1$; $L = Null$.

  While $\Delta IC_T^* > 0$
    Set $T^*_{(old)} =: T$.
    Most One-In Step
      For each k-mer k in R_D
        If($R(k, M_T) \leq RMMS(T)$)
          Add k into L;
      For each k-mer k in L
        Calculate α-ratio of k;
        Define $k_{min} = arg \min_{k \in L}\{\alpha(k)\}$;
      Add the k_{min} into T temporarily;
      $\Delta IC_T =: IC_{T(new)} - IC_{T(old)}$;
      If $\Delta IC_T \geq 0$
        Keep the k_{min} in T;
      Else remove the k_{min} from T;
      Empty L; $\Delta IC_T =: 0$;

    Most One-Out Step
      Rank α-ratio values for all k in T;
      Define $k_{max} = arg \max_{k \in T}\{\alpha(k)\}$;
      Remove the k_{max} from T temporarily;
      $\Delta IC_T =: IC_{T(new)} - IC_{T(old)}$;
      If $\Delta IC_T \geq 0$
        The k_{max} will be removed permanently;
      Else retain the k_{max} in T;
    Set $T^*_{(new)} =: T; \Delta IC_T^* =: IC_{T^*_{(new)}} - IC_{T^*_{(old)}}$.
END

After Merging, we introduce an "Most One-In-Out" (MOIO) process to further refine the candidates by employing the α-ratio. Given a candidate T, the "Most One-In" step intends to collect the weak true binding sites from the reduced dataset R_D that are missed out during the evolution process, while the "Most One-Out" step follows the "Most One-In" step immediately, aiming to eliminate some false-positives. Our proposed MOIO strategy is summarized in the form of pseudo code.

Model refinement will be terminated when the iterative "Most One-In-Out" process is applied to each merged model. The final models are then ranked according to their RMMS scores. The model with the smallest RMMS score will be selected as a motif candidate for performance evaluation.

4 Performance Evaluation

4.1 Datasets

Eight benchmark datasets used in both [9] and [10] were employed in this study to evaluate the performance of our iGAPK algorithm. By applying the same datasets, it makes the comparisons more reasonable and convincing. Also the datasets have been well purified, we assume no unknown binding sites existing in the sequences. Statistical details of the datasets are shown in Table 1.

Table 1. Statistics of the 8 datasets

Prop.	CREB	CRP	ERE	E2F	MEF2	MYOD	SRF	TBP
N	17	18	25	25	17	17	20	95
l	200	105	200	200	200	200	200	200
w	8	22	13	11	7	6	10	6
n	19	23	25	27	17	21	36	95
p	1.12	1.28	1	1.08	1	1.23	1.8	1
pk	16	212	12	10	58	15	46	21

N is num. of input sequences., l is the length of sequences., w is the expected width of binding site (BS), n is the number of BSs in each dataset, p is the average number of BSs per sequence, and pk is the number of known BSs from public domains.

4.2 Results with Comparisons

iGAPK is evaluated by carrying out the experimental comparisons with GAME [9], GALF-P [10] and GAPK [11]. With the purpose of making a fair comparison of the four applications, we first kept using their default parameters in the first 5 runs. Then, for each of them, we adjusted the key parameters of GA: probabilities of genetic operators (from 0.1 to 0.9 with the interval of 0.1), the population size (from 100 to 500), the number of generation (from 100 to 1000) across another 15 runs. In [10], authors compared the prediction results between GALF-P and GAME out of 20 runs. The promising strategy convinced us that running each algorithm 20 times with some parameter adjustments could provide sufficient evidences to support the performance comparisons. F-measure is applied here to measure the prediction accuracy [11]. The average precision, recall and F-measure along with the standard deviation (with the $\pm$ symbol) of 20-runs across 8 datasets are shown in Table 2.

It can be seen that both iGAPK and GAPK shows comparable performance against the other two GA tools. From the results, GAME obtains 2 best recalls and F-measures, while GALF-P has 3 best recalls, and 2 best F-measures respectively. Comparing with them, iGAPK achieves 5 best precision rates and 4 F-measures as well as 3 comparable best recalls. Also, the improvements against GAPK are obvious. In average, iGAPK outperforms the other tools in terms of precision, recall and F-measure. It is observed that both iGAPK and GAPK have relatively low standard deviation rates in most of the datasets, which shows our proposed framework can provide reliable performance with the changing of parameters. We also notice that dataset "MYOD" contains subtle binding sites along with many repetitive sequence patterns, which brings the poor prediction results from both GAME and GALF-P. With the help of prior knowledge and RMMS, iGAPK obtains very impressive improvements on MYOD. As a result, iGAPK shows its advantage on prediction accuracy as well as stable performance over the eight datasets.

Table 2. Comparisons among iGAPK, GAPK, GAME and GALF-P for 20 runs

Datasets	iGAPK			GAPK		
	Precision	Recall	F-measure	Precision	Recall	F-measure
CREB	**0.68** ± 0.06	**0.65** ± 0.06	**0.66** ± 0.06	0.65 ± 0.04	**0.68** ± 0.07	**0.66** ± 0.04
CRP	0.90 ± 0.05	0.84 ± 0.03	0.87 ± 0.02	**0.96** ± 0.06	0.80 ± 0.04	0.86 ± 0.04
ERE	**0.73** ± 0.15	**0.88** ± 0.03	**0.79** ± 0.11	0.71 ± 0.15	**0.89** ± 0.06	0.78 ± 0.10
E2F	**0.69** ± 0.02	0.83 ± 0.06	0.75 ± 0.03	0.66 ± 0.02	0.92 ± 0.05	0.77 ± 0.02
MEF	0.87 ± 0.10	0.92 ± 0.04	0.89 ± 0.06	**1.00** ± 0.00	0.77 ± 0.05	0.87 ± 0.03
MYOD	**0.83** ± 0.06	**0.92** ± 0.08	**0.87** ± 0.05	0.68 ± 0.14	0.82 ± 0.06	0.74 ± 0.09
SRF	**0.75** ± 0.04	0.81 ± 0.05	**0.78** ± 0.03	0.63 ± 0.05	0.74 ± 0.05	0.67 ± 0.04
TBP	0.73 ± 0.10	0.83 ± 0.04	0.77 ± 0.06	**0.83** ± 0.06	0.83 ± 0.06	**0.83** ± 0.06
Average	**0.77**	**0.84**	**0.80**	**0.77**	0.81	0.79

Datasets	GAME			GALF-P		
	Precision	Recall	F-measure	Precision	Recall	F-measure
CREB	0.44 ± 0.31	0.43 ± 0.30	0.43 ± 0.32	0.47 ± 0.24	0.60 ± 0.29	0.53 ± 0.26
CRP	0.93 ± 0.05	0.84 ± 0.03	0.88 ± 0.03	0.95 ± 0.02	**0.88** ± 0.05	**0.91** ± 0.04
ERE	0.63 ± 0.07	0.84 ± 0.06	0.72 ± 0.06	0.65 ± 0.15	0.84 ± 0.04	0.72 ± 0.10
E2F	0.62 ± 0.05	0.86 ± 0.09	0.72 ± 0.06	0.67 ± 0.08	**0.93** ± 0.05	**0.78** ± 0.07
MEF	0.90 ± 0.05	**0.96** ± 0.06	**0.93** ± 0.04	0.85 ± 0.16	0.94 ± 0.06	0.89 ± 0.11
MYOD	0.24 ± 0.17	0.24 ± 0.16	0.24 ± 0.16	0.28 ± 0.24	0.51 ± 0.45	0.36 ± 0.32
SRF	0.67 ± 0.06	**0.92** ± 0.06	**0.78** ± 0.06	0.68 ± 0.12	0.88 ± 0.06	0.76 ± 0.09
TBP	0.67 ± 0.28	0.58 ± 0.24	0.62 ± 0.25	0.74 ± 0.12	**0.86** ± 0.02	0.80 ± 0.09
Average	0.64	0.71	0.67	0.66	0.81	0.73

Values of standard deviation are symbolled as ±.

5 Conclusion

This paper further develops our previously proposed GAPK framework for computational motif discovery. Our technical contributions mainly include: (i) Define a new fitness function for evolutionary computations, which is indeed a novel metric for measuring the quality of motif models; (ii) With the help of prior knowledge, a ratio-based filtering system is developed to discard irrelevant kmers so that the search space reduction can be achieved; (iii) An effective model refinement strategy, termed as MOIO scheme, is proposed. In the proposed

MOIO scheme, one parameter is used for merging models. This greatly improves the original GAPK algorithm in terms of the system reliability, although the prediction performance has slight improvement.

Our further research includes a comprehensive analysis on performance robustness with respect to the system parameter setting. In addition, it is significant to look into some ways to maximize the utilization of various prior knowledge in algorithm development.

References

1. Hu, J., Li, B., Kihara, D.: Limitations and potentials of current motif discovery algorithms. Nucleic Acids Res. 33, 4899–4913 (2005)
2. Neuwald, A.F., Liu, J.S., Lawrence, C.E.: Gibbs motif sampling: detection of bacterial outer membrane protein repeats. Protein Science 4, 1618–1632 (1995)
3. Bailey, T.L., Elkan, C.: Unsupervised learning of multiple motifs in biopolymers using EM. Machine Learning 21, 51–80 (1995)
4. Tompa, M., Li, N., Bailey, T.L., et al.: Assessing computational tools for the discovery of transcription factor binding sites. Nature Biotechnology 23, 137–144 (2005)
5. Bailey, T.L., Elkan, C.P.: The value of prior knowledge in discovering motifs with MEME. Intell. Sys. Mol. Bilo. 3, 21–29 (1995)
6. Li, L.P., Liang, Y., Bass, R.L.L.: GAPWM: a genetic algorithm method for optimizing a position weight matrix. Bioinformatics 23, 1188–1194 (2007)
7. Wang, T., Stormo, G.D.: Combining phylogenetic data with co-regulated genes to identify regulatory motifs. Bioinformatics 19, 2369–2380 (2003)
8. Narang, V., Mittal, A., Sung, W.-K.: Localized motif discovery in gene regulatory sequences. Bioinformatics 26, 1152–1159 (2010)
9. Wei, Z., Jensen, S.T.: GAME: detecting cis-regulatory elements using a genetic algorithm. Bioinformatics 22, 1577–1584 (2006)
10. Chan, T.-M., Leung, K.-S., Lee, K.-H.: TFBS identification based on genetic algorithm with combined representations and adaptive post-processing. Bioinformatics 24, 341–349 (2008)
11. Wang, D.H., Li, X.: GAPK: Genetic algorithms with prior knowledge for motif discovery in DNA sequences. In: CEC 2009: IEEE Congress on Evolutionary Computation 2009, Trondheim, Norway, pp. 277–284 (2009)
12. Wang, D.H., Lee, N.K.: MISCORE: mismatch-based matrix similarity scores for DNA motif detection. In: Köppen, M., Kasabov, N., Coghill, G. (eds.) ICONIP 2008. LNCS, vol. 5506, pp. 478–485. Springer, Heidelberg (2009)
13. Wang, D.H.: Characterization of regulatory motif models. Technical Report, La Trobe University, Australia (October 2009)
14. Stormo, G.D., Fields, D.S.: Specificity, free energy and information content in protein-DNA interactions. Trends in Biochemical Sciences 23, 109–113 (1998)
15. Thijs, G., Lescot, M., Marchal, K., Rombauts, S., De Moor, B., Rouzé, P., Moreau, Y.: A higher-order background model improves the detection of promoter regulatory elements by Gibbs sampling. Bioinformatics 17, 1113–1122 (2001)

A Computer-Aided Detection System for Automatic Mammography Mass Identification

Hussein Samma[1], Chee Peng Lim[1,*], and Ali Samma[2]

[1] School of Electrical and Electronic Engineering, University of Science Malaysia, Malaysia
[2] School of Computer Science, University of Science Malaysia, Malaysia
cplim@eng.usm.my

Abstract. Automatic detection and identification of mammography masses is important for breast cancer diagnosis. However, it is challenging to differentiate masses from normal breast regions because they usually have low contrast and a poor boundary. In this study, we present a Computer-Aided Detection (CAD) system for automatic breast mass identification. A four-stage region-based procedure is adopted for processing the mammogram images, i.e. localization, segmentation, feature extraction, and feature selection and classification. The proposed CAD system is evaluated using selected mammogram images from the Mammographic Image Analysis Society (MIAS) database. The experimental results demonstrate that the proposed CAD system is able to identify mammography masses in an automated manner, and is useful as a decision support system for breast cancer diagnosis.

1 Introduction

Breast cancer is one of the major causes of mortality among women. According to statistics from the American Cancer Society (ACS), it is expected that more than 200 thousand new cases to occur among women in US during 2010 [1]. Early detection of breast cancer increases the chances of survival. The earliest sign of breast cancer is an abnormality detected on a mammogram screening image. In this aspect, a Computer-Aided Detection (CAD) system can be developed as a useful decision support system for automatic detection of breast cancer, whereby the predictions from the CAD systems can be employed to enhance diagnoses from radiologists [2]. There are two types of CAD systems for breast cancer detection: one for microcalcifications detection and another for mass detection. In this study, we focus on developing a CAD system for automatic mass detection. A good review on automatic mass detection and segmentation in mammographic images with a quantitative performance comparison can be found in [3].

A general CAD system for automatic mass detection normally comprises two stages: detection of suspicious regions and classification of suspicious regions as a mass or a normal tissue. In the first stage, pixel-based or region-based methods are normally adopted [2]. In pixel-based methods, image features are extracted from each pixel, and are used for classification as either a mass or a normal tissue. For example, Liu *et al.* [4] proposed a multi-resolution scheme for lesion detection. The image is

* Corresponding author.

K.W. Wong et al. (Eds.): ICONIP 2010, Part II, LNCS 6444, pp. 226–233, 2010.

first decomposed using wavelet transform. Four features are then extracted for each pixel at each decomposition level. A total of 19 mammograms containing masses are selected from the MIAS database [5] for evaluation. The results show 84.2% true positive detection at less than 1 false positive per image, and 100% true positive detection at 2.2 false positives per image.

The second method for mass detection is based on Region Of Interest (ROI). In this method, an ROI is first segmented. Then, the features are extracted from each ROI. A classifier is used to classify the region as a suspicious or a normal region. Again, many region-based approaches have been proposed in the literature [6-14]. Mudigonda *et al.* [6] presented an approach which uses both gradient and texture features to distinguish normal masses from malignant ones. The Mahalanobis distance is employed to classify breast masses as benign or malignant. An accuracy rate of 73.7% with 38 MIAS cases is reported. Hupse *et al* [7] used contextual information to identify suspicious regions. A neural network classifier is employed to detect the suspicious regions. .Classification performance has been evaluated using a database of 3262 normal mammograms (with 9688 extracted images) and 636 abnormal mammograms (2180 images). The results show that the mean sensitivity is in the interval of 0.05–0.5 false positives/image, and increases to more than 6% with the use of context features.

Recently, Gao *et al* [8] presented an automated method for detecting mammographic masses. The detection scheme is based on Morphological Component Analysis (MCA). First, a mammogram is decomposed into two: piecewise-smooth and texture components. The smooth component is used for extracting independent regions at different intensity layers. Four morphological features are calculated for each ROI. If the selected ROI has more than one concentric layer in the successive lower intensity layers, it is considered as a mass. The system has been evaluated with a total of 200 mammograms selected from the Digital Database for Screening Mammography (DDSM) [15], and 95.3% sensitivity and 2.83 false positive per image has been reported.

In this study, we aim to develop an automatic region-based CAD system. The proposed CAD system consists of four stages: (i) localization of candidate ROI mass from a mammogram; (ii) segmentation of ROI masse; (iii) feature extraction of each ROI; and (iv) feature selection and classification using computational intelligence techniques. In the localization stage, a hybrid morphological erosion and reconstruction method is employed to mark the best 10 candidate regions. Next, a level set segmentation algorithm, which is based on area minimization, is employed to extract the marked candidate regions. After that, wavelet transform is applied to extract a set of multi-resolution features. Finally, a hybrid intelligent model comprising the Genetic Algorithm (GA) and Support Vector Machine (SVM) is deployed for feature selection and classification.

The organization of this paper is as follows. An overview of the proposed methodology is given in Section 2. The experimental results are presented and discussed in Section 3. Concluding remarks are provided in Section 4.

2 Methodology

The CAD system developed in this work consists of the following stages: localization, segmentation, features extraction, as well as feature selection and classification, as shown in Fig.1.

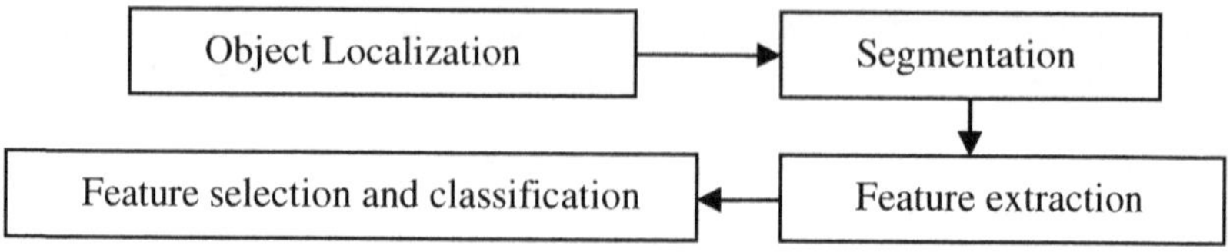

Fig. 1. The proposed methodology of the CAD system

Localization: The aim of this stage is to identify possible locations of candidate masses from the mammogram image. First, the intensity threshold value that separates the foreground regions from the background is identified using Otsu's algorithm [16] as shown in Fig.2. The maximum foreground object is selected as the target area for further processing. The next step of localization is marking candidate masses. First, a morphological erosion operation is applied with an incremental disk structure element (start by 5 pixels and increment by 1 each iteration). Then, a morphological reconstruction algorithm guided by the result of erosion as a marker is used [17]. The maximum regional area is identified from the result of morphological reconstruction (i.e. using the *imregionalmax* matlab function). This operation is repeated until a maximum of ten objects have been marked, as shown in Fig.3.

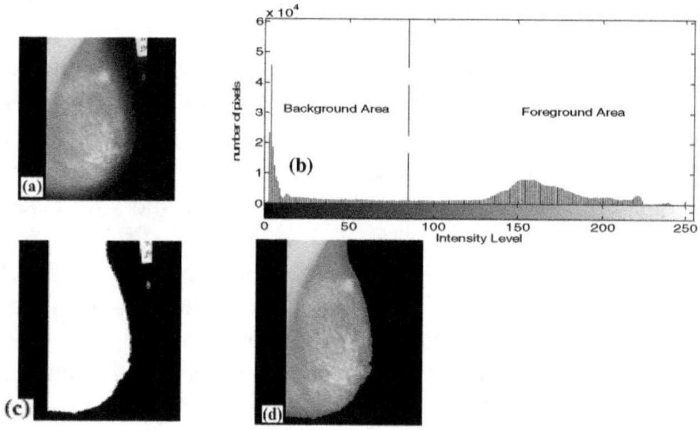

Fig. 2. (a). Original image, (b). Threshold value (c). After background removal, (d). Selection of the maximum foreground object

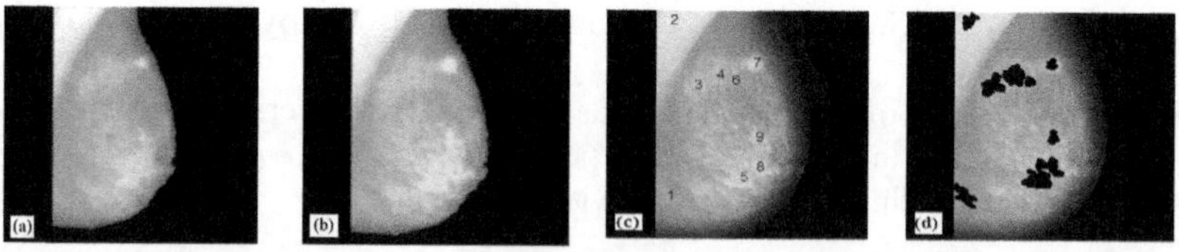

Fig. 3. (a). After erosion operation, (b). After reconstruction operation, (c-d). The resulting marked regions.

Segmentation: The aim of this stage is to extract the candidate masses from the mammogram image. A window around the candidate mass, which has to be large enough to include the whole mass boundary, is first taken. After many tests with several window sizes with the available MIAS images, the window size has been empirically chosen to be 900% with respect to the original mark size. Boxes are placed around each marked candidate masses. A schematic diagram of the segmentation stage is given in Fig. 4 and Fig.5.

Fig. 4. (a).Original image, (b) Candidate mass locations are marked

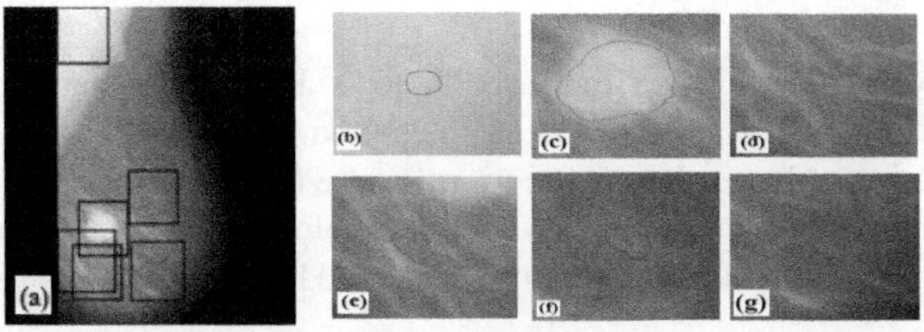

Fig. 5. (a) Boxes are placed around the localized candidate masses, (b-g). Each candidate mass is identified.

Next, Chunming's algorithm [18] is applied to extract the candidate mass within the window. The method is based on the active contour models, as it has the advantage of working in low homogeneity regions. The scale parameter, σ, and the number of iterations (N) govern the growing of the contour. For example, a large value of σ or N results in leakages in the mass border while a small value of σ or N results in a poor boundary. These effects are shown in Fig.6.

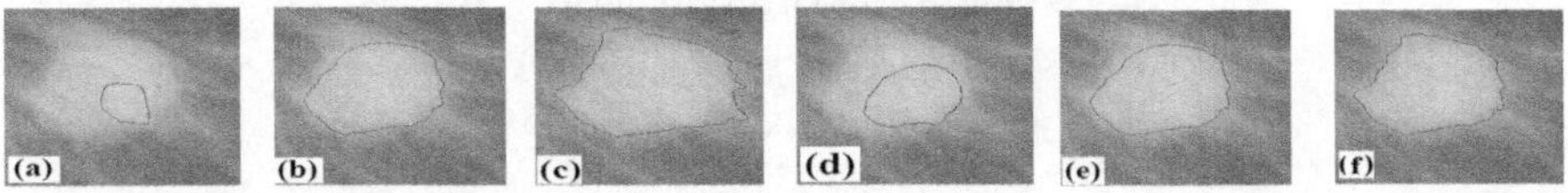

Fig. 6. Results of mass boundary with different scale parameter (σ) and number of iterations (N). (a). σ=5, N=100; (b) σ=15, N=100; (c) σ=30, N=100; (d) σ=15, N=30; (e) σ=15, N=100; (f) σ=15, N=300

Feature Extraction: A series of shape and wavelet energy related features is computed for each ROI candidate mass. The shape features comprise area and boundary of the ROI. The boundary is calculated based on the Euclidean distance from the mass center to the boundary points, as shown in Fig.7(a). The radial Euclidean distance for the whole boundary is shown in Fig.7(b).

The stationary wavelet transform is computed for the 1-D radial distance for two levels of decomposition, and the statistical mean and standard deviation values are computed for each decomposition level. A total of 10 features are computed from the ROI boundary. Moreover, two decomposition levels of ROI candidate masses are computed using a 2-D quadratic spline wavelet transform. Then, the statistical energy mean and standard deviation values are calculated for each wavelet sub-band. As such, a total of 12 2-D wavelet features are extracted for each ROI candidate mass.

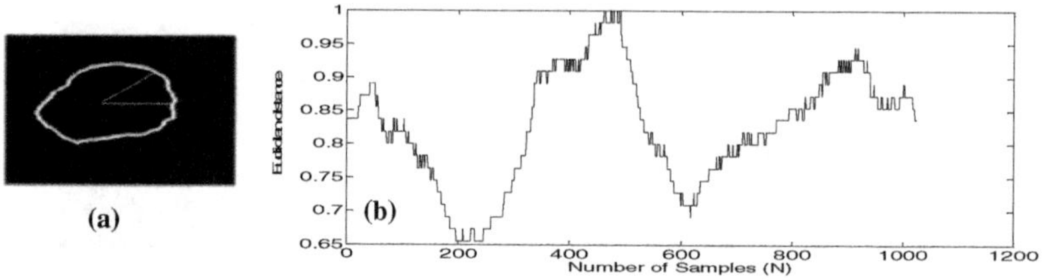

Fig. 7. (a). The Euclidean distance from the mass center to the boundary; (b) 1-D equivalent boundary signal with 1024 boundary points

Feature Selection and Classification: A hybrid GA-SVM model is employed for feature selection and classification, as depicted in Fig.8. The SVM classification accuracy rate is used as a fitness function for the GA. The length of the chromosome is equal to the length of the feature vector. A binary representation is used for each chromosome: 1 if the associated feature is selected and 0 if otherwise.

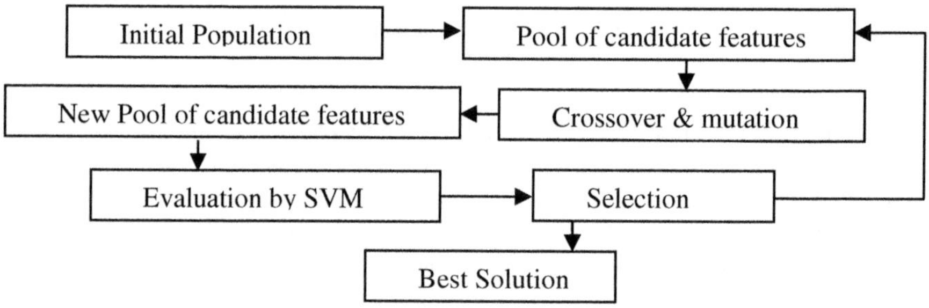

Fig. 8. The GA-SVM model

3 Results

The proposed CAD system is evaluated using 19 mammogram images containing speculated lesions and 19 normal mammograms images selected from the MIAS database. The images are in an 8-bit gray resolution format, and the size is 1024 by 1024 pixels. In the localization stage, a maximum of 10 ROIs are marked as the candidate masses. After localization, a total of 281 ROIs, with 19 mass ROIs and 262 normal ROIs, are available.

In the segmentation stage, after several empirical trials, the parameters of Chunming's algorithm are set as follows: scale parameter, σ=15, and number of iterations, N=100. The GA is used with the following parameters: population size=50, number of generation=50, probability of crossover=0.8, and probability of mutation=0.2. The non-linear kernel-based SVM was deployed with radial basis kernel functions. It was trained with a total of 38 ROIs, with 19 mass ROIs and 19 normal ROIs (randomly selected from 262 normal ROIs). The SVM was trained using the 10-fold cross-validation technique. Table 1 shows the results with and without GA-based feature selection and SVM classification.

Table 1. Eexperimental results

	No. of Features	Accuracy (%)	Sensitivity (%)	Specificity (%)
Without feature selection	23	74.94	82.10	67.95
With feature selection	6	91.57	87.65	95.54

With the use of the GA for feature selection, the number of features is reduced from 23 to 6. All three performance indicators, accuracy, sensitivity, and specificity, increase with the use of the reduced feature set. This implies that most of the features are not useful in discriminating between normal and abnormal ROIs. Note that the sensitivity rate is low because it is difficult to discriminate between normal and abnormal ROIs in some images. An example is shown in Fig.9. Fig. 9(a, b) shows two abnormal ROI masses. The candidate masses detected by the CAD system are shown in Fig. 9(c, d). Fig. 9(e) shows the border of the abnormal mass of Fig. 9(a). It appears to have similar border and intensity features like the normal candidate mass (no. 2) detected in Fig. 9(d), as shown in Fig. 9(f).

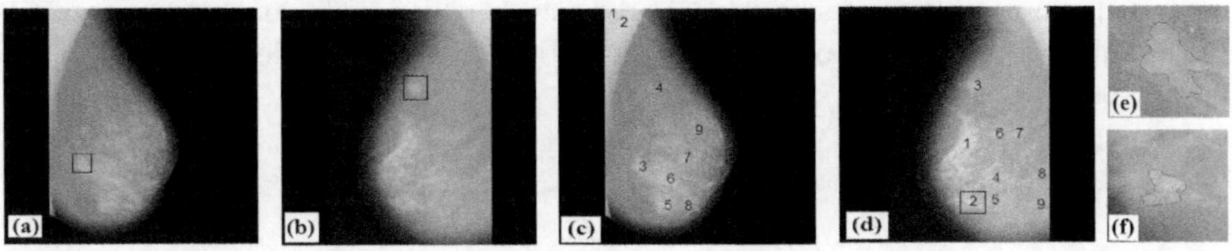

Fig. 9. (a, b) Two abnormal masses; (c, d) Candidate masses marked by the CAD system; (e) The border of the actual abnormal mass from (a); (f) The border for the normal mass (no. 2) marked by the CAD system from (d).

In another experiment, the CAD system was evaluated again with the same parameters except the maximum number of boxes in the localization stage was 3 (instead of 10) and the window size was empirically chosen to be 400% with respect to the original mark size. After localization, a total of 93 ROIs, with 19 mass ROIs and 74 normal ROIs, were available. The SVM was trained with a total of 38 ROIs, with 19 mass ROIs and 19 normal ROIs (randomly selected from the 74 normal ROIs). Table 2 shows the results with and without GA-based feature selection and SVM classification.

With a maximum of 3 boxes per image, only 39 images (from a total of 87 images that contain speculated masses) have been identified in the localization stage. In the first experiment with a maximum of 10 boxes per image, 82 images that contain speculated masses were identified. Moreover, the window size was reduced from 900% to 400% because the erosion structure element size and the marker size increased, as shown in Fig.10.

Table 2. Eexperimental results

	No. of Features	Accuracy (%)	Sensitivity (%)	Specificity (%)
Without feature selection	23	73.78	96.90	48.96
With feature selection	5	90.99	92.14	89.40

Fig. 10. Illustration of the marker size (a) Maximum of 10 candidate mass locations are marked; (b) Maximum of 3 candidate mass locations are marked

To compare the results with other approaches, the trained SVM model from the second experiment was evaluated with 19 new images that contain masses. The results show 88.9% true positive detection at 0.79 false positives per image with GA-based feature selection, and with a maximum of 3 boxes per image. Liu et al. [4] reported comparable results on the MIAS database with 84.2% true positive detection at less than 1 false positive per image, and 100% true positive detection at 2.2 false positives per image.

4 Summary

In this paper, we have presented a CAD system for automatic mammography mass identification. The proposed method consists of four stages, i.e., localization using hybrid of morphological operations, segmentation using Chunming's algorithm, feature extraction using wavelet transform, and feature selection and classification using a hybrid GA-SVM model. Encouraging results have been obtained. However, there is room for improvement. Future work will be focused on enhancing localization and segmentation of masses, as well as improving the classification performance with other computational intelligence models.

References

[1] American Cancer Society, http://www.cancer.org (access date: 2010-08-13)

[2] Sampat, M.P., Markey, M.K., Bovik, A.C.: Computer-aided detection and diagnosis in mammography. In: Bovik, A.C. (ed.) Handbook of Image and Video Processing, 2nd edn., pp. 1195–1217. Academic, New York (2005)

[3] Olivera, A., Freixeneta, J., Martía, J., Pérezb, E., Pontb, J., Dentonc, E.R.E., Zwiggelaard, R.: A review of automatic mass detection and segmentation in mammographic images. Med. Imag. Anal. 14(2), 87–110 (2010)
[4] Liu, S., Babbs, C.F., Delp, E.J.: Multiresolution detection of speculated lesions in digital mammograms. IEEE Trans Image Process 10(6), 874–884 (2001)
[5] The Mini-MIAS Database of Mammograms, http://peipa.essex.ac.uk (access date: 2010-08-13)
[6] Mudigonda, N.R., Rangayyan, R.M., Desautels, J.E.L.: Gradient and texture analysis for the classification of mammographic masses. IEEE Trans. Med. Imag. 19(10), 1032–1043 (2000)
[7] Hupse, R., Karssemeijer, N.: Use of normal tissue context in computer-aided detection of masses in mammograms. IEEE Trans. Med. Imag. 28(2), 2033–2041 (2009)
[8] Gao, X., Wang, Y., Li, X., Tao, D.: On combining morphological component analysis and concentric morphology model for mammographic mass detection. IEEE Trans. Info Tech. in Biomed. 14(2), 266–273 (2010)
[9] Shi, J., Sahiner, B., Chan, H.P., Ge, J., Hadjiiski, L.M., Helvie, M.A., Nees, A., Wu, Y.T., Wei, J., Zhou, C., Zhang, Y., Cui, J.: Characterization of mammographic masses based on level set segmentation with new image features and patient information. Med. Phys. 35(1), 280–290 (2008)
[10] Pu, J., Zheng, B., Leader, J.K., Gur, D.: An ellipse-fitting based method for efficient registration of breast masses on two mammographic views. Med. Phys. 35(2), 487–494 (2008)
[11] Velikova, M., Samulski, M., Lucas, P.J., Karssemeijer, N.: Improved mammographic CAD performance using multi-view information: a Bayesian network frameworks. Phys. Med. Biol. 54(5), 1131–1147 (2009)
[12] Varela, C., Tahoces, P.G., Mendez, A.J., Souto, M., Vidal, J.J.: Computerized detection of breast masses in digitized mammograms. Comput. Biol. Med. 37(2), 214–226 (2007)
[13] Rojas, A., Nandi, A.: Detection of masses in mammograms via statistically based enhancement, multilevel-thresholding segmentation, and region selection. Comput Med. Imag. Graph 32(4), 304–315 (2008)
[14] Qian, W., Song, D., Lei, M., Sankar, R., Eikman, E.: Computer-aided mass detection based on ipsilateral multiview mammograms. Acad. Radiol. 14(5), 530–538 (2007)
[15] Digital Database for Screening Mammography, http://marathon.csee.usf.edu (access date: 2010-08-13)
[16] Otsu, N.: A threshold selection method from gray-level histograms. IEEE Trans. Syst. Man, Cybern. 9(1), 62–66 (1979)
[17] Vincent, L.: Morphological grayscale reconstruction in image analysis: applications and efficient algorithms. IEEE Trans. Image Process 2(2), 176–201 (1993)
[18] Li, C., Kao, C., Gore, J.C., Ding, Z.: Minimization of region-scalable fitting energy for image segmentation. IEEE Trans. Image Process 17(10), 1940–1949 (2008)

Exploring Features and Classifiers to Classify MicroRNA Expression Profiles of Human Cancer

Kyung-Joong Kim[1] and Sung-Bae Cho[2]

[1] Dept. of Computer Engineering, Sejong University, Seoul, South Korea
[2] Dept. of Computer Science, Yonsei University, Seoul, South Korea
kimkj@sejong.ac.kr, sbcho@cs.yonsei.ac.kr

Abstract. Recently, some non-coding small RNAs, known as microRNAs (miRNA), have drawn a lot of attention to identify their role in gene regulation and various biological processes. The miRNA profiles are surprisingly informative, reflecting the malignancy state of the tissues. In this paper, we attempt to explore extensive features and classifiers through a comparative study of the most promising feature selection methods and machine learning classifiers. Here we use the expression profile of 217 miRNAs from 186 samples, including multiple human cancers. Pearson's and Spearman's correlation coefficients, Euclidean distance, cosine coefficient, information gain, mutual information and signal to noise ratio have been used for feature selection. Backpropagation neural network, support vector machine, and *k*-nearest neighbor have been used for classification. Experimental results indicate that *k*-nearest neighbor with cosine coefficient produces the best result, 95.0% of recognition rate on the test data.

Keywords: microRNA, Human Cancer, Classification, Feature Selection, Machine Learning.

1 Introduction

High-throughput messenger RNA (mRNA) expression profiling with microarray has produced huge amount of information useful for cancer diagnosis and treatment [1]. It has also promoted the development of techniques to analyze the large amount of information using statistical and machine learning approaches [2]. Computational methods selects relevant subsets of thousands genes and classify samples into normal or tumor tissues. Clustering technology reveals the relevant modules of co-expressed genes that show similar behavioral patterns in gene regulation process [3]. There are several microarray databases accessible by public [4][5].

Recently, small-non-coding RNAs, named microRNAs (miRNA) have drawn a lot of attention to identify their functional roles in biological processes [6][7]. Especially, researchers have investigated that the abnormal expression of miRNAs may indicate human diseases, such as cancers. Lu *et al.* collected 217 miRNAs expression profiles from 334 human and mouse samples using a bead based flow cytometric method [8]. They reported a down-regulation of miRNAs in cancer tissues compared with normal ones. In addition to the observation, they applied simple classification algorithms to

K.W. Wong et al. (Eds.): ICONIP 2010, Part II, LNCS 6444, pp. 234–241, 2010.

the samples which are not easily discriminated with mRNA expression profiles. Some researchers have been attempting to propose the optimal classification technique to work out this problem, especially dealing with predictive discrimination of multiple cancers [9][10].

Although there have been several comprehensive works to compare the possible methods with different feature selection and classification techniques for mRNA expression profiles [11], there have been still no work on the miRNA data. Like mRNA classification problems, there are a lot of possible choices on the combination of feature selection methods and classification algorithms resulting in different recognition accuracy. A through effort helps to find the best possible methods to classify human cancer using miRNA expression profiles. Also, it reveals the superiority of specific feature selection method and classification algorithm over alternatives for the problem.

In this paper, we attempt to explore the features and classifiers that efficiently detect the malignancy status (normal or cancer) of the tissues. We have adopted seven feature selection methods widely used in pattern recognition fields: Pearson's and Spearman's correlations, Euclidean distance, cosine coefficient, information gain and mutual information and signal-to-noise ratio. We have also utilized four *k*-nearest neighbor methods with different similarity measures (Euclidean, Pearson and Spearman correlation, and cosine coefficient), multilayer perceptrons, and support vector machines with linear kernel.

2 MicroRNA

Recently, hundreds of small, non-coding miRNAs have been discovered [7] which are averaging approximately 22 nucleotides in length (Table 1). They are involved with cell proliferation and death, gene regulatory networks, RNA metabolism, auxin signaling and neuronal synapse formation [6][7]. Especially, the expression of miRNAs indicates human diseases such as cancers [8]. Lu *et al.* used *k*-nearest neighbor and probabilistic neural network to classify human cancer using miRNA expression profiles. In their work, they used human miRNA expression data for multiple cancers as training samples to predict the mouse lung cancer's malignancy. They reported 100% accuracy for 12 mouse lung cancer tissues.

Table 1. Examples of miRNA expression profiles [8]

Description	Sample 1	Sample 2
hsa-miR-124a:UUAAGGCACGCGGUGAAUGCCA:bead_101-A	7.4204	6.931
hsa-miR-125b:UCCCUGAGACCCUAACUUGUGA:bead_102-A	10.8391	11.7231
hsa-miR-7:UGGAAGACUAGUGAUUUUGUU:bead_103-A	6.64631	6.78163
hsa-let-7g:UGAGGUAGUAGUUUGUACAGU:bead_104-A	9.86267	10.4861
hsa-miR-16:UAGCAGCACGUAAAUAUUGGCG:bead_105-A	10.6879	11.5479
hsa-miR-99a:AACCCGUAGAUCCGAUCUUGUG:bead_107-A	8.39361	8.88749
hsa-miR-92:UAUUGCACUUGUCCCGGCCUGU:bead_108-A	8.63981	9.06636

Xu *et al.* applied a neural based classifier, Default ARTMAP, to classify broad types of cancers based on their miRNA expression profiles [9]. In their work, particle swarm optimization (PSO) was used for selecting important miRNAs that contribute to the discrimination of different cancer types. Zheng *et al.* reported that discrete function learning algorithm (DFL) obtains better prediction performance than C4.5 (decision tree) and RIP algorithms.

3 MicroRNA Expression Classification

Generally, the miRNA expression profile has high dimensionality with small number of samples because of the limitation of sample availability, cost, or other reasons. After acquiring the miRNA expression profile, prediction systems goes through two stages: feature selection and pattern classification stages. The high dimensionality is one of the major challenges to analyze the miRNA expression profiles decreasing classification accuracy. There are a lot of feature selection methods proposed based on statistical similarity, information theory and signal-to-noise ratio [12]. Feature selection methods select relevant miRNAs which contribute to the discrimination of the malignancy type. At prediction stage, classification algorithms learn models with the selected miRNAs to predict the category of each sample. Finally, their goodness of the classifier is evaluated on unseen samples, called test data.

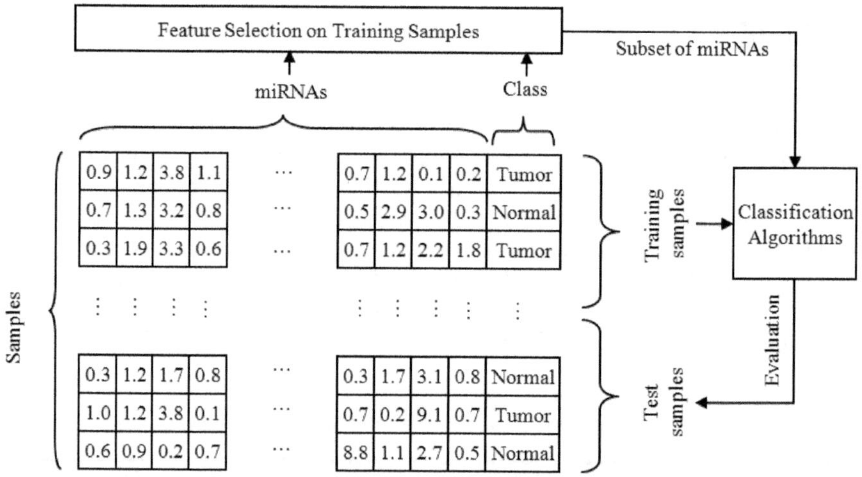

Fig. 1. Overview of miRNA classification system

3.1 Features

3.1.1 Similarity-Based Methods

In these methods, the value of each miRNA is evaluated based on the similarity to ideal vectors. In case of positive ideal vectors, the value is 1 if the training sample is cancer and vice versa. On the other hand, in the case of negative ideal vectors, the value is 1 if the training sample is normal. If there is a miRNA that shows the same behavior with the ideal vectors, this means that we can classify the training samples correctly with only the single miRNA. Because it is not common to classify samples correctly using only single miRNA, this vector is called as "ideal" one.

We can sort the miRNAs in accordance with the similarity between the miRNA's values for training samples and ideal vectors. Because we have the two ideal vectors, there are two different rankings based on positive and negative ideal vectors. Finally, half of the miRNAs are chosen from the rankings by the positive ideal vector, and others are from the one by the negative ideal vector. For example, if we decide to select 20 miRNAs, 10 miRNAs are very close to the positive ideal vectors and 10 miRNAs are very close to the negative ones. There are four different similarity measures used: inverse of Euclidean distance measure, Pearson correlation, cosine coefficient and Spearman correlation.

3.1.2 Information Gain

In the following formula, k is the total number of classes, n_l is the number of values in the left partition, n_r is the number of values in the right partition, l_i is the number of values that belong to class i in the left partition, and r_i is the number of values that belong to class i in the right partition. The information gain of a miRNA is defined as follows. The threshold for the portioning is a value to minimize class entropy. TN is the number of training samples.

$$IG(g_i) = \sum_{i=1}^{k} \left(\frac{l_i}{TN} \log \frac{l_i}{n_l} + \frac{r_i}{TN} \log \frac{r_i}{n_r} \right) - \sum_{i=1}^{k} \left(\frac{l_i + r_i}{TN} \right) \log \left(\frac{l_i + r_i}{TN} \right)$$

3.1.3 Mutual Information

Mutual information provides information on the dependency relationship between two probabilistic variables of events. If two events are completely independent, the mutual information is 0. The more they are related, the higher the mutual information is.

$$\begin{aligned} MI(g_i) = {} & MI(g_i \geq \overline{g_i}, t_i = NORMAL) + MI(g_i \geq \overline{g_i}, t_i = TUMOR) \\ & + MI(g_i < \overline{g_i}, t_i = NORMAL) + MI(g_i < \overline{g_i}, t_i = TUMOR) \end{aligned}$$

$$\overline{g_i} = \frac{1}{TN} \sum_{j=1}^{TN} g_{ji}$$

$$MI(g_i > \overline{g_i}, t_i = NORMAL) = P(g_i > \overline{g_i}, t_i = NORMAL) \log_{10} \frac{P(g_i > \overline{g_i}, t_i = NORMAL)}{P(g_i > \overline{g_i}) \times P(t_i = NORMAL)}$$

3.1.4 Signal-to-Noise Ratio

If we calculate the mean μ and standard deviation σ from the distribution of miRNA expressions within their classes, the signal-to-noise ratio (SN) of miRNA g_i is defined as follows:

$$SN(g_i) = \frac{|\mu_{NORMAL}(g_i) - \mu_{TUMOR}(g_i)|}{\sigma_{NORMAL}(g_i) + \sigma_{TUMOR}(g_i)}$$

3.2 Classifiers

3.2.1 K-Nearest Neighbor (KNN)

This is one of the most common methods for instance-based induction. Given an input vector, KNN extracts the *k* closest vectors in the reference set based on similarity measures, and makes a decision for the label of the input vector by using the labels of the *k* nearest neighbors. In this paper, many similarity measures were used such as the inverse of Euclidean distance (KNNE), Pearson correlation (KNNP), cosine coefficients (KNNC) and Spearman correlation (KNNS). If the *k* is not 1, the final outcome is based on the majority voting of the *k* nearest neighbors.

3.2.2 Multi-Layer Perceptron (MLP)

A feed-forward multilayer perceptron is an error backpropagation neural network that can be applied to pattern recognition problems. It requires engineering regarding the architecture of the model (the number of hidden layers, hidden neurons, and so on). In this classification problem, the number of output nodes is two (normal and tumor nodes). If the output from the normal node is larger than that from the tumor node, the sample is classified as normal.

3.2.3 Support Vector Machine (SVM)

This method classifies the data into two classes. SVM builds up a hyperplane as the decision surface in such a way as to maximize the margin of separation between positive and negative samples. In this paper, linear kernel (SVML) is used.

4 Experimental Results

We have used miRNA samples from Lu *et al.*'s work [8]. It contains expression values of 217 miRNAs from 186 samples including multiple cancer types (Table 2). In this work, we did binary classifications which classify samples as one of tumor or normal.

The expression level of each miRNA is normalized to 0~1. For miRNAs, we found the maximum and minimum expression values. The miRNA expression value is adjusted to (g-min)/(max-min). In the feature selection, the number of top-ranked miRNAs is 25. There is no report on the optimal number of miRNAs, but our previous study on mRNA expression profiles indicates that 25 is reasonable [2]. For Information Gain feature selection, we implemented it based on the RANKGENE source code and our IG method showed the same results with the RANKGENE [13]. We used LIBSVM for the SVM classification [14]. The parameters of classification algorithms are summarized in Table 3. The final results are an average of 10 runs. For each run, the miRNA expression data are randomly separated to the training dataset (2/3) and test dataset (1/3).

Table 2. The number of samples for each cancer type

Cancer	Normal	Tumor
Stomach	6	0
Colon	5	10
Pancreas	1	9
Liver	3	0
Kidney	3	5
Bladder	2	7
Prostate	8	6
Ovary	0	7
Uterus	9	10
Human Lung	4	6
Mesothelioma	8	0
Melanoma	0	3
Breast	3	6
Brain	2	0
B Cell ALL	0	26
T Cell ALL	0	18
Follicular Cleaved Lymphoma	0	8
Large B Cell Lymphoma	0	8
Mycosis Fungoidis	0	3
Sum	54	132

Table 3. Parameters of classification algorithms

Classifier	Parameter	Value
MLP	# of input nodes	25
	# of hidden nodes	8
	# of output nodes	2
	Learning rate	0.05
	Momentum	0.7
	Learning algorithm	Back propagation
KNN	*k*	3
SVM	Kernel function	Linear

Table 4 shows the comparison of accuracy on test data for the 42 combinations of feature selection and classifications. It shows that the KNNS-CC combination is the best accuracy 95% among them. Figure 2 shows the comparison of average performance of feature selection and classification methods. In the feature selection methods, CC is the best one. However, in the classification algorithm, KNNE is the best one. This means that it is important to find the appropriate combination of feature selection and classification algorithm to get the best accuracy from the miRNA expression profiles. Table 5 shows relevant miRNAs selected by CC methods.

Table 4. Accuracy on test data

	ED	PC	CC	SC	IG	MI	SN
KNNE	92.7	92.4	91.7	90.0	91.7	93.0	90.8
KNNP	93.3	86.4	94.1	87.5	92.9	91.7	93.2
KNNC	92.2	85.9	93.2	87.7	90.6	90.6	90.9
KNNS	93.0	85.3	**95.0**	87.9	89.5	90.9	91.2
MLP	92.0	92.5	94.0	91.2	89.3	91.1	89.5
SVML	91.6	91.7	92.2	91.7	90.9	92.9	90.6

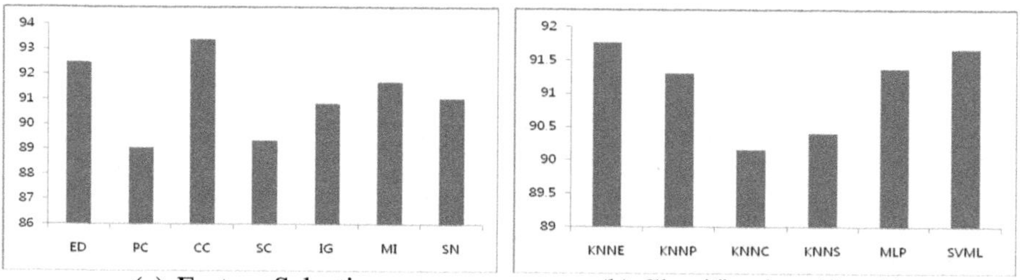

(a) Feature Selection (b) Classification Algorithms

Fig. 2. Comparison of average performance of feature selection and classification methods

Table 5. Relevant miRNAs selected by cosine coefficient

Value	Description
0.814328	hsa-miR-146:UGAGAACUGAAUUCCAUGGGUU:bead_109-A
0.812209	hsa-miR-296:AGGGCCCCCCCUCAAUCCUGU:bead_105-C
0.808118	hsa-miR-21:UAGCUUAUCAGACUGAUGUUGA:bead_119-B
0.805954	hsa-let-7a:UGAGGUAGUAGGUUGUAUAGUU:bead_159-B
0.803176	hsa-miR-16:UAGCAGCACGUAAAUAUUGGCG:bead_105-A
0.799869	hsa-let-7c:UGAGGUAGUAGGUUGUAUGGUU:bead_110-A

5 Conclusions

In this paper, we explore the feature selection and classification algorithms for miRNA expression profiles to classify human cancer. Compared to mRNA expression profile, there are few works using machine learning tools for miRNA data. In this work, we applied seven feature selection methods and six classification algorithms to find the best combination of them. Experimental results show that KNNS + CC method records the best accuracy 95%. For feature selection method, cosine coefficient is the best method. For classification algorithm, KNNE is the superior method. In conclusion, it is important to choose the proper combination of feature selection and classification algorithm to get the high accuracy for miRNA expression profiles.

Acknowledgements

This research was supported by Basic Science Research Program and the Original Technology Research Program for Brain Science through the National Research Foundation of Korea (NRF) funded by the Ministry of Education, Science and Technology (2010-0012876) (2010-0018948).

References

1. Golub, T., Slonim, D., Tamayo, P., Huard, C., Gaasenbeek, M., Mesirov, J., et al.: Moecular classification of cancer: Class discovery and class prediction by gene expression monitoring. Science 286, 531–537 (1999)
2. Cho, S.B., Won, H.H.: Machine learning in DNA microarray analysis for cancer classification. In: The First Asia Pacific Bioinformatics Conference (2003)
3. Segal, E., Shapira, M., Regev, A., Pe'er, D., Botstein, D., Koller, D., et al.: Module networks: Identifying regulatory modules and their condition-specific regulators from gene expression data. Nature Genetics 34, 166–176 (2003)
4. Stanford Microarray Database, `http://smd.stanford.edu/`
5. Gene Expression Omnibus, `http://www.ncbi.nlm.nih.gov/geo/`
6. Ambros, V.: The functions of animal microRNAs. Nature 431, 350–355 (2004)
7. Bartel, D.: MicroRNAs: Genomics, biogenesis, mechanism, and function. Cell 116, 281–297 (2004)
8. Lu, J., Getz, G., Miska, E.A., Alvarez-Saavedra, E., Lamb, J., Peck, D., Sweet-Cordero, A., et al.: MicroRNA expression profiles classify human cancers. Nature 435, 834–838 (2005)
9. Xu, R., Xu, J., Wunsch II, D.C.: MicroRNA expression profile based cancer classification using Default ARTMAP. Neural Networks 22, 774–780 (2009)
10. Zheng, Y., Kwoh, C.K.: Informative MicroRNA expression patterns for cancer classification. In: Li, J., Yang, Q., Tan, A.-H. (eds.) BioDM 2006. LNCS (LNBI), vol. 3916, pp. 143–154. Springer, Heidelberg (2006)
11. Cho, S.B.: Exploring features and classifiers to classify gene expression profiles of acute leukemia. International Journal of Pattern Recognition and Artificial Intelligence 16(7), 831–844 (2002)
12. Saeys, Y., Inza, I., Larranaga, P.: A review of feature selection techniques in bioinformatics. Bioinformatics 23, 2507–2517 (2007)
13. Su, Y., Murali, T.M., Pavlovic, V., Schaffer, M., Kasif, S.: RankGene: Identification of diagnostic genes based on expression data. Bioinformatics 19, 1578–1579 (2003)
14. `http://www.csie.ntu.edu.tw/~cjlin/libsvm/`

SOMIX: Motifs Discovery in Gene Regulatory Sequences Using Self-Organizing Maps

Nung Kion Lee and Dianhui Wang*

Department of Computer Science and Computer Engineering,
La Trobe University, Melbourne, Victoria 3086, Australia
dh.wang@latrobe.edu.au

Abstract. We present a clustering algorithm called Self-organizing Map Neural Network with mixed signals discrimination (SOMIX), to discover binding sites in a set of regulatory regions. Our framework integrates a novel intra-node soft competitive procedure in each node model to achieve maximum discrimination of motif from background signals. The intra-node competition is based on an adaptive weighting technique on two different signal models: position specific scoring matrix and markov chain. Simulations on real and artificial datasets showed that, SOMIX could achieve significant performance improvement in terms of sensitivity and specificity over SOMBRERO, which is a well-known SOM based motif discovery tool. SOMIX has also been found promising comparing against other popular motif discovery tools.

Keywords: self-organizing map, regulatory elements discovery, hybrid model.

1 Introduction

Identification of transcription factor binding sites (TFBS) is fundamental to understand gene regulation. The binding sites or motif instances are typically $10 \sim 15bp$ and degenerated in some positions. They are often buried in a large amount of non-functional background sequences which cause low motif signal-to-noise ratio. Computational discovery of the TFBS (that bind with common transcription protein) from the upstream DNA sequences of co-regulated genes, is regarded as computational motif discovery. Fundamental of these approaches is to search for motifs that are over-represented (over-abundance) in the input sequences compared to the background sequences. Algorithms based on various search strategies have been proposed to discover those over-represented motifs. They include MEME [1], ALIGNACE[2] and SOMBRERO[3]. In this paper, we aim to develop a self-organizing map (SOM) neural network with a customized hybrid node's model for motif discovery.

Standard SOM with weight vector as node model representation has been widely used in biological sequences clustering[4,5]. This representation is inappropriate for our purpose because it requires the input DNA sequences

* Corresponding author.

K.W. Wong et al. (Eds.): ICONIP 2010, Part II, LNCS 6444, pp. 242–249, 2010.

to be encoded into numerical values, which might cause loss of essential features and also no meaningful interpretation can be made after computations. Kohonen [6] proposed SOM with a string based node model for protein family clustering. Despite useful, the string representation has limited expressive power to exemplify the probabilistic nature of protein binding sites recognition. As a result, its discrimination ability is rather weak. Recently it is proposed to replace the standard weight vector with position- specific-scoring-matrix (PSSM) which is a signal based model in [3]. Despite some success in higher recall rates, it suffers from higher false positive rates.

The critical weakness of the SOM based traditional methods for (DNA) motif discovery is that they share a common assumption that, the motif and the background signals can be analogously and efficiently modelled by using a homogeneous node's model. This assumption is rather weak because the two classes of signals have some distinct statistical properties[7]. Forcing these incompatible signals to be represented using a homogeneous model usually produces high false positive rates as can be seen in [3]. Hence, it is necessary to consider the use of node model that takes into consideration these two classes of signals (i.e., motif and background) separately.

In this paper, we present a novel clustering framework based on SOM neural network, termed SOMIX, for the identification of motifs in DNA sequences. We propose a *hybrid node model* by combining/hybridizing PSSM[8] and markov chain (MC)[9] model to address some of the limitations of current SOM approaches. Then an adaptive weighting scheme is applied to control the soft-competition of those two model components in representing the mixture of signals in a node. We hypothesize that the fitness of each model's component with respect to the sequences in a node is a fuzzy indication of its signal class composition. We evaluate our proposed algorithm with several motif discovery tools using real and artificial DNA datasets.

The remainder of this paper is organized as follows: Section 2 presents our signal discrimination system framework; Section 3 presents the learning algorithm and data assignment; Section 4 reports the datasets used and results from the comparative studies; Section 5 discusses some issues related to our framework and offers a number of future directions.

2 System and Methods

2.1 Basic Concepts and Problem Formulation

The main idea of our system, called Self-organizing Map with mixed signals discrimination (SOMIX), is to use a hybrid node's model composed of heterogeneous models:Position Specific Scoring Matrix(PSSM) and Markov Chain(MC). We first give some notations used in this paper and then describe the SOMIX structure. Denoted $D = \{S_1, S_2, \ldots, S_N\}$, a DNA dataset with N sequences. Let a kmer $K_i = (b_1 b_2 \ldots b_k) \in \{A, C, G, T\}^k$ be a continuous subsequence of length k in a DNA sequence and $i = 1, \ldots, Z$, with Z is the total number of kmers. For a length L DNA sequence, there are $L - k + 1$ kmers in one of its strands. We also define a motif M as a collection of binding sites that characterizes a specific type of DNA binding domain or module and background B as non-functional random sequences.

A SOM neural network has two layers:input layer and output layer. The input layer has $4 \times k$ nodes that accept the encoded kmer as follows: $[(a_{1A}, a_{2C}, a_{2G}, a_{2T})$,..., $(a_{kA}, a_{kC}, a_{kG}, a_{kT})]$. The output layer is a 2-dimensional (2d) lattice of $R \times C$ nodes, where R, C is the number of rows and columns respectively. Each node V_{ij}, $i = 1, \ldots, R$ and $j = 1, \ldots, C$, has a parametrized model Θ, representing the data points assigned to it. For convenience, we use the notation V_l to represent a node, where $1 \leq l \leq (R \times C)$. The coordinate of a node V_l in the lattice is expressed as $z_l = (i, j)$.

2.2 Modeling the Motif Signals

We use the Position Specific Scoring Matrices (PSSM) to model the motif signals. The PSSM of a motif M is a matrix $W = [f(b, i)]_{4 \times k}$, $b \in \{A, C, G, T\}$ and $i = 1, ..., k$, with each entry $f(b, i)$ represents the probability of observing nucleotide b in position i. Hence, each column adds up to a total of one. The PSSM's entries can be estimated from the kmers in a node using the maximum likelihood principle, with a pseudo-count value added as under sample correction to the probabilistic model. The PSSM entries are computed as follows:

$$f(b, i) = (c(b, i) + g(b)) / (N + 1), \qquad (1)$$

where N is the number of kmers, $c(b, i)$ is the number of times nucleotide b occurs at position i of a set of kmers in a node, $g(b) = [n(b) + 0.25]/(N \times k + 1)$ and $n(b) = \sum_{i=1}^{k} c(b, i)$.

2.3 Modeling the Background Signals

The MC model [9] assumes that the probability of occurrence of a nucleotide b_i at position i in a DNA sequence is dependent only on the occurrences of m (i.e. the markov order) previous nucleotides. In our approach, the first order MC (i.e. $m = 1$) is used because higher order model usually requires more input data to avoid over-fitting. The maximum likelihood estimation of the conditional probability $p(b_i|b_{i-m}...b_{i-1})$ is given by[9]

$$p(b_i|b_{i-m}...b_{i-1}) = \frac{c'(b_{i-m}...b_{i-1}b_i)}{\sum_{w=A}^{T} c'(b_{i-m}...b_{i-1}w)}, \qquad (2)$$

where $c'(x)$ is the number of times sub-sequence x found in a set of kmers in a node.

Denoted $\pi(a, b)$, the conditional probability $p(b|a)$ of the first order MC, where $a, b \in \{A, C, G, T\}$. The MC transition matrix is given by $T = [\pi(a, b)]_{4 \times 4}$, where $\sum_{b=\{A,C,G,T\}} \pi(a, b) = 1$.

2.4 Similarity Metric

A similarity metric is needed for kmers assignment to the SOM's nodes during the learning process. The score of a kmer $K_j = (b_1 b_2 \ldots b_k)$ to the PSSM W_l of a node is computed as follows [10]:

$$W_l(K_j) = -\log(\prod_{i=1}^{k} f(b_i, i)). \tag{3}$$

Whereas, the score of a kmer K_j to the MC model with 1st-order transition matrix T_l is computed as [9]:

$$T_l(K_j) = -\log\left(p(b_1)\prod_{i=2}^{k} \pi(b_{i-1}, b_i)\right), \tag{4}$$

where $p(x)$ is the probability of observing nucleotide x, estimated from the set of kmers in the node.

2.5 Hybrid Node Model

We use a simple linear weighting scheme to combine the PSSM and MC as follows: Let W^N denote the PSSM constructed from N kmers in a node and T^N is the 1st-order MC transition matrix constructed using the same set of kmers. Define Θ to be $\beta W^N + (1-\beta)T^N$ for $\beta \in [0,1]$. Θ is an entry-wise weighted average of the matrices W^N and T^N, not a mixture of these two matrices. The weight value β corresponds to the fitness of the constituent models to represent kmers in a node. It is the likelihood of a node to be motif or background. The score of a kmer to a node's hybrid model Θ_l is given by

$$\Theta_l(K_j) = \sqrt{\frac{\beta}{W_l^N(K_j)} + \frac{1-\beta}{T_l^N(K_j)}}. \tag{5}$$

Note that, when $\beta = 1$, the hybrid model is simply a PSSM model; whereas when $\beta = 0$, it is a MC model.

3 Algorithm

3.1 Adaptation Process

We associate each node with three computing components including: two matrices $\Delta W, \Delta T$ and a counter r. Let V_l^* be BMU of an input kmer K. Denoted $\Delta W = [\Delta f(b,i)]_{4\times k}$ for $b \in \{A,C,G,T\}$ and $i = 1,\ldots,k$. Similarly, let $\Delta T = [\Delta\pi(m,n)]_{4\times 4}$ for $m,n \in \{A,C,G,T\}$. We initialize all entries in both matrices ΔW and ΔT as 0. Also let $r = 0$. Once a winning node for a kmer K is found, the matrices of a node V_l^* are updated as follows.

$$\Delta f(b,i) = \Delta f(b,i) + h(z_l^*, z_j, \sigma)a_{bi}, \tag{6}$$

$$\Delta\pi(m,n) = \Delta\pi(m,n) + h(z_l^*, z_j, \sigma)count(m,n)/(k-1), \quad (7)$$

where a_{ib} is an entry of the binary matrix $e(K)$ as defined in Sub-section 2.1, $count(m,n)$ is the frequency of di-nucleotide (mn) in kmer K and h is a Gaussian neighbourhood function. We also update $r = r+1$. Upon completion of an epoch, all nodes' model parameters will be updated as follows:

$$f(b,i)_{new} = f(b,i) + \eta\frac{\Delta f(b,i)}{r}, \quad (8)$$

$$\pi(m,n)_{new} = \pi(m,n) + \eta\frac{\Delta\pi(m,n)}{r}, \quad (9)$$

where η is the learning rate and $f(b,i)$ and $\pi(m,n)$ is defined in Eq. (1) and Eq. (2) respectively. Note that, in the computation of Eq. (8) and Eq. (9), we first compute $f(b,i)$ and $\pi(m,n)$ using the current set of kmers assigned to the node. It is also necessary to update the weighting parameters β. Assuming a set of N_l kmers $\{K_1,\ldots,K_{N_l}\}$ is assigned to a node V_l at the end of an epoch, the weighting parameters update equation is as follows:

$$\beta_{new} = \frac{\sum_{i=1}^{N_l} W_l(K_i)}{\sum_{i=1}^{N_l} (W_l(K_i) + T_l(K_i))}. \quad (10)$$

4 Results

For comparisons purposes, we employ the precision(P), recall(R) and F-measure(F) measure which are computed as [11]: P = TP/(TP + FP), R = TP/Y, F = 2/(1/P + 1/R), where TP, FP, and Y are the numbers of true positives, false positives and true sites in the dataset respectively. We consider a predicted site a true hit if it overlapped the true site for x nucleotides; where x depends on the length of the true motif consensus. To rank the motifs (i.e. clusters) from the final map, we use the MAP score [12].

4.1 Performance on Real Datasets

We compared the performances of SOMIX with four popular motif discovery tools, MEME, Weeder, SOMBRERO and AlignACE, on real biological datasets with experimentally verified motif locations. The eight test datasets are composed of seven datasets used in [13] and a dataset downloaded from the Promoter Database of *S. cerevisiae*. Each sequence contains at least one true binding site. SOMIX is run with map sizes that are arbitrarily selected as $s \times s$, with $s \in \{6, 8, ..., 20\}$. In each case, the SOMIX is trained for 100 epochs with a motif length value in $[l-3, l+3]$, where l is the expected known motif length.

Table 1. Evaluation results with comparisons

	SOMIX			SOMBRERO			MEME			ALIGNACE			WEEDER		
	R	P	F	R	P	F	R	P	F	R	P	F	R	P	F
CRP	**0.91**	0.89	**0.9**	0.83	0.43	0.56	0.59	0.88	0.69	0.83	0.98	**0.9**	0.75	0.83	0.79
GCN4	0.69	0.45	0.54	**0.8**	0.41	0.53	0.52	0.52	0.52	0.61	0.62	0.6	0.64	**0.87**	**0.73**
ERE	0.74	0.58	0.65	0.8	0.59	0.67	0.72	**0.82**	0.77	0.75	0.77	0.76	**0.76**	0.54	0.63
MEF2	0.81	**0.99**	**0.89**	0.35	0.22	0.27	**0.92**	0.8	0.85	0.86	0.87	0.86	0.88	0.88	0.88
SRF	0.84	0.74	**0.79**	0.67	**0.83**	0.74	**0.87**	0.72	**0.79**	0.83	0.71	0.77	0.83	0.71	0.76
CREB	**0.89**	0.67	**0.77**	0.83	0.43	0.56	0.59	**0.88**	0.69	0.52	0.66	0.57	0.79	0.71	0.75
E2F	**0.82**	0.64	0.71	0.76	0.67	0.71	0.68	0.64	0.65	0.75	**0.68**	**0.71**	0.89	0.67	**0.76**
MYOD	**0.66**	0.39	**0.49**	0.5	0.32	0.39	0.23	0.38	0.27	0.34	0.31	0.32	0.43	**0.5**	0.46
Average	**0.80**	0.67	**0.72**	0.69	0.49	0.55	0.64	**0.71**	0.65	0.69	0.70	0.69	0.75	**0.71**	0.72

The top 10 highest ranked motifs according to their MAP score are saved for evaluation purpose. The background model used in the MAP scores computation is a third-order markov chain model taken from [3]. The learning rate parameter is fixed at 0.005 in all the experiments. Weeder, MEME and ALIGNACE were rans online. Whereas, SOMBRERO is downloaded from the authors' website.

Table 1 shows the results of the comparative study, showing recall (R), precision(P) and F-measure (F) rates. The values in the table are computed from an average of ten runs returned by each program. The map sizes used in each dataset are: GCN:12x12-16x16, CREB:10x10-12x12, CRP:6x6-10x10, E2F:12x12-20x20, ERE:12x12-16x16, MYOD:12x12-16x16, MEF2:12x12-16x16, and SRF:12x12-14x14. In terms of recall rates, SOMIX performs better than or equally to other tools in four of the eight (8) datasets. Compared with SOMBRERO, SOMIX performs better in terms of recall rates in six (6) of the datasets. Also, SOMIX has higher precision and F-measure (except ERE) rates in six (6) and seven (7) of the test datasets respectively. Notably, for the MEF2 dataset, SOMIX obtained a much higher precision rate (0.99 vs 0.32) in comparison with SOMBRERO. The performances on all datasets show that SOMIX achieves significant improvements in the average precision (26.9%) and recall rate (13.8%) in comparison with SOMBRERO. This clearly shows that, SOMIX with heterogeneous node model can represents the true distribution of the DNA sequences better than homogeneous model.

It can be noticed that SOMIX performance is comparable or better than ALIGNACE, MEME and Weeder. For example, in terms of F-measure rates, SOMIX produces the best results for five (5) of the eight (8) datasets due to its higher precision rates (it is to be noted that both SOMIX and ALIGNACE achieve the same F-measure value for the CRP dataset). SOMIX's average F-measure value for all datasets (i.e. 0.72) is better than MEME, ALIGNACE and SOMBRERO and equally good to Weeder.

4.2 Performance on Artificial Datasets with Planted Multiple Motifs

We prepared five artificial DNA datasets with three planted distinct motifs in each dataset, generated from Annotated regulatory Binding Sites (ABS, v1.0)

Table 2. Evaluation results with comparisons for multiple motifs datasets

		SOMIX			SOMBRERO			MEME			WEEDER		
		R	P	F	R	P	R	R	P	F	R	P	F
Dataset1	CREB	0.43	0.26	**0.33**	**0.44**	0.26	**0.33**	0.20	**1.00**	**0.33**	0.00	0.00	0.00
	MYOD	**0.48**	**0.23**	**0.31**	0.20	0.08	0.11	0.00	0.00	0.00	0.00	0.00	0.00
	TBP	**0.36**	0.21	**0.26**	0.20	0.12	0.15	0.07	**0.50**	0.12	0.00	0.00	0.00
	Avg	**0.42**	0.23	**0.30**	0.28	0.15	0.20	0.09	**0.50**	0.15	0.00	0.00	0.00
Dataset2	NFAT	0.39	0.27	0.31	0.36	0.21	0.26	**0.44**	**0.78**	**0.56**	0.00	0.00	0.00
	HNF4	0.57	0.40	0.47	**0.63**	0.39	0.48	0.60	0.82	**0.69**	0.40	**1.00**	0.57
	SP1	0.50	0.53	**0.50**	**0.53**	0.35	0.42	0.38	**0.54**	0.44	0.00	0.00	0.00
	Avg	0.49	0.40	0.43	**0.51**	0.32	0.39	0.47	**0.71**	**0.56**	0.13	0.33	0.19
Dataset3	CAAT	**0.43**	0.21	0.25	0.32	0.17	0.22	0.29	**0.80**	**0.42**	0.00	0.00	0.00
	SRF	**0.70**	0.40	**0.50**	0.59	0.28	0.38	0.29	**0.57**	0.38	0.00	0.00	0.00
	MEF2	0.79	0.45	0.57	0.65	0.31	0.27	**0.80**	0.57	**0.67**	0.27	**1.00**	0.42
	Avg	**0.64**	0.35	0.44	0.52	0.25	0.29	0.46	**0.65**	**0.49**	0.09	0.33	0.14
Dataset4	USF	0.68	0.39	0.48	**0.73**	0.48	**0.57**	0.41	**0.88**	0.56	0.00	0.00	0.00
	HNF3B	**0.47**	0.25	**0.31**	0.26	0.13	0.17	0.15	**1.00**	0.27	0.00	0.00	0.00
	NFKB	0.71	0.47	0.56	0.66	0.46	0.54	**0.80**	0.57	**0.67**	0.33	**1.00**	0.50
	Avg	**0.62**	0.37	0.45	0.55	0.36	0.43	0.45	**0.82**	**0.50**	0.11	0.33	0.17
Dataset5	GATA3	**0.61**	0.37	0.46	0.49	0.33	0.36	0.40	0.75	0.52	0.40	**1.00**	**0.57**
	CMYC	0.74	0.47	0.57	**0.89**	0.70	0.84	0.75	**1.00**	**0.86**	0.19	0.75	0.30
	EGR1	**0.66**	0.36	0.47	0.47	0.26	0.33	0.64	**0.81**	**0.72**	0.00	0.00	0.00
	Avg	**0.67**	0.40	0.50	0.62	0.43	0.51	0.60	**0.85**	**0.70**	0.20	0.58	0.29

Note: The best value for each evaluation measure is bolded. The motif names in each dataset are listed in the second column.

database [14]. Every DNA dataset has twenty sequences and each sequence is 500 base-pairs in length. We run MEME, Weeder, SOMIX and SOMBRERO five times on each dataset. We asked SOMIX and MEME to return the top 20 motifs for the evaluation purposes. Again, we evaluate all best motifs returned by Weeder and SOMBRERO. Both SOMIX and SOMBRERO used the map size 40x20 for all datasets.

Table 2 shows the comparison between four algorithms on datasets with multiple motifs. On overall, SOMIX has the best recall rates in seven(7) out of fifteen(15) of the motifs. However, such higher recall rates come at the price of having lower precision rates compared with MEME and Weeder. Compared with SOMBRERO, SOMIX performs significantly better in most of the datasets in all performance measures. For example, in terms of recall rates, SOMIX is higher in ten(10) of the motifs; whereas, in terms of F-measure values, SOMIX has better results in twelve(12) of the motifs. Hence, we can conclude that the SOMIX hybrid model has better signal discrimination ability than SOMBRERO. In terms of the average F-Measure, MEME performs slightly better than SOMIX in four(4) out of five(5) of the datasets. Nevertheless, SOMIX has higher average recall rates in all of the datasets compared to MEME. Weeder performs poorly in most of the test datasets. This could be due to the weakness of its scoring function used to rank more potential motifs.

5 Conclusion

In this paper, we proposed a self-organizing map neural network clustering algorithm for simultaneous identification of multiple-motifs in DNA dataset.

Unlike existing works, our method uses a hybrid node model for motif and background signals discrimination based on their distinctive properties. Simulation results have demonstrated that, this representation has improved the false positive rates compared against SOMBRERO performances. Also, it obtained average higher recall rates as compared with other motif discovery tools. These results revealed that SOM with hybrid node's model has better representation of the kmers distribution in the input space. However, reducing the false positive rates using appropriate map sizes, requires further investigations. Some post-processing can be performed to reduce the false positive rates due to the non-optimal map sizes.

References

1. Bailey, T.L., Elkan, C.: Unsupervised learning of multiple motifs in biopolymers using expectation maximization. Machine Learning 21(1), 51–80 (1995)
2. Roth, F.P., Hughes, J.D., Estep, P.W., Church, G.M.: Finding dna regulatory motifs within unaligned noncoding sequences clustered by whole-genome mrna quantitation. Nat. Biotech. 16(10), 939–945 (1998)
3. Mahony, S., Hendrix, D., Golden, A., Smith, T.J., Rokhsar, D.S.: Transcription factor binding site identification using the self-organizing map. Bioinformatics 21(9), 1807–1814 (2005)
4. Ferrn, E.A., Ferrara, P.: Clustering proteins into families using artificial neural networks. Comput. Appl. Biosci. 8(1), 39–44 (1992)
5. Liu, D., Xiong, X., DasGupta, B., Zhang, H.: Motif discoveries in unaligned molecular sequences using self-organizing neural networks. IEEE Transactions on Neural Networks 17(4), 919–928 (2006)
6. Kohonen, T., Somervuo, P.: How to make large self-organizing maps for nonvectorial data. Neural Networks 15(8-9), 945–952 (2002)
7. Gunewardena, S., Zhang, Z.: A hybrid model for robust detection of transcription factor binding sites. Bioinformatics 24(4), 484–491 (2008)
8. Stormo, G.D.: Dna binding sites: representation and discovery. Bioinformatics 16(1), 16–23 (2000)
9. Robin, S., Rodolphe, F., Schbath, S.: DNA, Words and Models. Cambridge University Press, New York (2005)
10. Pavesi, G., Mauri, G., Pesole, G.: An algorithm for finding signals of unknown length in dna sequences. Bioinformatics 17(suppl. 1), S207–S214 (2001)
11. Fawcett, T.: An introduction to roc analysis. Pattern Recognition Letters 27(8), 861–874 (2006)
12. Liu, X.S., Brutlag, D.L., Liu, J.S.: An algorithm for finding protein-dna binding sites with applications to chromatin-immunoprecipitation microarray experiments. Nat. Biotech. 20(8), 835–839 (2002)
13. Wei, Z., Jensen, S.T.: GAME: detecting cis-regulatory elements using a genetic algorithm. Bioinformatics 22(13), 1577–1584 (2006)
14. Blanco, E., Farre, D., Alba, M.M., Messeguer, X., Guigo, R.: ABS: a database of Annotated regulatory Binding Sites from orthologous promoters. Nucl. Acids Res. 34(suppl. 1), D63–D67 (2006)

Microarray-Based Disease Classification Using Pathway Activities with Negatively Correlated Feature Sets

Pitak Sootanan[1], Santitham Prom-on[2], Asawin Meechai[3], and Jonathan H. Chan[4,*]

[1] Individual Based Program (Bioinformatics)
[2] Department of Computer Engineering
[3] Department of Chemical Engineering
[4] School of Information Technology
King Mongkut's University of Technology Thonburi, Bangkok, Thailand
jonathan@sit.kmutt.ac.th

Abstract. The vast amount of data on gene expression that is now available through high-throughput measurement of mRNA abundance has provided a new basis for disease diagnosis. Microarray-based classification of disease states is based on gene expression profiles of patients. A large number of methods have been proposed to identify diagnostic markers that can accurately discriminate between different classes of a disease. Using only a subset of genes in the pathway, such as so-called condition-responsive genes (CORGs), may not fully represent the two classification boundaries for Case and Control classes. Negatively correlated feature sets (NCFS) for identifying CORGs and inferring pathway activities are proposed in this study. Our two proposed methods (NCFS-i and NCFS-c) achieve higher accuracy in disease classification and can identify more phenotype-correlated genes in each pathway when comparing to several existing pathway activity inference methods.

Keywords: Microarray-based classification, pathway activity, negatively correlated feature sets, CORG-based, phenotype-correlated genes.

1 Introduction

Microarray technology is a powerful approach for genomics research. It is a general approach in gene expression profiling which offers a mean to study the molecular activities underlying a disease [1, 2]. As a result, microarray-based classification has become a widespread technique for identifying diagnostic markers of various disease states, outcomes, or responses to treatment especially in cancer [1, 3-6]. These markers are typically selected by scoring each individual gene for how well its expression pattern can discriminate between different classes of disease. However, finding reliable gene markers is a challenging problem, and several recent studies have questioned the reliability of many classifiers based on individual gene markers [7-9].

---

* Corresponding author.

K.W. Wong et al. (Eds.): ICONIP 2010, Part II, LNCS 6444, pp. 250–258, 2010.

There are usually many pathways involved in the mechanism of complex diseases such as cancers. Many genes have been proposed to be involved in the mechanism of cancer but a far lower number of pathways have been determined to link with cancer [10]. KEGG, Kyoto Encyclopedia of Genes and Genomes, is a one of the most common public database which stores the information of pathways [11]. These pathways can be used to help determine the biological-relevant gene expression profile from microarray data. In order to utilize this information, pathway-based analysis has been developed to perform disease classification of expression profiles for more precision than using individual genes [12]. There are many ways to infer the activity of a given pathway based on expression levels of the constituent genes to be the pathway markers in classification [12-15]. Most of them would use all of the member genes in each pathway [12-13, 15]. One recent method proposed by Lee *et al.* [14] infers pathway activity using only a subset of genes in the pathway. This subset contains so-called condition-responsive genes (CORGs). The CORG-based method can effectively incorporate pathway information into expression-based diagnosis of disease to build more biologically defensible models to accurately discriminate the phenotype of interest.

As an improvement of CORG-based method, the use of negatively correlated feature sets (NCFS) for identifying CORG set was incorporated in our previous study to increase the discriminative power of disease detection [16]. However, a potential shortcoming of this approach is that it only employs a small number of member genes in pathways for inferring its activity. It is possible that these member genes may not fully represent the differentiation between the two different aspects of classification boundary. In this paper, we propose a novel modification by using NCFS to separately identify CORG sets. Instead of using only one negatively correlated feature subset which represented only one differentiation pattern, two different gene subsets are used to infer pathway activities to ensure the maximization of difference in both up- and down-regulated fashion. We demonstrate the effectiveness of our proposed method by applying it to classify breast cancer metastasis and have compared its classification accuracy to several other pathway-based approaches.

2 Previous Work

There are a number of methods for inferring pathway activity that have been proposed recently [12-16]. The basic assumption sharing among these methods is that the classification system performs better if the markers are biologically relevant to the disease under study. All of these studies typically yield more reliable results compared to traditional gene-based classifiers. However, there are several shortcomings of the previous pathway activity inference methods. For example, Guo *et al.* [12] proposed methods to estimate the pathway activity by taking the mean or median of the gene expression values of the member genes. These methods cannot effectively capture the coherent gene expression patterns that may be present within a pathway. That is, much of the discriminative information contained in the respective gene expression values may be lost if we average them out. The PCA-based inference method can somewhat relieve this problem [13]. In the PCA approach, the first basis vector captures the average expression pattern of the member genes, and the first principal component can estimate the presence and the strength of this pattern in a gene expression profile. However, not all the member genes in a perturbed pathway are typically altered at the mRNA level under different phenotypes in a consistent manner. In fact, some genes may have expression changes that are irrelevant to the change of phenotype of interest. To address this problem, Lee *et al.* [14] proposed a new pathway activity inference method that uses only a subset of member genes, called CORGs (condition-responsive genes). Pathway activities which inferred by this method are highly discriminative of the phenotypes. However, the CORG-based method may disregard member genes that have consistent, but not large, expression changes under different phenotypes.

Instead of using only one subset of genes which represent one differentiation pattern, two different gene subsets are used to infer pathway activities to ensure the maximization of difference in both up- and down-regulated fashion. Sootanan *et al.* [16] proposed a pathway activity inferring method which used NCFS based on ideal markers – this is referred to as NCFS-i in this paper. The use of NCFS was adopted from the work of Kim and Cho [17]. They used distance-based feature selection methods with NCFS to determine the genes that have cancer-related functions. Based on the ideal feature vectors for case and control, the feature selection measure can choose different feature subsets.

3 Material and Method

3.1 Dataset

We obtained breast cancer datasets from large-scale gene expression studies by Wang *et al.* [18]. This dataset is chosen for comparison purposes because it is used in a number of previous studies [13, 15]. It contains the gene expression profiles of 286 breast cancer patients from the USA, where metastasis was detected in 107 of them (referred as relapse class) while the remaining 179 were metastasis-free (referred as non-relapse class). In this study, we did not consider the follow-up time or the occurrence of distant metastasis. We retrieved this breast dataset ID GSE2034 from the public database of Gene Expression Omnibus (GEO) [19]. This dataset can be analyzed readily with their expression levels. For genes with more than one probe in one platform, we chose the probe with the highest mean expression value.

To obtain the set of known biological pathways, we referred to the pathway information from KEGG (Kyoto Encyclopedia of Genes and Genomes) database [11]. We downloaded manually curated pathways containing 204 gene sets. 169 pathways with more than 10 member genes are selected. These gene sets are compiled by domain experts and they provide canonical representations of biological processes.

3.2 Method

Expression data is normalized to z-score form before mapping onto the member genes in each pathway. These data are then used to identify CORGs and their pathway activities with CORG-, NegTCORG- and the proposed NegSCORG-based methods.

Also, pathway activities (PACs) are obtained using mean- and median-based methods. A comparison of the classification performance with the above-mentioned inferring methods is then made.

3.2.1 Normalizing Expression Data and Matching onto the Member Genes in Each Pathway

Expression values g_{ij} are normalized to z-transformed score z_{ij} for each gene i and each sample j. To integrate the expression and pathway datasets, we overlay the expression values of each gene on its corresponding protein in each pathway.

3.2.2 Inferring Pathway Activities

Pathway activities (PACs) are the combined expression levels of all member genes or subset of genes of each pathway across the sample. Several inferring pathway activity methods have been proposed in previous studies. Five inferring methods are chosen for comparison in this study. First three inferring methods are based on CORG set, and the other two are based on simple statistics like mean and median [12]. Within each pathway, a subset of condition-responsive genes (CORGs) is searched. PACs inferred by using all member genes like mean and median approaches used in Guo *et al.* [12] are used to compare to those inferred by using subsets of member genes in pathway. A schematic diagram summarizing the steps to compute the pathway activities from CORGs using the three different methods is shown in Fig. 1.

CORG-based method

To identify the CORG set, member genes are first ranked by their t-test scores, in descending order if the average t-score among all member genes is positive, and in ascending order otherwise. For a given pathway, a greedy search is performed to identify a subset of member genes in the pathway for which the Discriminative Scores (DS, $S(G)$) is locally maximal. The individual z_{ij} of each member gene in the gene set are averaged into a combined z-score which is designated as the activity a_j [14]. The activity vector a of the final CORG set is regarded as the pathway activity across the samples (see Fig. 1A).

NCFS-i method

In our previous work, we found that the results of using Pearson correlation coefficient as the feature selection method are comparable to the use of t-test for the correlation of negatively correlated ideal marker 1 (1,1,…,1,0,0,…,0) and ideal marker 2 (0,0,…,0,1,1,…,1) [16]. However, in the NCFS-i method, t-scores are used to determine the correlation with ideal markers (hence the denotation of "*i*" in the method name) to enable direct comparison to the CORG-based method. All member genes in each pathway are first ranked by their t-scores in descending and ascending orders if the average t-score among all member genes is positive and negative, respectively. Then, within each pathway, top ranked genes in these two different gene subsets are used to search for a subset of CORGs (see Fig. 1B). To derive activity a_j, the activity a_{j1} is subtracted by activity a_{j2}. The activity vector a of the final CORG set is regarded as the pathway activity across the samples.

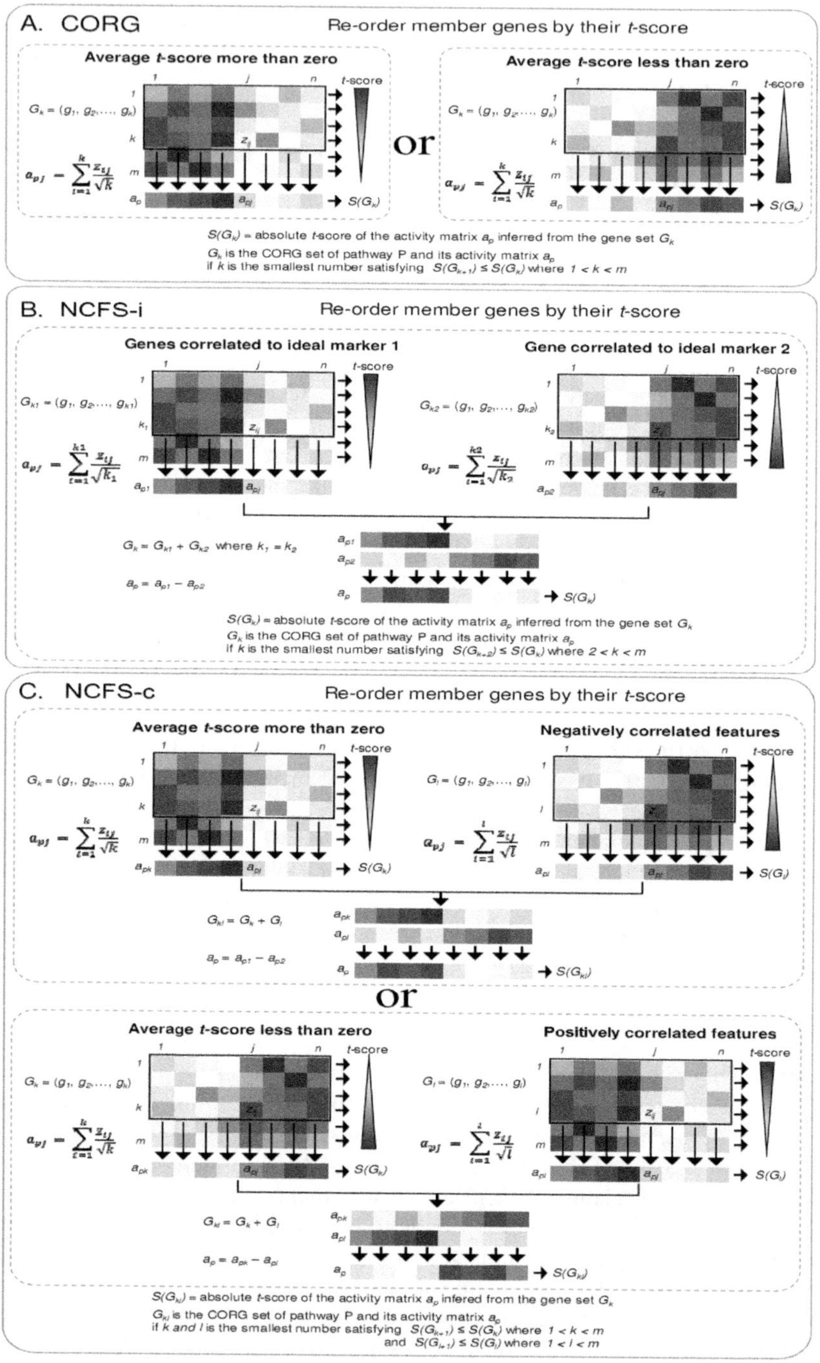

Fig. 1. Schematic diagram of key gene identification and pathway activity inference steps

NCFS-c method

This method is another modification from CORG-based method [14] by incorporating NCFS. The first set of CORGs is identified and the activity of the gene set is inferred using the CORG-based method. Then a second set of CORGs which are negatively correlated to the first set is identified and its activity is inferred. To derive activity a_j for a given pathway, the activity a_j inferred from the first CORG set is subtracted by the activity a_j inferred from the second CORG set to contrast the extreme difference between the two gene sets. (The "c" in the method name denotes CORG-based.) This final activity vector a is regarded as the pathway activity across the samples (see Fig. 1C).

Mean- and median-based method

All of member genes for a given pathway are inferred activity vector a by measuring mean and median of their z-transformed score in each sample. This final activity vector a is regarded as the pathway activity across the samples.

3.2.3 Classification Performance Evaluation Measure

In this work, we evaluated the performance of a classifier based on the Area Under Curve (AUC) of the Receiver Operating Characteristic (ROC) curve. A final classification performance was reported as the ROC area using ten folds cross validation. WEKA (Waikato Environment for Knowledge Analysis) version 3.6.2 [20] is used to build the classifier by using logistic regression (Logistic). The results in ROC Area are used to demonstrate the classification performance of different pathway activity inferring methods.

4 Results and Discussion

We computed the actual activity scores of all 169 pathways based on each pathway activity inference scheme, and ranked the pathways according to their discriminative score. The range of member genes in each pathway varies from 10 to 866, with the average being 62. Table 1 shows the discriminative score using different percentages of top pathways. We compared five pathway activity inference methods, namely, CORG-based method [14], NCFS-i method [16], mean and median methods [12], and NCFS-c method proposed in this paper. All member genes are used in the mean and medium methods. For the three different methods which used CORG set to infer the pathway activities, the average numbers are 5, 9, and 10, and the ranges are 1-12, 2-24, and 2-20 for CORG, NCFS-i and NCFS-c, respectively.

As shown in Table 1, the two pathway activity inference schemes which use NCFS significantly improved the power of pathway markers to discriminate between metastatic samples (relapse) and non-metastatic samples (non-relapse). That is, the gene sets obtained using negative correlation resulted in higher correlation with the two phenotype classes of *relapse* and *non-relapse*. The improvement is consistent for all cases of pathways used.

Table 1. Discriminative score (DS) and classification performance (ROC Area) of different pathway activity inferring methods of all pathway markers as a function of the top percentage of pathways used

Percentage	Pathway activity inferring methods									
	CORG		NCFS-i		NCFS-c		Mean		Median	
	DS	ROC Area	DS	ROC Area	DS	ROC Area	DS	ROC Area	DS	ROC Area
Top	7.547	0.752	9.586	0.806	9.628	0.792	4.318	0.643	4.005	0.638
20%	6.009	0.697	7.523	0.741	7.739	0.746	2.871	0.593	3.043	0.596
40%	5.608	0.684	7.003	0.725	7.197	0.730	2.350	0.575	2.536	0.582
60%	5.288	0.674	6.567	0.711	6.747	0.716	2.020	0.561	2.214	0.570
80%	4.956	0.662	6.145	0.698	6.284	0.702	1.718	0.548	1.899	0.557
100%	4.574	0.649	5.632	0.681	5.752	0.685	1.425	0.527	1.589	0.538

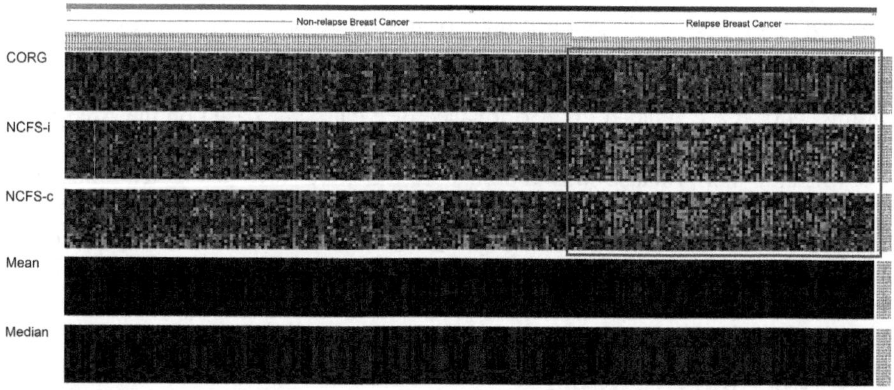

Fig. 2. Heat map of pathway activities of the five different inferring methods – based on 20 highest-ranked pathways by their discriminative score inferred by the CORG-based method

Heat map of pathway activities inferred by the five different methods are shown in Fig. 2. The 20 highest ranked pathways by their Discriminative Score (DS) inferred by the CORG-based methods are selected and clustered before plotting on this figure. Our newly proposed NCFS-c method can determine these PACs similar to CORG-based method but with higher discriminative power between the two classes. Both methods are different from NCFS-i which contains only up-regulated pathway markers in phenotype *non-relapse* group and down-regulated ones in phenotype *relapse* group, instead of a mixture of both up- and down-regulated ones.

We also used the proposed pathway activity inference scheme for classification of breast cancer metastasis to evaluate its usefulness in discriminating different cancer phenotypes. For this breast dataset, we performed ten-fold cross-validation experiments. We evaluated all pathway activity inference methods using logistic regression (Logistic) and assessed the classification performance using the ROC area. Table 1 shows that our methods are better than CORG-based method of Lee *et al.* [14] which used only one characterized set of gene to infer pathway activities and methods of Guo *et al.* [12]

which used simple statistic like mean and median. Classification performance of our newly proposed NCFS-c method is also better than our previously proposed NCFS-i method except when only the top pathway is used to obtain the ROC area. Mean and median approaches for inferring pathway activities are not appropriate to discriminate phenotype of interest because of their low classification performance. These results show that our two proposed methods which use NCFS clearly outperform other pathway-based classifiers in terms of both ROC area and discriminative score (Table 1). Note that a Naïve Bayes (NB) classifier has also been used with very similar results to the Logistic classifier. The results from NB are not shown because it is somewhat redundant and resulted in lower classification performance. Also, additional performance measures are not shown for a similar reason. That is, these results are all similar and show that the proposed methods outperform the other methods.

5 Conclusions

In this study, a novel method for inferring pathway activities and identifying CORG sets with negatively correlated feature sets NCFS) is proposed. We have demonstrated that effectively incorporating pathway information into expression-based disease diagnosis and using NCFS can provide better discriminative and more biologically defensible models. Our new proposed method which uses NCFS is promising for disease classification and identifying more meaningful CORG sets which can be used to search for genes in pathways that are relevant to a disease of interest. However, a limitation of our method is that it is not applicable for pathway gene sets that do not contain NCFS. Further work is underway to address this issue.

Acknowledgments. The main author (PS) gratefully acknowledges the financial support from Burapha University during his current doctorate study at King Mongkut's University of Technology Thonburi.

References

1. Golub, T.R., Slonim, D.K., Tamayo, P., Huard, C., Gassenbeek, M., Mesirov, J.P., Coller, H., Loh, M.L., Downing, J.R., Caligiuri, M.A., Bloomfield, C.D., Lander, E.S.: Molecular Classification of Cancer: Class Discovery and Class Prediction by Gene Expression Monitoring. Science 286, 531–537 (1999)
2. Young, R.A.: Biomedical Discovery with DNA Arrays: Cell 102, pp. 9–15 (2000)
3. Lakhai, S., Ashworth, A.: Microarray and Histopathological Analysis of Tumours: The Future the Past? Nat. Rev. Can. 1, 151–157 (2001)
4. Berns, A.: Cancer: Gene Expression Diagnosis. Nature 403, 491–492 (2000)
5. Su, A.I., Welsh, J.B., Sapinoso, L.M., Kern, S.G., Dimitrov, P., Lapp, H., Schultz, P.G., Powell, S.M., Moskaluk, C.A., Frierson Jr., H.F., Hampton, G.M.: Molecular Classification of Human Carcinomas by Use of Gene Expression Signatures. Cancer Res. 61, 7388–7393 (2001)
6. Lu, Y., Han, J.: Cancer Classification Using Gene Expression Data. Inform. Systems 28, 243–268 (2008)

7. Ein-Dor, L., Suk, O., Domany, E.: Thousands of Samples Are Needed to Generate a Robust Gene List for Predicting Outcome in Cancer. Proc. Natl. Acad. Sci. USA 103, 5923–5928 (2006)
8. Dupuy, A., Simon, R.M.: Critical Review of Published Miroarray Studies for Cancer Outcome and Guidelines on Statistical Analysis and Reporting. J. Natl. Cancer Inst. 99, 147–157 (2007)
9. Michiels, S., Koscielny, S., Hill, C.: Prediction of Cancer Outcome with Microarrays: A Multiple Random Validation Strategy. Lancet 365, 488–492 (2005)
10. Vogelstein, B., Kinzler, K.W.: Cancer Genes and the Pathways They Control. Nat. Med. 10, 789–799 (2004)
11. Kanehisa, M., Araki, M., Goto, S., Hattori, M., Hirakawa, M., Itoh, M., Katayama, T., Kawashima, S., Okuda, S., Tokimatsu, T., Yamanishi, Y.: KEGG for Linking Genomes to Life and the Environment. Nucleic Acids Res. 36, D480–D484 (2008)
12. Guo, Z., Zhang, T., Li, X., Wang, Q., Xu, J., Yu, H., Zhu, J., Wang, H., Wang, C., Topol, E.J., Wang, Q., Rao, S.: Towards Precise Classification of Cancers Based on Robust Gene Functional Expression Profiles. BMC Bioinformatics 6 (2005)
13. Bild, A.H., Yao, G., Chang, J.T., Wang, Q., Potti, A., Chasse, D., Joshi, M.-B., Harpole, D., Lancaster, J.M., Berchuck, A., Olson Jr, J.A., Marks, J.R., Dressman, H.K., West, M., Nevins, J.R.: Oncogenic pathway signatures in human cancers as a guide to targeted therapies. Nature 439, 353–357 (2006)
14. Lee, E., Chuang, H.-Y., Kim, J.-W., Ideker, T., Lee, D.: Inferring Pathway Activity Toward Precise Disease Classification. PLoS. Comput. Biol. 4, e1000217 (2008)
15. Su, J., Yoon, B.-J., Dougherty, E.R.: Accurate and Reliable Cancer Classification Based On Probabilistic Inference of Pathway Activity. PLoS ONE 4 (2009)
16. Sootanan, P., Meechai, A., Prom-on, S., Chan, J.H.: Pathway Activity Inferences with Negatively Correlated Features for Pancreatic Cancer Classification. In: 2nd International Conference on BioMedical Engineering and Informatics (BMEI 2009), pp. 1888–1892. IEEE Press, China (2009)
17. Kim, K.-J., Cho, S.-B.: Ensemble Classifiers Based on Correlation Analysis for DNA Microarray Classification. Neurocomputing 70, 187–199 (2006)
18. Wang, Y., Klijn, J.G., Zhang, Y., Sieuwerts, A.M., Look, M.P., Yang, F., Talantov, D., Timmermans, M., Meijer-van Gelder, M.E., Yu, J., Jatkoe, T., Berns, E.M.J.J., Atkins, D., Foekens, J.A.: Gene-expression profiles to predict distant metastasis of lymph-node-negative primary breast cancer. Lancet 365, 671–679 (2005)
19. Edgar, R., Domrachev, M., Lash, A.E.: Gene Expression Omnibus: NCBI Gene Expression and Hybridization Array Data Repository. Nucl. Acids Res. 30, 207–210 (2002)
20. Hall, M., Frank, E., Holmes, G., Pfahringer, B., Reutemann, P., Witten, I.H.: The WEKA data mining software: An update. SIGKDD Explorations 11 (2009)

A Malware Detection Algorithm Based on Multi-view Fusion

Shanqing Guo[1], Qixia Yuan[1], Fengbo Lin[1], Fengyu Wang[1], and Tao Ban[2]

[1] Shandong University Jinan 250101, Shandong, China
[2] Information Security Research Center, National Institute of Information and Communications Technology
Tokyo 184-8795, Japan
{guoshanqing,linfb}@sdu.edu.cn, bantao@nict.go.jp

Abstract. One of the major problems concerning information assurance is malicious code. In order to detect them, many existing run-time intrusion or malware detection techniques utilize information available in Application Programming Interface (API) call sequences to discriminate between benign and malicious processes. Although some great progresses have been made, the new research results of ensemble learning make it possible to design better malware detection algorithm. This paper present a novel approach of detecting malwares using API call sequences. Basing on the fact that the API call sequences of a software show local property when doing network, file IO and other operations, we first divide the API call sequences of a malware into seven subsequences, and then use each subsequence to build a classification model. After these building models are used to classify software, their outputs are combined by using BKS and the final fusion results will be used to label whether a software is malicious or not. Experiments show that our algorithm can detect known malware effectively.

Keywords: Malware Detection; API Call Sequences; Multi-view Fusion; BKS Algorithm.

1 Introduction

With the improvement of the techniques used by malwares like virus, trojan and worm, traditional malware detection approaches based on signatures, were not effectively enough to detect variables of known malwares. In order to solve this problem, some researchers started to use the API call sequences of a program for detecting malwares and made much progress in this field[1]. Recently, the local property, a phenomenon that described the affinity between API functions in the network, file IO or other operations, has been discovered, but not taken into consideration to design better detection method. Using this property, we design a novel malware detection algorithm based on ensemble leaning methods and the experiments show that our algorithm has a better classification result.

K.W. Wong et al. (Eds.): ICONIP 2010, Part II, LNCS 6444, pp. 259–266, 2010.

The rest of the paper is organized as follows: in part 2, we briefly review previous approaches closely related to this paper; In part 3 , we deliver our core technical contribution include how to extract features from the API call sequences of a program using information gain algorithm and design a malware detection algorithm based on a multi-modal fusion; in part 4, the experiment shows that our algorithm have a better classification results and can identify a malware type effectively whilst part 5 contains our concluding remarks.

2 Related Work

Forrest et al. proposed to use fixed-length system calls for distinguishing benign and malicious UNIX process [1]. Later on, Wepsiet al. improved the method using variable-length system calls[2]. In a recent work, the authors proposed to extract semantics information by annotating call sequences for malware detection [3]. Using flow graphs to model temporal information of system calls has been proposed in [4]. Faraz Ahmed proposed Using Spatio-Temporal Information for Malware Detection and Analysis [5], but the extracting method of temporal information leaved a backdoor for insidious malware author to fake API call sequences to cover up temporal information.

The technology used by the malwares is improving continuously. The statistics showed that most of the malwares came from the modification of the old ones [6] and traditional signature based anti-malware technology is not effective enough to detect varieties of the known malwares. In order to solve this problem, some researcher had used machine learning technology for detecting the polymorphic malwares, for example, Konrad Rieck used SVM to classify the behavior of malwares [7]. Similar method went into finite automaton, HMM [8], data mining [9] and neural network [10]. Although simple and well performed, different classification method performed large differences to different type of malwares. Each classification method had its own emphasis. As a result, there was not a classification method which performs well for every kind of malwares. In [5], fusion of multiple classification method showed well performance. However, on the chosen of the fusion method, there is still improvement space.

3 Multi-view Malware Detection Method Basing on API Call Sequences

The system is divided into two parts: offline analysis and online analysis. The offline analysis part is responsible for constructing classification modal by training with the historical data. The online analysis part is responsible for three functions: getting the API call sequences of an unknown software, extracting features from the API call sequences, and verifying whether it is a malicious software. We are now going into details with the analysis.

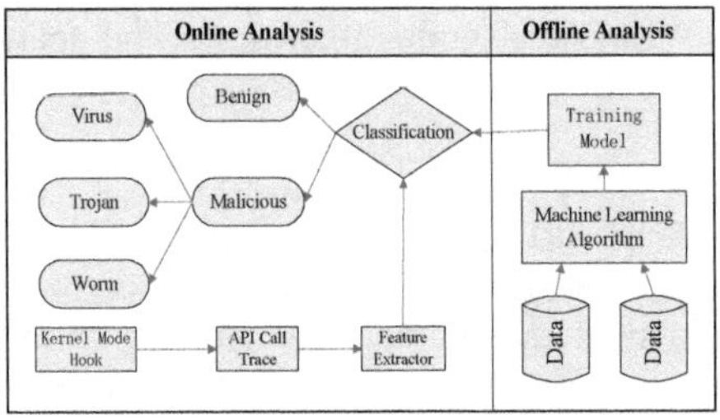

Fig. 1. System Design Figure

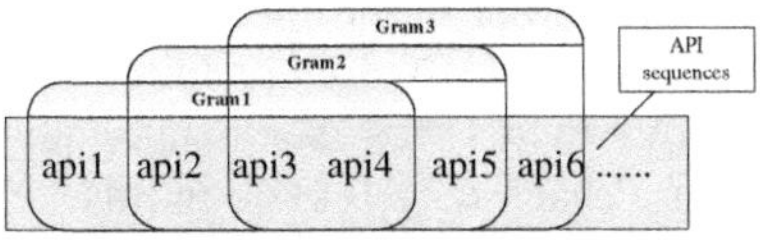

Fig. 2. 4-Gram Sequence

3.1 Getting API Call Sequences

APIs are interface functions that the Microsoft Windows operating system provides for users in dynamic link libraries. It can be run in user space or kernel space. Although there are many APIs in Microsoft Windows operating system, Naïve APIs are enough to show the characteristic of a program. As a result, we focus on the naïve APIs in this paper. There are over 900 naïve APIs in Microsoft windows operating system and we only consider the most important 229 ones.

Table 1. API Category [5]

API Category	Numbers	Description
File Input/Output (I/O)	50	Interface functions of file input and output
Dynamic-Link Libraries	7	Interface functions of DLL operations
Network Management	16	Interface functions for managing network
Memory Management	35	Interface functions for managing memory
Processes & Threads	39	Interface functions of process operations
Registry	41	Interface functions of registry operations
Windows Socket	41	Interface functions of Windows Socket

In order to catch the API call sequences produced by a running program, we design a hook program resident in memory to monitor the running of a program to catch the critical APIs listed above, which output the API call sequences.

3.2 Characterization of API Call Sequences

The API call sequences got from the hook program cannot be directly used as the input of a classification method. As a result, we design a feature extraction algorithm using N-Gram and information gain [11]. According to experience and experiment, we set the value of n as 2-6 in n-gram algorithm. After several comparative experiments, we find that 4-gram algorithm performs well and is reasonable in time and resource. For the above reasons, we choose 4-gram algorithm. The feature extraction algorithm includes the following process:

(1) Extracting the gram of every API call sequence.

In 4-gram algorithm, the probability of an API appears only depends on the 3 APIs before it. By doing sliding window operation in every API call sequence, we get a series of 4 length segment sequences. Each of the segment sequence is called a gram. What we need to do in this step is to count the probability of each gram appears in an API call sequence. After doing this operation to all the API call sequence, we get a series of 4-gram recording the probability of a 4-gram appears in its corresponding API call sequence file. Figure 2 shows the 4-gram API sequences.

(2) Calculating the value of information gain for each Gram

The gram we get in step 1 can be treated as a feature. However, there are too many grams, normally over ten thousands and not all the grams are a valuable feature. As a result, we need to filter out the useful ones. We treat every gram as a feature t and calculate its information gain [15] using the formula

$$IG(t) = -\sum_{i=1}^{n} P(C_i)\log_2 P(C_i) + P(t)\sum_{i=1}^{n} P(C_i \mid t)\log_2 P(C_i \mid t) + P(\bar{t})\sum_{i=1}^{n} P(C_i \mid \bar{t})\log_2 P(C_i \mid \bar{t})$$

The C_i represents a category. In our paper, there are 4 different categories, namely benign, virus, trojan and worm, so we have C_1, C_2, C_3 and C_4. P (C_i) means the probability of C_i appears. Similarly, P(t) represents the probability of feature t appears in all files.

(3) Generating the final signature vector

To sort the features from step 1 in the order of its information gain calculated in step 2. Choose the biggest n ones as signature vector S, namely: $S = \{gram_1, gram_2 ... gram_n\}$,Then every API call sequence file can be represent by a One-dimensional feature vector R of size n, namely:

$$R = \{r_1, r_2 ... r_n\}, r_i = \begin{cases} 1, \text{ if } gram_i \text{ exsits in file} \\ 0, \text{ if } gram_i \text{ not exsit in file} \end{cases}$$

3.3 Classification Model

We designed a multiple classification fusion algorithm (Figure 3) basing on the BKS (behavior knowledge space) proposed by Huang [12].It can be used to discriminate the benign and malicious programs, and it is also well enough for deeply classifying malicious program into virus, Trojan and worm. This algorithm's process procedures were listed bellows:

(1) Dividing the API call sequences of a software API call sequence into seven different sub-sequences according to the API category mentioned in 3.1.

(2) Classifying them using machine learning algorithm and seven results can be got.

(3) Fusing them with BKS algorithm or other ensemble methods, and using its fusion result as this software's label.

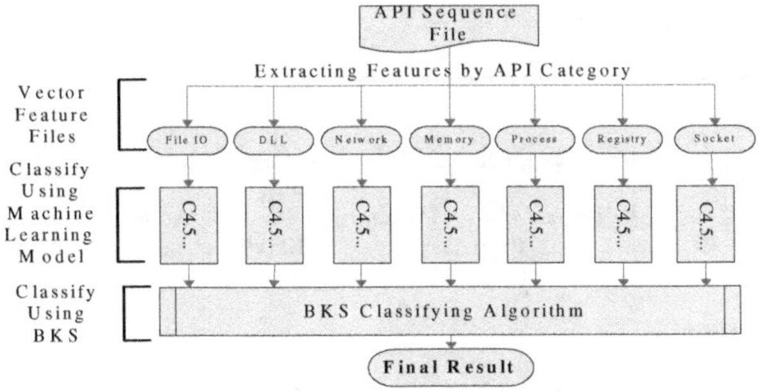

Fig. 3. Multi-View Classification Algorithm

4 Experiment

4.1 Experiment Samples

Totally, there are 817 sample programs in our experiment, including benign program, trojan, virus and worm. These programs can be downloaded from web[13]. We run Microsoft windows XP in VmWare and use API Monitor [14] reference as a "hook" program to catch API call sequences of a program. The constitution of the samples is shown in Table 2.

Table 2. Constitution of Experiment Samples

Program Category	Numbers of Samples
Benign	100
Trojan	193
Virus	289
Worm	235
Total	817

4.2 Performance Index

We use TP (true positive) Rate, FP (false positive) Rate, Precision, and Accuracy (number of correctly recognized samples/total number of samples) as the performance index to judge the classification modal. The terms true positives, true negatives, false positives and false negatives are used to compare the given classification of an item (the class label assigned to the item by a classifier) with the desired correct classification (the class the item actually belongs to). This is illustrated by Table 3.

Table 3. Shown of classification performance index

		correct result / classification	
		E1	**E2**
obtained result / classification	**E1**	**tp**(true positive)	fp(false positive)
	E2	**fn**(false negative)	tn(true negative)

The TP rate, FP rate, Precision, Recall, F-measure and Accuracy are then defined as: TP Rate=$\frac{tp}{tp+fn}$, FP Rate=$\frac{fp}{fp+tn}$, Precision=$\frac{tp}{tp+fp}$, Accuracy=$\frac{tp+tn}{tp+tn+fp+fn}$.

4.3 Performance

We use LibSVM, Id3, J48, NaiveBayes and SMO to carry out our experiment. For we use 10-fold cross-validation, we don't need to distinguish training samples and testing samples manually. First we check out the capability of these classifications modal to correctly distinguish a program as benign and malicious. The results are shown in Table 4. The "Five" in row API means that five types of api sequences, namely net, process, memory, reg and socket. The result shows that after fusion by BKS, most of the classification performance (the only exception is NaiveBayes) is improved. Specially, the Id3 and SMO reach a perfect result.

Table 4. Result of Malicious-benign Classification

Testing method	API	TP Rate	FP Rate	Precision	Accuracy (%)	Class
LibSVM	All	0.998	0.104	0.984	98.424	malicious
		0.896	0.002	0.986		benign
BKS (LibSVM)	Five	1.000	0.091	0.986	98.774	malicious
		0.909	0.000	1.000		benign
Id3	All	0.984	0.117	0.982	97.023	malicious
		0.883	0.016	0.895		benign
BKS (Id3)	Five	1.000	0.000	1.000	100	malicious
		1.000	0.000	1.000		benign
J48	All	0.99	0.091	0.986	97.898	malicious
		0.909	0.01	0.933		benign
BKS (J48)	Five	0.994	0.039	0.994	98.949	malicious
		0.961	0.006	0.961		benign

Table 4. (*continued*)

Testing method	API	TP Rate	FP Rate	Precision	Accuracy (%)	Class
SMO	All	1.000	0.065	0.99	99.124	malicious
		0.935	0.000	1.000		benign
BKS (SMO)	Five	1.000	0.000	1.000	100	malicious
		1.000	0.000	1.000		benign

We also carry out experiments for other fusion method, namely Vote. The results are shown in Table 5. The performance of BKS is better than that of Vote except for using NaiveBayes as the base classifier.

Table 5. Results of Different Fusion Method

Base Classifier	Fusion	TP Rate	FP Rate	Precision	Accuracy (%)	Class
LibSVM	Vote	0.998	0.104	0.984	98.424	malicious
		0.896	0.002	0.986		benign
	BKS	1.000	0.091	0.986	98.774	malicious
		0.909	0.000	1.000		benign
Id3	Vote	0.994	0.104	0.984	98.074	malicious
		0.896	0.006	0.958		benign
	BKS	1.000	0.000	1.000	100	malicious
		1.000	0.000	1.000		benign
J48	Vote	0.986	0.117	0.982	97.198	malicious
		0.883	0.014	0.907		benign
	BKS	0.994	0.039	0.994	98.949	malicious
		0.961	0.006	0.961		benign
SMO	Vote	1.000	0.078	0.988	98.949	malicious
		0.922	0.000	1.000		benign
	BKS	1.000	0.000	1.000	100	malicious
		1.000	0.000	1.000		benign

5 Conclusion and Future Work

In this paper, we propose the multi-view classification algorithm basing on t the local property that described the affinity between API functions in the network, file IO or other operations, which was evaluated on a large corpus of malicious code . The experiment results show that our multi-view classification algorithm can effectively discriminate a program into benign and malicious and can improve most of the classifier. Future work will extend to extend the techniques described here to improve the detection of malicious code, especially the novel malicious code.

Acknowledgments. Supported by Specialized Research Fund for the Doctoral Program of Higher Education (Grant No. 20090131120009), Natural Science Foundation of Shandong Province for Youths (Grant No. Q2008G01), Outstanding

Young Scientists Foundation Grant of Shandong Province(Grant No. BS2009DX018), Independent Innovation Foundation of Shandong University(Grant No. 2009TS031), The Key Science-Technology Project of Shandong Province of Shandong(Grant No. 2010GGX10117).

References

1. Forrest, S., et al.: A Sense of Self for Unix Processes. In: IEEE Symposium on Security and Privacy (S&P), pp. 120–128. IEEE Press, USA (1996)
2. Wespi, A., et al.: Intrusion Detection Using Variable-Length Audit Trail Patterns. In: Debar, H., Mé, L., Wu, S.F., et al. (eds.) RAID 2000. LNCS, vol. 1907, pp. 110–129. Springer, Heidelberg (2000)
3. Christodorescu, M., et al.: Semantics-Aware Malware Detection. In: IEEE Symposium on Security and Privacy (S&P). IEEE Press, USA (2005)
4. Beaucamps, P., Marion, J.-Y.: Optimized control flow graph construction for malware detection. In: International Workshop on the Theory of Computer Viruses (TCV), France (2008)
5. Ahmed, F., et al.: Using Spatio-Temporal Information in API Calls with Machine Learning Algorithms for Malware Detection and Analysis. In: Proceedings of the 2nd ACM Workshop on Security and Artificial Intelligence, Chicago, Illinois, USA, pp. 55–62 (2009)
6. Turner, D., et al.: Symantec Internet security thread report trends for January06-june 06. Symaatec Corporation Cupertino, CA, USA, Tech Rep: Volume X (2006)
7. Rieck, K., et al.: Learning and Classification of Malware Behavior. In: Proceedings of the 5th International Conference on Detection of Intrusions and Malware, and Vulnerability Assessment, Paris, France, pp. 108–125 (2008)
8. Ye, N.: A markov chain model of temporal behavior for anomaly detection. In: Workshop on Information Assurance and Security, West Point, NY (June 2000)
9. Lee, W., et al.: Data Mining Approaches for Intrusion Detection. In: 7th USENIX Security Symposium, San Antonio, TX (1998)
10. Han, S.-J., Cho, S.-B.: Evolutionary neural networks for anomaly detection based on the behavior of a program. IEEE Trans. Syst. Man Cybern. B Cybern., 559–570 (June 2006)
11. Mitchell, T.M.: Machine Learning. The McGraw-Hill Companies, Inc., New York (1997)
12. Huang, Y.S., et al.: The behavior-knowledge space method for combination of multiple classifiers. In: Computer Vision and Pattern Recognition, pp. 347–352 (1993)
13. `http://vx.netlux.org/vl.php`
14. `http://www.APImonitor.com/`

A Fast Kernel on Hierarchial Tree Structures and Its Application to Windows Application Behavior Analysis

Tao Ban, Ruo Ando, and Youki Kadobayashi

National Institute of Information and Communications Technology, Japan
bantao@nict.go.jp, ruo@nict.go.jp, youki-k@is.aist-nara.ac.jp

Abstract. System calls have been proved to be important evidence for analyzing the behavior of running applications. However, application behavior analyzers which investigate the majority of system calls usually suffer from severe system performance deterioration or frequent system crashes. In the presented study, a light weighted analyzer is approached by two avenues. On the one hand, the computation load to monitor the system calls are considerably reduced by limiting the target functions to two specific groups: file accesses and Windows Registry accesses. On the other hand, analytical accuracy is achieved by deep inspection into the string parameters of the function calls, where the proximity of the programs are evaluated by the newly proposed kernel functions. The efficacy of the proposed approach is evaluated on real world datasets with promising results reported.

Keywords: FTree, kernel on structured data, registry analysis, sequence analysis, system call analysis.

1 Introduction

Information security are receiving more and more concern because of the increasing iniquitousness and inter-connectedness of computer systems. Conventional security solution systems, e.g., anti-virus software, SPAM email filters, etc., usually rely on extensive human effort to identity new threats, extract particular characteristics, and update the system to treat with new threats [1,2,3]. Recent research shows that this labor-intensive process can be more efficient by applying machine learning algorithms.

In this paper, we present a study on application of pattern classification techniques for identifying the status of a target operating system (OS). In particular, we look into the activity of the running threads, which are the smallest unit of processing that can be scheduled by an OS. According to recent research [4,5], the behavior of a program can be represented by the sequence of system calls during its life-time. However, because of its versatility, monitoring all system calls will not only impose unnecessary cost but could give rise to unpredictable conflicts and system crashes as well.

K.W. Wong et al. (Eds.): ICONIP 2010, Part II, LNCS 6444, pp. 267–274, 2010.

Aiming at a lightweight monitoring and analysis system, we focus on a specific type of system call – file access. File access has the following features that render it very promising for program behavior characterizing. First, most programs will make file access during their life time. Unlike other system resources the status of which will get lost after a reset of the OS, e.g., memory block, socket buffer, and network interface buffer, file modification is permanent and thus it is employed by most programs to put on long term modification to the system. Second, essential information with respect to the program identity and behavior is presented as string parameters to the file access routine. Collective information on the accessed files can significantly reduce the probability of miss identifying a program and thus is of particular interest in this study. Finally, file-access related routines are considerably fewer than all the system calls, which makes the monitoring more light-weighted and the analysis more efficient.

In this study Windows OSes are of special interest because of their large market-share and many disclosed system vulnerabilities, which motivated the presented study – to detect suspicious running programs in the system by behavior analysis. Another interesting feature of Windows OS is the Registry, which is a special file that shares similar characteristic of a file system, so it can be treated similarly with an improvement on the system performance.

The rest of the paper is organized as follows. In Section 2, we introduce the characteristics of the tree structures composed of file paths or Windows registry keys. In Section 3, we describe the proposed kernels on the tree structures. In Section 4, we evaluate the proposed approach on real world datasets. We conclude the paper in Section 5.

2 Windows File System and Registry

In Windows File Systems (WFS), a drive letter abstraction is used to distinguish different partitions. For example, a path 'C:\WINDOWS\control.ini' represents a file 'control.ini' in directory 'WINDOWS' on the partition represented by letter 'C', with subsidiary directories divided by a backslash. Multiple paths can be presented in a hierarchial structure as shown in Fig. 1(a).

A file of special interest in a Windows OS is the Windows Registry (WR). WR is a hierarchical database that stores configuration settings and options on MS Windows OSes. See an example of the WR in Fig. 1(b). Since its first introduction with Windows 3.1, the WR's primary purpose have extended from storing configuration information for COM-based components to store configuration settings for all Windows programs. For now, most of the Windows programs make use of the registry to store respective configuration settings.

Files, especially WR, are one of the major sources for a program to read configuration settings or application data. Because the paths of the files (registry item) tend to vary little for different versions of the same program but much for different programs, they can be considered as signatures of the program and thus can exploited to distinguish one program from another. As seen from Fig. 1, both file paths and WR paths can be organized in a hierarchical tree structure. With

(a) Windows Registry (b) Windows File System

Fig. 1. Hierarchical structures of Windows file system and registry

duplicated leading paths merged, such a tree structure is conceptually much easier to show the structure of the file system than a list of paths. Hereinafter, we make no difference between a tree of file paths and a tree of registry keys, and call such a structure a File Tree (FTree).

3 Kernels on Trees

To facilitate machine learning algorithms operating on structured data such as the tree composed of file paths, we take the well known kernel based approach. The key to this approach is the kernel function which defines the proximity between the tree structures. With such a kernel function, most popular learning algorithms such as SVM, Linear discriminant analysis, c-means clustering, can be readily applied. In the following, we discuss how to operate efficiently on the tree structures for such a kernel function.

3.1 Kernels on FTrees

In Figures 2(a), 2(b), and 2(c), we show three FTrees, each of which is generated from a sequence of paths. In these figures, a shaded node denotes a path/file appeared in the sequence and all the nodes, shaded or open, compose a complete FTree associated with the superset of three sequences. Intuitively, FTree $\mathcal{A}$ is closer to FTree $\mathcal{B}$ than it is to FTree $\mathcal{C}$: as show in Fig. 2(d), $\mathcal{A}$ shares four nodes with $\mathcal{B}$, but it only has two common nodes with $\mathcal{C}$. Then, the distance from $\mathcal{A}$ to $\mathcal{C}$ is comparable with that from $\mathcal{B}$ to $\mathcal{C}$: $\mathcal{A}$ and $\mathcal{C}$ have two common nodes and so do $\mathcal{B}$ and $\mathcal{C}$. Interestingly, this common-subtree based similarity measure, agrees with our intuition of difference between tree structures. This gives rise to a our proposed kernel functions on FTrees.

Definition 1. A *FTree*, denoted by $\mathcal{T}$, for a sequence S of n unique file paths, is a rooted directed multi-way tree. Each node, other than the root, is labelled with a nonempty segment of the file paths, i.e., a substring of the key in between two consequent backslashes. No two child nodes of a node can have the same label. For any leaf node i, the concatenation of the node-labels on the path from root to leaf together with necessary back-slashes, spells out the corresponding file path. The *depth* of a node i, $d(i)$, is the minimum number of edges along

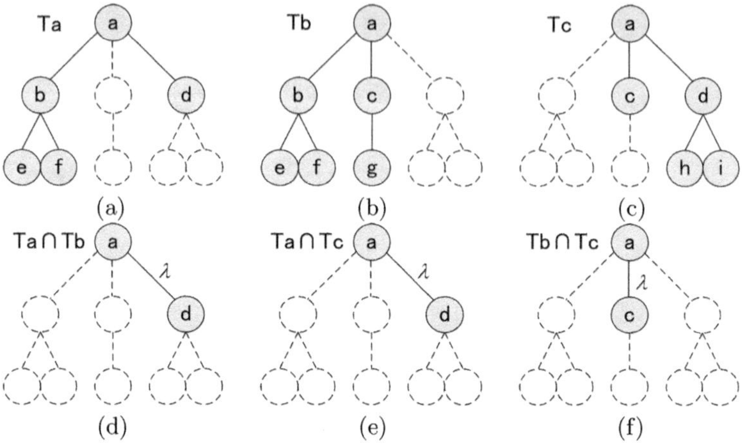

Fig. 2. FTrees and maximum common sub-trees

any path from this node to the root. A *weighted FTree* is a FTree that each of its nodes is associated with a counter $c(i)$ that counts the number of nodes in the subtree underneath.

FTree is simply a rigid definition of the tree structures common found in Windows OS, see the trees in Figures 1 and 2 for an example. The concept can be simply perceived by noticing the fact that in Windows OS, it is illegitimate to have two or more identical sub-folders under any folder.

Definition 2. A *generalized FTree* $\mathcal{T}_g$ of $\mathcal{T}_1$ and $\mathcal{T}_2$ is the FTree created from the superset of the two file path sequences, S_g, where $S_g = S_1 \cup S_2$. A *maximum common sub-tree* $\mathcal{T}_c$ between $\mathcal{T}_1$ and $\mathcal{T}_2$ is a FTree created from the intersection of the two file path sequences, S_c, where $S_c = S_1 \cap S_2$.

Here, a sequence of file paths are treated as a set, where the order of the paths are not take into consideration. Apparently, generalized FTree and maximum common sub-tree can be extended for cases of more than two FTrees. Based on the above definitions, we are ready to define the kernel between two FTrees.

Definition 3. The *FTree kernel* between two FTrees $\mathcal{T}_1$ and $\mathcal{T}_2$, is computed as

$$k(\mathcal{T}_1, \mathcal{T}_2) = \sum\nolimits_{i \in \mathcal{T}_c(\mathcal{T}_1, \mathcal{T}_2)} c_1(i) \cdot c_2(i), \tag{1}$$

where $c_1(i)$ and $c_2(i)$ are the counters of $\mathcal{T}_1$ and $\mathcal{T}_2$ associated with leaf i in $\mathcal{T}_c$. Similarly, we define the *plain (FTree) kernel* between two FTrees $\mathcal{T}_1$ and $\mathcal{T}_2$ as

$$k_p(\mathcal{T}_1, \mathcal{T}_2) = \sum\nolimits_{i \in \mathcal{T}_c(\mathcal{T}_1, \mathcal{T}_2)} 1. \tag{2}$$

From the definition, the plain kernel between $\mathcal{T}_1$ and $\mathcal{T}_2$ equal the number of nodes in $\mathcal{T}_c$, i.e., the number of common nodes between $\mathcal{T}_1$ and $\mathcal{T}_2$. The weighted kernel also takes the number of appearances of the node labels in the file path sequence into consideration.

When computing the kernel between two trees, sometimes it is natural to take the depth of a shared node into consideration. For example, for the two nodes a and e in Fig. 2(d), $\mathcal{T}_a$ and $\mathcal{T}_b$ are sharing e indicates that a is also in the common sub-tree of $\mathcal{T}_a$ and $\mathcal{T}_b$; however, it is not true, vice versa. To assign larger weights to nodes that are located deeper in the FTree, we introduce a decay parameter, λ, and define the *depth weighted kernel* between $\mathcal{T}_1$ and $\mathcal{T}_2$ as

$$k_{dw}(\mathcal{T}_1, \mathcal{T}_2) = \sum\nolimits_{i \in \mathcal{T}_c(\mathcal{T}_1, \mathcal{T}_2)} c_1(i) \cdot c_2(i) \lambda^{d(i)}, \tag{3}$$

where $d(i)$ is the depth of node i. Finally, we add kernel normalization to prevent the imbalance caused by FTrees created from too many or too few file paths. Thus, we define the *normalized depth weighted FTree kernel* as

$$k'_{dw}(\mathcal{T}_1, \mathcal{T}_2) = \frac{\sum_{i \in \mathcal{T}_c(\mathcal{T}_1, \mathcal{T}_2)} c_1(i) \cdot c_2(i) \lambda^{d(i)}}{\sqrt{k_{dw}(\mathcal{T}_1, \mathcal{T}_1) \cdot k_{dw}(\mathcal{T}_2, \mathcal{T}_2)}}, \tag{4}$$

Similarly, we define the *depth weighted plain kernel* between $\mathcal{T}_1$ and $\mathcal{T}_2$ as

$$k_{dwp}(\mathcal{T}_1, \mathcal{T}_2) = \sum\nolimits_{i \in \mathcal{T}_c(\mathcal{T}_1, \mathcal{T}_2)} \lambda^{l(i)}, \tag{5}$$

and *normalized depth weighted plain kernel* between $\mathcal{T}_1$ and $\mathcal{T}_2$ as

$$k'_{dwp}(\mathcal{T}_1, \mathcal{T}_2) = \frac{\sum_{i \in \mathcal{T}_c(\mathcal{T}_1, \mathcal{T}_2)} \lambda^{d(i)}}{\sqrt{k_{dwp}(\mathcal{T}_1, \mathcal{T}_1) \cdot k_{dwp}(\mathcal{T}_2, \mathcal{T}_2)}}. \tag{6}$$

3.2 Properties of the Kernels

To justify the correctness of the kernel function definitions we have to show whether they satisfy the positive semi-definite criterion. This is answered by the following proposition.

Proposition 1. The weighted kernel defined in Eq. (1) is positive semi-definite.

Proof: Let Σ be the set of all possible file paths with length $m < C$, where C is an arbitrarily large natural number. Let Ψ be the FTree generated by Σ. Then, as long as the alphabet is limited and C is limited, Ψ is comprised of a finite set of nodes, H. Denote the number of nodes in Ψ as N. Then we can create a dictionary from the H, so that each node in H is assigned a unique index in the dictionary. Because $\mathcal{T}_i \subset \Psi$ and $\mathcal{T}_i \cap \Psi = \mathcal{T}_i$, $\mathcal{T}_i$ can be presented as an N-dimensional vector, where the weight of the hth dimension is assigned to $c_i(h)$ if node h is in $\mathcal{T}_i$ and to zero otherwise. In this formulation, $k(\mathcal{T}_i, \mathcal{T}_j)$ is equivalent to the inner product between the two associated N-dimensional vectors. Because

Table 1. Algorithm to Create a (Generalized) FTree

```
function [T] = ExpandGFTree(T, S, j = 0)
Inputs: T: A generalize FTree previously obtained;
        S: Sequence of registry keys;
        j: index of the FTree in the generalized FTree;
Outputs: T: Expanded generalized FTree;
1    for m = 1 : |S|
2        L = Split(S_i); // split a registry key into list of node labels;
3        Insert(T, L, j, 0); // insert the node labels into the tree;
4    return T;
-----------------------------------------------------------------------
function [T] = Insert(T, L, j, l)
Inputs: L: List of node labels;
        l: level of the current tree node;
5    T.c[j] += 1; // increment the counter for the j-th FTree;
6    if ( l > len(L) ) return T; // return when reach the leaf;
7    [f, p] = Locate(L[l]); // locate the insert position of the label;
8    if (f) // find an identical node with L[l];
9        Insert(child[p], L, j, l + 1); // insert into the pth brach;
10   else // the labels does not exist, create a new branch;
11       child[len(child)] = new node;
12       Insert(child[len(child)], L, j, l + 1); // insert into the new branch;
13   return T;
```

the inner product of N-dimensional vectors are positive semi-definite, $k(\cdot,\cdot)$ is positive semi-definite. □

Corollary 1. The depth weighted kernel in Eq. (4) and the normalized depth weighted kernel in Eq. (3) are positive semi-definite.

Corollary 2. The plain kernel defined in Eq. (2), depth weighted plain kernel in Eq. (5), and normalized depth weighted kernel in Eq. (6) are positive semi-definite.

3.3 Algorithm

The algorithms to create a FTree from a list of file paths are shown in Table 1. To create a FTree, simply input an empty root node to function *ExpandGFTree* together with the sequence of registry keys S. The j parameter is to make difference between individual FTrees in the generalized FTree and is set to zero for a single FTree. To expand an existing FTree to a generalized FTree, use parameter j to represent the identity of the inserted FTree.

3.4 Computational Costs

Suppose the location of child node label in the list of all existing labels of a parent node (line 7 in Table 1) can be solved in $O(1)$ time, the computational cost to create a FTree is linear to the weighted number of nodes in the tree, i.e., $O(N)$, where N is the number of segments of file paths in S. To compute the

Table 2. Benchmark datasets information

Dataset	max depth	Training	Test	Classes
Registry	12	1200	1330	15
File I/O	8	800	785	15

kernel between two FTrees, we need to first build a generalized FTree T_g from the two registry key sequences and then traverse T_g to compute the kernel. The time complexity is $O(N_1 + N_2)$.

Assume sufficient storage, a time-efficient way for computing a kernel matrix between a fixed number of FTrees is to implement a generalized FTree for all the FTrees such that for each node in the generalized FTree a vector is deployed to store all the counters for every sequence. In this way, a sequence is examined only once for creating the generalized FTree. Then the cost to compute a row/column of the kernel matrix, i.e., all kernels between a fixed FTree $\mathcal{T}_i$ with all the other FTrees, will be $O(M|\mathcal{T}_i|)$, where M is the number of FTrees embedded in the generalized FTree and $|\mathcal{T}_i|$ is the number of nodes in $\mathcal{T}_i$.

3.5 Experiments

In this section, we evaluate the efficacy of proposed kernels on a Windows application classification task. Records of WR and file accesses are exploited to classify the Windows processes into known categories. The dataset is collected by the memento system [6], which is a virtualization based Windows API monitor. To record the API sequences, the programs are run within the guest Windows system that is installed upon a virtual machine. In the guest OS, DLL injection and filter drive insertion are employed to record all registry and file accesses during the life time of the target processes. The captured data are instantly forwarded to the host OS where processing and analysis are done on real time basis. The category of a process is determined by its program/process name. Each process are divided into multiple data instances by shifting time windows with fixed size. See more information on the classification task in Table 2.

Because of its popularity and good generalization performance, SVM is chosen as the classifier to incorporate with the proposed kernels. We compare the generalization ability between the proposed methods. All the necessary parameters, namely, width parameter γ, margin parameter C, and the gap parameter λ for FTree based kernels, are selected by 5-fold cross validation on the training set. We use the LibSVM toolbox [7] to train the SVM classifiers. For comparison, we also report the result for TF-IDF based features.

In Table 3, we compare the recognition rate of the TF-IDF representation and the FTree kernel approach. As seen in the table, the proposed FTree based string kernels for most of the cases have outperformed the TF-IDF feature based SVM. For the registry dataset, normalized gap weighted kernel shows the best accuracy of about 95%. And for the File access dataset, the normalize gap plain kernel show the best accuracy. Although not significant, it seems that normalization

Table 3. Recognition Rate Comparison

Dataset	TFIDF-SVM	Nor. W.FTree	W.FTree	Nor. U.FTree	U.FTree
Registry	93.31	**95.17**	94.28	94.72	93.44
File I/O	87.52	89.35	90.07	**90.63**	89.42

of the kernels can help to improve the recognition rate to some extent. Since there is no significant difference between the generalization performance of a weighted kernel and a plain kernel, it may be more convenient to neglect weight information in the generalized FTree. And thus can lead to a considerable saving on the storage to retain the counter for all FTrees.

From another point of view, it is quite promising that an accuracy above 90% can be achieved using only such a small subset of the whole system call library. By collecting more information on other categories of system calls could possibly increase the accuracy to some extent and make system call analysis an powerful tool for OS status monitoring.

4 Conclusion and Future Research

In this paper, we have presented a study on how to apply machine learning techniques for realtime monitoring and analyzing application behaviors in a Windows guest OS. A group of computationally efficient kernels function are proposed to measure the proximity between programs and promising generalization performance is achieved on two real world datasets with moderate size.

For future research, the proposed framework will be evaluated on larger datasets collected from different OSes. And we will also explore the possibility to treat Windows security problems with this approach.

References

1. Barbar, D., Jajodia, S. (eds.): Applications of Data Mining in Computer Security. Kluwer, Dordrecht (2002)
2. Chan, P.K., Lippmann, R.P.: Machine learning for computer security. Journal of Machine Learning Research 7, 2669–2672 (2006)
3. Maloof, M. (ed.): Machine Learning and Data Mining for Computer Security. Springer, Heidelberg (2006)
4. Mazeroff, G., Knoxville, T., Thomason, M., Ford, R.: Probabilistic suffix models for API sequence analysis of Windows XP applications. Pattern Recognition 41(1), 90–101 (2008)
5. Wang, C., Pang, J., Zhao, R., Liu, X.: Using API sequence and Bayes algorithm to detect suspicious behavior. In: 2009 International Conference on Communication Software and Networks, Macau, China (February 27-28, 2009)
6. Ando, R.: A Visualization of anomaly memory behavior of full-virtuallized windows OS using virtual machine introspection (to appear, 2010)
7. Chang, C., Lin, C.: LIBSVM: a library for support vector machines (2001), Software available at `http://www.csie.ntu.edu.tw/~cjlin/libsvm`

Evolution of Information Retrieval in Cloud Computing by Redesigning Data Management Architecture from a Scalable Associative Computing Perspective

Amir H. Basirat and Asad I. Khan

Clayton School of Informationn Technology
Monash University, Victoria, Australia
{Amir.Basirat,Asad.Khan}@monash.edu

Abstract. The new surge of interest in cloud computing is accompanied with the exponential growth of data sizes generated by digital media (images/audio/video), web authoring, scientific instruments, and physical simulations. Thus the question, how to effectively process these immense data sets is becoming increasingly urgent. Also, the opportunities for parallelization and distribution of data in clouds make storage and retrieval processes very complex, especially in facing with real-time data processing. Loosely-coupled associative computing techniques, which have so far not been considered, can provide the break through needed for cloud-based data management. Thus, a novel distributed data access scheme is introduced that enables data storage and retrieval by association, and thereby circumvents the partitioning issue experienced within referential data access mechanisms. In our model, data records are treated as patterns. As a result, data storage and retrieval can be performed using a distributed pattern recognition approach that is implemented through the integration of loosely-coupled computational networks, followed by a divide-and-distribute approach that allows distribution of these networks within the cloud dynamically.

Keywords: Pattern Recognition, Neural Networks, Associative Computing, Single-Cycle Learning, Distributed Hierarchical Graph Neuron.

1 Introduction

Cloud computing encompasses a pay-per-use paradigm for providing services over the Internet in a scalable manner. Supporting data intensive applications is an essential requirement for the clouds. However, dynamic and distributed nature of cloud computing environments makes data management processes very complicated, especially in the case of real-time data processing/database updating. According to Shiers [1], "it is hard to understand how data intensive applications, such as those that exploit today's production grid infrastructures, could achieve adequate performance through the very high-level interfaces that are exposed in clouds". In addition to this complexity, there are other underlying issues that need to be addressed properly by any data management scheme deployed for clouds. Some of these concerns are highlighted by Abadi [2] including: capability to parallelize data workload, security concerns as a result of storing data at an untrusted host, and data replication

K.W. Wong et al. (Eds.): ICONIP 2010, Part II, LNCS 6444, pp. 275–282, 2010.

functionality. Hence the existing data management schemes do not work well when data is partitioned among numerous available nodes dynamically [3].

An approach towards application virtualization in cloud, which offers greater portability, manageability and compatibility of applications and data, has yet to be fully-explored. In this regard, the concept of universal data access for cloud computing can act as an alternative towards database-application integration while it can provide a common access mechanism through the use of fast and scalable data access framework within cloud. With this in mind, in this research paper we would like to explore new possibilities to evolve a novel virtualization scheme that can efficiently partition and distribute data for clouds. For this matter, loosely-coupled associative techniques, not considered so far, can be the key to effectively partitioning and distributing data in the clouds. Doing so will improve elastic scaling of system resources and remove one of the main obstacles in provisioning data centric software-as-a-service (SaaS) for the clouds. Our approach will entail two-fold benefit. On one hand, applications based on associative computing models will efficiently utilize the underlying hardware to scale up and down the system resources dynamically and on the other hand, the main hurdle towards providing scalable partitioning and distribution of data in the clouds will be removed, bringing forth a vastly superior solution for virtualizing data intensive applications and the system infrastructure to support pay on per-use basis.

2 Cloud Data Access Scheme

The efficiency of the cloud system in dealing with data intensive applications through parallel processing, essentially lies in how complex data is partitioned among nodes, and how collaboration among nodes is handled to accomplish a specific task. Data access schemes for cloud infrastructure should be able to distribute data across different networks and provide data services for remote clients. As a result, and to address the aforementioned concerns in relation to data storage and retrieval in cloud, any data access scheme should aim to handle partitioning between processing nodes, as well as node collaborations in a robust manner. These two features are still lacking in the current data access mechanisms. Hence, new data management approaches need to be investigated for cloud computing environments. In this paper, a distributed neural network technique is proposed by redesigning data management architecture from a scalable associative computing perspective for creating a database-like functionality that can scale up or down over the available infrastructure without interruption or degradation, dynamically. It eliminates data imbalances and completes transition to cloud by replacing referential data access mechanisms with fast and highly distributable associative memory.

3 Efficient Model Using Single-Cycle Learning

Our proposal is based on a special type of Associative Memory (AM) model, which is readily implemented within distributed architectures. Our aim is to apply a data access scheme that enables data retrieval to be conducted across multiple records and data segments within a single-cycle utilizing a parallel approach.

3.1 Distributed Hierarchical Graph Neuron (DHGN)

DHGN is a novel distributed associative memory (AM) algorithm for pattern recognition. The main idea behind this algorithm is that common pattern recognition approaches for various kinds of patterns would be able to be conducted within the body of a network. DHGN shifts the recognition algorithm paradigm from employing CPU-centric processing towards network-centric processing approach. It also adopts single-cycle learning with in-network processing capability for fast and accurate recognition scheme. The basic foundation of DHGN algorithm is based upon the functionalities and capabilities of two other associative memory algorithms known as Graph Neuron (GN) [4] and Hierarchical Graph Neuron (HGN) [5]. It eliminates the crosstalk issue in GN implementation, as well as reduces the complexity of HGN algorithm by reducing the number of processors required for its execution. DHGN is also a lightweight pattern recognizer that supports adaptive granularity of the computational network, ranging from fine-grained networks such as WSN to coarse-grained networks including computational grid. Figure 1 depicts a GN array, which is capable of converting the spatial/temporal patterns into a simple graph-based representation in which input patterns are compared with the edges in the graph for memorization or recall operations. Recognition process within DHGN involves a single-cycle learning of patterns on a distributed processing manner. Among various applications of the GN based AM, an input pattern in GN pattern recognition may represent bit elements of an image [6] or a stimulus/signal spike produced within a network intrusion detection application [7]. In order to solve the issue of the crosstalk in GN model due to the limited perspective of GNs, the capabilities of perceiving GN neighbours in each GN was expanded in Hierarchical Graph Neuron (HGN) to prevent pattern interference [5]. We then extended the HGN by dividing and distributing the recognition processes over the network (See Figure 2) [6].

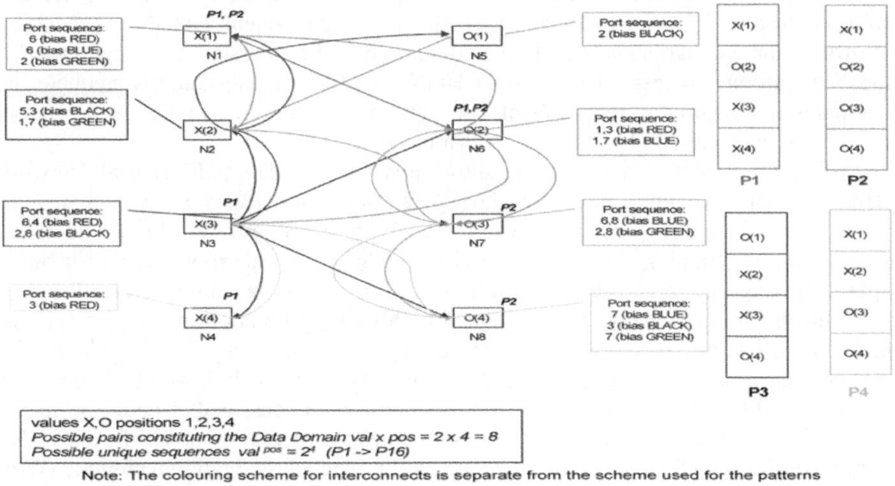

Fig. 1. An eight node GN is in the process of storing patterns. P1 (RED), P2 (BLUE), P3 (BLACK), and P4 (GREEN)

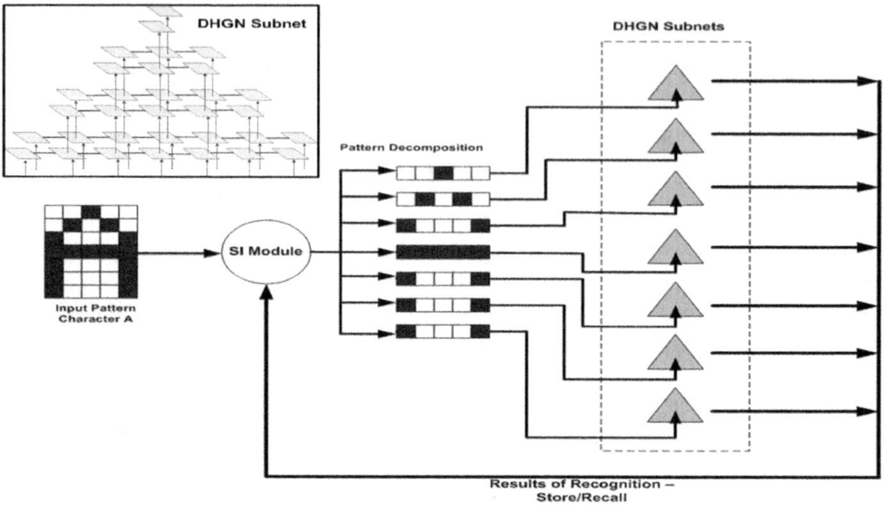

Fig. 2. DHGN distributed pattern recognition architecture

At macro level, DHGN pattern recognition algorithm works by applying a divide-and-distribute approach to the input patterns. It involves a process of dividing a pattern into a number of sub-patterns and the distribution of these sub-patterns within the DHGN network as shown in Figure 2. This distributed scheme minimizes the number of processing nodes by reducing the number of levels within the HGN. Figure 3 shows the divide-and-distribute transformation from a monolithic HGN composition (top) to a DHGN configuration for processing the same 35-bit patterns (bottom). The base of the HGN structure in Figure 3 represents the size of the pattern. Note that the base of HGN structure is equivalent to the cumulative base of all the DHGN subnets/clusters. This transformation of HGN into equivalent DHGN composition allows on the average 80% reduction in the number of processing nodes required for the recognition process. Therefore, DHGN is able to substantially reduce the computational resource requirement for pattern recognition process – from 648 processing nodes to 126 for the case shown in Figure 3.

Unlike other pattern recognition algorithms such as Hopfield Neural Network (HNN) [8] and Kohonen SOM [9], DHGN employs in-network processing feature within the recognition process. The test reveals that DHGN offers higher accuracy with minimum training data, in comparison to SOM. Furthermore, our distributed approach requires no training iteration, as it adopts a single-cycle learning mechanism. Comparatively, SOM requires high training iteration in order to achieve high classification accuracy. This processing capability of DHGN allows the recognition process to be performed by a collection of lightweight processors (referred to PEs). PE is an abstract representation of processor that could be in the form of a specific memory location or a single processing node. DHGN also eliminates the need for complex computations for event classification technique. With the adoption of single-cycle learning and adjacency comparison approaches, DHGN implements a non-iterative and lightweight computational mechanism for event recognition and classification.

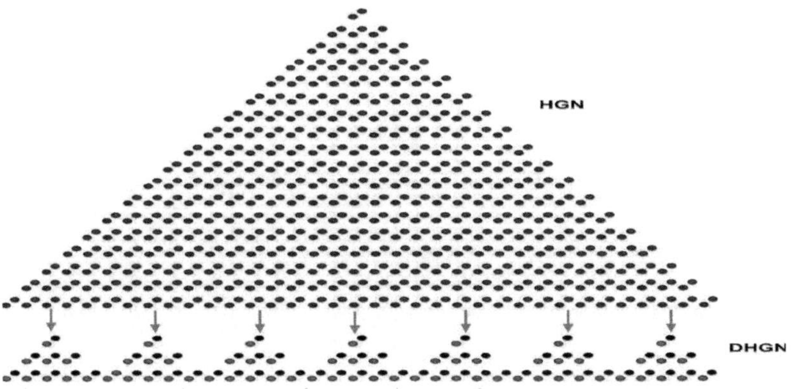

Fig. 3. Transformation of HGN structure (top) into an equivalent DHGN structure (bottom)

The results of our performance analysis have also shown that DHGN recognition time increases linearly with an increase in the number of processing elements (PEs) within the network. This simply reveals that DHGN's computational complexity is also scalable with an increase in the size of the sub-patterns. DHGN allows the recognition process to be conducted in a smaller sub-pattern domain, hence minimizing the number of processing nodes which in turn reduces the complexity of pattern analysis. In addition, the recognition process performed using DHGN algorithm is unique in a way that each subnet is only responsible for memorizing a portion of the pattern (rather than the entire pattern). A collection of these subnets is able to form a distributed memory structure for the entire pattern. This feature enables recognition to be performed in parallel and independently. The decoupled nature of the sub-domains is the key feature that brings dynamic scalability to data management within cloud computing.

4 Tests and Results

An important contribution of our research work is the identification of novel DHGN usages for distributed information processing in clouds. For that purpose, a web-based DHGN is implemented to illustrate the fact that dynamic scalability of DHGN has the potential to remarkably empower virtualization in clouds and drive the future of data centre networking. Using this web-based DHGN, users can start drawing images based on provided image templates. For simplicity of our model, users are provided with three sample pattern templates of Apple, Banana and Leaf as depicted in Figure 4. As a result, the user drawn image will be compared with these three master patterns using DHGN algorithm. The approach is not only capable of detecting the type of image but it also can provide users with various data analysis based on their drawn input patterns (See Figure 5).

Having an image as an input pattern, the bitmap object built from the input pattern is re-sized to form a static size of 128x128. Then, DHGN model for this object is formed by using 128 HGNs. In fact, each individual HGN is assigned the task of processing a single line of image. Following that, a vertex of size 16K is constructed which includes total number of HGN nodes (16384 nodes).

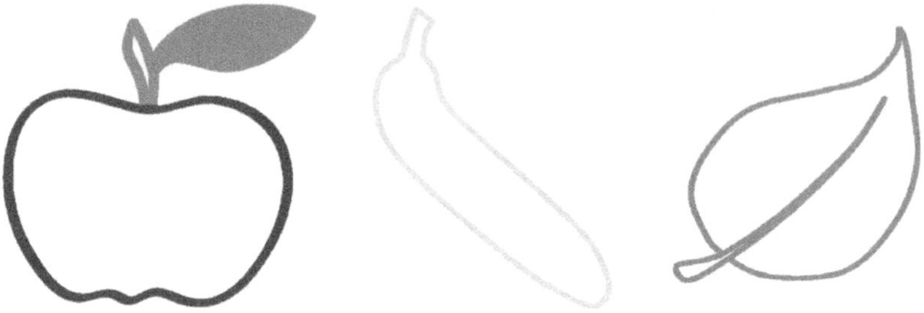

Fig. 4. Three provided master patterns as image templates (Apple, Banana, and Leaf)

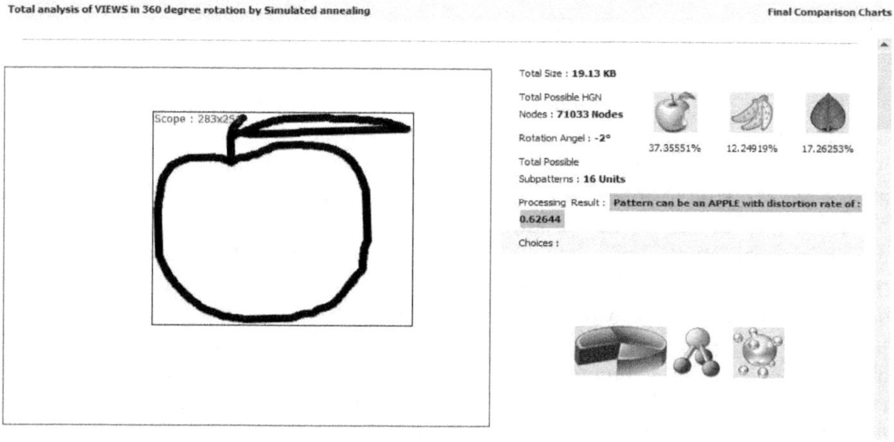

Fig. 5. User drawn image is detected as an Apple with 62.6% distortion rate

As it is clearly depicted in Figure 5, the input pattern is heavily distorted compared with the provided apple base pattern. However, the algorithm exhibits its remarkable strength by detecting the image as an apple, while its distortion rate is almost 62.6%. Comparing the input pattern with banana and leaf pattern templates, results in distortion rates of 87.8% and 82.7% respectively.

Figure 6(a) illustrates total number of positive and negative matches between the input pattern nodes and all of the three pattern templates. Figure 6(b) represents distortion rates for each individual line of image (each HGN) when compared against three provided base patterns. Using simulated annealing technique and to minimize distortion rate, input image is also rotated within a pre-defined range of values. The rotation range for the example case depicted in Figure 5 is set to -2 to 2 degrees with a step of 1 degree. As a result, the image is rotated in steps of 1-degree and each time it is compared with all three master patterns using DHGN pattern matching technique. As can be seen from Figure 7, the minimum distortion rate is achieved with rotation degree of -2 (2 degrees anticlockwise).

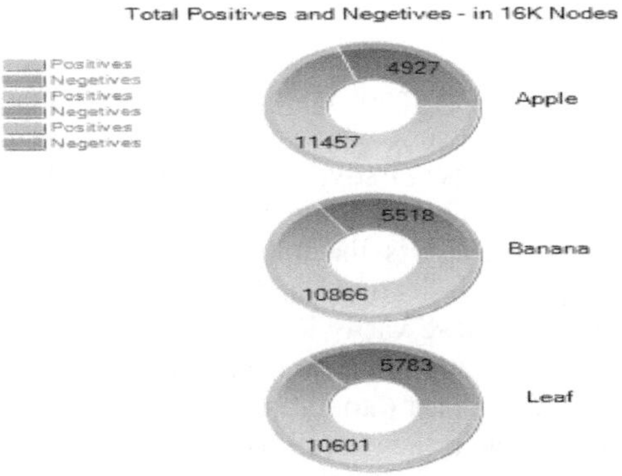

(a)

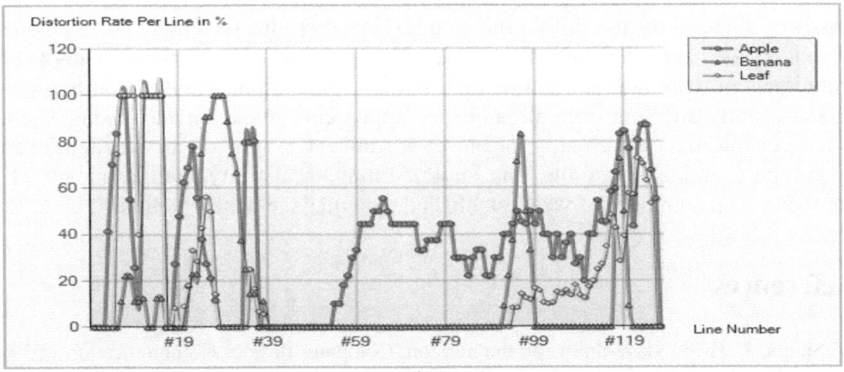

(b)

Fig. 6. **(a)** Total number of positive and negative matches. **(b)** Distortion rates for each line of image (each constructed HGN).

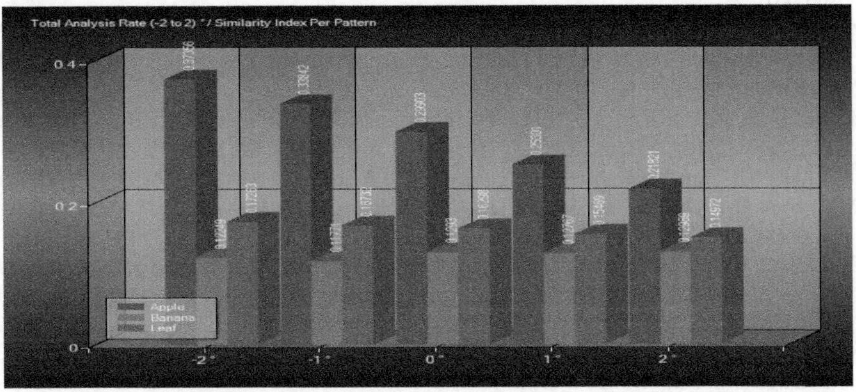

Fig. 7. Image distortion rates vs. rotation degrees

5 Remarks and Conclusion

This paper heralds a new breed of innovative cloud applications by making data universally available within the network. The work presented here enables innovative cloud applications on two accounts. Firstly, it brings a content-based (associative) searching mechanism to Clouds where in complex data types such as images may be specified as keys. Secondly, DHGN is the only distributable associative memory approach that provides single-cycle learning and entails a large number of loosely coupled parallel operations to achieve vastly improved performance. In fact, strength of DHGN lies in the processing of non-uniform data patterns as it implements a finely distributable framework at the smallest (atomic) logical sub-pattern level. The results are easily obtained by summation at the overall pattern level. DHGN is also a highly scalable algorithm that incorporates content addressable memory within a clustered framework. This one-shot approach will allow real-time in-network data manipulation; essential feature for online processing. Our proposed technique is primarily focused for use within the clouds. However, this technique has the potential of wider use provided we can compare its characteristics with state-of-the-art techniques in data management using a pattern recognition scheme. Our approach is fundamentally different from all published approaches in data management. The large heterogeneous datasets created for our case studies can provide an excellent resource to compare and contrast the one shot learning, scalability, and accuracy of our approach with a number of well-established data management techniques.

References

1. Shiers, J.: Grid today, clouds on the horizon. Computer Physics Communications, 559–563 (2009)
2. Abadi, D.J.: Data Management in the Cloud: Limitations and Opportunities. Bulletin of the Technical Committee on Data Engineering, 3–12 (2009)
3. Szalay, A., Bunn, A., Gray, J., Foster, I., Raicu, I.: The Importance of Data Locality in Distributed Computing Applications. In: Proceedings of the NSF Workflow Workshop (2006)
4. Khan, A.I., Mihailescu, P.: Parallel Pattern Recognition Computations within a Wireless Sensor Network. In: Proc. of 17th Intl. Conf. on Pattern Recognition, United Kingdom (2004)
5. Nasution, B.B., Khan, A.I.: A Hierarchical Graph Neuron Scheme for Real-Time Pattern Recognition. IEEE Transactions on Neural Networks, 212–229 (2008)
6. Khan, A.I., Muhamad Amin, A.H.: One Shot Associative Memory Method for Distorted Pattern Recognition. In: Orgun, M.A., Thornton, J. (eds.) AI 2007. LNCS (LNAI), vol. 4830, pp. 705–709. Springer, Heidelberg (2007)
7. Baig, Z.A., Baqer, M., Khan, A.I.: A pattern recognition scheme for distributed denial of service (DDOS) attacks in wireless sensor networks. In: Proc. of the 18th International Conference on Pattern Recognition (2006)
8. Hopfield, J.J., Tank, D.W.: Neural Computation of Decisions in Optimization Problems. Biological Cybernetics 52, 141–152 (1985)
9. Kohonen, T.: Self-Organizing Maps. Springer, Heidelberg (2007)

Factorizing Class Characteristics via Group MEBs Construction

Ye Chen, Shaoning Pang, and Nikola Kasabov

KEDRI, Auckland University of Technology, New Zealand & NICT, Japan

Abstract. Classic MEB (minimum enclosing ball) models characteristics of each class for classification by extracting core vectors through a $(1+\varepsilon)$-approximation problem solving. In this paper, we develop a new MEB system learning the core vectors set in a group manner, called group MEB (g-MEB). The g-MEB factorizes class characteristic in 3 aspects such as, reducing the sparseness in MEB by decomposing data space based on data distribution density, discriminating core vectors on class interaction hyperplanes, and enabling outliers detection to decrease noise affection. Experimental results show that the factorized core set from g-MEB delivers often apparently higher classification accuracies than the classic MEB.

Keywords: Minimum Enclosing Ball, Core Vector Machine, Group Minimum Enclosing Ball.

1 Introduction

The minimum enclosing ball (MEB) problem is to compute a ball of minimum radius enclosing a given set of objects (points, balls, etc) in $\mathcal{R}^d$. It has been wildly implemented for clustering applications, such as support vector clustering [1]; classification applications, such as area gap tolerant classifiers [2], and core vector machine (CVM) [3]; as well as approximation applications, such as 1-cylinder problem approximation [4].

Classic MEB for classification computes a $(1+\varepsilon)$-approximation [4] for a minimum radius ball learning, and extracts those data points located at the outer area of a MEB for classification modelling. The set of those extracted data points characterize the given entire dataset, thus are called core vector set or core set. For classification modelling, MEB can be used to approximate each class data distribution, so that one class can be distinguished from another by core set computing.

However in practice, such classic MEB has the following difficulties (1) MEB encloses often sparseness together with data. To enclose an isolated outlier points, a huge MEB is required, which makes the MEB include actually more sparseness than the data occupation (2) MEB keens on enclosing data, thus disables the detection of any outliers despite that outliers produce the sparseness of MEB.

K.W. Wong et al. (Eds.): ICONIP 2010, Part II, LNCS 6444, pp. 283–290, 2010.

To mitigate the above problems, this paper proposes a novel group MEB (g-MEB) approach to learning core set in a group manner. g-MEB sets MEBs in different data distribution area, reducing the sparseness in MEB by decomposing data space based on data distribution density, discriminating core vectors on class interaction hyperplanes, and enabling outliers detection to decrease noise affection. The rest of paper is structured as follows: Section 2 reviews the original MEB algorithm. Section 3 presents the proposed g-MEB learning. In Section 4, we cover experimentation and algorithm evaluation. Lastly, in Section 5 we draw our conclusion and state future directions.

2 Classic MEB

Given a data matrix $\mathbf{X} \in \mathbb{R}^{n \times d}$ in l categories, $\mathbf{X} = [\mathbf{X}_1, \mathbf{X}_2, \ldots, \mathbf{X}_l]$ where $\mathbf{X}_i \in \mathbb{R}^{n_i \times d}$ and $\sum_{i=1}^{l} n_i = n$. An exact MEB over $\mathbf{X}$ is modelled as a smallest hypersphere $B_{\mathbf{X}} = \{\mathbf{c}, r, \mathbf{X}\}$, where $\mathbf{c}, r$ are the center and radius of $B_{\mathbf{X}}$, respectively. Alternatively, the MEB can be computed by $(1+\varepsilon)$-approximation [5] as a $(1+\varepsilon)$ MEB $B_{\mathbf{X}} = \{c, (1+\varepsilon)r, \mathbf{S}\}$, where $\varepsilon > 0$, and $\mathbf{S} \subset \mathbf{X}$, is the core set of $\mathbf{X}$ that contains the instances located at the outer area. Alternatively, kernel $(1+\varepsilon)$-approximation MEB computing is conducted by 2 steps.

The first step is Kernel MEB initialization. Given data matrix $\mathbf{X} = \cup_{i=1}^{l} \mathbf{X}_i$, and $0 < \varepsilon < 1$, we select subset $\mathbf{S} = \{\mathbf{x}_1, \mathbf{x}_z | \arg\max_{\mathbf{x}_z \in \mathbf{X}_i} ||\mathbf{x}_z - \mathbf{x}_1||, \mathbf{x}_1 \in \mathbf{X}_i\}$ and have the $(1+\varepsilon)$-*approximation* MEB over $\mathbf{S}$ as, $B_{\mathbf{S}} = \{\mathbf{c}_i, (1+\varepsilon)r_i, \mathbf{S}\}$. To adopt $B_{\mathbf{S}}$ for kernel computing, we map $\mathbf{S}$ with an associated embedding φ using kernel κ satisfying $\kappa(\mathbf{x}, \mathbf{z}) = \langle \varphi(\mathbf{x}_a), \varphi(\mathbf{x}_b) \rangle$. Then, we obtain $\mathbf{c}_i$ and r_i by solving the following optimization problem: $\min_{\mathbf{c}_i \in \mathbb{R}^d} \max_{\mathbf{x}_z \in \mathbf{S}} ||\mathbf{c}_i - \varphi(\mathbf{x}_z)||$, or more precisely

$$\begin{array}{ll} \min_{\mathbf{c}_i, r_i} & r_i^2 \\ \textit{subject to} & ||\mathbf{c}_i - \varphi(\mathbf{x}_z)||^2 = (\varphi(\mathbf{x}_z) - \mathbf{c}_i)'(\varphi(\mathbf{x}_z) - \mathbf{c}_i) \le r^2 , \\ & \mathbf{x}_z = \mathbf{S} \end{array} \tag{1}$$

which can be solved by introducing a Lagrange multiplier $\alpha_z \ge 0$ for each constraint

$$L(\mathbf{c}_i, r_i, \alpha) = r_i^2 + \sum_{z=1}^{|\mathbf{S}|} \alpha_z \left[||\varphi(\mathbf{x}_z) - \mathbf{c}_i||^2 - r_i^2 \right] . \tag{2}$$

Calculating the derivative of $\mathbf{c}_i$ and r_i from Eq. (2),

$$\begin{array}{l} \frac{\partial L(\mathbf{c}_i, r_i, \alpha)}{\partial \mathbf{c}_i} = 2 \sum_{z=1}^{|\mathbf{S}|} \alpha_z (\varphi(\mathbf{x}_z) - \mathbf{c}_i) = 0, \textit{ and} \\ \frac{\partial L(\mathbf{c}_i, r_i, \alpha)}{\partial r_i} = 2r_i \left(1 - \sum_{z=1}^{|\mathbf{S}|} \alpha_z\right) = 0, \end{array} \tag{3}$$

we have $\sum_{z=1}^{|\mathbf{S}|} \alpha_z = 1$, and

$$\mathbf{c}_i = \sum_{z=1}^{|\mathbf{S}|} \alpha_z \varphi(\mathbf{x}_z), \quad r_i = \sqrt{\alpha' diag(K) - \alpha' K \alpha}. \tag{4}$$

Thus, with kernel definition, the above $(1+\varepsilon)$-*approximation* MEB is renovated to kernel MEB as $B_{\mathbf{S}} = \{\mathbf{c}_i, (1+\varepsilon)r_i, \varphi(\mathbf{S})\}$ in which $\mathbf{c}_i$ and r_i are obtained from Eq. 4.

The second step is MEB expansion. Based on the above initialized $\mathbf{S}$, we update $B_{\mathbf{S}}$ by adding the furthest point to $B_{\mathbf{S}}$, $\mathbf{S} = \mathbf{S} \cup \mathbf{x}^*$, $\mathbf{x}^* = \arg\max_{\mathbf{x}\in\mathbf{X}_i} ||\varphi(\mathbf{x}) - \mathbf{c}_i||$, where the distance of $\mathbf{x}$ to $B_{\mathbf{S}}$ is evaluated in kernel space as:

$$\|\mathbf{c}_i - \varphi(\mathbf{x})\|^2 = \sum_{\mathbf{x}_a,\mathbf{x}_b\in\mathbf{S}} \alpha_a\alpha_b\kappa(\mathbf{x}_a,\mathbf{x}_b) - 2\sum_{\mathbf{x}_a\in\mathbf{S}} \alpha_a\kappa(\mathbf{x}_a,\mathbf{x}) + \kappa(\mathbf{x},\mathbf{x}). \qquad (5)$$

The MEB expansion continues until $\mathbf{X}_i \subset B_{\mathbf{S}}$. As a result, we obtain kernel MEB for ith class data $\mathbf{X}_i$.

For the overall dataset $\mathbf{X} = \cup_{i=1}^{l}\mathbf{X}_i$, let $\mathbf{Q}_i = \varphi(\mathbf{S})$ be the core set, then we have the general MEB model Ω_{MEB} as a set of kernel MEB,

$$\Omega_{MEB} = \begin{pmatrix} \mathbf{c}_1, & (1+\varepsilon)r_1, & \mathbf{Q}_1 \\ \mathbf{c}_2, & (1+\varepsilon)r_2, & \mathbf{Q}_2 \\ \vdots & \vdots & \vdots \\ \mathbf{c}_l, & (1+\varepsilon)r_l, & \mathbf{Q}_l \end{pmatrix}, \qquad (6)$$

and summarize the computation of MEB learning as Algorithm 1.

Algorithm 1. Kernel MEBs algorithm

Input: Set of points $\mathbf{X} = \cup_{i=1}^{l}\mathbf{X}_i \in \mathbb{R}^d$; parameter $\varepsilon = 2^{-m}$
Output: Kernel MEB learning model Ω_{MEB}

for each $\mathbf{X}_i \subset \mathbf{X}$ **do**
  Initialize $\mathbf{Q}_i^0 \leftarrow \mathbf{x}_1, \mathbf{p}$, where $\mathbf{p} \in \mathbf{X}_i$ is the furthest instance to $\mathbf{x}_1$
  Computer the initial MEB B_i^0 and its radius r_i^0 and center $\mathbf{c}_i^0$ using equation 4 on $\mathbf{Q}_i^0$
  for $u \leftarrow 0$ to $m-1$ **do**
    if $\mathbf{X}_i \subseteq B_i^u$ **then**
      find $\mathbf{x} \leftarrow \arg\max_{\mathbf{x}\in\mathbf{X}_i} ||\mathbf{x} - \mathbf{c}_i^u||$
      $\mathbf{Q}_i^u \leftarrow \{\mathbf{Q}_i^u, \varphi(\mathbf{x})\}$
      Computer MEB B_i^u and its radius r_i^u and center $\mathbf{c}_i^u$ using equation 4 on $\mathbf{Q}_i^u$
    else
      $\mathbf{c}_i \leftarrow \mathbf{c}_i^u$, $r_i \leftarrow r_i^u$, and $\mathbf{Q}_i \leftarrow \mathbf{Q}_i^u$
      Break for loop
    end if
  end for
  $\Omega_{MEB} \leftarrow \Omega_{MEB} \bigcup \{\mathbf{c}_i, (1+\varepsilon)r_i, \mathbf{Q}_i\}$
end for

3 The Proposed g-MEB Learning

In terms of complex data distribution, a large MEB follows the sparseness of the ball, thus is often unable to approximate a data distribution accurately. A single MEB for one class data is likely to encloses wrongly almost all data points from another class, especially when data is zonally distributed. As a solution, a number of smaller MEBs are able to drill into the details of any data distribution, apparently allowing a more accurate approximation.

Motivated by this, the above *kernel* MEB is renovated for group manner MEB computing (g-MEB). Instead of addressing a whole class data $\mathbf{X}_i$ with one MEB $B_{\mathbf{Q}_i}$, g-MEB models class data using with a set of MEBs $B_{\mathbf{Q}_i} = \cup_{j=1}^{k} B_{\mathbf{Q}_{i,j}^u}$, where k is the number of MEBs. Consider that MEB learning is an iterative learning process, we represent an individual *kernel* MEB here as $B_{\mathbf{Q}_{i,j}^u} = \{\mathbf{c}_{i,j}^u, (1+\varepsilon)r_{i,j}^u, \varphi(\mathbf{Q}_{i,j}^u)\}$ with j as the index of MEB, and u as the iteration number of MEB updating. In this way, given $\mathbf{X} = \cup_{i=1}^{l}\mathbf{X}_i$ as the training dataset, the proposed g-MEB learning is described as follows.

Similar to *kernel* MEB, we initialize one MEB (i.e. $B_{\mathbf{Q}_{i,k}^0}$, $k = 1$) on one class data $\mathbf{X}_i$ at very beginning. Here, we initialize the core set as $\mathbf{Q}_{i,1}^0 = \{\varphi(\mathbf{x}_a), \varphi(\mathbf{x}_b)\}$, where $\mathbf{x}_a$ is the furthest data point to a random $\mathbf{x} \in \mathbf{X}_i$ and $\mathbf{x}_b$ is calculated as:

$$\mathbf{x}_b = \mathbf{x}_a + \frac{\mathbf{x}_a - \arg\max_{\mathbf{x}\in\mathbf{X}} ||\mathbf{x}_a - \mathbf{x}||}{\lambda}, \quad \lambda > 1. \tag{7}$$

Then, we obtain $B_{\mathbf{Q}_{i,1}^0} = \{\mathbf{c}_{i,1}^0, (1+\varepsilon)r_{i,1}^0, \mathbf{Q}_{i,1}^0\}$ using Eq. 4.

Theorem 1. *The first initialized MEB's radius* $r_{i,1}^0 \approx \frac{\Delta}{2\lambda}$, *where* Δ *denotes the diameter of* $\mathbf{X}_i$.

Proof. Science $\mathbf{Q}_{i,1}^0 = \varphi(\mathbf{x}_a, \mathbf{x}_b)$, $\mathbf{x}_a$ is the furthest point to $\mathbf{x}_1$. Clearly, the distance from $\mathbf{x}_a$ to it's furthest data point approximate to δ. Then, $\mathbf{x}_b$ locates about Δ/λ away from $\mathbf{x}_a$. According to the definition $r_{i,1}^0$ is the radius of $B_{\mathbf{Q}_{i,1}^0}$, we have $r_{i,1}^0 \approx \frac{\Delta}{2\lambda}$.

If there has $\mathbf{x} \in \mathbf{X}_i$ not contained in any MEB $B_{\mathbf{Q}_{i,j}^{u-1}}$, $(j = \{1, 2, \ldots, k\})$. It results in one of the following 3 g-MEB updating cases.

In the first case, if MEB $B_{\mathbf{Q}_{i,j}^u}$ over the existed core set $\mathbf{Q}_{i,j}^{u-1}$ with $\mathbf{x}$ (i.e. $\mathbf{Q}_{i,j}^u = \mathbf{Q}_{i,j}^{u-1} \cup \varphi(\mathbf{x})$) has radius $r_{i,j}^u$, which is less or equal to the upper bound value as $1 + \frac{\varepsilon^2 \times \eta}{16}$, $(\eta > 1)$ times of the existed radius $r_{i,j}^{u-1}$. The existed MEB $B_{\mathbf{Q}_{i,j}^{u-1}}$ is updated by replacing with $B_{\mathbf{Q}_{i,j}^u}$. Note that, $\mathbf{c}_{i,j}^{u-1}$ is the closest MEB center to $\mathbf{x}$.

Theorem 2. *Given an upper bound of radius increment as* $r_{i,j}^u \leq (1+\frac{\varepsilon^2 \times \eta}{16})r_{i,j}^{u-1}$, *the g-MEB expansion is between* $32\lambda/\varepsilon^2$ *to* $128\lambda\delta/\varepsilon^2$*times.*

Proof. Science $r_{i,j}^0 \geq \Delta/2\lambda$, and each step we increase the radius by at least $(\Delta/4)\varepsilon^2/16 = \Delta\varepsilon/64$, it follows that we cannot encounter this case more than $64/\varepsilon$ times, as δ is an upper bound of the radius of the minimum enclosing ball of $\mathbf{X}_i$.

In the second case, if radius $r_{i,j}^{u}$ is greater than the upper bound value $(1 + \frac{\varepsilon^2 \times \eta}{16}) r_{i,j}^{u-1}$, a new fragmentary core set $\mathbf{Q}_{i,k+1}^{0} = \{\varphi(\mathbf{x})\}$ is created as a completed core set which has at least 2 vectors.

In the third case, if the distance from $\mathbf{x}$ to the closest fragmentary core set $\mathbf{Q}_{i,j}^{0}$ is less than $\frac{\Delta}{\lambda}$, we add $\varphi(\mathbf{x})$ into the fragmentary core set $\mathbf{Q}_{i,j}^{0}$ as $\mathbf{Q}_{i,j}^{0} = \{\mathbf{Q}_{i,j}^{0}, \varphi(\mathbf{x})\}$, in this way, the fragmentary core set $\mathbf{Q}_{i,j}^{0}$ becomes completed. In addition, a new MEB $B_{\mathbf{Q}_{i,k}^{0}} = \{\mathbf{c}_{i,k}^{0}, (1+\varepsilon) r_{i,k}^{0}, \mathbf{Q}_{i,k}^{0}\}$, $(k = k+1)$ is created using Eq. 4. The threshold $\frac{\Delta}{\lambda}$ is also considered as the outliers threshold, as if a single data point away from the rest data farther than the threshold, this data point is treated as an outlier by g-MEB.

Theorem 3. *The total number of g-MEB k equals to approximately* $\lambda/2$.

Proof. Science we guarantee that the radius of each firstborn g-MEB $r_{i,j}^{0}$, $(j = \{1, 2, \ldots, k\})$ is less than $\frac{1}{2\lambda}$ diameter of the given data $\mathbf{X}_i$. Thus, the total number of the g-MEB k approximates to $\lambda/2$.

The g-MEB updating is terminated once $\mathbf{X}_i \subset \cup_{j=1}^{k} B_{\mathbf{Q}_{i,j}^{k}}$. For the overall dataset $\mathbf{X} = \cup_{i=1}^{l} \mathbf{X}_i$, let k_i be the number of MEBs of $i\text{-}th$ class, then we factorize the core sets by abandoning these core vectors contained in just one MEB. As a result, we have the g-MEB model Ω_{gMEB} as the set of MEBs that constructed by the above 3 g-MEB updating cases.

$$\Omega_{gMEB} = \begin{pmatrix} \cup_{j=1}^{k_1} \mathbf{c}_{1,j} & \cup_{j=1}^{k_1} (1+\varepsilon) r_{1,j} & \cup_{j=1}^{k_1} \mathbf{Q}_{1,j} \\ \cup_{j=1}^{k_2} \mathbf{c}_{2,j} & \cup_{j=1}^{k_2} (1+\varepsilon) r_{2,j} & \cup_{j=1}^{k_2} \mathbf{Q}_{2,j} \\ \vdots & \vdots & \vdots \\ \cup_{j=1}^{k_l} \mathbf{c}_{l,j} & \cup_{j=1}^{k_l} (1+\varepsilon) r_{l,j} & \cup_{j=1}^{k_l} \mathbf{Q}_{l,j} \end{pmatrix}, \quad (8)$$

and summarize the computation of MEB learning as Algorithm 2.

4 Experiments and Discussions

In this section, we give 2 experiments where we used g-MEB for benchmark UCI data classification, and Face Membership Authentication (FMA) [6].

4.1 Classification Accuracy Comparison

To evaluate class factorization ability of g-MEB, MEB and g-MEB are compared in terms of the classification performance achieved with 4 conventional classification methods on 5 two-class and 5 multi-class benchmark datasets. For each dataset, we conducted K folds cross validation, we set $K = 10$ as 10-fold cross-validation is commonly used.

Table 1 presents the comparison results. For two-class datasets, MEB outperforms other 4 conventional classification methods for 3 out of 5 datasets, but

Algorithm 2. g-MEB algorithm

Input: Set of points $\mathbf{X} \in \mathbb{R}^d$; parameter ε, λ, and η
Output: A g-MEB learning model Ω_{gMEB}

for each $\mathbf{X}_i \subset \mathbf{X}$ **do**
  Initialize $k \leftarrow 1$, $T = \emptyset$, and Δ which is the diameter of $\mathbf{x}$
  Initialize $\mathbf{Q}^0_{i,j}$ by equation 7
  Computer the initial MEB $B^0_{i,1}$ and its radius $r^0_{i,1}$ and center $\mathbf{c}^0_{i,1}$ using equation 4 on $\mathbf{Q}^0_{i,1}$
  for each $\mathbf{x} \in \mathbf{X}_i$ **do**
    if $\mathbf{x} \notin \cup_{j=1}^{k} B_{i,j}$ **then**
      Find $\mathbf{t} \leftarrow \arg\min_{\mathbf{t}\in\mathbf{T}} ||\mathbf{x} - \mathbf{t}||$
      if $||\mathbf{x} - \mathbf{t}|| < \frac{\Delta}{2\lambda}$ **then**
        Remove $\mathbf{t}$ from $\mathbf{T}$
        $k \leftarrow k + 1$
        $\mathbf{Q}^0_{i,k} = \{\varphi(\mathbf{t}), \varphi(\mathbf{x})\}$
        Computer a new MEB $B^0_{i,k}$ and its radius $r^0_{i,k}$ and center $\mathbf{c}^0_{i,k}$ using equation 4 on $\mathbf{Q}^0_{i,k}$
      else
        Find $\mathbf{c}^u_{i,j} \leftarrow \arg\min_{j\in\{1:k\}} ||\mathbf{x} - \mathbf{c}^u_{i,j}||$
        $\mathbf{Q}^{u+1}_{i,j} \leftarrow \{\mathbf{Q}^u_{i,j}, \varphi(\mathbf{x})\}$
        Update MEB $B^{u+1}_{i,j}$ and its radius $r^{u+1}_{i,j}$ and center $\mathbf{c}^{u+1}_{i,j}$ using equation 4 on $\mathbf{Q}^{u+1}_{i,j}$
        if $r^{u+1}_{i,j} > r^u_{i,j} \times (1 + \frac{\varepsilon^2 \times \eta}{16})$ **then**
          $\mathbf{T} \leftarrow \mathbf{T} \cup \mathbf{x}$
          Undo this update
        end if
      end if
    end if
  end for
  $\Omega_{gMEB} \leftarrow \Omega_{gMEB} \bigcup \{\cup_{j=1}^{k} \mathbf{c}_{i,j}, \cup_{j=1}^{k}(1+\varepsilon) r_{i,j}, \cup_{j=1}^{k} \mathbf{Q}_{i,j}\}$
end for

MEB is defeated by g-MEB for all dataset except for ionosphere. For multi-class datasets, again g-MEB wins for 4 out of 5 datasets. In comparison to MEB, the g-MEB exhibits a more stable behavior in that g-MEB wins MEB in both two-class and multi-class categories. The results indicate that g-MEB is worthy to be noted as a perfect method for most of the selected datasets.

4.2 Class Factorization Evaluation

To evaluate class factorization ability of g-MEB, again we study the FMA problem [6] which is to distinguish the membership class from the non-membership in a total group through a binary class classification. FMA involving different levels of class overlapping which involves most discriminative class characteristics [7] because class overlapping increases while the membership size is closing to the non-membership size. The size of the membership group can be dynamically changed which makes class characteristic of membership and non-membership

Table 1. Classification accuracy comparison on 10 selected data sets

Dataset	KNN	Bayes	MLP	SVM	MEB	g-MEB
liver-disorder	64.92%	63.17%	**71.28%**	64.92%	59.13%	65.13%
breast-cancer	96.63%	95.97%	96.65%	**97.07%**	95.43%	**97.07%**
heart	82.59%	75.65%	83.33%	83.19%	83.67%	**84.11%**
ionosphere	64.11%	65.32%	89.16%	92.06%	**92.73%**	92.72%
web Spam	93.32%	84.13%	94.62%	94.60%	94.64%	**94.89%**
iris	94.67%	90.35%	95.97%	96.00%	**96.66%**	**96.66%**
wine	**97.16%**	71.31%	93.91%	74.85%	73.76%	83.12%
vehicle	71.04%	66.24%	77.78%	71.61%	71.61%	**78.13%**
vowel	96.89%	91.37%	98.08%	97.47%	97.45%	**98.56%**
KDD99	89.32%	75.21%	90.53%	90.42%	90.23%	**94.95%**

manually adjustable, the smaller size of membership group, the less discriminative class characteristics involved.

FMA is performed on MPEG-7 face dataset, which consists of 1355 face images of 271 persons (5 different face images per person are taken). We set the membership size ranging from 35 to 230 with a 10 persons interval to achieve datasets with dynamic class characteristics, and compared the proposed g-MEB with MEB under the condition of dynamic distinctive class characteristics in FMA.

(a) (b)

Fig. 1. (a) and (b) shows the comparison results of MEB and g-MEB on classification accuracy and number of core vectors under the condition of different membership group size, respectively

Fig. 1a and Fig. 1b illustrate the number of core vectors and classification accuracy for MEB and g-MEB under the condition of different membership group size respectively. As seen in Fig. 1a, the number of core vectors from MEB stays constantly around 80, while the membership group size is growing from 35 to 135 (equals to 50% of total group size). However, the number of core vectors from g-MEB has a spiking increase, but never going above the number of core vectors from MEB. On the other hand, Fig. 1b shows that g-MEB achieves general higher FMA accuracy than MEB, and the difference becomes as significant as 8% when the member group size is ranged from 50 to 135. Recall that the number of g-MEB core vectors is always smaller than the number of MEB core vectors, which indicates that g-MEB core vectors are more discriminative than MEB for FMA. In other words, g-MEB is more capable than MEB on factorizing the class characteristics of membership and non-membership as fewer g-MEB core vectors delivers often better FMA.

5 Conclusions and Future Work

In this paper, a novel g-MEB is proposed that learns characteristics of class (i.e. core vectors) in a group manner. g-MEB factorizes class characteristics by reducing the sparseness area, discriminating core vectors on class interaction hyperplanes, and enabling outliers detection. g-MEB is evaluated by conducting a comparison of classification accuracy on 10 data sets with 5 conventional classification methods. g-MEB obtains 7 out of 10 highest classification accuracy and compared to classic MEB, g-MEB wins on all the data sets except ionosphere. To evaluate class factorization ability of g-MEB, we compare g-MEB with classic MEB on FMA, the result shows that g-MEB is more capable than MEB on factorizing the class characteristics of membership and non-membership as fewer g-MEB core vectors delivers often better FMA.

In our further work, we will address the optimization problem of the total number of g-MEB required for optimal solution. It can only be determined using cross validation method and by investigating on measuring the density of the MEB, and further exploiting new models based on 'group minimum enclosing ball'.

References

1. Ben-Hur, A., Horn, D., Siegelmann, H.T., Vapnik, V.: Support vector clustering. J. Mach. Learn. Res. 2, 125–137 (2002)
2. Burges, C.J.: A tutorial on support vector machines for pattern recognition. Data Mining and Knowledge Discovery 2, 121–167 (1998)
3. Tsang, I.W., Kwok, J.T., Cheung, P.M.: Core vector machines: Fast svm training on very large data sets. Journal of Machine Learning Research 6, 363–392 (2005)
4. Bādoiu, M., Har-Peled, S., Indyk, P.: Approximate clustering via core-sets. In: STOC 2002: Proceedings of the Thiry-Fourth Annual ACM Symposium on Theory of Computing, pp. 250–257. ACM, New York (2002)
5. Welzl, E.: Smallest enclosing disks (balls and ellipsoids). In: Maurer, H.A. (ed.) New Results and New Trends in Computer Science. LNCS, vol. 555, pp. 359–370. Springer, Heidelberg (1991)
6. Pang, S., Kim, D., Bang, S.Y.: Face membership authentication using svm classification tree generated by membership-based lle data partition. IEEE Transactions on Neural Networks 16(2), 436–446 (2005)
7. Garcia, V., Alejo, R., Sinchez, J.S., Sotoca, J.M., Mollineda, R.A.: Combined effects of class imbalance and class overlap on instance-based classification (2008)

A Hybrid Fuzzy-Genetic Colour Classification System with Best Colour Space Selection under Dynamically-Changing Illumination

Heesang Shin, Napoleon H. Reyes, and Andre L. Barczak

Institute of Information and Mathematical Sciences, Massey University, Auckland, New Zealand

H.Shin@massey.ac.nz, N.H.Reyes@massey.ac.nz, A.L.Barczak@massey.ac.nz,
http://www.massey.ac.nz/~hshin

Abstract. This paper contributes in colour classification under dynamically changing illumination, extending further the capabilities of our previous works on Fuzzy Colour Contrast Fusion (FCCF), FCCF-Heuristic Assisted Genetic Algorithm (HAGA) for automatic colour classifier calibration and Variable Colour Depth (VCD). All the aforementioned algorithms were proven to accurately in real-time with a pie-slice technique. However, the pie-slice classifier is the accuracy-limiting factor in these systems. Although it is possible to address this problem by using a more complex shape for specifying the colour decision region, this would only increase the chances of overfitting. We propose a hybrid colour classification system that automatically searches for the best colour space for classifying any target colour. Moreover, this paper also investigates the general selection of training sets to get a better understanding of the generalisation capability of FCCF-HAGA. The experiments used a professional Munsell ColorChecker Chart with extreme illumination conditions where the colour channels start hitting their dynamic range limits.

1 Introduction

An exemplary colour classifier in colour-based object recognition system generally compensates for the effects of illumination changes in the exploratory environment to accurately identify colours comprising an object and recognising it. Real-time operation and automatic calibration are also one of the preferred capabilities, and these are all embodied in the combination of FCCF [10], FCCF-HAGA [13], VCD and VCD-LUT [14]. This work proposes an extension of these previous algorithms aggregating into an FCCF-HAGA-VCD-VCD LUT system, with automatic best colour space identification. The inclusion of the best colour space identification appears almost trivial, but this easily complicates the colour classifier extraction task as the search space grew seven times bigger. Per target colour, there are 12 colour descriptors [13] to calibrate (classification angles and radii, contrast angles, colour contrast operations and variable colour depth sub ranges), 7 colour spaces to explore and 10 different illumination settings. It also requires tweaking of the fitness function to accurately pick the best colour

K.W. Wong et al. (Eds.): ICONIP 2010, Part II, LNCS 6444, pp. 291–299, 2010.

classifiers across the many colour spaces. Due to the inherent limitations of the camera, and the extreme illumination settings, the training sets are plagued with duplicate colour pixels representing different colour classes. As depicted in Figure 1, a camera would capture a pink colour patch, under dark illumination, with some red colour pixels along its border. While a red colour patch, under bright illumination would be rendered with pink colour pixels along its border. Such colour pixel duplications makes the colour classifier extraction task extremely difficult.

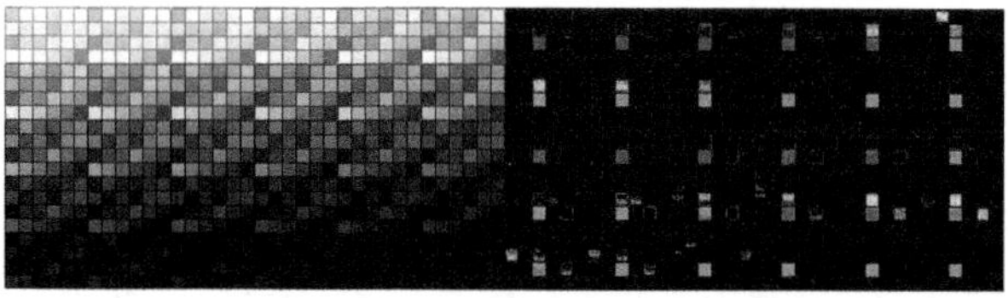

Fig. 1. Example of Duplicate Colour Pixels. The left pane shows the test set image, while the right pane shows the colour classification results for the Brute Force LUT algorithm for target colour Red. On the right pane, the red pixels depict the correct classifications for the red target colour. The green pixels denote the masks for the red targets but was missed. The yellow pixels are duplicate colour pixels of the red targets but they are actually representing another shade of red called 'moderate red'. These duplications are due to changes in illumination conditions.

An earlier attempt at using a Fuzzy-Genetic approach to colour classification [12] is using 4-inputs denoting proportional rgb colour values with a component involving intensity for each input. This approach has the effect of simple white-balancing on the inputs. In contrast to this research, a prototype colour sensor was used in their system, and not a typical colour CCD camera. Their classification system is using a conventional fuzzy inference system with 4 fuzzy sets per rule, implemented by trapezoidal membership functions. Moreover, 4 parameters were used to describe each membership function. The Genetic Algorithm (GA) was used solely for optimising these membership functions. On the other hand, in our proposed research we are using the GA to calibrate not just the membership functions. In the robot soccer framework [6] [5], there are lots of evidences in the literature suggesting that working under uncontrolled illumination settings is extremely difficult. In fact, in [8] it was reported that in 2004, 24 robot teams around the world attempted to solve the problem, but their approach failed to cope up when lights turned dimmer. It was also noted that traditional colour segmentation usually works well only under one illumination condition and would become increasingly inaccurate at other illumination settings [8]. In order to compensate for the effects of dynamically changing illumination conditions, others propose a GA-assisted hardware-calibration approach [4]. This approach, however, is hardware-specific as it alters the exposure, iris, gain, saturation, white balance, etc. and requires full-access to these hardware parameters. Therefore, this technique is limited by the hardware parameter boundaries.

2 Colour Spaces

Seven different colour spaces were transformed to work with the pie-slice colour decision region [15]. By transformation here means that we had to extract only the chromaticity components and disregard the brightness or lightness component. In addition, we moved the origin at the position of an achromatic colour (e.g. white, black, gray). Table 1 shows the colour spaces and transformation formulae used.

Table 1. Colour Spaces Used in the Experiments

Colour Space Identifier	Colour Space Name	Base Colour Space	x Component Transformation	y Component Transformation	Origin Position
0	RG	RGB	$\frac{R}{R+G+B}$	$\frac{G}{R+G+B}$	$\frac{1}{3}x, \frac{1}{3}y$
1	CH	L^*a^*b	$\frac{^*a}{127}$	$\frac{^*b}{127}$	0,0
2	CbCr	$Y'CbCr$	C_R	C_B	0,0
3	$C'_{yb}C'_{rg}$	oRGB	C'_{yb}	C'_{rg}	0,0
4	CM	CMY(K)	$\frac{C}{C+M+Y}$	$\frac{M}{C+M+Y}$	$\frac{1}{3}x, \frac{1}{3}y$
5	$C'_1C'_2$	oRGB	C'_1	C'_2	0,0
6	HS	HSV	$\cos(H\frac{\pi}{180})S$	$\sin(H\frac{\pi}{180})S$	0,0

Previously, we tested the modified RG chromaticity [10], CbCr chromaticity based on $Y'CbCr$ (YUV) [11] and the LCH chromaticity based on L^*a^*b colour space [7]. Three more new colour spaces were used in this work, apart from those tested previously: the CM chromaticity, based on the CMY(K) colour space, the $C'_1C'_2$ and the $C'_{yb}C'_{rg}$ chromaticity colour spaces, based on the opponent colour space called oRGB [1].

3 FCCF-HAGA Algorithm

FCCF, introduced in [10] is a colour correction and classification algorithm for colour-based object recognition. It uses an unconventional fuzzy inference system, not relying on a centroid formula for defuzzification. It is utilising a set of novel fuzzy colour contrast operators, and rules, in conjunction with a pie-slice decision region technique, and is proven to be better than just using the pure pie-slice technique [11] . On the other hand, FCCF-HAGA, introduced in [13], is an extension of FCCF, and is capable of fully-automatic colour classifier calibration, utilising a heuristic-assisted Genetic algorithm. Further improvements in colour classification, came from VCD and VCD LUT [14], proposing a variable colour depth representation of the colour tri-stimulus. This technique has the effect of channel-level colour averaging, while conserving memory space.

3.1 Fitness Function

FCCF-HAGA adaptively evolves the colour classifiers, as represented by chromosomes. They are automatically graded using a fitness function that is a slight modification of the fitness described in [13]. The new fitness function(Eqn. (1)) adaptively forgives false positive classifications and encourages finding classifiers that return high true positives. Moreover, it tries to avoid getting trapped in local maxima by reducing rewards in cases where true positives and false positives are both very low. Two constants parameters in the fitness function were revised from -10 to -7 and 0.5 to 0.7 in order to encourage more true positive results.

$$
\begin{aligned}
x &= \frac{\mathit{true\ positive\ pixels\ count\ within\ the\ target\ area}}{\mathit{total\ pixels\ in\ the\ target\ area}} \\
y &= \frac{\mathit{false\ positive\ pixels\ outside\ target\ area}}{\mathit{total\ pixels\ outside\ the\ target\ area}} \\
fitness &= \frac{1}{2}\left[\frac{1}{e^{-7(y-0.7)}} + \left(\frac{1-\frac{1}{1+e^{-75(x-0.05)}}}{1+e^{-10(y-0.4)}}\right)\right]
\end{aligned}
\qquad (1)
$$

4 Experiment Set-Up

A Munsell x-rite ColorChecker Image Reproduction Target was used for colour classification. Figure 2 shows the image acquisition system, the source illuminants and the Munsell colour targets.

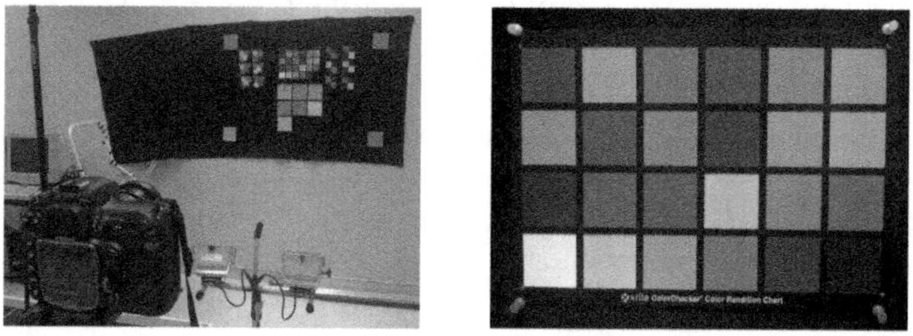

Fig. 2. The image acquisition system, source illuminants and the Munsell ColorChecker Image Reproduction Target chart

The light source is comprised of a combination of ambient fluorescent illuminants from the ceiling of the room, and two 500w halogen lamps. The exact combined illumination condition was measured at 5620 lux, 0.2499u', 0.5213v', 0.4360x, 0.4043y, 6061X, 5620Y, 2220Z by a Minolta CL-200 Chroma Metre. For training and testing the colour classifiers at varying illumination settings, 30 images of the same colour chart were taken at varying shutter speeds, from 2ms. (1/500 sec.) to 1600ms. (1.6 sec.). The training images were divided into 7 groups, each containing 10 different illumination settings, by changing the shutter speed settings of the camera. Each training set generally represents a certain illumination condition. For example, training set 0 generally represents the medium illumination condition range. Table 2 summarises the different groups of training images. On the other hand, Figure 3 shows all of the 30 training images put together as one. Any training set used at any one time is a subset of this image. It is worth noting that at the extreme illumination sub ranges (topmost and bottom rows), most of the colours are no longer distinguishable from each other through the naked eye.

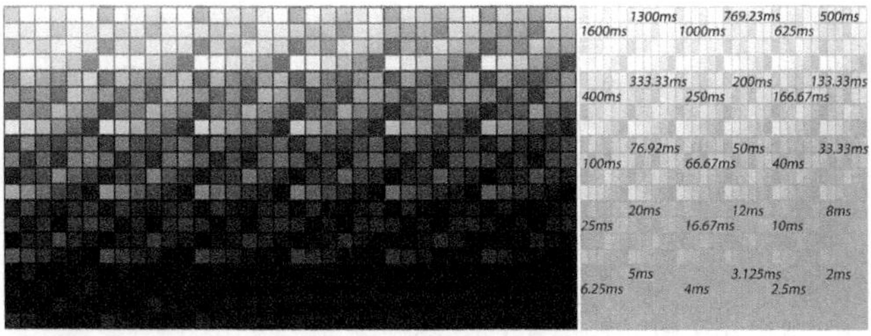

Fig. 3. Test target image containing 30 different illumination settings. The numbers (in msec.) correspond to the shutter speed settings for producing the illumination condition.

As can be viewed in Figure 2, there are 18 unique target colour patches in the Munsell colour rendition chart. We trained the system to recognise these 18 target colours individually as well as to recognise 5 families of colours (i.e. blue shades, green shades, skin colours ,etc.). Colour masks were used to define the target colour regions. However, the 6 grey shades at the bottom of the colour chart were all excluded because they are all achromatic. Table 3 lists all the colours used for colour classifier extraction.

5 Results and Discussion

5.1 Colour Classification Results

Table 4 shows the best classification results for each target colour. A classification score of more than 0.5 generally indicates a 'good' classifier that is effective for colour classification tasks on spatially varying illumination settings. It can be

Table 2. Images used for traning

Training Set	Shutter Speeds (ms)	Remarks
0	40, 50, 66.67, 76.92, 100, 166.67, 200, 250, 333.33, 400	Medium illumination
1	2.5, 5, 10, 20, 40, 76.92, 166.67, 133.33, 625, 1300	Mostly dark with few bright
2	2, 4, 8, 16.67, 33.33, 66.67, 125, 250, 500, 1000	Mostly dark with few medium
3	3.125, 6.25, 12, 25, 50, 100, 200, 400, 769.23, 1600	Balanced sampling across all illuminations
4	40, 50, 66.67, 76.92, 100, 625, 769.23, 1000, 1300, 1600	From medium to bright
5	10, 12, 16.67, 20, 25, 625, 769.23, 1000, 1300, 1600	Mostly bright with few dark
6	10, 12, 16.67, 20, 25, 166.67, 200, 250, 333.33, 400	From dark to medium
Test Set	All of the above	
Alternative Test Set	All of the above excluding 2, 2.5, 3.125, 4, 5 and 6.25	Excluding the darkest illumination range

Table 3. Target colours used for classification. Patch numbers begin from the top-left corner of the colour chart

Colour Name	Patch Numbers	Colour Name	Patch Numbers	Colour Name	Patch Numbers
Dark Skin	1	Light Skin	2	Blue Sky	3
Foilage	4	Blue Flower	5	Blue Green	6
Orange	7	Puplish Blue	8	Moderate Red	9
Purple	10	Yellow Green	11	Orange Yellow	12
Blue	13	Green	14	Red	15
Yellow	16	Magenta	17	Cyan	18
Blue Shades	3, 8, 13, 18 as (19)	Green Shades	4, 11 as (20)	Orange Shades	7, 12, 16 as (21)
Red Shades	9, 15 as (22)	Skin Colours	1, 2, 7 as (23)		

observed by inspecting both Tables 4 and Table 2 that the majority of the best colour classifiers were trained using Training sets 2, 1 and 3, which are generally representing darker illumination settings.

Table 4. Colour Classification Results of the FCCF-HAGA using 'ALL' 30 illuminations. The colour names marked in **bold** scored more than 0.5. 12 out of 23 colours scored more than 0.5.

Colour Name	Colour Space(s)	Training Set(s)	Classification Score	Colour Name	Colour Space(s)	Training Set(s)	Classification Score
Dark Skin	1,2,4,5,6	1,2,3,5	0.385116	Light Skin	All	1,2,3,4	0.445318
Blue Sky	1	2	0.387193	**Foilage**	**0**	**1**	**0.540411**
Blue Flower	5	6	0.452554	**Blue Green**	**5**	**2**	**0.742187**
Orange	All	1,2,3,4,5,6	0.444499	**Puplish Blue**	**1**	**1**	**0.763247**
Moderate Red	All	1,2,3,5,6	0.44112	Purple	1	2	0.430284
Yellow Green	**6**	**2**	**0.668154**	Orange Yellow	All	1,2,3,4,5	0.445107
Blue	**1**	**3**	**0.672016**	**Green**	**6**	**1**	**0.767966**
Red	**3**	**2**	**0.545756**	Yellow	All	1,2,3,4,5,6	0.445099
Magenta	**5**	**2**	**0.612172**	**Cyan**	**1**	**1**	**0.791258**
Blue Shades	**1**	**2**	**0.598744**	**Green Shades**	**0**	**2**	**0.667565**
Orange Shades	All	1,2,3,5,6	0.444901	**Red Shades**	**3**	**1**	**0.569657**
Skin Colours	0,1,2,3,4,5	1,2,3,5	0.429195				

In contrast, interestingly, classifiers trained on relatively brighter training sets did not do very well on the broader illumination test ranges. As an alternate validation of the findings, we examined the classifiers on an alternative test set which contains 24 different illumination settings derived from the original test set. In this alternate set, we excluded the darker illumination settings used during training (2, 2.5, 3.125, 4, 5 and 6.25ms). Consequently, we found that the classifiers trained using darker illumination settings performed well on test sets under bright illumination settings. Table 5 shows the classification results. Due to the absence of the extreme dark illumination settings, it is noticeable that we now have 3 more colour classifiers that yield higher than 0.5 classification scores than the classifiers performance in the original test set(Table 4).

5.2 Comparison of the Different Colour Classification Algorithms

We have compared the performance of our proposed algorithm against two other colour classifiers. The first one is simply exhaustively mapping all the pixels of

Table 5. Colour Classification Results of FCCF-HAGA Algorithm using the Alternate Test Set. Colour names marked in **bold** scored more than 0.5. 15 out of 23 colours scored more than 0.5.

Colour Name	Colour Space(s)	Training Set(s)	Classification Score	Colour Name	Colour Space(s)	Training Set(s)	Classification Score
Dark Skin	All	1,2,3,5	0.445445	Light Skin	0,1,2,4,5	0,1,2,3,4	0.445452
Blue Sky	0,2,4	2,5	0.44533	**Foilage**	**0**	**1**	**0.695956**
Blue Flower	**5**	**6**	**0.596275**	**Blue Green**	**5**	**2**	**0.824066**
Orange	All	1,2,3,4,5,6	0.445452	**Puplish Blue**	**1**	**1**	**0.899813**
Moderate Red	**3**	**1**	**0.574498**	**Purple**	**1**	**2**	**0.612651**
Yellow Green	**6**	**1**	**0.751064**	Orange Yellow	All	1,2,3,4,5	0.445452
Blue	**1**	**3**	**0.835202**	**Green**	**6**	**1**	**0.862613**
Red	**3**	**2**	**0.683617**	Yellow	All	1,2,3,4,5,6	0.445452
Magenta	**5**	**2**	**0.761289**	**Cyan**	**1**	**6**	**0.910595**
Blue Shades	**1**	**2**	**0.759713**	**Green Shades**	**0**	**2**	**0.757176**
Orange Shades	All	1,2,3,5,6	0.445452	**Red Shades**	**3**	**1**	**0.699453**
Skin Colours	0,1,2,3,4	1,2,3,5	0.445449				

Table 6. Colour Classification Score Comparisons Between Brute Force Colour Classification Algorithm, Adaboost Classification Algorithm and FCCF-HAGA Algorithm. Classifiers were tested against the original test set. **Bold** numbers denote best classification scores.

Colour Name	Brute Force Classification Score	Adaboost Classification Score	FCCF-HAGA Classification Score	Colour Name	Brute Force Classification Score	Adaboost Classification Score	FCCF-HAGA Classification Score
Dark Skin	0.243974	**0.436888**	0.385116	Light Skin	0.230752	0.44346	**0.445318**
Blue Sky	0.248425	**0.438708**	0.387181	Foilage	0.370085	0.427861	**0.540411**
Blue Flower	0.227394	0.445371	**0.452554**	Blue Green	0.20612	0.444845	**0.742188**
Orange	0.360171	0.440933	**0.444499**	Puplish Blue	0.211463	0.444388	**0.763247**
Moderate Red	0.24944	**0.445158**	0.44112	Purple	0.21912	**0.432815**	0.430284
Yellow Green	0.325943	0.473285	**0.66338**	Orange Yellow	0.247082	0.442122	**0.445107**
Blue	0.210872	0.444174	**0.672016**	Green	0.216792	0.613775	**0.767966**
Red	0.314519	0.445063	**0.545756**	Yellow	0.289477	0.444601	**0.445099**
Magenta	0.205332	0.21068	**0.612172**	Cyan	0.237986	0.345114	**0.791358**
Blue Shades	0.255146	0.25223	**0.598744**	Green Shades	0.352426	0.149648	**0.667565**
Orange Shades	0.318534	0.0552074	**0.444901**	Red Shades	0.362675	0.104209	**0.569657**
Skin Colours	0.327321	0.0952789	**0.429195**				

the colours from the training set and then storing its colour classification into a look-up table, which we called a brute force approach. The second one used a cascaded AdaBoost [3] [2], known to train faster than Neural Networks and good at multi-dimensional feature space classification problems [9]. The training targets were 98% of hit ratios with the lowest possible false detection, using a maximum of 30 cascades, limiting the number of weak classifiers used by each one of the final classifiers.

The comparison of the results can be viewed in Table 6. As evidenced by the table, the brute force approach did not yield accurate results. AdaBoost only outperformed FCCF-HAGA in 4 cases, and just by a small margin. This is an indication that multiple duplicate colour pixels are representing different colour categories (some training points appear as true positives as well as false positives in the same training set). These duplicate colour pixels occur mostly in the extreme dark and extreme bright settings, where colour clipping occurs.

6 Conclusions

This research further broadens the colour classification capability of FCCF-HAGA with the inclusion of the automatic best colour space selection. A professional standard Munsell colour test chart was used and results prove the proposed algorithms efficacy, as compared to an AdaBoost colour classifier and a simple brute force classifier.

This work also proposes a training set selection strategy for FCCF-HAGA. Empirical results show that FCCF-HAGA performed better colour classification when it was fed with training sets that were under relatively dark illuminations.

References

1. Bratkova, M., Boulos, S., Shirley, P.: orgb: A practical opponent color space for computer graphics. IEEE. M. CGA 29(1), 42–55 (2009)
2. Freund, Y., Schapire, R.E.: A short introduction to boosting. Journal of Japanese Society for Artificial Intelligence 14(5), 771–780 (1999)
3. Freund, Y., Schapire, R.E.: Experiments with a new boosting algorithm. In: Proceedings of the 13th International Conference in Machine Learning, Bari, Italy, pp. 148–156 (1996)
4. Grillo, E., Matteucci, M., Sorrenti, D.G.: Getting the most from your color camera in a color-coded world, pp. 221–235 (2005), http://dx.doi.org/10.1007/978-3-540-32256-6_18
5. Kim, J.H., Seow, K.T.: Soccer Robotics. Springer, Heidelberg (2004)
6. Kitano, H., Asada, M., Kuniyoshi, Y., Noda, I., Osawa, E., Matsubara, H.: Robocup: A challenging problem for ai. AI Magazine 18, 73–85 (1997)
7. Kloss, G.K., Shin, H., Reyes, N.H.: Dynamic colour adaptation for colour object tracking. In: Proceedings of the 24th International Image and Vision Computing New Zealand 2009 (IVCNZ 2009), Wellington, New Zealand, pp. 340–345 (November 2009)
8. Lovell, N.: Illumination independent object recognition. In: Bredenfeld, A., Jacoff, A., Noda, I., Takahashi, Y. (eds.) RoboCup 2005. LNCS (LNAI), vol. 4020, pp. 384–395. Springer, Heidelberg (2006)
9. Reyes, N., Barczak, A.L.C., Messom, C.H.: Fast colour classification for real-time colour object identification: Adaboost training of classifiers. In: International Conference on Autonomous Robots and Agents (ICARA 2006), Palmerston North, NZ, pp. 611–616 (December 2006)
10. Reyes, N.H., Dadios, P.E.: Dynamic color object recognition using fuzzy logic. Journal of Advanced Computational Intelligence and Intelligent Informatics 8, 29–38 (2004)
11. Reyes, N.H., Messom, C.: Identifying colour objects with fuzzy colour contrast fusion. In: 3rd International Conference on Computational Intelligence, Robotics and Autonomous Systems, and FIRA RoboWorld Congress (2005)
12. Sakurai, M., Kurihara, Y., Karasawa, S.: Color classification using fuzzy inference and genetic algorithm. In: Proceedings of the Third IEEE Conference on Fuzzy Systems 1994, IEEE World Congress on Computational Intelligence, vol. 3, pp. 1975–1978 (26-29, 1994)

13. Shin, H., Reyes, N.: Finding near optimum colour classifiers: genetic algorithm-assisted fuzzy colour contrast fusion using variable colour depth. Memetic Computing Journal, 1–18 (2009), http://dx.doi.org/10.1007/s12293-009-0025-8
14. Shin, H., Reyes, H.: N.: Variable colour depth look-up table based on fuzzy colour processing. In: Köppen, M., Kasabov, N., Coghill, G. (eds.) ICONIP 2008. LNCS, vol. 5506, pp. 1071–1078. Springer, Heidelberg (2009)
15. Thomas, P., Stonier, R., Wolfs, P.: Robustness of color detection for robot soccer. In: Proceedings of the Seventh International Conference on Control, Automation, Robotics and Vision, pp. 1245–1249 (2002)

Identifier Based Graph Neuron: A Light Weight Event Classification Scheme for WSN

Nomica Imran[1] and Asad Khan[2]

School of information Technology
Clayton 3168, Victoria, Australia
nmcho1@student.monash.edu.au, asad.khan@monash.edu.au

Abstract. Large-scale wireless sensor networks (WSNs) require significant resources for event recognition and classification. We present a light-weight event classification scheme, called Identifier based Graph Neuron (IGN). This scheme is based on highly distributed associative memory which enables the objects to memorize some of its internal critical states for a real time comparison with those induced by transient external conditions. The proposed approach not only conserves the power resources of sensor nodes but is also effectively scalable to large scale WSNs. In addition, our scheme overcomes the issue of *false-positive detection* -(which existing associated memory based solutions suffers from) and hence promises to deliver accurate results. We compare Identifier based Graph Neuron with two of the existing associated memory based event classification schemes and the results show that IGN correctly recognizes and classifies the incoming events in comparative amount of time and messages.

1 Introduction

A wireless sensor network may contains tens to thousands of wireless transducer nodes capable of sensing, computing, storing and communicating various environmental properties such as temperature, humidity or pressure; hence sensory data are generally the measurements taken by the onboard sensors of the sensor nodes. Global information from the sensed region is collected and analysed. The problem of analysing real time sensory data occurs in many important wireless sensor network (WSN) applications such as security, surveillance, climatic-change studies, and structural health monitoring. The most predominant model used for sensor network applications involves sending sensory data to a base-station for analysis. Two fundamental problems arise here. First, communicating sensory data from the physical environment to centralized server is an expensive task. Secondly, sending streams of raw data from each sensor node to the centralized server may overwhelm the processing capacity of the server. Moreover, in both the cases, sensory data may encounter delays that could diminish its temporal value and significance. An efficient way to process this huge data is through collaborative in-network processing where sensory data are quasi-processed within the sensor network before reaching their destination.

K.W. Wong et al. (Eds.): ICONIP 2010, Part II, LNCS 6444, pp. 300–309, 2010.

Sensor networks present unique challenges: individual nodes are resource restricted, radio communication between nodes is loss prone and a major energy consumer [12], and the networks are tightly coupled to the physical environment. That's why sensor network application designs need to take into consideration the application energy requirement, and incorporate energy awareness and energy conservation approaches. One of the main motives driving sensor network research is the ability to operate sensor networks in an untethered and unattended manner for long-term outdoor deployment. A sensor node continues to provide sensory data from the physical environment until it depletes its own on-board battery. Replacing or recharging batteries is not possible in sensor networks owing to the sheer number of deployed sensor nodes and the inaccessibility of the physical environment. Therefore, incorporating energy saving techniques is central to the design of sensor network applications.

Traditional pattern recognition methods [6] are not considered feasible on wireless ad hoc infrastructure. These focus on minimizing the error rate by finding the probability distribution of sensory data and creating feature vectors for high dimensional sensory data, both of which consumes resources. Traditional signal processing based event detection schemes also proves to be too complex to use in WSN. Recognizing event patterns in a distributed manner is a challenging problem. We need to develop solution to transform the local pattern match results into an accurate global pattern recognition through sensor node collaboration while conserving the sensor nodes energy. This has actually been our general motivation to carry out the research.

In this paper, we present a light-weight event classification scheme, called Identifier based Graph Neuron (IGN). Through in-network processing, the sets of GN nodes concurrently compare their sub patterns with the historical information, which is locally available within each GN via associative memory. The local state of an event is recognize through *locally* assigned identifiers. These nodes run an iterative algorithm to coordinate with other nodes to reach a consensus about the global state of the event. The main contribution of this paper is in providing a solution on how a group of resource-limited sensor nodes can recognize physical world events in a distributed, real-time, and energy efficient way. The proposed scheme accurately recognizes large-scale event patterns within the network, thus reducing the dependency on base-stations. The proposed approach not only conserves the power resources of sensor nodes but is also effectively scalable to large scale WSNs. We have also solved the low false positive rate issue (also called as crosstalk problem) faced by earlier publications in an efficient and simple way. We have also compared IGN with Graph Neuron (GN) and hierarchical Graph Neuron (HGN). We found that IGN correctly resolves the crosstalk problem in comparative amount of time and messages.

2 Related Work

The main characteristic of an intelligent system is the capability of mimicking human intelligence, such as memorizing, recognizing, understanding, and reasoning. These characteristics of AI have attracted researchers in WSN to use

AI in their systems. Many systems, such as home care [13], target surveillance system [14], and access control, have now been developed with such intelligent capabilities. AI researchers hope that the focus on reduced energy consumption by using AI inspired techniques [8] can improve the quality of WSN system. Not only accurate, the approaches using AI can also work quickly and consume less memory resources [4].

A well known application of AI is pattern recognition [5]. Not only quality and performance is the target to improvement in pattern recognition, the focus is also shifted to some new perspectives, algorithms, and architectures. As an example, unsupervised learning methods for categorization of the sesory inputs are presented by Kulakov and Davcev who proposed two possible implementations of the ART and FuzzyART neural-networks algorithms. They are tested on a data obtained from a set of several motes, equipped with several sensors each. Results from simulations of purposefully faulty sensors show the data robustness of these architectures. The proposed neural-networks classifiers have distributed short and long-term memory of the sensory inputs and can function as security alert when unusual sensor inputs are detected.

Away from those recognising strategies, the development of Hopfields associative memory [11] has given a new opportunity in recognizing patterns. Associative memory applications provide means to recall patterns using similar or incomplete patterns. Associative memory is different to conventional computer memory. In conventional computer memory contents are stored and recalled based on the address. In associative memory architecture, on the other hand, contents are recalled based on how closely they are associated with the input (probe) data. Khan [1] have replaced the term content-addressable, introduced by Hopfield [9] as associative memory (AM). Hence, content addressable memory is interchangeable with AM.

The spatio-temporal encoded, or spiking, neurons developed by Hopfield [11] draw upon the properties of loosely coupled oscillators. However, the problem of scaling up the AM capacity still remains. Izhikevich [7] states, though some new capabilities at differentiating between similar inputs have been revealed, however, there is no clear evidence to suggest their superiority over the classical Hopfield model in terms of an overall increase in the associative memory capacity. Hence, one of the motivations behind GN development was to create an AM mechanism, which can scale better than the contemporary methods. Furthermore, we needed to create an architecture which would not be dependent for its accuracy on preprocessing steps such as patterns segmentation [3] or training [1]. From the discussion of several pattern recognition techniques, it seems that GN [2] is a promising AI architecture for recognising patterns in WSN.

3 System Model

In this section, we model the wireless sensor network. Table 1 provides the summary of notations used in this paper.

Let there are N sensor nodes in the wireless sensor network. Each sensor senses a particular data from its surroundings. Let $\mathcal{E} = \{e_1, e_2, \ldots, e_E\}$ be a non-empty

Table 1. Notation of proposed system model

Symbol	Description
N	number of sensors nodes
$\mathcal{E}$	set of all elements $\{e_1, e_2, \ldots, e_E\}$
E	size of set $\mathcal{E}$
L	pattern length
$\mathcal{P}$	set of all patterns of length L over $\mathcal{E}$
P	size of set $\mathcal{P}$ and is equal to E^L
$\mathcal{G}$	GN overlay, a two-dimentional array of $E \times L$
$n(e_i, j)$	a node at i-th row and j-th column in $\mathcal{G}$
L_s	Local State of active node $n(e_i, j)$

finite set of such data elements sensors sense from their surroundings. We find it convenient to describe input data in the form of *patterns*. A pattern over $\mathcal{E}$ is a finite sequence of elements from $\mathcal{E}$. The length of a pattern is the number of sensors in the system. We define $\mathcal{P}$ as a set of all possible patterns of length L over $\mathcal{E}$:

$$\mathcal{P} = \{p_1, p_2, \ldots, p_P\},$$

where P is the total number of possible unique patterns in $\mathcal{P}$ and can be computed as:

$$P = E^L.$$

For example, if $\mathcal{E} = \{x, y\}$, then $\mathcal{P}$ of length, say, 3 over $\mathcal{E}$ is:

$$\mathcal{P} = \{xxx, xxy, xyx, xyy, yxx, yxy, yyx, yyy\}.$$

We model GN as a structured overlay $\mathcal{G} = \{(\mathcal{E} \times \mathcal{L})\}$ where $\mathcal{L} = \{1, 2, \ldots, L\}$:

$$\mathcal{G} = \{n(e_i, j)\} \quad \text{forall} \quad e_i \in \mathcal{E}, j \in \mathcal{L},$$

where $n(e_i, j)$ is a node in $\mathcal{G}$ at i-th row and j-th column. GN can be visualized as a two dimensional array of L rows and E columns. Total number of nodes in the $\mathcal{G}$ are $E \times L$.

We refer all the nodes in the $(j-1)$ column as the left neighbors of any node $n(*, j)$ in j-th column. Similarly, all the nodes in the $(j+1)$ column are called as the right neighbors of $n(*, j)$.

4 Proposed Scheme

In a GN-based classifier, each node $n(e_i, j)$ is programmed to respond to only a specific element e_i at a particular position j in a pattern. That is, node $n(e_i, j)$ can only process all those patterns in $\mathcal{P}$ such that e_i is at the j-th position in that pattern.

Each node maintains an *active/inactive* state flag to identify whether it is processing the incoming pattern or not. Initially all nodes are in inactive state. Upon arrival of a pattern, if a node finds its programmed element e_i at the given position in the pattern, it switches its state to active otherwise it remains inactive. Only active nodes participate in our algorithm and inactive nodes remain idle.

At any time, there can be exactly L active nodes in a GN. Hence, there are exactly one active left-neighbor and exactly one active right-neighbor of a node $n(e_i, j)$ where $j \neq 0, l$. Whereas terminal nodes $n(e_i, 0)$ and $n(e_i, L)$ has only one active left and right neighbor respectively. Fig.1 shows how two characters are being mapped into GN array and a simple GN-based classifier is explained in Algorithm 1.

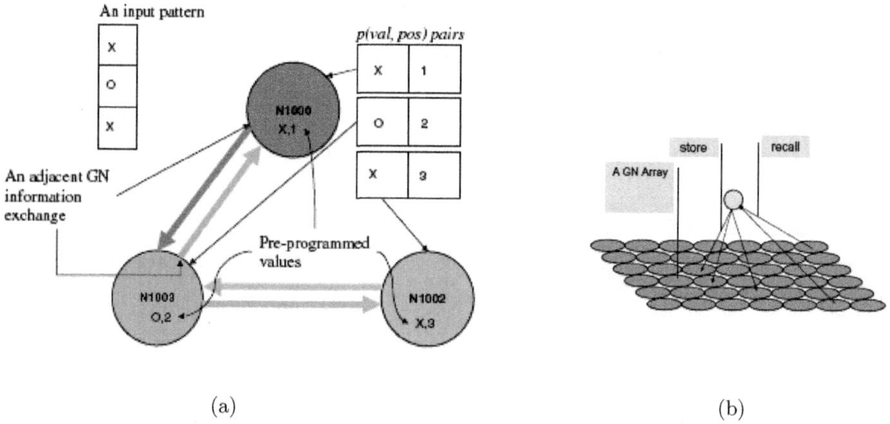

Fig. 1. Execution time for selected patterns - (a) Character "x" and "O" mapped to the GN nodes, (b) GN communication Scheme: Pattern store and recall in a typical GN-based classifier.

5 Proposed Protocol

On arrival of an input pattern $\mathcal{P}$, each active node $n(e_i, j)$ store e_i in its j_{th} position. Each node $n(e_i, j)$ sends its matched element e_i to its active neighbors $(j+1)$ and $(j-1)$. The GNs at the edges will send there matched elements to there penultimate neigbours only. Upon receipt of the message, the active neighboring nodes update there bais array. Each active node $n(e_i, j)$ will assign a local state L_s to the received $(e_i,)$ value. The generated local state L_s will be *Recall* if the the added value is already present in the bais array of the active node and it will be a store if in-case its new value. An $< ID >$ will be generated against each added value. There are certain rules that need to be considered by each active node $n(e_i, j)$ while creating states and assigning $< IDS >$ against those states. The rules are as under:

Rule 1: *Store*$_{(Si)}$ > *Recall*$_{(Ri)}$. *Store* S_i has a natural superiority over *Recall* R_i i-e $Store_{(Si)} > Recall_{(Ri)}$. If node $n(e_i, j)$ self local state L_s is *Recall* but it receives a *Store* command from any of its neighbors, $(j+1)$ or $(j-1)$, it will update its own state from $Recall_{(Ri)}$ to $Store_{(Si)}$.

Rule 2: All New Elements. If any of the elements presented to $\mathcal{G}$ is not existing in the bais array of any of the active nodes $n(e_i, j)$ suggests that its a new pattern. Each active node $n(e_i, j)$ will create a new <ID> by incrementing the already stored maximum <ID> in there bias array by 1.

Rule 3: All Recalls with Same ID. If e_i presented to $\mathcal{G}$ is the recall of previously stored pattern with same <ID>, means that its a repetitive pattern. The same <ID> will be allocated to this pattern.

Rule 4 : All Recalls with Different IDs. If all the e_i of the pattern $\mathcal{P}$ presented to $\mathcal{G}$ are the recall of already stored patterns with different <IDs> indicates that it's a new pattern. Each active node $n(e_i, j)$ will find out the $max(ID)$ in there bias array and will increment it by 1.

Rule 5: Mix of Store and Recall. If the end decision is to *Store* due to mix of *Store* and *Recall*, each active node $n(e_i, j)$ will again find out the $max(ID)$ in there bais array and will increament it by 1.

After generating local <IDS> against generated states each active node $n(e_i, j)$ will undergo phase transition mode. During first round of phase transition mode, all the active nodes $n(e_i, j)$ will share locally generated <IDS> and L_s with there $(j+1)$ and $(j-1)$ neighbors. On receiving the values all the $n(e_i, j)$ will compare the received values with there local values. If received value and self value is same there won't be any change in there state. If received value $\neq$ local value, the node will upgrade its local value according to the rules listed below:

Transition Rule 1. If the active node $n(e_i, j)$ has received a greater value from its left neighbor $(j+1)$, it will upgrade its local state and transfer this updated value to its right $(j-1)$ neighbor only.

Transition Rule 2. Incase if the received value from both the neighbors $(j+1)$ and $(j-1)$ are smaller than the local value, node $n(e_i, j)$will upgrade its value and transfer this new value to both of its neighbors.

When the pattern is resolved, the generated <ID> will be stored in the bias array of each active GN. Bias array wont be having duplicated <IDs>. It is also not necessary that the <ID> in the bais array will be in ordered form. Once the pattern has been stored, a signal is sent out within the network informing all the IGN nodes that the pattern has been stored. This is called the pattern resolution phase.

5.1 Discussion

The basic motivation behind *GN* development was to create an associative memory network that work more faster and is scalable compared to existing schemes.

The GN array converts the spatial/ temporal patterns into a graph based representation and then compares the elements of the graphs for memorization and recall operations. The advantage of having a graph-like representation is that it provides a mechanism for placing the spatial/ temporal information in a context. Hence not only can we compare the individual data points but we may also compare the order in which they occur. Also, the proposed graph based approach allows us to explicitly model relationships between different objects, or parts of the objects under consideration. This is a clear advantage over pattern classification based on feature vectors, which are restricted to using unary feature values. Moreover, in our graph based approach we can use a variable number of entities, i.e. nodes and edges, depending on the complexity of the object under considerations.

It is a simple and light scheme that best suits the resource constrained nature of WSN. Our proposed scheme is base station independent. Through in-network processing we are efficiently utilizing the resources of sensor nodes. It is also not dependent on preprocessing steps such as patterns segmentation or training for its processing. Through parallel processing, the scalability issues in WSN are catered well. GN provides an emergent property of real-time processing regardless the size of the network.

6 Simulation Work

For the purpose of ensuring the accuracy of our proposed IGN algorithm for pattern recognition, we have conducted a series of tests involving random patterns. We develop an implementation of IGN in Eclipse Java. We have performed several runs of experiments each time choosing random patterns. We have constructed a database of stored patterns in which patterns can be added manually and stored in the database too. The performance metrics are accuracy and time.

Fig. 2 shows the number of entries in the bias array of IGNs while processing a database of 1000 patterns, in which each pattern is consisted of 7 elements in 5 positions However, the percentage of recalls for each of the GNs is also illustrated.

To estimate the communication overhead for HGN two factors are considered. 1). Bias array size and 2). time required by the HGN array to execute the request. Incase, of IGN the number of bias array entries are significant but they are not as high as in HGN as it doesnt have to create hierarchies and secondly the execution time is also less as compared to HGN as it only has to communicate to its immediate neigbours. In IGN the decision is not taken by the top node as a result lot of processing and communication cost is saved and the storage requirements within the bias array are not significantly extended with the increase in the number of stored patterns from the scalability point of view. It should be mentioned that the implementation of HGN requires asynchronous communication and mutually exclusive processes while these features can be easily implemented in Java, compared with other real-time programming languages such as C or Ada.

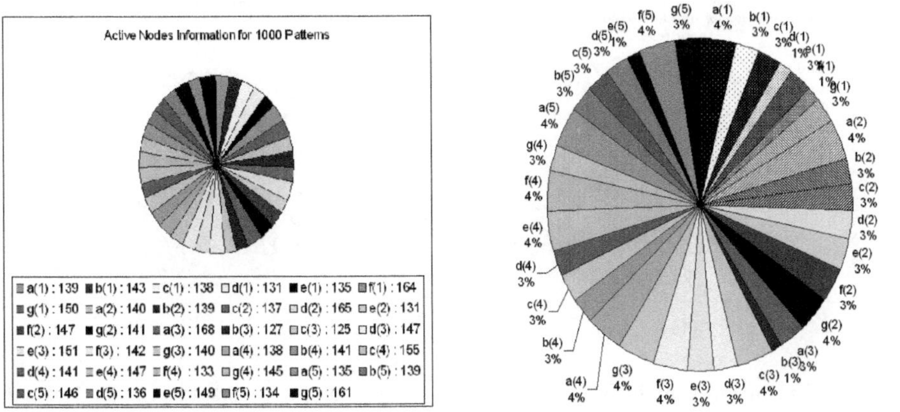

(a) (b)

Fig. 2. (a) Information about all the active nodes in IGN (b) Recall percentage of active nodes in IGN

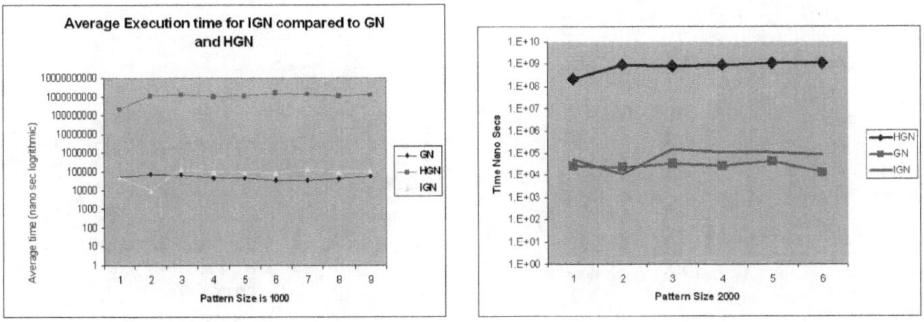

(a) (b)

Fig. 3. Execution time for selected patterns - (a) Average execution time for 1000 patterns in IGN, GN and HGN, (b) Average execution time for 2000 patterns in IGN and GN.

The results in Fig. 3 also show that the pattern detection time in IGN is much less as compared to HGN. This is mainly due to the fact that by increasing the number of stored patterns, the search will be accomplished through higher layers of HGN hierarchy which results in more delays in delivering the final outcome.

To investigate the impact of response time,Fig. 4 illustrates a comparison between GN, IGN and HGN. It may be observed from the figures that the response time for GN remains almost consistent however the HGN response time increases due to fact that by increasing the pattern size, the number of GNs within the composition will be increased drastically which in turn results in excessive overheads owing to GNs intercommunicating. The response time for IGN is also not consistent because of the overhead of message communication that varies from pattern to pattern. A comparison between the response time of GN and IGN for selected patterns have been made it Fig. 5.

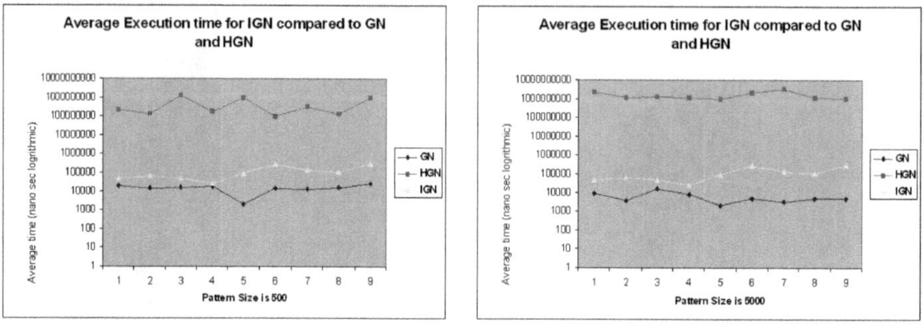

(a) (b)

Fig. 4. Execution time for selected patterns - (a) Average execution time for 500 patterns in IGN, GN and HGN, (b) Average execution time for 5000 patterns in IGN and GN.

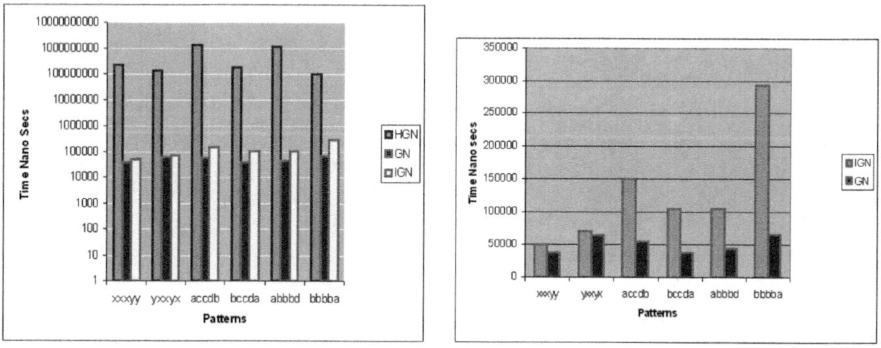

(a) (b)

Fig. 5. Execution time for selected patterns - (a) Average execution time for six selected patterns in IGN, GN and HGN, (b) Average execution time for six selected patterns in IGN and GN.

7 Conclusion and Future Work

In this paper we have proposed an in-network based pattern matching approach for providing scalable and energy efficient pattern recognition within a sensor network. IGN algorithm works by locally generating Ids and states against each incoming pattern element according to the set rules and then share these ids to reach a common consensus. IGN algorithm not only provides a single-cycle learning model which is remarkably suitable for real time applications but also overcome the issue of crosstalk in normal GN approach by delivering accurate results.

References

[1] Amin, A.H.M., Mahmood, R.A.R., Khan, A.I.: Analysis of pattern recognition algorithms using associative memory approach: A comparative study between the hopfield network and distributed hierarchical graph neuron (dhgn), pp. 153–158. IEEE, Los Alamitos (2008)

[2] Basirat, A.H., Khan, A.I.: Building context aware network of wireless sensors using a novel pattern recognition scheme called hierarchical graph neuron. In: IEEE International Conference on Semantic Computing, ICSC 2009 (2009)

[3] Basirat, A.H., Khan, A.I.: Building context aware network of wireless sensors using a novel pattern recognition scheme called hierarchical graph neuron. In: ICSC 2009: Proceedings of the 2009 IEEE International Conference on Semantic Computing, Washington, DC, USA, pp. 487–494. IEEE Computer Society, Los Alamitos (2009)

[4] Brandl, M., Kellner, K.H., Posnicek, T., Kos, A., Mayerhofer, C., Fabian, C.: An efficient source initiated on-demand data forwarding scheme for wireless sensor networks. In: ICICS 2009: Proceedings of the 7th International Conference on Information, Communications and Signal Processing, Piscataway, NJ, USA, pp. 1349–1355. IEEE Press, Los Alamitos (2009)

[5] Frank, J., Mda-c, N.U.: Artificial intelligence and intrusion detection: Current and future directions. In: Proceedings of the 17th National Computer Security Conference (1994)

[6] Gu, L., Jia, D., Vicaire, P., Yan, T., Luo, L., Tirumala, A., Cao, Q., He, T., Stankovic, J.A., Abdelzaher, T., Krogh, B.H.: Lightweight detection and classification for wireless sensor networks in realistic environments. In: SenSys 2005: Proceedings of the 3rd International Conference on Embedded Networked Sensor Systems, pp. 205–217. ACM, New York (2005), `http://doi.acm.org/10.1145/1098918.1098941`

[7] Izhikevich, E.M.: Dynamical Systems in Neuroscience: The Geometry of Excitability and Bursting (Computational Neuroscience), 1st edn. The MIT Press, Cambridge (November 2006)

[8] Kim, J., Lim, J.S., Friedman, J., Lee, U., Vieira, L., Rosso, D., Gerla, M., Srivastava, M.B.: Sewersnort: a drifting sensor for in-situ sewer gas monitoring. In: SECON 2009: Proceedings of the 6th Annual IEEE Communications Society Conference on Sensor, Mesh and Ad Hoc Communications and Networks, Piscataway, NJ, USA, pp. 691–699. IEEE Press, Los Alamitos (2009)

[9] Kim, J., Hopfield, J.J., Winfree, E.: Neural network computation by in vitro transcriptional circuits (2004). In: Saul, L.K., Weiss, Y., Bottou, L. (eds.) Advances in Neural Information Processing Systems, vol. 17, pp. 681–688 (2007)

[10] Kulakov, A., Davcev, D.: Tracking of unusual events in wireless sensor networks based on artificial neural-networks algorithms, vol. 2, pp. 534–539. IEEE Computer Society, Los Alamitos

[11] McEliece, R.J., Posner, E.C., Rodemich, E.R., Venkatesh, S.S.: The capacity of the hopfield associative memory. IEEE Trans. Inf. Theor. 33(4), 461–482 (1987)

[12] Pantazis, N.A., Vergados, D.D.: A survey on power control issues in wireless sensor networks, vol. 9, pp. 86–107 (2007)

[13] Chung, N.H., Gu, T., Xue, W.: Context-aware middleware for pervasive elderly homecare. IEEE Journal on Selected Areas in Communications, 510–524, IEEE (2009)

[14] Wang, X., Wang, S., Bi, D.: Distributed visual-target-surveillance system in wireless sensor networks, vol. 39, pp. 1134–1146. IEEE Press, Los Alamitos (2009)

Clustering Categorical Data Using an Extended Modularity Measure

Lazhar Labiod, Nistor Grozavu, and Younès Bennani

LIPN-UMR 7030, Université Paris 13,
99, av. J-B Clément, 93430 Villetaneuse, France
{firstname.secondname}@lipn.univ-paris13.fr

Abstract. Newman and Girvan [12] recently proposed an objective function for graph clustering called the Modularity function which allows automatic selection of the number of clusters. Empirically, higher values of the Modularity function have been shown to correlate well with good graph clustering. In this paper we propose an extended Modularity measure for categorical data clustering; first, we establish the connection with the Relational Analysis criterion. The proposed Modularity measure introduces an automatic weighting scheme which takes in consideration the profile of each data object. A modified Relational Analysis algorithm is then presented to search for the partitions maximizing the criterion. This algorithm deals linearly with large data set and allows natural clusters identification, i.e. doesn't require fixing the number of clusters and size of each cluster. Experimental results indicate that the new algorithm is efficient and effective at finding both good clustering and the appropriate number of clusters across a variety of real-world data sets.

1 Introduction

In the exploratory data analysis of high dimensional data, one of the main tasks is the formation of a simplified, usually visual, overview of data sets. This can be achieved through simplified description or summaries, which should provide the possibility to discover most relevant features or patterns. Clustering is among the examples of useful methods to achieve this task; classical clustering algorithms produce a grouping of the data according to a chosen criterion. Most algorithms use similarity measures based on Euclidean distance. However there are several types of data where the use of this measure is not adequate. This is the case when using categorical data since; generally, there is no known ordering between the feature values. If the data vectors contain categorical variables, geometric approaches are inappropriate and other strategies must be developed [4]. The problem of clustering becomes more challenging when the data is categorical, that is, when there is no inherent distance measures between data values. This is often the case in many applications where data is described by a set of descriptive or binary attributes, many of which are not numerical. Examples of such include the country of origin and the color of eyes in demographic data. Many algorithms

K.W. Wong et al. (Eds.): ICONIP 2010, Part II, LNCS 6444, pp. 310–320, 2010.

have been developed for clustering categorical data, e.g., (Barbara et al [3]., 2002; Gibson et al [7]., 1998; Huang [9], 1998).

Modularity measure have been used recently for graph clustering [12] [13] [1]. In this paper, we show that the Modularity clustering criterion can be formally extended for categorical data clustering. We also establish the connections between the Modularity criterion and the Relational Analysis (RA) approach [10][11] which is based on Condorcet's criterion. We then develop an efficient procedure inspired from the RA heuristic to find the optimal partition for maximizing the Modularity criterion. Experiments demonstrate the efficacy and effectiveness of our approach.

The rest of the paper is organized as follows: Section 2 introduces some notations and definitions; Section 3 presents the Relational Analysis approach (RA), Section 4 provides the extended modularity measure and its connection with the RA criterion. Section 5 discusses the proposed optimization procedure; Section 6 shows our experimental results and finally, Section 7 presents our conclusions.

2 Definitions and Notations

Let be D a dataset with a set I of N objects $(O_1, O_2, ..., O_N)$ described by the set V of M categorical attributes (or variables) $V^1, V^2., V^m, .., V^M$ each one having $p_1, .., p_m, .., p_M$ categories respectively and let $P = \sum_{m=1}^{M} p_m$ denote the full number of categories of all variables. Each categorical variable can be decomposed into a collection of indicator variables. For each variable V^m, let the p_m values naturally correspond to the numbers from 1 to p_m and let $V_1^m, V_2^m, ..., V_{p_m}^m$ be the binary variables such that for each j, $1 \leq j \leq p_m$, $V_k^m = 1$ if and only if the V^m takes the j-th value. Then the data set can be expressed as a collection of M $N \times p_m$ matrices K^m, $(m = 1, .., M)$ of general term k_{ij}^m such as:

$$k_{ij}^m = \begin{cases} 1 \text{ if the object } i \text{ takes the categorie } j \text{ of } V^m \\ 0 \text{ otherwise} \end{cases} \tag{1}$$

which gives the N by P binary disjunctive matrix $K = (K^1|K^2|...|K^m|...|K^M)$.

Let us recall that each variable V^m provides a similarity matrix S^m which can be expressed as $S^m = K^m\,{}^tK^m$ and the global similarity matrix $S = K\,{}^tK$ where ${}^tK^m$ and ${}^tK$ are the transposed K^m matrix respectively K matrix.

2.1 Undirect Graph and Data Matrices

An interesting connection between data matrices and graph theory can be established,. A data matrix can be viewed as a weighted undirect graph $G = (V, E)$, where $V = I$ is the set of vertices and E is the set of edges. The data matrix S can be viewed as a weighted undirect graph where each node i in I corresponds to a row. The edge between i and i' has weight $s_{ii'}$, denoting the element of the matrix in the intersection between row i and column i'.

2.2 Equivalence Relation

We set down some notations. Suppose that a set of N by P-dimensional binary data vectors, K, represented as an N by P matrix, is partitioned into L classes $C = \{C_1, C_2, ...C_L\}$ and we want the points within each class are similar to each other. We view C as an equivalence relation X which models a partition in a relational space, and must respect the following properties:

$$\begin{cases} x_{ii} = 1, \ \forall ii & \text{reflexivity} \\ x_{ii'} - x_{i'i} = 0, \forall (i, i') & \text{symetry} \\ x_{ii'} + x_{i'i''} - x_{ii''} \leq 1, \forall (i, i', i'') & \text{transitivity} \\ x_{ii'} \in \{0, 1\}, \forall (i, i') & \text{binarity} \end{cases} \quad (2)$$

3 The Relational Analysis Approach

The relational analysis theory is a data analysis technique that has been initiated and developed at IBM in the 1970s, by F. Marcotorchino and P. Michaud [10] [11]. This technique is used to resolve many problems that occur in fields like: preferences, voting systems, clustering, etc. The Relational Analysis approach is a clustering model that automatically provides the suitable number of clusters, this approach takes as input a similarity matrix. In our context, since we want to cluster the objects of the set I, the similarity matrix S is given, then we want to maximize the following clustering function

$$\mathcal{R}_{RA}(S, X) = \sum_{i} \sum_{i'} (s_{ii'} - m_{ii'}) x_{ii'} \quad (3)$$

Where $M = [m_{ii'} = \frac{s_{ii} + s_{i'i'}}{4}]_{i,i'=1,...,N}$ is the matrix of threshold values. In others words, objects i and i' have chances to be in the same cluster providing their similarity measure $s_{ii'}$, is greater or equal to their threshold value of majority $m_{ii'}$. X is the solution we are looking for, it is a binary relational matrix with general term $x_{ii'} = 1$ if object i is in the same cluster as object i'; and $x_{ii'} = 0$, otherwise. X represents an equivalence relation, thus it must respect the properties in (2).

4 Extensions of the Modularity Measure

This section shows how to adapt the Modularity measure for categorical data clustering

4.1 Modularity and Graphs

Modularity is a recently quality measure for graph clusterings, it has immediately received a considerable attention in several disciplines [12] [1]. As for the RA clustering problem, maximizing the modularity measure can be expressed in the form of an integer linear programming. Given the graph $G = (V, E)$, let A be a binary, symmetric matrix whose (i, j) entry, $a_{ij} = 1$ if there is edge between nodes i and j. If there is no edge between nodes i and j, a_{ij} is zero. We note that in our problem setting, A is an input having all information on the given graph G and is often called an adjacency matrix. Finding a partition of the set nodes V into homogeneous subsets leads to the resolution of the following integer linear programming:

$$\max_X Q(A, X) \tag{4}$$

where

$$Q(A, X) = \frac{1}{2|E|} \sum_{i,i'=1}^{n} (a_{ii'} - \frac{a_{i.}a_{i'.}}{2|E|})x_{ii'} \tag{5}$$

is the modularity measure, $2|E| = \sum_{i,i'} a_{ii'} = a_{..}$ is the total number of edges and $a_{i.} = \sum_{i'} a_{ii'}$ the degree of i. X is the solution we looking for wich must satisfies the properties of an equivalence relation defined on $I \times I$.

4.2 First Extension: Early Integration

The early integration consist in a direct combination of graphs from all variables into a single dataset (graph) before applying the learning algorithm. Let us consider the Condorcet's matrix S where each entry $s_{ii'} = \sum_{m=1}^{M} s_{ii'}^{m}$, which can be viewed as weight matrix associated to graph $G = (I, E)$, where each edge $e_{ii'}$ have the weight $s_{ii'}$. Similarly to the classical Modularity measure, we define the extension $Q_1(S, X)$ as follow:

$$Q_1(S, X) = \frac{1}{2|E|} \sum_{i,i'=1}^{n} (s_{ii'} - \frac{s_{i.}s_{i'.}}{2|E|})x_{ii'} \tag{6}$$

where $2|E| = \sum_{i,i'} s_{ii'} = s_{..}$ is the total number of edges and $s_{i.} = \sum_{i'} s_{ii'}$ the degree of i.

4.3 Modularity Extensions as a Modified RA Criterion

This subsection shows the theoretical connection between the RA criterion and the proposed extension of the modularity measure. We can establish a relationship between the Modularity measure extension and the RA criterion, indeed the function $Q_1(S, X)$ can be expressed as a modified RA criterion in the following way:

$Q_1(S, X)$ as a modified RA criterion:

$$Q_1(S, X) = \frac{1}{2|E|}(\mathcal{R}_{RA}(S, X) + \psi_1(S, X)) \tag{7}$$

where

$$\psi_1(S,X) = \frac{1}{2|E|}\sum_{i=1}^{n}\sum_{i'=1}^{n}(m_{ii'} - \frac{s_{i.}s_{i'.}}{2|E|})x_{ii'} \quad (8)$$

is the weighting term that depends on the profile of each pair of objects (i, i'). This extension of the modularity measure allows to introduce a weighting scheme depending on the profile of each data object.

5 Optimization Procedure

As the objective function is linear with respect to X and as the constraints that X must respect are linear equations, theoretically wa can solve the problem using an integer linear programming solver. However, this problem is NP-hard. As a result in practice, we use heuristics for dealing with large data sets.

5.1 Modularity Decomposition

The extension of the modularity measure can be decomposed in terms of the contribution of each object i in each clusters C_l of the searched partition as follows:

$$Q_1(S,X) = \sum_{l=1}^{L}\sum_{i=1}^{N} cont(i, C_l) \quad (9)$$

where

$$cont_{Q_1}(i, C_l) = \frac{1}{2|E|}\sum_{i' \in C_l}(s_{ii'} - \frac{s_{i.}s_{i'.}}{2|E|}) \quad (10)$$

Using the transformations, $s_{ii'} =< K_i, K_{i'} >$ and $s_{i.} = \sum_{i''} < K_i, K_{i''} >$, the contribution expression becomes[1],

$$\begin{aligned} cont_{Q_1}(i, C_l) &= \frac{1}{2|E|}\sum_{i' \in C_l}(< K_i, K_{i'} > \\ &\quad - \frac{\sum_{i''} < K_i, K_{i''} > \sum_{i''} < K_{i'}, K_{i''} >}{2|E|}) \quad (11) \\ &= \frac{1}{2|E|} < K_i, P_l > - \sum_{i' \in C_l} \delta_{ii'} \quad (12) \end{aligned}$$

where

$$\delta_{ii'} = \sum_{i' \in C_l} \frac{\sum_{i''} < K_i, K_{i''} > \sum_{i''} < K_{i'}, K_{i''} >}{2|E|} \quad (13)$$

---

[1] Let us recall that this new writing of the contribution formula allows to reduce considerably the computational cost related to the square similarity matrix S and to characterize each cluster C_l using his prototype P_l.

The new contribution formula introduces an automatic weighting scheme, the value of the new formula will be great, let or equal to the RA contribution depending on the weight $\delta_{ii'}$. The contribution formula $cont_{Q_1}$ can be written in term of the RA contribution $cont_{RA}$ by adding a weighting term depending on the profile of each objects pairwise (i, i'):

$$\begin{aligned} cont_{Q_1}(i, C_l) &= \frac{1}{2|E|}[(< K_i, P_l > - \sum_{i' \in C_l} m_{ii'}) \\ &+ \sum_{i' \in C_l} (m_{ii'} - \delta_{ii'})] \qquad (14) \\ &= \frac{1}{2|E|}[cont_{RA}(i, C_l) + \sum_{i' \in C_l} (m_{ii'} - \delta_{ii'})] \qquad (15) \end{aligned}$$

The change in the contribution formula is interesting because it introduces a weighting relative to the profiles of the data objects automatically, without requiring the presence of an expert. There are three scenarios;

1. Taking $\delta_{ii'} = m_{ii'}, \ \ \forall i, i'$ we find the case of the RA algorithm.
2. If the weight $\delta_{ii'}$ is less than $m_{ii'}, \ \ \forall i, i'$ the contribution formula $cont_{Q_1}$ is greater than the old contribution $cont_{RA}$, and therefore it is more likely to be positive than $cont_{RA}$, the observation i is then found aggregated to an existing cluster. the number of clusters will be smaller.
3. If the weight $\delta_{ii'}$ is great than $m_{ii'}, \ \ \forall i, i'$, the contribution formula $cont_{Q_1}$ is greater than the old contribution $cont_{RA}$, and therefore it is more likely to be negative than $cont_{RA}$, the observation i is then found in a new cluster. The number of clusters will be more important.

5.2 Relational Analysis Heuristic

The heuristic process starts from an initial cluster (a singleton cluster) and build in an incremental way, a partition of the set I by increasing the value of Condorcet's criterion $\mathcal{R}_{RA}(S, X)$ at each assignment. We give in (**Algorithm1**) the description of the relational analysis algorithm which was used by the Relational Analysis methodology (see Marcotorchino and Michaud for further details). The presented algorithm aims at maximizing the criteria ($\mathcal{R}_{RA}$ and Q_1) given above, based on the contribution computation.

We have to fix a number of iterations in order to have an approximate solution in a reasonable processing time. Besides, it is also required a maximum number of clusters, but since we don't need to fix this parameter, we put by default $L_{max} = N$. Basically, this algorithm has $O(N_{iter} \times L_{max} \times N)$ computation cost. In general term, we can assume that $N_{iter} << N$, but not $L_{max} << N$. Thus, in the worst case, the algorithm has $O(L_{max} \times N)$ computation cost.

Algorithm 1. RA heuristic

Inputs:
L_{max}= maximal number of clusters, N_{iter}= number of iterations, N= number of examples (objects)
- Compute the threshold matrix (δ or M)
- Take the first object as the first element of the first cluster.
- $l = 1$ where l is the current number of clusters

for t=1 to N_{iter} **do**
    for $i = 1$ to N **do**
        for $j = 1$ to l **do**
            Compute the contribution of object i : $cont(i, j)$
        end for
        $l^* = arg \max_j cont(i, j)$,
where l^* is the cluster id which has the highest contribution with the object i
        $cont(i, l^*) \leftarrow$ the computed contribution
        if $cont(i, l^*) < 0$ **and** $l < L_{max}$ **then**
            create a new cluster where the object i is the first element
                $l \leftarrow l + 1$
        else
            assign object i to cluster C_{l^*}
        endif
    endfor
endfor

Output:
at most L_{max} clusters

6 Experiments

A performance study has been conducted to evaluate our method. In this section, we describe those experiments and the results. We ran our algorithm on real-life datasets obtained from the UCI Machine Learning Repository to test its clustering performance against the RA algorithm.

6.1 Performance Measures

There are many ways to measure the accuracy of clustering algorithm.

Cluster purity: One of the ways of measuring the quality of a clustering solution is the cluster purity. Let there be L clusters of the dataset I and size of cluster C_l be $|C_l|$. The purity of this cluster is given by purity(C_l)= $\frac{1}{|C_l|} \max_k(|C_l|_{cluster=k})$ where $|C_l|_{cluster=k}$ denote the number of items for the cluster k assigned to cluster l. The overall purity of a clustering solution could be expressed as a weighted sum of individual cluster purities

$$purity = \sum_{l=1}^{L} \frac{|\mathcal{C}_l|}{|I|} purity(\mathcal{C}_l) \tag{16}$$

In general, if the values of purity are larger, the clustering solution is better.

Rand Index. This index (Rand, 1971) simply measures the number of pairwise agreements between a clustering K and a set of class labels C, normalized so that the value lies between 0 and 1:

$$RI(U, V) = \frac{a + b}{a + b + c + d} \tag{17}$$

where a denotes the number of pairs of points with the same label in U and assigned to the same cluster in V, b denotes the number of pairs with the same label, but in different clusters, c denotes the number of pairs in the same cluster, but with different class labels and d denotes the number of pairs with a different label in U that were assigned to a different cluster in V. The index produces a result in the range [0,1], where a value of 1.0 indicates that U and V are identical. A high value for this measure generally indicates a high level of agreement between a clustering and the annotated natural classes.

Jaccard Index. In this index (Jaccard, 1912), which has been commonly applied to assess the similarity between different partitions of the same dataset, the level of agreement between a set of class labels U and a clustering result V is determined by the number of pairs of points assigned to the same cluster in both partitions:

$$JI(U, V) = \frac{a}{a + b + c} \tag{18}$$

The index produces a result in the range [0,1], where a value of 1.0 indicates that U and V are identical.

Tanimoto Index. Similarity between different partitions of the same data set can be measured as the ratio of their common elements to the number of all different elements,

$$TI(U, V) = \frac{\frac{1}{2}(a + d)}{\frac{1}{2}(a + d) + b + c} \tag{19}$$

The index produces a result in the range [0,1].

6.2 The Datasets for Validation

In this section, we evaluate the performance of the RA heuristic on several databases available at the UC Irvine Machine Learning Repository [2]. The description of the used data sets is given in Table 1:

Table 1. Description of the data set

Data set	# of Objects	# of Attributes	Classes
Soybean small	47	21	4
Zoo	101	16	7
Soybean large	307	35	19
SPECTF	267	22	2
Post-Operative	90	8	3
Balance Scale	625	4	3
Audiology Normalized	226	69	24

6.3 Results for the Early Integration

The proposed method is tested with data sets obtained from the UCI machine learning data repository. As the proposed method is a modified RA approach, we have compared the performance of the proposed algorithm with RA algorithm. From the Table 2 and the Figure (Fig1: Purity measure), it is clear that the performance of the proposed method based the extended modularity measure is the best then the RA approach for all data sets. This means that the introduced weighting scheme improves the purity clustering.

Table 2. Purity measure for $\mathcal{R}_{RA}(S,X)$ criterion and the extended Modularity $Q_1(S,X)$

DB	DB size	$\mathcal{R}_{RA}(S,X)$	$Q_1(S,X)$
Soybean small	47x21	78 %	100 %
Zoo	101x16	83.17%	88.12 %
Soybean large	307x35	70 %	72.31 %
SPECTF	267x22	61.25 %	85 %
Post-Operative	90x8	71.11 %	73.33% %
Balance Scale	625x4	63.52 %	63.52 %
Audiology Normalized	226x69	50.50 %	58 %

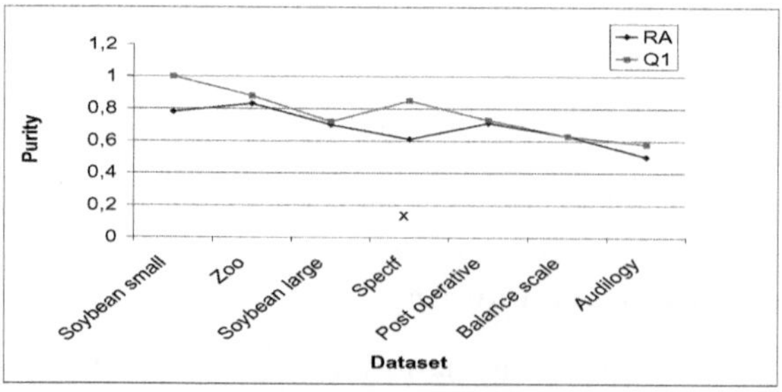

Fig. 1. Purity measure

Table 3. Results on different datasets using $Q_1(S, X)$

DB	DB size	RI	JI	TI
Soybean small	47x21	100 %	100 %	100 %
Zoo	101x16	94.2%	79.9 %	89.2 %
Soybean large	307x35	91.2 %	26.6 %	83.9 %
SPECTF	267x22	60.98 %	38.28 %	43.86 %
Post-Operative	90x8	50.75%	37.37% %	34.01%
Balance Scale	625x4	58 %	20 %	40 %
Audiology Normalized	226x69	82 %	20 %	69%

Table 4. Results on different datasets using $\mathcal{R}_{RA}(S, X)$

DB	DB size	RI	JI	TI
Soybean small	47x21	86.66 %	45.88 %	76.47 %
Zoo	101x16	72.9%	46.09 %	57.37 %
Soybean large	307x35	85.03 %	25.7 %	73.97 %
SPECTF	267x22	55.74 %	38.69 %	38.64 %
Post-Operative	90x8	54.44%	41.17% %	37.4%
Balance Scale	625x4	57 %	19 %	39 %
Audiology Normalized	226x69	82 %	20 %	69 %

In order to show the good performance of the proposed approach we use several categorical data sets of different sizes and we indicate in the table 4 the RI, JI and TI index of clustering using the classical RA criterion and in Table 3 the RI, JI and TI index using the extended modularity measure. The results illustrate that the proposed technique increase the index value compared to the classical RA and allows to introduce an automatic weighting scheme which relates to the object profile in the data set.

7 Conclusions

In this paper, we studied the extensions of the Modularity-based criterion for the categorical data clustering and we illustrate its relations with Condorcet's criterion. An efficient, iterative procedure for optimization is presented. The experimental results indicate the effectiveness of the proposed early integration method compared to the classical Relational Analysis approach. In future work, we will propose a second extension called **intermediate integration**, the main idea is to compute a combined Modularity objective measure from separate Modularity on each variable, and then pass to the learning algorithm.

References

1. Agarwal, G., Kempe, D.: Modularity-maximizing graph communities via mathematical programming. The European Physical Journal B 66(33), 409–418 (2008)
2. Asuncion, A., Newman, D.J.: UCI Machine Learning Repository. University of California, School of Information and Computer Science, Irvine, CA (2007), http://www.ics.uci.edu/~mlearn/MLRepository.html
3. Barbara, D., Couto, J., Li, Y.: COOLCAT: an entropy-based algorithm for categorical clustering. In: Proceedings of the Eleventh ACM CIKM Conference, pp. 582–589 (2002)
4. Bock, H.-H.: Probabilistic aspects in cluster analysis. In: Opitz, O. (ed.) Conceptual and Numerical Analysis of Data, pp. 12–44. Springer, Berlin (1989)
5. Celeux, G., Govaert, G.: Clustering criteria for discrete data and latent class models. Journal of Classification 8, 157–176 (1991)
6. Ganti, V., Gehrke, J., Ramakrishnan, R.: CACTUS - clustering categorical data using summaries. In: Proceedings of the Fifth ACM SIGKDD Conference, pp. 73–83 (1999)
7. Gibson, D., Kleinberg, J., Raghavan, P.: Clustering categorical data: An approach based on dynamical systems. In: Proceedings of the 24rd VLDB Conference, pp. 311–322 (1998)
8. Guha, S., Rastogi, R., Shim, K.: ROCK: A robust clustering algorithm for categorical attributes. Information Systems 25, 345–366 (2000)
9. Huang, Z.: Extensions to the k-means algorithm for clustering large data sets with categorical values. Data Mining and Knowledge Discovery 2, 283–304 (1998)
10. Marcotorchino, J.F.: Relational analysis theory as a general approach to data analysis and data fusion. In: Cognitive Systems with Interactive Sensors (2006)
11. Marcotorchino, J.F., Michaud, P.: Optimisation en analyse ordinale des données (1978) (in Masson)
12. Newman, M., Girvan, M.: Finding and evaluating community structure in networks. Physical Review E 69, 26113 (2004)
13. White, S., Smyth, P.: A spectral clustering approach to finding communities in graphs. In: SDM, pp. 76–84 (2005)

A Morphological Associative Memory Employing a Reverse Recall

Hidetaka Harada and Tsutomu Miki

Graduate School of Life Science and Systems Engineering,
Kyushu Institute of Technology, Kitakyushu 808-0196, Japan
harada-hidetaka@edu.brain.kyutech.ac.jp,
miki@brain.kyutech.ac.jp

Abstract. Recently the morphological associative memory proposed by Ritter attracts researcher's attention. The model is superior to other models in terms of memory capacity and perfect recall rate. However the conventional MAM has a problem that the correct pattern cannot be recalled if a pattern has inclusive relation to other stored pattern. In this paper, as one of the solutions, an effective MAM employing a reverse recall is proposed. In the proposed method, candidate patterns of an input can be estimated by reverse recall from the kernel image recalled by a given inclusion input pattern, and then the plausible recall pattern can be determined by comparing the candidates with input pattern. We confirm the validity of the proposed method through hetero association experiments for twenty six alphabet patterns with inclusion patterns.

Keywords: morphological associative memory, reverse recall, perfect recall, pattern with inclusive relation.

1 Introduction

Associative memory is one of crucial brain functions. The researches of the associative memory have been studied especially from early in 80's [1-4]. Hopfield network [4] is well known as the traditional model, which is used as not only the associative memory but also an optimization tool. However it has not been used for practical functional memory devices, despite the attractive features. It is resulting from the drawbacks that are low memory capacity in contrast to the number of memory units and instability in recall caused by the local minimum.

As one of the improving models, Ritter proposed a morphological associative memory (MAM) using a concept of the morphology [5]. The MAM is superior to other models in terms of the memory capacity and perfect recall rate. The MAM, on the other hand, has a drawback that the kernel image used for a recall index becomes hard to design as the number of stored patterns increase. To overcome the problem, Hattori *et al.* proposed a high speed kernel designing method [6], and Ida *et al.* developed a MAM employing additional kernel images [7]. We proposed the block splitting type MAM (BMAM) which avoids spreading noises by splitting a pattern in several blocks [8]. In this method, the perfect recall rate was improved by about 8-10 % compared to

K.W. Wong et al. (Eds.): ICONIP 2010, Part II, LNCS 6444, pp. 321–328, 2010.

the MAM without the kernel image. We also proposed an effective kernel design method employing the kernel images independent to the corresponding stored pattern [9]. In this proposed method, the plausible kernel image can be determined by applying the BMAM method to the kernel images. The method employing the independent kernel image facilitated the design of the kernel image. However it remains as a problem that an associative memory is hard to recall the correct pattern if the recall pattern is inclusion pattern (e.g., "C and G", "E and F").

In this paper, we propose the effective method employing reverse recall to solve this problem. In the method, the kernel image can be represented with unique one bit by using the kernel design method that does not depend on the stored patterns. Finally, in the proposed MAM, the plausible recall pattern can be determined by comparing input pattern with patterns obtained by reverse recall. The correct recall pattern can be obtained by comparing input pattern with patterns obtained by the reverse recall. We confirm the validity of the proposed method through hetero association experiments for twenty six alphabet characters including patterns with inclusion relation to other stored patterns.

2 Morphological Associative Memory: MAM

2.1 Ritter's MAM

The MAM proposed by Ritter [5] has two-stage recall process using memory matrices "*M*" and "*W*" in the stages as shown Fig.1. In the recall process, a kernel image is used as an intermediate image. Here, let assume *S* pattern pairs $(X^1,Y^1),\dots,(X^S,Y^S)$ as the stored patterns. Respectively, $X^r=(x_1^r,\cdots,x_n^r)$, $Y^r=(y_1^r,\cdots,y_m^r)$, *S* is the total number of the pattern pairs. The kernel image Z^r works as the index for recalling the stored pattern Y^r and consists of partial units of the stored pattern X^r. The memory matrices "*M*" and "*W*" are given as;

$$m_{ij}=\bigvee_{r=1}^{S}\left(z_i^r-z_j^r\right), \tag{1}$$

$$w_{ij}=\bigwedge_{r=1}^{S}\left(y_i^r-z_j^r\right), \tag{2}$$

where m_{ij} and w_{ij} are *(i, j)-th* element of memory matrices "*M*" and "*W*", respectively. The symbols $\wedge$ and $\vee$ denote minimum and maximum operators, respectively. When an input pattern X^r is fed into the MAM, the output is obtained by two-stage recall process given by follows;

$$z_i^r=\bigwedge_{j=1}^{n}\left(m_{ij}+x_j^r\right) \qquad i=1,\cdots,n, \tag{3}$$

$$y_i^r=\bigvee_{j=1}^{n}\left(w_{ij}+z_j^r\right) \qquad i=1,\cdots,m. \tag{4}$$

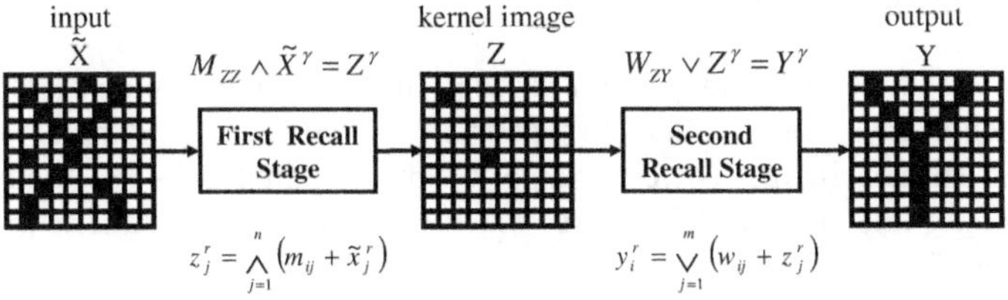

Fig. 1. Two-stage recall process in Ritter's MAM

2.2 MAM Using the Stored Pattern Independent Kernel Image

Ritter's MAM has a problem that the design of the kernel image becomes difficult as the number of stored patterns increase because the kernel image is created using a part of the stored pattern. In order to overcome the problem, we proposed a MAM using a stored pattern independent kernel image [9].

The recall process of the MAM using the independent kernel image as shown Fig.2 processes through the following steps;

step1 An input pattern is divided evenly into sub blocks,
step2 the first recall is executed every sub block independently,
step3 the recalled patterns of all sub blocks are summed up,
step4 the kernel image is determined by a majority logic for the kernel pattern obtained in step3,
step5 finally the output pattern is recalled using the kernel image in the second recall stage.

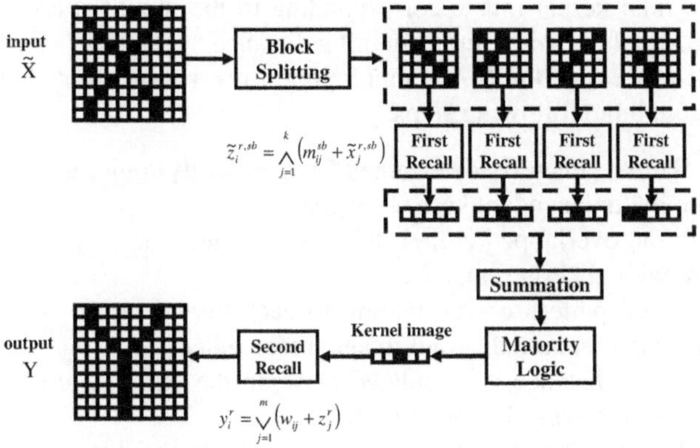

Fig. 2. Recall process of the MAM using the independent kernel image

3 MAM Employing a Reverse Recall

In the MAM using the independent kernel image, when a pattern is completely included with other stored pattern (a pattern in the pair is called "inclusion pattern"), overlapped kernel image corresponding to those patterns is recalled in hetero-association for the pattern. Therefore, the overlapped output pattern is recalled as shown in Fig.3.

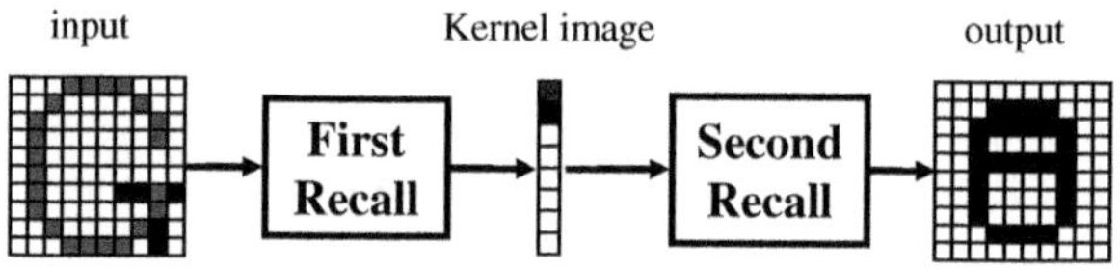

Fig. 3. Recall process of the MAM using the independent kernel image for the inclusion pattern in hetero-association. ■ denotes the pattern "C" completely included with the pattern "G".

As one of solutions, we propose an effective method employing a reverse recall. In the proposed method, overlapped kernel image is separated to individual kernel images. The stored pattern corresponding to each kernel image is reversely recalled for each separated kernel image, independently. Here, we introduce the feedback scheme proposed by Ritter [10] into the proposed method. The stored pattern $X^r = (x_1^r, \ \cdots \ x_n^r)$ reversely recalled for kernel images $Z^r = (z_1^r, \ \cdots \ z_m^r)$ is given as:

$$x_j^r = \bigvee_{i=1}^{m} \left(z_i^r - m_{ij}\right), \tag{5}$$

where, x_j^r is *j-th* unit of the stored pattern x^r, z_i^r is *i-th* unit of the kernel image z^r.

Hamming distance is calculated between an input pattern and patterns reversely recalled. The final kernel image corresponding to the input pattern is determined by the kernel image of the minimum Hamming distance.

The recall process of the MAM employing a reverse recall is shown in Fig.4 and processes through the following steps:

- step1 The kernel image is determined by a majority logic, as same as the MAM using the independent kernel image,
- step2 when the overlapped kernel image is recalled, the kernel image is separated to individual kernel images,
- step3 the stored pattern corresponding to each kernel image is reversely recalled using each separated kernel image, independently,
- step4 Hamming distance is calculated between the input pattern and the reversely recalled patterns obtained in step3,
- step5 The kernel image having the minimum Hamming distance is selected for the plausible kernel image,
- step6 finally, in the second recall stage, the output pattern is recalled using the kernel image determined in step5.

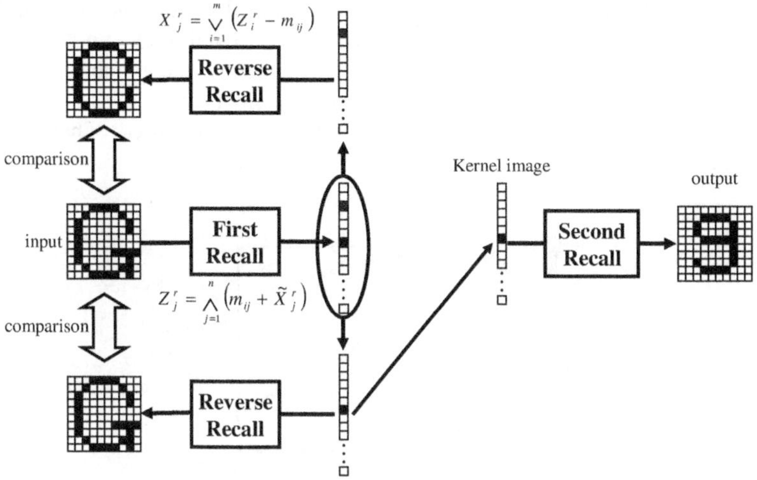

Fig. 4. Recall process of the MAM employing a reverse recall

4 Experimental Results

In order to evaluate the validity of the proposed method, hetero-association experiments are examined. The performance is evaluated with the perfect recall rate. Here the perfect recall rate is evaluated by an average of 10,000 trials in the simulation. The perfect recall is defined as recalling the correct pattern which is the pattern recalled by the noiseless input pattern.

In the experiments, each pattern consists of 10 x 10 = 100 binary units. The unit of pattern takes '0' or '1', the '1' represents black and '0' white. The noise is assigned to change '1' into '0' at random and vice versa.

4.1 Hetero Association Using Twenty-Six Alphabet Characters

Firstly, we investigate the perfect recall rate of the MAM using the independent kernel image for twenty-one alphabet patterns excluding the inclusion patterns as shown in Fig.5.

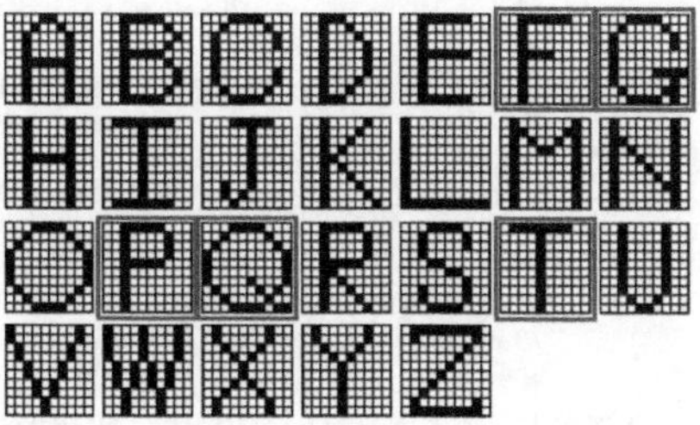

Fig. 5. Stored pattern: twenty-six alphabet capital letter patterns. Patterns surrounded by red square represent the inclusion patterns.

Fig.6 shows the noise tolerance of the MAM using the independent kernel image for twenty-one patterns illustrated in Fig.5. The perfect recall rate is investigated with changing the number of block splits. Here, *sb* represents the number of the block splits, and *sb* = 1 is the special case, which means no block split.

In the case, pattern excluding the inclusion patterns are used. Therefore, good performance can be achieved.

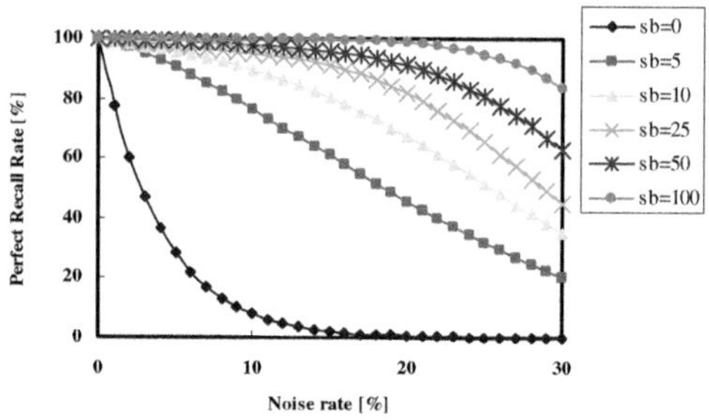

Fig. 6. Noise tolerance of the MAM using the independent kernel image in hetero-association for twenty-one alphabet patterns

Next, we investigate the perfect recall rate of the MAM using the proposed method and the MAM using the independent kernel image for twenty-six alphabet patterns including the inclusion patterns as shown in Fig.5.

Fig.7 shows the noise tolerance of the MAM using the independent kernel image and the proposed method for twenty-six alphabet patterns including the inclusion patterns. In ordinary MAM using the independent kernel image, as shown in Fig.7 (a), the perfect recalling can not be achieved even if the input pattern does not include any noise.

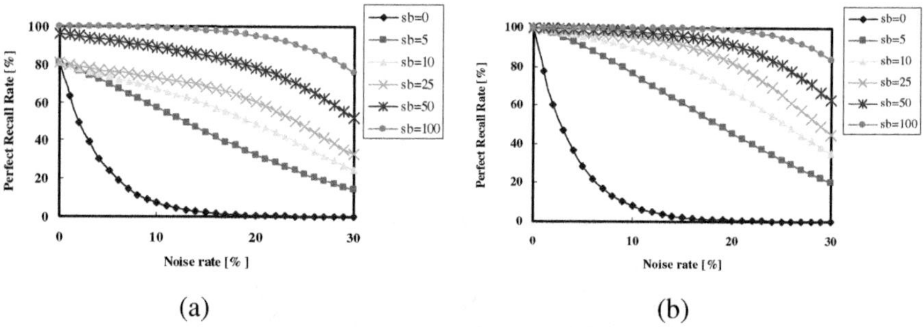

Fig. 7. Noise tolerance in hetero-association for twenty-six alphabet patterns. (a) is the result of the MAM using the independent kernel image, and (b) the proposed method.

On the other hand, as shown in Fig.7 (b), in the proposed method, the perfect recalling can be achieved when the input pattern does not include any noise, differently from the result of ordinary MAM using the independent kernel image.

4.2 Hetero Association Using Sets of Threefold Inclusion Patterns

Thirty patterns as shown in Fig.8 are used in the experiment.

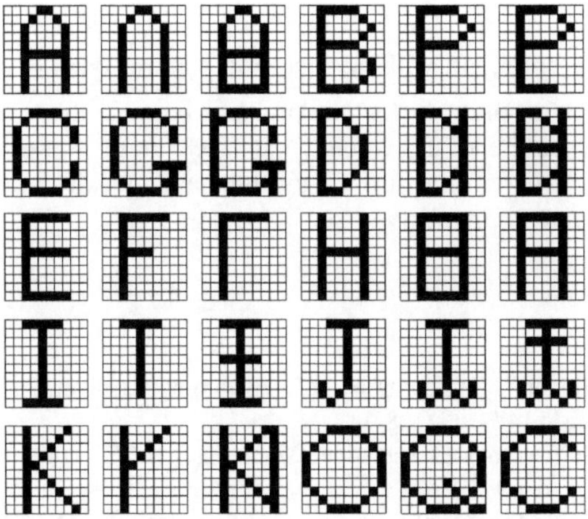

Fig. 8. Stored patterns: thirty patterns that consist of only sets of threefold inclusion patterns

Fig.9 shows the noise tolerance of the MAM using the independent kernel image and the proposed method for thirty patterns in hetero-association. Even in the case of threefold inclusion patterns, Fig.9 (b) shows that the perfect recalling can be achieved when the input pattern does not include any noise.

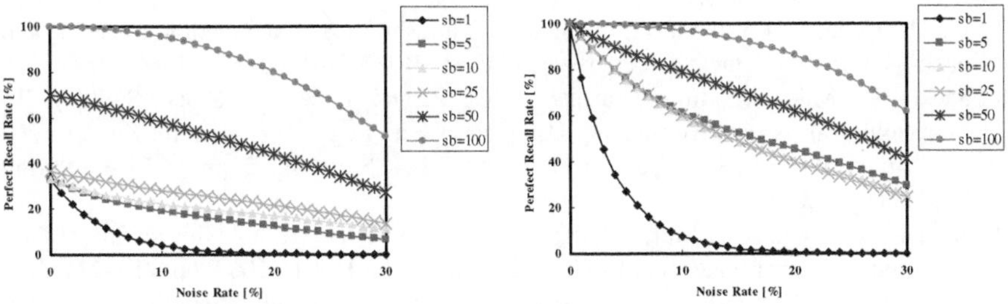

(a) (b)

Fig. 9. Noise tolerance in hetero-association for thirty patterns. (a) is the result of the MAM using the independent kernel image, and (b) the proposed method.

5 Conclusion

Conventional MAMs cannot achieve the perfect recall even if it is noiseless pattern when a pattern is completely included by other stored pattern. In order to overcome this problem, we proposed the MAM employing the reverse recall method available for inclusive patterns. The validity of the proposed method was confirmed by hetero-association experiments. Furthermore, we showed that the proposed method can handle even threefold inclusion patterns. On the other hand, the proposed method needs an additional calculation for the reverse recall. The additional time is approximately equal to $0.5 \times S_{ip} \times t_{recall}$. Here S_{ip} is the number of included patterns and t_{recall} is the time of recall with no inclusive pattern. Parallel computation can be expected as one of the solutions for this problem because the reverse recall on each kernel image can be done individually. In the future works, we will improve calculation speed by refining the algorithm and tackle practical applications employing an associative memory.

References

1. Ackley, D.H., Hinton, G.E., Sejnowski, T.J.: A learning algorithm for Boltzmann machines. Cognitive Science 9, 147–169 (1985)
2. Kosko, B.: Bidirectional Associative Memories. IEEE Trans. Systems, Man and Cybernetics 8(1), 46–60 (1988)
3. Hagiwara, M.: Multidirectional associative memory. In: International Joint Conference on Neural Networks, vol. I, pp. 3–6 (1990)
4. Hopfield, J.J.: Neural network and physical systems with emergent collective computational abilities. Proc. Of the National Academy of Sciences of the United States of America 79(8), 2554–2558 (1982)
5. Ritter, G.X., Sussner, P., Diaz-de-Leon, J.L.: Morphological associative memory. IEEE Trans. Neural Networks 9(2), 281–293 (1998)
6. Hattori, M., Fukui, A., Ito, H.: A fast method to decide kernel patterns for Morphological Associative memory. Transactions of the Institute of Electrical Engineers of Japan. C 123(10), 1830–1838 (2003)
7. Ida, T., Ueda, S., Kashima, M., Fuchida, T., Murashima, S.: On a method to decide kernel patterns of morphological associative memory. D-II J83-D-II(5), 1372–1380 (2000)
8. Saeki, T., Miki, T.: Improvement of the Perfect Recall of Block Splitting Type Morphological Associative Memory Using a Majority Logic Approach. In: King, I., Wang, J., Chan, L.-W., Wang, D. (eds.) ICONIP 2006. LNCS, vol. 4232, pp. 352–360. Springer, Heidelberg (2006)
9. Harada, H., Miki, T.: A Morphological Associative Memory Employing A Stored Pattern Independent Kernel Image and Its Hardware Model. In: IWCIA 2009, pp. 219–224 (2009)
10. Ritter, G.X., Diaz-de-Leon, J.L., Sussner, P.: Morphological bidirectional associative memories. Neural Networks 12, 851–867 (1999)

Analysis of Packet Traffics and Detection of Abnormal Traffics Using Pareto Learning Self Organizing Maps

Hiroshi Dozono[1], Masanori Nakakuni[2], Takaru Kabashima[1], and Shigeomi Hara[1]

[1] Faculty of Science and Engineering, Saga University, 1-Honjyo Saga 840-8502, Japan
hiro@dna.ec.saga-u.ac.jp

[2] Information Technology Center, Fukuoka University, 8-19-1, Nanakuma, Jonan-ku, Fukuoka 814-0180, Japan
nak@fukuoka-u.ac.jp

Abstract. Recently, the spread of the Internet makes familiar to the incident concerning the Internet, such as a DoS attack and a DDoS attack. Some methods which detect the abnormal traffics in the network using the information from headers and payloads of IP-packets transmitted in the networks are proposed. In this research, we propose a method of Pareto Learning SOM (Self Organizing Map) for IP packet flow analysis in which the occurrence rate is used for SOM computing. The flow of the packets can be visualized on the map and it can be used for detecting the abnormal flows of packets.

1 Introduction

As for the popularization of computers and development of internet connections, the incidents on the network are increasing. The purpose of this research is as follows. From the network which uses the packet communication, the packets are captured. After preprocessing, the occurrence rate of the each element in the packets is calculated, and is learned by Pareto learning Self Organizing Map. From, the result of learning, the traffics of the IP packets are visualized on the map. Using this map, the changes of traffics are also visualized and it will be applicable to detect network incidents.

Self Organizing map is the feed forward type neural network which consists of 2 layers, competitive layer and input layer without hidden layers. The learning method is unsupervised learning. After learning SOM can map the multi-dimensional data on the 2 dimensional plane. Pareto learning SOM (P-SOM) is a variation of SOM which is proposed for the learning of multi-modal vectors. The packet data consists of some types of attributes, so it will be effective to apply P-SOM for this problem. As the previous work using SOM for network traffic analysis, in [1] the large amount of logs from IDS system were analyzed and the information of each packet was learned using SOM. In [2], the occurrence rate of the IP address in the list of IP address was analyzed by using SOM. Our approach also uses the occurrence rate, however the occurrence rate

K.W. Wong et al. (Eds.): ICONIP 2010, Part II, LNCS 6444, pp. 329–336, 2010.

of each attributes of IP packets is counted, and not only IP address, but also other attributes in IP-header and payload information are used for the learning. The occurrence rates are often used for the learning of statistical information of the large amounts of input data. For example, In [3], the occurrence rates of the words in the documents are used for clustering the large amounts of document data, and in [4], the occurrence rates of the tuples of the DNA sequences are used for discovering the hidden properties of DNA sequences. The traffics of IP packets also will be large amounts, so it will be effective to use occurrence rate for learning the statistical information of the IP-traffics.

After learning the map, the packet traffics can be visualized and it can also used for detecting the abnormal traffics. In this paper, the experimental results of the detection of abnormal traffics, using supervised learning and using unsupervised learning are shown.

2 Analysis of Packet Traffics Using Pareto Learning Self Organizing Maps

2.1 Pareto Learning Self Organizing Maps

For the learning of multi-modal vectors using conventional Self Organizing Map (SOM)s, the simply concatenated vectors are often used. But, the resulting maps are dominated by the largely scaled vectors and are easily affected by unreliable vectors. For this problem, the concatenated vectors with weight values are used. But, the resulting maps heavily depend on the setting of weight values and it is difficult to select optimal weight value for each vector. For this problem, we proposed Pareto learning Self Organizing Map(P-SOM)s[5][6]. P-SOM organizes the input data composed of the multiple independent vectors based on the Pareto optimal concept[5]. Additionally, we proposed Supervised Pareto learning SOM(SP-SOM) s to improve the accuracy of classification by adding the supervised learning of category vector as feature vectors[5]. We applied P-SOM and SP-SOM to the authentication problem using multi-modal behavior vectors such as key typing features and pen drawing features on touch screen[5]. The algorithm of P-SOM is as follows.

P-SOM Algorithm

1. Initialization of the map
   Initialize the vector $\mathbf{m^{ij}}$ which are assigned to unit U^{ij} on the map using the 1st and 2nd principal components as base vectors of 2-dimensional map.
2. Batch learning phase
   (1) Clear all learning buffer of units U^{ij}.
   (2) For each vector x^i, search for the pareto optimal set of the units $P = \{U_p^{ab}\}$. U_p^{ab} is an element of pareto optimal set P, if for all units $U_{kl} \in P - U_p^{ab}$, existing h such that $e_h^{ab} \leq e_h^{kl}$ where

$$e_h^{kl} = |\mathbf{x_h^i} - \mathbf{m_h^{kl}}| \quad (1)$$

   (3) Add x^i to the learning buffer of all units $U_p^{ab} \in P$.

3. Batch update phase
For each unit U^{ij} update the associated vector $\mathbf{m}^{ij}$ using the weighted average of the vectors recorded in the buffer of U^{ij} and its neighboring units as follows.
(1)For all vectors x recorded in the buffer of U^{ij} and its neighboring units in distance $d \leq Sn$, calculate weighted sum $\mathbf{S}$ of the updates and the sum of weight values W.

$$\mathbf{S} = \mathbf{S} + \eta fn(d)(\mathbf{x} - \mathbf{m}^{\mathbf{i}'\mathbf{j}'}) \tag{2}$$

$$W = W + fn(d) \tag{3}$$

where $U^{i'j'}$s are neighbors of U^{ij} including U^{ij} itself, η is learning rate, $fn(d)$ is the neighborhood function which becomes 1 for d=0 and decrease with increment of d.
(2) Set the vector $\mathbf{m}^{\mathbf{ij}} = \mathbf{m}^{\mathbf{ij}} + \mathbf{S}/W$.

Repeat 2. and 3. with decreasing the size of neighbors Sn for pre-defined iterations.

As shown in step 2 of this algorithm, Pareto winner set for the integrated input vector $\mathbf{x}$ are searched for based on the concept of Pareto Optimality using the distance defined by (6) as objective function $f_h(\mathbf{x})$ for each element $\mathbf{x_h}$ in $\mathbf{x}$. Thus, the multiple units become winners. The winners and their neighboring units are modified in the update process in step 3. Overlapped neighbors are updated multiply and the overlapped region will contribute to generalization ability and integration ability of P-SOM.

2.2 Learning IP-Packets Using Occurrence Rate

As the input vector for learning IP-packets using P-SOM, the occurrence rates of the elements in the packets are used as the statistical information of the group of packets. This method is considered to be effective for handling large number of packets. In our experiments, IP packets can be captured only in out laboratory which is inside of the firewall. So, the packets concerning the attacks did not appear. So, we develop the algorithm which can visualize the time flow of the packets. This algorithm will be applicable for detecting illegal attacks which will be mapped to the specific regions on the map. The calculation method of occurrence rate is as follow. The occurrence rate of source and destination IP address is counted for each 8 bits digit in 256 dimensions of vector. Fig. 1 shows an example of counting IP address. For 192.168.36.18, the 192th element of the 1st vector, 168th element of 2nd vector, 36th element of 3rd vector and 18th element of 4th vector are counted up. The port number is described in 16 bits digit. The port numbers which are often used for illegal accesses are concentrated to the well known port numbers, TCP/25 for mail service, TCP/80 for web service and UDP/53 for DNS(Domain Mail Service). The port numbers are counted in 5 elements, which represents these 3 port numbers and other well-known ports and other ports which are temporally allocated for communication. The packet length is counted in 2 elements which denote less than 100 and greater than 100.

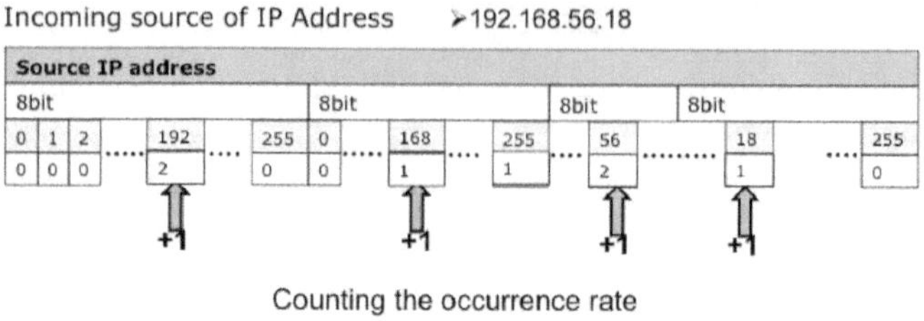

Fig. 1. Counting of IP Addresses

The payload data is counted in 256 elements which denote the occurrence of 8 bits numbers in payload data. Fig.2 shows the configuration of input vectors. As shown in Fig.2, the dimension and scale of the input vectors are different each other. P-SOM can integrate such vectors naturally based on the concept of Pareto optimality.

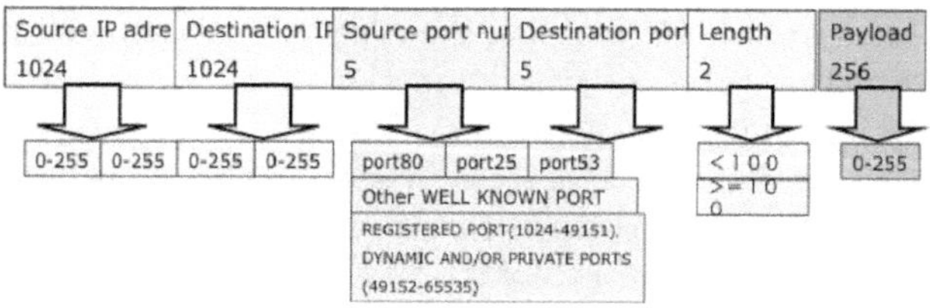

Fig. 2. Configuration of input vector

2.3 Experiment of Mapping the Traffics of Packets

At first, we made the experiments using the small number of packet data captured in the laboratory. As mentioned before, the packets are already filtered by firewall of university, so the packets concerning illegal access are not contained. In this experiments, we examine how the traffics of the packets are mapped using P-SOM algorithm. We made some experiments with changing the number of packets in a group. Number of the groups is set 100 for all cases. For each group, the input vector is configured as shown in Fig.2 and the category vector which represents the sequential number of the group is added. Fig.3 shows the result for the size of the group set as 500 packets. This map is organized as torus map. The number in each cell denotes the sequence number of the group of packets which is derived from the category vector learned on the map. The gray scale level also represents the sequence number. If the group of IP Packets are mapped continuously depending on the sequence number of the groups, the change of the color will be continuous on the map and it reflects the changes of the traffics of IP packets in the time flow on the map. The IP communications are performed using group of packets. Thus, the occurrence rate of the group of

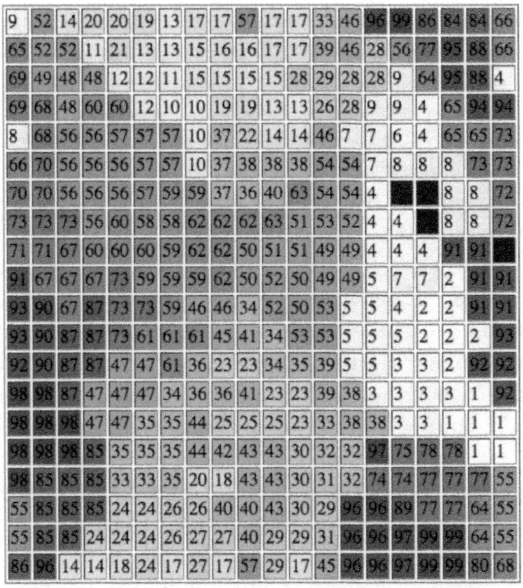

Fig. 3. Map of the occurrence rate vectors of 500 IP packets

the packets which are close in time scale become similar and mapped closely by P-SOM and this map reflects the changes of the traffics of IP Packets in local network.

3 Experiments for Detecting Abnormal Packets

3.1 Supervised Learning Method

Next, we examine the availability for detecting abnormal packets using supervised learning. For this experiments, the abnormal packet data including illegal attacks is artificially generated from the captured data. An attack consists of group of packets and the IP address fields and Port number fields of the packets are modified as follows.

Random IP scan (IP-R). Assuming the IP scan to local network, the destination IP address is set to random IP address in local network for each packet.

Fixed IP attack (IP-F). Assuming the attack to a computer in local network, the destination IP address is set to identical IP address in local network during an attack and is set random for each attack.

Random Port scan (Port-R). Assuming the port scan to the computers in local network, the destination port number is set to random number for each packet.

Fixed Port attack (Port-F). Assuming the attack to a fixed Port number of the computers in local network, the destination Port number is set to identical number during an attack and is set random for each attack.

For all cases, the source IP addresses and source port numbers are set identical during an attack and is set random for each attack.

We made experiments with changing the combinations of the modifications of IP and Port number. 500 groups of packets, which include 100 attacks, are used for learning SP-SOM and 5000 groups of packets, which include 1000 attacks, are used for test. Each attack consists of 100 packets. Fig.4 shows the map for case4 (IP-R, Port-R) and Table 1 shows the sensibility and specificity for classifying the groups including abnormal packets. Sensibility and specificity means the rates for classifying the group of packets including abnormal packets and the group not including abnormal packets correctly. In Fig.4, the groups of packets including attacks are colored in red and clustered mostly in two clusters. For the test data, specificity, which means the detection rate of non-attacking packets, is high enough for all cases. Sensibility, which means the detection rate of attacking packets, is about over 80% except in case 2. The sensibility can be improved at the expense of specificity with changing the threshold of discrimination of SP-SIM, however, the specificity is considered to be more important than sensibility is this experiments to avoid too sensitive detections which annoy the administrator of the network.

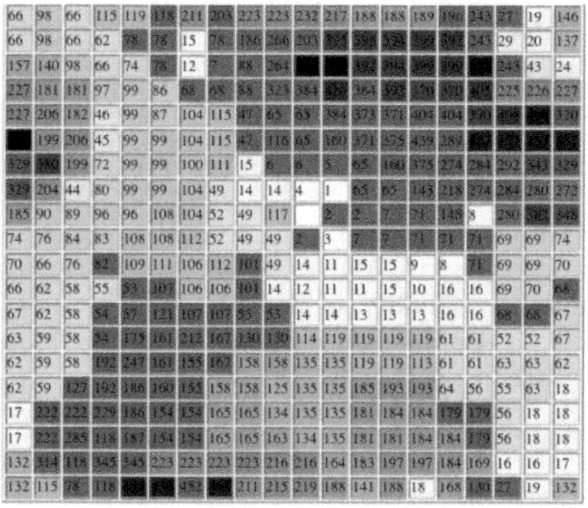

Fig. 4. Map of the group of packets including attacks

3.2 Unsupervised Learning Method

In the previous subsection, the results of the experiments of detecting abnormal packets using supervised learning is mentioned. But, for supervised learning, the pattern of the attacks should be known and the patterns which are different from learned one can not be detected. In the recalling process of Pareto learning SOM, the size of Pareto set becomes larger for the un-learned input vectors. In the authentication system using SP-SOM, the unregistered users can be detected using this feature. For the detection of abnormal packets, the size of the Pareto set becomes larger for the group of packets including abnormal packets after learning the map using the group of normal packets. Thus, the group of packets including attacks can be detected with setting the threshold value for Pareto set. The size of the Pareto set decreases and is adaptively adjusted during the learning process of P-SOM. Thus, the average size of the Pareto set in the last learning step(ps_{last}) is used as the reference and the threshold value is set as $1.5 \times ps_{last}$.

Table 1. Sensibility and specificity of attacks using supervised learning

	Attacks	Sensibility	Specificity
Case 1	IP-F, Port-F	86.5 %	92.0 %
Case 2	IP-F, Port-R	68.9 %	96.2 %
Case 3	IP-R, Port-F	79.9 %	96.3 %
Case 4	IP-R, Port-R	96.2 %	98.9 %

The setting of the experiments is almost same with that mentioned in previous section except that the map is learned using the 500 groups of normal packets. Under this settings, both of the sensibility and specificity becomes 1.0 for all cases of the attack patterns. The detection method based on the size of Pareto set shows much superior results compared with those in the previous subsection. To make severe the condition, the number of packets in an attack is decreased. Fig.5 shows the results for sensibility. Specificity is almost 1.0 for all cases. From these results, the packets concerning the attacks can be detected if more than 15

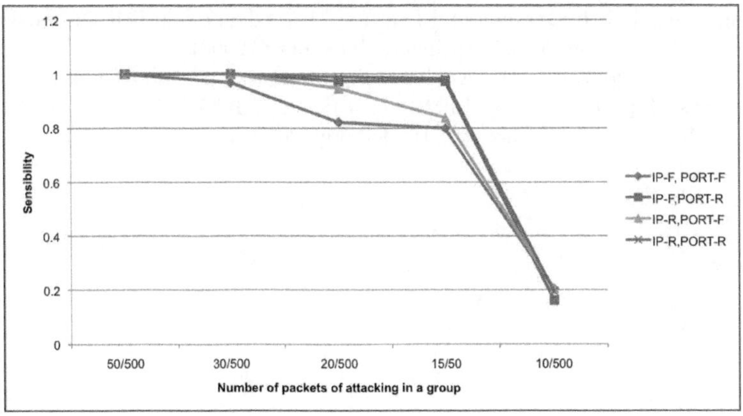

Fig. 5. Change of the sensibility with decreasing number of packets in an attack

packets concerning the attacks are contained in a group sized 500 packets, thus this detection method is considered to be very sensitive to abnormal traffics.

4 Conclusions

We propose a method for analysis of IP traffics based on Pareto learning SOM (P-SOM). The map which reflects the change of IP traffics can be organized by P-SOM and the simulated attacks can be detected using Supervised P-SOM. This method will be available for monitoring IP traffics and detecting the attacks from outside of local network. For detecting attacks, the unsupervised learning method based on the size of Pareto set, which does not need the prior knowledge of the attacks, shows much better results than those of supervised learning method. As the feature works, we must examine this system in larger scale network and in the environment of outside firewalls.

References

1. Ohkouchi, K., Rikitake, K., Nakao, K.: A Study on Network Incident Analysis Using Self Organizing Maps. In: Proceedings of the 2006 Symposium on Cryptography and Information Security (2006)
2. Kanenishi, K., Togawa, S., Matsuura, K., Mitsuhara, H., Yano, Y.: Aberrant Detection from Behavior of Campus Network Traffic. Journal of Academic Computing and Networking (13), 74–83 (2009)
3. Lagus, K., Kaski, S., Kohonen, T.: Mining Massive Document Collections by the WEB- SOM method. Information Sciences 163/1-3, 135–156 (2004)
4. Abe, T., Ikemura, T., et al.: A Novel Bioinfor- matics Strategy for Phylogenetic Study of Genomic sequence Fragments: Self Organizing Map (SOM) of Oligonucleotide Frequencies. In: Proceedings of 5th Workshop on Self Organizing Maps, pp. 669–676 (2005)
5. Dozono, H., Nakakuni, M.: Application of Supervised Pareto Learning Self Organizing Maps amd Its Incremental Learning. In: Advances in Self Organizing Maps. LNCS, vol. 5629, pp. 54–62. Springer, Heidelberg (2009)
6. Dozono, H., Nakakuni, M.: Analysis of Robustness of Pareto Learning SOM to Variances of Input Vectors. In: Chan, J.H. (ed.) ICONIP 2009, Part II. LNCS, vol. 5864, pp. 836–844. Springer, Heidelberg (2009)

Log Analysis of Exploitation in Cloud Computing Environment Using Automated Reasoning

Ruo Ando[1], Kang Byung[2], and Youki Kadobayashi[3]

[1] National Institute of Information and Communication Technology,
4-2-1 Nukui-Kitamachi, Koganei,
Tokyo 184-8795, Japan
[2] Korea Advanced Institute of Science and Technology
335 Gwahak-ro(373-1 Guseong-dong), Yuseong-gu, Daejeon 305-701,
Republic of Korea
[3] Guraduate School of Information Science,
Nara Institute of Science and Technology, Takayama-cho 8916-5,
Ikoma, Nara, 630-0192, Japan

Abstract. Recently server consolidation using virtualization leverages cloud computing. In cloud computing, we can apply centralized logging system using server consolidation. In this paper we propose a log analysis method in cloud computing environment using automated reasoning. On cloud computing providers, VM (virtual machine) monitoring is important to detect security incident. We discuss how to monitor VM, formatting and analyzing logs. Automated reasoning is more effective to retrieves information from large amount of log string. In proposed system, VM log is represented as clausal form and processed by FoL (First order Logic) theorem prover. We also present the numerical output of proposed system.

Keywords: Cloud computing, log analysis, automated reasoning, IaaS, Internet Explorer.

1 Introduction

With the rapid advance of high speed Internet, high performance CPU and virtualization technologies, cloud computing is possible. PaaS, IaaS and HaaS is possible to be provided with by reasonable computing resources and price. At the same time, our computing environment become sophisticated, diversified and complicated. To detect security incidents, conventional signature matching and stateful inspection is not enough. More fine-grained monitoring and sophisticated analysis is necessary to cope with complicated cloud computing management. In this paper we propose the application of automated reasoning processing log strings from VM monitoring.

K.W. Wong et al. (Eds.): ICONIP 2010, Part II, LNCS 6444, pp. 337–343, 2010.

2 Could Computing Environment

2.1 Cloud Computing

Cloud computing is a concept of Internet-based computing, which becomes feasible by the rapid improvement of high performance network and virtualization technologies. Cloud computing provides high usability thin client, more flexible provisioning and effective server consolidation. On the other hand, cloud computing yields complicated VM infrastructure where more sophisticated tracking and analyzing system to achieve availability and security.

2.2 PaaS, SaaS and IaaS

Cloud computing is divided into three styles: PaaS, SaaS and IaaS. PaaS is platform as a service to provide computing environment. SaaS is software as a service to deliver application without installing on client side. IaaS is infrastructure as a service to provide platform using virtualization technologies. In this paper we cope with IaaS to obtain VM logs and apply automated reasoning for log strings.

2.3 Virtual Machine Monitor

VMM (virtual machine monitor) is a thin layer of software between the physical hardware and the guest operating system. The rapid increase of CPU performance enables VMM to run several operating system as virtual machine, multiplexing CPU, memory and I/O devices in reasonable processing time. Recent VMM is a successful implementation of microkernels. Under the guest OS, VMM runs directly on the hardware of a machine which means that VMM can provides the useful inspection and interposition of guest OS.

3 Mechanized Reasoning

Mechanized reasoning is also called as automated reasoning in which fields researchers cope with the creation of software which makes computers "reason" in the sense of mathematical aspects such as solving puzzle and proving theorems. In this field, software such as FoL (first-order logic) or HoL (higher order logic) theorem prover, SAT solver and model generator is created to automated the mathematical process. In this paper we apply automated deduction system for automated log analysis of infected Windows OS using mechanized reasoning.

3.1 Resolution

Basically, proposed system is based on resolution. If clause Cls1 and Cls2 have literal L1 and L2, the clause C_R is resolved below.

$$\boxed{C_R = (C_1\sigma \setminus L_1\sigma) \cup (C_2\sigma \setminus \bar{L_2}\sigma)}$$

where σ is unifier which L1 and L2 equal. σ is sometimes could be most general unifier. The resolution in $Lit_1 \in Cls_1$ is also possible as $Lit_n \in Cls_n$. According to the above formation, hyper resolution several clauses.

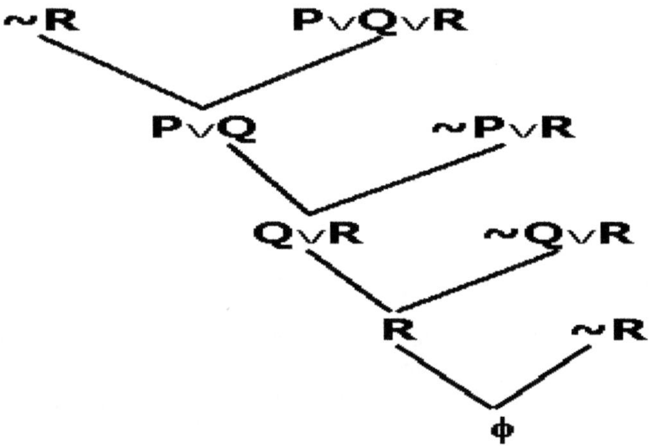

Fig. 1. Figure1 show the resolution process in set of support strategy, where S={P and Q and R, P and R, Q and R, R}.The restriction imposes the reasoning so that the program do not apply an inference rule to a set of clauses that are not the complement of set of support.

3.2 Set of Support Strategy

Set of support was introduced by L.Wos, S.Robinson and Carson in 1965[1]. If the clause T is retrieved from S, SOS is possible with the satisfiability of S-T. Set of support strategy enable the researcher to select one clause characterizing the searching to be placed in the initializing list called SOS. For the searching to be feasible and more effective, the resolution of more than one clauses not in SOS is inhibited in order to prevent the prover go into abundant searching place. Definition. H is satisfiable subset of S. T is set of support where

$$\boxed{S = \cup H \text{ and } H \cap T = \phi}$$

Figure1 show the resolution process in set of support strategy, where S=P and Q and R, P and R, Q and R, R. The restriction imposes the reasoning so that the program do not apply an inference rule to a set of clauses that are not the complement of set of support.

3.3 Hyperresolution

In generating encoder, we apply the inference rule called hyper resolution[2], which is a kind of resolution that can do resolutions at once compared with several steps in another rules. For hyperresolution, these must be the negative or mixed clause with the remaining clauses equal to the number of literals in the negative or mixed clause. The positive clause are described as satellites, the negative clause nucleus. "Hyper" means that in this resolution more process has occur than another resolution such as binary resolution.

3.4 Subsumption

Subsumption[4] is the process of discarding a specific statement. The clause that duplicated or is less general is discarded in the already-existing information. As a result, subsumption prevents a reasoning program from retaining clauses that is obviously redundant, especially is logically captured by more general clauses. Definition. The clause A subsumes the clause B when B is the instance that is logically captured by B.

```
The clause P(X) Subsumes the clause P(a).
```

There are several paths and axioms that could be applied. Subsumption strategy is effective when the same or more specific clause in the present of already-existing clause is generated. The clause is crossed and the generated clauses on the process of resolution is also eliminated. The effectiveness of this strategy is presented in experimental results.

4 Logging Techniques of Windows OS

Windows modification OS consists of three steps: [1]inserting DLL into user process, [2]inserting filter driver into kernel space. In this section we discuss library insertion (DLL injection) and filter driver injection.

4.1 DLL Injection

DLL injection is the insertion of original library on userland. We apply DLL injection for inspecting illegal resource access of malicious process. DLL injection is debugging technology to hook API call of target process. Windows executable applies some functions from DLL such as kernel32. dll. Executable has import table to use the linked DLL. This table is called as import section. Among some techniques of DLL injection, modifying import table is useful because this technique is CPU-architecture independent. Figure 5 show the modification of import table. Address of function A on left side is changed to the address of inserted function on right side. In code table, some original functions are appended to executable. Modified address is pointed to code of inserted function. By doing this, when the function A is invoked, the inserted function is executed.

4.2 Filter Driver

Filter driver is an intermediate driver which runs between kernel and device driver. By using filter driver, we can hook events on lower level compared with library insertion technique on userland. In detail, System call table is modified to insert additional routine for original native API. In proposed system, filter driver is implemented and inserted for hooking events on file system.

```
505 [] file(type(PreRead),pid(1040),processName(IEXPLORE_EX),
filename(C__DOCUMENTSANDSETTINGS_ADMINISTRATOR_DESKTOP_WAR_FTPD_SYSMSG9_TXT)).
646 [] registry(pid(1040),processName(IEXPLORE_EXE),operation(SetValue),
regName(HKEY_LOCAL_MACHINE_SOFTWARE_MICROSOFT_CRYPTOGRAPHY_RNG)).
669 [] registry(pid(1040),processName(IEXPLORE_EXE),operation(EnumValue),
regName(HKEY_LOCAL_MACHINE_SOFTWARE_MICROSOFT_WINDOWSNT_
CURRENTVERSION_LANGUAGEPACK)).
733 [] registry(pid(1040),processName(IEXPLORE_EXE),operation(QueryValue),
regName(HKEY_LOCAL_MACHINE_SYSTEM_CONTROLSET001_CONTROL_NLS_LANGUAGEGROUPS)).
734 [] registry(pid(1040),processName(IEXPLORE_EXE),operation(QueryKey),
regName(HKEY_LOCAL_MACHINE_SOFTWARE_MICROSOFT_WINDOWSNT_
CURRENTVERSION_FONTLINK_SYSTEMLINK)).
2675 [] -file(type(PreRead),pid(x1),processName(x2),filename(C__DOCUMENTSANDSETTINGS_
ADMINISTRATOR_DESKTOP_WAR_FTPD_SYSMSG9_TXT))|
-registry(pid(x1),processName(y2),operation(QueryValue),regName(y4))|access1(pid(x1),
processName(y2),operation(QueryValue),regName(y4)).
2676 [] -file(type(PreRead),pid(x1),processName(x2),filename(C__DOCUMENTSANDSETTINGS_
ADMINISTRATOR_DESKTOP_WAR_FTPD_SYSMSG9_TXT))|
-registry(pid(x1),processName(y2),operation(EnumValue),regName(y4))|access2(pid(x1),
processName(y2),operation(EnumValue),regName(y4)).
2677 [] -file(type(PreRead),pid(x1),processName(x2),filename(C__DOCUMENTSANDSETTINGS_
ADMINISTRATOR_DESKTOP_WAR_FTPD_SYSMSG9_TXT))|
-registry(pid(x1),processName(y2),operation(SetValue),regName(y4))|access3(pid(x1),processName(y2),
operation(SetValue),regName(y4)).
2678 [] -file(type(PreRead),pid(x1),processName(x2),filename(C__DOCUMENTSANDSETTINGS_
ADMINISTRATOR_DESKTOP_WAR_FTPD_SYSMSG9_TXT))|
-registry(pid(x1),processName(y2),operation(QueryKey),regName(y4))|access4(pid(x1),
processName(y2),operation(QueryKey),regName(y4)).
2679 [] -access1(pid(x1),processName(x2),operation(x3),regName(x4))| -access2(pid(x1),
processName(x2),operation(y3),regName(y4))|
-access3(pid(x1),processName(x2),operation(y5),regName(y6))| -access4(pid(x1),
processName(x2),operation(y7),regName(y8))|ok.
2681 [] -ok.
2691 [hyper,646,2677,505] access3(pid(1040),processName(IEXPLORE_EXE),
operation(SetValue),regName(HKEY_LOCAL_MACHINE_SOFTWARE_MICROSOFT_
CRYPTOGRAPHY_RNG)).
2693 [hyper,669,2676,505] access2(pid(1040),processName(IEXPLORE_EXE),
operation(EnumValue),regName(HKEY_LOCAL_MACHINE_SOFTWARE_MICROSOFT_
WINDOWSNT_CURRENTVERSION_LANGUAGEPACK)).
2704 [hyper,733,2675,505] access1(pid(1040),processName(IEXPLORE_EXE),
operation(QueryValue),regName(HKEY_LOCAL_MACHINE_SYSTEM_CONTROLSET001_
CONTROL_NLS_LANGUAGEGROUPS)).
2705 [hyper,734,2678,505] access4(pid(1040),processName(IEXPLORE_EXE),
operation(QueryKey),regName(HKEY_LOCAL_MACHINE_SOFTWARE_MICROSOFT_
WINDOWSNT_CURRENTVERSION_FONTLINK_SYSTEMLINK)).
2832 [hyper,2705,2679,2704,2693,2691] ok.
2834 [binary,2832.1,2681.1] $F.
```

Fig. 2. Sample log output of proposed system. Internet Explore which is compromised accessed illegal directory of FTP server. Also, related registry accesses are shown in line 646, 669, 733 and 734.

4.3 Centralized Logging Architecture

One of the thrusts of virtualizaton technologies including VMM is the ability of consolidation. Consolidation makes it possible for logging several virtualized servers at the same time, on the same physical machine. In proposed system, victim VM and attacker VM is running on the same physical machine with the consolidation technologies.

	hyper	binary	ur		hyper	binary	ur
clauses given	2654	2973	2642	unit deletions	0	5592	0
clauses generated	6293	6760	6281	factor simplifications	0	297	0
hyper_res generated	3640	0	2641	clauses kept	151	0	151
binary_res generated	0	3789	0	new demodulators	0	474	0
factors generated	0	0	0	empty clauses	2	2	2
ur_res generated	0	0	3640	clauses back demodulated	0	0	0
demod_inf generated	2653	2971	2641	clauses back subsumed	0	156	0
demod & eval rewrites	0	2971	0	usable size	2660	2978	2648
clauses wt,lit,sk delete	0	0	0	sos size	171	20	183
tautologies deleted	0	0	0	demodulators size	0	0	0
clauses forward subsumed	6142	0	6130	passive size	1	1	1
(subsumed by sos)	5730	6286	5730	hot size	0	0	0

Fig. 3. Comparison of three resolution methods. About the number of clauses generated, UR (unit reference) is best effective with 2642.

5 Experimtal Results

5.1 Internet Explorer Aurora Attack: MS979352

In this paper we cope with an exploitation of the vulnerability of Internet explorer which is called Aurora attack. Aurora attack, Microsoft Security Advisory (979352), is implemented for the vulnerability in Internet explorer which could allow remote code execution. Reproduction of aurora attack is done by Java script with attack vector on server side and Intenet explorer connecting port 8080, resulting in the shell code operation with port 4444.

5.2 Experimental Results

Figure 2 Sample log output of proposed system. Internet Explore which is compromised accessed illegal directory of FTP server. Also, related registry accesses are shown in line 646, 669, 733 and 734. Figure 3 shows Comparison of three resolution methods. About the number of clauses generated, UR (unit reference) is best effective with 2642.

6 Conclusion

With the rapid advance of monitoring and filtering technologies of IT system, we can cope with a variety of access logs to retrieve information of security incidents. Particularly, recently server consolidation using virtualization leverages cloud computing. In cloud computing, we can apply centralized logging system using server consolidation. In this paper we have proposed the log analysis method in cloud computing environment using automated reasoning. On cloud computing providers, VM (virtual machine) monitoring is important to detect security incident. We have discussed how to monitor VM, formatting and analyzing logs. Automated reasoning is more effective to retrieves information from large amount of log string. In proposed system, VM log has been represented as clausal form and processed by FoL (First order Logic) theorem prover. We also have presented the numerical output of proposed system.

References

1. Goth, G.: Virtualization: Old Technology Offers Huge New Potential. IEEE Distributed Systems Online 8(2) (2007)
2. Wos, L.: The Problem of Explaining the Disparate Performance of Hyperresolution and Paramodulation. J. Autom. Reasoning 4(2), 215–217 (1988)
3. Wos, L.: The Problem of Self-Analytically Choosing the Weights. J. Autom. Reasoning 4(4), 463–464 (1988)
4. Wos, L.: The Problem of Choosing the Type of Subsumption to Use. J. Autom. Reasoning 7(3), 435–438 (1991)
5. Wos, L., Robinson, G.A., Carson, D.F., Shalla, L.: The Concept of Demodulation in Theorem Proving. Journal of Automated Reasoning (1967)
6. Ando, R.: Automated Log Analysis of Infected Windows OS Using Mechanized Reasoning. In: Chan, J.H. (ed.) ICONIP 2009, Part II. LNCS, vol. 5864, pp. 540–547. Springer, Heidelberg (2009)
7. A Virtual Machine Introspection Based Architecture for Intrusion Detection Tal Garfinkel and Mendel Rosenblum in the Internet Society's 2003 Symposium on Network and Distributed System Security (NDSS), pp. 191–206 (February 2003)

A Multidirectional Associative Memory Based on Self-organizing Incremental Neural Network

Hui Yu[1,2], Furao Shen[1,2], and Osamu Hasegawa[3]

[1] National Key Laboratory for Novel Software Technology, Nanjing University, China
frshen@nju.edu.cn
http://cs.nju.edu.cn/rinc/
[2] Jiangyin Information Technology Research Institute, Nanjing University, China
[3] Imaging Science and Engineering Lab., Tokyo Institute of Technology, Japan
hasegawa.o.aa@m.titech.ac.jp

Abstract. A multidirectional associative memory (AM) is proposed. It is constructed with three layer networks: an input layer, a memory layer, and an associate layer. The proposed method is able to realize many-to-many associations with no predefined conditions, and the association can be incrementally added to the network without destruction of old associations. Experiments show that the proposed AM works well for real tasks.

Keywords: Incremental learning; Self-organizing incremental neural network; Many-to-many association.

1 Introduction

An associative memory (AM) is a memory that stores data in a distributed fashion and which is addressed through its contents. Traditional methods such as the Hopfield network [1], the bidirectional associative memory (BAM) [2] and their variants realize one-to-one association, i.e., according to one key vector, only one stored vector is recalled. However, with a stimuli, we human beings usually remember much things rather than one. We hope AM can simulate the memory of human beings by realizing many-to-many association.

Another challenge is incremental learning of associations. We human beings are capable of learning new knowledge without destruction of learned knowledge. Therefore, AM should incrementally memorize new key-response information without destroying stored key-response information.

Some neural models have been proposed for multidirectional AM or incremental learning. Recently published self-organizing incremental associative memory (SOIAM) [3] incrementally stores new key-response pairs without destruction of memorized information, however, to realize incremental learning, SOIAM spends plenty of storage and computation cost. M. Hagiwara proposed a multidirectional AM [4] for many-to-many association, but the association was not flexible. It was necessary to preset the number of layers with the predetermined number of associations.

K.W. Wong et al. (Eds.): ICONIP 2010, Part II, LNCS 6444, pp. 344–351, 2010.

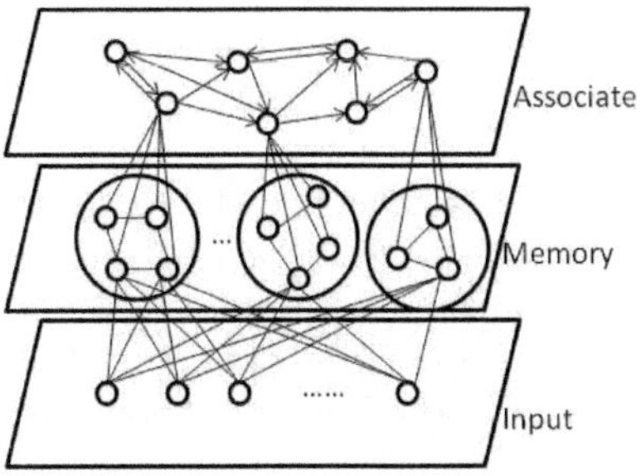

Fig. 1. Network structure of the proposed multidirectional AM

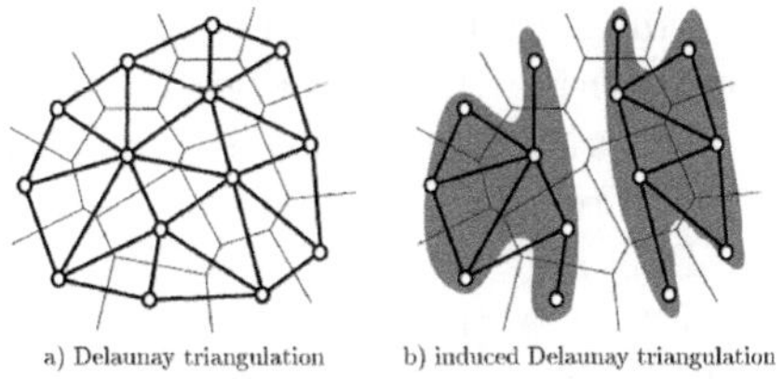

Fig. 2. a) *Delaunay triangulation* (thick lines) connects points having neighboring *Voronoi polygons* (thin lines). b) *Induced Delaunay triangulation* (thick lines): masking Delaunay triangulation with a data distribution (shaded)

In this paper, we propose a multidirectional AM to realize many-to-many association and incremental learning. A three-layer network is adopted for our targets (Fig. 1). The input layer inputs key vector, response vector, and association into the AM system. The memory layer stores information coming from the input layer. The associate layer builds the associative relation between the key vector and the response vector. In associate layer, we incrementally construct many-to-many associations.

2 Learning Algorithms

2.1 Memory Layer

Herein, we adopt a self-organizing incremental neural network (SOINN) [5] to build the memory layer. SOINN is based on competitive learning. Neural nodes are used to represent the data distribution of input data. The weights of such nodes are used to store the input patterns. The memory layer comprises some sub-networks, and each sub-network is used to represent one class. For each class we adopt one SOINN to represent the distribution of that class. Algorithm 1 shows the learning algorithm of memory layer.

Algorithm 1. Learning of the memory layer

1: Initialize the memory layer network: node set A, sub-network set S, and connection set C, $C \subset A \times A$ to the empty set: $A = \emptyset$, $S = \emptyset$, $C = \emptyset$.
2: Input a pattern $x \in R^n$ to the memory layer, the class name of x is c_x.
3: **if** There is no sub-network with name c_x **then**
4: Add a sub-network c_x to memory layer. This sub-network has a node c_x^1, the weight of c_x^1 is set as x.
5: Add node c_x^1 to the node set A, i.e., $S = S \cup \{c_x\}$, $A = A \cup \{c_x^1\}$.
6: **else**
7: Update the sub-network c_x with SOINN (Algorithm 2.1 in [5]).
8: **end if**

According to [6], to build connections among neural nodes, SOINN adopts the competitive Hebbian rule [7]: for each input signal, connect the two closest nodes with an edge. This rule forms a network whose edges are in the area suggested by input data distributions. The network represents a subgraph of the Delaunay triangulation (Fig. 2-a). Using the competitive Hebbian rule, the resultant graph approximates the shape of the input data distributions (Fig. 2-b). Hence, Algorithm 1 is capable of representing the input data distribution well. The nodes of sub-networks are the centers of Voronoi regions. Such nodes serve as the attractors for future recalling phase, the Voronoi regions form the basins of attraction.

In Algorithm 1, each class is allocated a sub-network. It shows that, for different classes, the dimension of vectors might be different. On other words, the memory layer is capable of memorizing data of different types (patterns with any different dimensions).

Algorithm 2. Learning of the associate layer

Initialize the associate layer network: node set B, arrow edge set $D \subset B \times B$ to the empty set: $B = \emptyset$, $D = \emptyset$
Input a key vector $x \in R^n$, the class name of x is c_x.
Use Algorithm 1 to memorize key vector x in the memory layer.
if No node b exists in the associate layer representing class c_x **then**
  Insert a new node b representing class c_x into the associate layer: $B = B \cup \{b\}$, $c_b = c_x$, $m_b = 0$, $W_b = x$.
else
  Increment the associative index of b: $m_b \leftarrow m_b + 1$;
  Find node i that is most frequently being winner in sub-network c_x.
  Update the weight of node b in associate layer: $W_b = W_{c_x^i}$.
end if
Input the response vector $y \in R^m$, the class name of y is c_y.
Use Algorithm 1 to memorize the response vector y in the memory layer.
if No node d representing class c_y in the associate layer **then**
  Insert a new node d representing class c_y into the associate layer: $B = B \cup \{d\}$, $c_d = c_y$, $m_d = 0$, $W_d = y$.
else
  Find node i which is most frequently being winner in sub-network c_y.
  Update the weight of node d in associate layer: $W_d = W_{c_y^i}$.
end if
if There is no arrow between node b and d **then**
  Connect node b and d with an arrow edge.
  Add arrow (b, d) to connection set D: $D = D \cup \{(b, d)\}$,
  Set the m_bth response class of b as c_d: $RC_b[m_b] = c_d$,
  Set the weight of arrow (b, d) as 1: $W_{(b,d)} = 1$.
else
  Set the m_bth response class of b as c_d: $RC_b[m_b] = c_d$,
  Increment the weight of arrow (b, d) with 1: $W_{(b,d)} \leftarrow W_{(b,d)} + 1$.
end if

2.2 Associate Layer

Associate layer is to build association between key vectors and response vectors. We designate the class a "key class", to which the key vector belongs, and call the class the "response class", to which the response vector belongs. In the associate layer, nodes are connected with arrow edges. Each node represents one class: the beginning of the arrow means the key class; the end of the arrow means the response class.

For training of the associate layer, firstly, Algorithm 1 is used to memorize information of both the key vector and the response vector. Then, the class name of the key class and response class are sent to the associate layer. In the associate layer, if there already exist nodes representing the key class and response class, we connect the nodes of the key class and response class with an arrow edge. If no node represents the key class (or response class) within the associate layer, we add a node to the associate layer and use that node to express the new class and then we build an arrow edge between the key class and response class.

Algorithm 2 gives details of learning associate layer. The building of the associate layer with Algorithm 2 discloses that it can realize many-to-many associations. The third layer of Fig. 1 presents an example of a many-to-many association network.

In the associate layer, weight vector of every node is selected from the corresponding sub-network of memory layer. Step 8, 9, 16, and 17 in Algorithm 2 show that the node that is most frequently being winner is chosen as the typical node of the sub-network in memory layer, and the weight vector of the typical node is set as the weight of that class node in associate layer.

Algorithm 2 is able to realize incremental learning. For example, we presume that Algorithm 2 has built the association of $x_1 \rightarrow y_1$. We want to build $x_2 \rightarrow y_2$ association incrementally. If c_{x_2} and c_{y_2} differ from class c_{x_1} and c_{y_1}, we need only build a new arrow edge from class c_{x_2} to class c_{y_2}. This new arrow edge has no influence to the arrow edge (c_{x_1}, c_{y_1}). If one of c_{x_2} and c_{y_2} is the same as c_{x_1} or c_{y_1}, for example, $c_{x_2} = c_{x_1}$, and $c_{y_2} \neq c_{y_1}$, then Algorithm 1 memorizes the pattern x_2 in sub-network c_{x_1} incrementally, and Algorithm 2 updates the weight and associative index of node c_{x_1} in the associate layer. Then Algorithm 2 finds or generates a node c_{y_2} in the associate layer and build an arrow edge from c_{x_1} to c_{y_2}, which differs from arrow edge (c_{x_1}, c_{y_1}). In this situation, the pair $x_2 \rightarrow y_2$ is learned incrementally. For the situation $c_{x_2} \neq c_{x_1}$, $c_{y_2} = c_{y_1}$, we can give a similar analysis.

3 Recall and Associate

3.1 Recall in Auto-associative Mode

Figure 3 shows the basic idea for auto-associative task. There exists some attractor, and every attractor has an attraction basin. If the key vector is located in an attraction basin, the corresponding attractor will be the associated result.

According to Fig. 2, the memory layer separates input patterns to different Voronoi regions, every Voronoi region acts as attraction basin for associative

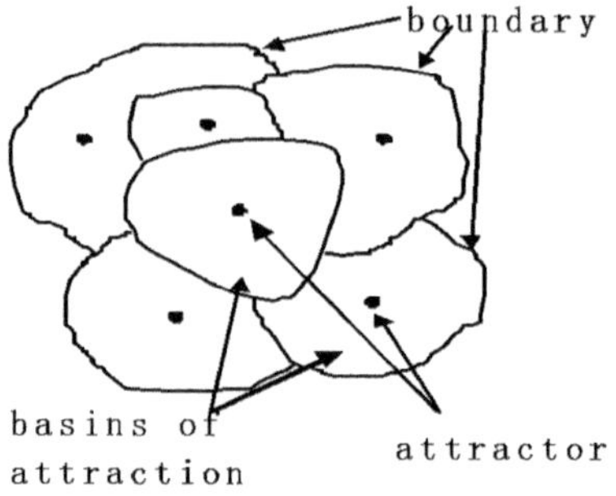

Fig. 3. Every attractor has an attraction basin. If key vector is located in an attraction basin, the corresponding attractor is the associated result.

Fig. 4. Humanoid robot HOAP-3. After hearing the bell sound, HOAP-3 turns its head to the bell and watches the bell; then it points to the bell with its finger.

process, and the node in the Voronoi region acts as attractor. For the associative process, if an input key vector lies in one Voronoi region V_i, we give the weight vector W_i of the corresponding node i as the associative result. Algorithm 3 gives the detail for auto-associative recalling process.

Algorithm 3. Auto-associative: recall the stored pattern with a key vector

1: Assume there are n nodes in the memory layer, input a key vector x.
2: **for** $i = 1, 2, ..., n$ **do**
3: Calculate the weight sum of input vector, and $\frac{1}{2}||W_i||^2$ is a bias.

$$g_i(x) = W_i^T x - \frac{1}{2}||W_i||^2 \tag{1}$$

4: **end for**
5: Find the maximum $g_k(x) = \max_{i=1,2,\ldots,n} g_i(x)$
6: Output W_k as the recalling pattern.
7: Output the class of node k as the class of x.

By step 2-5, Algorithm 3 judges to which Voronoi region the input x most likely belongs. It is because

$$||x - W_i||^2 = ||x||^2 - 2W_i^T x + ||W_i||^2 \tag{2}$$

$||x||$ is the common item for all nodes, thus minimize $||x - W_i||^2$ is equivalent to maximize $W_i^T x - \frac{1}{2}||W_i||^2$, which is calculated in step 3 of Algorithm 3.

3.2 Associate in Hetero-associative Mode

With Algorithm 2, the proposed AM memorizes the $x \to y$ pair. To associate y from x, firstly we use Algorithm 3 to recall the stored key class c_x of key vector x, the corresponding node for class c_x in the associate layer is b_x; then, we use

Algorithm 4. Hetero-associative: associate stored patterns with a key vector

Input a key vector x.
Using Algorithm 3 to classify x to class c_x.
In associate layer, find node b_x corresponding to sub-network c_x.
for $k = 1, 2, ..., m_{b_x}$ **do**
  Find the response classes $c_y[k]$: $c_y[k] = RC_{b_x}[k]$.
  Sort $c_y[k]$ with the order of $W_{(c_x, c_y[k])}$.
end for
for $k = 1, 2, ..., m_{b_x}$ **do**
  Find node $b_y[k]$ in the associate layer corresponding to sub-network $c_y[k]$.
  Output weight $W_{b_y[k]}$ as the associated result of key vector x.
end for

$RC_{b_x}[k], k = 1, ..., m_{b_x}$ to obtain the response class c_y and corresponding node b_y. Finally, we output all W_{b_y} as the hetero-associative results for key vector x. Algorithm 4 shows details of associating y from key vector x.

4 Experiment

4.1 Binary (Bipolar) Data

Here we use a binary text character dataset taken from the IBM PC CGA character font. This dataset is adopted by some methods such as SOIAM [3], BAM with PRLAB [8], Kohonen feature map associative memory (KFMAM) [9], and KFMAM-FW [10] to test their performance. There are 26 capital letters and 26 small letters. Each letter is a 7×7 pixel image, and every pixel has only -1 (black) or 1 (white) value. During memorization, capital letters are used as the key vectors, and small letters are used as the response vectors, i.e., $A \rightarrow a$, $B \rightarrow b$, ..., $Z \rightarrow z$.

Firstly, we consider incremental learning. The patterns of $A \rightarrow a$, $B \rightarrow b$, ..., $Z \rightarrow z$ are input into the system sequentially. At the first stage, only $A \rightarrow a$ are memorized, then $B \rightarrow b$ are input into the system and memorized, and so on. This environment is non-stationary, new patterns and new classes are incrementally input to the system. Table 1 shows comparison results between the proposed AM and other methods. For the proposed AM, 94 nodes in all are needed for memorization. The correct recall rate is 100%. It is difficult for BAM and KFMAM to realize incremental learning. Later input patterns will destroy the memorized patterns. For SOIAM, it needs 99 nodes to represent the association pairs; it recalls the associated patterns with a 100% correct recall rate. For KFMAM-FW, if we adopt sufficient nodes (more than 36), it can achieve perfect recalling results. We must mention that if the maximum number of patterns to be learned is not revealed in advance, we do not know how to give the total number of nodes for KFMAM-FW [3].

Then, we consider the many-to-many association. The BAM based and KF-MAM based methods are unsuitable for this task. In fact, SOIAM can realize many-to-many association. However, for SOIAM, if it incrementally learns a new

Table 1. Comparison: recalling results of the proposed AM and other methods under an incremental environment

Method	Number of nodes	Recall rate
Proposed AM	94	100%
SOIAM	99	100%
BAM with PRLAB	-	3.8%
KFMAM	64	31%
	81	38%
	100	42%
KFMAM-FW	16	infinite loop
	25	infinite loop
	36	100%
	64	100%

association pair, SOIAM will put together the key vector and response vector of new pair to one combination vector and send it to SOIAM for clustering. In [3], pairs such as (A, a), (A, b), (C, c), (C, d) (C, e), (F, f), (F, g), (F, h), and (F, i) are used to test the one-to-many association. To realize this target, SOIAM puts together A and b to produce vector $A+b$, C and d to produce vector $C+d$, and so on, then clusters such combination vectors with new nodes: new nodes different from $A+a$, $C+c$, and $F+f$ are added into the system to represent the associative relation between $A \to b$, $C \to d$, etc. With the proposed AM, we need only add new associative relation (arrow edge) between nodes in the associate layer without adding new nodes in both the memory layer and associate layer. For example, to realize $A \to b$ association, we need only add an arrow edge from node A to node b in the associate layer: no new nodes are generated. Both the proposed AM and SOIAM can recall old associated patterns and new added response vectors well (100% correct recall rate). However, SOIAM spends new storage and computation time to cluster new association pairs and adds 81 new nodes. The proposed AM requires no new storage and nearly no additional computation for building new associations, and it saves much more storage and computation time than SOIAM.

4.2 Real Task for Robot with GAM

This experiment uses a humanoid robot with an image sensor and sound sensor to test whether the proposed AM is applicable to real tasks. A humanoid robot HOAP-3 (as depicted in Fig. 4) (Fujitsu Ltd.) is adopted for this experiment.

We show a bell to HOAP-3 and then push the button of the bell to present the bell sound to HOAP-3. Sounds of the bell are served as key vectors and images of the bell are served as response vectors. The associative action is set as the instruction for HOAP-3 to point to the bell with its finger. The association pairs are presented to the proposed AM, which is built in the brain of HOAP-3, to build association between key-response pair vectors. For features of images collected by the image sensor of HOAP-3, we do grayscale transformation and adopt 36-dimension low-frequency DCT coefficients as the feature vector. For features of the sounds collected by sound sensor of HOAP-3, we extract the 15-dimensional spectrum feature on the 20 kHz – 50 ms sampling rate.

After the AM is trained, we show HOAP-3 the sound of the bell, and the sound of the bell serves as the key vector. HOAP-3 recalls the image of the bell, turns its head to the bell and watches the bell, then it points to the bell with its finger, as depicted in Fig. 4.

This experiment demonstrates that the proposed AM is able to realize real tasks with good performance. It is also noteworthy that, in this experiment, the dimensions of image and sound are different and the proposed AM builds an association between vectors with different dimensions quite well.

Acknowledgement

This work was supported in part by 973 Program (2010CB327903). It is also supported in part by China NSF grant (#60975047, #60723003, #60721002), Jiangsu NSF grant (#BK2009080).

References

1. Hopfield, J.J.: Neural networks and physical systems with emergent collective computational abilities. Proc. Nat. Acad. Sci. USA 79, 2554–2588 (1982)
2. Kosko, B.: Bidirectional associative memories. IEEE Trans. Systems, Man and Cybernetics 18(1), 49–60 (1988)
3. Sudo, A., Sato, A., Hasegawa, O.: Associative memory for online learning in noisy environments using self-organizing incremental neural network. IEEE Trans. on Neural Networks 20(6), 964–972 (2009)
4. Hagiwara, M.: Multidirectional associative memory. In: Proc. of the 1990 International Joint Conference on Neural Networks, pp. 3–6 (1990)
5. Shen, F., Hasegawa, O.: An incremental network for on-line unsupervised classification and topology learning. Neural Networks 19, 90–106 (2006)
6. Shen, F., Hasegawa, O.: An enhanced self-organizing incremental neural network for online unsupervised learning. Neural Networks 20, 893–903 (2007)
7. Martinetz, T.M., Berkovich, S.G., Schulten, K.J.: "neural-gas" network for vector quantization and its application to time-series prediction. IEEE Trans. On Neural Networks 4(4), 558–569 (1996)
8. Oh, H., Kothari, S.C.: Adaptation of the relaxation method for learning in bidirectional associative memory. IEEE Trans. On Neural Networks 5(4), 576–583 (1994)
9. Kohonon, T.: Self-organization and associative memory. Springer, Berlin (1984)
10. Yamada, T., Hattori, M., Morisawa, M., Ito, H.: Sequential learning for associative memory using kohonen feature map. In: Proc. of the 1999 International Joint Conference on Neural Networks, pp. 1920–1923 (1999)

Range Image Registration Using Particle Filter and Competitive Associative Nets

Shuichi Kurogi, Tomokazu Nagi, and Takeshi Nishida

Kyusyu Institute of technology, Tobata, Kitakyushu, Fukuoka 804-8550, Japan
{kuro@,nagi@kurolab.,nishida@}cntl.kyutech.ac.jp
http://kurolab.cntl.kyutech.ac.jp/

Abstract. This paper presents a method using a particle filter (PF) and competitive associative nets (CAN2s) for range image registration to fuse 3D surfaces on range images taken from around an object by the laser range finder (LRF). The method uses the CAN2 for learning to provide a piecewise linear approximation of the LRF data involving various noise, and obtaining a coarse but fast pair-wise registration. The PF is used for reducing the cumulative error of the consecutive pair-wise registration. The effectiveness is shown by using the real LRF data of a rectangular box.

Keywords: Range Image Registration, Particle Filter, Competitive Associative Nets, Reduction of Cumulative Error by Loop Closing.

1 Introduction

This paper describes a method for range image registration to fuse three-dimensional (3D) surfaces on range images taken from around an object by the laser range finder (LRF). As shown in a survey of range image registration [1], the most common registration methods employ pair-wise registration, such as ICP (iterative closest point), and they have the problem of propagation or cumulative error. To solve this problem, we introduce the particle filter (PF), known as a sequential Monte Carlo method to estimate hidden states from the observations [2]. Since the PF estimates the states sequentially, the cumulative error also occurs, but we introduce a loop closing technique for reducing the error. Here, the loop closing means that the state comes back to a previous state after experiencing other states and it is a problem to be solved in SLAM (simultaneous localization and mapping) applications (see e.g. [3]).

On the other hand, the range images obtained by the LRF are characterized as involving lack of data called black spots, quantization errors owing to the range (distance) resolution (e.g. 10mm), and a large number of data owing to high angular resolution (e.g. 0.25°). To deal with such data, we have developed a pair-wise registration method using the CAN2 (competitive associative net) [4], where the CAN2 learns to extract piecewise planner surfaces. This paper shows an improved method utilizing a distance measure of range points and the piecewise planes extracted by the CAN2. Note that the method using a distance measure of points and planes has shown a very good performance in the comparative experiments shown in [1], and the present measure is a simpler version.

K.W. Wong et al. (Eds.): ICONIP 2010, Part II, LNCS 6444, pp. 352–359, 2010.

In the next section, we show the registration method using a PF and CAN2s, and then the effectiveness of the method is evaluated in Sect. 3.

2 Range Image Registration Using PF and CAN2

We use the SICK LMS200 as a LRF for scanning the horizontal 2D plane to measure the distance to an object and a suspension unit for rotating the LRF vertically by means of a geared stepping motor (see Fig. 1(a)). The yaw and pitch angle resolutions are 0.25° and 0.05°, respectively, and the range (distance) resolution is 10mm. Let $Z_{[\mathrm{polar}]t} = \{\boldsymbol{p}^{(i)}_{[\mathrm{polar}]t} = (\theta^{(i)}_t, \phi^{(i)}_t, r^{(i)}_t) \mid i = 1, 2, \cdots\}$ denote the tth range image taken from around an object, where $\theta^{(i)}_t$, $\phi^{(i)}_t$ and $r^{(i)}_t$, respectively, are the yaw and pitch angles and the range of the ith scan data (see Fig. 1(b)), and $t = 0, 1, 2, \cdots, T-1$. From $Z_{[\mathrm{polar}]t}$, we have the Cartesian data $\boldsymbol{p}^{(i)}_{[\mathrm{s}]t}$ as $\boldsymbol{p}^{(i)}_{[\mathrm{s}]t} = (x^{(i)}_{[\mathrm{s}]t}, y^{(i)}_{[\mathrm{s}]t}, z^{(i)}_{[\mathrm{s}]t})^T = r^{(i)}_t(\sin\theta^{(i)}_t, \cos\theta^{(i)}_t \sin\phi^{(i)}_t, \cos\theta^{(i)}_t \cos\phi^{(i)}_t)^T$, where the subscript [s] indicates the scan center coordinate because the z-axis of $\boldsymbol{p}^{(i)}_{[\mathrm{s}]t}$ directs to the center of the LRF scan. We denote this dataset as $Z_{[\mathrm{s}]t} = \{\boldsymbol{p}^{(i)}_{[\mathrm{s}]t} = (x^{(i)}_{[\mathrm{s}]t}, y^{(i)}_{[\mathrm{s}]t}, z^{(i)}_{[\mathrm{s}]t})^T \mid i = 1, 2, \cdots\}$.

The registration from the tth image to the $(t-1)$th image is executed by the transform $\boldsymbol{p}^{(i)}_{[\mathrm{s}]t-1,t} = \boldsymbol{R}_{[\mathrm{s}]t-1,t}\boldsymbol{p}^{(i)}_{[\mathrm{s}]t} + \boldsymbol{t}_{[\mathrm{s}]t-1,t}$, where the parameter $u_t = (\boldsymbol{R}_{[\mathrm{s}]t-1,t}, \boldsymbol{t}_{[\mathrm{s}]t-1,t})$ consists of the rotation matrix $\boldsymbol{R}_{[\mathrm{s}]t-1,t}$ and the translation vector $\boldsymbol{t}_{[\mathrm{s}]t-1,t}$. By means of applying this relation recursively, we can transform $\boldsymbol{p}^{(i)}_{[\mathrm{s}]t}$ to the 0th LRF coordinate as $\boldsymbol{p}^{(i)}_{[\mathrm{s}]0,t} = \boldsymbol{R}_{[\mathrm{s}]0,t}\boldsymbol{p}^{(i)}_{[\mathrm{s}]t} + \boldsymbol{t}_{[\mathrm{s}]0,t}$, where

$$x_t = (\boldsymbol{R}_{[\mathrm{s}]0,t}, \boldsymbol{t}_{[\mathrm{s}]0,t}) = (\boldsymbol{R}_{[\mathrm{s}]0,t-1}\boldsymbol{R}_{[\mathrm{s}]t-1,t}, \boldsymbol{t}_{[\mathrm{s}]0,t-1} + \boldsymbol{R}_{[\mathrm{s}]0,t-1}\boldsymbol{t}_{[\mathrm{s}]t-1,t}). \quad (1)$$

Note that x_t indicates the pose, or the orientation $\boldsymbol{R}_{[\mathrm{s}]0,t}$ and the position $\boldsymbol{t}_{[\mathrm{s}]0,t}$, of the tth LRF w.r.t. the 0th LRF coordinate, and u_t indicates the movement of the LRF.

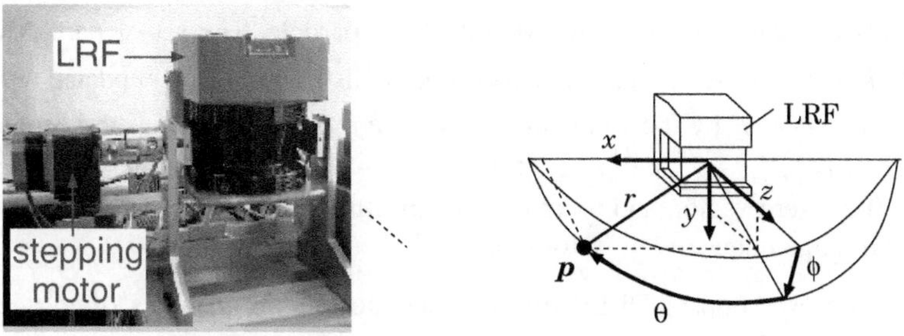

(a) (b)

Fig. 1. (a) The LRF with the stepping motor and (b) the LRF coordinate system

2.1 Modeling of Range Image Registration for the PF

Mathematical formulation: Let x_t be the current state or the pose of the LRF, u_t the control motion or the movement of the LRF, and $z_t = Z_{[s]t}$ the measurement or the scan dataset. In order to use PF, we employ two assumptions:

(A1) the probability of the current state x_t only depends on the previous state x_{t-1} and the control u_t; or x_t is a Markov process, and
(A2) the probability of the measurement $z_t = Z_{[s]t}$ only depends on x_t.

Then, the following recursive target distribution for PF becomes reasonable:

$$\begin{aligned}
p(x_{0:t}|u_{0:t}, z_{0:t}) &= \underbrace{p(x_t|x_{0:t-1}, u_{0:t}, z_{0:t})p(x_{0:t-1}|u_{0:t}, z_{0:t})}_{\text{conditional probability for } p(x_{0:t-1}, x_t)} \\
&= \underbrace{p(x_t|x_{t-1}, u_t, z_t)}_{\text{(A1) and (A2)}}\ \underbrace{\eta\, p(z_t|x_{0:t-1}, u_{0:t}, z_{0:t-1})p(x_{0:t-1}|u_{0:t}, z_{0:t-1})}_{\text{Bayes rule for } p(x_{0:t-1}|z_t)} \\
&= \eta\, p(x_t|x_{t-1}, u_t, z_t)\ p(z_t|x_{0:t-1}, u_{0:t}, z_{0:t-1})\underbrace{p(x_{0:t-1}|u_{0:t-1}, z_{0:t-1})}_{x_{0:t-1} \text{ is independent to } u_t}, \qquad (2)
\end{aligned}$$

where η is the normalization coefficient and the initial distribution $p(x_0)$ for $t = 0$ is supposed to be given. We make the recursive proposal distribution of particles, $q(x_t)$, and the weight, w_t, as

$$q(x_t) := p(x_t|x_{t-1}, u_t, z_t)\ q(x_{t-1}) \qquad (3)$$

$$w_t := p(z_t|x_{0:t-1}, u_{0:t}, z_{0:t-1})\ w_{t-1} \qquad (4)$$

for $t = 1, 2, \cdots$ from the initial $q(x_0) = p(x_0)$ and $w_0 = 1$. Then, $w_t q(x_t)$ is proportional to the target distribution given by the right-hand side of Eq.(2). Here, note that the proposal distribution $q(x_t)$ involving the factor $p(x_t|x_{t-1}, u_t, z_t)$ is not so usual but formulated in several applications such as FastSLAM2.0 [2] for obtaining more accurate state than using $p(x_t|x_{t-1}, u_t)$.

Algorithm: Let $X_t^{[k]} = (x_t^{[k]}, w_t^{[k]})$ be the kth particle, where $x_t^{[k]} = (\boldsymbol{R}_{[s]0,t}^{[k]}, \boldsymbol{t}_{[s]0,t}^{[k]})$ represents the state, and $w_t^{[k]}$ the weight of the particle. The set of particles, $X_t = \left\{X_t^{[k]} \mid k = 1, \cdots, K\right\}$, represents the proposal distribution $q(x_t)$ updated recursively for $t = 1, 2, \cdots, T$ by the following steps, where the initial values are set as $X_0 := \{X_0^{[k]} = (\boldsymbol{R}_{[s]0,t}^{[k]}, \boldsymbol{t}_{[s]0,t}^{[k]}, w_1^{[k]}) = (\boldsymbol{I}, \boldsymbol{0}, 1) \mid k = 1, 2, \cdots, K\}$, where $\boldsymbol{I}$ is the unit matrix, $\boldsymbol{0}$ the zero vector, and $:=$ indicates substitution. Furthermore, we use the initial or the 0th range image at $t = T$ for loop closing.

1. (Update of Proposal Distribution) Generate random samples of $u_{[s]t-1,t}^{[k]}$ from the pair-wise registration parameter, $(\widehat{\boldsymbol{R}}_{[s]t-1,t}, \widehat{\boldsymbol{t}}_{[s]t-1,t})$, obtained via the method shown below (see Sect. 2.2),

$$u_{[s]t-1,t}^{[k]} = (\boldsymbol{R}_{[s]t-1,t}^{[k]}, \boldsymbol{t}_{[s]t-1,t}^{[k]}) := (\widehat{\boldsymbol{R}}_{[s]t-1,t}\delta\boldsymbol{R}^{[k]},\ \widehat{\boldsymbol{\xi}}_{[s]t-1} - \boldsymbol{R}_{[s]t-1,t}^{[k]}\widehat{\boldsymbol{\xi}}_{[s]t} + \delta\boldsymbol{t}^{[k]}) \qquad (5)$$

Here, $\widehat{\boldsymbol{\xi}}_{[\mathrm{s}]t}$ indicates the center of rotation, and $\delta\boldsymbol{R}^{[k]}$ is a random rotation matrix made via Rodrigues' formula from the rotation axis $\delta\boldsymbol{r}^{[k]} = (\delta r_x^{[k]}, 1, \delta r_z^{[k]})^T$ consisting of $\delta r_x^{[k]}$ and $\delta r_z^{[k]}$ sampled randomly from the normal distribution $N(0, \sigma_r^2)$, and the rotation angle $\delta\varphi^{[k]} \sim N(0, \sigma_\varphi^2)$. The translation vector $\delta\boldsymbol{t}^{[k]}$ is also with the elements sampled from $N(0, \sigma_t^2)$. From X_{t-1} and Eq.(1), we have X_t involving

$$x_{[\mathrm{s}]0,t}^{[k]} = (\boldsymbol{R}_{[\mathrm{s}]0,t}^{[k]}, \boldsymbol{t}_{[\mathrm{s}]0,t}^{[k]}) := (\boldsymbol{R}_{[\mathrm{s}]0,t-1}^{[k]} \boldsymbol{R}_{[\mathrm{s}]t-1,t}^{[k]}, \ \boldsymbol{R}_{[\mathrm{s}]0,t-1}^{[k]} \boldsymbol{t}_{[\mathrm{s}]t-1,t}^{[k]} + \boldsymbol{t}_{[\mathrm{s}]0,t-1}^{[k]}). \tag{6}$$

2. (Weight Update) Let $Z_{[\mathrm{s}]t}^{\mathrm{ROI}}$ be the dataset in the ROI (see Sect. 2.2) generated from $z_t = Z_{[\mathrm{s}]t}$, and $Z_{[\mathrm{s}]t-1,t}^{\mathrm{ROI},[k]}$ be the dataset transformed from $Z_{[\mathrm{s}]t}^{\mathrm{ROI}}$ by $u_{[\mathrm{s}]t-1,t}^{[k]}$. Then, a square distance $(\Delta Z_{t-1,t}^{[k]})^2$ between two images, $Z_{[\mathrm{s}]t-1,t}^{\mathrm{ROI},[k]}$ and $Z_{[\mathrm{s}]t-1}^{\mathrm{ROI}}$, is obtained by Eq.(14) shown below. Assuming $p(\Delta Z_{t-1,t}^{[k]})$ is Gaussian $N(0, \sigma_z^2)$ and the likelihood $p(z_t \mid x_{0:t-1}^{[k]}, u_{0:t}^{[k]}, z_{0:t-1}^{[k]})$ is proportional to $p(\Delta Z_{t-1,t}^{[k]})$, then we have

$$w_t^{[k]} := w_{t-1}^{[k]} \exp\left(-\frac{(\Delta Z_{t-1,t}^{[k]})^2}{2\sigma_z^2}\right). \tag{7}$$

3. (Loop Closing) Since the initial image is used at $t = T$, it is desirable that the joint distribution $p(x_{0:T}|u_{0:T}, z_{0:T}) \propto p(x_T|x_{0:T-1}, u_{0:T}, z_{0:T})$ has the factor with the mean at $x_T^{[k]} = (\boldsymbol{R}_{[\mathrm{s}]0,T}^{[k]}, \boldsymbol{t}_{[\mathrm{s}]0,T}^{[k]}) = x_0^{[k]} = (\boldsymbol{I}, \boldsymbol{0})$. Thus, we modify the weight as

$$\widetilde{w}_t^{[k]} := w_t^{[k]} \exp\left(-\frac{\left\|\boldsymbol{R}_{[\mathrm{s}]0,T}^{[k]} - \boldsymbol{I}\right\|^2}{2\sigma_R^2}\right) \exp\left(-\frac{\left\|\boldsymbol{t}_{[\mathrm{s}]0,T}^{[k]}\right\|^2}{2\sigma_t^2}\right). \tag{8}$$

Here, note that this equation indicates a loop closing, where the loop closing in the SLAM applications is not so easy as above, while it becomes easier in this application owing that we can set the loop closing poses of the LRF offline.

After executing the above three steps, we transform the scan data $\boldsymbol{p}_{[\mathrm{s}]t}^{(j)} \in Z_{[\mathrm{s}]t}$ weighted by $\widetilde{w}_t^{[k]}$ at each $t = 1, 2, \cdots, T$ to the initial LRF coordinate by using the particles $X_t^{[k]}$. The obtained weighted data represent the reconstructed shape of the object.

2.2 Pair-Wise Registration by the CAN2

In order to obtain the pair-wise registration parameter $(\widehat{\boldsymbol{R}}_{[\mathrm{s}]t-1,t}, \widehat{\boldsymbol{t}}_{[\mathrm{s}]t-1,t})$ and the square distance $(\Delta Z_{t-1,t}^{[k]})^2$ of two images for the above steps, we utilize the piecewise planes extracted by the CAN2 as follows.

Plane extraction by the CAN2: Since the relation $y = f(\boldsymbol{x})$ for $y = z_{[\mathrm{s}]t}^{(i)}$ and $\boldsymbol{x} = (x_{[\mathrm{s}]t}^{(i)}, y_{[\mathrm{s}]t}^{(i)})^T$ cannot be a single valued function for the points on the floor plane parallel to the z-axis, we rotate $\boldsymbol{p}_{[\mathrm{s}]t}^{(i)}$ so that the new z-axis directs to the object center as

$$\boldsymbol{p}_{[\mathrm{o}]t}^{(i)} = \boldsymbol{R}_P(-\phi_t^{(0)})\boldsymbol{R}_Y(-\theta_t^{(0)})\boldsymbol{p}_{[\mathrm{s}]t}^{(i)} = \boldsymbol{R}_{[\mathrm{o,s}]t}\boldsymbol{p}_{[\mathrm{s}]t}^{(i)} \tag{9}$$

where $\phi_t^{(0)}$ and $\theta_t^{(0)}$ are the yaw and pitch angles of the center of the object,

$$\boldsymbol{R}_Y(\theta) = \begin{bmatrix} \cos\theta & 0 & -\sin\phi \\ 0 & 1 & 0 \\ \sin\theta & 0 & \cos\theta \end{bmatrix} \quad \text{and} \quad \boldsymbol{R}_P(\phi) = \begin{bmatrix} 1 & 0 & 0 \\ 0 & \cos\phi & -\sin\phi \\ 0 & \sin\phi & \cos\phi \end{bmatrix} \tag{10}$$

are the yaw and pitch rotation matrices, and $\boldsymbol{R}_{[\mathrm{o,s}]t}$ is the rotation matrix from the scan center to the object center coordinate, where the subscript [o] indicates the object center coordinate. Note that the reverse transform is given by $\boldsymbol{p}_{[\mathrm{s}]t}^{(i)} = \boldsymbol{R}_{[\mathrm{s,o}]t}\boldsymbol{p}_{[\mathrm{o}]t}^{(i)} = \boldsymbol{R}_{[\mathrm{o,s}]t}^T\boldsymbol{p}_{[\mathrm{o}]t}^{(i)}$, so that we can easily transform each other.

Let us prepare the CAN2 with N units for each tth image. The jth unit has a weight vector $\boldsymbol{w}_{[\mathrm{o}]t}^{(j)} = (w_{[\mathrm{o}]t,1}^{(j)}, w_{[\mathrm{o}]t,2}^{(j)})^T$ and an associative matrix (row vector) $\boldsymbol{M}_{[\mathrm{o}]t}^{(j)} = (M_{[\mathrm{o}]t,0}^{(j)}, M_{[\mathrm{o}]t,1}^{(j)}, M_{[\mathrm{o}]t,2}^{(j)})$. After learning $Z_{[\mathrm{o}]t} = \{\boldsymbol{p}_{[\mathrm{o}]t}^{(i)} = (x_{[\mathrm{o}]t}^{(i)}, y_{[\mathrm{o}]t}^{(i)}, z_{[\mathrm{o}]t}^{(i)})^T | i = 1, 2, \cdots\}$ as a function $y = f(\boldsymbol{x})$ for $\boldsymbol{x} = (x_{[\mathrm{o}]t}^{(i)}, y_{[\mathrm{o}]t}^{(i)})^T$ and $y = z_{[\mathrm{o}]t}^{(i)}$, the CAN2 divide the input space into Voronoi regions $V_{[\mathrm{o}]t}^{(j)} = \{\boldsymbol{x} | j = \underset{i}{\operatorname{argmin}} \{\|\boldsymbol{x} - \boldsymbol{w}_{[\mathrm{o}]t}^{(i)}\|\}$ for $j = 1, 2, \cdots, N$, and performs linear approximation $y = \boldsymbol{M}_{[\mathrm{o}]t}^{(j)}\boldsymbol{x}$ in each region. As a result, the range image is divided into piecewise planes given by $(\boldsymbol{n}_{[\mathrm{o}]t}^{(j)})^T\boldsymbol{p}_{[\mathrm{o}]t} = \alpha_{[\mathrm{o}]t}^{(j)}$, where the normal vector $\boldsymbol{n}_{[\mathrm{o}]t}^{(j)} = (n_{[\mathrm{o}]t,x}^{(j)}, n_{[\mathrm{o}]t,y}^{(j)}, n_{[\mathrm{o}]t,z}^{(j)})^T$ and the distance to the origin, $\alpha_{[\mathrm{o}]t}^{(j)}$, are given by

$$((\boldsymbol{n}_{[\mathrm{o}]t}^{(j)})^T, \alpha_{[\mathrm{o}]t}^{(j)}) = \frac{(-M_{[\mathrm{o}]t,1}^{(j)}, -M_{[\mathrm{o}]t,2}^{(j)}, 1, M_{[\mathrm{o}]t,0}^{(j)})}{\sqrt{(M_{[\mathrm{o}]t,1}^{(j)})^2 + (M_{[\mathrm{o}]t,2}^{(j)})^2 + 1}}. \tag{11}$$

Here, note that $n_{[\mathrm{o}]t,z}^{(j)} > 0$ or the normal vector directs forward from the origin or the tth LRF. We use $Z_{[\mathrm{o}]t}^{\mathrm{CAN2}} = \{\boldsymbol{q}_{[\mathrm{o}]t}^{(j)} = (\boldsymbol{w}_{[\mathrm{o}]t}^{(j)}, \boldsymbol{M}_{[\mathrm{o}]t}^{(j)} \widetilde{\boldsymbol{w}}_{[\mathrm{o}]t}^{(j)})^T \; | i \in I^N\}$ for rough registration because it approximates $Z_{[\mathrm{o}]t}$ and the normal vectors $\boldsymbol{n}_{[\mathrm{o}]t}^{(j)}$ of $\boldsymbol{q}_{[\mathrm{o}]t}^{(j)}$ can be utilized in several ways as shown below.

ROI for registration: From $Z_{[\mathrm{o}]t}^{\mathrm{CAN2}}$, we remove the following data and obtain the ROI (Region of Interest) dataset $Z_{[\mathrm{o}]t}^{\mathrm{ROI}} = \{\boldsymbol{q}_{[\mathrm{o}]t}^{(j)} | j \in I_{[\mathrm{o}]t}^{\mathrm{ROI}}\}$.

(i) (Remove floor) By means of the plane extraction method using the CAN2 [5], we extract the floor plane from $Z_{[\mathrm{o}]t}^{\mathrm{CAN2}}$, and remove the data within the distance θ_h(= 30mm) to the floor.

(ii) (Remove jump edge) The data on the jump edge hold $(\boldsymbol{n}_{[\mathrm{o}]t}^{(j)})^T\boldsymbol{q}_{[\mathrm{o}]t}^{(j)} = 0$. So, we remove the data with $|(\boldsymbol{n}_{[\mathrm{o}]t}^{(j)})^T\boldsymbol{q}_{[\mathrm{o}]t}^{(j)}|/\|\boldsymbol{q}_{[\mathrm{o}]t}^{(j)}\| < \cos(\pi/2 - \theta_e)$, where $\theta_e(= 5°)$ indicates allowable error.

(iii) (Remove unreliable piecewise planes) remove the data in the Voronoi region of the unit which involves less than 4 data because the plane is unreliable.

Let us consider the registration of the $c(= t)$th image $Z_{[\mathrm{o}]c}^{\mathrm{ROI}}$ to the $r(= t-1)$th image $Z_{[\mathrm{o}]r}^{\mathrm{ROI}}$, where c and r represent the current and reference images, respectively.

Registration using planes extracted by the CAN2: Suppose that the plane extraction method [5] applied above has extracted the centers $\boldsymbol{\xi}^{(j)}_{[\mathrm{o}]t}$ and the normal vectors $\boldsymbol{\nu}^{(j)}_{[\mathrm{o}]t}$ of plane surfaces on the object for $t = c$ and r, and there is a common plane surface on both cth and rth images whose normal vector is not orthogonal to the rotation plane of the LRF. Then, we can estimate the registration transform by the following steps, where s_j represents a correspondence of the planes such that the jth plane in the cth image corresponds to the s_jth plane in the rth image.

1. (Obtain registration parameters) For a correspondence s_j, we obtain the yaw angle of the corresponding planes by $\theta^{(s_j)}_{[\mathrm{s}]r,c} = \mathrm{atan2}(\nu^{(s_j)}_{[\mathrm{s}]r,z}, \nu^{(s_j)}_{[\mathrm{s}]r,x}) - \mathrm{atan2}(\nu^{(j)}_{[\mathrm{s}]c,z}, \nu^{(j)}_{[\mathrm{s}]c,x})$. where $\boldsymbol{\nu}^{(j)}_{[\mathrm{s}]t} = (\nu^{(j)}_{[\mathrm{s}]t,x}, \nu^{(j)}_{[\mathrm{s}]t,y}, \nu^{(j)}_{[\mathrm{s}]t,z})$ is the normal vector represented by the scan center coordinate, and atan2(z,x) gives the angle of the point (x,z) from the positive x-axis. Then, the registration or the transform of the points in the cth image to the rth image, $\boldsymbol{q}^{(j)}_{[\mathrm{o}]r,c} = \boldsymbol{R}_{[\mathrm{o}]r,c}\boldsymbol{q}^{(j)}_{[\mathrm{o}]c} + \boldsymbol{t}_{[\mathrm{o}]r,c}$, is obtained, where $(\boldsymbol{R}_{[\mathrm{o}]r,c}, \boldsymbol{t}_{[\mathrm{o}]r,c}) = (\boldsymbol{R}_{[\mathrm{o,s}]r}\boldsymbol{R}_Y(\theta^{(s_j)}_{[\mathrm{s}]r,c}), \boldsymbol{\xi}^{(s_j)}_{[\mathrm{o}]r} - \boldsymbol{R}_{[\mathrm{o}]r,c}\boldsymbol{\xi}^{(j)}_{[\mathrm{o}]c})$.
2. (Transform and remove hidden data) Transform the cth data to the rth coordinate by $\boldsymbol{q}^{(j)}_{[\mathrm{o}]r,c} = \boldsymbol{R}_{[\mathrm{o}]r,c}\boldsymbol{q}^{(j)}_{[\mathrm{o}]c} + \boldsymbol{t}^{(j)}_{[\mathrm{o}]c}$, and let $Z^{\mathrm{ROI}}_{[\mathrm{o}]r,c} = \{\boldsymbol{q}^{(j)}_{[\mathrm{o}]r,c} | j \in I^{\mathrm{ROI}}_{r,c}\}$ be the set of all transformed data except the ones whose third elements of the normal vectors $\boldsymbol{n}^{(j)}_{[\mathrm{o}]r,c} = \boldsymbol{R}_{[\mathrm{o}]r,c}\boldsymbol{n}^{(j)}_{[\mathrm{o}]c}$ are negative, because they are invisible from the origin of the rth LRF.
3. (Evaluate the registration) Let $\boldsymbol{q}^{(l_j)}_{[\mathrm{o}]r} \in Z^{\mathrm{ROI}}_{[\mathrm{o}]r}$ be the closest point to $\boldsymbol{q}^{(j)}_{[\mathrm{o}]r,c} \in Z^{\mathrm{ROI}}_{[\mathrm{o}]r,c}$, and $\boldsymbol{\xi}^{(l_j)}_{[\mathrm{o}]r}$ be the closest point on the tangent plane involving $\boldsymbol{q}^{(l_j)}_{[\mathrm{o}]r}$, namely,

$$\boldsymbol{q}^{(l_j)}_{[\mathrm{o}]r} = \underset{\boldsymbol{q}^{(l)}_{[\mathrm{o}]r} \in Z^{\mathrm{ROI}}_{[\mathrm{o}]r}}{\mathrm{argmin}} \left\{ \|\boldsymbol{q}^{(j)}_{[\mathrm{o}]r,c} - \boldsymbol{q}^{(l)}_{[\mathrm{o}]r}\| \;\middle|\; (\boldsymbol{n}^{(j)}_{[\mathrm{o}]r,c})^T \boldsymbol{n}^{(l)}_{[\mathrm{o}]r} > 0 \right\} \tag{12}$$

$$\boldsymbol{\xi}^{(l_j)}_{[\mathrm{o}]r} = \boldsymbol{q}^{(l_j)}_{[\mathrm{o}]r} + \left(\alpha^{(l_j)}_{[\mathrm{o}]r} - (\boldsymbol{n}^{(l_j)}_{[\mathrm{o}]r})^T \boldsymbol{q}^{(j)}_{[\mathrm{o}]r,c}\right) \boldsymbol{n}^{(l_j)}_{[\mathrm{o}]r}. \tag{13}$$

Then, a distance of $Z^{\mathrm{ROI}}_{[\mathrm{o}]r,c}$ and $Z^{\mathrm{ROI}}_{[\mathrm{o}]r}$ is given by

$$(\Delta Z_{r,c})^2 = \frac{1}{|Z^{\mathrm{ROI}}_{[\mathrm{o}]r,c}|} \sum_{\boldsymbol{q}^{(j)}_{[\mathrm{o}]r,c} \in Z^{\mathrm{ROI}}_{[\mathrm{o}]r,c}} \|\boldsymbol{q}^{(j)}_{[\mathrm{o}]r,c} - \boldsymbol{\xi}^{(l_j)}_{[\mathrm{o}]r}\|^2. \tag{14}$$

For all correspondences s_j, let $(\widehat{\boldsymbol{R}}_{[\mathrm{o}]r,c}, \widehat{\boldsymbol{t}}_{[\mathrm{o}]r,c})$ be the parameter which achieves the smallest $(\Delta Z_{r,c})^2$. Then, from Eq.(9) we have the registration parameters for the scan center data as follows,

$$(\widehat{\boldsymbol{R}}_{[\mathrm{s}]r,c}, \widehat{\boldsymbol{t}}_{[\mathrm{s}]r,c}) = \left(\boldsymbol{R}_{[\mathrm{s,o}]r}\widehat{\boldsymbol{R}}_{[\mathrm{o}]r,c}\boldsymbol{R}_{[\mathrm{o,s}]c},\; \boldsymbol{R}_{[\mathrm{s,o}]r}\widehat{\boldsymbol{t}}_{[\mathrm{o}]r,c}\right). \tag{15}$$

3 Experimental Results

In order to examine the effectiveness of the present method, we have conducted experiments using four range images taken from around a rectangular box on the floor. The

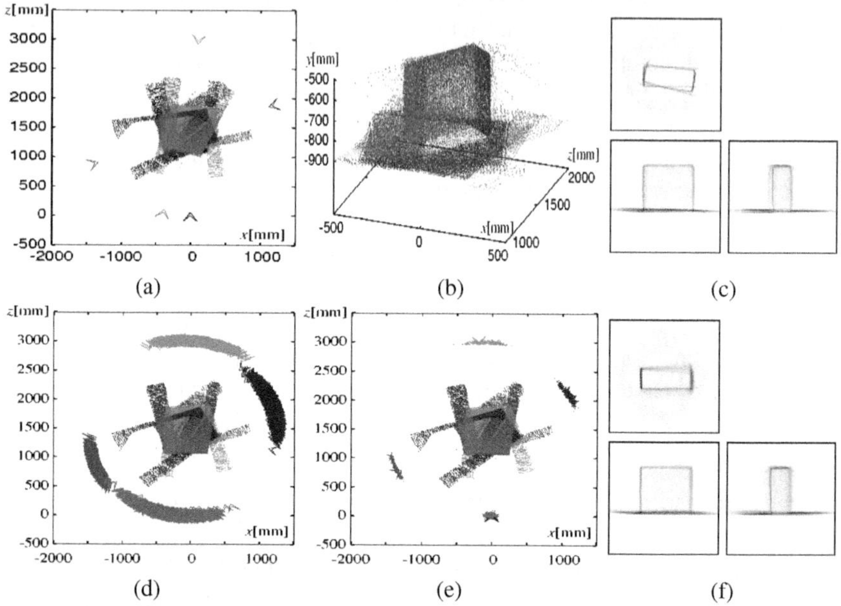

Fig. 2. Experimental result using four range images taken from around a rectangular box, the CAN2 with $N = 2000$ units and the PF with $K = 5000$ particles. (a),(b) and (c) indicate the result of the pair-wise registration. (a) shows the over-head view, where the arc-heads colored black, red, green, blue and pink indicate the registered poses (orientations and the positions) of the tth LRF for $t = 0, 1, 2, 3, 4$, respectively, and the dots indicate the LRF data. (b) shows the perspective view. (c) shows the top, front and side view of the registered object, where the range data are digitized into 10mm^3 cubic volumes in the 1m^3 cubic area involving the object, and the numbers of the data in the volumes are summed up orthogonally to the top, front and side planes, respectively. (d) and (e) show the overhead view of the registration by the PF before and after the loop closing, respectively. The arc-heads indicate the poses of the registered LRFs corresponding to particles, and the dots represent the range data for the particle with the biggest weight. (f) shows three side views of the registered and weighted range data.

result is shown in Fig. 2. From (a), (b) and (c), we can see the performance of the pair-wise registration by the CAN2. Especially, from (a), we can see the cumulative error as the difference of the LRF poses for $t = 0$ and $t = 4$ or the black and the pink arc-heads. From (c), we can also see the cumulative error in the top view.

The registration by the PF is shown in (d), (e) and (f). From the over-head view before (d) and after (e) the loop closing, we can see the effectiveness of the loop closing. Namely, the broad distribution of the poses in (d) is reduced in (e) and the pink poses for $t = T(= 4)$ distributes around the initial pose at the origin $(x, y, z) = (0, 0, 0)$. We can also see the reduction of the cumulative error in the top view in (f).

The actual size of the box is $485 \times 175 \times 396$ [mm^3] in width×depth×height. From the thick edges in (f), we obtain the size as $47 \times 17 \times 40$ [$(\text{pixel}/10\text{mm})^3$]. Considering that every surface of the rectangular box may have 10mm resolution error, the estimated values seem to be very accurate.

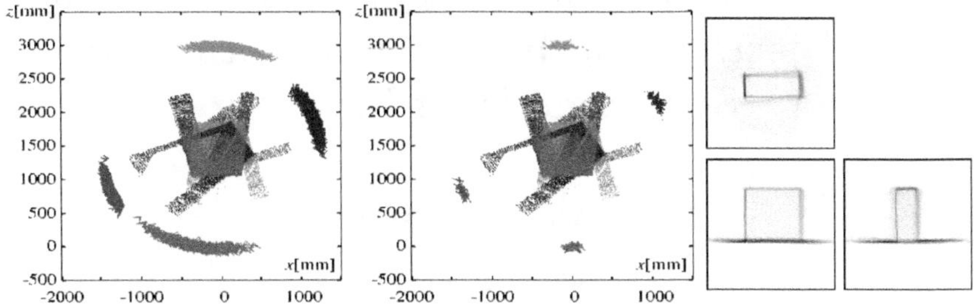

(a) (b) (c)

Fig. 3. Experimental result using $(N, K) = (500, 500)$. (a) and (b) show the overhead view of the registration before and after the loop closing, respectively, and (c) shows the three side views of the registered range data.

The above result is obtained with the CAN2 using $N = 2000$ units and the PF using $K = 5000$ particles for an accurate registration and the computational cost was 904s by Intel Core2Duo 2.67GHz CPU. The result using $(N, K) = (500, 500)$ is shown in Fig. 3, where the computational cost was 107s, and we have obtained the sise 46×18×41 [(pixel/10mm)^3], which seems not so accurate but not so bad.

Incidentally, the distance measure of two images shown in Eq.(14) using the piecewise planes extracted by the CAN2 has largely contributed to reducing the computational cost from the simple distance measure of closest points on two images. Namely, the reduction rate of the computational cost for the above two examples using $(N, K) = (2000, 5000)$ and $(500, 500)$, respectively, is 0.087$\simeq$ 904s/10332s and 0.094$\simeq$107s/1143s.

4 Conclusion

We have presented a method using a PF and CAN2s for range image registration to fuse 3D surfaces of range images taken from around an object by the LRF. The effectiveness is shown and examined through the experimental results.

This work was partially supported by the Grant-in Aid for Scientific Research (C) 21500217 of the Japanese Ministry of Education, Science, Sports and Culture.

References

1. Salvi, J., Matabosh, C., Foli, D., Forest, J.: A review of recent range image registration methods. Image and Vision Computing 25, 578–596 (2007)
2. Doucet, A., Feitas, N., Gordon, N. (eds.): Sequential Monte Carlo methods in particle. Springer, Heidelberg (2001)
3. Thrun, S., Burgard, W., Fox, D.: Probabilistic robotics. MIT Press, Cambridge (2005)
4. Kurogi, S., Koya, H., Nagashima, R., Wakeyama, D., Nishida, T.: Range image registration using plane extraction by the CAN2. In: Proc. CIRA 2009 (2009)
5. Kurogi, S.: Plane extraction from range data using competitive associative nets. The Brain & Neural Networks (Journal of Japanese Neural Network Society) 14(4), 273–281 (2007)

Rotation Invariant Categorization of Visual Objects Using Radon Transform and Self-Organizing Modules

Andrew P. Papliński

Monash University, Australia
Clayton School of IT
Andrew.Paplinski@monash.edu

Abstract. The Radon transform in combination with self-organizing maps is used to build the rotation invariant systems for categorization of visual objects. The first system has one SOM per the Radon transform direction. The outputs from these directional SOMs that represent positions of the winners and related post-synaptic activities, form the input to the final categorizing SOM. Such a network delivers robust rotation invariant categorization of images rotated by angles up to around 12°. In the second network the angular Radon transform vectors are combined together and form the input to the categorizing SOM. This network can correctly categorized visual stimuli rotated by up to 30°. The rotation invariance can be improved by increasing the number of Radon transform angle, which has been equal to six in our initial experiments.

Keywords: Radon transform, Self-organizing maps, Rotation invariant vision.

1 Introduction

Radon transform has a long history of application in computer tomography, and relatively recently has been applied in a variety of image processing problems. Typically, Radon transform is used in conjunction with other transforms, wavelet and Fourier included. Magli et al. [1] and Warrick and Delaney [2] seem to initiate the use of Radon transform in combination with wavelet transform. More recently, a similar combination of transforms has been used in rotation invariant texture analysis [3,4], and in shape representation [5]. Other approach to rotation invariant texture analysis uses Radon transform in combination with Fourier transform [6]. Chen and Kégl [7] consider feature extraction using combination of three transforms: Radon, wavelet and Fourier. In [8], texture classification is performed by using a feature descriptor based on Radon transform and an affine invariant transform. Miciak [9] describes a character recognition system based on Radon transform and Principal Component Analysis. Hejazi et al. [10] present discrete Radon transform in rotation invariant image analysis. Close to our considerations are object identification problems discussed by Hjouj and Kammler in [11].

K.W. Wong et al. (Eds.): ICONIP 2010, Part II, LNCS 6444, pp. 360–366, 2010.

In the above papers the reader can find many variants of detailed description of Radon transform and its properties. Here, we can only reiterate the basic fact that Radon transform, $R(\theta, r)$, is composed of sums of pixels along the line that crosses the visual object under the angle θ at the distance r from the origin. It can be noted that Radon transform of the image rotated be a known angle, θ, can be easily inferred from the transform of the un-rotated image. This property makes Radon transform attractive in rotation invariant vision systems.

In this paper we use combination of Radon transform and Self-Organizing maps [12]. Our original idea was related to the way in which human vision could possibly recognize rotated characters as in a process of reading. However, the presented solutions can be used in a variety of systems of rotation invariant categorization of visual objects. We discuss two networks of self-organizing modules that perform the above task in different way.

2 One SOM Per Direction

We start with the system presented in Figure 1in which there is one dedicated self-organizing module, Dir, per Radon transform direction. The image is presented at the receptive field, RF, and is randomly sampled at the points symbolically indicated as green dots. Each line crossing the receptive field symbolizes the 'dendritic' summation of image pixels implementing a single point of Radon transform for a given line, (θ, r). In the example of Figure 1 Radon transform is calculated for $m = 6$ angles, along the $n = 8$ lines, hence, dimensionality of each vector $\mathbf{x}_D$ is $n = 8$, whereas the number of self-organizing modules, Dir, is equal to $m = 6$, that is, the number of Radon transform directions, θ.

In our particular computational examples presented below the diameter of the receptive field, RF, is $n = 75$ pixels normalised into a unity circle. We use letters of the Latin alphabet in 28-point font as the set of the test images. Each directional self-organizing module contain a randomly generated number of neurons approximately equal to πr^2, where r is selected to be equal to 16. Hence that number of neurons varies around 804. Each module produces a 3-dimensional output $\mathbf{y}_D$:

$$\mathbf{y}_D = g(\mathbf{x}_D) \tag{1}$$

where $\mathbf{x}_D$ and $\mathbf{y}_D$ represent input and output signals, respectively, and $g(\cdot)$ describes the Winner-Takes-All function of $\mathbf{W_D x_D}$ which produces a 2-D positional vector $\mathbf{v}$ and related postsynaptic activity d. Such 3-D outputs can be thought of as low dimensional signatures, or labels, specific for each input to the self-organizing modules. In this we follow our other works [13,14,15].

In Figure 2A we show the result of training one of the directional maps, Dir, namely, the 60° map. Each map is excited with n-dimensional vectors ($n = 75$ in our example) representing the value of Radon transform for a given direction, θ. Each Dir map encapsulates directional similarities of the letters capturing features characteristic for a given direction. The 3-dimensional signatures, or labels, $\mathbf{y}_D$, generated by directional maps are then applied to the combined map, TrImg. An input vector $\mathbf{x}_T$ is of dimensionality $3m$, where m is the number of

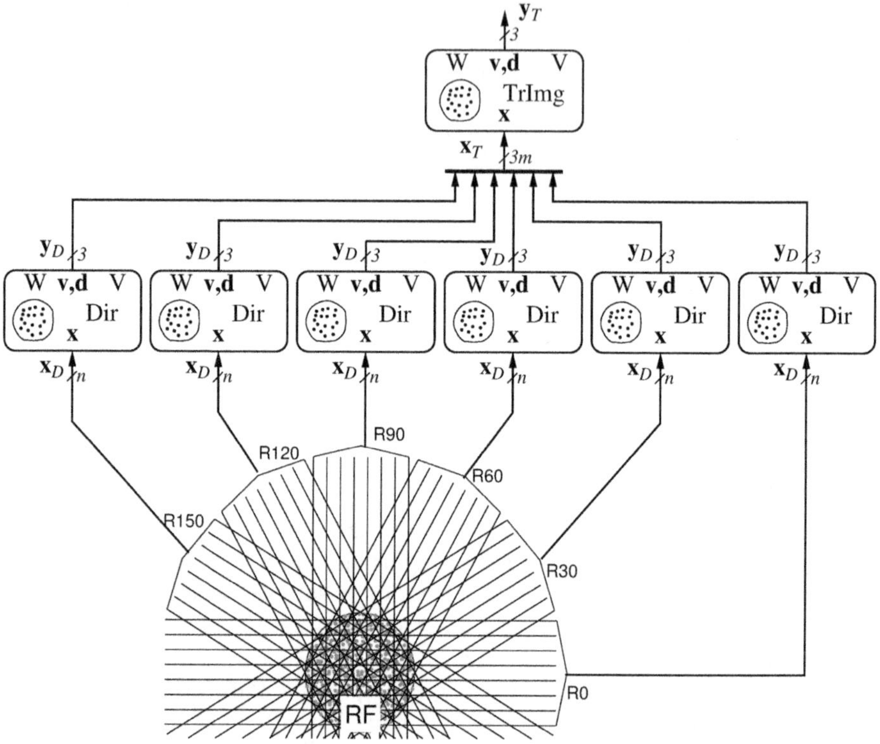

Fig. 1. The categorization network with one self-organizing module per Radon transform direction

Radon transform directions. More formally, for the combined module, TrImg, we can write:

$$\mathbf{y}_T = g(\sum_{i=1}^{m} W_{Ti}\mathbf{y}_{Di}) \tag{2}$$

After training we obtain a combined map as presented in Figure 2B. It can be observed that the combined map captures, as expected, visual similarities between letter.

Now we test the responses of the trained network to the rotated images of letters. The results are presented in Figure 3. We rotate the letters by 2° angles varying from -12° to 12° as indicated in the map. Firstly, it can be noted that majority of the rotated letters are correctly clustered. The quality of the clustering is indicated by the right-hand side plots. The upper plot gives a relative confidence level as measured by the inner products of respective weights and input vectors (see sec. 4 for details). Since these are unity vectors, the maximum of the inner products is equal to 1. The bottom-left plot gives the average size of the rotated letter clusters. Again the radius of the neuronal circle is unity, which gives an idea about the size of the clusters.

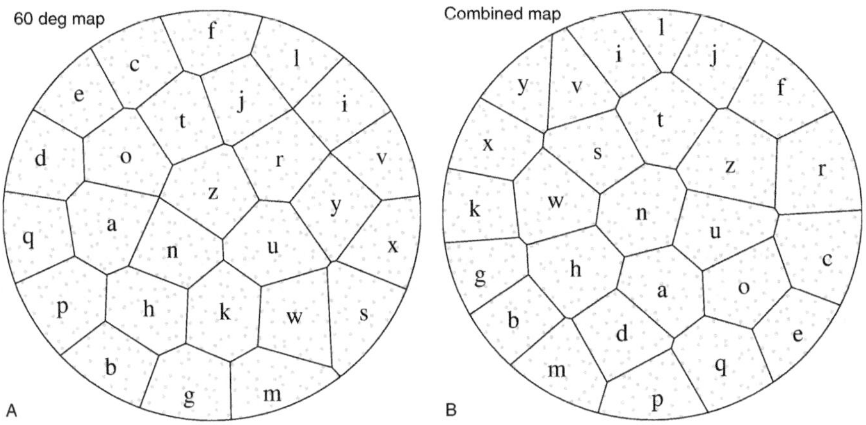

Fig. 2. The result of training maps. A: A directional 60° Dir map. B: A combined categorization map.

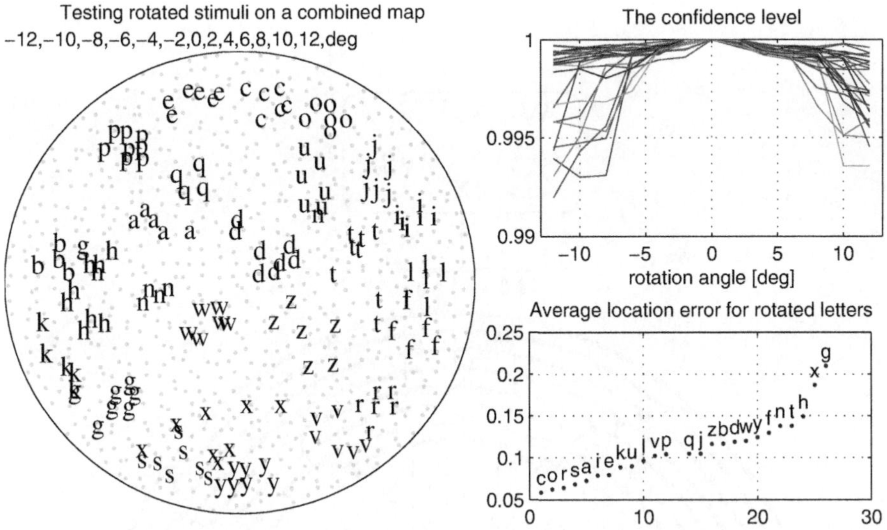

Fig. 3. Categorization of rotated letter by the first network

Although it is encouraging that the above network deliverers categorization invariance for relatively small angles, it would be interesting to find a solution that was invariant to relatively large rotation angles. One possibility is presented in the next section. With reference to eqn (2) and Figure 1 we note that if we rotate the image by the Radon transform angle, it is equivalent to shifting responses, $\mathbf{y}_{Di}$, between the directional self-organizing modules. This gives the potential of the perfect rotation invariant behaviour. This potential is lost, however, when we multiply $\mathbf{y}_{Di}$ by the segments of the weight matrix W_{Ti} that have been trained for non-rotated images.

3 The Single-SOM Network

In this solution we have replaced the directional SOMs by simple summations, as shown in Figure 4. The circular 'dendrites' symbolize the summations of the respective Radon transform rays. The resulting vectors $\mathbf{x}_S$ are of dimensionality $n = 75$ (8 in Figure 4), and are inputs to a single categorizing SOM, TrImg. More formally, we can write:

$$\mathbf{x}_S = \sum_{i=1}^{m} R(\theta_i, \mathbf{r}) \; , \quad \mathbf{y}_S = g(W_T \mathbf{x}_S) \tag{3}$$

where $R(\boldsymbol{\theta}, \mathbf{r})$ is a Radon transform matrix for a given image and vectors $\boldsymbol{\theta}, \mathbf{r}$ of all possible angles and lines, respectively. Summation over all Radon transform angles does remove directional sensitivity, but, unfortunately, ignores the

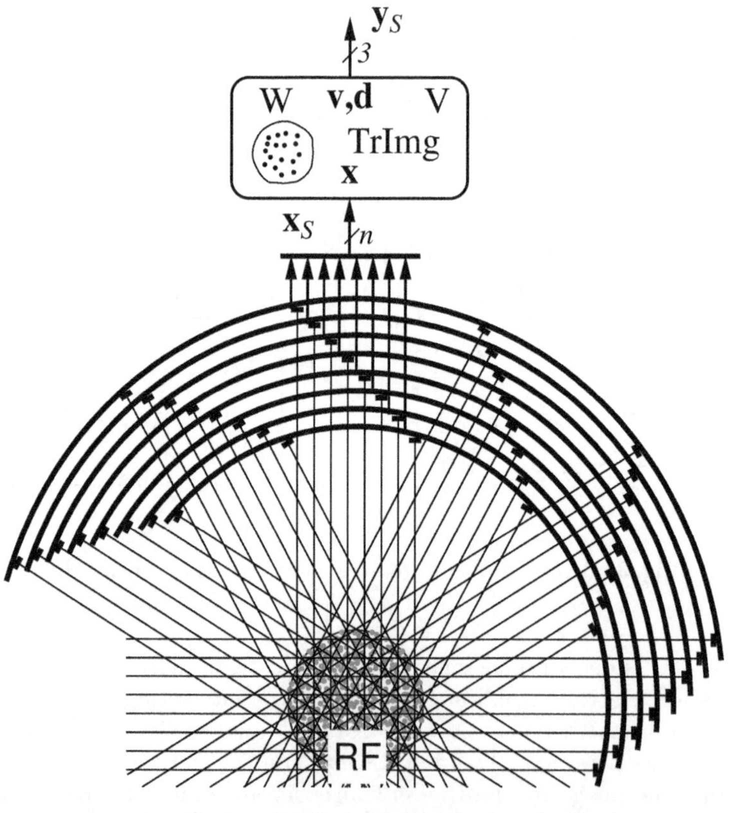

Fig. 4. The categorization network with summed Radon transform and a single SOM

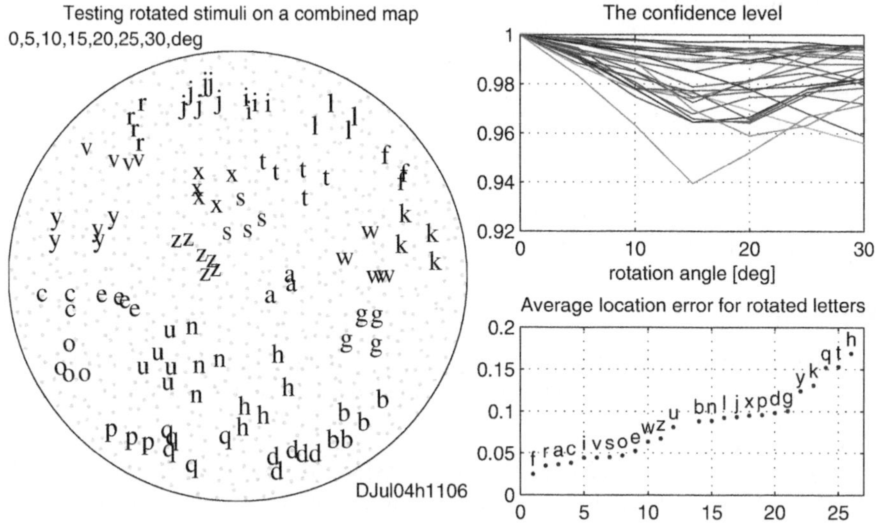

Fig. 5. Categorization of rotated letter by the first network

richness of the image details contained in the full Radon transform. It is, however, expected that the network will correctly classify images rotated by larger angles that in the network of Figure 1. The results of testing the behaviour of the network of Figure 4 are presented in Figure 5. This times we rotate the test images by angles varying from $\mathbf{5^o}$ to $\mathbf{30^o}$. Comparing with the results presented in Figure 3 we note that visually the letters are also well clustered despite of larger rotation angles. This is reflected in the left hand side bottom plot of the average location error, and is confirmed by the higher values of the confidence level presented in the upper plot.

4 Some Implementation Remarks

All input vectors applied to the self-organizing modules, e.g., $\mathbf{x}_D$ in eqn (1), or $\mathbf{x}_S$ in eqn (3), are normalised and projected on the unity hypersphere by adding one additional dimension. Similarly, all weight vectors are kept on the unity hypersphere. As a result of such an arrangement the inner products of weight and input vectors are equal to the cosine of the angle between such vectors. Working with unity vectors makes it possible to use the dot-product learning law [12] which speeds up the training.

The neuronal lattice is organized in such a way that each neuron is assigned a random position inside a unity circle (see Figure 2). By adding third dimension the position vectors are projected on a 3-D unity sphere. All position vectors are stored in the position matrix $\boldsymbol{V}$ of dimension $\boldsymbol{N}\times 3$, where $\boldsymbol{N}$ is the total number of neurons.

5 Conclusion

We have described preliminary investigation of two networks combining Radon transform and self-organizing maps that are used in categorization of rotated images. Radon transform is easy to implement since it involves only summation of image pixel values along the set of parallel lines crossing the image under a specified set of angles. We have shown that such networks can produce a degree of rotation invariance that can be attractive both in image processing tasks and in analysis of aspects of human vision.

References

1. Magli, E., Presti, L.L., Olmo, G.: A pattern detection and compression algorithm based on the joint wavelet and Radon transform. In: Proc. IEEE 13th Int. Conf. Dig. Sig. Proc., pp. 559–562 (1997)
2. Warrick, A., Delaney, P.A.: Detection of linear features using a localized Radon transform with a wavelet filter. In: Proc. ICASSP, pp. 2769–2772 (1997)
3. Jafari-Khouzani, K., Soltanian-Zadeh, H.: Rotation-invariant multiresolution texture analysis using Radon and wavelet transforms. IEEE Trans. Img. Proc. 14(6), 783–795 (2005)
4. Yu, G., Cao, W., Li, Z.: Rotation and scale invariant for texture analysis based on Radon transform and wavelet transform. In: Proc. 3rd ICPCA, pp. 704–708 (2008)
5. Yao, W., Pun, C.M.: Invariant shape representation by Radon and wavelet transforms for complex inner shapes. In: Proc. IEEE Int. Conf. Inform. Autom., pp. 1144–1149 (2009)
6. Xiao, S.S., Wu, Y.X.: Rotation-invariant texture analysis using Radon and Fourier transforms. J. Phys.: Conf. Ser. 48, 1459–1464 (2007)
7. Chen, G., Kégl, B.: Feature extraction using Radon, wavelet and Fourier transform. In: Proc. IEEE Int. Conf. Syst. Man and Cybernetics, pp. 1020–1025 (2007)
8. Liu, G., Lin, Z., Yu, Y.: Radon representation-based feature descriptor for texture classification. IEEE Trans. Img. Proc. 18(5), 921–928 (2009)
9. Miciak, M.: Character recognition using Radon transformation and principal component analysis in postal applications. In: Proc. Int. Multiconf. Comp. Sci. Info. Tech., pp. 495–500 (2008)
10. Hejazi, M., Shevlyakov, G., Ho, Y.S.: Modified discrete Radon transforms and their application to rotation-invariant image analysis. In: Proc. IEEE Workshop Mult. Sig. Proc., pp. 429–434 (2006)
11. Hjouj, F., Kammler, D.W.: Identification of reflected, scaled, translated, and rotated objects from their Radon projections. IEEE Trans. Img. Proc. 17(3), 301–310 (2008)
12. Kohonen, T.: Self-Organising Maps, 3rd edn. Springer, Berlin (2001)
13. Papliński, A.P., Gustafsson, L., Mount, W.M.: A model of binding concepts to spoken names. In: Proc. 17th Int. Conf. Neural Inf. Proc., Sydney (submitted, 2010)
14. Chou, S., Papliński, A.P., Gustafsson, L.: Speaker-dependent bimodal integration of Chinese phonemes and letters using multimodal self-organizing networks. In: Proc. Int. Joint Conf. Neural Networks, Orlando, Florida (2007)
15. Papliński, A.P., Gustafsson, L.: Feedback in multimodal self-organizing networks enhances perception of corrupted stimuli. In: Sattar, A., Kang, B.-h. (eds.) AI 2006. LNCS (LNAI), vol. 4304, pp. 19–28. Springer, Heidelberg (2006)

Learning Topological Constraints in Self-Organizing Map

Guénaël Cabanes and Younès Bennani

LIPN-CNRS, UMR 7030, Université de Paris 13
99, Avenue J-B. Clément
93430 Villetaneuse, France
cabanes@lipn.univ-paris13.fr

Abstract. The Self-Organizing Map (SOM) is a popular algorithm to analyze the structure of a dataset. However, some topological constraints of the SOM are fixed before the learning and may not be relevant regarding to the data structure. In this paper we propose to improve the SOM performance with a new algorithm which learn the topological constraints of the map using data structure information. Experiments on artificial and real databases show that algorithm achieve better results than SOM. This is not the case with trivial topological constraint relaxation because of the high increase of the Topological error.

1 Introduction

The Self-Organizing Map (SOM) [1] is a popular algorithm to analyze the structure of a dataset. A SOM consist in a set of artificial neurons that represent the data structure. Neurons are connected with topological (or neighborhood) connections to form a two dimensional grid. Two connected neurons should represent the same type of data, two distant neurons (according to the grid) should represent different data. These properties are insured during the learning process by using neighborhood information as topological constraints, i.e. each neuron is activated by data that are represented by this neuron, but each neuron also responds in a less degree to data represented by its neighbors.

However, in the SOM algorithm, the topological information is fixed before the learning process and may not be relevant regarding to the data structure. To solve this problem, some works have been done in order to adapt the number of neurons during the learning process, using informations from the database to analyze [2]. Results have shown that the quality of the model is improved when the number of neurons is learned from the data.

Despite of these results, there is very few works that address the problem of learning the topological constraints from the data structure. However, at the end of the learning process, some "neighbors" neurons may not represent the same data [3,4]. In this paper we propose to improve the SOM performance with a new algorithm able to learn the topology of the map using data structure information: the Data-Driven Relaxation – SOM algorithm (DDR-SOM). The main idea is to associate to each topological connection of the map a value indicate how well the two connected neurons represent the same type of data, then to use this values to reduce some topological constraint between neurons that represent different data. These constraint relaxation are expected to improve

K.W. Wong et al. (Eds.): ICONIP 2010, Part II, LNCS 6444, pp. 367–374, 2010.

the quality of the SOM, especially by reducing the quantization error of the map and increasing the number of neurons that really participate to the data representation.

The remainder of the paper is organized as follow. Section 2 presents the Self-Organizing Map algorithm. Section 3 describes the DDR-SOM algorithm based on SOM. Section 4 shows experimental validations and discuss the obtained results. Finally, a conclusion is given in Section 5.

2 Self-Organizing Maps

A SOM consists in a two dimensional map of neurons which are connected to n inputs according to a set of prototypes vectors and to their neighbors with topological connection [5,1].. The training set is used to organize these maps under topological constraints of the input space. Thus, a mapping between the input space and the network space is constructed ; two close observations in the input space would activate two close neurons of the SOM. When an observation is recognized, the activation of an output neuron inhibits the activation of other neurons and reinforce itself. It is said that it follows the so called "Winner Takes All" rule. The winner neuron updates its prototype vector, making it more sensitive for later presentation of that type of input. To achieve a topological mapping, the neighbors of the winner neuron can adjust their prototype vector towards the input vector as well, but at a lesser degree, depending on how far away they are from the winner.

The connectionist learning is often presented as a minimization of a cost function. In most case, it will be carried out by the minimization of the distance between the input samples and the map prototypes, weighted by the neighborhood function K_{ij}. The cost function to be minimized is defined by:

$$\tilde{R}(w) = \frac{1}{N} \sum_{k=1}^{N} \sum_{i=1}^{M} K_{iu^*(x^{(k)})} \parallel x^{(k)} - w_i \parallel^2$$

N represents the number of data in the database, M the number of neurons in the map, $u^*(x^{(k)})$ is the neuron having the closest weight vector to the input data $x^{(k)}$. The relative importance of a neuron i compared to a neuron j is weighted by the value of the kernel function K_{ij} which can be defined as:

$$K_{ij} = \frac{1}{\lambda(t)} \times e^{-\frac{d_M^2(i,j)}{\lambda^2(t)}}$$

Where $\lambda(t)$ is the temperature function modeling the topological neighborhood extent. $d_M(i,j)$ is the Manhattan distance defined between two neurons i and j on the map grid, i.e. the minimal number of topological connection between i and j.

3 Data-Driven Relaxation in Self-Organizing Map

3.1 Principle

In the DDR-SOM algorithm, we propose to associate each neighborhood connection a real value v which indicates the relevance of the connected neurons. Given the organization constraint of the SOM, both closest neurons of each data must be connected by a neighborhood connection. A pair of neighbor neurons that are together a good representative of a set of data should be strongly connected, whereas a pair of neighbor neurons that don't represent the same type of data should be weakly connected. Each neighborhood connection is then associated to a value varying from 1 (strong connection) to 2 (week connection) using a logistic function depending on the number of data well represented by both the two connected neurons.

These values are used to estimate a weighted Manhattan distance $d_{WM}(i, j)$ between two neurons i and j. This distance is the minimal number of connections between i and j weighted by the value associated to each connection (see fig. 1 for an example).

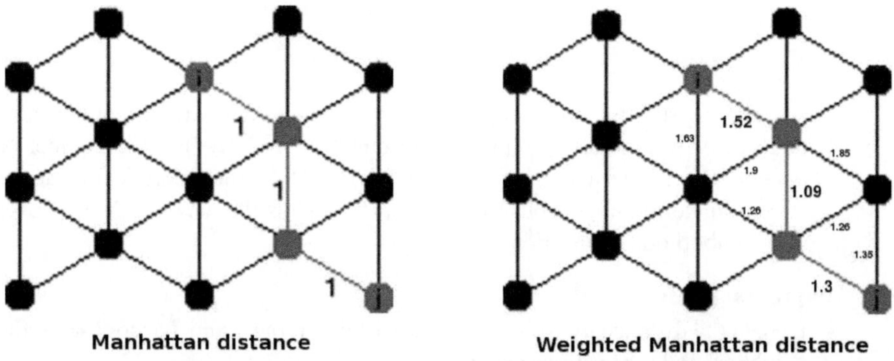

Fig. 1. Comparison of Manhattan distance and Weighted Manhattan distance between neurons i and j for hexagonal topology. Here $d_M(i, j) = 3$, it's the minimal number of topological connection between i and j, whereas $d_{WM}(i, j) = 1.52 + 1.09 + 1.3 = 3.91$, it's the minimal path between i and j according to the connection values v.

For that purpose we use Johnson's algorithm [6] to find:

- the shortest paths along neighborhood connection between all pair of neurons.
- the length of this path according to the values v associated to each connection.

In this way the distance between two neurons is expected to reflect the "true" neighborhood of these neurons.

Connection values and distances between neurons are updated during the learning of the map. Thus, the final quality of the SOM should be improved.

3.2 Algorithm

The DDR-SOM algorithm proceeds in tree steps :

1. **Initialization step:**
   - Define the topology of the SOM.
   - Initialize all prototypes vectors of the map w.
   - Initialize all neighborhood connections values v to zero.

2. **Competition step:**
   - Find the first BMU u^* and the second BMU u^{**} for each input data $x^{(k)}$:

$$u^*(x^{(k)}) = \underset{1 \leq i \leq M}{Argmin} \parallel x^{(k)} - w_i \parallel^2$$

   and

$$u^{**}(x^{(k)}) = \underset{1 \leq i \leq M,\ i \neq u^*}{Argmin} \parallel x^{(k)} - w_i \parallel^2$$

   - Update the neighborhood connections values v according to the following rule:

$$v_{i,j} = \frac{1 + 2e^{\frac{Nth - N(i,j)}{\sigma Nth}}}{1 + e^{\frac{Nth - N(i,j)}{\sigma Nth}}}$$

   $v_{i,j}$ is a logistic function that grown from 1 (when i and j represent the same data) to 2 (when i and j represent different data). $N(i,j)$ is the number of data (x) that have i and j in $\{u^*(x), u^{**}(x)\}$. Nth is the theoretical value for $N(i,j)$ under homogeneous hypothesis, i.e. Nth is the mean of $N(i,j)$ over all neighborhood connection.

3. **Adaptation phase:**
   - Update the distance $d_{WM}(i,j)$ for all pair of neurons i and j according to the neighborhood connections values v.
   - Update the kernel function K according to the temperature $\lambda(t)$ and d_{WM}.
   - Update prototypes vectors w_i of each neuron i in order to minimize the cost function:

$$w_i = \frac{\sum_{k=1}^{N} K_{iu^*(x^{(k)})}.x^{(k)}}{\sum_{k=1}^{N} K_{iu^*(x^{(k)})}}$$

4. **Repeat steps 2 and 3** until $t = t_{max}$

4 Experimental Results

4.1 Databases Description

In order to test the validity of the new algorithm we used 10 artificial and real databases with different number of data and features.

Databases "Target", "TwoDiamonds", "Tetra" and "Hepta" come from Fundamental Clustering Problem Suite (FCPS) [7]. They are artificial data in low dimensional space with well known structure. They are often used as benchmark for clustering algorithms [3,4,7]. Databases "Housing", "Harot", "Iris" and "Wine" are well known databases in various dimensional spaces from the UCI repository [8]. Finally, "Cockroach" and "Chromato" are very noisy real databases from biological experiments.

These databases are expected to reflect the diversity of the modeling problems that are encountered by SOM's users.

4.2 Estimation of the Quality of the SOM

We use the following three usual quality indexes to evaluate the training performance of SOM-based algorithms:

Quantization Error Qe:
This measures the average distance between each data vector and its BMU [1]. The smaller is the value of Qe, the better is the algorithm.

$$Qe = \frac{1}{N}\sum_{k=1}^{N} \| x^{(k)} - w_{u^*(x^{(k)})} \|^2$$

Topographic Error Te:
Te describes how well the SOM preserves the topology of the studied data set [9]. It's the proportion of all data vectors for which first and second BMUs are not adjacent neurons (i.e. are not connected with a topological connection). A small value of Te is more desirable. Unlike the quantization error, it considers the structure of the map.

Neuron Utilization Ne:
Ne measures the percentage of neurons that are not BMU of any data in the database [10]. A good SOM should have a small Ne, i.e. all neurons must be used to represent the data.

In all following experiments, all indexes are normalized in order to compare efficiently results on different databases. To represent the gain or the loss in comparisons to SOM, each error index is divided to the value obtained with the SOM algorithm (the SOM's error is then alway equal to 1). For each experiments in this Section we used the SOM-Toolbox [11] package, all parameters of the SOM have been set to default values (in particular we use hexagonal grid as initial topology of the map).

4.3 Topological Relaxation in SOM

Basically, the main principle of the new algorithm is to decrease the topological constraint of the SOM by increasing the distance between neurons. These modifications are data-driven to optimize the final quality of the SOM.

The first step in our experiment is to analyze how behave a SOM with a trivial relaxation of the topological constraint. We expect that smaller constraint leads to a better modeling (i.e. smaller Quantification and Neuron Utilization errors), but also leads to worst topological error, as the topological constraint's function is to reduce the Topological error of the SOM.

We calculate the Quantification, Neuron Utilization and Topological error for each databases from results of different versions of the SOM algorithm where each distance between two neurons is multiplied by a constant value (see Fig. 2). The higher is this value, the weaker is the topological constraints. For example, SOM(α) is similar to the SOM algorithm, but $d(i,j) = \alpha \times d_M(i,j)$. We tested different values for this constant from SOM(1), similar to SOM, to SOM(10), where neurons are almost independent (in that case the algorithm behavior is similar to a K-means algorithm).

As the gain in Ne and Qe is associated to a loss in Te, we propose to define a General error that reflect the trade-off between Ne, Qe and Te:

$$Ge = Te^2 \times Ne \times Qe$$

Ge is the product of two trade-off: Ne vs. Te and Qe vs. Te. The value of Ge is smaller when the gain in Ne and Qe is higher than the loss of Te in comparison to the SOM algorithm. Ge is bigger in the other case.

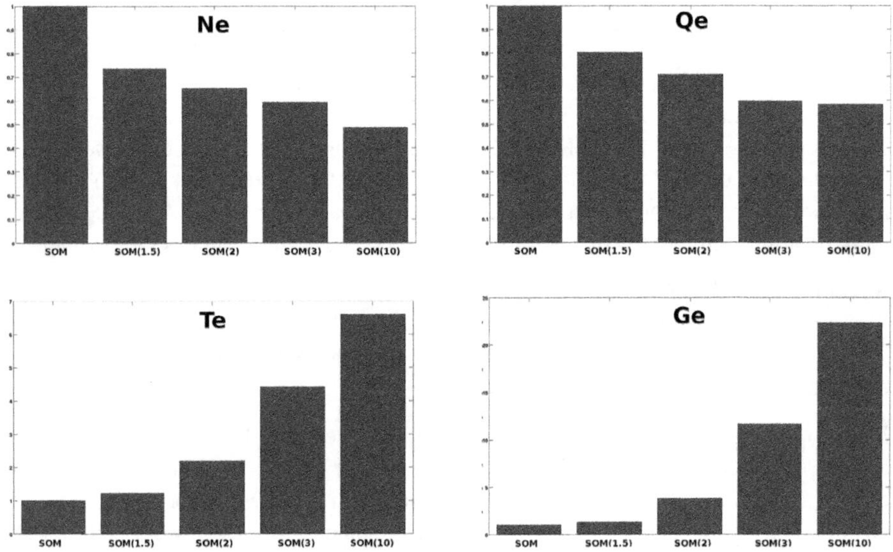

Fig. 2. Visualization of the mean value of Qe, Ne, Te and Ge over all databases

Result are summarized in Fig. 2, here we show the mean value of Qe, Ne, Te and Ge over all databases for different values of α. As expected, Ne and Qe decrease when topological constraint decrease, whereas Te highly increase. But the value of Ge shows that under relaxation of the topological constraint the gain of Ne and Qe don't overcome the loss of Te. Thus, the best trade-off is to use the classical SOM algorithm!

Now the question is: can we use data to find some topological constraint relaxations which are a better trade-off than the SOM.

4.4 Evaluation of DDR-SOM

To evaluate the quality of DDR-SOM, we compare SOM and different versions of DDR-SOM with different values of the parameter σ. Fig. 3 shows means values of Qe, Ne, Te and Ge over all databases. Table 1 show the Ge value for each databases and each algorithms.

We can note the DDR-SOM mean topological constraint is similar to SOM(1.5). But as we can see, the DDR-SOM algorithm performance is much better than SOM(1.5)

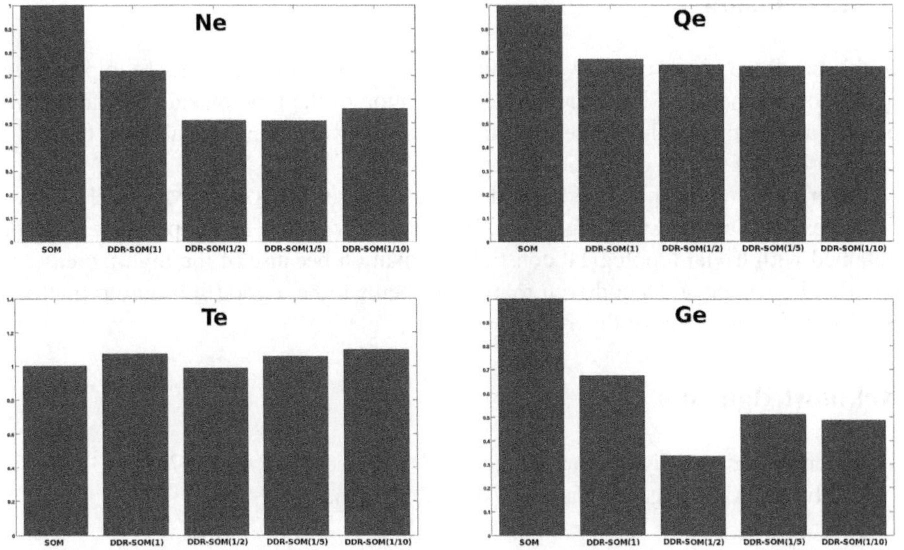

Fig. 3. Visualization of the mean value of Qe, Ne, Te and Ge over all databases for different values of σ in DDR-SOM

and SOM, i.e. the gain in Ne and Qe is higher than the loss in Te. Actually, with DDR-SOM Qe is similar that with SOM(2) and Ne is very low, similar that with SOM(10), whereas Te is similar that with the classical SOM algorithm.

Table 1. Ge value for each database and each algorithm

	DDR(1)	DDR(1/2)	DDR(1/5)	DDR(1/10)
Target	0,57	0,39	0,87	0,89
TwoDiamonds	0,22	0,27	0,22	0,11
Hepta	1,82	0,62	0,97	0,57
Tetra	1,17	0,53	1,61	1,12
Iris	0,09	0,21	0,29	0,28
Harot	0,79	0,17	0,06	0,38
Housing	0,83	0,54	0,35	0,55
Wine	0,51	0,14	0,13	0,33
Cockroach	0,62	0,42	0,52	0,54
Chromato	0,12	0,08	0,09	0,09

These results lead to two remarks:

1. The DDR-SOM quality is better than SOM for all values of σ, although a value of $\sigma = 1/2$ seems to give better results for those databases.
2. The gain in Ge in comparison to SOM tend to be higher for databases in high dimensional space (e.g. "Chromato", "Wine", etc ...).

5 Conclusion

In this paper we propose a new algorithm adapted from SOM, in order to improve the quality of the model, using a data-driven relaxation of the topological constraints. We defined a Global error that represent the trade-off between Topological error, Quantification error and Neural Utilization error.

Experiments on artificial and real databases show that DDR-SOM model achieve better results than the SOM algorithm. We also showed that this improvement is not obtained with trivial topological constraint relaxation because of the high increase of the Topological error. Data-driven relaxation seems to be a good solution to improve the $NeQe/Te$ trade-off of the SOM.

Acknowledgments

This work was supported in part by the $CADI$ project (N^o ANR-07 TLOG 003) financed by the ANR (Agence Nationale de la Recherche).

References

1. Kohonen, T.: Self-Organizing Maps. Springer, Berlin (2001)
2. Fritzke, B.: Growing grid - a self-organizing network with constant neighborhood range and adaptation strength. Neural Processing Letters 2(5), 9–13 (1995)
3. Cabanes, G., Bennani, Y.: A local density-based simultaneous two-level algorithm for topographic clustering. In: Proceeding of the International Joint Conference on Neural Networks, pp. 1176–1182 (2008)
4. Matsushita, H., Nishio, Y.: Self-Organizing Map with False-Neighbor Degree between Neurons for Effective Self-Organization. IEICE Transactions on Fundamentals E91-A(6), 1463–1469 (2008)
5. Kohonen, T.: Self-Organization and Associative Memory. Springer, Berlin (1984)
6. Johnson, D.: Efficient algorithms for shortest paths in sparse networks. Journal of the ACM 24(1), 1–13 (1977)
7. Ultsch, A.: Clustering with SOM: U*C. In: Procedings of the Workshop on Self-Organizing Maps, Paris, France, pp. 75–82 (2005)
8. Frank, A., Asuncion, A.: UCI machine learning repository (2010)
9. Kiviluoto, K.: Topology Preservation in Self-Organizing Maps. In: International Conference on Neural Networks, pp. 294–299 (1996)
10. Cheung, Y., Law, L.: Rival-Model Penalized Self-Organizing Map. IEEE Trans. Neural Networks 18(1), 289–295 (2007)
11. Vesanto, J., Himberg, J., Alhoniemi, E., Parhankangas, J.: Self-Organizing Map in Matlab: the SOM Toolbox. In: Proceedings of the Matlab DSP Conference, pp. 35–40 (1999)

Pseudo-network Growing for Gradual Interpretation of Input Patterns

Ryotaro Kamimura

IT Education Center,
1117 Kitakaname Hiratsuka Kanagawa 259-1292, Japan
ryo@keyaki.cc.u-tokai.ac.jp

Abstract. In this paper, we propose a new information-theoretic method to interpret competitive learning. The method is called "pseudo-network growing," because a network re-grows gradually after learning, taking into account the importance of components. In particular, we try to apply the method to clarify the class structure of self-organizing maps. First, the importance of input units is computed, and then input units are gradually added, according to their importance. We can expect that the corresponding number of competitive units will be gradually increased, showing the main characteristics of network configurations and input patterns. We applied the method to the well-known Senate data with two distinct classes. By using the conventional SOM, explicit class boundaries could not be obtained, due to the inappropriate map size imposed in the experiment. However, with the pseudo-network growing, a clear boundary could be observed in the first growing stage, and gradually the detailed class structure could be reproduced.

1 Introduction

In this paper, we propose a new type of information-theoretic competitive learning method called "pseudo-network growing." The method is called "pseudo" because we try to reproduce a growing process after learning is finished. The main objective is to interpret network configurations as well as input patterns clearly by inspecting the reproduction of growing processes.

Neural networks have been applied to many practical problems, but one of the major problems lies in the difficulty in interpreting final network configurations [1]. There have been many methods to simplify network configurations and to extract explicit rules for interpretation [2]. Unlike these conventional methods, our method aims to reproduce as many network configurations as possible. The method begins with the generation of the simplest configuration, with one input unit, and gradually increases the complexity. This property enables us to interpret final network configurations at different levels. This means that the overall characteristics of input patterns can be extracted at the beginning, and detailed characteristics can be detected gradually.

The pseudo-growing method can be applied to the self-organizing maps (SOM) in particular, because it is easy to visually demonstrate the good performance of

K.W. Wong et al. (Eds.): ICONIP 2010, Part II, LNCS 6444, pp. 375–382, 2010.

our method with the SOM. In addition, we consider the SOM as one of the main attempts to visually interpret internal representations. From our point of view, one of the main contributions of the SOM lies in creating interpretable network configurations with many visualization techniques. However, as is well known, the conventional self-organizing maps have several shortcomings, for example, the problem of predetermined map structure and class boundaries [3]. In the self-organizing maps, the number of competitive units and map structure must be fixed before learning. In addition, links among competitive units should also be determined, which may degrade the extraction of class boundaries. When the predetermined map structure is not appropriate to given input patterns, obtained class structures may be degraded.

To overcome these shortcomings, network growing algorithms [4] have been proposed. In those growing methods, it is not necessary to determine the map size before learning. In addition, links between competitive units are gradually generated in the course of learning, and class boundaries are gradually formed. However, in growing networks, computational complexity is increased where additional parameters should be introduced to insert or delete units [5]. This means that, though the network growing seems to be promising, the process of network growing needs many computational techniques to grow and stabilize learning with additional parameters.

In our new method, we limit ourselves to the interpretation of obtained network configurations by using the growing method. Then, the growing process is reproduced when learning has already been finished. This assures computationally inexpensive and stable learning processes. First, we compute the importance of input units by using mutual information between competitive units and input patterns. Thus, we produce feature maps by adding new input units according to the importance of the input units by mutual information. We can expect that the important features or class boundaries generated by the important input units will be generated in a course of network growing.

2 Theory and Computational Methods

2.1 Concept of Pseudo-network Growing

The pseudo-network growing aims to reproduce learning processes after learning in order to clearly show class boundaries and the main and detailed characteristics of network configurations and input patterns. At the initial stage, competitive learning is used to train connection weights. After learning is finished, a network is decomposed and the importance of input units is evaluated. Then, the network grows by recruiting a new input unit according to its importance. In addition, we can expect that the number of competitive units will be increased as the number of input units is increased.

2.2 Mutual Information and Information Enhancement

We have used information-theoretic competitive learning to realize competition [6]. We have demonstrated that mutual information between competitive units and input patterns can be used to describe the competition processes of networks. As the mutual information is increased, competition becomes more accentuated. When the mutual information is completely maximized, only one competitive unit is turned on, while all the other units are off. Thus, competitive processes are realized by maximizing mutual information between competitive units and input patterns. The output from the jth competitive unit can be computed by

$$u_j^s \propto \exp\left\{-\frac{1}{2}(\mathbf{x}^s - \mathbf{w}_j)^T \boldsymbol{\Sigma}(\mathbf{x}^s - \mathbf{w}_j)\right\}, \tag{1}$$

where $\mathbf{x}^s$ and $\mathbf{w}_j$ are supposed to represent L-dimensional input and weight vectors. The klth element of the $L \times L$ scaling matrix $[\boldsymbol{\Sigma}]_{kl}$ is defined by

$$[\boldsymbol{\Sigma}]_{kl} = \delta_{kl}\frac{p(k)}{\sigma^2}, \tag{2}$$

where σ is a spread parameter. The probability $p(k)$ denotes the firing probability of the kth input unit, and at the initial stage,

$$p(k) = \frac{1}{L}, \tag{3}$$

because we have no preference for input units at the initial stage.

To estimate the information of each input unit, we introduce the concept of attention." Attention has been one of the central themes in psychology [7], and in neural computing, many neural models have been proposed so far [8], [9]. When attention is paid to the tth input unit, we have the firing probabilities of input units

$$p(k;t) = \delta_{kt}. \tag{4}$$

Thus, we have competitive unit outputs focusing upon the tth input unit

$$u_j^s(t,\sigma) \propto \exp\left\{-\frac{1}{2}(\mathbf{x}^s - \mathbf{w}_j)^T \boldsymbol{\Sigma}_{(t)}(\mathbf{x}^s - \mathbf{w}_j)\right\}, \tag{5}$$

where the klth element of the scaling matrix $\left[\boldsymbol{\Sigma}_{(t)}\right]_{kl}$ is defined by

$$\left[\boldsymbol{\Sigma}_{(t,\sigma)}\right]_{kl} = \delta_{kl}\frac{p(k;t)}{\sigma^2}. \tag{6}$$

Because we use the self-organizing maps in the training mode, in the output layer, all output units should cooperate with each other. To realize this cooperation, we introduce a neighborhood kernel h between two units

$$h_{jc} = \exp\left(-\frac{\parallel \mathbf{r}_c - \mathbf{r}_j \parallel^2}{2\sigma_{rd}^2}\right), \tag{7}$$

where $\mathbf{r}_c$ and $\mathbf{r}_j$, respectively, denote vectors representing the position of the corresponding competitive unit. The spread parameter σ_{rd} is always set to one, because the final value of σ_{rd} in the training mode with the conventional SOM is one. Then, cooperative outputs can be defined by

$$v_j^s(t,\sigma) \propto \sum_{c=1}^{M} h_{jc} u_c^s(t,\sigma), \tag{8}$$

where M is the number of competitive units. Normalizing these outputs, we have

$$p(j \mid s; t, \sigma) = \frac{v_j^s(t,\sigma)}{\sum_{m=1}^{M} v_m^s(t,\sigma)}. \tag{9}$$

The firing probability of the jth competitive unit is defined by

$$p(j; t, \sigma) = \sum_{s=1}^{S} p(s) p(j \mid s; t, \sigma). \tag{10}$$

By using these probabilities, we have the mutual information of competitive units I when attention is paid to the tth input unit:

$$I(t,\sigma) = \sum_{s=1}^{S} \sum_{j=1}^{M} p(s) p(j \mid s; t, \sigma) \log \frac{p(j \mid s; t, \sigma)}{p(j; t, \sigma)}. \tag{11}$$

This mutual information shows how well the tth input unit contributes to a process of competition among competitive units [6]. As this information gets larger, the tth input unit plays a more essential role in realizing competitive processes. Using this information, we estimate the firing probability of the kth input unit

$$p(k; \sigma) = \frac{I(k; \sigma)}{\sum_{t=1}^{L} I(t; \sigma)}. \tag{12}$$

Then, we define information in input units by

$$I(\sigma) = \sum_{k=1}^{L} p(k; \sigma) \log L p(k). \tag{13}$$

We try to increase this information in input units as much as possible to make the number of important input units as small as possible.

2.3 Pseudo-network Growing

Then, we reproduce self-organizing maps by increasing the number of input units gradually. A new input unit is added according to the information or importance of the input unit by changing the spread parameter σ. Now, let us formulate the gradual change of mutual information by adding a new input unit to the existing network. For this, we define Φ_m^M to denote a set of m important units of total M units, and defined by

$$q(k; \Phi_m^M, \sigma) = \begin{cases} p(k; \sigma) & \text{if } k \in \Phi_m^M, \\ 0 & \text{otherwise.} \end{cases}$$

The firing probabilities of the kth input unit are defined

$$p(k; \Phi_m^M, \sigma) = \frac{q(k; \Phi_m^M, \sigma)}{\sum_{t \in \Phi_m^M} q(t; \Phi_m^M, \sigma).} \tag{14}$$

The competitive unit output is defined by

$$u_j^s(\Phi_m^M, \sigma) \propto \exp\left\{ -\frac{1}{2}(\mathbf{x}^s - \mathbf{w}_j)^T \mathbf{\Sigma}_{(\Phi_m^M)} (\mathbf{x}^s - \mathbf{w}_j) \right\}, \tag{15}$$

where the klth element of the matrix $\left[\mathbf{\Sigma}_{(\Phi_m^M)}\right]_{kl}$ is defined by

$$\left[\mathbf{\Sigma}_{(\Phi_m^M)}\right]_{kl} = \delta_{kl} \frac{p(k; \Phi_m^M, \sigma)}{\sigma^2}. \tag{16}$$

The cooperative firing probabilites can be defined by

$$p(j \mid s; \Phi_m^M, \sigma) = \frac{\sum_{c=1}^{M} h_{jc} u_c^s(\Phi_m^M, \sigma)}{\sum_{l=1}^{M} \sum_{c=1}^{M} h_{lc} u_c^s(\Phi_m^M, \sigma)}. \tag{17}$$

By using these probabilities, the information with the first m important input units is computed by

$$I(\Phi_m^M, \sigma) = \sum_{s=1}^{S} \sum_{j=1}^{M} p(s) p(j \mid s; \Phi_m^M, \sigma) \log \frac{p(j \mid s; \Phi_m^M, \sigma)}{p(j; \Phi_m^M, \sigma)}. \tag{18}$$

3 Results and Discussion

We present experimental results on the Senate data [10] to show the good performance of our method. We used the SOM toolbox developed by Vesanto et al. [11], because experimental results presented in this paper can thus be easily reproduced . The value of the spread parameter σ was determined so as to maximize the information of input units (equation 13).

Figures 1(a), (b) and (c) show quantization errors, topographic errors and mutual information, respectively, without any specific attention paid to input units (equation 11) as a function of the number of input units. Figure 1(a) shows quantization errors as a function of the number of input units. The quantization errors are increased as the number of input units is increased. For comparison, we used the random method in which a new input unit to be added was randomly chosen. The final values of the random method are the averages over ten different runs. With this random method, the final error becomes equivalent to that of the conventional SOM, namely, 0.049, shown in Figure 1(a). On the other hand, the network growing method produces the value of 0.042, which is lower than that produced by the conventional method, again shown in Figure 1(a). Figure 1(b) shows the topographic errors as a function of the number of input units. When

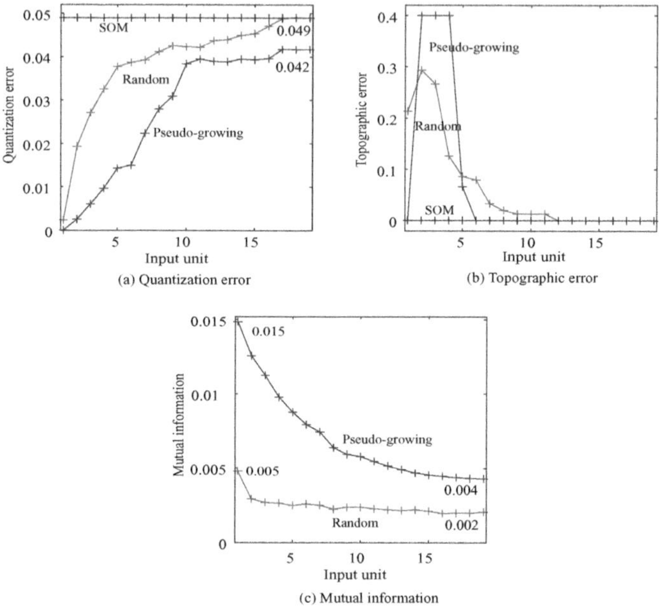

Fig. 1. Quantification errors (a), topographic errors (b) and mutual information as a function of the number of input unit (c).

the network growing method is used, at the initial phase, the error is zero, and then it jumps to 0.4. Finally, the error becomes zero. Compared with the errors produced by the random method, the errors produced by network growing are reduced to zero faster. Figure 1(c) shows mutual information as a function of the number of input units. At the initial stage, the mutual information is 0.015 with the network growing method, while with the random method, the information is less than 0.005. Even at the final stage, the mutual information is 0.004 with the network growing method, while with the conventional method, the information is 0.002.

Figure 2 shows a U-matrix (1) and labels (2) obtained by the conventional SOM (a) and the network growing method (b)-(f). As already mentioned, this data can be clearly divided into two classes by any conventional statistical method, such as cluster analysis and principal component analysis. However, as shown in Figure 2(a), the U-matrix obtained by the conventional SOM fails to show the main class boundaries, because the map size is too large for extracting appropriate class boundaries. On the other hand, with the network growing, the number of input units is increased from one in Figure 2(b) to 19 in Figure 2 (f). In addition, the number of competitive units responding to input patterns is increased from two in Figure 2(a) to 14 in Figure 2(f). These figures clearly show that the pseudo-network growing method can realize a process of growing that increases the number of input units as well as competitive units. When the number of units is one, in Figure 2(b), a clear boundary in warmer color in the

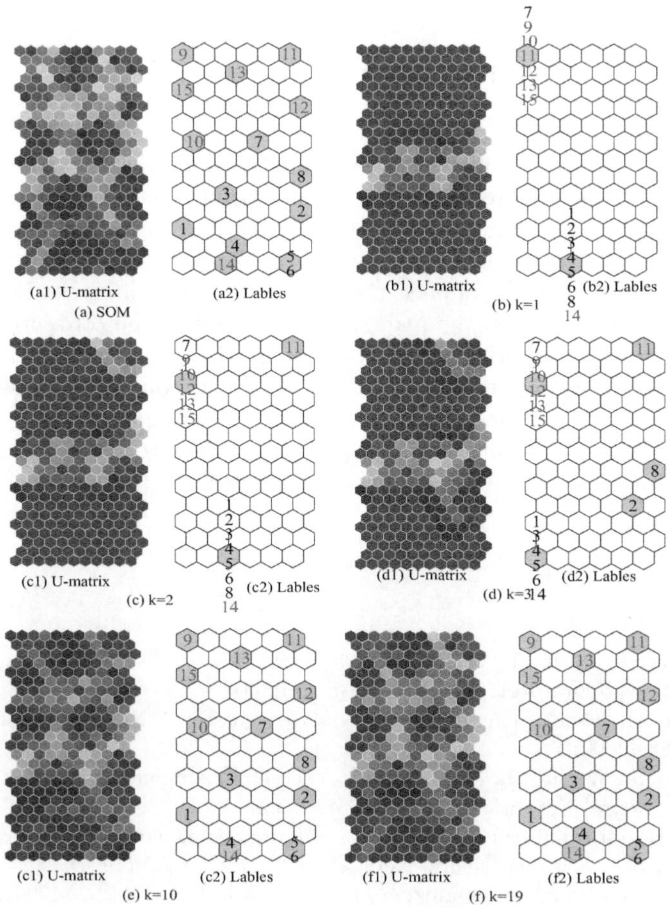

Fig. 2. U-matrix and labels obtained by the conventional SOM (a) and the network growing method (b)-(f). Republicans and Democrats are represented in black and red, respectively.

middle of the map can be seen. This means that input patterns can be classified into two groups by this boundary in the middle of map. When the number of input units is increased to two, shown in in Figure 2(c), the number of competitive units is increased to three, of which one unit, corresponding to the 11th input pattern, is separated from the others. When the number of input units is increased to three, Figure 2(d) shows that two additional competitive units corresponding to input patterns No. 2 and No. 8 are separated. In the course of growing, the number of competitive units to be separated is gradually increased, as shown in Figure 2(e). Finally, when the number of input units is 19, namely, when all input units are used, we can still see a class boundary clearly in the middle of the map in Figure 2(f).

4 Conclusion

In this paper, we have proposed a new type of information-theoretic method to interpret network configurations. We applied the method to the self-organizing maps in particular, because it is thus easy to demonstrate the good performance of our method visually. Contrary to the conventional approach to network growing, in our method a network re-grows after learning has been finished. In the growing process a new input unit is added according to its importance or information content. In addition, the number of competitive units is expected to be gradually increased. We applied the method to the well-known Senate data with distinct classes. When the map size was too large, the conventional SOM failed to produce appropriate class boundaries. However, the pseudo-network growing method was able to gradually produce main boundaries classifying input patterns into two groups and, later, detailed class boundaries. Though there are some problems with this new method, such as how to stop growing in this pseudo-growing, it certainly contributes to the interpretation problem of neural networks.

References

1. Alexander, J.A., Mozer, M.C.: Template-based procedures for neural network interpretation. Neural Networks 12, 479–498 (1999)
2. Ishikawa, M.: Rule extraction by successive regularization. Neural Networks 13, 1171–1183 (2000)
3. Marsland, S., Shapiro, J., Nehmzow, U.: A self-organizing network that grows when required. Neural Networks 15, 1041–1058 (2002)
4. Fritzke, B.: Growing cell structures – a self-organizing network for unsupervised and supervised learning. Neural Networks 7(9), 1441–1460 (1994)
5. Fritzke, B.: Growing self-organizing networks – why? In: ESANN 1996: European Symposium on Artificial Neural Networks, pp. 61–72 (1996)
6. Kamimura, R.: Information-theoretic competitive learning with inverse Euclidean distance output units. Neural Processing Letters 18, 163–184 (2003)
7. Anderson, J.R.: Cognitive Psychology and its Implication. Worth Publishers, New York (1980)
8. Korsten, N.J.H., Fragopanagos, N., Hartle, M., Taylor, N., Taylor, J.G.: Attention as a controller. Neural Networks 19, 1408–1421 (2006)
9. Hamker, F.H., Zirnsak, M.: V4 receptive field dynamics as predicted by a systems-level model of visual attention using feedback from the frontal eye field. Neural Networks 19, 1371–1382 (2006)
10. Romesburg, H.C.: Cluster Analysis for Researchers. Krieger Publishing Company, Florida (1984)
11. Vesanto, J., Himberg, J., Alhoniemi, E., Parhankangas, J.: SOM toolbox for Matlab. tech. rep., Laboratory of Computer and Information Science, Helsinki University of Technology (2000)

The Adaptive Authentication System for Behavior Biometrics Using Pareto Learning Self Organizing Maps

Hiroshi Dozono[1], Masanori Nakakuni[2], Shinsuke Itou[1], and Shigeomi Hara[1]

[1] Faculty of Science and Engineering, Saga University,
1-Honjyo Saga 840-8502, Japan
hiro@dna.ec.saga-u.ac.jp
[2] Information Technology Center, Fukuoka University,
8-19-1, Nanakuma, Jonan-ku, Fukuoka 814-0180, Japan
nak@fukuoka-u.ac.jp

Abstract. In this paper, we propose an authentication system which can adapt to the temporal changes of the behavior biometrics with accustoming to the system. We proposed the multi-modal authentication system using Supervised Pareto learning Self Organizing Maps. In this paper, the adaptive authentication system with incremental learning which is applied as the feature of neural networks is developed.

1 Introduction

The most basic part of the computer security is authentication. The most popular method for authentication is password authentication. But, password authentication has some issues. At first, anyone can login to the system if the password is known because the password is simple text information. It may be possible to obtain the password with peeping the login operation, guessing from the personal information like birthday or family name, getting the memo with password. Furthermore, because identical password tends to be used for some different systems, the secure system can be logged in using the password obtained from week system. For this problem, biometric authentication systems are often used[1].

Biometric authentication systems are classified to the system which uses biological characteristics and the system which uses behavior characteristics. As the biological characteristics, the fingerprint, iris patterns and the vein patterns are often used, and the accuracy of authentication is very high. But, the special hardware for measuring the characteristics is required and it raises the cost of computer system. Furthermore, it is possible to spoof a registered user using the dummy of body parts because the biological characteristics is static information. As the behavior characteristics, keystroke timings[2], sign and handwritten symbols are often used. Some of these characteristics can be measured using the devices equipped to the computer without adding special hardwares. It is difficult to spoof because the behavior characteristics is dynamic information even if the authentication process is seen by illegal user. However, the accuracy

K.W. Wong et al. (Eds.): ICONIP 2010, Part II, LNCS 6444, pp. 383–390, 2010.

of authentication is lower than that using biological information, because the behavior characteristics varies with the variation in each authentication action, noises and changes by time with accustoming to the authentication action. Some methods which can improve the accuracy of authentication with integrating multiple behavior characteristics are proposed[3].

For the integration of multiple behavior characteristics, Self Organizing Map (SOM)s can be applied. Conventional SOM can integrate the multiple input vectors with concatenating them with weight values. Using these features, SOM can be applied to find the behavior characteristics effective for authentication and to implement the authentication system using multiple behavior characteristics[5]. As an variation of SOM, Supervised Pareto learning SOM (SP-SOM), which can integrate the multiple vectors without weight values and can perform supervised learning, is proposed to improve the accuracy of authentication and applied for the authentication system using multiple behavior characteristics[6].

As mentioned before, the behavior characteristics may change by time while the users accustom to authentication method. Thus, the authentication system should adapt the change of behavior characteristics by time. Neural networks can adapt the change of environment with learning. The authentication systems using SOM and SP-SOM, which are mentioned before, learn the behavior characteristics while they register initial data. The adaptive authentication system using the incremental learning ability of SP-SOM was proposed[7].

On the other hand, behavior characteristics is easily affected by variation of behaviors and noises, and it decreases the accuracy of authentication. Neural networks considered to be robust to the variation of input data. However, the accuracy may be affected with incremental learning the input vectors including noises. The affects of the noises to the learning using SP-SOM were examined in [8].

In this paper, considering the implementation the authentication system, some modifications and experimental results of adaptive authentication system which is based on SP-SOM is mentioned. The experimental results under the environment adding both of the changes by time and noises in behavior characteristics are shown. Furthermore, for the implementation of authentication system, the algorithm of SP-SOM is modified. At first, the algorithm of incremental learning is modified as to use the input vector which are successfully authenticated because it is impossible to distinguish the legal user who is failed to be authenticated from illegal user who intend to spoof the legal user for authentication system. However, this modification may affects the accuracy of authentication and the affects are examined experimentally. Secondly, the algorithm of recalling process is modified as to detect the unregistered user. The original SP-SOM algorithm classifies the any input vector to one of the learned class, so the input vector of unregistered user is also classified to one of the registered user. The recalling algorithm is modified as to detect the unlearned data with introducing the threshold values to the size of Pareto set and the magnitude of category value. However, it may also affect the adaptation ability of incremental learning and the affects are examined experimentally.

2 Pareto Learning Self Organizing Maps

The Pareto learning SOM is proposed for the learning of multi-modal vectors, such as multi-modal biometric data. Pareto learning SOM(P-SOM) uses the concept of Pareto Optimality in the competitive phase of learning and the members in the Pareto set becomes winners and the region of Pareto winners and their neighbors are updated simultaneously. Fig.1 shows the difference of the learning algorithm. With this updating method, the multi-modal vectors are organized

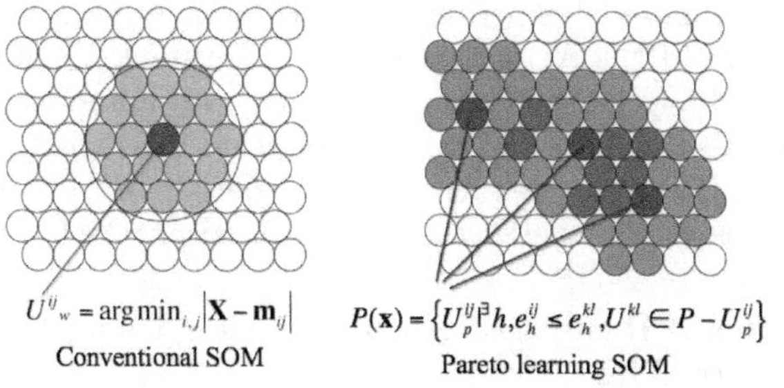

Fig. 1. Difference of the algorithm between SOM and P-SOM

naturally regardless to the magnitude of each vector in multi-modal vector. P-SOM can integrate not only the vectors but also the objects for which metric can be defined, for example graph models, images and other neural networks. As the extensions of P-SOM, Supervised Pareto learning SOM (SP-SOM) , which uses the input vectors with category vectors, is proposed to improve the accuracy of classification using the category vectors for organizing the better clusters on the map. And the adaptive learning algorithm for tuning the size of Pareto sets is introduced. The details of the algorithm is mentioned in [7] and [8].

3 Experimental Results of Authentication Using Keystroke Timings and Key Typing Sounds

The purpose of this paper is the development of authentication system which can adapt to the changes of behavior characteristics by time. However, it is almost impossible to obtain the behavior characteristics data changing by time because it will need very long term and heavy load to the examinees to obtain the behavior characteristics changing by time experimentally. Thus, the artificial data changing by time are generated from measured data. The artificial data will not make the experimental results any less justified because very severe settings of the changes in data compared with actual changes of users are selected and enough randomness is introduced in the change to avoid the advantages of machine learning systems to the steady changing data. In this experiment, the keystroke timings and key typing sounds data, which are obtained from 10 users, typing a phrase "kirakira" in ten times for each user, are used. As a behavior characteristics, each data contains 15 keystroke timing data, which is the key pushing times and intervals between keys, and 8 key typing sound data, which is the maximum amplitude of key typing sound for each key, and they are composed as 2 feature vectors of behavior characteristics.

3.1 Experimental Results of Authentication Using Original Data

In this experiments, 5 data are used for learning and remaining 5 data are used for testing for each user. All combinations of learning data and testing data were examined, so the number of experiments is ${}_{10}C_5 = 2520$, and the number of test data for each user is 1260. As the authentication system, SP-SOM is used. The size of map is 16x16, size of initial neighbor is 4, and initial learning coefficient is 0.5. FRR and FAR denote the rate for rejecting the registered user and the rate for accepting the wrong user respectively, and both of them should be small enough. As the result, the averages are FRR=0.108 and FAR=0.012.

3.2 Experimental Results of Authentication Using the Data Changing by Time

In this sub-sectuon, the authentication experiments using the data changing by time are mentioned. As mentioned before, it is considered to be difficult to obtain data for behavior characteristics changing by time, artificial data was used for the experiments. In this experiment, N_{ci} elements of the vector $\mathbf{x_i}$ in test data are randomly selected and multiplied by R_{ci} before authentication and substituted. where $\mathbf{x_1}$ is keystroke timing vector and $\mathbf{x_2}$ is key typing sound. Because SP-SOM has no dynamics, constant multiply rate R_{ci} does not give advantage to machine learning process. All of the data are learned and and are used as test data with change. For all data, the authentication experiments are made in 20 times. Only average FRR is shown in the following graphs and average FAR is 1/9 of FRR because the threshold for rejecting wrong user is not used in this experiment. Fig.2 shows the results for changing key stroke timings using the parameters $N_{c1} = 4, R_{c1} = 0.8$(right side) and for changing both of the keystroke timings and key typing sounds using the parameters $N_{c1} = 4, N_{c2} = 2, R_{c1} = R_{c2} = 0.8$. 2 modes of incremental learning is examined, supervised incremental learning which uses both the input vector with the user label and unsupervised incremental learning which uses the input vector without user label. These experiments simulate the situation that the key typing speed becomes faster for each authentication and latter one also simulate the changes of key typing sounds by time. For the case of changing key stroke timings, FRR becomes worse as repeating the tests without incremental learning, because the test data is changing. With incremental learnings, FRR does not becomes so worse becausethe authentication system can adapt the change of test data. The supervised incremental learning showsthe best result and the FRR is almost kept to 0. For the case of changing both of keystroke timings and key typing sound, FRR increases rapidly and becomes 0.9, which means the random selection without incremental learning because both of keystroke timing data and

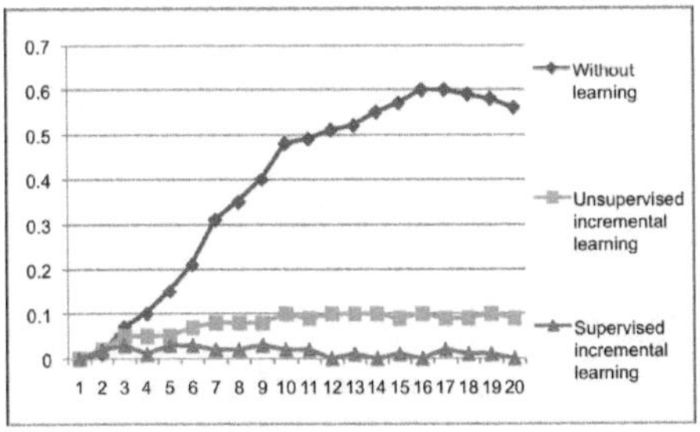

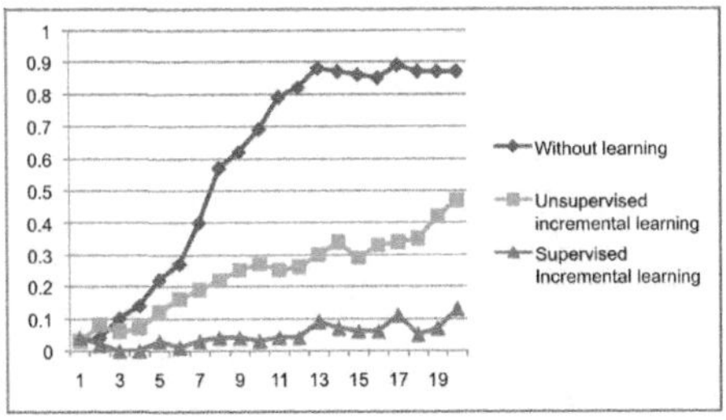

Fig. 2. Experimental result of FRR for the case of changing the features by time. Left:($N_{c1} = 4$, $R_{c1} = 0.8$), Right:($N_{c1} = 4$,$N_{c2} = 2$, $R_{c1} = R_{c2} = 0.8$).

key typing sounds becomes unusable for authentication. WIth supervised incremental learning, FRR kept low enough, so the system is considered to adapt the changes of the data by time.

3.3 Experimental Results of Authentication Using the Test Data with Variations and Noises

As mentioned before, behavior characteristics may vary in each authentication and it may be affected by noises especially for the case using key typing sounds. With incremental learning, the noises and variations may affect more to the accuracy of authentication. Both noises and variations vary the test data temporally, thus both of them can be modeled as noises. In this experiments, N_{ni} elements of the vector $\mathbf{x_i}$ in test data are randomly selected and added the uniform random noises which amplitudes are R_{ni} times of current values in maximum. the test data. The difference from the change by time is that the noise is temporal and test data is not substituted. Fig.3 shows the results of authentication experiments with adding the changes by time and noises simultaneously. In the left figure, FRR becomes worse(0.2) even in the case with supervised incremental learning because both changes by time and noises are too large. However, this setting is too severe considering the actual variations. In the right figure, which shows the results for adding the half amounts of variations, FRR is kept low because the system can adapt the changes by time despite of adding noises. The adaptation ability is affected by noises, however it can adapt to the changes by time to some degree.

3.4 Experimental Results of Incremental Learning of Authenticated User Data

As mentioned before, the incremental supervised learning is effective as the adaptive authentication system. However, some system may use the policy that the user who can not be authenticated by behavior characteristics can not login to the system using other authentication method. In this case, the data of failed user is not used for incremental learning and it may lead to the deterioration of

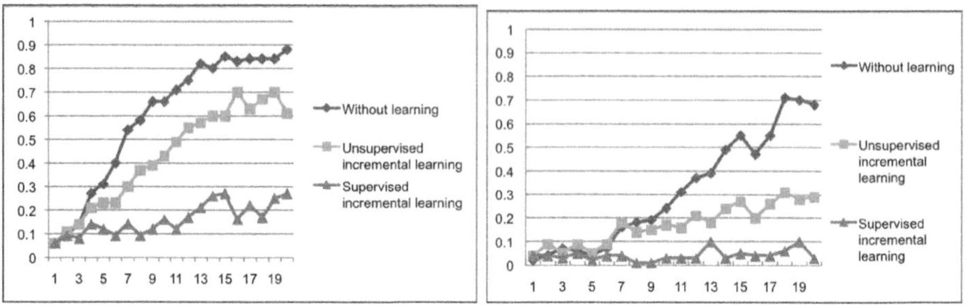

Fig. 3. Experimental result of FRR with changing inputs and adding noises(Left: $N_{c1} = 4, N_{c2} = 2, R_{c1} = R_{c2} = 0.8, N_{n1} = 8, N_{n2} = 4, R_{n1} = R_{n2} = 0.5$, Right: $N_{c1} = N_{n1} = 4, N_{c2} = N_{n2} = 2, R_{c1} = R_{c2} = 0.9, R_{n1} = R_{n2} = 0.5$)

adaptation. Fig.4 shows the results of incremental learning of the authenticated user data. The other parameters are same as those of the right figure in the right figure of Fig.3. Compared with the case of learning all data(all time), FRR

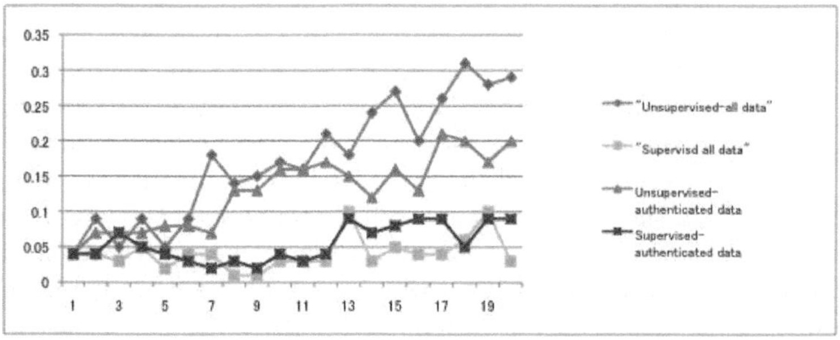

Fig. 4. Experimental result of FRR for the case of incremental learning at successful authentication ($N_{c1} = N_{n1} = 4, N_{c2} = N_{n2} = 2, R_{c1} = R_{c2} = 0.9, R_{n1} = R_{n2} = 0.5$)

becomes worse with unsupervised incremental learning using authenticated data only. With supervised incremental learning, FRR is almost same as that of the case of learning all data, and this system can adapt the changes by time using authenticated data only. The authentication system can adapt the changes by time and robust to the noises for the case with the incremental learning of authenticated used data.

3.5 Experimental Results of Detecting Unregistered User Using Threshold

In the experiments mentioned before, the threshold values which are used for detecting unregistered users are not introduced. The unregistered user can be detected using the threshold of size of Pareto set and magnitude of category

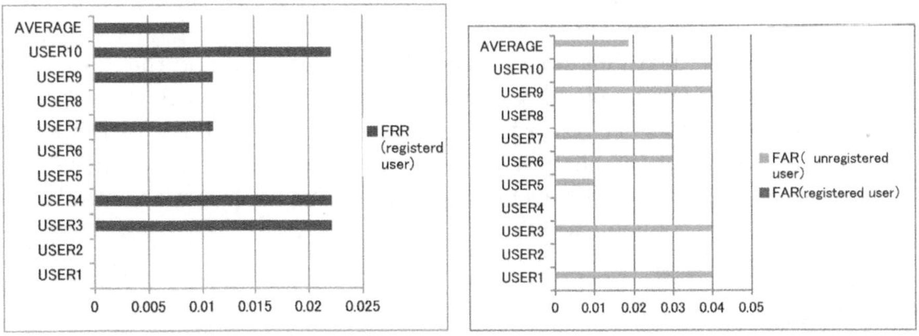

Fig. 5. Result of the authentication experiments for the case rejecting unregistered users

value. For the unlearned data, the size of Pareto set and magnitude of category value tend to be large and small respectively. At first, the authentication experiment using 9 of 10 users as registered user and remaining 1 user as unregistered user without incremental learning is made. Fig.5 shows FRR and FAR of this case. The Thresholds of Pareto size and category value are set as 10 and 0.5 respectively. FRR of the registered user is 2% in maximum and FAR is almost 0. FAR for unregistered user is 2% in average and 4% in maximum. The unregistered users are detected well without changes by time and noises. Fig.6 shows averages of FRR and FAR with supervised incremental learning adding changes by time and noises. FRR becomes gradually worse, however it kept under 10%. FAR is kept low for both registered user and unregistered user. The authentication system can adapt the changes by time and robust to the noises for the case of detecting the unregistered user.

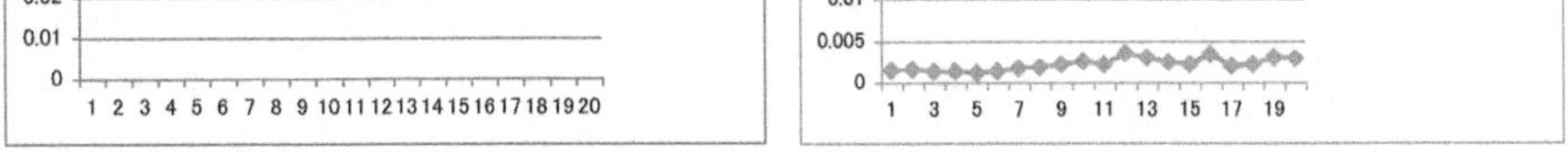

Fig. 6. Result of the authentication experiments for the case rejecting unregistered users with changing inputs and noises

4 Conclusions

In this paper, an adaptive authentication system which can adapt to the changes of the behavior characteristics by time using incremental learning of Supervised

Pareto Learning SOM is developed and the robustness of the incremental learning to the noises is examined. For the implementation of actual authentication system, incremental learning of the authenticated user data and the authentication system using the threshold to detect the unregistered users are examined and confirm the effectiveness of supervised incremental learning. In the experiments, the amount of the changes by time and noises are set extremely large, so it will be expected that the accuracy will be better than those shown in this paper in the actual implementation.

References

1. Bolle, R., Connell, J., Pankanti, S., Ratha, N., Senior, A.: Guide to Biometrics. Springer, Heidelberg (2004)
2. Monrose, F., Rubin, A.D.: Keystroke Dynamics as a Biometric for Authentication. Future Generation Computer Systems (March 2000)
3. Dokic, S., Kulesh, A., et al.: An Overview of Multi-modal Biometrics for Authentication. In: Proceedings of The 2007 International Conference on Security and Management, pp. 39–44 (2007)
4. Kohonen, T.: Self Organizing Maps. Springer, Heidelberg, ISBN 3-540-67921-9
5. Nakakuni, M., Dozono, H., et al.: Application of Self Organizing Maps for the Integrated Authentication using Keystroke Timings and Handwritten Symbols. Wseas Transactions On Information Science & Applications 2-4, 413–420 (2007)
6. Dozono, H., Nakakuni, M.: An Integration Method of Multi-Modal Biometrics Using Supervised Pareto Learning Self Organizing Maps. In: Proceedings of 2008 International Joint Conference on Neural Networks, pp. 603–607. IEEE, Los Alamitos (2008)
7. Dozono, H., Nakakuni, M.: Application of Supervised Pareto Learning Self Organizing Maps amd Its Incremental Learning. In: Advances in Self Organizing Maps. LNCS, vol. 5629, pp. 54–62. Springer, Heidelberg (2009)
8. Dozono, H., Nakakuni, M.: Analysis of robustness of pareto learning SOM to variances of input vectors. In: Chan, J.H. (ed.) ICONIP 2009, Part II. LNCS, vol. 5864, pp. 836–844. Springer, Heidelberg (2009)

Human Action Recognition by SOM Considering the Probability of Spatio-temporal Features

Yanli Ji, Atsushi Shimada, and Rin-ichiro Taniguchi

Department of Advanced Information Technology
Kyushu University, Fukuoka, Japan
{yanli,atsushi,rin}@limu.ait.kyushu-u.ac.jp
http://limu.ait.kyushu-u.ac.jp

Abstract. In this paper, an action recognition system was invented by proposing a compact 3D descriptor to represent action information, and employing self-organizing map (SOM) to learn and recognize actions. Histogram Of Gradient 3D (HOG3D) performed better among currently used descriptors for action recognition. However, the calculation of the descriptor is quite complex. Furthermore, it used a vector with 960 elements to describe one interest point. Therefore, we proposed a compact descriptor, which shortened the support region of interest points, combined symmetric bins after orientation quantization. In addition, the top value bin of quantized vector was kept instead of setting threshold experimentally. Comparing with HOG3D, our descriptor used 80 bins to describe a point, which reduced much computation complexity. The compact descriptor was used to learn and recognize actions considering the probability of local features in SOM, and the results showed that our system outperformed others both on KTH and Hollywood datasets.

Keywords: Computer vision, Human action recognition, SOM.

1 Introduction

Generally, a variety of studies on action recognition concentrate on two important issues. One is how to extract useful information from raw video data. While the other is how to model different actions, then their similarities is measured for recognition.

When extracting useful information from a video, there are mainly two categories, global feature and local feature extraction. We mainly consider local feature extraction in this paper. For local feature, many spatial-temporal feature detectors[1,2,3,4,6,7] and descriptors[5,8,9,10] have been proposed in the past few years. The detectors usually differ in the type and the sparsity of selected points. While feature descriptors capture shape and motion information in neighbor of the selected point by various image measurements, such as spatial or spatio-temporal image gradients, optical flow, etc.

However, complicated descriptors were employed in previous methods. They would suffer from high computation complexity and high memory requirement.

K.W. Wong et al. (Eds.): ICONIP 2010, Part II, LNCS 6444, pp. 391–398, 2010.

For instance, although HOG3D[9] outperforms other currently used descriptors, it employs a quite complex algorithm to calculate descriptor, and uses a vector with 960 bins to describe an interest point. It takes long time to compute descriptors if there are a quantity of interest points. Generally, a considerable quantity of points are required to represent human actions exactly. Thus it is necessary to construct a simpler system with a compact descriptor.

In recognizing actions using local features, the representative learning algorithms are Support Vector Machines(SVM)[8,9], Hidden Markov Model(HMM), Near Neighbor(NN), Beyesian classifier, Fern classifier, etc. SOM[12] is a type of artificial neural network that is trained using unsupervised learning to produce a low-dimensional, discretized representation of the input space of the training samples. It is certified to perform well in [11] for action recognition.

In this paper, inspired by [13], we detect corners spatially and temporally to obtain information of action shape and motion. Features from Accelerated Segment Test(FAST) corner detector[4] is extended to 3D space for detecting spatio-temporal corners. Furthermore, we modify the HOG3D descriptor to be a compact descriptor, which shortens the support region of interest points. In orientation quantization, it combines symmetric bins after orientation quantization by icosahedron, and keeps the top value bin of quantized vectors. In addition, suitable parameters for the descriptor are determined experimentally. And the first order gradient is chosen for descriptor calculation. The compact descriptor reduces much calculation time, occupies less memory. In recognition, SOM is adopted to learn and recognize different actions. In addition, the probability of local features of each action class is considered in SOM to recognize actions correctly.

The remaining parts of this paper are organized as follows. Detailed information of 3D corner detection in ROI is shown in section 2. Then the modified 3D feature descriptor is introduced in detail in section 3. Section 4 is learning and recognizing by SOM. The experiment results and comparison with other researches are shown in section 5.

2 3D FAST Corner Detection in ROI

FAST[4] is a faster and more stable feature detector. Here, it is extended to 3D space, detecting features spatially and temporally, to obtain shape and motion information of human actions.

2.1 3D FAST Corner Detection

If x, y represent axes of spatial plane, and t is temporal axis, a video can be regarded as a 3D space with axes x, y and t. The spatio-temporal corners in videos can be obtained by detecting corners in xy, xt, yt planes. In addition, xt planes in video compose xt channel, while yt planes compose yt channel. xy plane is each frame in video.

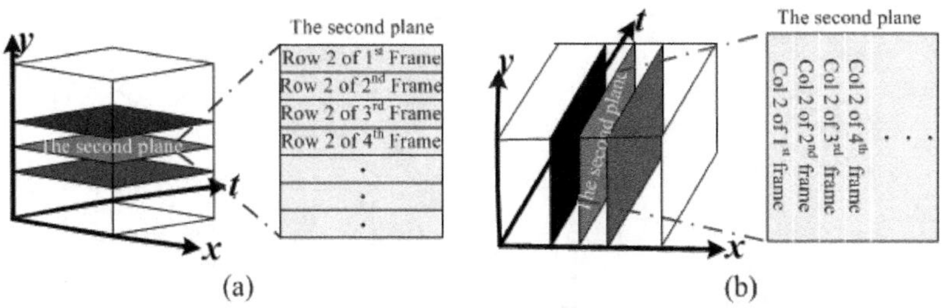

Fig. 1. (a) xt channel and its composition. The left part of (a) shows positions of xt tangent planes in video, and the right part of (a) is the detailed structure of the second xt plane, which is composed of the second row pixels of every frame in video; (b) yt channel and its composition. The left part of (b) shows the positions of yt planes, and the right part of (b) shows the structure of the second yt plane, which is composed of the second column pixels of each frame.

xt planes are serial tangent planes in horizontal orientation of video as shown in Fig.1(a), and axes of each plane are row x of frame and temporal axis t. yt channel is also composed of serial tangent planes, but yt planes are vertical tangent planes in video. The axes of every plane are the column y of frame and temporal axis t. The sketch figure of yt channel is shown in Fig.1(b).

In this paper, FAST corner detection can be operated in two or three channels to obtain 3D corners. At last, detection in two channels is chosen based on our experiments. Furthermore, to achieve more efficient and representative corners, 3D corner detection is applied in original video and down sampled video. The corners which are detected twice in these two steps are regarded as interest corners and they are kept for the following processing.

2.2 ROI Extraction

Generally, human actions occur in rather complex surroundings than in experimental backgrounds. When corner detection is operated on frames with a complex background, many corners in the background will be included. It will largely influence the following steps. To avoid the sad effect from them and to reduce computation complexity, region of interest(ROI) extraction is performed.

The centroid of the detected corners and its neighbor are used to extract the ROI of each frame in video. On each frame, the centroid of corner points is calculated by obtaining the mean position of all corners in xt, yt channels.

And ROI is determined considering the distance from the centroid to the ROI boundary. Denote that T is a threshold of distance from centroid to boundary of ROI, which is experimentally decided. It is compared with the maximum and minimum of corner coordinates. If the distance from centroid to the maximum, minimum is less than T, ROI is established based on the maximum and minimum coordinates. Otherwise, $centroid \pm T$ is decided to be the boundary of ROI.

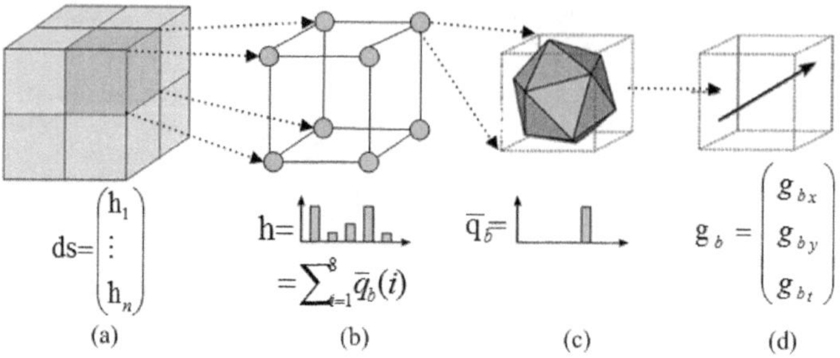

Fig. 2. Compact descriptor calculation. (a) the support region ($2 \times 2 \times 2$ cells) and descriptor of one interesting point; (b) histogram calculation of one cell ($2 \times 2 \times 2$ points); (c) orientation quantization; (d) gradient calculation.

3 Compact Descriptor

Based on the original HOG3D[9] ,we have improved it to save computation time and to provide more exact descriptors. First of all, the method of support region division is changed to a simpler way. The support region of an interest point is defined as a cuboid with the size of $4 \times 4 \times 4$ pixels around the point, and the cuboid is divided into 8 cells again, each with the size of $2 \times 2 \times 2$ pixels. In this case, each cuboid contains 8 cells, and each cell contains 8 neighbor points. Our descriptor calculation is shown in Fig.2, and detailed explanation is in the following parts.

3.1 3D Gradients and Orientation Quantization

In section 2.1, xt, yt planes are separated from videos. The gradients of x, t orientations (g_x, g_t) can be obtained from calculating the gradients on xt planes, while (g_y, g_t) can be obtained from calculating the gradients on yt planes. Then each point is given a gradient vector of x, y, t orientations, $(g_x, g_y, g_t)^T$.

In [9], average gradient of each point calculated in support region was employed for orientation quantization. In this paper, we try the average gradient, the second order gradient, and the original gradient $(g_x, g_y, g_t)^T$ for orientation quantization in the experiments. It is found that the original gradient performs best. Denote g_{bx} is gradient in x orientation of point b, the gradient of the point for orientation quantization is $\mathbf{g_b} = (g_{bx}, g_{by}, g_{bt})$, which is shown in Fig.2(d).

In the step of orientation quantization, we keep half elements of orientation quantized gradient since elements of it are positive-negative symmetry. And we keep the maximum element of quantized vectors obtained in above step instead of setting threshold experimentally in [9]. After that, the quantized gradient vectors are recorded as $\bar{\mathbf{q}}_\mathbf{b}$, as shown in Fig.2(c).

For each cell, the histogram $\mathbf{h}$ for the cell region is obtained by summing the quantized gradient vectors $\bar{\mathbf{q}}_\mathbf{b}$ of all elements. When $\bar{\mathbf{q}}_\mathbf{b}(i, j)$ is the jth bin in the quantized gradient of element i, and there are total s elements in the cell, each bin $\mathbf{h}(j)$ of $\mathbf{h}$ is calculated as shown in Fig.2(b):

$$\mathbf{h}(j) = \sum_{i=1}^{s} \bar{\mathbf{q}}_{\mathbf{b}}(i, j) \tag{1}$$

3.2 Descriptor Computation

As section 3.1 introduced, the support region of one interesting point is divided into 8 cells. Then each cell is described by one histogram. The same with [9], all histograms in support region of one interest point are finally concatenated to one feature vector, as shown in Fig.2(a).

$$\mathbf{ds} = (\mathbf{h}_1, \mathbf{h}_2, \cdots, \mathbf{h}_8) \tag{2}$$

It is the final descriptor for one interesting point. Then the feature vectors for all interesting points of one action class are given an action label *action*. So the descriptor for one action class becomes

$$ds\_action = \{\mathbf{ds}_1, \mathbf{ds}_2, \cdots, \mathbf{ds}_C, action\} \tag{3}$$

where C is the interesting point numbers of the action, and it varies on different action classes because the quantity of corners detected from different videos is always different.

4 Recognition by SOM

In our experiments, the basic theory of Batch learning[12] is employed to train a map using descriptors of all action classes.

Supposing that $\{\mathbf{m}_i\}$ are model vectors of neuron nodes in SOM map, where i refers to neuron number. $\mathbf{ds}_{lk}(l = 1, \cdots, L, k = 1, \cdots, C_l)(=\{ds\_action_l\})$ are input descriptors of all actions for training, where L refers to action classes and C_l refers to interesting point numbers of action class l. $\mathbf{m}_i$ has the same dimension with feature vector $\mathbf{ds}_{lk}$. In the first step, the map with model vectors $\{\mathbf{m}_i\}$ is trained based on Batch learning[12] using all descriptors $\mathbf{ds}_{lk}$. Supposing that $\mathbf{R_i}$ is the neighborhood set of nodes that lie up to a certain radius from node i in the map, the procedure can be described as:

(1) Initialize the values of the $\{\mathbf{m}_i\}$ in some proper way. (2) Input all the $\mathbf{ds}_{lk}$, one by one, and list each of them under the model $\mathbf{m}_i$ that is closest to $\mathbf{ds}_{lk}$ according to some distance, generally Euclidean distance. (3) Let U_i denote the union of descriptors matched with model $\mathbf{m}_i$ and those matched with nodes in $\mathbf{R_i}$. Compute the means of the vectors $\mathbf{ds}_{lk}$ in each U_i, and replace the old values of $\mathbf{m}_i$ by the respective means by formula (4). (4) Repeat from the second step a few times until the solutions can be regarded as steady.

$$\mathbf{m}_i^* = \frac{\sum_{\mathbf{ds}_{lk} \in U_i} \mathbf{ds}_{lk}}{n(U_i)} \tag{4}$$

where, $n(U_i)$ means the number of descriptors $\mathbf{ds}_{lk}$ that belong to U_i.

Following that, all the descriptors $\{\mathbf{ds}_{lk}\}$ are input to the trained map one by one again to match with $\mathbf{m}_i$. The number of $\mathbf{ds}_{lk}$ matching with $\mathbf{m}_i$ is counted and recorded as N_{il}. Then N_{il} is normalized by interesting point numbers C_l of each action class using formula (5). The normalized Nc_{il} represents the contribution on neuron i from action class l in the map, which is denoted as contribution probability of action l.

$$Nc_{il} = N_{il}/C_l = N_{il}/\sum_i N_{il} \tag{5}$$

So the trained map contains 2 components, and it is described as $\{\mathbf{m}_i, Nc_{il}\}$.

The trained SOM map is employed to recognize actions of videos. Supposing that there is a video with frame number F to be tested, and we detect corners and calculate descriptors as section 2 and 3 introduced. The descriptor for the video is recorded as $\{\mathbf{ds}_{fk}, f = 1, \cdots, F, k = 1, \cdots, K\}$, where K is number of interesting points on frame f and it varies on different frame.

To recognize the action class of current video, the action of each frame is determined firstly. On the fth frame, if $\mathbf{ds}_{fk}$ matches with $\mathbf{m}_i$ best, $\max(Nc_{il}, l \in 1, \cdots, L)$ determines the action class of $\mathbf{ds}_{fk}$. Then statistic of labels of $\{\mathbf{ds}_{fk}\}$ on the frame f is recorded as N_{fl}, and l, corresponding to the $\max(N_{fl})$, gives us the action class of the frame. This process is repeated for all frames. The maximum of the statistic of labels of all frames shows us the action class of the tested video.

5 Experiment Results

The recognition in this paper are operated on KTH[14] and Hollywood[8] datasets. Some sample figures are shown in Fig.3, where the top row figures are from KTH, while the bottom row figures are from Hollywood.

As other researchers always do for KTH dataset, we divide the dataset samples into training/validation set (8+8 persons) and test set (9 persons). With the purpose of testing our descriptors in a living context, 4 types actions ("stand up", "sit down", "hug" and "hand shake") in the Hollywood dataset are selected for recognition, and 9 video samples of each action for training and 9 samples for testing are chosen.

In table 1, we compared the average recognition accuracy of our result with the results of correlative algorithms. [15] gave an evaluation of local spatio-temporal features on currently used detectors and descriptors. Based on it, algorithms of HoF, HoG/HoF and HOG3D performed better than other algorithms, and the average accuracy of HoF and HoG/HoF were 92.1%, 91.8%, respectively. In [9], the average accuracy of HoF and HOG3D were recorded as 86.7% and 91.4%. Here we compare our result with the best result of each algorithm. Results show that our algorithm outperforms others. The first reason is that the compact descriptor describes local features exactly. When descriptor is calculated, the top value bin of quantized gradient is kept, which corresponds feature of a point. It

Fig. 3. Sample figures of KTH and Hollywood dataset

Table 1. Average recognition accuracy comparison

Algorithm	HoG[8]	HoF[8]	HoG/HoF[15]	3DHOG[9]	Ours
KTH (%)	81.6	86.7	91.8	91.4	93
Hollywood(%)	30.55	22.55	-	25.52	36.1

makes descriptor representing feature of a cell more exactly. Frame match experiment is performed using compact descriptor, and matching ratio reaches 88.9%. Furthermore, we have tried recognition by SVM and SOM with the compact descriptors, and result indicates that SOM provides better recognition result. It can be said that the recognition system based on the probability of local features in SOM provides more correct distinguish of action classes.

About Hollywood, our algorithm also performs better. However the accuracy is much lower comparing with KTH because scene cut, scale change and luminance change always exist in a movie. In addition, the compact descriptor works not so well when the action is in large size, for instance actions in Hollywood. Another reason is that the actions in Hollywood are always incomplete because of close shot.

6 Conclusion

A compact 3D descriptor and a new human action recognition system based on the probability of local features in SOM are proposed in this paper, and our results are compared with other researcher's. As it is described above, our system performs better in recognizing actions in both KTH and Hollywood. However, our algorithm have some limitations. Firstly, the quantity of necessary local features is not easy to determine. Then compact descriptor is not suitable for close shot action, for instance, "hug" in Hollywood. In addition, the system performs bad in describing actions of a few person interaction.

In our future research, we will try to look for efficient characters of actions, and try to use less information to describe actions. At the same time, we will devote on recognizing actions of a few person interaction.

References

1. Harris, C., Stephens, M.: A combined corner and edge detector. In: 4th Alvey Vision Conference. Elsevier North-Holland, The Netherlands (1988)
2. Dollar, P., Rabaud, V., Cottrell, G., Belongie, S.: Behavior recognition via sparse spatio-temporal features. In: 2nd Joint IEEE International Workshop on Visual Surveillance and Performance Evaluation of Tracking and Surveillance, pp. 65–72 (2005)
3. Laptev, I., Lindeberg, T.: On Space-time interest points. In: 6th IEEE International Conference on Computer Vision, pp. 432–439 (2003)
4. Rosten, E., Drummond, T.: Machine learning for high-speed corner detection. In: Leonardis, A., Bischof, H., Pinz, A. (eds.) ECCV 2006. LNCS, vol. 3951, pp. 430–443. Springer, Heidelberg (2006)
5. Willems, G., Tuytelaars, T., Gool, L.V.: An efficient dense and scaleinvariant spatio-temporal interest point detector. In: Forsyth, D., Torr, P., Zisserman, A. (eds.) ECCV 2008, Part II. LNCS, vol. 5303, pp. 650–663. Springer, Heidelberg (2008)
6. FeiFei, L., Perona, P.: A bayesian hierarchical model for learning natural scene categories. In: 15th IEEE Computer Society Conference on Computer Vision and Pattern Recognition, pp. 524–531 (2005)
7. Jurie, F., Triggs, B.: Creating efficient codebooks for visual recognition. In: 8th IEEE International Conference on Computer Vision, pp. 604–610 (2005)
8. Laptev, I., Marszalek, M., Schmid, C., Rozenfeld, B.: Learning realistic human actions from movies. In: 18th IEEE Computer Society Conference on Computer Vision and Pattern Recognition (2008)
9. Klaser, A., Marszalek, M., Schmid, C.: A spatio-temporal descriptor based on 3D gradients. In: 19th British Machine Vision Conference, pp. 995–1004. British Machine Vision Association, Worcs (2008)
10. Scovanner, P., Ali, S., Shah, M.: A 3-dimensional sift descriptor and its application to action recognition. In: 15th ACM International Conference on Multimedia, pp. 357–360. ACM, New York (2007)
11. Shimada, A., Taniguchi, R.: Gesture recognition using sparse code of hierarchical SOM. In: 18th International Conference on Pattern Recognition (2008)
12. Kohonen, T.: Self-Organizing Maps. Springer, Berlin (1995)
13. Gilbert, A., Illingworth, J., Bowden, R.: Fast realistic multi-action recognition using mined dense spatio-temporal features. In: 12th IEEE International Conference on computer Vision (2009)
14. Schuldt, C., Laptev, I., Caputo, B.: Recognizing human actions: A local SVM approach. In: 14th International Conference on Pattern Recognition, pp. 32–36 (2004)
15. Heng, W., Muhammad, M.U., Klaser, A., Laptev, I., Schmid, C.: Evaluation of local spatio-temporal features for action recognition. In: British Machine Vision Conference, pp. 127–137 (2009)

On Generalization Error of Self-Organizing Map

Fumiaki Saitoh[1] and Sumio Watanabe[2]

[1] Imagine Science and Engineering Laboratory, Tokyo Institute of Techinology, R2-52, 4259, Nagatsuta-chou, Midori-ku, Yokohama, 226-8503, Japan
saitoh.f.aa@m.titech.ac.jp
[2] Precision and Intelligent Laboratory, Tokyo Institute of Techinology, R2-5, 4259, Nagatsuta-chou, Midori-ku, Yokohama, 226-8503, Japan
swatanab@pi.titech.ac.jp

Abstract. Self-organizing map is usually used for estimation of a low dimensional manifold in a high dimensional space. The main purpose of applying it is to extract the hidden structure from samples, hence it has not been clarified how accurate the estimation of the low dimensional manifold is. In this paper, in order to study the accuracy of the statistial estimation using the self-organizing map, we define the generalization error, and show its behavior experimentally. Based on experiments, it is shown that the learning curve of the self-organizing map is determined by the order that are smaller than dimensions of parameter. We consider that the topology of self-organizing map contributed to abatement of the order.

Keywords: Self-organizing Map, Generalization Error, Statistical Learning, Information Extraction.

1 Introduction

The self-organizing map(SOM)[1]is an artificial neural network model proposed by kohonen. There is information that corresponds to the topology of mathematics between each node of SOM. This information shows whether each node is near or is far.Because topology is defined, SOM can extract low dimensional manifold from the high dimensional data space as information.

On the other hand, recent studies have suggested that Statistical models and neural network models are different in a mathematical character;

(1) Competitive learning models and layered models aren't regular models. Hence, neural network models cannot be discussed by Fisher's statistical asymptotic theory.
(2) Learning models that have layered structure and hidden state like neural network models have many singularities. Learning models like neural network that have layered structure and hidden state have many singularities. These models can learn accurately and stably by bayesian learning.

These mathematical characters were derived by considering the behavior of the generalization error.

K.W. Wong et al. (Eds.): ICONIP 2010, Part II, LNCS 6444, pp. 399–406, 2010.

Therefore, investigation of the generalization error is important for not only model design and model selection but also showing that the neural network models are significant in the information engineering. The behavior of the generalization error of SOM has not been researched enough. In this paper, we define the generalization error, and demonstrate its behavior experimentally. We consider that research of generalization error plays an important role in optimum design of SOM and character clarification of SOM as statistical model.

2 Definition of Generalization Error

Because the self orgaizing map is not uniquely decided as the probabilistic model, generalization error is not uniquely decided. Furthermore, various learning algorithm of self-organizing map are proposed. Henceforth in this paper, we use 1-dimentional SOM to simplify description. But, this theory can be applied similarly to multi-dimensional SOM. The following cases are assumed in this paper.

2.1 Distribution of Data

Let,(x, y) is element of euclidean space$\mathbf{R}^1 \times \mathbf{R}^1$. The value x is a score from the probability distribution $q(x)$. The value y comes from conditional probability distribution

$$q(y|x) = \frac{1}{\sqrt{2\pi\sigma^2}} \exp\Big(-\frac{1}{2\sigma^2}|y - f(x)|^2\Big)$$

when the value x is given.where $\sigma > 0$ is standard deviation,$f(x)$ is continuous function. Let generate samples $z_i = (x_i, y_i)(i = 1, 2...n)$ independently from joint probability distribution

$$q(z) = q(x)q(y|x),$$

where $z = (x, y) \in \mathbf{R}^2$. Let us call these samples training data. Here, we considers one dimensional manifold

$$y - f(x) = 0$$

in two dimensional euclidean space to simplify description.

2.2 Learning Algorithms of Batch-Learning SOM (BL-SOM)

We present the definition of Self-organizing Map(SOM) and batch-learning algorithm here. Let m be a natural number and let $\boldsymbol{w}_1, \boldsymbol{w}_2, ..., \boldsymbol{w}_m \in \mathbf{R}^2$ be weight vectors of all nodes. These vectors are optimized to input data by learning algorithms. The equation that show the winner node and the update rule are described as follows:

$$c = \arg\max_i \| \boldsymbol{w}_i - \boldsymbol{z}_j \| \tag{1}$$

$$w_i(t+1) = \frac{\sum_{j=1}^{j_{\max}} h_{ci} z_j}{\sum_{j=1}^{j_{\max}} h_{ci}} \tag{2}$$

where h_{ci} is neighborhood function,$j_{\max}$is the number of samples, $z_j$is input vector.

The equation of neighborhood function are described as follows:

$$h_{ci}(t) = \alpha(t) \cdot \exp\left(-\frac{\| r_i - r_c \|^2}{2\sigma^2(t)}\right) \tag{3}$$

$$\sigma(t) = \sigma_f + (\sigma_i - \sigma_f)\exp\left(-\frac{t}{T}\right) \tag{4}$$

where $\alpha(t)$is learning rate, $\| r_i - r_c \|$is euclidean distance between winner node c and $i-th$ node defined by topology, $\sigma(t)$is a Neighborhood radius, $\sigma_i$is a initial value of neighborhood radius, $\sigma_f$is a last value of neighborhood radius.

The algorithms are described as follows:

(1) Weight vectors in self-organizing map($w_1, w_2, ..., w_m$) are initialized to random values.

(2) This process is repeated for a number of cycles γ(usually large).

(2.1) Input vectors $\{(z_1, z_2, ..., z_{j_{max}})\}$ are inputted to all the nodes at the same time in parallel.

(2.1.1) Distances between z_j and all the weight vectors are calculated. The winner node, denoted by c, is the node with the weight vector closest to the input vector z_j(Eq.1).

(2.1.2) Neighborhood functions of all nodes, denoted by h_{cj},are calculated to dataz_j

(2.2) $z_1, z_2, ..., z_{j_{max}}$ is calculated by the update rule(Eq.2).

---

in this study, we consider that probability distribution of self-organizing map is represented by

$$p(z|b) = \frac{1}{m}\sum_{k=1}^{m} \frac{1}{(2\pi b^2)} \exp\Big(-\frac{1}{2b^2}|z - w_k|^2\Big), \tag{5}$$

where b is a control parameter used when the generalization error is measured.

The definition of the distribution estimated by self-organizing map is not unique. In this paper, we use Eq.5 as the distribution estimated by self-organizing map. Information on topology is not included in this definition. It is future tasks to define probability distribution including information of topology of SOM.

2.3 Generalization Error

The generalization error is defined by Kullback Leibler distance

$$G = \int q(z) \log \frac{q(z)}{p(z|b)} dz,$$

where $q(z)$ is true probability distribution, $p(z|b)$ is probability distribution estimated by self-organizing map. G is approximately obtained by the empirical Kullback information

$$G \approx \frac{1}{N} \sum_{j=1}^{N} \log \frac{q(z_j)}{p(z_j|b)},$$

where $\{z_1, z_2, ..., z_N\}$independent and identical samples from the true distribution $q(z)$.

2.4 Learning Curve

In general, it is known well that Learning Curve becomes

$$G(n) = L_0 + \frac{\lambda}{n} + o(\frac{1}{n}), \tag{6}$$

where $G(n)$ is the average generalization error,and is the function of N that is size of training data, L_0 and λ are constants decided by true distribution, model and algorithm. In the learning theory, it is a key problem to clarify the relation

$$(TrueDistribution, Model, Algorithm) \mapsto (L_0, \lambda). \tag{7}$$

For example, If true distribution is included in the learning model and fisher information matrix is regular, L_0 is entropy of true distribution and $\lambda = d/2$ where d is dimention of parameter. And, it doesn't depend on the learning algorithm. If true distribution isn't included in the learning model and fisher information matrix isn't regular, L_0 and λ are depend on true distribution, model and algorithm.Nevertheless, whether fisher information matrix is regular or not, Eq.(6) is approved.

In this paper, we study the learning curve of self-organizing map that has not been researched until now. In the data space of the p-dimension, self-organizing map has parameter of mp-dimension. However, we predicted that because $\boldsymbol{w}_k (k = 1, 2, ..., m)$ is a parameter that mutually influences when it is learning, degree of freedom is lower than parameter of dimension.

3 Experiments

This section presents results of several experiments to examine behavior of generalization error of self-organizing map.

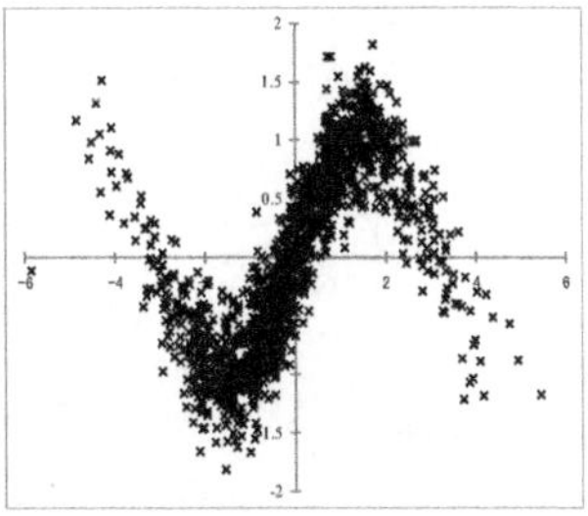

Fig. 1. An example of data set sample

3.1 Experimental Setting

We conducted our experiment on the artificial data set generated by

$$q(x) = \frac{1}{\sqrt{2\pi\rho^2}} \exp\Big(-\frac{x^2}{2\rho^2}\Big)$$

$$q(y|x) = \frac{1}{\sqrt{2\pi\sigma^2}} \exp\Big(-\frac{1}{2\sigma^2}|y - \sin(x)|^2\Big).$$

In this experiment,we use next parameter set.

$$\begin{aligned} \rho &= \pi/2 \\ \sigma &= 0.3 \\ n &= 50, 100, 150, ..., 950, 1000 \\ m &= 10, 20, 30 \end{aligned}$$

The schematic diagram of an example of data set sample(n=1000) is shown in Fig.1.

3.2 Experimental Result

Learning result of SOM. An example of learning result of self-organizing map is shown in Fig.2.

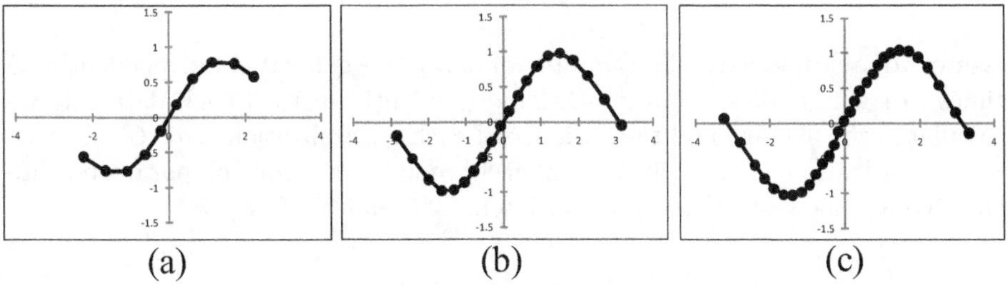

Fig. 2. Learning results of Batch-Learning Self-Organizing Map for Target data. (a)m=10. (b)m=20. (c)m=30.

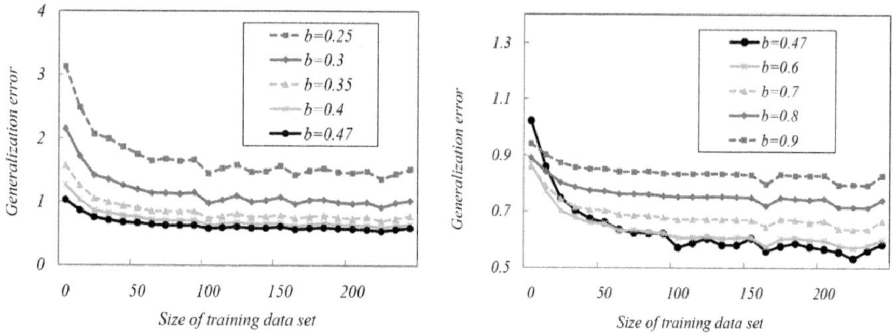

Fig. 3. Generalization error(m=10)

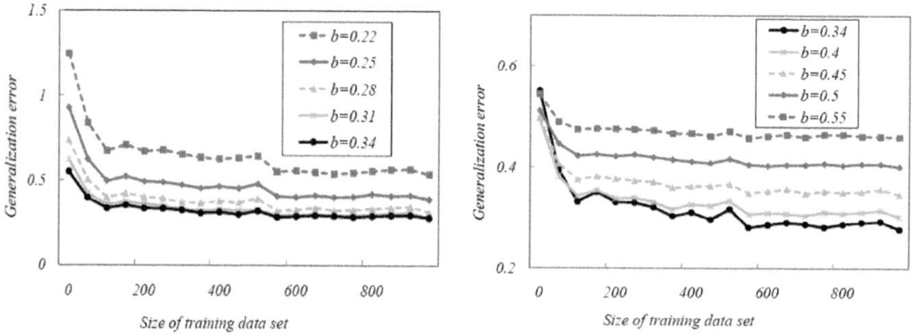

Fig. 4. Generalization error(m=20)

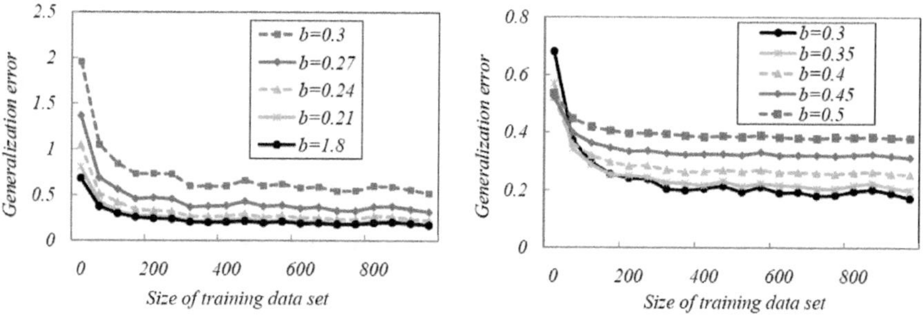

Fig. 5. Generalization error(m=30)

Generalization Error. The training data set was generated independently 25 times in each condition. After SOM independently learned each data set, we calculated the average and the variance of each generalization error G.

Genrearization errors when number of nodes and control parameter are changed are shown in Fig.3(m=10),Fig4(m=20) and Fig.5(m=30).

$$G = A + \frac{B}{n} + noise$$

where A corresponds to L_0,B corresponds to λ.

Table 1. Coefficient of learning curve obtained from experimental result

m	10	20	30
A	0.549	0.271	0.164
B	5.192	12.949	18.304

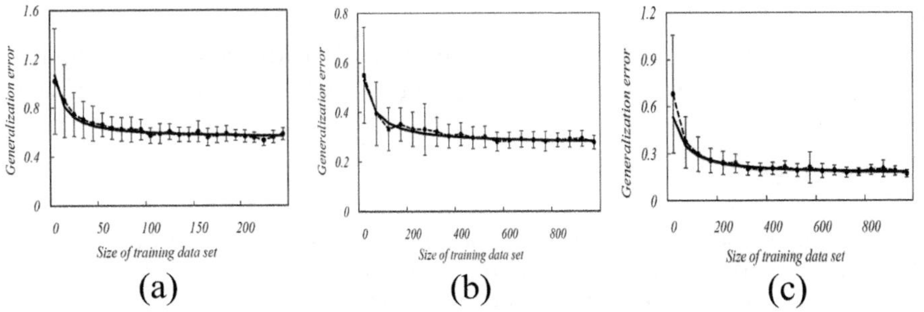

Fig. 6. Generalization error. (a)$m = 10$, optimal parameter $b = 0.47$. (b)$m = 20$, optimal parameter $b = 0.34$. (a)$m = 30$, optimal parameter $b = 0.3$.

G was fitted to the learning curveg that uses value of optimum b by least squares method.

Table1 shows the calculation value of A and B obtained for $m = 10, 20, 30$. fitted learning curves and generalization error is shown in Fig.6.

3.3 Discussion

In this experiment,when the number of nodes is m, the dimension of the parameter becomes $2m$ because SOM exists in two dimension space.

If self-organizinng map is statistical regular model, the relation between constant m and λ is as follows.

$$m = 10, 20, 30,$$
$$\lambda = 10, 20, 30.$$

The relation between constant m and B estimated from the outcome of an experimentis as follows(see Table1).

$$m = 10, 20, 30,$$
$$B = 5.192, 12.949, 18.304.$$

We consider that because the degree of freedom of SOM is lower than the dimension of the parameter, B has become smaller than λ. Therefore, topological structue of SOM influences the estimation of those parameters. Here,the predictions of section2 were confirmed.

4 Conclusions

In this paper,we defined the generalization error of self-organizing map, and showed its behavior experimentally. The experimental results demonstrated that the learning curve of the self-organizing map is determined by the order that are smaller than dimensions of parameter. We consider that the topology of self-organizing map contributed to abatement of the order.

Acknowledgments. This work was supported by the Ministry of Education, Science, Sports, and Culture in Japan, Grand-in-aid for scientific research 18079007.

References

1. Kohonen, T.: Self Organizing Maps, 2nd edn. Springer, Berlin (1995)
2. Aoyagi, M., Watanabe, S.: Stochastic complexities of reduced rank regression in Bayesian estimation. Neural Networks 18(7), 924–933 (2005)
3. Hartigan, J.A.: A Failure of likelihood asymptotics for normal mixtures. In: Proceedings of the Berkeley Conference in Honor of J.Neyman and J.Kiefer, pp. 807–810 (1985)
4. McLachlan, G., Peel, D.: Finite Mixture Models. Wiley, New York (2000)
5. Watanabe, S.: Algebraic analysis for nonidentifiable learning machines. Neural Computation 13(4), 899–933 (2001)
6. Watanabe, S.: Algebraic geometry and statistical learning theory. Cambridge University Press, Cambridge (2009)
7. Yamazaki, K., Watanabe, S.: Singularities in mixture models and upper bounds of stochastic complexity. International Journal of Neural Networks 16(7), 1029–1038 (2003)
8. Yamazaki, K., Aoyagi, M., Watanabe, S.: Asymptotic Analysis of Bayesian Generalization Error with Newton Diagram (to appear in Neural Networks)

A Novel Approach for Sound Approaching Detection

Hirofumi Tsuzuki, Mauricio Kugler, Susumu Kuroyanagi, and Akira Iwata

Department of Computer Science and Engineering
Nagoya Institute of Technology
Showa-ku, Gokiso-cho, 466-8555, Nagoya, Japan
hrfm.tsuzuki@gmail.com, mauricio@kugler.com, {bw,iwata}@nitech.ac.jp

Abstract. The detection of approaching vehicles is a very important topic on the development of complementary traffic safety systems. However, the majority of the proposed approaches are very complex and not suitable for embedded applications. This paper proposes a new sound approaching detection algorithm specifically intended for hardware implementation. Experimental results show higher accuracy and earlier detection when comparing to other methods.

Keywords: approaching detection, time before arrival, hardware implementation.

1 Introduction

Driving safety is one of the major concerns of the automotive industry nowadays. Video cameras and movement sensors are used in order to improve the driver's perception of the environment surrounding the automobile [2][9]. These methods present good performance when detecting objects (e.g., cars, bicycles, and people) which are in line of sight of the sensor, but fail in case of obstruction or dead angles. Moreover, the use of multiple cameras or sensors for handling dead angles increases the size and cost of the safety system. The human being auditory system plays a major role in the surrounding environment perception. Not only the recognition of potentially hazardous events but also the localization and movement judgment (distance and speed) of such sounds is a very advanced ability [8]. If such ability could be reproduced by artificial systems, it would enable the development of new security systems, man-machine interface and interactive devices. In particular, to detect the approach of a sound emitting object without depending on cameras or more complex sensors is a topic that has been studied by several researches, usually aiming complementary traffic safety systems.

For instance, Hoshino [3] proposed an approaching detection method based on heuristical relations of the level variation between time windows of sound taken from two microphones. This algorithm can detect an approaching vehicle 2.2 seconds before its arrival, in average, although real experiments showed an

K.W. Wong et al. (Eds.): ICONIP 2010, Part II, LNCS 6444, pp. 407–414, 2010.

average of 1.4 seconds with a large variance. Kodera, Itai and Yasukawa [4] introduced a method of speed and arrival time estimation of vehicles using the signals of a 4-microphone array. Their method calculates the average and medians of a spectrogram constructed from the beam-forming of the Short-Term Fourier Transform of each microphone signal. A Support Vector Machine then classifies the extracted features in several categories, representing all combinations of possible speeds and arrival times, with great accuracy.

These methods, however, fail in two important aspects: they are either too complex to be implemented in embedded systems or only detect the approaching shortly before the source arrival. This paper proposes a sound approaching strategy specifically intended for hardware implementation. The new approach is suitable for embedded applications, while presenting good performance on real-environment experiments.

2 Proposed Method

Kugler *et. al.* [5,6] demonstrated a sound recognition system specifically developed for hardware implementation. This system can be divided in three main blocks: signal processing, feature extraction and classification. The signal processing block, inspired by the human hearing system, is composed by a bandpass filter bank, a hair-cell non-linear function and a spike generator. The classification consists of a Learning Vector Quantization (LVQ) neural network followed by a integration layer named time potentials. These modules form a robust framework for classifying real-time sound-related data, with the requirement that the generated features must be binary. The proposed system is based on the same framework, except for the feature generation module, and its main structure is shown in Fig. 1.

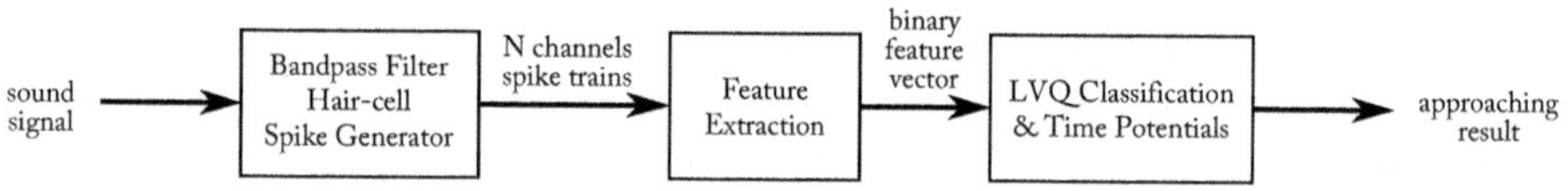

Fig. 1. Proposed sound approaching system structure

The sound signal is sampled at 48kHz, converted to single-precision floating-point representation and sent to the filter bank module, which divides it in N frequency channels. For sound approaching detection, 5 channels between 1 kHz and 2 kHz where used. The signals are compressed by a non-linear function corresponding to the inner hair-cells of the hearing system. The hair-cells function[7] is defined as:

$$f(x) = \begin{cases} x^{\frac{1}{3}} & x \geq 0 \\ \frac{1}{4}x^{\frac{1}{3}} & x < 0 \end{cases} \quad (1)$$

where x is the input value. The spikes are generated from the compressed signal. The inter-spike period T is calculated as follow:

$$T = K\frac{x_{\max} - x_{\min}}{x - x_{\min}} \tag{2}$$

where $x_{\max}$ and $x_{\min}$ are limiting factors for the value of x and K is a constant. After the period is calculated, the equivalent integer value it is compared with the correspondent timer and, if it overflows the calculated period, a spike is generated on the channel. All spike trains $p_n(t)$ $(n = 1 \ldots N)$ become the input data of the feature extraction module.

In order to detect an approaching sound, it is necessary to measure the variation of the sound signal energy along time. If the variation is positive, the sound source is approaching the subject, otherwise, it is either stopped or getting away of it. Both later cases are not important for the current application and will be considered as the negative category. This has to be performed for each of the frequency channels generated by the signal processing module.

As the sound is already converted to spikes, the signal energy for each nth frequency channel can be measured simply by counting the number of spikes in a time window of W samples:

$$x_n(t) = \sum_{i=0}^{W-1} p_n(t - i) \tag{3}$$

where $p_n(t)$ is the spike train value on time t, $n = 1 \ldots N$.

A naive approach for determining if the sound is approaching would be just calculating the difference of two consequent vectors $x_n(t)$ and $x_n(t + W)$. However, in real situations, the sound signal contains noise from the environment (e.g. wind) and also secondary sounds possibly moving in different directions. Thus, S windows are used and the feature vector is calculated as follow:

$$F_n^s(t) = \begin{cases} 1 & if \;\; x_n(t + sW) - x_n(t + (s-1)W) > 0 \\ 0 & otherwise \end{cases} \tag{4}$$

where F_n^s represents the sth binary feature of the nth frequency channel and $s = 1 \ldots S$. If the difference of the two consecutive windows is too small, the feature should represent a stationary sound. Hence, another vector with the same number of features is defined as:

$$V_n^s(t) = \begin{cases} 1 & if \;\; |x_n(t + sW) - x_n(t + (s-1)W)| > \beta \\ 0 & otherwise \end{cases} \tag{5}$$

where β is the minimal level threshold. Finally, the complete features vector is formed by the concatenation of the two vectors $\mathbf{F}$ and $\mathbf{V}$.

As stated before, the classification is performed by a standard LVQ neural network [1]. As the patterns were reduced to simple binary vectors, they can be compared by Hamming distance:

$$d(\mathbf{z},\omega) = \sum_{i=1}^{R} |z_i - \omega_i| \qquad (6) \qquad\qquad \omega_n^b = \begin{cases} 0 & if \;\; \omega_n < 0.5 \\ 1 & otherwise \end{cases} \qquad (7)$$

where $\mathbf{z}$ represents the samples formed by the $\mathbf{F}$ and $\mathbf{V}$ vectors, ω are the reference vectors and $R = 2NS$ is the number of dimensions of the final feature vector. The elements of ω, during the training process, are converted to binary values only for distance calculation.

It can be assumed that the sound sources will not present instant changes on speed and direction, i.e. they keep the movement direction for periods of time much larger than the size of the time windows. Thus, by the use of potentials similar to the membrane potential of spiking neurons, one can remove the instant errors from the LVQ neural network without modifying the training process. The time potentials are defined as:

$$u_k(t) = \begin{cases} \min(u_{\max}, u_k(t-1) + \gamma) & if \;\; k = y(t) \\ \max(0, u_k(t-1) - 1) & if \;\; k \neq y(t) \end{cases} \qquad (8)$$

where u_k is the potential of the kth category, γ is the increment for the winner category $u_{\max}$ is the maximal potential and $y(t)$ is the LVQ's classification. Hence, the winner category at time t is the one with higher $u_k(t)$ value.

3 Experiments

Fig. 2 shows experimental conditions. The microphone was fixed on the right bottom car's rear windshield. When recording the sound of approaching vehicles, the reference car with the microphone was idling on the side lane. The feature extraction parameters were tuned experimentally. The LVQ neural network labels the input vectors as "approaching" or "nothing" (stopped or departing categories).

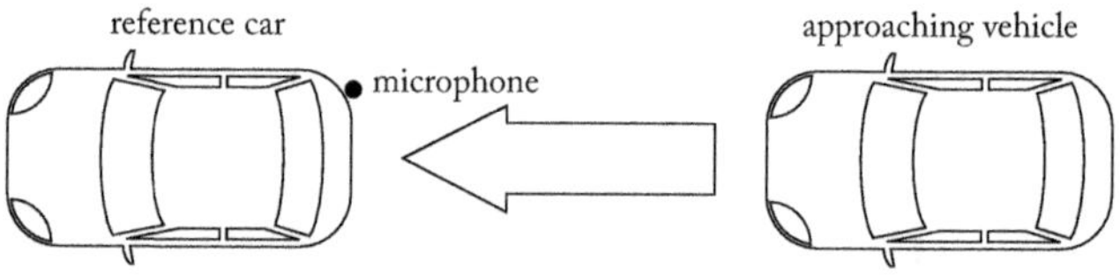

sampling frequency	48kHz
frequency channels (N)	5
frequency range	1kHz ~ 2kHz
window length (W)	15000
window number(S)	5
threshold (β)	5
LVQ codebook	50 clusters/class

Fig. 2. Experimental conditions

3.1 Parameter Decision

The feature extraction parameters window length (W),window number (S) and threshold (β) were tuned in order to maximize the accuracy of the LVQ classifier. Table 1 shows parameter candidates of the feature extraction process. Sounds from an approaching scooter were taken in a test area and several signals from approaching cars were taken in normal transit in the streets of Nagoya.

Table 1. Parameter candidate

window length (W)	5000, 7500, 10000, 12500, 15000, 17500
window number (S)	2, 3, 4, 5, 6
threshold (β)	2, 3, 4, 5, 6, 7

Fig. 3 shows the accuracy of the LVQ neural network. Accuracy is directly proportional and very dependent on the window length and number of windows. However, large number of wide windows increase system's latency. The threshold parameter is not critical when using wide windows.

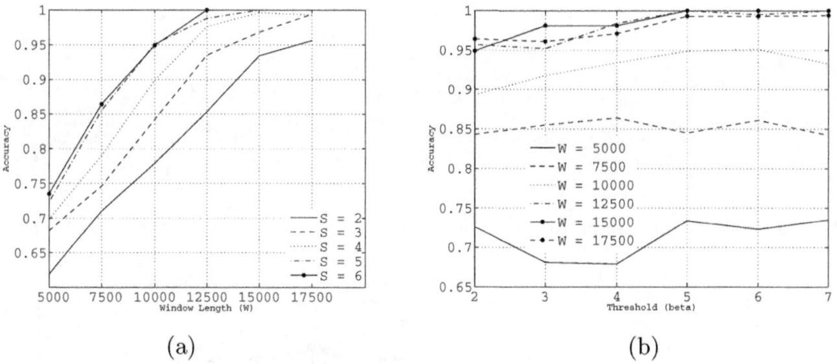

Fig. 3. Detection accuracy: (a) window length and window number, (b) threshold and window length

From Fig. 3, the best parameter set was selected as shown in Fig. 2. For the number of clusters greater than 50, accuracy does not change significantly.

3.2 Performance Evaluation

The experimental results of the scooter approaching at 30km/h and 40km/h are shown, respectively, in Fig. 4(a) and 4(b). In the 30km/h case, the model detected the approaching vehicle 4.3 seconds before the arrival, while in the 40km/h case, the approach was detected 3.6 seconds in advance. For these experiments, all approaches were detected successfully.

Results from the sounds taken in normal traffic for a single approaching car are shown in Fig. 5. The approaching vehicle could be detected 4.7 seconds before the arrival. The approaching vehicles were manually verified from a video stream taken from the back of the car containing the microphones. Fig. 6 shows the accuracy of detection as a function of time before arrival. For these measurements, a margin of 5 seconds before and after each vehicle arrival was used, i.e. only vehicles separated by more than 5 seconds from others were considered in the accuracy calculation. Also, due to the use of non-directional microphones, only cars on the same lane or on the immediate neighbor lanes were considered.

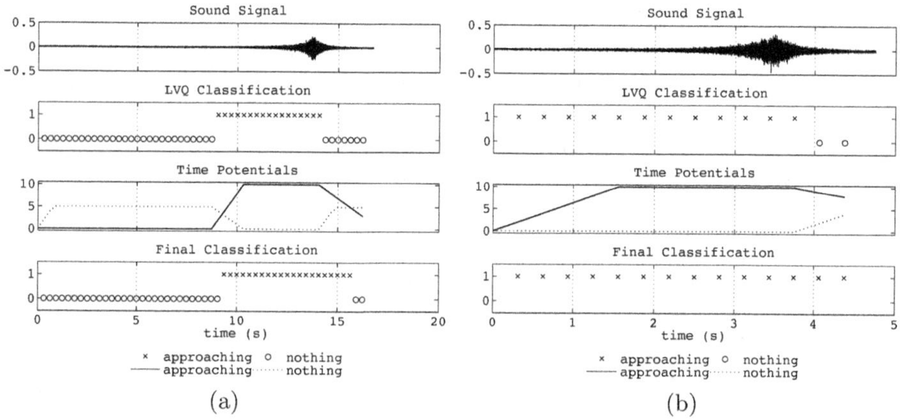

Fig. 4. Scooter approaching detection result(test area): (a) 30 km/h, (b) 40 km/h

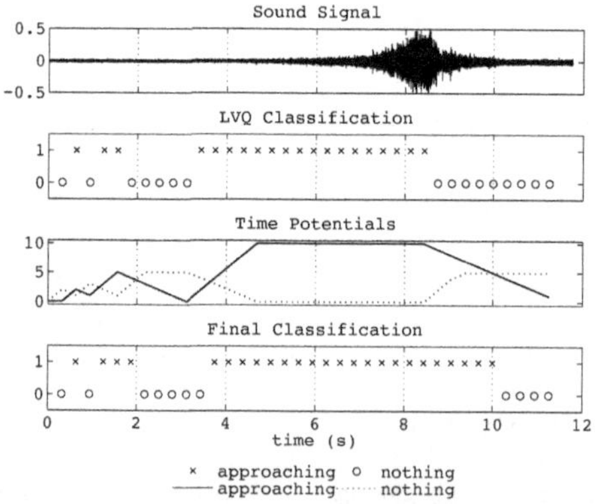

Fig. 5. Scooter approaching detection result(city streets)

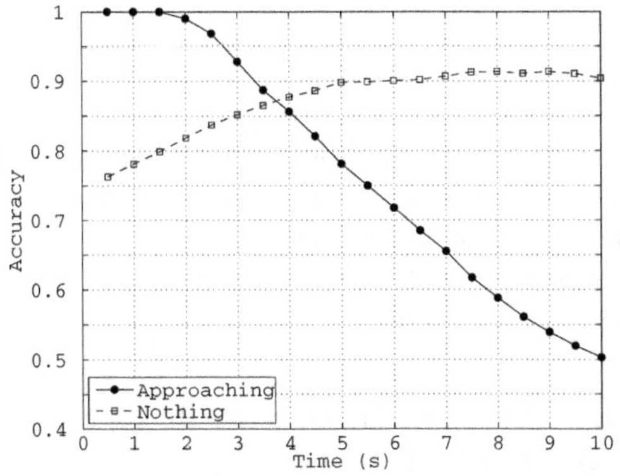

Fig. 6. Detection accuracy for single vehicle in normal traffic

3.3 Performance Comparison

The proposed method was compared with the approach described by Hoshino[3]. In this method, the features are extracted from sound level difference and classified based on heuristic rules. The sound signal is filtered by an octave-bandpass filter which center frequency is 2 kHz. After the filtering, the average sound pressure level of every 200 ms interval is calculated. The approaching vehicle detection is defined as the three section increase of the level or two section continuous increase larger than 1.0 dB of the level.

Fig. 7 shows the accuracy of detection as a function of time before arrival processed by this method. This experiments were performed using the same data of the experimental results in Fig. 6.

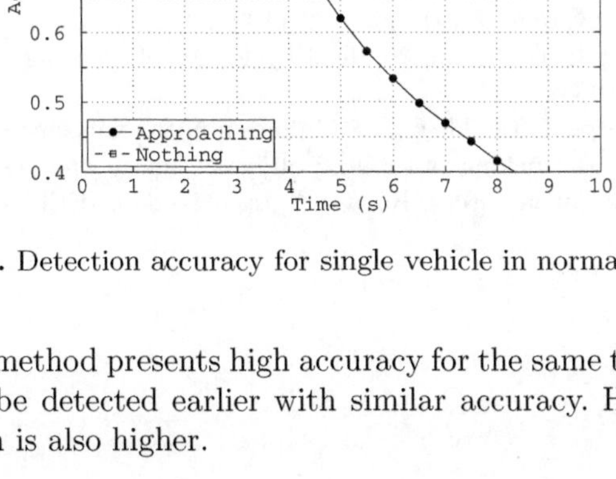

Fig. 7. Detection accuracy for single vehicle in normal traffic

The proposed method presents high accuracy for the same time before arrival, i.e. vehicles can be detected earlier with similar accuracy. However, the false-positive detection is also higher.

4 Conclusions

This paper presented a new method for approaching vehicles detection. The obtained algorithm is very robust, presenting high accuracy and a time detection before arrival larger than other methods. The classifier consists of a standard LVQ neural network, with simple learning and no critical parameters.

However, although the results for single vehicle approaching detection were encouraging, multiples vehicles still presents a challenge and are the main focus of the future steps of this research.

Acknowledgments. This research was partially supported by Toyota Motor Corporation and Aisin Seiki Corporation.

References

1. Fausett, L.: Neural Networks Based on Competition. In: Fundamentals of Neural Networks: architectures, algorithms and applications, Fundamentals of Neural Networks, 1st edn., New Jersey, pp. 156–217 (1994)
2. Gupte, S., Masoud, O., Martin, R.F.K., Papanikolopoulos, N.P.: Detection and classification of vehicles. IEEE Transactions on Intelligent Transportation Systems 3(1), 37–47 (2002)
3. Hoshino, H.: Approaching vehicles detection system by using vehicle noise for driver support. Journal of the Acoustical Society of Japan 62(3), 265–274 (2006)
4. Kodera, K., Itai, A., Yasukawa, H.: Estimation of speed and arrival time of approaching vehicles using sound. IEICE Technical Report, 13–18 (2007)
5. Kugler, M., Benso, V.A.P., Kuroyanagi, S., Iwata, A.: A novel approach for hardware based sound classification. In: Köppen, M., Kasabov, N., Coghill, G. (eds.) ICONIP 2008. LNCS, vol. 5507, pp. 859–866. Springer, Heidelberg (2009)
6. Kugler, M., Iwasa, K., Benso, V.A.P., Kuroyanagi, S., Iwata, A.: A complete hardware implementation of an integrated sound localization and classification system based on spiking neural networks. In: Ishikawa, M., Doya, K., Miyamoto, H., Yamakawa, T. (eds.) ICONIP 2007, Part II. LNCS, vol. 4985, pp. 577–587. Springer, Heidelberg (2008)
7. Kuroyanagi, S., Iwata, A.: Auditory pulse neural network model to extract the inter-aural time and level difference for sound localization. IEICE Transactions on Information and Systems 77(4), 466–474 (1994)
8. Pickles, J.O.: An Intruduction to the Physiology of Hearing, 2nd edn. Academic Press, London (1988)
9. Wang, C.C., Thorpe, C.E., Thrun, S.: Online simultaneous localization and mapping with detection and tracking of moving objects: theory and results from a ground vehicle in crowded urban areas. In: ICRA, pp. 842–849. IEEE, Los Alamitos (2003)

Ground Penetrating Radar System with Integration of Mutimodal Information Based on Mutual Information among Multiple Self-Organizing Maps

Akira Hirose, Ayato Ejiri, and Kunio Kitahara

Department of Electrical Engineering and Inform. Sys., The University of Tokyo
7–3–1 Hongo, Bunkyo-ku, Tokyo 113–8656, Japan
ahirose@ee.t.u-tokyo.ac.jp, {ejiri,kitahara}@eis.t.u-tokyo.ac.jp
http://www.eis.t.u-tokyo.ac.jp/

Abstract. We propose a ground penetrating radar system to integrate mutimodal information of space- and frequency- domain textural features in self-organization that is modulated by mutual information. We use the MuSOM (mutual-information-based self-organizing map) architecture we proposed previously, in which the mutual information among the data fed to multiple SOMs modulates the SOM dynamics. Experiments demonstrate that our system makes meaningful clusters of landmine features clearer than a conventional non-MuSOM system does.

1 Introduction

Many ground penetrating radar (GPR) systems have been proposed for visualization of plastic landmines [1] . We previously proposed an adaptive visualizer based on a complex-valued self-organizing map (Complex-valued SOM, CSOM) in which we focus on the textural features in the space and frequency domains [2] [3] . We have been successful in visualization even for small plastic landmines buried in heavily wet laterite soil where both the permeability and the permittivity are very high [4] [5] . In the system, however, we had room of improvement in the information integration of the textural features in the two different domains.

In this paper, we propose a method to integrate the two different types of information and generate a concept of "plastic landmine" in a self-organizing manner based on the observation data. The method modulates the bottom-up dynamics of self-organization in multiple (complex-valued) SOMs in a top-down manner.

Note that we may call the present network "mutual complex-valued self-organizing map (MuCSOM)." Here in this paper, however, we call it simply "MuSOM" since the dynamics modulation based on mutual information is not limited to CSOM but applicable widely to SOM in general.

K.W. Wong et al. (Eds.): ICONIP 2010, Part II, LNCS 6444, pp. 415–422, 2010.

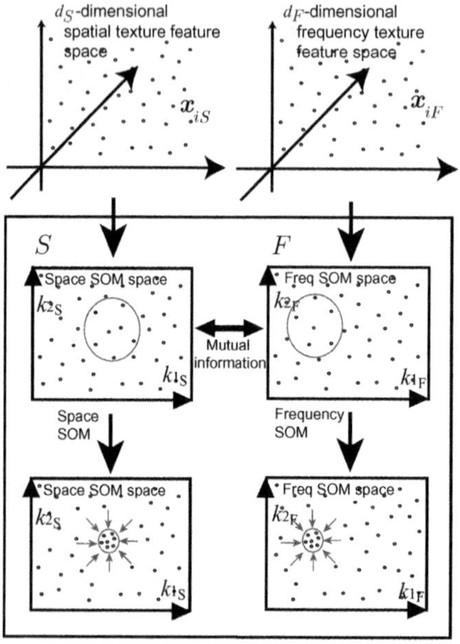

Fig. 1. Schematic diagram illustrating the concept generation by modulation of self-organization dynamics based on the mutual information [6]

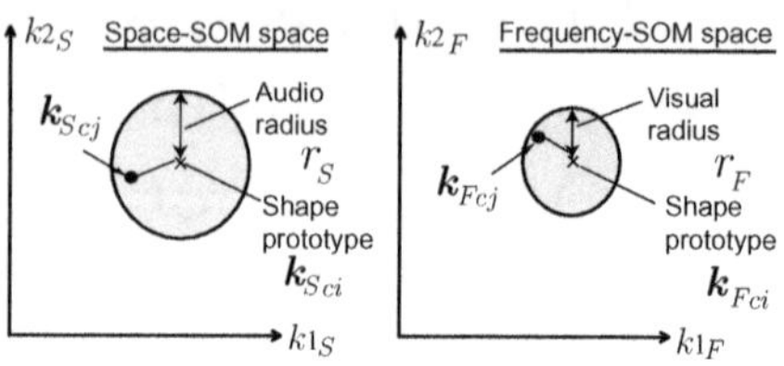

Fig. 2. Illustration showing how to determine the region that maximizes mutual information [6]

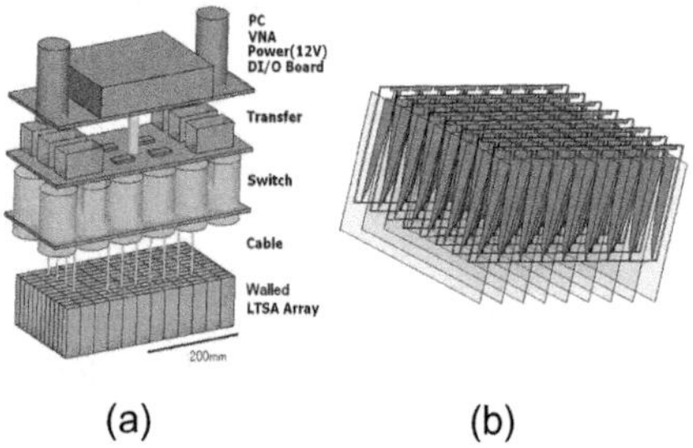

Fig. 3. (a) Front-end and (b) array antenna of the plastic landmine visualization system [3]

2 Integration of Multimodal Information and Concept Generation in the MuSOM

2.1 Basic Idea

The authors previously proposed a basic method of information integration and concept generation based on mutual information among multiple self-organizing maps (SOMs) by modulating SOM parameters in respective SOMs [6]. We named the system architecture the MuSOM, meaning mutual-information-based SOM. The MuSOM is applicable widely to various applications independent of the types of information or situations under consideration.

In this paper, we apply the idea of MuSOM to our adaptive radar system by showing the qualitative improvement of the performance to visualize plastic landmines. In our recent ground penetrating radar (GPR) systems, we basically extract two textural features, i.e., space- and frequency-domain textural features, as a concatenation into a single feature vector, and feed them to a complex-valued self-organizing map. However, since the two sets of features represent different information essentially, there has been room for improvement in its integration. This paper proposes the application of MuSOM to the integration of these two qualitatively different types of information so that the system generates the concept of landmine self-organizingly. We construct a radar system to demonstrate the meaningful clustering of features of plastic landmines based on the mutual information among multiple self-organizing maps, i.e., in the MuSOM.

Figure 1 is the schematic diagram showing the mutual-information-based self-organizing process in the MuSOM, in which each SOM is basically a conventional one that maps input information into low dimensional information space. The typical SOM dynamics is expressed as the weight updating process in terms of the weights of neurons $\boldsymbol{w}_{\boldsymbol{k}}(t)$, at position $\boldsymbol{k}$ in the SOM space at time t (or learning iteration number t), and a set of signal input vector $\boldsymbol{x}_i$ fed to the SOM as

$$\boldsymbol{w}_{\boldsymbol{k}}(t+1) = \boldsymbol{w}_{\boldsymbol{k}}(t) + \Theta(\boldsymbol{k}, \boldsymbol{k}_c; t)\ \alpha(t)\ \left(\boldsymbol{x}_i - \boldsymbol{w}_{\boldsymbol{k}}(t)\right) \tag{1}$$

where

$$\boldsymbol{k}_c = \arg\min_{\boldsymbol{k}} \left|\boldsymbol{x}(t) - \boldsymbol{w}_{\boldsymbol{k}}(t)\right| \qquad \text{(Winner neuron position)} \tag{2}$$

$$\alpha(t) \equiv \alpha(0)\left(1 - \frac{t}{T}\right) \qquad \text{(Learning coefficient)} \tag{3}$$

$$\Theta(\boldsymbol{k}, \boldsymbol{k}_c; t) = \exp\left(-\frac{|\boldsymbol{k} - \boldsymbol{k}_c|^2}{2\delta^2(t)}\right) \qquad \text{(Neighborhood function)} \tag{4}$$

$$\delta(t) \equiv \delta(0)\left(1 - \frac{t}{T}\right) \qquad \text{(Inverse of sharpness of } \Theta\text{)} \tag{5}$$

and T is the maximum time (maximum number of iteration) in the self-organization.

However, in our MuSOM [6], the dynamics is modulated in accordance with mutual information among input signals fed to a set of SOMs. That is, a SOM finds correspondence of input information with another input signal fed to another SOM by paying attention to mutual information. Each SOM changes the stretch of its neighborhood according to the mutual information so that a set of corresponding data makes a cluster through self-organization. For example, let's assume visual and audio data streams existing simultaneously. By referring to the mutual information between the visual and audio data, we can segment the visual stream commutatively with the audio data, and vice versa.

2.2 MuSOM: A Set of SOMs Modulated by Mutual Information

The MuSOM is a set of SOMs that realizes a clustering process taking into account the relationship among information in multiple modes [6]. The dynamics in each SOM inside is modulated by mutual information quantity to realize a mapping that takes the relationship among multiple-mode data sets into account. The detail is described as follows. We consider only two modes of data for simplicity here, e.g., visual and audio data streams. We then prepare two SOMs. Each SOM self-organizes by referring to the mutual information and changing the neighborhood function used in the self-organization. We calculate the mutual information between the visual and audio SOM neurons $I(A;V)$ as

$$I(S;F) = \sum_{d_S=0,1} \sum_{d_F=0,1} p(d_S,d_F) \log \frac{p(d_S,d_F)}{p(d_S)p(d_F)} \tag{6}$$

$$p(d_S) = \frac{m(d_S)}{M} \;, \quad p(d_F) = \frac{m(d_F)}{M} \;, \quad p(d_S,d_F) = \frac{m(d_S,d_F)}{M^2} \tag{7}$$

$$d_S = \begin{cases} 1, & \text{if } \left|\boldsymbol{k}_{Sci} - \boldsymbol{k}_{Scj}\right| \le r_S \\ 0, & \text{otherwise} \end{cases} \;, \quad d_F = \begin{cases} 1, & \text{if } \left|\boldsymbol{k}_{Fci} - \boldsymbol{k}_{Fcj}\right| \le r_F \\ 0, & \text{otherwise} \end{cases} \tag{8}$$

$d_S(\boldsymbol{k}, \boldsymbol{k}_c)\mathrm{A}d_F(\boldsymbol{k}, \boldsymbol{k}_c)$ where $n(d_A)$ and $n(d_V)$ (or $n(d_{A/V})$ in short) denote the number of the data that gives $d_{A/V}$, N is the total number of the data in the set (i.e., $\boldsymbol{x}_1 = [\boldsymbol{x}_{S1}, \boldsymbol{x}_{F1}], \boldsymbol{x}_2 = [\boldsymbol{x}_{S2}, \boldsymbol{x}_{F2}], \cdots, \boldsymbol{x}_N = [\boldsymbol{x}_{SN}, \boldsymbol{x}_{FN}]$), $\left|\boldsymbol{k}_{A/Vcj} - \boldsymbol{k}_{A/Vci}\right|$ is the distance between the winners for data $\boldsymbol{x}_i$ and $\boldsymbol{x}_j$ in the audio / visual SOM space, and $r_{A/V}$ denotes radius of a variable circle in the SOM space that gives a boundary of the category represented by $\boldsymbol{k}_{A/Vci}$ as shown in Fig.2. We vary the values r_A and r_V to find a pair of r_A and r_V that maximize the mutual information I_{max}. Then we modify the neighborhood function in accordance with the I_{max} value as follows.

If the mutual information I_{max} is larger than a threshold H, we employ a modified $\delta(t)$ in (5), namely $\tilde{\delta}(t)$ expressed as

$$\tilde{\delta}(t) = \begin{cases} \delta(0)\ \left(1 - \frac{t}{T}\right)^2 \\ \qquad \text{for} \quad \boldsymbol{k}_{S/Fcj} \in \text{the circle of } \boldsymbol{k}_{S/Fci} \\ \delta(0)\ \left(1 - \frac{t}{T}\right) \\ \qquad \text{otherwise} \end{cases} \tag{9}$$

That is, in the upper case in (9), the neighborhood is sharper. In other words, if a winner position $\boldsymbol{k}_{A/Vcj}$ for a data $\boldsymbol{x}_j$ falls within the circle, we employ a smaller $\tilde{\delta}(t)$ in the neighborhood function $\Theta(\boldsymbol{k}, \boldsymbol{k}_c, t)$ when the maximum mutual information I_{max} exceeds the threshold H, which indicates a strong correlation between the audio and video data $\boldsymbol{x}_A$ and $\boldsymbol{x}_V$. Though a locally adaptive threshold may improve the dynamics, we employ a fixed H for simplicity in the experiment.

3 Experiment

3.1 Construction of the Adaptive Radar System to Visualize Plastic Landmines

Figures 3(a) and (b) illustrate the constructions of our front-end and array antenna of the radar system to visualize plastic landmines. It is a stepped-frequency radar covering 8-12GHz band. We obtain complex-amplitude images of scattering / reflection at 10 frequency points with 0.4GHz interval. The antenna element is the so-called walled linearly tapered slot array (walled LTSA) we proposed in Ref. [3]. 12×12 elements form an array. The elements is so wideband and high gain that the system needs no aperture synthesis, and suitable for near field imaging.

A vector network analyzer (VNA) generates microwave. Layered radio-frequency (RF) switches lead the microwave to one of the antenna elements (transmitter: Tx). Another element (receiver: Rx) receives scattered or reflected wave. That is, they form a bistatic radar configuration. We obtain the amplitude and phase of the received wave at the VNA. The switches realize various combinations of Tx and Rx [4].

A personal computer (PC) controls the switches and the VNA, and performs all the preprocessing such as signal normalization, feature extraction, and SOM clustering as shown below. The data obtained at the VNA represent the amplitude and phase values at position (l_x, l_y) and frequency f_n. First we conduct the preprocessing and the feature extraction inspired by the human early vision system as follows.

3.2 Preprocessing and Feature Extraction

The preprocessing consists of the following four processes in series. 1)Subtraction of the direct coupling between Tx and Rx by using calibration data obtained in advance. 2)Subtraction of rotation phase value (in the phase domain) corresponding to the rotation caused by the direct coupling. 3)Transformation of the complex-amplitude itself into that in decibel (dB) representation, just like in the way the human beings feel. 4)Normalization of the amplitude in dB in such a manner that the minimum and maximum values in a single data-set acquisition are transformed linearly to 0 and 1, respectively. Finally we obtained a set of preprocessed data $z(l_x, l_y, f_n)$ ($\in \boldsymbol{C}$: complex domain).

Figure 4 shows the flowchart of the whole signal processing performed in the landmine visualization system. Out of the data $z(l_x, l_y, f_n)$ obtained by the front-end, we extract the local textural features as a feature vector $\boldsymbol{x}$ in a small local window of $L \times L$ pixels.

$$\boldsymbol{x} = [m, \boldsymbol{x}_s, \boldsymbol{x}_f] \tag{10}$$

$$m = \frac{1}{L^2 N} \sum_{l_x=1}^{L} \sum_{l_y=1}^{L} \sum_{n=1}^{N} z(l_x, l_y, f_n) \tag{11}$$

$$\boldsymbol{x}_s = [x_s(0,0), x_s(1,0), x_s(0,1), \cdots] \tag{12}$$

$$x_s(i,j) = \frac{1}{L^2 N} \sum_{l_x=1}^{L} \sum_{l_y=1}^{L} \sum_{n=1}^{N} z(l_x, l_y, f_n) z^*(l_x + i, l_y + j, f_n) \tag{13}$$

$$\boldsymbol{x}_f = [x_f(f_1), \cdots, x_f(f_{N-1})] \tag{14}$$

$$x_f(f_n) = \frac{1}{L^2} \sum_{l_x=1}^{L} \sum_{l_y=1}^{L} z(l_x, l_y, f_n) z^*(l_x, l_y, f_{n+1}) \tag{15}$$

where $(\cdot)^*$ denotes the Hermite conjugate of $(\cdot)$. That is, we represent the texture by the mean m, space domain correlation $\boldsymbol{x}_s$, and frequency domain correlation

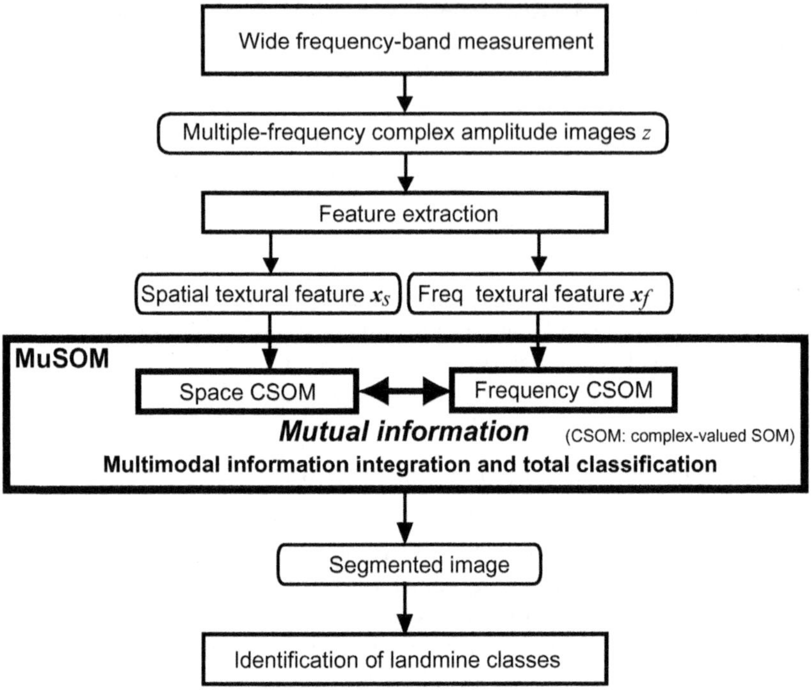

Fig. 4. Flowchart of MuSOM signal processing for plastic landmine visualization

$\boldsymbol{x}_f$. Then we feed $[m, \boldsymbol{x}_s]$ and $[m, \boldsymbol{x}_f]$ to the space and frequency SOMs, respectively, in the MuSOM.

3.3 Results

The number of the feature vectors is $\ell_{\max}$=360, which is also the number of the local windows which are set in an overlapping manner in the observation land area. A feature vector $\boldsymbol{x}_\ell = [m, \boldsymbol{x}_{S\ell}, \boldsymbol{x}_{F\ell}]$ represents the texture in a local window. Among the elements of $\boldsymbol{x}_\ell$, we feed the space components with the mean $[m, \boldsymbol{x}_{S\ell}]$ to a space SOM that deal with the space domain information, while we do the frequency components with the mean $[m, \boldsymbol{x}_{F\ell}]$ to the frequency SOM to treat the frequency domain information. During the self-organization in the respective SOMs, we modulate the dynamics if we find correlation between the two types of information.

In our previous system contrarily, we dealt with a single joint vector $\boldsymbol{x}_\ell = [m, \boldsymbol{x}_{S\ell}, \boldsymbol{x}_{F\ell}]$ in a single SOM. Even in such a case, we find meaningful self-organization, and we are able to visualize landmines by the textural classification. On the other hand, it is obvious that $\boldsymbol{x}_{S\ell}$ is qualitatively different from $\boldsymbol{x}_{F\ell}$ and that the simple concatenation may be inappropriate. Rather the information should be integrated naturally based on the information itself. This paper insists on this point.

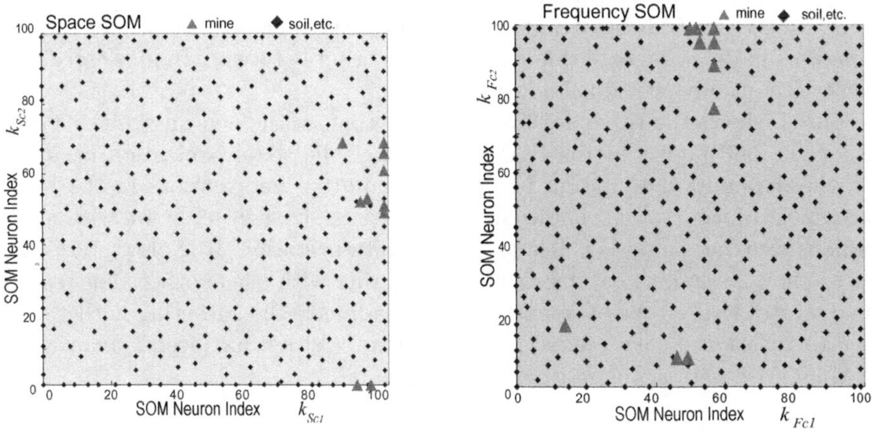

(a) (b)

Fig. 5. Winner neuron distributions in conventional (a)space and (b)frequency SOMs (separate and non-modulated SOMs): △ shows the neurons firing for landmine area, while • for other areas

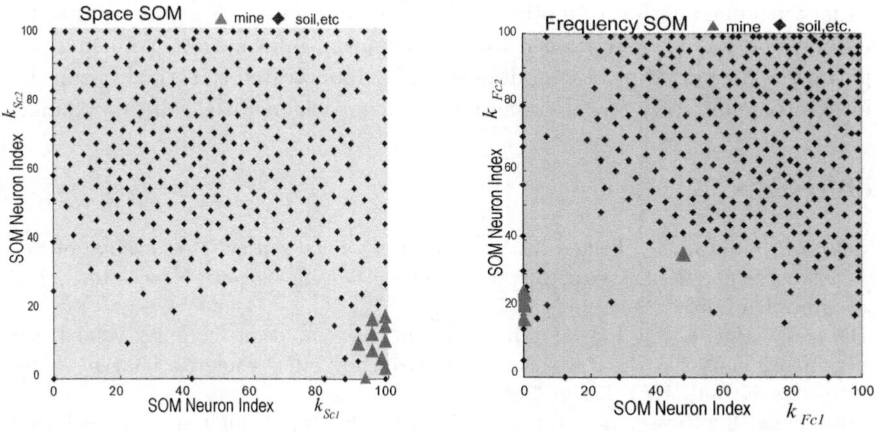

(a) (b)

Fig. 6. Winner neuron distributions in the MuSOM (a)space and (b)frequency SOMs interacting with each other based on the mutual information: △ shows the neurons firing for landmine area, while • for other areas

Figure 5 displays the positions of the winner neurons that fire for the 360 input vectors in the SOM space after the self-organization completed conventionally without the MuSOM configuration, i.e., no modulation in the dynamics. The data used was obtained for soil including a buried landmine but a little difficult for our conventional system to visualize [3]. In the present experiment, we checked the landmine position beforehand, and we know the landmine position.

In Fig.5, △ shows the neuron position firing for a feature vector of the landmine region, whereas • indicates that for another area such as just soil or stones. Though △ gathers, there is difficulty in distinction between landmine area and others without any previous knowledge.

Figure 6 shows the result for the MuSOM, that is, the neuron positions that fire for the 360 inputs in the SOM space after the self-organization occurred with the dynamics modulation based on the mutual information. In the SOM space, most of the △ are separate from • positions, though some are within the • regions. Further analysis is needed for the interpretation. It is also interesting that the non-landmine neurons also tend to gather with one another. The results demonstrates that a type of concept, corresponding to the gathering in the SOM space to reflect the characteristic set of feature vectors, has been generated in the MuSOM based on the mutual information.

4 Summary

We proposed a MuSOM-based method to integrate the multimodal information of space- and frequency- domain textural features and to generate a concept of landmines in a self-organization modulated by the mutual information. In the system, multiple self-organizing maps dealing with different modes of information interact with one another to modulate their dynamics in such a manner that the neighborhood region is sharpened for larger mutual information. We conducted experiments for buried plastic landmines and demonstrated that the information is integrated effectively and the landmine concept is generated self-organizingly.

References

1. Sato, M., Fujiwara, J., Fenga, X., Kobayashi, T.: Dual sensor ALIS evaluation test in Afghanistan. IEEE Geoscience and Remote Sensing Society Newsletter, 22–27 (Septermber 2005)
2. Hara, T., Hirose, A.: Plastic mine detecting system using complex-valued self-organizing map that deals with multiple-frequency interferometric images. Neural Networks 17(8-9), 1201–1210 (2004)
3. Masuyama, S., Hirose, A.: Walled LTSA array for rapid, high spatial resolution, and phase sensitive imaging to visualize plastic landmines. IEEE Transactions on Geoscience and Remote Sensing 45(8), 2536–2543 (2007)
4. Masuyama, S., Yasuda, K., Hirose, A.: Multiple mode selection of walled-ltsa array elements for high resolution imaging to visualize antipersonnel plastic landmines. IEEE Geoscience and Remote Sensing Letters 5(4), 745–749 (2008)
5. Nakano, Y., Hirose, A.: Improvement of plastic landmine visualization performance by use of ring-CSOM and frequency-domain local correlation. IEICE Transactions on Electronics E92-C(1), 102–108 (2009)
6. Kitahara, K., Hirose, A.: A concept generation method based on mutual information quantity among multiple self-organizing maps. In: Proceedings of the International Conference on Neural Inform. Processing (ICONIP) 2009 Bangkok, pp. 263–272 (2009)

Information-Theoretic Competitive and Cooperative Learning for Self-Organizing Maps

Ryotaro Kamimura

IT Education Center,
1117 Kitakaname Hiratsuka Kanagawa 259-1292, Japan
ryo@keyaki.cc.u-tokai.ac.jp

Abstract. In this paper, we propose a new type of information-theoretic method for competitive learning based, upon mutual information between competitive units and input patterns. In addition, we extend this method to a case where cooperation between competitive units exists to realize self-organizing maps. In computational methods, free energy is introduced to simplify the computation of mutual information. We applied our method to two problems, namely, the Senate data and ionosphere data problems. In both, experimental results confirmed that better performance could be obtained in terms of quantization and topographic errors. We also found that the information-theoretic methods tended to produce more equi-probable distribution of competitive units.

1 Introduction

In this paper, we propose a new information-theoretic method to realize competitive learning, based upon our finding that mutual information between input patterns and competitive units can be used to realize competitive processes [1]. Information-theoretic methods have been successfully introduced in neural networks and machine learning, because of their ability to deal with higher-order statistics and non-linear problems [2]. One of the main shortcomings of the information-theoretic method lies in its computational complexity when we must compute mutual information. Though there have been several attempts [3], [2] to reduce the complexity, the problem remain serious, even at the present stage of research [3], [2]. To simplify the computation of mutual information, we introduce a type of free energy by which we can replace the computation of mutual information with that of the simple partition function [4].

The method can easily be extended to the generation of self-organizing maps [5], where the good performance of our method can be shown visually and quantitatively. To realize the self-organizing maps, we introduce cooperation between competitive units in addition to competition between the units. The cooperation is realized by the weighted sum of all competitive units. Thus, one of the main characteristics of our method is soft competition, where we can flexibly control the degree of competition, between competitive units, while in the conventional

K.W. Wong et al. (Eds.): ICONIP 2010, Part II, LNCS 6444, pp. 423–430, 2010.

SOM, the rigid winner-take-all algorithm is used. The winner-take-all operation can be considered to be a specific case of our soft competition. Because this flexibility in the control of competition is surely related to the performance of competitive learning, we try to show in this paper that this flexible control of competition is directly related to the better performance of our method.

2 Theory and Computational Methods

2.1 Two Types of Information

Our method presented here is based upon information-theoretic competitive learning. Thus, we should summarize the results of information-theoretic competitive learning [6], [1]. In this method, competition processes are supposed to be realized by maximizing mutual information between competitive units and input patterns.

Let us compute mutual information for a network shown in Figure 1(a). The jth competitive unit output can be computed by

$$v_j^s \propto \exp\left\{-\frac{1}{2}(\mathbf{x}^s - \mathbf{w}_j)^T \mathbf{\Sigma}^{-1}(\mathbf{x}^s - \mathbf{w}_j)\right\}, \tag{1}$$

where $\mathbf{x}^s$ and $\mathbf{w}_j$ are supposed to represent L-dimensional input and weight column vectors, where L denotes the number of input units. The $L \times L$ matrix $\mathbf{\Sigma}$ is called a "scaling matrix," and the klth element of the matrix denoted by $(\mathbf{\Sigma})_{kl}$ is defined by

$$(\mathbf{\Sigma})_{kl} = \delta_{kl}\frac{\sigma^2}{p(k)}, \tag{2}$$

where $p(k)$ is greater than zero and initially set to $1/L$, because we have no preference in input units. The spread parameter σ is computed by

$$\sigma = \frac{1}{\alpha}, \tag{3}$$

where $\alpha > 1$. The output is increased when connection weights become closer to input patterns. The conditional probability of the firing of the jth competitive unit, given the sth input pattern of the S input patterns, can be obtained by

$$p(j \mid s) = \frac{\exp\left\{-\frac{1}{2}(\mathbf{x}^s - \mathbf{w}_j)^T \mathbf{\Sigma}^{-1}(\mathbf{x}^s - \mathbf{w}_j)\right\}}{\sum_{m=1}^{M}\exp\left\{-\frac{1}{2}(\mathbf{x}^s - \mathbf{w}_m)^T \mathbf{\Sigma}^{-1}(\mathbf{x}^s - \mathbf{w}_m)\right\}}, \tag{4}$$

where M is the number of competitive units. The probability of the firing of the jth competitive unit is computed by

$$p(j) = \sum_{s=1}^{S} p(s)p(j \mid s). \tag{5}$$

With these probabilities, we can compute two types of information, namely, the first order and the second order information (mutual information) [7]. The first order information is defined by

$$I_1 = \sum_{j=1}^{M} p(j) \log Mp(j). \tag{6}$$

The first order information shows how far the distribution of competitive units is from the equi-probable one. The second order information, or mutual information, is defined by

$$I_2 = \sum_{s=1}^{S} \sum_{j=1}^{M} p(s) p(j \mid s) \log \frac{p(j \mid s)}{p(j)}. \tag{7}$$

When this second order information, or mutual information, is maximized, just one competitive unit fires, while all the other competitive units cease to do so.

Finally, we should note that one of the main properties of this mutual information is that it is dependent upon the scaling matrix, or more concretely, the spread parameter σ. As the spread parameter is decreased, the mutual information between competitive units and input patterns tends to be increased.

2.2 Free Energy Minimization

We can differentiate the mutual information and obtain update rules, but direct computation of mutual information is accompanied by computational complexity. To simplify the computation, we introduce free energy [4]. The free energy F can be defined by

$$F = -2\sigma^2 \sum_{s=1}^{S} p(s) \log \sum_{j=1}^{M} p(j) \exp\left\{ -\frac{1}{2} (\mathbf{x}^s - \mathbf{w}_j)^T \mathbf{\Sigma}^{-1} (\mathbf{x}^s - \mathbf{w}_j) \right\}. \tag{8}$$

Here, we suppose the following equation

$$p(j \mid s) = \frac{p(j) \exp\left\{ -\frac{1}{2} (\mathbf{x}^s - \mathbf{w}_j)^T \mathbf{\Sigma}^{-1} (\mathbf{x}^s - \mathbf{w}_j) \right\}}{\sum_{m=1}^{M} p(j) \exp\left\{ -\frac{1}{2} (\mathbf{x}^s - \mathbf{w}_m)^T \mathbf{\Sigma}^{-1} (\mathbf{x}^s - \mathbf{w}_m) \right\}}. \tag{9}$$

Then, the free energy can be expanded as

$$\begin{aligned} F &= \sum_{s=1}^{S} p(s) \sum_{j=1}^{M} p(j \mid s) \|\mathbf{x}^s - \mathbf{w}_j\|^2 \\ &\quad + 2\sigma^2 \sum_{s=1}^{S} p(s) \sum_{j=1}^{M} p(j \mid s) \log \frac{p(j \mid s)}{p(j)}. \end{aligned} \tag{10}$$

This equation shows that, by minimizing the free energy, we can decrease mutual information as well as quantization errors. As already noted, the mutual information can be increased by decreasing the spread parameter σ. We usually set $p(j)$ into $1/M$ for simplification. Then, by differentiating the free energy, we have

$$\mathbf{w}_j = \frac{\sum_{s=1}^{S} p(j \mid s)\mathbf{x}^s}{\sum_{s=1}^{S} p(j \mid s)}, \tag{11}$$

where

$$p(j \mid s) = \frac{\exp\left\{-\frac{1}{2}(\mathbf{x}^s - \mathbf{w}_j)^T \mathbf{\Sigma}^{-1}(\mathbf{x}^s - \mathbf{w}_j)\right\}}{\sum_{m=1}^{M} \exp\left\{-\frac{1}{2}(\mathbf{x}^s - \mathbf{w}_m)^T \mathbf{\Sigma}^{-1}(\mathbf{x}^s - \mathbf{w}_m)\right\}}. \tag{12}$$

2.3 Competitive and Cooperative Learning

We can easily extend the information-theoretic competitive learning to a case where cooperation between competitive units must be taken into account, namely, self-organizing maps.

In the training mode, we try to borrow the computational methods developed for the conventional self-organizing maps, and then we use the ordinary neighborhood kernel used for SOM, namely,

$$h_{jc} = \exp\left(\frac{\|\mathbf{r}_j - \mathbf{r}_c\|^2}{2\sigma_{nh}^2}\right), \tag{13}$$

where $\mathbf{r}_j$ and $\mathbf{r}_c$ denotes the position of the jth unit on the output space. The cooperative outputs can be defined by the summation of all neighboring competitive units

$$y_j^s = \sum_{c=1}^{M} h_{jc} \exp\left\{-\frac{1}{2}(\mathbf{x}^s - \mathbf{w}_j)^T \mathbf{\Sigma}_{coop}^{-1}(\mathbf{x}^s - \mathbf{w}_j)\right\}, \tag{14}$$

where the klth element of the scaling matrix $(\mathbf{\Sigma}_{coop})_{kl}$ is given by

$$(\mathbf{\Sigma})_{kl} = \delta_{kl}\frac{\sigma_{coop}^2}{p(k)}. \tag{15}$$

The conditional probability of the firing of the jth competitive unit, given the sth input pattern, can be obtained by

$$q(j \mid s) = \frac{y_j^s}{\sum_{m=1}^{M} y_m^s}. \tag{16}$$

The free energy can be defined by

$$F = -2\sigma^2 \sum_{s=1}^{S} p(s) \log \sum_{j=1}^{M} q(j|s) \exp\left\{-\frac{1}{2}(\mathbf{x}^s - \mathbf{w}_j)^T \mathbf{\Sigma}^{-1}(\mathbf{x}^s - \mathbf{w}_j)\right\}. \tag{17}$$

Now, it is easy to differentiate the free energy to have update rules. Suppose that input patterns are given with the same probabilities, and then we have

$$\mathbf{w}_j = \frac{\sum_{s=1}^{S} p(j \mid s)\mathbf{x}^s}{\sum_{s=1}^{S} p(j \mid s)}, \tag{18}$$

where

$$p(j \mid s) = \frac{q(j \mid s) \exp\left\{-\frac{1}{2}(\mathbf{x}^s - \mathbf{w}_j)^T \boldsymbol{\Sigma}^{-1}(\mathbf{x}^s - \mathbf{w}_j)\right\}}{\sum_{m=1}^{M} q(j \mid s) \exp\left\{-\frac{1}{2}(\mathbf{x}^s - \mathbf{w}_m)^T \boldsymbol{\Sigma}^{-1}(\mathbf{x}^s - \mathbf{w}_m)\right\}}. \tag{19}$$

In this paper, the most simplified version of the computational methods is used. This means that we suppose that the probability $p(j \mid s)$ is already equivalent to $q(j \mid s)$ before learning, and we have

$$p(j \mid s) = q(j \mid s). \tag{20}$$

Then, update rules are simplified into

$$\mathbf{w}_j = \frac{\sum_{s=1}^{S} q(j \mid s)\mathbf{x}^s}{\sum_{s=1}^{S} q(j \mid s)}. \tag{21}$$

We use this simplified version because the final equation is very close to the equation of the Batch learning in the conventional SOM [5]. With this simplified computational method, we can easily use the values of parameters tuned for the self-organizing maps.

3 Results and Discussion

We present two experiments, namely, the Senate data [8] and the ionosphere data from the machine learning database[1] to show the good performance of our method. We use the SOM toolbox developed by Vesanto et al. [9], because it is easy to reproduce final results in the present paper by using this package. In the SOM, the Batch method is used, which has shown better performance than the popular real-time method in terms of visualization, quantization and topographic errors. The quantization error is simply the average distance from each data vector to its BMU(best-matching unit). The topographic error is the percentage of data vectors for which the BMU and the second-BMU are not neighboring units [10]. Weights are linearly initialized along the greatest eigenvectors of the covariance matrix of the data. The number of competitive units is heuristically determined, and the normal size N is determined roughly by $5\sqrt{S}$, where S is the number of input patterns. The ratio of side lengths is determined by the ratio between the two largest eigenvalues and vectors of the data. The product of two side lengths is adjusted so as to be equal to the expected number of competitive units.

3.1 Senate Data

First, we apply the method to the well-known Senate data [8]. The numbers of input units and patterns are 19 and 15, respectively. With the use of the conventional SOM, the quantization error is 0.218, while with the information-theoretic

[1] http://archive.ics.uci.edu/ml/

method, the error decreases to 0.193. With the conventional SOM, the topographic error is 0.133, while with the information-theoretic method, the error decreases to zero. Figures 1(a) and (b) show the U-matrix and labels obtained by the conventional SOM and the information-theoretic method, respectively. Two characteristics can be pointed out to show the difference between the two methods. First, the slim class boundary in warmer colors obtained by the conventional SOM, shown in Figure 1(a1), becomes wider and stronger than the one in Figure 1(b1), obtained by the information-theoretic method. In addition, we can see that the number of competitive units responding to input patterns increases from eight with the conventional method (Figure 1(a2) to 12 with the information-theoretic method (Figure 1(b2)). Finally, we should note that the mutual information is almost equal with the two methods, while the first order information by the conventional SOM is larger than that by the information-theoretic method.

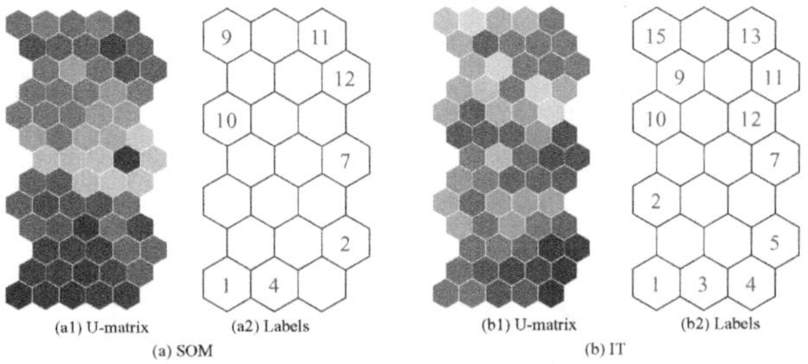

Fig. 1. U-matrices (1) and labels (2) obtained by the conventional SOM (a) and the information-theoretic method (b)

3.2 Ionosphere Data

The second example is the ionosphere data from the well-known machine learning database. The radar data are classified as good and bad returns. The number of input variables and patterns are 34 and 351, respectively. The data are normalized to range between zero and one.

First, we examine quantization and topographic errors. The quantization error gradually decreases to 0.131, while with the conventional SOM, the quantization error is 0.134. The topographic error with the conventional SOM is 0.009, while with the information-theoretic method, the quantization error decreases to 0.005. Figure 2 shows the U-matrix and labels obtained by the conventional SOM (a) and the information-theoretic method (b). Compared with the class boundaries in warmer colors produced by the conventional SOM, those by the information-theoretic method are clearer and stronger. Figure 3 shows the first and the second order information obtained by the conventional SOM and by the information-theoretic method. As can be seen in Figure 3(a), the second order information is 1.622 with the conventional SOM, while with the information-theoretic method, the information is 1.564, slightly below the level obtained by the conventional SOM. Figure 3(b) shows the first order information as a function of the parameter α. The first order information obtained by the conventional SOM increases greatly and then decreases. In the later stage, the information again increases. On the other hand, with the information-theoretic method, the values of the first order information remain relatively constant and small.

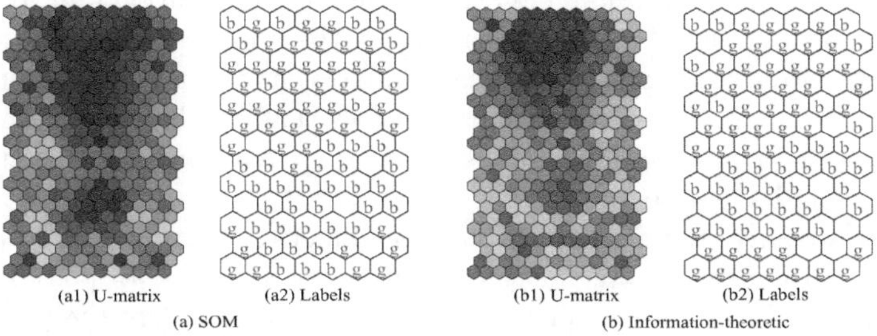

Fig. 2. U-matrices (1) and labels (2) obtained by the conventional SOM (a) and the information-theoretic method (b)

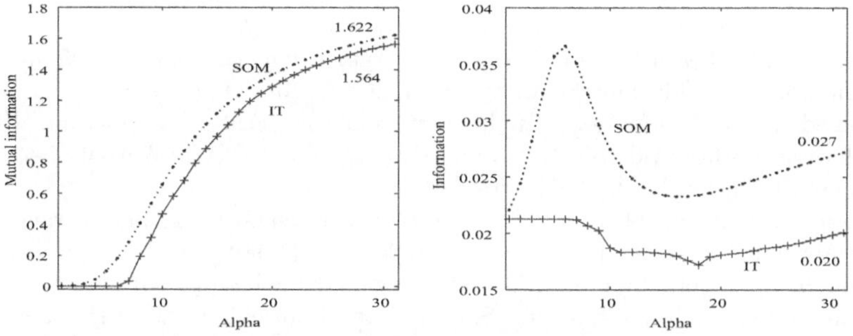

Fig. 3. Second order information (mutual information) and first order information obtained by the conventional SOM (a) and the information-theoretic method (b)

4 Conclusion

We have proposed a new information-theoretic competitive learning method that includes a computational method for free energy. In particular, we have applied the method to the self-organizing maps. One of the main differences is that the degree of competition between competitive units is controlled flexibly by the parameter in our model, while in the conventional method, the winner-take-all algorithm is used. In our method, the winner-take-all operation is obtained as an extreme case of our soft competition. Then, to simplify the computation of mutual information, we have introduced a free energy concept that corresponds to minimizing mutual information as well as quantization errors. In addition, we simplify the computation of the free energy so as to make the final update rules closer to those of the conventional SOM to profit from the sophisticated computational techniques developed for the conventional SOM.

We have applied the information-theoretic method to two well-known problems, namely, the Senate data and ionosphere data. In both problems, experimental results showed that the information-theoretic method had better performance than the conventional SOM in terms of quantization and topographic errors. We have explored the main reason why the information-theoretic method has shown better performance. We have found that the first order information obtained by our method is much smaller than that by the conventional SOM. As the first order information grows smaller, more equi-probable distribution of competitive units can be obtained.

Though we have tried to use computational methods close to the conventional SOM to facilitate the parameter tuning, there are a number of possible computational methods for the free energy. Thus, we should explore the possibility of the free energy as a computational method more exactly.

References

1. Kamimura, R.: Information-theoretic competitive learning with inverse Euclidean distance output units. Neural Processing Letters 18, 163–184 (2003)
2. Torkkola, K.: Feature extraction by non-parametric mutual information maximization. Journal of Machine Learning Research 3, 1415–1438 (2003)
3. Nenadic, Z.: Information discriminant analysis: Feature extraction with an information-theoretic objective. IEEE Transactions on Pattern Analysis and Machine Intelligence 29(8), 1394–1407 (2007)
4. Rose, K., Gurewitz, E., Fox, G.C.: Statistical mechanics and phase transition in clustering. Physical review letters 65(8), 945–948 (1990)
5. Kohonen, T.: Self-Organizing Maps. Springer, Heidelberg (1995)
6. Kamimura, R., Kamimura, T., Shultz, T.R.: Information theoretic competitive learning and linguistic rule acquisition. Transactions of the Japanese Society for Artificial Intelligence 16(2), 287–298 (2001)
7. Kamimura, R., Kamimura, T., Uchida, O.: Flexible feature discovery and structural information control. Connection Science 13(4), 323–347 (2001)
8. Romesburg, H.C.: Cluster Analysis for Researchers. Krieger Publishing Company, Florida (1984)
9. Vesanto, J., Himberg, J., Alhoniemi, E., Parhankangas, J.: SOM toolbox for Matlab. tech. rep., Laboratory of Computer and Information Science, Helsinki University of Technology (2000)
10. Kiviluoto, K.: Topology preservation in self-organizing maps. In: Proceedings of the IEEE International Conference on Neural Networks, pp. 294–299 (1996)

Early Recognition Based on Co-occurrence of Gesture Patterns

Atsushi Shimada, Manabu Kawashima, and Rin-ichiro Taniguchi

Department of Advanced Information Technology, Kyushu University
744 Motooka, Nishi-ku, Fukuoka, Japan
{atsushi,kawashima,rin}@limu.ait.kyushu-u.ac.jp
http://limu.ait.kyushu-u.ac.jp/

Abstract. We propose an approach to achieve early recognition of gesture patterns. We assume that there are two people who interact with a machine, a robot or something. In such a situation, a gesture of a person often has a relationship with a gesture of another person. We exploit such a relationship to realize early recognition of gesture patterns. Early recognition is a method to recognize sequential patterns at their beginning parts. Therefore, in the case of gesture recognition, we can get a recognition result of human gestures before the gestures have finished. Recent years, some approaches have been proposed. In this paper, we expand the application range of early recognition to multiple people based on the co-occurrence of gesture patterns. In our approach, we use Self-Organizing Map to represent gesture patterns of each person, and associative memory based approach learns the relationship between co-occurring gestures. In the experiments, we have found that our proposed method achieved the early recognition more accurately and earlier than the traditional approach.

Keywords: Gesture Recognition, Early Recognition, Co-occurring Gesture, Self-Organizing Map.

1 Introduction

A man-machine seamless interaction is an important tool for various interactive systems such as virtual reality systems, video game consoles, human-robot communication, and so on[5,6]. To realize such a interaction, the system has to estimate human gestures in real-time. Generally, a gesture recognition result is acquired after the gesture has finished. Therefore, if a long gesture is observed, we have to wait for the response until the recognition result is determined. This is a problem to realize a "real-time" man-machine interaction.

Recent years, a new approach called "early recognition" has been proposed for gesture recognition[4,8,1]. The early recognition means that a system outputs a recognition result before a gesture has finished. It is a very useful technique to realize a real-time interaction. The most difficult problem of early recognition is that when the system determines the recognition result. In other words, the system has to ensure the recognition result before the observing gesture has finished. Most traditional approaches suffer from this problem since the gestures

K.W. Wong et al. (Eds.): ICONIP 2010, Part II, LNCS 6444, pp. 431–438, 2010.

comprehend ambiguity. Especially at the beginning part of them, it is very difficult to determinate the recognition result since enough input data has not been observed yet. To solve this problem, we propose a new approach. The biggest difference between traditional approaches and our approach is that we target not only an individual person but also two or more people in the environment. It means that there are two or more people who interact with a machine, a robot, or so on, simultaneously. In such a situation, a gesture of a person is often related to a gesture of another person. We call such a relationship "co-occurring gesture", and we use the information of co-occurrence for realizing early recognition.

Our approach uses Self-Organizing Map (SOM) and its sparse codes to represent gesture patterns. This approach is based on the approach proposed by Shimada *et al.*[7]. In this research, we have modified their approach to adapt for co-occurring gesture recognition. In addition, we introduce an associative memory to describe a relationship between co-occurring gestures.

2 Definition of Early Recognition of Gesture Patterns

In this section, we give conceptual explanation about early recognition of individual gesture and co-occurring gesture.

2.1 Typical Gesture Recognition

Let $\boldsymbol{C}^i = \{c_1^i, \ldots, c_n^i\}$ be a training gesture pattern which belongs to gesture class $i \in \boldsymbol{L}$. The $\boldsymbol{L}$ is a set of class labels. A gesture can be represented in a sequential n-long posture patterns. Therefore, c_n^i means the n-th posture of the gesture. When an unknown gesture $\boldsymbol{X} = \{x_1 \ldots, x_l\}$ is observed, the typical gesture recognition problem is to find the most similar gesture from training patterns by

$$p = \underset{i}{\operatorname{argmin}} \{f(\boldsymbol{X}, \boldsymbol{C}^i)\} \tag{1}$$

where p is the class label and $f()$ is a distance function which evaluate the similarity between the gesture pattern $\boldsymbol{X}$ and $\boldsymbol{C}^i$.

2.2 Early Recognition of Individual Gesture Patterns

The key issue of early recognition is to output a recognition result before acquiring complete input pattern. In the case of gesture recognition, especially individual gesture patterns, it corresponds to the following problem. When a part of gesture pattern (unfinished gesture) $\boldsymbol{X}' = \{x_1 \ldots, x_k\}, (k < l)$ is observed, the recognition result is determined by

$$p = \underset{i}{\operatorname{argmin}} \{f(\boldsymbol{X}', \boldsymbol{C}^i) < TH_I\} \tag{2}$$

where TH_I is a threshold of distance which adjusts the timing of recognition result. If the threshold is not introduced, a recognition result will be output without concrete proof. Therefore, we set a threshold to ensure reliability for the recognition result.

2.3 Early Recognition of Co-occurring Gesture Patterns

Unlike the other recognition strategies mentioned above, the system has to observe gestures of two people simultaneously. Let $\boldsymbol{Y}^{'}$ be a gesture pattern of another person (the gesture has not finished yet). The output of the early recognition of co-occurring gesture patterns can be defined as follows.

$$(p,q) = \underset{(i,j)\in \boldsymbol{M}}{\operatorname{argmin}} \{f(\boldsymbol{X}^{'}, \boldsymbol{C}_A^i) + f(\boldsymbol{Y}^{'}, \boldsymbol{C}_B^j) < TH_C\} \tag{3}$$

where $\boldsymbol{M}$ is a subset of $\boldsymbol{L} \times \boldsymbol{L}$. The subscripts of $\boldsymbol{C}$ denote the person labels, i.e, person A and person B. Note that the $\boldsymbol{L} \times \boldsymbol{L}$ is a set of all combination of co-occurring gestures. Actually, the combination is restricted by the application, environment or so on, and the co-occurrence is not always the all combination of gestures. This is why we introduce the subset $\boldsymbol{M}$. Fig. 1 shows an example of the relationship between $\boldsymbol{M}$ and $\boldsymbol{L}\times\boldsymbol{L}$. The rows denote the label of gesture of person A, and the columns denote the one of person B. The “circle mark” indicates that the corresponding gestures will be observed simultaneously. In the case of Fig. 1(a), all gestures of person A would be observed at the same time with the all gestures of person B. On the other hand, in the case of Fig. 1(b), some cells are “blank”, which means that such co-occurrence will not be observed. Therefore, the possible co-occurring gestures between person A and B are a subset of all combination $\boldsymbol{L} \times \boldsymbol{L}$, i.e., $\boldsymbol{M} = \{(g1, g1), (g1, g2), (g2, g1), (g2, g3), (g3, g1)\}$.

The TH_C is a threshold which controls the timing of early gesture recognition. The difference between Eq. 2 and Eq. 3 is that the latter determines the output timing based on two distance functions, i.e, $f(\boldsymbol{X}^{'}, \boldsymbol{C}_A^i)$ and $f(\boldsymbol{Y}^{'}, \boldsymbol{C}_B^j)$. Therefore, even if the system does not have high confidence in one person's gesture recognition, it can output the result when another person's gesture is recognized with higher confidence (i.e, very smaller distance of $f(\boldsymbol{Y}^{'}, \boldsymbol{C}_B^j)$).

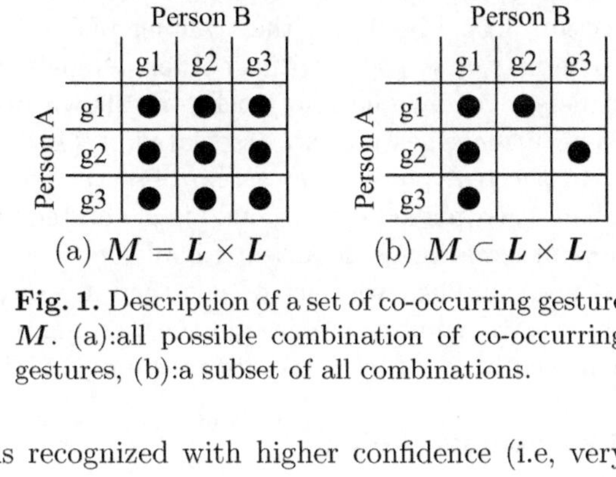

(a) $\boldsymbol{M} = \boldsymbol{L} \times \boldsymbol{L}$ (b) $\boldsymbol{M} \subset \boldsymbol{L} \times \boldsymbol{L}$

Fig. 1. Description of a set of co-occurring gesture $\boldsymbol{M}$. (a):all possible combination of co-occurring gestures, (b):a subset of all combinations.

3 Early Recognition Strategy

3.1 System Overview

First of all, we show the system overview in Fig. 2. The process can be divided into two phases; training phase and test phase. In the training phase, Self-Organizing Map (SOM) is used to learn postures, which are elements of all gestures. The advantages of using SOM are 1) to reduce dimensionality of

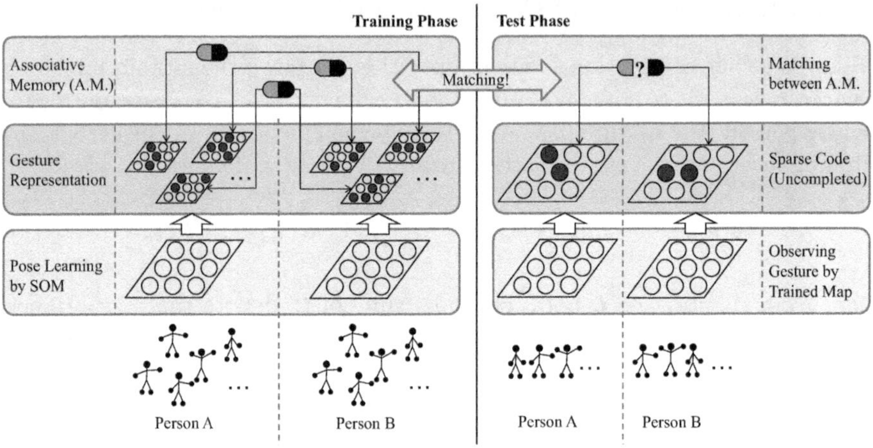

Fig. 2. Processing Flow of Training/Recognition of Co-occurring Gesture Patterns

gesture patterns, 2) to reduce some redundant postures, 3) to represent a gesture pattern by combination of smaller number of neurons and so on. Due to space limitation, we skip the detailed explanation about SOM and how to learn the postures (refer to the literature [7] for detail). After the training of all postures, element postures of each gesture are input to the map again. And then, we can get a "Sparse Code" which represent a gesture pattern on the SOM (see section 3.2). Finally, in the training phase, we associate one person's gesture pattern (sparse code) with another person's gesture pattern based on teacher signals given by the relationship as shown in Fig. 1. In this way, all possible co-occurrence gestures are associated by "Associative Memory"(see section 3.3).

In the test phase, the system observes two people's gesture simultaneously. Then, each person's parse code is generated/updated immediately whenever a new observation is acquired. Finally, two sparse codes (person A and person B) are examined whether or not they are co-occurring gesture by referring to associative memory acquired in the training phase. Actually, the examination is achieved by measuring the distance between sparse codes(see section 3.4).

3.2 Sparse Coding

When a posture x_k is input to the SOM, one neuron will be selected as winner. When a set of postures which consist of a gesture is sequentially input to the SOM, some neurons will be activated. We regard such an activation pattern as "Sparse Code", which represents an input gesture. Here, we define the notation of a sparse code. Let $\boldsymbol{S}$ be a sparse code which means a set of activated neuron s. In the training phase, all training gestures are represented by using sparse code $\boldsymbol{C}^i$. Meanwhile, in the test phase, a sparse code of observing gesture is represented by $\boldsymbol{X}^{'}$ or $\boldsymbol{Y}^{'}$, which corresponds to Eq. 3.

Note that the sparse code described here has not an ability to distinguish the gesture patterns whose elements are the same but the sequences are different.

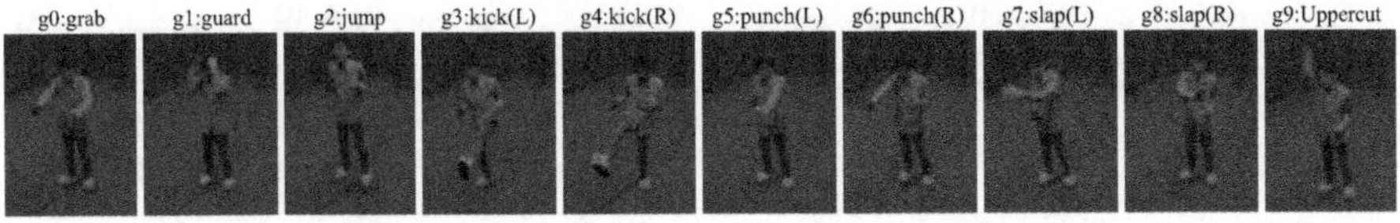

Fig. 3. Gestures used in our experiments

However, we can easily improve introduce temporal information into the system by our previous study[7].

3.3 Associative Memory

To realize early recognition of co-occurring gesture patterns, we introduce an associative memory. After the training of all gestures(actually, the training of sequential postures by SOM), we get sparse code $\boldsymbol{C}_A^i$ and $\boldsymbol{C}_B^j$ which co-occur with each other. The combination of i and j is restricted by $(i, j) \in \boldsymbol{M}$ which is defined by application(see section 4 for our configuration). Our system memorize these relationships between co-occurring gestures as "associative memory". In other words, the system has several combinations between $\boldsymbol{C}_A^i$ and $\boldsymbol{C}_B^j$, which will be observed as co-occurring gestures. Actually, in our implementation, we stored each pair of sparse codes, which indicates the list of winner neurons' indices for the corresponding gesture, in the memory storage of the computer and used them as associative memory.

3.4 Similarity Measure

The number of elements in sparse code $\boldsymbol{S}$ is different from each other since the number of activated neurons depend on the gesture length and the gesture pattern. Therefore, we introduce the Hausdorff distance to measure the similarity between two sets of sparse code. Let X and Y be two non-empty subsets of a metric space. The Hausdorff distance $f(X, Y)$, which corresponds to the distance function in Eq. 1, 2 and 3, is defined by

$$f(X, Y) = \max_i \{ \sup_{x \in \boldsymbol{X}} \inf_{y \in \boldsymbol{Y}} d(x, y), \sup_{y \in \boldsymbol{Y}} \inf_{x \in \boldsymbol{X}} d(x, y) \} \quad (4)$$

where $d(x, y)$ is the distance function. In our research, we use L2-distance between the coordinates of activated neurons.

4 Experimental Results

4.1 Condition

We demonstrate proposed early recognition of gesture patterns using motion data prepared by ourselves. Each gesture consists of a sequence of postures, and each posture is represented by 5 measured markers. Each marker is composed

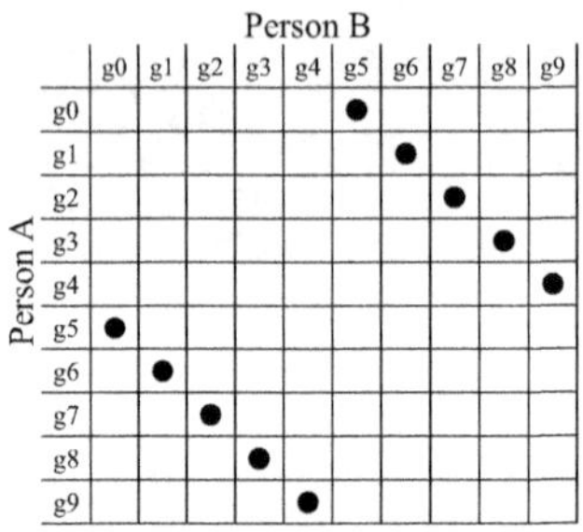

(a) case 1: one-to-one

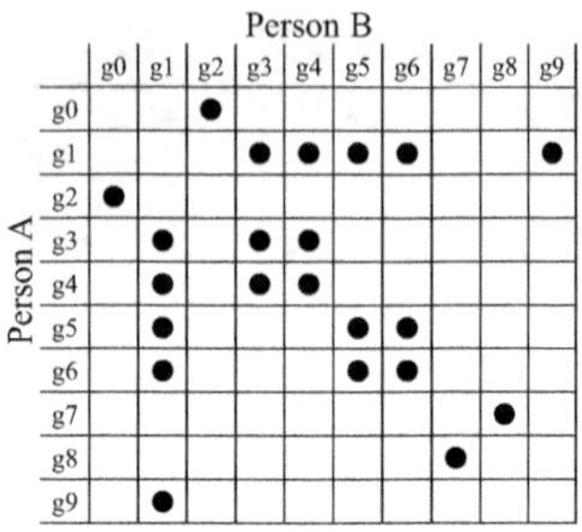

(b) case 2: one-to-many

Fig. 4. Configuration of co-occurring gestures

of data of (x, y, z)-axis. We prepared 10 kinds of gesture patterns ($|\boldsymbol{L}| = 10$, see Fig. 3) from 7 examinees. Each person did each gesture 40 times. We used 20 patterns for training and other 20 patterns for test. We conducted the experiment through cross-validation among examinees.

The co-occurring gestures used in the experiments are shown in Fig. 4. The Fig. 4(a) shows the simple configuration that each gesture of person A corresponds to unique gesture of person B. Meanwhile, in the case of Fig. 4(b), the problem becomes more difficult since there are some gesture candidates(one-to-five correspondence at maximum) which occurs at the same time. We can investigate how the co-occurrence information is effective and helpful to determine the recognition result.

4.2 Early Recognition Result of Individual Gesture Patterns

Fig. 5 shows the result of early recognition for an individual person. The horizontial axis denotes the complete ratio of observing gesture pattern, and the vertical axis denotes the recognition accuracy. The bold curve indicates the average ratio of accuracy. For example, the recognition ratio exceeded 90% when more than 50% long gestures had been observed on average. We regards this result as baseline in the following experiments.

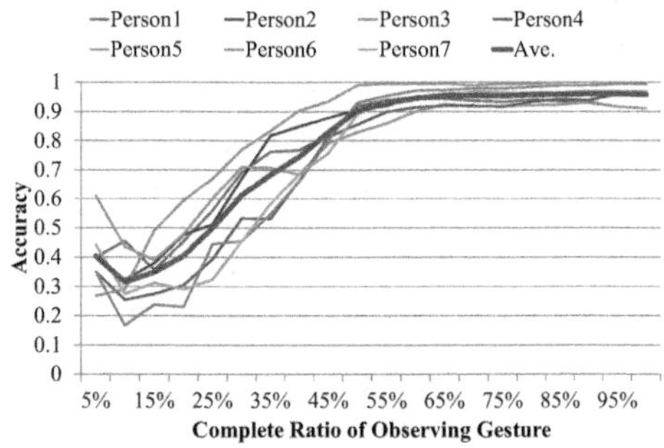

Fig. 5. Early Recognition Result of Individual Gesture Patterns

4.3 Early Recognition Result of Co-occurring Gesture Patterns

As mentioned above, we investigated the recognition accuracy of co-occurring gestures under two conditions(see Fig. 4). First, we examined the case 1 in

Complete Ratio of Observing Gesture of Person B

Complete Ratio of Observing Gesture of Person A	5	10	15	20	25	30	35	40	45	50	55	60	65	70	75	80	85	90	95	100
5					1.00	1.00	0.67	0.80	0.67	0.61	0.69	0.71	0.71	0.71	0.63	0.63	0.63	0.64	0.64	0.66
10					1.00	1.00	0.67	0.65	0.66	0.69	0.70	0.69	0.69	0.65	0.61	0.58	0.58	0.58	0.62	0.60
15					1.00	0.86	0.55	0.60	0.64	0.60	0.61	0.57	0.59	0.56	0.53	0.48	0.51	0.49	0.51	0.51
20					0.83	0.72	0.55	0.65	0.73	0.71	0.72	0.74	0.74	0.73	0.74	0.72	0.73	0.73	0.73	0.73
25			1.00	1.00	0.98	0.96	0.90	0.87	0.89	0.88	0.89	0.89	0.88	0.87	0.86	0.84	0.85	0.85	0.84	0.83
30	1.00	1.00	1.00	0.65	0.82	0.95	0.92	0.91	0.91	0.92	0.92	0.92	0.93	0.92	0.91	0.92	0.91	0.91	0.90	0.89
35	1.00	1.00	0.90	0.69	0.80	0.92	0.96	0.95	0.96	0.97	0.97	0.97	0.97	0.97	0.96	0.95	0.96	0.95	0.96	0.95
40	1.00	1.00	0.88	0.66	0.86	0.93	0.95	0.95	0.97	0.97	0.97	0.97	0.97	0.97	0.97	0.96	0.97	0.96	0.97	0.97
45	1.00	1.00	0.92	0.72	0.89	0.94	0.96	0.94	0.97	0.97	0.98	0.97	0.97	0.98	0.98	0.97	0.98	0.98	0.98	0.98
50	1.00	1.00	0.92	0.67	0.84	0.89	0.94	0.93	0.97	0.98	0.98	0.98	0.98	0.98	0.98	0.98	0.98	0.98	0.98	0.98
55	0.86	0.94	0.89	0.67	0.85	0.91	0.95	0.94	0.97	0.98	0.97	0.97	0.97	0.98	0.98	0.97	0.98	0.98	0.98	0.98
60	0.86	0.94	0.89	0.69	0.85	0.91	0.95	0.94	0.97	0.98	0.97	0.97	0.97	0.98	0.98	0.97	0.98	0.98	0.98	0.98
65	0.86	0.94	0.89	0.69	0.86	0.91	0.95	0.94	0.97	0.98	0.97	0.97	0.97	0.98	0.98	0.97	0.98	0.98	0.98	0.98
70	0.86	0.94	0.89	0.69	0.86	0.92	0.95	0.94	0.97	0.97	0.97	0.97	0.97	0.98	0.98	0.98	0.98	0.98	0.98	0.98
75	0.86	0.95	0.85	0.71	0.85	0.92	0.95	0.94	0.97	0.97	0.97	0.97	0.97	0.98	0.98	0.97	0.98	0.98	0.98	0.98
80	0.89	0.95	0.80	0.65	0.83	0.92	0.95	0.94	0.98	0.97	0.97	0.97	0.97	0.97	0.98	0.97	0.98	0.98	0.98	0.98
85	0.90	0.95	0.83	0.70	0.85	0.91	0.95	0.94	0.97	0.98	0.97	0.97	0.97	0.98	0.98	0.97	0.98	0.98	0.98	0.98
90	0.90	0.96	0.83	0.70	0.84	0.90	0.95	0.94	0.97	0.98	0.97	0.98	0.97	0.98	0.97	0.97	0.98	0.97	0.98	0.97
95	0.89	0.94	0.82	0.70	0.83	0.90	0.95	0.94	0.97	0.98	0.97	0.98	0.97	0.98	0.97	0.97	0.98	0.97	0.98	0.98
100	0.90	0.95	0.82	0.70	0.82	0.90	0.96	0.93	0.97	0.98	0.97	0.97	0.97	0.98	0.97	0.97	0.98	0.97	0.98	0.98

Fig. 6. Early Recognition Result of Co-occurring Gesture Patterns

Fig. 4(a). We got about 100% accuracy when two people's gesture had been observed at least 15%. Due to limitations of space, we left out the detailed result here, but we consider that we could get such good results because of the simple co-occurrence rule between two people's gesture patterns. In other words, the system could narrow the recognition result easily with the help of co-occurrence information.

Second, we examined the case 2 in Fig. 4(b). The early recognition results are shown in Fig. 6. The vertical line shows the complete ratio of observing gesture patterns of person A and the horizontal line shows the one of person B. Each cell in the figure shows each recognition result. The blank cell denotes that the system didn't output the recognition result because the condition of early recognition was not satisfied in Eq. 3. We gave each cell a color based on the accuracy ratio. The red or yellow cell means a good result which exceeds 95% accuracy. Most cells have a red-like or yellow-like color, which indicates higher accuracy. For example, it was enough for the system to determine the recognition result when the one person's gesture patterns had been observed at least 25%.

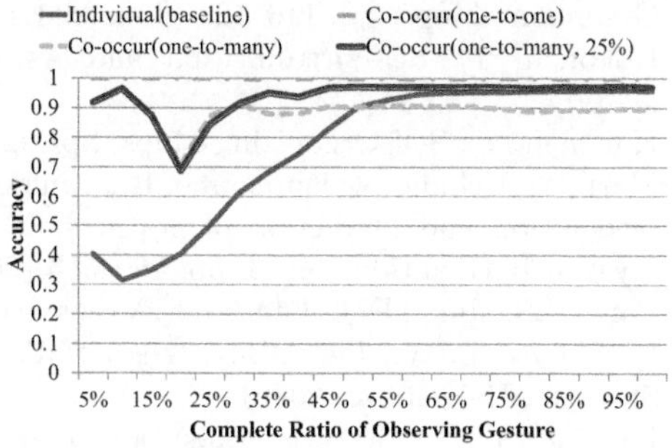

Fig. 7. Comparison between Individual and Co-occurrence Recognition

To compare the early recognition accuracy between individual and co-occurrence, we draw the accuracy curve as shown in Fig. 7. The bold red curve is referred from the average accuracy in Fig. 5. In fact, though there are two complete ratio axes in the case of co-occurring gesture patterns(i.e., person A and person B), we marginalized with respect to one person's complete ratio to represent in the same format with Fig. 5. We can see that the system achieved gesture recognition much earlier than the baseline method. In particular, if one person's gesture had been observed at least 25%, the performance was very high compared with the baseline method(see the bold blue line in Fig. 7).

5 Conclusion

We have proposed a new approach for early recognition of gesture patterns which targets two people. When there is co-occurrence of gestures between two people, the system can recognize the recognition result using its co-occurring information. We have developed prototype of early recognition system using SOM and the associative memory. Through experiments, we confirmed that our proposed method performs well.

In a future work, we are going to use our proposed early recognition system for actual man-machine interaction and investigate its effectiveness. Before that, we will conduct further experiments; increasing the number of gesture classes, the number of people and so on.

References

1. Kawashima, M., Shimada, A., Taniguchi, R.: Early recognition of gesture patterns using sparse code of self-organizing map. In: 7th International Workshop On Self-Organizing Maps, pp. 116–123 (June 2009)
2. Kohonen, T.: Self-Organization and Associative Memory. Springer, Heidelberg (1989)
3. Kohonen, T.: Self-Organizing Maps. Springer Series in Information Science (1995)
4. Mori, A., Uchida, S., Kurazume, R., Taniguchi, R., Hasegawa, T., Sakoe, H.: Early recognition and prediction of gestures. In: Proc. of International Conference on Pattern Recognition, vol. 3, pp. 560–563 (2006)
5. Park, H.S., Jung, D.J., Kim, H.J.: Vision-based game interface using human gesture. In: Chang, L.-W., Lie, W.-N. (eds.) PSIVT 2006. LNCS, vol. 4319, pp. 662–671. Springer, Heidelberg (2006)
6. Park, J., Yi, J.: Gesture recognition based interactive boxing game. International Journal of Information Technology 12, 36–44 (2006)
7. Shimada, A., Taniguchi, R.: Gesture recognition using sparse code of hierarchical som. In: Proc. of International Conference on Pattern Recognition (2008)
8. Uchida, S., Amamoto, K.: Early recognition of sequential patterns by classifier combination. In: Proc. of International Conference on Pattern Recognition (2008)

A Dynamically Reconfigurable Platform for Self-Organizing Neural Network Hardware

Hakaru Tamukoh and Masatoshi Sekine

Institute of Engineering, Tokyo University of Agriculture and Technology

Abstract. In this paper, we propose a dynamically reconfigurable platform for self-organizing neural network hardware. In the proposed platform, a hardware unit can be handled as a hardware object in object-oriented design. The hardware object is loaded into the FPGA's virtual hardware circuit space, and accelerates the calculation of self-organizing neural networks. We design two types of the distance calculation, a winner-take-all and a rough-winner-take-all virtual hardware circuit as common parts of self-organizing neural networks. By combining them, we realize four types of self-organizing neural network. Experimental results show that the implemented self-organizing neural network hardware achieves about 100 times faster than the software implementation. Besides, the proposed platform can change its learning mode easily as well as the software implementation. Therefore, the proposed platform features both of the speed of hardware and the flexibility of software.

Keywords: Self-Organizing Map, Vector Quantization, hw/sw complex system, FPGA, Digital Hardware Implementation.

1 Introduction

Self-Organizing Map (SOM) is widely used to various application areas such as data analysis, classification and control tasks [1]. Besides, modified or advanced algorithms of SOM also have been proposed such as Neural Gas (NG) [2], Self-Organizing Relationship Network (SORN) [3] and so on.

A digital hardware implementation of self-organizing neural networks achieves high-performance calculation, low-power consumption, and small area implementation, which are important properties to realize real-time applications and embedded systems. Therefore, numerous special-purpose hardware architectures have been proposed. Porrmann and Rückert et al., who are a pioneer of the SOM hardware implementation, proposed a massively parallel architecture of SOM and implemented into an Application Specified Integrated Circuit (ASIC). Hikawa proposed SOM hardware in a pulse mode using a digital phase locked loop to conserve the Field Programmable Gate Array (FPGA) resources [5]. Rovetta et al. proposed an efficient learning strategy of NG and its ASIC implementation in an analog mode [6]. We also proposed a FPGA implementation of SOM [7] [8] and a parameter-less SOM with an adaptive learning rule [9], and applied them to an image processing and a mobile robot controlling. However,

K.W. Wong et al. (Eds.): ICONIP 2010, Part II, LNCS 6444, pp. 439–446, 2010.

these implementations are realized into ASICs or specified FPGA boards. Thus, a design-reuse or an applying the design to other applications is quite difficult.

In this paper, we propose a dynamically reconfigurable platform for self-organizing neural network hardware. In the proposed platform, a hardware unit can be handled as an object in object-oriented design. This hardware object can be dynamically constructed, executed, and destructed from a user application in software. The hardware object is loaded into the FPGA's virtual hardware circuit space, and accelerates the calculation of self-organizing neural networks. This mechanism facilitates a rapid application prototyping, which allows the design reuse and the combining existing hardware designs easily.

2 Self-Organizing Neural Networks

First, we introduce Kohonen's SOM which is the basic algorithm for the other self-organizing neural networks. The algorithm of SOM consists of the following three calculation steps.

1. Distance calculation: An input vector $\boldsymbol{x}$ is compared with all the reference vectors $\boldsymbol{w}_i$ by following function.

$$d_i = \| x - w_i \|, \tag{1}$$

d_i is the distance between the input vector and the i-th reference vector. Euclidean distance is usually used for the software implementation, and Manhattan distance is often employed for the digital hardware implementation to conserve the hardware resources.

2. Competition: A winner reference vector c is selected by winner-take-all (WTA).

$$c = arg \min_i (d_i). \tag{2}$$

3. Updating: The winner and its neighborhood reference vectors are updated by:

$$w_i^{new} = w_i^{old} + h_{ci} \cdot (x - w_i^{old}). \tag{3}$$

h_{ci} is called as the neighborhood function, and the detailes is shown in the [1].

The various self-organizing neural networks have been proposed which is the modified version of SOM. For instance, in order to enhance the vector quantization ability, NG [2] and rough WTA self-organizing neural network (RWTA SONN) [10] have been proposed. In the algorithm of NG, the sorting and the ranking function are used instead of the WTA and the neighborhood function. Similarly, the algorithm of RWTA SONN employed rough WTA as the substitute of the ordinary WTA. Moreover, combining the fuzzy inference unit as an output layer to the learned network, it can be used as an execution mode of self-organizing relationship network [3] or radial basis function network (RBFN).

Table 1 summarizes the algorithm of self-organizing neural networks and learning results shown in Fig.1. Among these algorithms, the distance calculation is used as the common component, i.e., SOM hardware's distance calculation

Table 1. Algorithm examples of self-organizing neural networks

Algorithm	SOM	NG	RWTA SONN	K-means
Distance calculation	Euclidean / Manhattan			
Competition	WTA	Sorting	Rough WTA	WTA
Updating manner	Neighborhood	Ranking	Winner only	Winner only

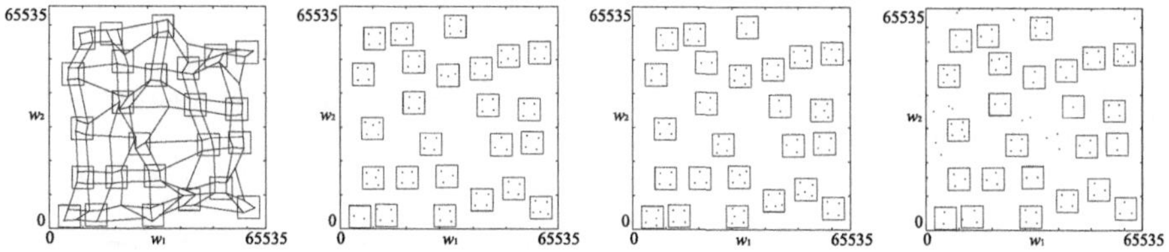

SOM (1358.10) NG (1292.08) RWTA SONN (1308.05) K-means (1508.24)

Fig. 1. Learning results of self-organizing neural networks [10]. Parenthetic values are the mean square error to evaluate the vector quantization ability. The squares and the dots represent the data clusters and reference vectors, respectively. The lines between reference vectors represents neighborhood in SOM. NG and RWTA SONN specialized to the vector quantization ability while SOM can make the topological map.

unit can be reused to the other self-organizing neural networks. Currently, the modified or advanced SOMs are continuously investigating. Therefore, the design reuse is important to develop new self-organizing neural network hardware.

3 Dynamically Reconfigurable FPGA Platform

In order to realize a rapid-prototyping of a new self-organizing neural network hardwareC we propose a hw/sw complex system based on a hardware object model [11] and its platform FPGA board named "hwModule". The abstract model of hw/sw complex system is shown in Fig.2 (a). In this system, an object processed by the Host Processor is termed "swObject", an object processed by the FPGA is termed "hwObject", and a virtual hardware circuit in the FPGA that processes a hwObject is termed "hwNet".

The hw/sw complex system is composed of two essential units. The hardware unit is constructed by the hwModule FPGA board which consists of a number of FPGAs and a Local Memory SDRAMs. The hwNet is separated from peripheral circuits through its corresponding I/F (Interface) circuits. It is also possible to connect to external devices such as LAN and USB through GPIO (General Purpose Input Output). By designing a hwNet that executes a very large scale computing, we can develop a high-performance hw/sw complex system which possesses both the speed of hardware and the flexibility of software. All of the hardware unit's components, excluding the hwNet space, are peripheral circuits which are standardized for all hwModule FPGA board series. These peripheral

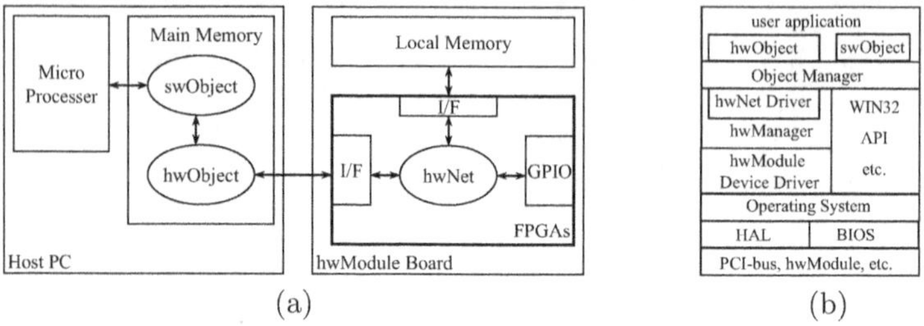

Fig. 2. The hw/sw complex system. (a) A conceptual model of hw/sw complex system. (b) A layers model from hwObject to hwModule Board.

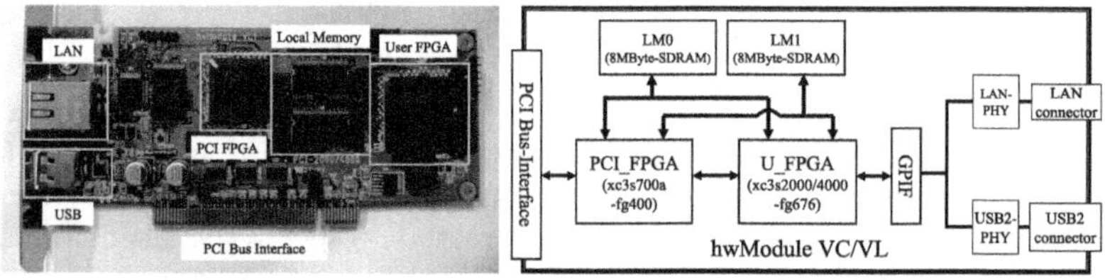

Fig. 3. The hwModule FPGA board. Left: Photgraph of hwModuleVL. Right: Block diagram of hwModule FPGA board.

circuits are provided in the system development kit to implement a new hardware function easily.

The software unit is specialized for software processing, and is running on the high-performance processor and the large-scaled main memory of the Host PC. Since the hardware unit handling is abstracted in the software unit, the software engineer accesses the hwNet through a hwObject. This achieves a tight connection between the hardware and software units, and a high-performance and flexible processing system.

The hierarchical diagram of the components supporting the hwObject's operation is shown in Fig.2 (b). The user application calls the hwObject in the same manner as it calls a swObject. The Object Manager includes an event-queue and the management table of the objects. It controls multiple accesses from the hwObjects to a hardware manager (hwManager). The hwNet Driver defines the I/O of the hwNet, and is dynamically attached to the hwManager when the hwObject is constructed in the user application. The hwManager and the hwModule Device Driver intermediates between a hwObject and the hwModule, in such way, the details of the hardware are separated from the user application. In particular, these two components translate the hwObject instruction to an I/O access command. This command is then forwarded into the hwModule Device Driver to be sent as physical signals to the corresponding hwNet via the PCI Bus. Currently, the hwModule Device Driver is designed according to the Windows Driver Model specifications and operates under Windows 2000/XP OS. By

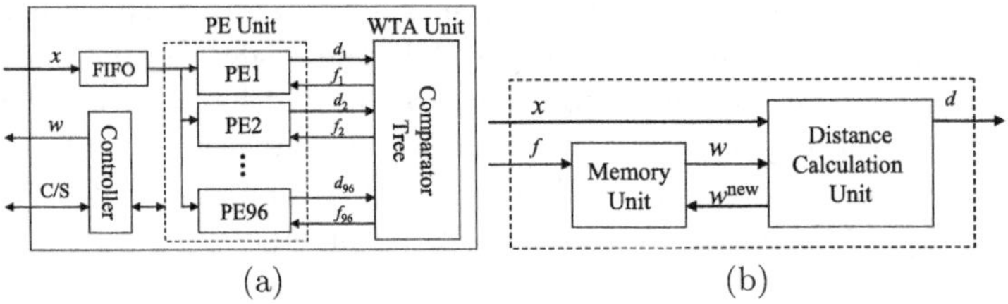

Fig. 4. Block diagram of the self-organizing neural network hardware. (a) Top view. C/S is control and status signals which are used for the communication between the host PC and the controller. (b) Processing Element.

providing these various components as a System Development Kit, it is possible to program the hwObject in the user application.

In order to realize the concept of hw/sw complex system, we have developed hwModule FPGA board series shown in Fig.3. We have already developed practical hw/sw complex systems such as an image processing and a web video streaming applications using this platform and hardware object model [11] [12]. We employ hwModuleVL as a dynamically reconfigurable platform for self-organizing neural network hardware in this paper.

4 Implementation of Self-Organizing Neural Network

Fig.4 (a) shows the block diagram of the developed hardware architecture which consists of two essential units. PE Unit consists of 96 processing elements which are executed in the massively parallel manner, and each PE calculates eqs. (1) and (3). WTA Unit consists of a binary comparator tree which calculates eq. (2). In this section, we introduce two types of PE Unit and two types of WTA Unit, and four types of self-organizing neural networks are presented by combining these units.

4.1 Distance Calculation Unit

Distance calculation unit calculates the distance d between the input vector x and the reference vector w, and updates reference vector if the winner flag $f = 1$ is obtained. We have designed following two types of calculation unit.

Type 1: Manhattan type. This architecture widely used for the digital hardware implementation of SOM because the required hardware resource is less than Euclidean distance.

$$d_i = \sum_j |x_j - w_{ij}|. \tag{4}$$

Type 2: Euclidean type. We employed the embedded multiplier of the FPGA and the square root operation is not applied.

$$d_i = \sum_j (x_j - w_{ij}) \times (x_j - w_{ij}). \tag{5}$$

Both types of the distance calculation unit takes n clock cycle to calculate n-dimensional vector's distance. The proposed distance calculation unit is designed as a parameterized verilog source code, i.e., the bit-width of vector's element and the vector's dimension can be easily changed by the hardware designer. This idea is similar to an automatic SOM VHDL code generation [13], which facilitates a development of SOM hardware with various parameters.

4.2 WTA and Rough WTA Unit

WTA Unit selects the winner which has a minimum d_i. After WTA, the flag $f = 1$ is assigned to only the winner. We employed a binary comparator tree for the WTA Unit. It takes 7 clock cycles to select the winner when the number of the input distance is 96. We have designed following two types of WTA unit.

Type1: Accurate type. This is an ordinary (normal) architecture which selects accurate winner by eq.(2).

Type 2: Rough type. This is called as a rough comparison winner-take-all (RWTA) [10] which selects winner by the following equation.

$$c = arg \min_i (d_i \gg r(t)). \tag{6}$$

'$\gg$' means the logical right shift operation and $r(t)$ represents a parameter of the selection accuracy. The large $r(t)$ causes the rough winner selection because the shifted lower bits becomes the quantization error. To realize a good vector quantization, $r(t)$ should decrease with the learning iteration t.

$$r(t) = D_a \cdot \frac{T - t}{T}, \tag{7}$$

where, D_a and T represent the number of the distance register bits and the total number of the learning steps, respectively. This strategy leads a high quality vector quantization as same as NG which shown in Fig.1.

4.3 Results and Discussion

We designed and synthesized four types of self-organizing neural networks using velilog HDL and Xilinx tools. Memory Unit was designed based on a BRAM (FPGA internal memory) and the neighborhood function h_{ci} was defined by a look up table. We configured the input and reference vector which has 2,048 dimensions with 8 bits calculation accuracy. From the implementation results shown in Table 2, all hardware can be implemented into the hwModuleVL and operate in 66MHz. As the result, a theoretical maximum performance of the designed hardware achieved 6,305 MCUPS (Mega Connection Update Per Second). On the other hand, with a software implementation on a state-of-the-art personal computer (Intel Core2Duo, 3.16GHz, Single thread programming) only a performance of 62 MCUPS can be achieved.

Table 2. Implementation results of four networks using xc3s4000 User FPGA

No.	Distance	WTA	Slice	BRAM	MULT.	Freq. (MHz)
1.	Manhattan	Accurate	5,442 (19.7 %)	96 (100 %)	0 (0 %)	74.43
2.	Euclidean	Accurate	6,665 (24.1 %)	96 (100 %)	96 (100 %)	74.43
3.	Manhattan	Rough	8,995 (32.5 %)	96 (100 %)	0 (0 %)	69.26
4.	Euclidean	Rough	12,681 (46.2 %)	96 (100 %)	96 (100 %)	69.26

Table 3. Comparison of the proposed platform with the other platforms

Reference	Proposed	[14]	[9]	[6]
Device	XC3S4000	XC2V10000	XC3S1600E	ASIC 0.8μm
Performance (MCUPS)	6,305	17,360	632	NA
Learning mode	SOM, VQ	SOM, RBFN, etc.	parameter-less SOM	NG
Reconfigurability	Yes	Yes	No	No

From the view point of application development which uses self-organizing neural networks, the proposed platform allows several learning modes. For instance, a data mining application requires topological map, thus, no.2 can be selected to a new implementation. On the other hand, an image compression application desires a high quality vector quantization and it not require the topological map. In such situation, the application designer can select no.4 hardware. Moreover, if the pre- or post-processing requires much FPGA area, the designer can select no.1 or no.3 to conserve hardware resources.

Table 3 summerizes the comparison result of the proposed platform. The almost all of the platforms such as [9] and [6] does not have a reconfigurability. On the other hand, both of the proposed and [14] have the reconfigurability which can operate several learning mode. The performance difference between the proposed and [14] platform was caused by the employed FPGA. The platform [14] used Xilinx Virtex series which is a high-end FPGA. In our platform, we employed Xilinx Spartan series, which is a low-end FPGA, to consider the system cost. Currently, to overcome the system performance, we are developing a new platform which consists of 128 XC3S4000 FPGAs. Further discussion of the system performance will be a future task.

5 Conclusions

In this paper, we propose a dynamically recofigurable platform for the self-organizing neural network hardware, which features both of the speed of hardware and the flexibility of software. Currently, the proposed platform can operate four learning mode. Building a useful hardware object library of the self-organizing neural network and a development of practical applications would be future tasks.

Acknowledgment

This work was supported by a Grant-in-Aid for Young Scientists by JSPS (Research Project Number: 20700207), and also supported by a Creation and Support Program for Start-ups from Universities by JST.

References

1. Kohonen, T.: Self-Organizing Maps, 3rd edn. Springer, Heidelberg (2001)
2. Martinetz, T.M., Berkovich, S.G., Schulten, K.J.: "Neural-Gas" Network for Vector Quantization and its Application to Time-Series Prediction. IEEE Trans. on Neural Networks 4(4), 558–569 (1993)
3. Yamakawa, T., Horio, K.: Self-Organizing Relationship (SOR) Network. IEICE Trans. on Fundamentals E82-A(8), 1674–1678 (1999)
4. Porrmann, M., Witkowski, U., Rückert, U.: A Massively Parallel Architecture for Self-Organizing Feature Maps. IEEE Trans. on Neural Networks 5, 1110–1121 (2003)
5. Hikawa, H.: FPGA Implementation of Self Organizing Map with Digital Phase Locked Loops. Neural Networks 18(5-6), 514–522 (2005)
6. Rovetta, S., Zunino, R.: Efficient Traning of Neural Gas Vector Quantizers with Analog Circuit Implementation. IEEE Trans. on Circuits and Systems II: Analog and Digital Signal Processing 46(6), 688–698 (1999)
7. Tamukoh, H., Aso, T., Horio, K., Yamakawa, T.: Self-Organizing Map Hardware Accelerator System and its Application to Realtime Image Enlargement. In: Proc. of International Joint Conference on Neural Networks, pp. 2683–2687 (2004)
8. Tamukoh, H., Horio, K., Yamakawa, T.: Fast Learning Algorithms for Self-Organizing Map employing Rough Comparison WTA and its Digital Hardware Implementation. IEICE Trans. Elect. E87-C(11), 1787–1794 (2004)
9. Sonoh, S., Aou, S., Horo, K., Tamukoh, H., Koga, T., Yamakawa, T.: A Human Robot Interaction by a Model of the Emotional Learning in the Brain. Journal of Automation, Mobile Robotics & Intelligent Systems 4(2), 48–54 (2010)
10. Tamukoh, H., Koga, T., Horio, K., Yamakawa, T.: Rough-Winner-Take-All Self-Organizing Neural Network for Hardware Oriented Vector Quantization Algorithm. In: Proc. of the 50th IEEE Int. Mid. Symp. on Circuits and Systems, pp. 349–352 (2007)
11. Kudo, K., Myokan, Y., Than, W.C., Akimoto, S., Kanamaru, T., Sekine, M.: Hardware Object Model and its Application to the Image Processing. IEICE Trans. Fund. E87-A(3), 547–558 (2004)
12. Tamukoh, H., et al.: Internet Booster: A Netwroked hw/sw Complex System and its Application to Hi-Performance WEB Application. In: Proc. of Wolrd Automation Congress (accepted, 2010)
13. Onoo, A., et al.: On Automatic Generation of VHDL Code for Self-Organizing Map. In: Proc. of Int. Joint Conference on Neural Networks, pp. 2366–2373 (2009)
14. Porrmann, M., Witkowski, U., Kalte, H., Rückert, U.: Implementation of Artificial Neural Networks on a Reconfigurable Hardware Accelerator. In: Proc. of the 10th Euromicro Workshop on Parallel, Distributed and Network-based Processing (2002)

Inversion of Many-to-one Mappings Using Self-Organising Maps

Anne O. Mus

College of Information Systems and Technology
University of Phoenix
anne1mus@ymail.com

Abstract. Bidirectionally trained neural networks would be very useful in many circumstances. Often, we have data available for a prediction problem, but prediction of properties for unknown or new situations is only part of the story. In many cases we know the effect we wish to achieve on the output, but what we do not know is how to modify the inputs to achieve this goal. A basic problem in this area is the inversion of many to one mappings. Our work is based on the popular backpropagation neural network to predict the GDP of developing countries. These networks are integrated with a Self-Organising Map to allow the inversion of many to one mappings.

Keywords: bidirectional training, backpropagation neural network, SOM.

1 Bidirectional Backprogration

Most neural networks can be applied to real world systems to perform classification, pattern recognition or prediction tasks on the basis of input data. Given the output data, however, these neural network models are not able to produce any plausible input data unless another network is trained specifically for that task. This is done easily by humans. For example, we can retrieve an image for an elephant from the word *elephant*, and when we see an elephant we will find the corresponding word.

Networks which can produce plausible input values for a given output value have many applications. Bidirectional associative memories [1, 2] and the bidirectional version of counterpropagation networks [3] have been developed to learn bidirectional mappings. The problem with these approaches is their low capacity, low efficiency and multiple learning. By multiple learning we mean the different learning methods which are used in the first and second layer of bidirectional version of counterpropagation networks.

1.1 Normal Backpropagation Training

This network has no backward, lateral or layer-skipping connections. All processing neurons have a bias implemented as an extra input always on. Following the usual convention, input neurons are drawn, however these are not processing neurons, merely switch boxes distributing the single input x_i to the hidden neurons.

K.W. Wong et al. (Eds.): ICONIP 2010, Part II, LNCS 6444, pp. 447–453, 2010.

1.2 Reverse Direction Training

We have applied the error back-propagation technique [4] in both reverse and forward directions to adjust the weight matrix of the network. In our experiments we did not need to use more hidden units in training a bidirectional network in comparison to the case of training a network in the traditional unidirectional way [5].

When trained from left to right in Figure 1, the weights on the connections between layers are used normally, along with biases. Note that here input neurons are processing neurons when used in the reverse direction.

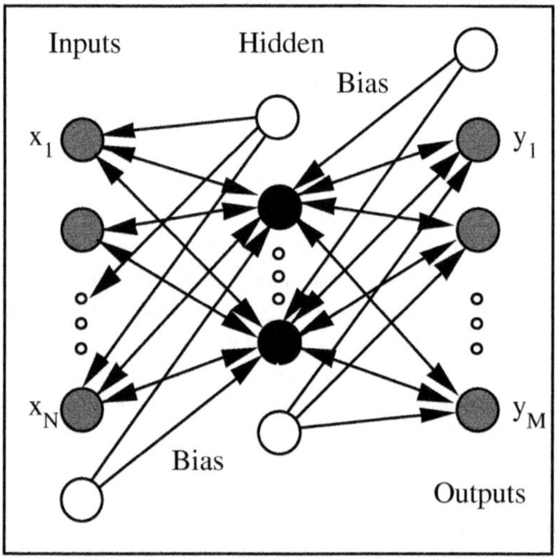

Fig. 1. Bidirectional network

When training in the reverse direction, different biases are used as shown. This would be the case for multiple hidden layers also. Due to a flatter search space, sometimes a higher number of epochs may be necessary for the network to converge in comparison to traditional unidirectional networks.

The remaining challenge is the case where the function relating inputs to outputs is not invertible, or there is a relation which is many-to-one between inputs and outputs.

2 Data

We use two data sets to demonstrate our techniques. These are a small synthetic data set, and a complex real world data set. These are described below.

2.1 dh46

This data set consists of 16 patterns, being all combinations of 4 binary elements [6, 7]. The single output is on when 3 out of 4 binary elements are on. This data set is of the class of data sets such as *xor* which are beyond the power of single perceptrons, and require a hidden layer trained by algorithms such as the back-propagation training algorithm described in §1.2. At the same time, the data set is complex enough that in a 4-6-1 back-propagation network, it shows a range of behaviour from immediate learning, learning after some time with little visible improvement (training plateau), and never learning (stuck on a plateau / local minimum).

2.2 gdp

The gdp data set is from the United Nations Conference on Trade and Development selected economic and social indicators of developing countries [8]. The data set includes the following 14 indicators: % population urban; size of total labour force; % of labour force in agriculture; % infant mortality; % public health expenditure; population per doctor; % access to water; illiteracy %; % public education expenditure; food production per capita; number of phones per capita; total population; average growth of population; density of population. The task is of course to predict the GDP of each country.

3 Interactive Competition Model

After training a back-propagation network, using as the input to the network the economic and social factors for a particular developing country, the network can quite accurately predict the GDP (given the quality of the data due to the difficulty of collection). But how is the network's conclusion useful?

In modelling the relationship between GDP and the selected socio-economic indicators, the real objective is not to predict GDP accurately as it can be measured directly, but to answer questions relating to change such as "If we wanted a higher GDP, how should (for example) general access to safe drinking water change," or (more likely) "given the economic and social factors we are committed to politically, what factors can we change to achieve an effect on GDP or health or ..." etc.

A predictive model such as the trained feed-forward network can be forced to this end by perturbing the input values into the network and examining the results. This is very inefficient, given the large number of inputs even in this relatively simple case.

The causal index of each input to the feed-forward network to the GDP output value provides a significance measure of that input's relationship to the output, and can be used as a weight in a simple interactive neural network model. That is, the rate of change of the output when the input is in the vicinity of its actual value in a pattern represents the significance in the context of the actual example.

3.1 Iac Model

The interactive activation and competition (iac) model consists of a collection of processing units organised into some number of competitive pools. There are excitatory connections among units in different pools and inhibitory connections between units within the same pool. The excitatory connections between pools are generally bi-directional, thereby making the processing interactive in the sense that processing in each pool both influences and is influenced by processing in other pools. Within a pool, the inhibitory connections run from each unit in the pool to every other unit in the pool implementing a kind of competition among the units such that the unit or units in the pool that receive the strongest activation tend to drive down the activation of the other units [9].

The network is operated by turning on a number of units and allowing the network to cycle until state has been reached. It is also possible to clamp some of the unit on to serve as a constant stimulus to the network.

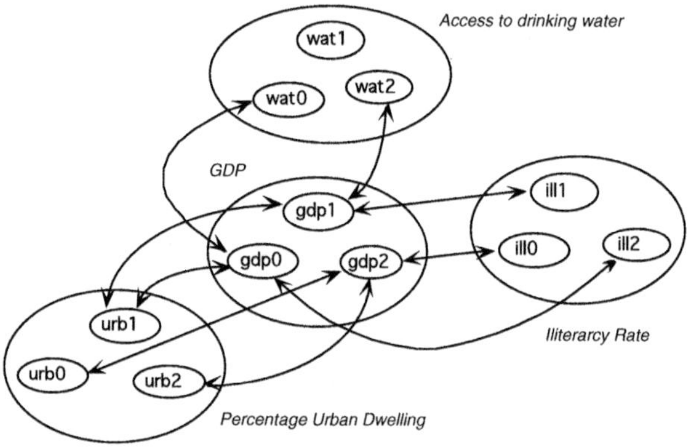

Fig. 2. Simplified iac diagram showing pools

A simplified diagram of an iac network is shown in Figure 2, with three sample input pools. Values within pools are sub-ranges of the input data to the back-propagation net [10]. Weights between pools are bi-directional, set to the sum of causal indexes for each input / output sub-range pair. Note that not all sub ranges need to be indexed.

3.2 Iac Results

Testing the interactive network in its response to change, for an example we clamp the gdp6 unit on, being two GDP sub-ranges higher than the actual GDP. All of the input values are clamped on except for the % public health expenditure, and % public education expenditure. After the stable state of the network is reached, the values of the variables which were not clamped provide the solution.

Validation of the solution was done by using the trained feed-forward network, by constructing a new input pattern consisting of the old values of the clamped variables, and the resulting values of the unclamped variables. The test is whether the value of GDP predicted by the feed-forward network based on the new pattern is similar to the value the GDP was clamped to in the interactive network. In the case of the prediction, the result of the new pattern is a GDP value of 0.47 which is an increase in the GDP of about 2% which is not insignificant. This confirms both the increase we have tried to produce, and at the same time the difficulty of increasing the GDP with only the flexibility allowed by the few variables left unclamped. This suggests there are one or more countries in the list which have similar relationships between their public health and education expenditures and GDP. There is, however, a component of this approach which is still too simplistic, that of clamping values. The above strategy also assumes that, for example, the illiteracy rate will not be affected by the reduced public education expenditure, since the value was clamped.

4 Hybrid SOM Model

The key notion in solving the inversion of many to one (many to many) mappings is to recognise the different categories on input which may give rise to a particular output. For this purpose we will use a self-organising map, described below.

4.1 The Hybrid Model

The graphical description of the model is below, in Figure 3, with only a few connections shown for clarity. The model consists of a standard back-propagation trained feedforward neural network, a single dimension self-organising map (SOM), the output of which is used to create the training set for a single layer back-propagation (or perceptron) network. This latter network uses (selected instances of) the activation values of the first network as its training input and produces the appropriate original input as its output. For use on unknown patterns, the output value is input to the SOM, mapped to the relevant SOM categories, and the appropriate SOM exemplars are used as inputs to the single layer to produce outputs.

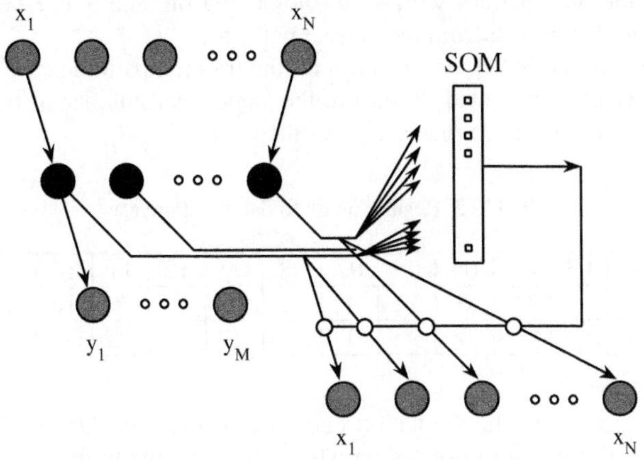

Fig. 3. The hybrid model during training

4.2 Input into SOM

The activation values produced by each hidden neuron in the original back-propagation network are used as input for the SOM.

These activation values were used to create the neuron behaviour graphs, and used to determine when to stop training the back-propagation network. In this case, the activation values from every tenth epoch were used. The SOM was trained using the algorithm described earlier, with an initial gain term of 0.9, initial neighbourhood of 30%, and 16 SOM neurons for both data sets. The SOM results for the *dh46* and *gdp* data sets are described below.

SOM results - dh46. For each of the 16 patterns (labelled *p00* to *p15*), the second row shows the index of the nearest SOM unit (ie its category). The third row shows the 4 patterns which satisfy the '3 out of 4' condition.

Table 1. Results on netork running the dh46 dataset

00	01	02	03	04	05	06	07	08	09	10	11	12	13	14	15
1	1	2	5	3	4	4	**8**	1	6	6	**13**	4	**11**	**9**	15
							1				1		1	1	

The four sets of patterns which have a 1 in the desired output in the original training set have been identified by a contiguous subset of the SOM indices. There is substantial redundancy in the indices for many of the remaining patterns. Note that the SOM did not have direct access to the original training set, only to the activations of neurons being trained on that data set. Thus, for example, the activations of the hidden neurons every ten epochs for a total of 2000 epochs on pattern *p07* had clearly contained some indication of the significance of this pattern and that it had some relationship to the patterns *p11*, *p13*, and *p14*. It is clear that *p15* is also somewhat related, being the only pattern with 4 out of 4 being on, and hence being somewhat more difficult to distinguish from the correct patterns.

For comparison, the SOM was also run on the training pattern set. Clearly, there is only one set, which is repeated, to ensure the same total number of presentations of inputs to the SOM have taken place. The results were:

Table 2. Results on dh46 dataset statically

00	01	02	03	04	05	06	07	08	09	10	11	12	13	14	15
12	7	13	8	15	6	14	**1**	11	10	12	**9**	16	**5**	**15**	3
							1				1		1	1	

There is no pattern to the 4 four on patterns, clearly the SOM on hidden neuron activations was discovering information which it is not able to discover directly from the original training set. This validates our use of the hidden neuron activations as the input to the SOM.

SOM results - gdp. There are 143 patterns in the original *gdp* training set, so we can not readily show the pattern in the same way. Also, the target output is not symbolic and as easy to interpret as the *dh46* data set.

Some qualitative comparisons are possible. The developing countries most familiar to many in the West are the richest developing country. A test of the success of our technique would be whether these countries are sorted together.

The SOM grouped the following countries in category 16: Brunei, Qatar, and United Arab Emirates; and in category 4: Chile, Colombia, Fiji, Hong Kong, Malta, Mauritius, Panama, Saudi Arabia, Singapore, Uruguay, and Venezuela. While these are surprises in the latter group, the three richest are grouped together, and clearly Hong Kong and Singapore we would have expected to end up in the same group.

Again we compare with the result of running the SOM on the original data set. The results are less good. The three richest countries are in three different groups. The second grouping identified above partially survives, with 3 different countries. The changes would be hard to justify, for example Saudi Arabia is now clustered with Ghana. Thus out method is significantly better than static analysis of the data set [11].

5 Conclusion

We have introduced our technique for training a back-propagation network in the reverse direction by the use of a Self-Organising Map. This hybrid structure bidirectional neural network solves the previously identified problems of inverting many to one or many to many mappings. The steps in the process have been individually justified. The ability of such networks to provide suggestions for modification of input parameters to achieve the desired results will be useful in a number of application areas.

References

1. Kosko, B.: Bidirectional Associative Memories. IEEE Transactions on Systems, Man and Cybernetics SMC-18, 49–60 (1988)
2. Kosko, B.: Neural Networks and Fuzzy Systems: A Dynamical Systems Approach to Machine Intelligence. Prince-Hall, Inc., Englewood Cliffs (1992)
3. Hecht-Nielsen, R.: Counterpropagation Networks. Applied Optics 26(3), 4979–4984 (1987)
4. Gedeon, T.D.: Stochastic bidirectional training. In: IEEE International Conference on System Man and Cybernetics (SMC 1998), San Diego, pp. 1968–1971 (1998)
5. Nejad, A.F., Gedeon, T.D.: BiDirectional MLP Neural Networks. In: Proceedings International Symposium on Artificial Neural Networks, Taiwan, pp. 308–313 (1994)
6. Gedeon, T.D., Harris, D.: Hidden Units in a Plateau. In: 1st International Conference on Intelligent Systems, Singapore, pp. 391–395 (1992)
7. Gedeon, T.D., Harris, D.: Network Reduction Techniques. In: International Conference on Neural Networks Methodologies and Applications, San Diego, vol. 1, pp. 119–126 (1991)
8. UNCTAD Statistical Pocket Book, United Nations, New York (1984)
9. McClelland, J.L., Rumelhart, D.E.: Explorations in Parallel Distributed Processing. MIT Press, Cambridge (1988)
10. Gedeon, T.D., Good, R.P.: Interactive modelling of a neural network model of GDP. In: International Conference on Modelling and Simulation, Perth, pp. 355–360 (1993)
11. Slade, P., Gedeon, T.D.: Bimodal Distribution Removal. In: Mira, J., Cabestany, J., Prieto, A.G. (eds.) IWANN 1993. LNCS, vol. 686, pp. 249–254. Springer, Heidelberg (1993)

Self-Organizing Hidden Markov Models

Nobuhiko Yamaguchi

Faculty of Science and Engineering, Saga University, Saga-shi, 840–8502 Japan

Abstract. The self-organizing mixture models (SOMMs) were proposed as an expectation-maximization (EM) algorithm that yields topology preserving maps of data based on probabilistic mixture models. Compared to self-organizing maps, the SOMM algorithm has a clear interpretation: it maximizes the sum of data log likelihood and a penalty term that enforces self-organization. The object of this paper is to extend the SOMM algorithm to deal with multivariate time series. The standard SOMM algorithm assumes that the data are independent and identically distributed samples. However, the i.i.d. assumption is clearly inappropriate for time series. In this paper we propose the extension of the SOMM algorithm for multivariate time series, which we call self-organizing hidden Markov models (SOHMMs), by assuming that the time series is generated by hidden Markov models (HMMs).

1 Introduction

In the field of time series analysis [1], [2], it is often necessary to deal with time series whose statistical properties are evolving over time. For example, the statistical series, such as average temperature and precipitation vary cyclically according to four seasons. In this paper, we propose a method for visualizing states of such time series.

In recent years, self-organizing mixture models (SOMMs) [3], [4] have been proposed as a multidimensional data visualization technique. The SOMM algorithm yields the topology preserving maps by applying a modified EM algorithm to mixture models [5]. The SOMM algorithm has the following advantages over self-organizing maps [6]. First, since the SOMM algorithm is based on the EM algorithm, it can be applied to any mixture models and can handle missing data in a natural way. Second, the objective function that the algorithm optimizes has a clear interpretation: it sums the data log likelihood and a penalty term that enforces the topology preservation.

The object of this paper is to extend the SOMM algorithm to deal with multivariate time series. The standard SOMM algorithm treats the data as independent and identically distributed samples. However, the i.i.d. assumption is clearly inappropriate for time series. In this paper we propose the extension of the SOMM algorithm for multivariate time series, which we call self-organizing hidden Markov models (SOHMMs), by assuming that the time series is generated by hidden Markov models (HMMs) [7].

K.W. Wong et al. (Eds.): ICONIP 2010, Part II, LNCS 6444, pp. 454–461, 2010.

2 EM Algorithm

We begin by reviewing the EM algorithm as the basis for the proposed algorithm. In this paper, we denote the set of all model parameters by $\boldsymbol{\theta}$, all of the observed variables by $\boldsymbol{X} = (\boldsymbol{x}_1', \cdots, \boldsymbol{x}_T')'$, and all of the latent variables by $\boldsymbol{Z} = (z_1, \cdots, z_T)'$. We call $(\boldsymbol{X}, \boldsymbol{Z})$ the complete data set. The object of the EM algorithm is to find the parameters $\boldsymbol{\theta}$ that maximize the log likelihood defined as

$$\mathcal{L}(\boldsymbol{\theta}) = \ln p(\boldsymbol{X}; \boldsymbol{\theta}) = \ln \sum_{\boldsymbol{Z}} p(\boldsymbol{X}, \boldsymbol{Z}; \boldsymbol{\theta}).$$

To find the parameters, the EM algorithm introduces a distribution $q(\boldsymbol{Z})$ defined over the latent variables and maximizes a lower bound of the log likelihood instead of the log likelihood itself. Considering the non-negativity of the KL divergence, we can obtain the lower bound $\mathcal{F}(q, \boldsymbol{\theta})$ as follows:

$$\begin{aligned}\mathcal{F}(q, \boldsymbol{\theta}) &= \mathcal{L}(\boldsymbol{\theta}) - \mathrm{KL}(q\|p) &(1)\\ &= \mathrm{E}_q\left[\ln p(\boldsymbol{X}, \boldsymbol{Z}; \boldsymbol{\theta})\right] + \mathcal{H}(q) &(2)\end{aligned}$$

where $\mathrm{KL}(q\|p)$ is the KL divergence between the distribution $q(\boldsymbol{Z})$ and the posterior distribution $p(\boldsymbol{Z}|\boldsymbol{X}; \boldsymbol{\theta})$ and $\mathcal{H}(q)$ is the entropy of the distribution $q(\boldsymbol{Z})$.

$$\mathrm{KL}(q\|p) = \mathrm{E}_q[\ln \frac{q(\boldsymbol{Z})}{p(\boldsymbol{Z}|\boldsymbol{X}; \boldsymbol{\theta})}]$$

$$\mathcal{H}(q) = -\mathrm{E}_q[\ln q(\boldsymbol{Z})]$$

The equality of (2) is due to $p(\boldsymbol{X}; \boldsymbol{\theta}) = p(\boldsymbol{X}, \boldsymbol{Z}; \boldsymbol{\theta})/p(\boldsymbol{Z}|\boldsymbol{X}; \boldsymbol{\theta})$.

The EM algorithm finds the maximum likelihood estimate by maximizing the lower bound $\mathcal{F}(q, \boldsymbol{\theta})$. In order to maximize the lower bound, the EM algorithm can be started by initializing the parameters $\boldsymbol{\theta}$ to random values, and then iterating E-step and M-step. Suppose that the parameters obtainded in the previous M-step are $\boldsymbol{\theta}^{\tau-1}$. In the E-step, the lower bound $\mathcal{F}(q, \boldsymbol{\theta}^{\tau-1})$ is maximized with respect to $q(\boldsymbol{Z})$ while holding $\boldsymbol{\theta}^{\tau-1}$ fixed. Because the log likelihood $\mathcal{L}(\boldsymbol{\theta})$ does not depend on $q(\boldsymbol{Z})$, the largest $\mathcal{F}(q, \boldsymbol{\theta}^{\tau-1})$ occurs when the KL divergence is equal to 0, that is, the distribution $q(\boldsymbol{Z})$ can be found by

$$\begin{aligned}\textbf{E-step}: q^{\tau} &= \arg\max_q \mathcal{F}(q, \boldsymbol{\theta}^{\tau-1}) = \arg\min_q \mathrm{KL}(q\|p(\boldsymbol{Z}|\boldsymbol{X}; \boldsymbol{\theta}^{\tau-1}))\\ &= P(\boldsymbol{Z}|\boldsymbol{X}; \boldsymbol{\theta}^{\tau-1}).\end{aligned}$$

In this case, the lower bound $\mathcal{F}(q^{\tau}, \boldsymbol{\theta}^{\tau-1})$ is equal to the log likelihood $\mathcal{L}(\boldsymbol{\theta}^{\tau-1})$ since the KL divergence is equal to 0.

Suppose that the distribution obtained in the previous E-step is q^{τ}. In the M-step, the lower bound $\mathcal{F}(q^{\tau}, \boldsymbol{\theta})$ is maximized with respect to the parameters $\boldsymbol{\theta}$ while holding q^{τ} fixed. Because the entropy $\mathcal{H}(q)$ does not depend on $\boldsymbol{\theta}$, the largest $\mathcal{F}(q^{\tau}, \boldsymbol{\theta})$ can be found by maximizing the expectation of the log likelihood of the complete data.

$$\textbf{M-step}: \boldsymbol{\theta}^{\tau} = \arg\max_{\boldsymbol{\theta}} \mathcal{F}(q^{\tau}, \boldsymbol{\theta}) = \arg\max_{\boldsymbol{\theta}} \mathrm{E}_{q^{\tau}} \left[\ln p(\boldsymbol{X}, \boldsymbol{Z}; \boldsymbol{\theta})\right]$$

In this case, the M-step increases the lower bound $\mathcal{F}(q^{\tau}, \boldsymbol{\theta})$ and increases the corresponding log likelihood $\mathcal{L}(\boldsymbol{\theta})$ since the the lower bound $\mathcal{F}(q^{\tau}, \boldsymbol{\theta}^{\tau-1})$ is set to equal the log likelihood $\mathcal{L}(\boldsymbol{\theta}^{\tau-1})$ in the E-step.

3 Self-Organizing Hidden Markov Models

3.1 Hidden Markov Models

In section 3.1 we briefly review HMMs, and then, in section 3.2 we will propose an extension of the SOMM algorithm that yields topology preserving maps of multivariate time series based on the HMMs. A HMM consists of a hidden state sequence $\boldsymbol{Z} = (z_1, \cdots, z_T)'$ and a corresponding observation sequence $\boldsymbol{X} = (\boldsymbol{x}_1, \cdots, \boldsymbol{x}_T)'$. Each state z_t is a discrete random variable with possible values $1, 2, \cdots, M$. Transitions between the states are governed by a first order Markov chain parameterized by the transition probabilities $p_{ij} = P\{z_t = j | z_{t-1} = i\}$, while the initial state probabilities are $\rho_i = P\{z_1 = i\}$. Of course, the transition probabilities and initial state probabilities must sum to one:

$$\sum_{j=1}^{M} p_{ij} = 1, \qquad i = 1, \cdots, M, \tag{3}$$

$$\sum_{i=1}^{M} \rho_i = 1. \tag{4}$$

On the other hand, each observation $\boldsymbol{x}_t$ is generated by a distribution corresponding to the state z_t. For example, in the SOHMM algorithm, we choose the distribution of $\boldsymbol{x}_t$ to be a radially-symmetric Gaussian centered on $\boldsymbol{\mu}_i$ having variance β^{-1} so that

$$p(\boldsymbol{x}_t | z_t = i; \boldsymbol{\theta}_i) = \left(\frac{\beta}{2\pi}\right)^{d/2} \exp\left(-\frac{\beta}{2}\|\boldsymbol{x}_t - \boldsymbol{\mu}_i\|^2\right)$$

where d is the dimension of the observation vector $\boldsymbol{x}_t$ and $\boldsymbol{\theta}$ is the adaptive vector composed of the $\boldsymbol{\mu}_i$, β, p_{ij}, and ρ_i.

3.2 Estimating HMM Parameters Using the Constrained EM Algorithm

In this section, we propose the SOHMM that visualizes states of multivariate time series. The SOHMM algorithm assumes that a time series is generated by a

HMM, that is, the value of a time series at time t, $\boldsymbol{x}_t$, is generated according to the normal distribution corresponding to the state z_t and the state transitions are governed by a first order Markov chains. In order to visualize the states, we prepare a regular grid of M points $\boldsymbol{g}_1, \cdots, \boldsymbol{g}_M$ in visualization space. In addition, calculating the probability $p(z_t = i|\boldsymbol{X}; \boldsymbol{\theta})$, we can visualize the state z_t by the corresponding point $\boldsymbol{g}_{z_t}$ in the visualization space.

To yield the topology preserving maps, the SOHMM algorithm estimates the HMM parameters by the constrained EM algorithm that uses a constrained distribution $\tilde{q}(\boldsymbol{Z}; \boldsymbol{W})$ instead of the distribution $q(\boldsymbol{Z})$ in the EM algorithm. By constraining the distribution $q(\boldsymbol{Z})$, we can enforce a topological ordering of the states in the sense that if the corresponding points $\boldsymbol{g}_i$ are close together the probabilities $p(z_t = i|z_{t-1}, \boldsymbol{X}; \boldsymbol{\theta})$ are also close together. The constrained distribution $\tilde{q}(\boldsymbol{Z}; \boldsymbol{W})$ is defined by using the normalized neighborhood function of Kohonen's SOM as follows:

$$\tilde{q}(\boldsymbol{Z}; \boldsymbol{W}) = \tilde{q}_1(z_1; w_1) \prod_{t=2}^{T} \tilde{q}_{t,t-1}(z_t|z_{t-1}; \boldsymbol{w}_t),$$

$$\tilde{q}_1(z_1 = i_1; w_1) = \frac{\exp(-\lambda \|\boldsymbol{g}_{i_1} - \boldsymbol{g}_{w_1}\|^2)}{\sum_{j_1=1}^{M} \exp(-\lambda \|\boldsymbol{g}_{j_1} - \boldsymbol{g}_{w_1}\|^2)},$$

$$\tilde{q}_{t,t-1}(z_t = i_t|z_{t-1} = i_{t-1}; \boldsymbol{w}_t) = \frac{\exp(-\lambda \|\boldsymbol{g}_{i_t} - \boldsymbol{g}_{w_{t,i_{t-1}}}\|^2)}{\sum_{j_t=1}^{M} \exp(-\lambda \|\boldsymbol{g}_{j_t} - \boldsymbol{g}_{w_{t,i_{t-1}}}\|^2)}$$

where w_1 and $w_{t,i}$ are discrete parameters with possible value $1, \cdots, M$, $\boldsymbol{w}_t = (w_{t,1}, \cdots, w_{t,M})'$, $\boldsymbol{W} = (w_1, \boldsymbol{w}_2', \cdots, \boldsymbol{w}_T')'$ and λ controls the width of the neighborhood function. The $\tilde{q}_{t,t-1}(z_t|z_{t-1}; \boldsymbol{w}_t)$ is a distribution which decreases as the distance between the point $\boldsymbol{g}_{z_t}$ and $\boldsymbol{g}_{w_{t,z_{t-1}}}$ increases, and the $\tilde{q}_1(z_1; w_1)$ is also a distribution which decreases as the distance increases.

The SOHMM algorithm estimates the parameters by maximizing the lower bound $\mathcal{F}(q, \boldsymbol{\theta})$ while constraining the $q(\boldsymbol{Z})$ to be the distribution $\tilde{q}(\boldsymbol{Z}; \boldsymbol{W})$. Specifically, the SOHMM algorithm estimates the parameters $\boldsymbol{\theta}$ by iterating the following E-step and M-step:

$$\begin{aligned} \textbf{E-step}: \boldsymbol{W}^{\tau} &= \arg\max_{\boldsymbol{W} \in Q} \mathcal{F}(\tilde{q}, \boldsymbol{\theta}^{\tau-1}) \\ &= \arg\min_{\boldsymbol{W} \in Q} \mathrm{KL}(\tilde{q}(\boldsymbol{Z}; \boldsymbol{W}) \| p(\boldsymbol{Z}|\boldsymbol{X}; \boldsymbol{\theta}^{\tau-1})), \end{aligned} \tag{5}$$

$$\textbf{M-step}: \boldsymbol{\theta}^{\tau} = \arg\max_{\boldsymbol{\theta}} \mathcal{F}(\tilde{q}^{\tau}, \boldsymbol{\theta}) = \arg\max_{\boldsymbol{\theta}} \mathrm{E}_{\tilde{q}(\boldsymbol{Z}; \boldsymbol{W}^{\tau})} [\ln p(\boldsymbol{X}, \boldsymbol{Z}; \boldsymbol{\theta})] \tag{6}$$

where $Q = \{1, \cdots, M\}^{M(T-1)+1}$ and $\tilde{q}^{\tau} = \tilde{q}(\boldsymbol{Z}; \boldsymbol{W}^{\tau})$.

By maximizing the lower bound of the log likelihood, we can fit the HMM parameters to the distribution of the time series. Further, according to (1), the SOHMM algorithm can be interpreted as the maximum penalized likelihood estimation in which the penalty term is $-\mathrm{KL}(\tilde{q}(\boldsymbol{Z};\boldsymbol{W})\|p(\boldsymbol{Z}|\boldsymbol{X};\boldsymbol{\theta}))$. The penalty term can be expanded as the average of the KL divergences at each time t:

$$\mathrm{KL}(\tilde{q}(\boldsymbol{Z};\boldsymbol{W})\|p(\boldsymbol{Z}|\boldsymbol{X};\boldsymbol{\theta})) = \mathrm{KL}(\tilde{q}_1(z_1;w_1)\|p(z_1|\boldsymbol{X};\boldsymbol{\theta}))$$

$$+\sum_{t=2}^{T}\sum_{i_{t-1}=1}^{M} \tilde{q}_{t-1}(i_{t-1};\boldsymbol{W}_t)\mathrm{KL}(\tilde{q}_{t,t-1}(z_t|i_{t-1};\boldsymbol{w}_t)\|p(z_t|i_{t-1},\boldsymbol{X};\boldsymbol{\theta})),$$

$$\tilde{q}_t(z_t=i_t;\boldsymbol{W}_t) = \sum_{i_1=1}^{M}\cdots\sum_{i_{t-1}=1}^{M} \tilde{q}_1(i_1;w_1)\prod_{u=2}^{t}\tilde{q}_{u,u-1}(i_u|i_{u-1};\boldsymbol{w}_u)$$

$$= \sum_{i_{t-1}=1}^{M} q_{t-1}(i_{t-1};\boldsymbol{W}_{t-1})q_{t,t-1}(i_t|i_{t-1};\boldsymbol{w}_t)$$

where $\boldsymbol{W}_t = (w_1, \boldsymbol{w}_2', \cdots, \boldsymbol{w}_t')'$. The objective function $\mathcal{F}(\tilde{q},\boldsymbol{\theta})$ prefers the $\tilde{q}$ for which the penalty term are the largest, i.e. the $\tilde{q}_{t,t-1}$ for which the distribution $p(z_t|z_{t-1}=i_{t-1},\boldsymbol{X};\boldsymbol{\theta})$ is most similar to the normalized neighborhood function. From this reason, we can enforce a topological ordering of the states in the sense that if the corresponding points $\boldsymbol{g}_i$ are close together the probabilities $p(z_t=i|z_{t-1},\boldsymbol{X};\boldsymbol{\theta})$ are also close together. And from the same reason, we can enforce that the probabilities $p(z_1=i|\boldsymbol{X};\boldsymbol{\theta})$ are also close together.

3.3 E-step

In the E-step we estimates the parameters $\boldsymbol{W}$ by minimizing the KL divergence in (5). However, there are $M^{M(T-1)+1}$ numbers of combinations of the parameters $\boldsymbol{W}$ and it is therefore difficult to find the optimal parameters from all the combinations. In this section, we propose the method for finding the optimal parameters $\boldsymbol{W}$ with less computational time.

First, we rewrite the KL divergence in (5) as the following recursive equations:

$$\mathrm{KL}(\tilde{q}(\boldsymbol{Z};\boldsymbol{W})\|p(\boldsymbol{Z}|\boldsymbol{X};\boldsymbol{\theta}^{\tau-1})) = r_1(w_1,\boldsymbol{w}_2,\cdots,\boldsymbol{w}_T),$$

$$r_1(w_1,\boldsymbol{w}_2,\cdots,\boldsymbol{w}_T) = \mathrm{KL}(\tilde{q}_1(z_1;w_1)\|p(z_1|\boldsymbol{X};\boldsymbol{\theta}^{\tau-1}))$$
$$+\sum_{i_1=1}^{M}\tilde{q}_1(i_1;w_1)r_2(i_1;\boldsymbol{w}_2,\cdots,\boldsymbol{w}_T),$$

$$r_t(i_{t-1};\boldsymbol{w}_t,\cdots,\boldsymbol{w}_T) = \mathrm{KL}(\tilde{q}_{t,t-1}(z_t|i_{t-1};\boldsymbol{w}_t)\|p(z_t|i_{t-1},\boldsymbol{X};\boldsymbol{\theta}^{\tau-1}))$$
$$+\sum_{i_t=1}^{M}\tilde{q}_{t,t-1}(i_t|i_{t-1};\boldsymbol{w}_t)r_{t+1}(i_t;\boldsymbol{w}_{t+1},\cdots,\boldsymbol{w}_T),$$

$$r_T(i_{T-1}; \boldsymbol{w}_T) = \mathrm{KL}(\tilde{q}_{T,T-1}(z_T|i_{T-1}; \boldsymbol{w}_T) \| p(z_T|i_{T-1}, \boldsymbol{X}; \boldsymbol{\theta}^{\tau-1})).$$

Suppose that the parameters that minimize the function $r_t(i_{t-1}; \boldsymbol{w}_t, \cdots, \boldsymbol{w}_T)$ for all i_{t-1} are $\boldsymbol{w}_t^\tau, \cdots, \boldsymbol{w}_T^\tau$. Since the probabilities $\tilde{q}_{t,t-1}(z_t = i_t|z_{t-1} = i_{t-1}; \boldsymbol{w}_t)$ are nonnegative, the optimal parameters $\boldsymbol{w}_{t+1}^\tau, \cdots, \boldsymbol{w}_T^\tau$ except for $\boldsymbol{w}_t^\tau$ are equal to the parameters that minimize the function $r_{t+1}(i_t; \boldsymbol{w}_{t+1}, \cdots, \boldsymbol{w}_T)$ for all i_t. Hence, the optimal parameters $\boldsymbol{w}_t^\tau, \cdots, \boldsymbol{w}_T^\tau$ can be found by finding the parameters that minimize the function $r_{t+1}(i_t; \boldsymbol{w}_{t+1}, \cdots, \boldsymbol{w}_T)$ for all i_t, and then finding the parameters $\boldsymbol{w}_t^\tau$ that minimize the function $r_t(i_{t-1}; \boldsymbol{w}_t, \boldsymbol{w}_{t+1}^\tau, \cdots, \boldsymbol{w}_T^\tau)$ for all i_{t-1}. In the E-step, the optimal parameters $\boldsymbol{W}^\tau$ can be found by sequentially optimizing the parameters $\boldsymbol{w}_t^\tau$ from $t = T$ to $t = 1$, that is, by sequentially iterating (7)C(8) and (9) from $t = T$ to $t = 1$.

$$\textbf{E-step}: w_1^\tau = \mathop{\arg\min}_{w_1 \in \{1,\cdots,M\}} r_1(w_1, \boldsymbol{w}_2^\tau, \cdots, \boldsymbol{w}_T^\tau) \tag{7}$$

$$w_{t,i_{t-1}}^\tau = \mathop{\arg\min}_{w_{t,i_{t-1}} \in \{1,\cdots,M\}} r_t(i_{t-1}; \boldsymbol{w}_t, \boldsymbol{w}_{t+1}^\tau, \cdots, \boldsymbol{w}_T^\tau) \tag{8}$$

$$w_{T,i_{T-1}}^\tau = \mathop{\arg\min}_{w_{T,i_{T-1}} \in \{1,\cdots,M\}} r_T(i_{T-1}; \boldsymbol{w}_T) \tag{9}$$

3.4 M-step

In the M-step, we update the parameters $\boldsymbol{\theta}$ by maximizing the expected log likelihood of the complete data in (6). The expected log likelihood of the complete data can be written as

$$\mathrm{E}_{\tilde{q}(\boldsymbol{Z};\boldsymbol{W}^\tau)}[\ln p(\boldsymbol{X}, \boldsymbol{Z}|\boldsymbol{\theta})] = \sum_{i_1=1}^{M} \cdots \sum_{i_T=1}^{M} \tilde{q}(z_1 = i_1, \cdots, z_T = i_T; \boldsymbol{W}^\tau)$$
$$\times \left\{ \ln p(z_1 = i_1; \boldsymbol{\theta}) + \sum_{t=1}^{T} \ln p(\boldsymbol{x}_t|z_t = i_t; \boldsymbol{\theta}) + \sum_{t=2}^{T} \ln p(z_t = i_t|z_{t-1} = i_{t-1}; \boldsymbol{\theta}) \right\}.$$

According to the Lagrange multiplier method, we can find the optimal parameters subject to the constraints in (3) and (4) as follows:

$$\rho_i^\tau = \frac{\tilde{q}_1(i; w_1^\tau)}{\sum_{j=1}^{M} \tilde{q}_1(j; w_1^\tau)}, \tag{10}$$

$$p_{ij}^\tau = \frac{\sum_{t=2}^{T} \tilde{q}_{t-1}(i; \boldsymbol{W}_{t-1}^\tau) \tilde{q}_{t,t-1}(j|i; \boldsymbol{w}_t^\tau)}{\sum_{t=2}^{T} \tilde{q}_{t-1}(i; \boldsymbol{W}_{t-1}^\tau)}, \tag{11}$$

$$\boldsymbol{\mu}_i^{\tau} = \left\{ \sum_{t=1}^{T} \tilde{q}_t(i; \boldsymbol{W}_t^{\tau}) \right\}^{-1} \sum_{t=1}^{T} \tilde{q}_t(i; \boldsymbol{W}_t^{\tau}) \boldsymbol{x}_t, \tag{12}$$

$$\beta^{\tau} = \frac{dT}{\sum_{t=1}^{T} \sum_{i=1}^{M} \tilde{q}_t(i; \boldsymbol{W}_t^{\tau}) \| \boldsymbol{x}_t - \boldsymbol{\mu}_i^{\tau-1} \|^2}. \tag{13}$$

In the M-step, we update the parameters $\boldsymbol{\theta}$ by (10)C (11)C (12) and (13).

3.5 State Visualization

In order to visualize state and state transitions, we use the probability $p(z_t = i|\boldsymbol{X}; \boldsymbol{\theta})$ computed in the Baum-Welch algortihm. The visualization is achieved by using the mean of the points $\boldsymbol{g}_i$ in visualization space:

$$\bar{\boldsymbol{g}}_t = \sum_{i=1}^{M} p(z_t = i|\boldsymbol{X}; \boldsymbol{\theta}) \boldsymbol{g}_i.$$

4 Experimental Results

In this experiment, we used the data collected at the Tokyo meteorological station in Japan during a period of 713 months from January 1951 to May 2010. The meteorological data is issued by the Meteorological Agency of Japan, which contains monthly mean temperature, precipitation, relative humidity, sea level air pressure, station level pressure, and vapor pressure. For comparison we also test the performance of SOMM. The SOMM and SOHMM models consist of a 8×8 grid of nodes in visualization space. The dimension of the visualization space is fixed to 2 and the width of neighborhood function is $\lambda = 0.3$.

Fig. 1 shows the visualization of state z_t using $\bar{\boldsymbol{g}}_t$. The SOHMM exhibits better separation of four seasons than the SOMM model. This is due to the fact that, in the representation of the SOHMM model, the four clusters of seasons are clustered. On the contrary, in the representation of the SOMM model, the four clusters of seasons are more diffuse than those of the SOHMM models and particularly the spring and autumn clusters have more overlap.

5 Conclusions

We have proposed an extension of the SOMM for multivariate time series by assuming that the time series is generated by a HMM. Using this algorithm, we can visualize the state of the multivariate time series. Experimental result shows that the proposed algorithm provides better visualization of meteorological data than the conventional SOMM algorithm.

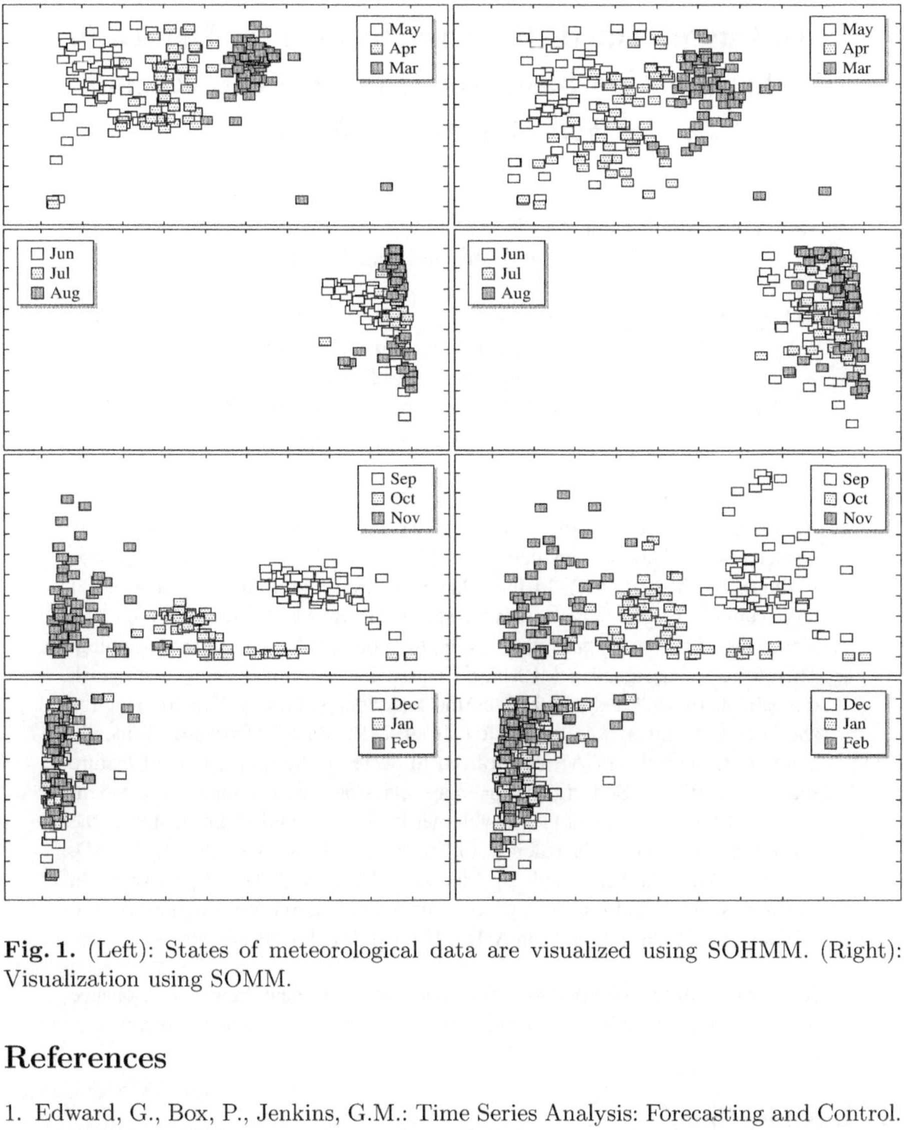

Fig. 1. (Left): States of meteorological data are visualized using SOHMM. (Right): Visualization using SOMM.

References

1. Edward, G., Box, P., Jenkins, G.M.: Time Series Analysis: Forecasting and Control. Prentice Hall PTR, Upper Saddle River (1994)
2. Hamilton, J.D.: Time Series Analysis. Princeton University Press, Princeton (1994)
3. Verbeek, J., Vlassis, N., Kröse, B.: Self-organizing mixture models. Neurocomputing 63, 99–123 (2005)
4. Verbeek, J.: Mixture Models for Clustering and Dimension Reduction. PhD thesis, University of Amsterdam (2004)
5. McLachlan, G.J., Peel, D.: Finite Mixture Models. John Wiley & Sons, Inc., New York (2000)
6. Kohonen, T., Schroeder, M.R., Huang, T.S. (eds.): Self-Organizing Maps. Springer, Heidelberg (2001)
7. Rabiner, L.R.: A tutorial on hidden Markov models and selected applications in speech recognition. Proc. IEEE 77(2), 257–286 (1989)

An Image-Aided Diagnosis System for Dementia Classification Based on Multiple Features and Self-Organizing Map

Shih-Ting Yang[1], Jiann-Der Lee[1], Chung-Hsien Huang[1], Jiun-Jie Wang[2], Wen-Chuin Hsu[3], and Yau-Yau Wai[3]

[1] Department of Electrical Engineering, Chang Gung University, Taiwan 333
[2] Department of Medical Imaging and Radiological Sciences, Chang Gung University, Taiwan 333
[3] Department of Neuroscience, Chang Gung Memorial Hospital, Taiwan 333
jdlee@mail.cgu.edu.tw

Abstract. Mild cognitive impairment (MCI) is considered as a transitional stage between normal aging and dementia. MCI has a high risk to convert into Alzheimer's disease (AD). In the related research, the volumetric analysis of hippocampus is the most extensive study. However, the segmentation and identification of the hippocampus are highly complicated and time-consuming. Therefore, we designed a MRI-based classification framework to distinguish the patients of MCI and AD from normal individuals. First, volumetric features and shape features were extracted from MRI data. Afterward, Principle component analysis (PCA) was utilized to decrease the dimensions of feature space. Finally, a Self-organizing map classifier was trained for patient classification. By combining the volumetric features and shape features, the classification accuracy is reached to 86.76%, 66.67%, and 46.67% in AD, amnestic MCI (aMCI), and dysexecutive MCI (dMCI), respectively. In addition, with the help of PCA process, the classification result is improved to 93.63%, 73.33%, and 53.33% in AD, aMCI and dMCI, respectively.

Keywords: Alzheimer's disease, Mild cognitive impairment, Magnetic resonance imaging, Shape descriptors, Self-organizing map, Principle component analysis.

1 Introduction

Mild cognitive impairment (MCI) is considered as a transitional stage between normal aging and dementia [1]. It can be divided into two categories: amnestic MCI (aMCI) and dysexecutive MCI (dMCI). The majority of MCI patients will develop Alzheimer's disease (AD) [2]. Recent reports in the treatment or prevention of AD leads to a growing concerns in the early diagnosis. A thorough understanding of the conversion process is, therefore, of clinical interest and importance.

Magnetic resonance imaging (MRI) is a very important tool in diagnosing MCI and AD. MRI demonstrats that volumetric atrophy appeared in the early stages of MCI and AD [3]. The image-based volumetric analysis of hippocampus draws a lot of attention in AD-related research in the past decade [4]. In addition, Pruessner *et al.* [5]

K.W. Wong et al. (Eds.): ICONIP 2010, Part II, LNCS 6444, pp. 462–469, 2010.

revealed that the medial temporal lobe structures, especially the entorhinal cortex and the hippocampus, have high relationship with AD.

However, the segmentation and identification of hippocampus are usually sensitive to the subjective opinion of the operator and also time-consuming. On the other hand, the enlargement of ventricles is also a significant characteristic of AD due to neuronal loss [6]. Ventricles are filled with cerebro-spinal fluid (CSF) surrounded by gray matter (GM) and white matter (WM). The coverage of GM and WM structures are often affected by dementia diseases. As a result, measuring the ventricular enlargement, hemispheric atrophy rate reveals significant variation between normal individuals and the subjects with MCI and AD [4].

In this study, we have designed a MRI-based classification framework to distinguish the patients of MCI and AD from normal individuals. Section 2 explains the proposed framework comprising system flowchart and the shape features selected. Statistical analysis and experimental results are revealed in Section 3. Finally, conclusions are included in Section 4.

2 Flow Chart and Feature Extraction

Figure 1 illustrates the flowchart of the proposed system. First, each individual's brain MRI is normalized to a T1-weighted MRI template for a spatially coherent purpose. Followed by a segmentation procedure, brain tissues are separated into GM, WM, CSF and cerebral ventricle. Volume-related and shape-related features are utilized for further classification. After extracting these features from a training data set, Mann-Whitney U test is adopted to filter out the features which are with low discriminative power. Afterwards, principle component analysis (PCA) is applied to reduce the dimensions of feature space and then a self-organized map (SOM) is adopted to classify testing subjects into four categories comprising normal individuals, dMCI, aMCI, and AD patients. Details are explained in the following sub-chapters.

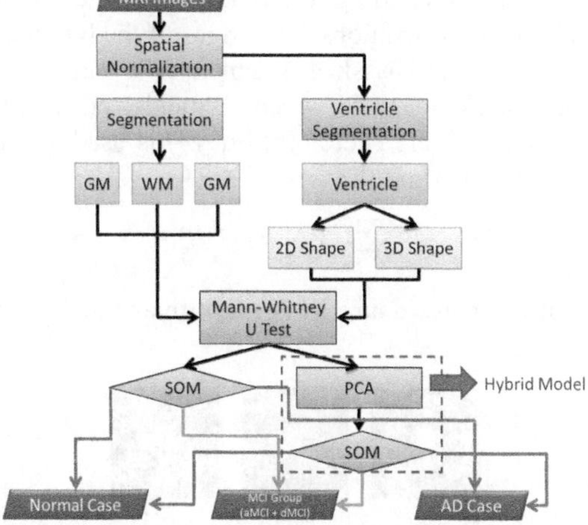

Fig. 1. Flowchart of the proposed image-aided diagnosis system

2.1 Spatial Normalization of MRI Data

Spatial normalization is a procedure to register a MRI data set to a standard coordinate system [7]. Therefore, each voxel is thus comparable with the other registered MRI or a reference template. The normalization herein was performed by using a 12-parameter affine transform and a Bayesian framework to a T1-weighted MRI template, provided by ICBM, NIH P-20 project [8].

2.2 Volume Features

The volumes of brain tissues such as GM, WM and CSF indicate important information, especially in brain degeneration diseases [9]. A clustering-based segmentation algorithm provided by SPM8 [10] is adopted to extract GM, WM and CSF probability maps from the original MRI data. The value of each voxel in the corresponding probability map indicates the posterior of the voxel belonging to the tissue by giving its gray intensity. As a result, we can calculate the volumes of GM, WM, CSF and the whole brain by the following equations:

$$\text{volume}_{\text{GM}} \approx \sum_{\forall i \in I} (P(C_{gray} \mid f(i)) > 0.5) \tag{1}$$

$$\text{volume}_{\text{WM}} \approx \sum_{\forall i \in I} (P(C_{white} \mid f(i)) > 0.5) \tag{2}$$

$$\text{volume}_{\text{CSF}} \approx \sum_{\forall i \in I} (P(C_{CSF} \mid f(i)) > 0.5) \tag{3}$$

$$\text{volume}_{\text{Whole}} \approx \sum_{\forall i \in I} (P(C_{GM \vee WM} \mid f(i)) > 0.5) \tag{4}$$

where i is any pixel of the MRI data and $f(i)$ stands for the gray level of i.

Binary ventricle volume data, $M(x, y, z)$, are extracted from MR images using a region growing algorithm with a threshold, which was estimated through a double threshold algorithm. After thresholding, the binary ventricle regions are obtained by using fill, erosion and dilation operations. The edges of the binary images are detected by Sobel operation on a slice-by-slice manner. The segmented region is then represented as a binary mask image M, where 1 stands for the ventricle pixel and 0 stands for the non-ventricle pixel. Therefore, Eq. (5) is used to measure the cerebral ventricle, as shown in Fig. 2 (a) and (b).

$$\text{volume}_{\text{Ventricle}} \approx \sum_{\forall i \in M} (P(C_{Ventricle} \mid f(i)) = 1) \tag{5}$$

where i is any pixel of the mask data , M is mask image and $f(i)$ stands for the gray level of i.

(a) (b) (c)

Fig. 2. (a) CSF binary map, (b) ventricle mask image, and (c) edges of (b)

2.3 Shape Features

Since the volume features extracted from the whole 3-D volume cannot capture the variation of the anatomical shape, Wang proposed a shape-based classification method [11]. The shape features comprises 3-D and 2-D shape features.

In the feature of 3-D shape, we used a leave-one-out method to construct training set and testing set. Four sets of probability map were then built by using Eq. (6) and as shown in Fig. 3.

$$P_t(x,y,z)=\frac{1}{M}\sum_{i=1}^{M} I_t^i(x,y,z) \tag{6}$$

Where t indicates the type of the subjects, comprising normal, AD, aMCI and dMCI. M is the number of training samples, and I stands for the grey level of the ventricular mask image. Next, we obtained a discriminate map - by subtracting the normal probability map from the patient probability map, as shown in Fig. 3 (c). Lastly, a matching coefficient (MC) between a testing input and the discriminate map can be calculated using Eq. (7). Where $D(x, y, z)$ is the discriminate map and T stands for the testing ventricular mask image.

$$MC^i_{Normal\ or\ patient} = \sum_{\forall x,y,z} D(x,y,z)T^i_{Normal\ or\ patient}(x,y,z) \tag{7}$$

(a) (b) (c)

Fig. 3. (a) Probability of normal controls, (b) probability of patients and (c) discriminate map

The 2-D shape features adopted herein are referred to the work of Lee *et al.* [12] and listed as follows:

(1) Area: Binary image's pixels belonging to the GM, WM, CSF, and ventricle.
(2) Perimeter: Sobel edge detection algorithm is used to extract the boundary.
(3) Compactness: Feature is calculated with the square of the perimeter.
(4) Elongation: The ratio of the height and the width of a rotated minimal bounding box, as shown in Fig. 4(a), which can fit the ventricle.
(5) Rectangularity: The ratios of an image object area and the area of the minimum bounding rectangle.
(6) Distances: There are four important corner points on the brain ventricle shape, i.e. the points A, B, C and D as shown in Fig. 4(b). The centroid M of the ventricle is computed. Then, six different distances, i.e. d (A,M), d (B,M), d (C,M), d (D,M), d (A,C) and d (B,D) are obtained, respectively.
(7) Minimum thickness: It is defined as the minimum distance between path (A, D) and path (B, C), as shown in Fig. 4(c).
(8) Mean signature value: the mean value of the distances from each boundary pixel to the centroid. It starts with the corner point A and follows in the clockwise direction.

(a) (b) (c)

Fig. 4. (a) Minimum bounding box, (b) corner and mass points, and (c) minimum thickness

2.4 Self-Organizing Map Architecture

A self-organizing map (SOM) is a type of artificial neural network that is trained by using unsupervised learning to produce a low-dimensional, discretized representation of the input space of the training samples. SOMs are different from other artificial neural networks in the sense that they use a neighborhood function to preserve the topological properties of the input space. SOMs divide into two parts: training and mapping. Training builds the map using input examples, called a Kohonen map [13]. Mapping automatically classifies a new input vector.

In this study, SOM was adopted as a classifier. A 2*2 map topology was used. Iterative times were set as 1000 epochs. Ordering phase learning rate = 0.9, tuning phase learning rate = 0.5, and tuning phase neighborhood distance = 0.5. In order to verify the stability of SOM to generalize the correct tendency, the classifier was trained in 10 times to get reliable results. Thirty cases were chosen (AD = 10, Normal = 10, aMCI = 5, dMCI = 5) to be the training set randomly. Scaling of variables is of special importance in our model since the SOM algorithm uses Euclidean metric to measure distances between vectors. In order to solve this problem, we achieved this by linearly scaling all variables so that their variances were equal to one.

3 Experimental Results

3.1 Material

According to the research [14], most patients with Alzheimer's disease are aged at 65 or older. Therefore, the whole data we choose is ranged over 65 years old. The image data used in this study were provided by Chang Gung Memorial Hospital, Lin-Kou, Taiwan. The whole dataset consists of four groups. Demographic information is provided in Table 1.

The whole-brain MRI scans were obtained by a 3T MR scanner T1 MPRage series with TR = 2000ms and TE = 2.63ms. The results were represented as a 224×256 matrix, and slice thickness = 1mm in 160 slices.

Table 1. Demographic data and cognitive scores

Group	Normal control	dMCI	aMCI	AD
Individuals (Male/Female)	28 (18/10)	15 (7/8)	17 (9/8)	24 (11/13)
Mean age (yrs)	67 ± 5.67	70 ± 5.01	73 ± 5.13	71 ± 7.37
Education time (yrs)	10 ± 4.8	8 ± 4.23	8 ± 5.24	6.96 ± 5.84
MMSE scores	28 ± 1.24	28 ± 1.63	25 ± 4.05	14.38 ± 6.55

3.2 Statistical Analysis and Classification

Mann-Whitney U test was performed on each feature to evaluate its discriminative power, as shown in Eq. (14). U_{obt} is the smaller one of the two calculated test statistics (U_1 & U_2), where n_1 and n_2 are the sample size for sample 1 and 2.

$$Z_U = \frac{U_{obt} - \left(n_1 n_2 / 2 \right)}{\sqrt{\frac{n_1 n_2 (n_1 + n_2 + 1)}{12}}} \tag{7}$$

The p-values obtained from the tests providing a generally known and comparable criterion. It rejects the null hypothesis of equal distributions when $p < 0.05$. Table 2 illustrates the statistical results of volume and shape features. In the experiment results, since circularity and rectangularity's *p-value* > 0.05, they were excluded (circularity = 0.871, rectangularity = 0.628) in the following steps of classification.

Table 2. Statistical analysis of features

Features	**Mean volume in [mm] ± S.D.**				
Volume	**Normal**	**dMCI**	**aMCI**	**AD**	**p-value**
V_{GM}	849.5 ± 62.1	820.9 ±113.2	801.5 ± 47.6	776.6 ± 114.3	0.009
V_{WM}	621.6 ± 57.3	610.5 ± 82.1	599.7 ± 11.7	534.5 ± 71.9	0.027
V_{CSF}	849.6 ± 137.1	861.4 ± 125.4	916.2 ± 146.5	969.8 ± 117.8	0.011
Shape	**Normal**	**dMCI**	**aMCI**	**AD**	**p-value**
Area	1581.1 ± 268.3	1732.4 ± 235.1	1968.5 ± 513.6	2206.4 ± 713.8	0.029
Area (PR)	614.4 ± 112.1	746.5 ± 62.0	831.9 ± 128.1	901.7 ± 211.6	0.011
Area (PL)	611.7 ± 118.4	739.2 ± 187.1	854.2 ± 199.4	907.9 ± 234.1	0.009
Area (FR)	132.8 ± 98.5	189.8 ± 77.6	217.5 ± 134.2	253.9 ± 176.1	0.013
Area (FL)	140.5 ± 76.9	201.2 ± 62.1	264.9 ± 164.3	276.4 ± 191.0	0.017
Perimeter	214.3 ± 18.9	244.5 ± 11.4	263.7 ± 21.3	283.8 ± 36.3	0.021
Circularity	43.9 ± 5.6	41.0 ± 2.1	38.5 ± 4.7	37.0 ± 3.1	0.013
Elongation	1.2 ± 0.7	1.2 ± 1.1	1.3 ± 0.7	1.3 ± 0.1	0.019
Rectangularity	0.5 ± 0.1	0.5 ± 0.8	0.6 ± 0.4	0.6 ± 0.1	0.020
d(A,G)	34.7 ± 3.1	36.8 ± 2.5	39.7 ± 5.1	39.8 ± 6.4	0.016
d(B,G)	35.1 ± 2.9	37.9 ± 6.7	41.1 ± 4.9	42.3 ± 5.8	0.030
d(C,G)	37.3 ± 2.1	38.7 ± 3.2	40.6 ± 3.2	42.6 ± 5.1	0.026
d(D,G)	35.1 ± 3.7	36.7 ± 3.1	39.2 ± 0.9	41.3 ± 4.6	0.021
d(A,C)	73.2 ± 5.1	78.6 ± 10.3	83.6 ± 15.3	82.4 ± 12.9	0.008
d(B,D)	69.5 ± 6.7	74.4 ± 6.1	79.5 ± 2.7	80.9 ± 10.4	0.004
Min thickness	25.9 ± 2.1	25.7 ± 0.7	28.9 ± 3.1	29.5 ± 3.7	0.016
Mean Sig.	24.5 ± 2.9	26.9 ± 1.8	27.4 ± 3.1	29.1 ± 2.8	0.007

In fact, some of features may be redundant or have highly correlation. Therefore, PCA [15] was introduced to reduce the dimensions of the feature space. The principal components which contribute 95% to the total variation in data set were chosen herein. To train a volume-feature-based classification, all the volume features were adopted. To train a shape-feature-based classification, only the first five principal components which convey a large amount of information quantified by 95% energy were adopted. In the case of using both shape and volume features, the first six principle components were used.

SOM was used to train a classifier. Table 3 shows the accuracy (proportion of all subjects correctly classified), sensitivity (proportion of individuals with a true positive result) and specificity (proportion of individuals with a true negative result) with using different features. Obviously, incorporating with shape features, volume features, and PCA shows excellent classification ability than others.

In AD, aMCI, and dMCI, the accuracy, sensitivity and specificity have been improved respectively. However, the classified results in aMCI and dMCI are not very well than AD. That is because aMCI's characteristics are very similar with AD, and dMCI's characteristics are very similar with normal control. Therefore, the results are relatively weak.

Table 3. Classification results

Proportion	Volume features	Shape features	Volume + Shape features	Volume + Shape features + PCA
		AD		
Accuracy	79.27%	72.43%	86.76%	93.63%
Sensitivity	74.43%	78.33%	79.37%	80.91%
Specificity	80.69%	74.28%	84.13%	88.63%
		aMCI		
Accuracy	53.33%	46.67%	66.67%	73.33%
Sensitivity	66.67%	46.67%	53.33%	72.63%
Specificity	73.33%	53.33%	53.33%	80.14%
		dMCI		
Accuracy	26.67%	40.00%	46.67%	53.33%
Sensitivity	46.67%	53.33%	53.33%	60.33%
Specificity	53.33%	46.67%	60.33%	72.67%

4 Conclusions

In this study, we design a classification framework for image-aided diagnosis for AD by using easy-extractable volume-related and shape-related features. The measurement of global GM, WM and CSF volumes and the local shape analysis on ventricle, especially in the properties of the ventricular area, perimeter and distances, provide atrophy information and show statistically discriminative power ($p < 0.05$). By combining the volumetric features and shape features, the classification accuracy is reached to 86.76%, 66.67% and 46.67% in AD, amnestic MCI (aMCI), and dysexecutive MCI (dMCI), respectively. In addition, with the help of PCA process, the classification result is improved to 93.63%, 73.33% and 53.33% in AD, aMCI and dMCI, respectively.

Based on the testing results, we conclude that the volume features and shape features can be selected together because of their low computational complexity and classification ability. For the future work, we will increase the size of our dataset to support the outcome of our experiment. In addition, according to the classification results, aMCI and dMCI are similar with each other. Res-fMRI shows a new light to the understanding of neural network connectivity. DTI can be used to assess the fiber integrity. The information will help us to improve the outcome of diagnosing the neurodegenerative diseases.

Acknowledgements

This work was supported by Ministry of Economic Affairs, Taiwan under Technology Development Program for Academia (TDPA) with Grant No. 98-EC-17-A-19-S1-035 and Chang Gung Memorial Hospital with Grant No. CMRPD270052.

References

1. Petersen, R.C.: Mild cognitive impairment as a diagnostic entity. Journal of Internal Medicine 256, 183–194 (2004)
2. Thal, L.J., Ferris, S.H., Kirby, L., et al.: A randomized, double-blind, study of rofecoxib in patients with mild cognitive impairment. Neuropsychopharmacology 30, 1204–1215 (2005)
3. Vemuri, P., Wiste, H.J., Weigand, S.D., Shaw, L.M., Trojanowski, J.Q., Weiner, M.W., Knopman, D.S., Petersen, R.C., Jack Jr., C.R.: MRI and CSF biomarkers in normal, MCI, and AD subjects: predicting future clinical change. Neurology 73(4), 294–301 (2009)
4. Schott, J.M., Price, S.L., Frost, C., Whitwell, J.L., Rossor, M.N., Fox, N.C.: Measuring atrophy in Alzheimer diseases: a serial MRI study over 6 and 12 months. Neurology 65, 119–124 (2005)
5. Pruessner, J.C., Collins, D.L., Pruessner, M., Evans, A.C.: Age and gender predict volume decline in the anterior and posterior hippocampus. Jornal of Neuroscience 21(1), 194–200 (2001)
6. Nestor, S., Rupsingh, R., Accomazzi, V., Borrie, M., Smith, M., Wells, J., Bartha, R.: Changes in brain ventricle volume associated with mild cognitive impairment and alzheimer disease in subjects participating in the alzheimer's disease neuroimaging initiative. Alzheimer's and Dementia 3, S114 (2007)
7. Talairach, J., Tournoux, P.: Co-Planar Stereotaxic Atlas of a Human Brain: Dimensional Proprotional System: An Approach to Cerebral Imaging. Georg Thieme Verlag (1988)
8. Mazziotta, J.C., Toga, A., Evans, A., Fox, P., Lancaster, J., Zilles, K.: A probabilistic atlas and reference system for the human brain: International Consortium for Brain Mapping (ICBM). The Royal Society 356(1412), 1293–1322 (2001)
9. Fritzsche, K.H., von Wangenheim, A., Abdala, D.D., Meinzer, H.P.: A computational method for the estimation of atrophic changes in Alzheimer's disease and mild cognitive impairment. Computerized Medical Imaging and Graphics 32, 294–303 (2008)
10. UCL Institute of Neurology, http://www.fil.ion.ucl.ac.uk/spm/
11. Wang, J., de Haan, G., Unay, D., Soldea, O., Ekin, A.: Voxel-based discriminant map classification on brain ventricles for Alzheimer's disease. Medical Imaging 7259 (2009)
12. Lee, J.D., Su, S.C., Huang, C.H., Wang, J.J., Xu, W.C., Wei, Y.Y.: Combination of Multiple Features for Alzheimer's Disease Diagnosis by SVM with PCA. In: 16th International Conference on Neural Information Processing (2009)
13. Kohonen, T.: Self-Organizing Maps, 3rd extended edition. Springer, Heidelberg (2001)
14. Alzheimer's disease facts and figures (2010), http://www.alz.org/alzheimers_disease_facts_figures.asp?type=homepage
15. Jolliffe, I.T.: Principal Component Analysis, 2nd edn. Springer Series in Statistics. Springer, NY (2002)

Parallel Batch Training of the Self-Organizing Map Using OpenCL

Masahiro Takatsuka and Michael Bui

ViSLAB, School of Information Technologies
The University of Sydney, NSW 2006, Australia
masa.takatsuka@sydney.edu.au

Abstract. The Self-Organizing Maps (SOMs) are popular artificial neural networks that are often used for data analyses through clustering and visualisation. SOM's mathematical model is inherently parallel. However, many implementations have not successfully exploited its parallelism because previous attempts often required cluster-like infrastructures. This article presents the parallel implementation of SOMs, particularly the batch map variant using Graphics Processing Units (GPUs) through the use of Open Computing Language (OpenCL).

1 Introduction

The Self-Organizing Map (SOM) is popular artificial neural networks that can produce a topology preserved mapping of a high-dimensional feature space. When a feature vector is presented to the network, a search for the Best Matching Unit (BMU) is carried out. Once found, the BMU c and neurons within its neighbourhood on the lattice update their connection weight vector w according to the following equation:

$$w_i(t+1) = w_i(t) + h_{ci}(t)[x(t) - w_i(t)], \tag{1}$$

where w_i is the weight vector of the neuron i, x is the input vector and t is a variable in the discrete time index. The neighbourhood function h_{ci} is typically a Gaussian function:

$$h_{ci}(t) = \alpha(t).exp(-\frac{\|r_c - r_i\|^2}{2\sigma^2(t)}), \tag{2}$$

where $\alpha(t)$ is the learning rate and $\sigma(t)$ is the neighbourhood radius and r_i denotes the physical location of neuron i on the neural lattice. Both $\alpha(t)$ and $\sigma(t)$ are monotonically decreasing values. Moreover, the decreasing neighbourhood has been argued to be a condition that is necessary for self-organizing to occur. After training the network with input data over several iterations, the algorithm produces a topologically ordered mapping of the data, such that dissimilar data samples are mapped to distant locations on lattice. This characteristic has allowed the SOM to be successfully used in a large number of applications, such as geospatial analysis [1] and content-based image retrieval [2].

K.W. Wong et al. (Eds.): ICONIP 2010, Part II, LNCS 6444, pp. 470–476, 2010.

While the above is a typical description of SOM's incremental learning algorithm, a batch training scheme also exists that can run faster and produce similar results. In the batch training, all the feature vectors are presented to the SOM and mapped to their BMU's sublist of points at each iteration. The weight vector of each neuron i is then updated to be the weighted mean of feature vectors in the neighbourhood as:

$$m_i^* = \frac{\sum_j^{N_i} n_j h_{ji} \bar{x}_j}{\sum_j^{N_i} n_j h_{ji}}, \tag{3}$$

where N_i is the set of neurons within the neighbourhood of neuron i, n_j is the number of feature vectors $x(t)$ in the Voronoi set V_j and $\bar{x}_j$ is the mean of these feature vectors.

This batch map variant is particularly suited to parallelization. However, efforts to parallelize the SOM have mainly focused on the incremental learning algorithm [3,4]. Despite the nature of parallelism in its mathematical model and previous implementation efforts, not many SOM tools fully utilize the parallelism. One of common reasons for this was the requirement of a cluster-like high-performance computing infrastructure.

Recent advances in Computer Graphics resulted in commodity high-performance Graphics Processing Units (GPUs). Although they are specially design to facilitate fast graphics processing, they have been used as general purpose computing devices (often called General Purpose GPU: GPGPU). There have been some attempts toleverage the power of GPUs for training the neural network [5,6]. The work by Zhongwen et al. relied on the use of computer graphics language (Cg) to implement their parallel SOM[5]. Since their implementation was based on Cg, they first needed to map SOM's data structure to data structures used in Computer Grahics. Moreover, they needed to employ multi-path algorithm due to the architecture of the graphics pipeline. Jang et al. implemented a multilayer perceptron on both GPU and multi-core CPU and evaluated the benefit of GPGPU programming for the neural information processing[6].

In this article, we present re-visit the parallel implementation of the SOM using GPGPUs. In particular, our focus is on the implementation of the batch training algorithm. We chose the Open Computing Language (OpenCL) [7] library for this implementation is used where the resultant code is similar to the C programming language and programmers would not require knowledge of computer graphics and shading languages.

2 OpenCL

OpenCL (Open Computing Language) [7] is a framework that facilitates parallel programming on GPUs as well as CPUs. An OpenCL application is executed on a host that is connected to one or more OpenCL devices (such as a graphics ship on a graphics card). An OpenCL device consists of compute units that are divided into processing elements as shown in Figure 1. Code that can be run in parallel are written as kernels that are executed on the OpenCL devices, more precisely on the processing elements. An instance of a kernel is known as a work-item, which are organized into work-groups. Each work-item can be thought of a thread, such that each work-item operates on a portion of the input data in parallel. Furthermore, there are different levels of memories:

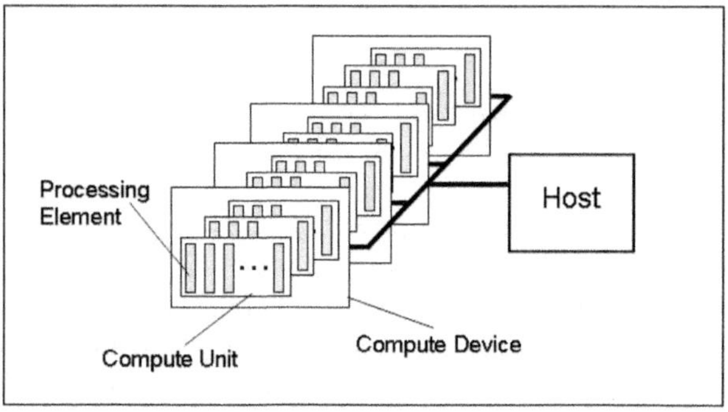

Fig. 1. Platform model of OpenCL architecture: one host and one or more compute devices each with one or more compute units each with one or more processing elements. While the host executes the main program, processing elements execute small kernel program in parallel. [7]

- Global memory: It can be read and written by all work-items in all work-groups.
- Constant memory: It is a read-only region of global memory that is constant. Allocated and initialized by host.
- Local memory: It is a memory visible to all work-items in a single workgroup.
- Private memory: It is a memory that is visible to a single work-item.

The access to the global memory is typically very expensive and has the largest latency while the access to the private memory is the fastest. It is strongly encouraged to use local memory when there are instances of code that perform many read/write operations to the same memory. Moreover, the transfer between host and device memory can be very expensive (eg. through PCI express bus). Consequently, such data transfers should be minimised. For more details on OpenCL's memory and execution model can be found in its specification[7].

3 Implementation of the Batch Training Process Using OpenCL

The SOM, like other neural networks, consists of elements that are inherently parallelisable. In this section, we describe how to parallelise the SOM algorithm

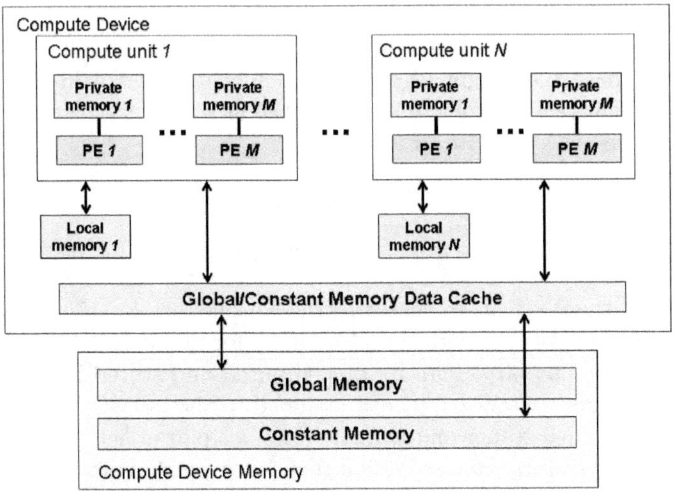

Fig. 2. Memory model of OpenCL architecture with respect to the compute units and device. There are four memory levels.[7]

for implementation with OpenCL. There are two main components of the batch training algorithm that can be parallelised:

1. the mapping/projection of each feature vector onto the SOM, and
2. the adaptation of neurons' connection weight vectors.

Essentially, our implementation uses data parallelism and the data sent from the host to the GPU are 1) all the feature vectors, 2) initial weight vector values, 3) dimensionality of the data set and 4) the current update radius. Therefore, data is only transferred between the host and GPU at most twice, once to send the values for training and, if needed, another transfer to retrieve the results for any processing that needs to be done on the host.

3.1 Finding the BMU

The implementation of the first component is rather trivial. The data set is partitioned into so that each workgroup operates on one portion. Each work-item deals with a single feature vector to find it's corresponding BMU by calculating the minimum distance between the feature vector and a neuron's weight vector on the SOM. As this computation requires accessing the values of the feature vector for each neuron, its values are copied from global memory to local memory. Each distance computation using this feature vector then reads the values from local memory space. Furthermore, during this process, the variable representing the current minimum distance value, and the index of the current BMU may be updated up to M times, where M is the number of neurons. These values are therefore also stored in local memory. Once the BMU has been found, its index is stored in a table of indices that resides in global memory.

3.2 Neuron Adaptation

Once the BMUs of all the feature vectors have been found, the algorithm can proceed to update the weight vectors by calculating the weighted mean as per Equation 3. However, in order to minimise the large latency memory access, we will compute the weighted mean as:

$$m_i^* = \frac{\sum_j^{N_i} h_{ji} \sum_k^{n_j} x(t)_k}{\sum_j^{N_i} n_j h_{ji}}, \tag{4}$$

where $x(t)_k$ is the kth feature vector in the Voronoi set V_j. Hence, rather than calculating the sum of the feature vectors in V_j for all neurons i that need to be updated, these sums and n_j can be precomputed and stored in global memory once all the feature vectors have been assigned to their BMUs.

Furthermore, the distance on the lattice for each pair of neurons i and j can be calculated before training and stored on the device's global memory as a distance matrix. Otherwise, the position of each neuron on the lattice would need to be stored. This would increase the number of memory access and increase the time of computation when determining if a neuron lies within the neighbourhood set N_i. Another optimisation that can be done to reduce memory access time is to store the m_i^* in local memory so that updates to the weighted mean writes to local memory. The final result can then be copied to global memory afterwards.

4 Results and Discussion

For our experiments, we implemented the Geodesic SOM [8] (a spherical SOM) in C++ and using OpenCL. To make the comparisons more fair, some of the optimisations discussed were also implemented in the C++ version. Namely, the pre-computation of the sum of feature vectors at each neuron and the distance matrix.

Table 1. The size of the data used: Iris, Ionosphere and Torus

Dataset	# of Instances	# of Attributes
Iris	150	4
Ionosphere	351	34
Torus	2000	3

We tested the performance of each SOM (C++ version and OpenCL version) by training them using three datasets (Iris, Ionosphere and Torus). The Iris and Ionoshphere data are from the UCI (University of California at Irvine) data repository (http://archive.ics.uci.edu/ml/index.html) and the Torus dataset is a set of 2000 three-dimensional points randomly sampled on the surface of a torus. For each dataset, both SOMs were trained 10 times and recorded the computation time required for the training process. Experiments were conducted on a commodity PC with an Intel Pentium 4 3.2 GHz processor, 1GB of RAM and a NVIDIA GTS 250 (128 CUDA cores) graphics card.

Table 2 shows the mean computation time for each data set and implementation of the Geodesic SOM. A three-frequency geodesic dome (92 neurons) was used, with an initial update radius of 9, initial learning rate of 1 and training occurred over 1000 epochs. The initial weight vectors were linearly initialized. It should be noted that the training time consists of the time took for finding the BMUs and updating neurons' connection weights.

Table 2. The mean time required to train the Geodesic SOM with each data and for each implementation

Dataset	Implementation	Avg. Time (sec)
Iris	C++	1.386
	OpenCL	0.316
Ionosphere	C++	21.314
	OpenCL	1.707
Torus	C++	12.508
	OpenCL	1.228

As shown in Table 2, when the small dataset (Iris: N=150, d=4) was used the performance gain was not significant. However, when the number of attribute is large or the number of instance is large, the computational cost saving on GPU became apparent. This indicates that the process of finding BMUs costs more than updating their weight vectors. The parallel implementation on the GPU does not involve inter-process (or inter-processing element) communication like often seen in the message-passing interface approach. Hence this approach seems to be suited for a large SOM or a large-scale dataset.

5 Conclusion

This paper proposed parallel implementation of SOM's batch training algorithm on commodity graphics hardware. The OpenCL was used for the implementation because it provides a C-like language and does not rely on data structures used in Computer Graphics libraries. For the process of finding BMUs, one processing element computes the distance between the given input vector against all connection weight vectors. Hence, this was done simply partitioning the input data so that they are mapped onto the array of processing elements. For updating the connection weight vectors of BMUs, each processing element computes the weighted mean for each BMU. In order to minimise the memory access to Host's memory space (PC's main memory), many intermediate values (such as the sum of feature vectors and distance matrices) are precomputed and stored in device's global memory. The experiments evaluated the proposed implementation through standard datasets, and confirmed a significant speed-up on the GPU compared to the execution of sequential algorithm on CPU.

References

1. Takatsuka, M., Gahegan, M.N.: Exploratory geospatial analysis using GeoVISTA studio: From a desktop to the web. In: Proceedings of the 1st International Workshop on Web Geographical Information Systems, WGIS 2001, pp. 446–455. IEEE Computer Society, Los Alamitos (December 2001)
2. Laaksonen, J., Koskela, M., Oja, E.: PicSOM-self-organizing image retrieval with MPEG-7 content descriptors. IEEE Transactions on Neural Networks 13(4), 841–853 (2002)
3. Strupl, D., Neruda, R.: Parallelizing self-organizing maps. In: Jeffery, K. (ed.) SOFSEM 1997. LNCS, vol. 1338, pp. 563–570. Springer, Heidelberg (1997)
4. Ozdzynski, P., Lin, A., Liljeholm, M., Beatty, J.: A parallel general implementation of kohonen's self-organizing map algorithm: performance and scalability. Neurocomputing 44-46, 567–571 (2002)
5. Luo, Z., Liu, H., Yang, Z., Wu, X.: Self-organizing maps computing on graphic process unit. In: Proceedings of the 13th European Symposium on Artificial Neural Networks, pp. 557–562 (2005)
6. Jang, H., Park, A., Jung, K.: Neural network implementation using cuda and openmp. In: Proceedings of the International Conference on Digital Image Computing: Techniques and Applications, pp. 155–161 (2008)
7. Khronos OpenCL Working Group: The OpenCL Specification. Verion 1.0 rev.43 edn. (May 2009)
8. Wu, Y., Takatsuka, M.: Spherical self-organizing map using efficient indexed geodesic data structure. Neural Networks 19(6-7), 900–910 (2006)

Fast Kohonen Feature Map Associative Memory Using Area Representation for Sequential Analog Patterns

Hiroki Midorikawa and Yuko Osana

Tokyo University of Technology,
1401-1 Katakura, Hachioji, Tokyo, Japan
osana@cs.teu.ac.jp

Abstract. In this paper, we propose a Fast Kohonen Feature Map Associative Memory with Area Representation for Sequential Analog Patterns (FKFMAM-AR-SAP). This model is based on the conventional Improved Kohonen Feature Map Associative Memory with Area Representation for Sequential Analog Patterns (IKFMAM-AR-SAP). The proposed model can realize the one-to-many associations even when the first patterns are same in the plural sequential patterns. And, it has enough robustness for noisy input and damaged neurons. Moreover, the learning speed of the proposed model is faster than that of the conventional model. We carried out a series of computer experiments and confirmed the effectiveness of the proposed model.

1 Introduction

Recently, neural networks are drawing much attention as a method to realize flexible information processing. In the field of neural networks, although a lot of models have been proposed, their learning and recall processes are divided, and therefore they need all information to learn in advance. However, in the real world, it is very difficult to get all information to learn in advance, so we need the model whose learning process and recall process are not divided. As such model, some models based on the Kohonen Feature Map (KFM) associative memory[1] have been proposed[2]-[6]. For example, the KFM associative memory with area representation for sequential analog patterns[5], the improved KFM associative memory with area representation for sequential analog patterns[6] and so on have been proposed. These models can deal with associations of sequential patterns including common terms, and has enough robustness for noisy input and damaged neurons. However, they can not realize the one-to-many associations when the first patters are same in the plural sequential patterns.

In this paper, we modify the winner neuron selection method and the connection weights update method of the conventional improved KFM associative memory with area representation for sequential analog patterns[6], and propose a Fast KFM Associative Memory with Area Representation for Sequential Analog Patterns (FKFMAM-AR-SAP). The proposed model can realize the one-to-many associations even when the first patterns are same in the plural sequential

K.W. Wong et al. (Eds.): ICONIP 2010, Part II, LNCS 6444, pp. 477–484, 2010.

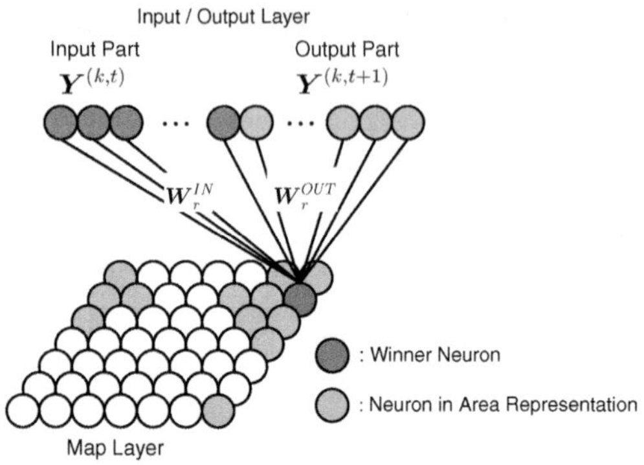

Fig. 1. Structure of Proposed Model

patterns. And, it has enough robustness for noisy input and damaged neurons. Moreover, the learning speed of the proposed model is faster than that of the conventional model.

2 Fast KFM Associative Memory with Area Representation for Sequential Analog Patterns

Here, we explain the proposed Fast Kohonen Feature Map Associative Memory with Area Representation for Sequential Analog Patterns (FKFMAM-AR-SAP). The proposed model is based on the conventional IKFMAM-AR-SAP[6]. In the proposed model, the winner neuron selection method and the connection weights update method are modified, and realize the one-to-many associations even when the first patterns are same in the plural sequential patterns. Moreover, the learning speed of the proposed model is faster than that of the conventional model.

2.1 Structure

Figure 1 shows the structure of the proposed model. As seen in Fig.1, it has two layers; (1) Input/Output Layer and (2) Map Layer, and the Input/Output Layer is divided into two parts; (1) Input Part and (2) Output Part as similar as the conventional model. In this model, as shown in Fig.1, the Map Layer is treated as torus.

2.2 Learning Process

Let $\boldsymbol{Y}^{(k,1)} \rightarrow \boldsymbol{Y}^{(k,2)} \rightarrow \cdots \rightarrow \boldsymbol{Y}^{(k,t_k)}$ be the kth temporal sequence to be stored, where t_k shows the length of the kth sequence.

In the sequential learning algorithm for the proposed FKFMAM-AR-SAP, the connection weights are learned as follows:

(1) The initial values of weights are chosen randomly and the recurrent difference vector is set to $\boldsymbol{y_i} = \boldsymbol{0}$.

(2) The recurrent difference vector of the neuron $\boldsymbol{i}$ in the Map Layer $\boldsymbol{y_i}(t)$ is calculate by

$$y_{ij}(t) = \begin{cases} \sum_{n=0}^{t}(1-\beta)^{t-n}X_j^{(k,n)} - \sum_{n=0}^{t}(1-\beta)^n W_{ij}(t), & (j \le M/2) \\ X_j^{(k,t)} - W_{ij}(t), & (\text{otherwise}) \end{cases} \quad (1)$$

where M is the number of neurons in the Input/Output Layer, and β ($0.5<\beta<1$) is the weighting factor determining the effect of the earlier difference vectors and the new input vector in the computation of $\boldsymbol{y_i}(t)$.

(3) The winner neuron $\boldsymbol{r}$ is determined as follows:

$$r = \underset{i:1-sH^{learn}(d_{ii*})<\theta^r}{\text{argmin}} \left(\|\boldsymbol{y_i}(t)\| \left(1 - sH^{learn}(d_{ii^*})\right)\right) \quad (2)$$

where θ^r is the threshold for the winner neuron decision, s ($0<s<1$) is the coefficient, d_{ii^*} is the distance between the neuron i and the nearest weights fixed neuron i^*.

In Eq.(2), $H^{learn}(d_{ii^*})$ is given by

$$H^{learn}(d_{ii^*}) = \frac{1}{1 + \exp\left(-\frac{d_{ii^*} - 2D}{\varepsilon^t}\right)} \quad (3)$$

where D is the constant which decides area size and ε^t is the steepness parameter. In the proposed model, the all area can be ensured as a circle with radius D.

(4) The connection weights between the neurons in the Input/Output Layer and the neuron i in the Map Layer except those of fixed neurons are updated by

$$W_{ij}(t+1) = \begin{cases} \dfrac{\sum_{n=0}^{t}(1-\beta)^{t-n}X_j^{(k,n)}}{\sum_{n=0}^{t}(1-\beta)^n}, & (\theta_1^{learn} \le H(d_{ri}) \text{ and } j \le M/2) \\ X_j^{(k,t)}, & (\theta_1^{learn} \le H(d_{ri}) \text{ and } M/2 < j) \\ W_{ij}(t) + H(d_{ri}) \left(\dfrac{\sum_{n=0}^{t}(1-\beta)^{t-n}X_j^{(k,n)}}{\sum_{n=0}^{t}(1-\beta)^n} \right), & \\ \quad (\theta_2^{learn} \le H(d_{ri}) < \theta_1^{learn} \text{ and } H(d_{ii^*}) < \theta_1^{learn} \text{ and } j \le M/2) & \\ W_{ij}(t) + H(d_{ri})X_j^{(k,t)}, & \\ \quad (\theta_2^{learn} \le H(d_{ri}) < \theta_1^{learn} \text{ and } H(d_{ii^*}) < \theta_1^{learn} \text{ and } M/2 < j) & \\ W_{ij}(t), & (\text{otherwise}) \end{cases} \quad (4)$$

where θ_1^{learn} and θ_2^{learn} are thresholds ($\theta_1^{learn}>\theta_2^{learn}$). And, $H(d_{ri})$ and $H(d_{ii^*})$ are given by Eq.(5).

$$H(d_{ij}) = \frac{1}{1 + \exp\left(\dfrac{d_{ij} - D}{\varepsilon}\right)} \tag{5}$$

If there is no weight-fixed neuron, we use

$$H(d_{ii^*}) = 0 \tag{6}$$

instead of Eq.(5).

(5) The connection weights of the winner neuron r are fixed.
(6) (2)~(5) are iterated until $t = t_k - 1$.
(7) (2)~(6) are repeated when a new pattern sequence is given.

2.3 Recall Process

When the pattern $\boldsymbol{X}^{(t)} (= (\boldsymbol{Y}^{(t)}, \boldsymbol{0})^T)$ is given, the output of the neuron i in the Map Layer at the time t, $x_i^{map}(t)$ is given by

$$x_i^{map}(t) = \begin{cases} 1, & (i = r) \\ 0, & (\text{otherwise}) \end{cases} \tag{7}$$

where r is selected randomly from the neurons which satisfy

$$\| \boldsymbol{y_i}(t) \| \leq \theta^{map} \tag{8}$$

where $\boldsymbol{y_i}(t)$ is the recurrent difference vector of the neuron i in the Map Layer at the time t which is given by

$$\boldsymbol{y_i}(t) = (1 - \beta)\boldsymbol{y_i}(t - 1) + \beta(\boldsymbol{X} - \boldsymbol{W_i}) \tag{9}$$

where θ^{map} is the threshold of the neuron in the Map Layer and is given by

$$\theta^{map} = y_{min} + a(y_{max} - y_{min}) \tag{10}$$

$$y_{min} = \min_i \| \boldsymbol{y_i}(t) \| \tag{11}$$

$$y_{max} = \max_i \| \boldsymbol{y_i}(t) \| \tag{12}$$

where a $(0 < a < 1)$ is a coefficient.

The output of the neuron j in the Input/Output Layer at the time t, $x_j^{io}(t)$ is given by

$$x_j^{io}(t) = W_{rj}. \tag{13}$$

3 Computer Experiment Results

Here, we show the computer experiment results to demonstrate the effectiveness of the proposed model.

3.1 Association Result

In this experiment, two sequential analog patterns shown in Fig.2 were memorized successively. Figure 3 shows the association result of the proposed model when the training set (1) was memorized. As shown in Fig.3, the proposed model could recall sequential analog patterns including a common term correctly.

We also examined the case when the first patterns are same in the two sequential patterns. Figure 4 (a) and (b) show the association results of the proposed model. As shown in Fig.4 (a) and (b), the proposed model could recall the desired patterns correctly. In contrast, the conventional IKFMAM-AR-SAP[6]

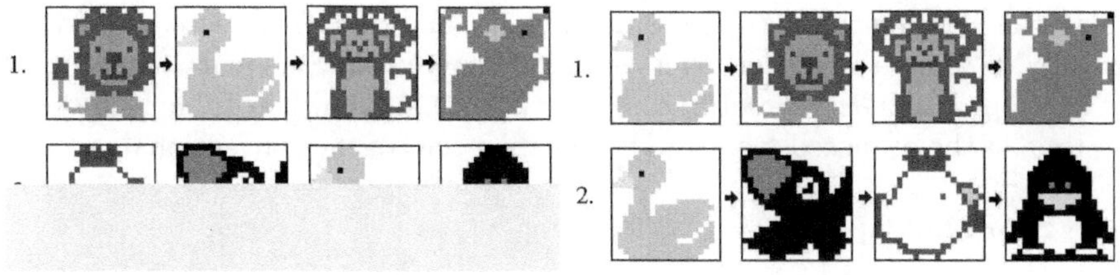

(a) Training Set (1) (b) Training Set (2)

Fig. 2. Stored Sequential Analog Patterns

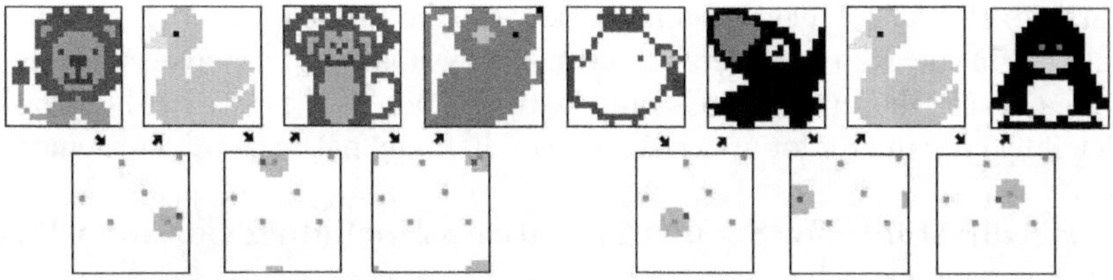

(a) When "lion" was Given (b) When "hen" was Given

Fig. 3. Association Result of Proposed Model

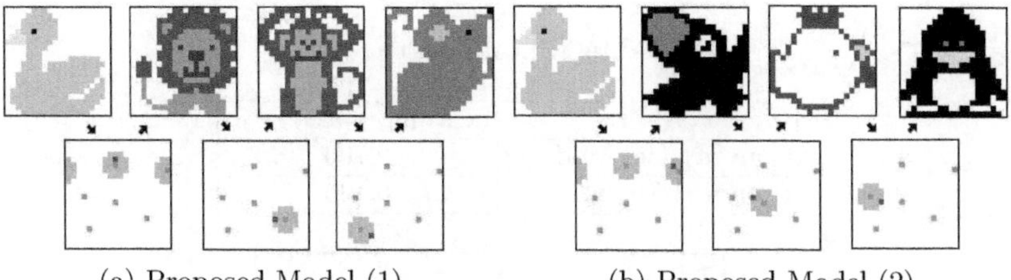

(a) Proposed Model (1) (b) Proposed Model (2)

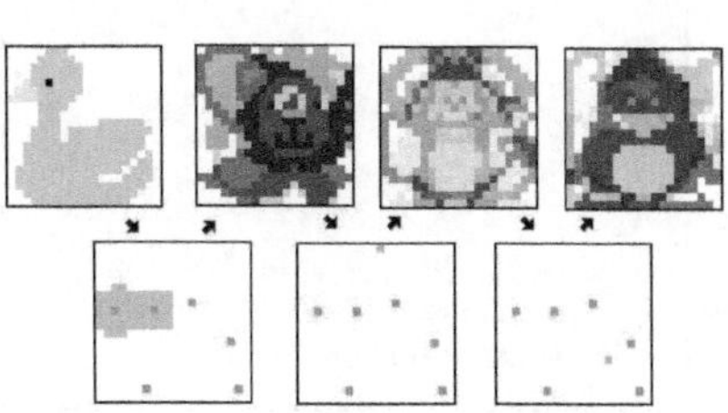

(c) Conventional Model[6]

Fig. 4. Association Result when "duck" was Given

could not recall correct patterns (Fig.4 (c)) because the superimposed patterns were recalled.

3.2 Learning Speed

Here, we examined the learning speed of the proposed model. In this experiment, one random pattern sequence composed of four patterns was memorized. Table 1 shows the learning time of the proposed model and the conventional IKFMAM-AR-SAP[6]. These results are average of 100 trials on the Personal Computer (Intel Pentium 4 (3.2GHz), FreeBSD 4.11, gcc 2.95.3). As shown in Table 1, the learning time of the proposed model is shorter than that of the conventional model.

3.3 Storage Capacity

Here, we examined the storage capacity of the proposed model. Figure 5 shows the storage capacities of the proposed model and the conventional IKFMAM-AR-SAP[6]. As shown in Fig.5, the storage capacity of the proposed model is smaller than that of the conventional model. This is because the area sizes for the training patterns in the Map Layer are almost same in the proposed model. In contrast, some areas in the Map Layer are sometimes very small in the conventional model, and as a result the number of stored patterns becomes large. In the proposed model, if the parameter for area size D is small, many patterns can be memorized.

3.4 Recall Ability for Sequential Patterns Including Common Terms

Here, we examined the recall ability for sequential patterns including common terms of the proposed model. Figure 6 (a) shows the recall ability of the proposed model when two sequential patterns composed of N patterns (only first and last patterns are different and the other $N-2$ patterns are common) were memorized

Table 1. Learning Time

	Learning Speed (sec)
Conventional Model[6]	2.706
Proposed Model	0.180

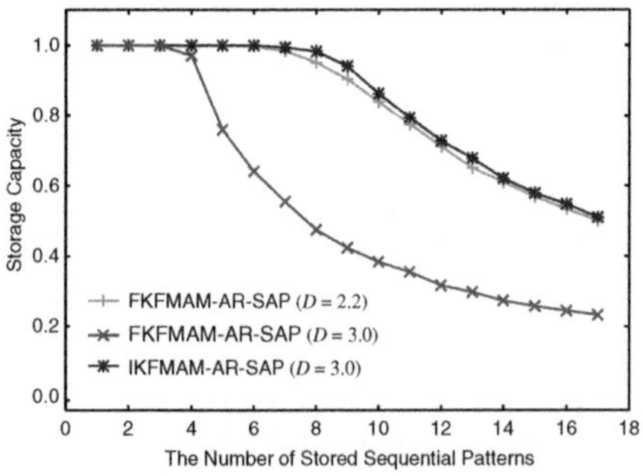

Fig. 5. Storage Capacity

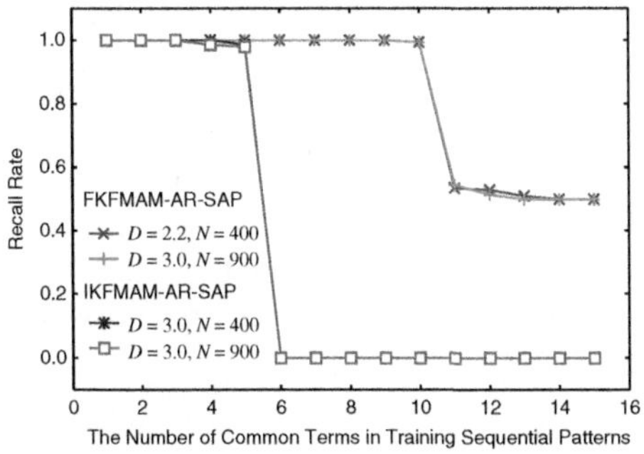

(a) When Two Sequential Patterns are Memorized

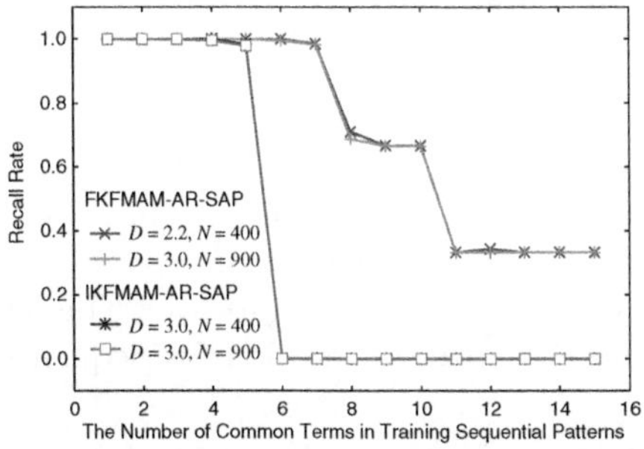

(b) When Three Sequential Patterns are Memorized

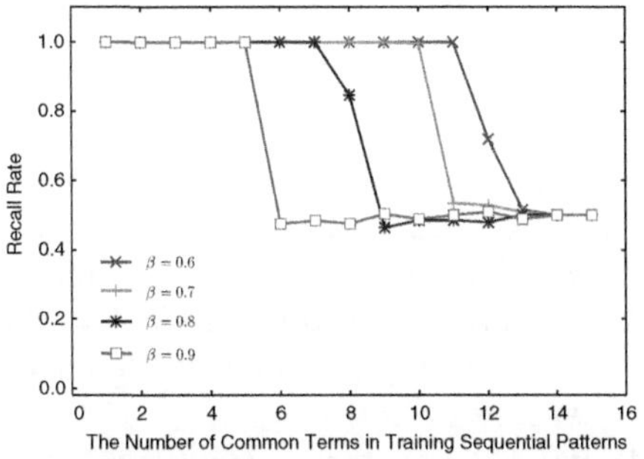

(c) Dependence Property to Weighting Coefficient β

Fig. 6. Recall Ability for Sequential Patterns including Common Terms

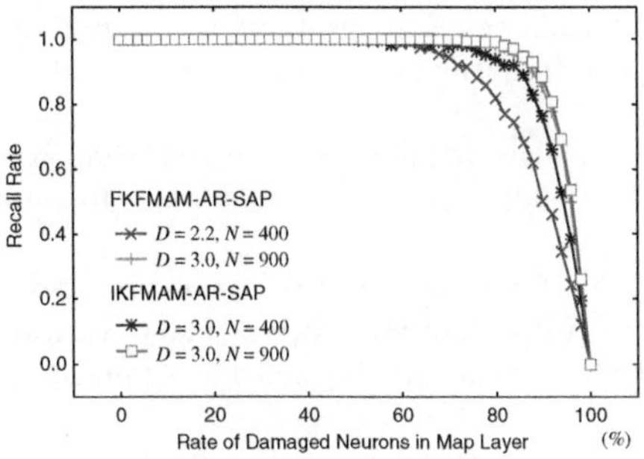

(a) Damaged Neurons

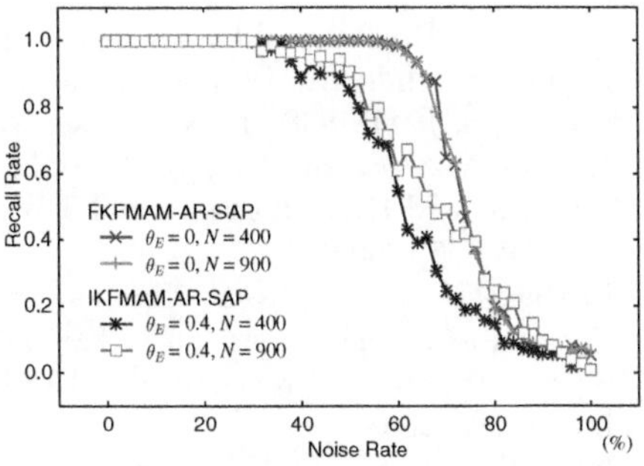

(b) Noisy Input

Fig. 7. Robustness for Damaged Neurons and Noisy Input

and the first pattern was given. Figure 6 (b) shows recall ability of the proposed model when the three sequential patterns were memorized.

Figure 6 (c) shows the recall ability of the proposed model for the weighting coefficient β. As shown in this figure, the proposed model whose weighting coefficient β is small can recall the sequential patterns including many common terms.

3.5 Robustness for Damaged Neurons and Noisy Input

Here, we examined the robustness for damaged neurons and noisy input of the proposed model and the conventional IKFMAM-AR-SAP[6]. In these experiments, five random pattern sequences composed of four patterns were memorized. Figure 7 shows the robustness of the proposed model and the conventional IKFMAM-AR-SAP. As shown in these figures, the proposed model has enough robustness for damaged neurons and noisy input as similar as the conventional model.

4 Conclusions

In this paper, we have proposed the Fast Kohonen Feature Map Associative Memory with Area Representation for Sequential Analog Patterns. We carried out a series of computer experiments and confirmed that the proposed model has following features.

(1) It can learn sequential patterns successively.
(2) It can deal with sequential analog patterns including common terms even when the first patterns are same in some sequential patterns.
(3) Its learning speed is higher than that of the conventional Improved Kohonen Feature Map Associative Memory with Area Representation for Sequential Analog Patterns.
(4) It has large storage capacity.
(5) It has robustness for damaged neurons.
(6) It has robustness for noisy input.

References

1. Ichiki, H., Hagiwara, M., Nakagawa, M.: Kohonen feature maps as a supervised learning machine. In: Proceedings of IEEE International Conference on Neural Networks, pp. 1944–1948 (1993)
2. Hattori, M., Arisumi, H., Ito, H.: SOM associative memory for temporal sequences. In: IEEE and INNS International Joint Conference on Neural Networks, Honolulu, pp. 950–955 (2002)
3. Imabayashi, T., Osana, Y.: Implementation of association of one-to-many associations and the analog pattern in Kohonen feature map associative emory with area representation. In: Proceedings of IASTED Artificial Intelligence and Applications, Innsbruck (2008)
4. Iwai, Y., Osana, Y.: Kohonen feature map associative memory with area representation for sequential patterns. In: Proceedings of International Symposium on Nonlinear Theory and its Applications, Vancouver (2007)
5. Shiratori, T., Osana, Y.: Kohonen feature map associative memory with area representation for sequential analog patterns. In: Proceedings of IEEE and INNS International Joint Conference on Neural Networks, Hong Kong (2008)
6. Shiratori, T., Osana, Y.: Improved Kohonen feature map associative memory with area representation for sequential analog patterns. In: Proceedings of International Conference on Artificial Neural Networks, Limassol (2009)

Facial Expression Based Automatic Album Creation

Abhinav Dhall[1], Akshay Asthana[2], and Roland Goecke[3,1]

[1] School of Computer Science, CECS, Australian National University, Canberra, Australia
[2] School of Engineering, CECS, Australian National University, Canberra, Australia
[3] Vision & Sensing, Faculty of Information Sciences and Engineering, University of Canberra, Australia
abhinav.dhall@anu.edu.au, aasthana@rsise.anu.edu.au, roland.goecke@ieee.org

Abstract. With simple cost effective imaging solutions being widely available these days, there has been an enormous rise in the number of images consumers have been taking. Due to this increase, searching, browsing and managing images in multi-media systems has become more complex. One solution to this problem is to divide images into albums for meaningful and effective browsing. We propose a novel automated, expression driven image album creation for consumer image management systems. The system groups images with faces having similar expressions into albums. Facial expressions of the subjects are grouped into albums by the Structural Similarity Index measure, which is based on the theory on how easily the human visual system can extract the shape information of a scene. We also propose a search by similar expression, in which the user can create albums by providing example facial expression images. A qualitative analysis of the performance of the system is presented on the basis of a user study.

Keywords: Automatic album creation, Facial expression analysis, Active Appearance Model, Structural Similarity Index, Image clustering.

1 Introduction

With the advent of low-cost and easy to use consumer level imaging solutions, the number of consumer images has grown incredibly. With these increasing numbers, the management of images has become increasingly cumbersome. Classification of images into albums is a potential solution to this problem. Semantics based albums can be very helpful for effective browsing and retrieval. We propose a method for generating automatic emotion based image albums for better image management and representation. Facial expressions convey powerful discriminating information in facial images and hence form a strong criteria for image clustering. The user can group images 'based' on the emotions the faces in the image convey, such as 'happy', 'excited' or 'neutral' albums. The proposed

K.W. Wong et al. (Eds.): ICONIP 2010, Part II, LNCS 6444, pp. 485–492, 2010.

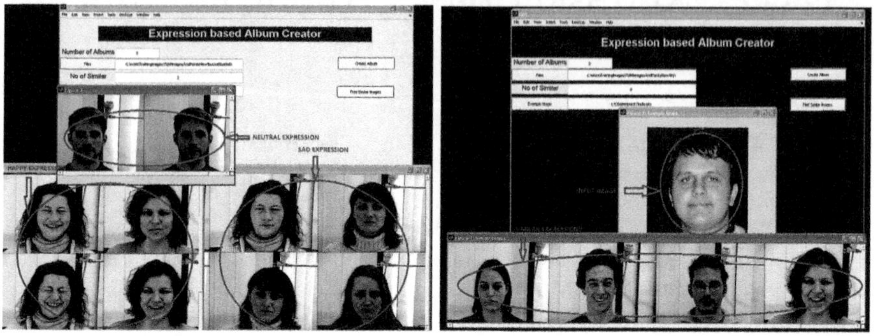

(a) Expression based albums (b) Album by similar expression

Fig. 1. Outputs of the system

system uses *Active Appearance Models* (AAM) [2], which have been widely used for facial expression recognition and related applications in recent years. The *Structural Similarity Index Measure* (SSIM) [15] is used to compare similar facial expressions. The SSIM is based on the theory of the human brain noticing slight changes in the structure of a scene easily and fast. The user decides on the number of clusters/albums. New images are then added to the existing albums via facial expression comparison to the mean expression shapes of the albums. Another option for the user is to input an example image, which depicts a specific facial expression. The system then searches for images, which match the expression of the input image. We term this as 'album creation by similar expression'. In the experiment section, we present a user study on the performance of the facial expression based album creation.

1.1 Related Work

Of the manual/semi-automatic techniques, labelling has been used for long. However, as the image databases grow, managing labels becomes a complex and time consuming task. In [7], time stamping techniques are used to link photographs for effective browsing. In [3], the *Media Browser* exploits the metadata information in the images for tagging faces. Face detection and automatic labelling are used in the *FotoFile* system [9]. In [17], faces are detected and name labels are suggested based on a Gaussian framework to the user to choose from. In [11], the *AutoAlbum* uses time based clustering followed by a hidden Markov model based probabilistic approach for content-based clustering.

Recently, image editing and management tools, such as Google Picasa [6], have been used to manage and group images. The software uses robust face detection and groups all the images of one specific person together, which are then labelled by the user. In [1], face models based on AdaBoost are used to extract facial features and semi-supervised clustering is used to group similar

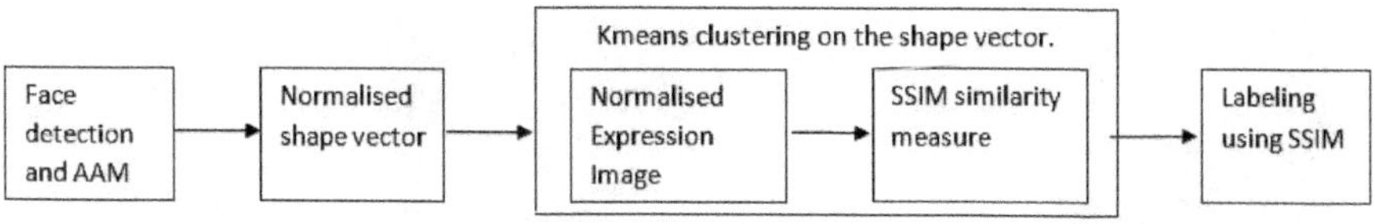

Fig. 2. Block diagram of the system

faces. [18] uses multiple representation spaces viz. faces, background and time of capture as input for mean-shift clustering.

Facial expression recognition is a well researched field. [5] present extensive surveys on facial expression recognition techniques. [10] use AAM for extracting facial features post fitting and machine learning techniques to classify emotions into FACS AU units [4].

Our contribution in this paper lies in creating albums from consumer images on the criteria of human expressions. Emotion based album creation is a useful feature for any image management system. Facial expressions are used to identify similarity among images. We assume that the majority of images contain faces. In existing systems, images of a specific person are grouped based on identity and labelled as an album. We want to explore the emotion/mood aspect of images, which can be a strong grouping criterion. The expression features can be used with the existing date, time and face criteria for album creation. For example a user may want to extract all the happy moments from a particular day's photographs or extracting the surprised expressions of people from an event.

Figure 1 depicts outputs of album by clustering similar expressions. The details of the technique are discussed in depth in the following sections. The paper is divided as follows: Section 2 describes the system and its component, Section 3 shows experimental results and Section 4 provides the conclusions.

2 System

The system constitutes of four major steps, which are described in the following sub sections. Figure 2 depicts the flow of the system.

2.1 Facial Feature Extraction

The face is localised using the Viola Jones [13] face detector, that gives the location of the face. This is used as initialisation for AAM [2] tracking. The AAM are a powerful generative class of methods for modelling and registering non-rigid deformable objects. Their real benefit comes from its compact representation of appearance, which comprises of shape and texture, as well as its rapid fitting to unseen images. We used the AAM fitting method described in [12] for its speed and accuracy.

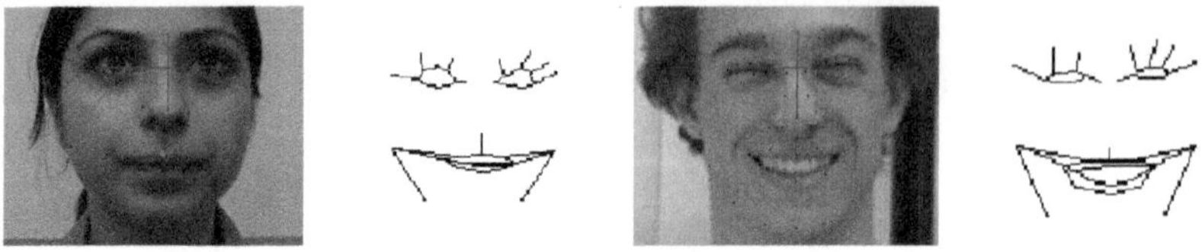

(a) Input (b) EI (c) Input (d) EI

Fig. 3. (a) and (c) are the input faces with red lines representing the landmark points, which are static with respect to local facial moment and donot contribute to the expression. (b) and (d) are the corresponding Expression Images (EI), which are compared for their structural similarity.

2.2 Expression Image Formation

The expression image is a visual map, which depicts the facial expression of the face. AAM fitting gives the shape vectors of the tracked faces, which constitute the landmark points. We then extract the eyes and mouth landmark positions from the shape vector. The new vectors obtained from the shape vectors of different images are then aligned to a common coordinate via translation, scaling and rotation for comparison.

The normalisation of the shape vector is performed by taking the horizontal Euclidean distance between the extreme end points of eye on the left and the right side. And the vertical distance is the Euclidean distance between the nose and upper eye brow. The choice for this normalisation is driven by the static nature of these points with respect to the expressions. The new EI is formed via drawing distance vectors among the new landmark points. The choice of specific landmark points and its corresponding distance vector image is derived from two motivations. One, choosing all points will bias the system towards similar faces. But our aim is different; we wish to find similar facial expressions rather than the images of the same person. Hence, a balanced number of landmark points, which represent enough information for representing the facial expression are chosen. The number and choice of landmark points was calculated with experimentation as on how much person independent the SSIM comparison can become. Two, SSIM works on images hence EI are created from the chosen landmark points. Figure 3 depicts two faces with red lines depicting the static nature of the points chosen for normalisation and their corresponding expression images.

2.3 Structural Similarity Index

We use Structural Similarity index (SSIM) [15] as the distance measure, it is a technique of calculating similarity among two images. SSIM is based on the theory that the human vision system is highly sensitive to changes in structure of the view. Hence, a measure for calculating the structural information change can provide valuable information. In our system, SSIM is used as a distance metric of similarity among EI images. The SSIM metric between two windows w_1 and w_2 on the same size $N \times N$ is given by:

$$SSIM(w_1, w_2) = \frac{(2\mu_{w_1}\mu_{w_2} + c_1)(2\sigma_{w_1 w_2} + c_2)}{({\mu_{w_1}}^2 + {\mu_{w_2}}^2 + c_1)({\sigma_{w_1}}^2 + {\sigma_{w_2}}^2 + c_2)} \quad (1)$$

where μ_{w_1} and μ_{w_2} are the average of w_1 and w_2 respectively. $\sigma^2_{w_1}$ and $\sigma^2_{w_2}$ are the variance of w_1 and w_2 respectively. $\sigma_{w_1 w_2}$ is the covariance between w_1 and w_2. $c_1 = (k_1 L)^2$ and $c_2 = (k_2 L)^2$ are the variables to stabilise the division with weak denominator. L is the dynamic range of the pixel-values. We use this as a comparison of two EI.

2.4 Album Creation and User Options

Semi-unsupervised Album Creation. For image album formation K-Means algorithm is calculated over the expression data. K-means clustering algorithm splits a set of observations into subsets by minimizing the intra-cluster variation. The numbers of image albums k serves as the initial number of clusters for K-Means clustering algorithm where the distance metric is SSIM. Therefore the clustering becomes:

$$\arg_{\mathbf{S}} \min \sum_{i=1}^{k} \sum_{x_j \in \mathbf{S}_i} \mathbf{SSIM}(\mathbf{x}_j, \mu_{\mathbf{i}}) \quad (2)$$

where $(x_1, x_2, .., x_n)$ are the Expression Images EI and μ_i in S_i is the mean EI. The clustering is done on the normalise landmark points of the shape vector, which are used to construct the distance vectors of EI, this is done to keep low dimensionality during clustering. Though the distance comparison is calculated on EI. The mean EI representing each clusters are then compared using the SSIM distance metric with pre-stored labeled EI. The pre-stored EI are labeled into four expressions (Happy, Neutral, Sad, and Excited). This leads to automatic labelling of the albums into the fundamental expression classes. Once a new image arrives it is added to the exiting albums via comparing its closeness to the mean image representing the respective albums.

Album by Example. A user may be interested in finding images, which have the expression similar to a specific facial expression. In this case, the user provides the system with one example image. The user also specifies the number of similar images, which decides the size of this album. The system extracts the EI for the example and the group of images. The example EI is then compared using SSIM with all other images in the group. The similarity distances are then sorted and the user desired number of similar images is selected as an album with respect to the relevance. Figure 4 depicts two examples of this function.

(a) Input (b) Similar Expressions

(c) Input (d) Similar Expressions

Fig. 4. "Album by similar expression" example, (a) and (c) are the input images. (b) and (d) are the corresponding similar expressions.

3 Empirical Experiment and Outputs

Since different users may have different perception about an expression hence analysing the correct clustering performance is a non trivial task. To validate the performance we created a test set of sixty images from the FEEDTUM [14] and LFW [8] databases. A total of fifteen human users were asked to judge the album creation performance by figuring out the the images, which seem to have a different expression and do not belong to the album created. The average total error classification rate came out to be 13.7%. We also compare our system with fuzzy clustering algorithm. Figure 5 shows the outputs of the systems. The Figure 1 displays the experimental GUI of the system. In 1(a) the images are from the FEEDTUM database [14]. The three sub windows in the figure are the albums created after SSIM based clustering. Please note that the system groups faces of similar facial expressions into one set.

Figure 4, is the experiment on images from the LFW database [8]. Figure 4(a)is the user example input with a happy expression. Figure 4(b) is the album of top matching expressions with decreasing relevance from left to right. Similarly, Figure 4(c) is face with smiling expression and Figure 4(d) are the similar expressions. Figure 1(b) depicts album by similar expression example. The user inputs an image, which contains an example expression. The user also specifies the number of similar images desired in the album. This input serves as the number of similar images to be presented. The larger eclipse shows the searched similar images and the upper one is the user input image.

Fig. 5. Sample result on images from FEEDTUM [14] database after executing our system and Fuzzy clustering algorithm in the upper and lower box respectively

4 Conclusions and Future Work

We propose a novel system, which can categorise images into albums on the basis of facial expression analysis, for effective image browsing and searching. It has applications in modern day image management systems such as Google Picasa [6] and Flickr [16]. The system uses AAM for facial feature extraction, a shape vector is extracted and normalised, and an EI is formed, which represents the facial expression of the image. Then, the SSIM is used as a distance metric for similarity, to cluster similar facial expression images together. The user also has the option to search for a particular image and form an album based on it ("creation by similar expression"). Future work is to add illumination invariance before AAM fitting, so as to have more robust fitting. Experimenting with a robust generic AAM tracker can also increase the performance of the system. Another potential area is exploring robust methods for unsupervised clustering.

References

1. Chu, W., Lee, Y., Yu, J.: Visual language model for face clustering in consumer photos. In: MM 2009: Proceedings of the seventeen ACM international conference on Multimedia. ACM, New York (2009)

2. Cootes, T.F., Edwards, G.J., Taylor, C.J.: Active Appearance Models. In: Burkhardt, H., Neumann, B. (eds.) ECCV 1998. LNCS, vol. 1407, p. 484. Springer, Heidelberg (1998)
3. Drucker, S.M., Wong, C., Roseway, A., Glenner, S., Mar, S.D.: MediaBrowser: reclaiming the shoebox. In: AVI 2004: Proceedings of the working conference on Advanced visual interfaces, pp. 433–436. ACM, New York (2004)
4. Ekman, P., Friesen, W.V.: The Facial Action Coding System: A Technique for the Measurement of Facial Movement. In: Consulting Psychologists (1978)
5. Fasel, B., Luettin, J.: Automatic Facial Expression Analysis: A Survey. PR 36(1) (2003)
6. Google. Google Picasa, `http://picasa.google.com/`
7. Graham, A., Garcia-Molina, H., Paepcke, A., Winograd, T.: Extreme Temporal Photo Browsing. In: Visual Interfaces to Digital Libraries [JCDL 2002 Workshop]. Springer, Heidelberg (2002)
8. Huang, G.B., Ramesh, M., Berg, T., Learned-Miller, E.: Labeled Faces in the Wild: A Database for Studying Face Recognition in Unconstrained Environments. University of Massachusetts, Amherst, Technical Report (2007)
9. Kuchinsky, A., Pering, C., Creech, M.L., Freeze, D., Serra, B., Gwizdka, J.: FotoFile: a consumer multimedia organization and retrieval system. In: CHI 1999: Proceedings of the SIGCHI conference on Human factors in computing systems. ACM, New York (1999)
10. Lucey, S., Matthews, I., Hu, C., Ambadar, Z., Torre, F.d.l., Cohn, J.: AAM Derived Face Representations for Robust Facial Action Recognition. In: FGR 2006: Proceedings of the 7th International Conference on Automatic Face and Gesture Recognition. IEEE Computer Society, Los Alamitos (2006)
11. Platt, J.C.: AutoAlbum: Clustering Digital Photographs using Probabilistic Model Merging. In: CBAIVL 2000: Proceedings of the IEEE Workshop on Content-based Access of Image and Video Libraries. IEEE Computer Society, Los Alamitos (2000)
12. Saragih, J., Göcke, R.: Learning AAM fitting through simulation. Pattern Recognition 42(11) (2009)
13. Viola, P.A., Jones, M.J.: Rapid Object Detection using a Boosted Cascade of Simple Features. In: CVPR (1) (2001)
14. Wallhoff, F.: Facial Expressions and Emotion Database (2006), `http://www.mmk.ei.tum.de/~waf/fgnet/feedtum.html`
15. Wang, Z., Bovik, A.C., Sheikh, H.R., Student Member, Simoncelli, E.P., and Senior Member.: Image quality assessment: From error visibility to structural similarity. IEEE Transactions on Image Processing, 600–612 (2004)
16. Yahoo. Flickr, `http://www.flickr.com`
17. Zhang, L., Chen, L., Li, M., Zhang, H.: Automated annotation of human faces in family albums. In: MULTIMEDIA 2003: Proceedings of the Eleventh ACM International Conference on Multimedia. ACM, New York (2003)
18. Zhang, T., Xiao, J., Wen, D., Ding, X.: Face based image navigation and search. In: MM 2009: Proceedings of the Seventeen ACM International Conference on Multimedia. ACM, New York (2009)

Age Classification Combining Contour and Texture Feature

Yan-Ming Tang[1] and Bao-Liang Lu[1,2]

[1] Center for Brain-Like Computing and Machine Intelligence
Department of Computer Science and Engineering
[2] MOE-Microsoft Key Laboratory for Intelligent Computing and Intelligent Systems Shanghai Jiao Tong University
800 Dong Chuan Rd., Shanghai 200240, China
bllu@sjtu.edu.cn

Abstract. Age classification based on computer vision has widespread applications. Most of previous works only utilize texture feature or use contour and texture feature separately. In this paper, we proposed an age classification system that integrate contour and texture information. Besides, we improve the traditional Local Binary Pattern(LBP) feature extraction method and get pure texture feature. Support Vector Machines with probabilistic output (SVM-PO) is used as individual classifiers. Then we use combination mechanism based on fuzzy integral to merge the output of different classifiers. The experiment results show pure texture feature outperforms other features and it can be well combined with contour feature.

Keywords: Age Classification, Contour Feature, Texture Feature, Located Local Binary Patterns, Fuzzy Integral.

1 Introduction

Age classification has a lot of applications, such as supervision of minors, demographics, commercial advertisement and so on. Most of previous researches only use texture feature [1,2] or use contour features and texture features separately [3,4]. We find both the shape of faces and skin roughness can help determine a person's age. Fig. 1(a) shows faces which can be discriminated by contour feature. All 8 images have soft skins, but the upper four faces are close to circles while the lower four faces are close to ovals. Fig. 1(b) are example of faces that can be distinguished by texture feature. We can see wrinkles on the forehead and at the corner of eyes clearly.

Seeing the above example, it is natural to expect better performance by combining contour feature and texture feature. Although [3] also use these two features, they didn't use them at the same time. First, contour feature is adopted to determine whether the facial image is a child or not, then it's classified as young people or elderly people according to texture feature. In this paper, we propose an age classification system that combing features together and achieve a performance which is comparable to humans.

The remaining part of the paper is organized as follows: in section 2, the age classification system we proposed is introduced in details. In section 3, we describe the

K.W. Wong et al. (Eds.): ICONIP 2010, Part II, LNCS 6444, pp. 493–500, 2010.

Fig. 1. Classify the facial image by different features: (a)Classify by contour feature; (b)Classify by texture feature

features extraction method we use in our system. Experiments and analysis are conducted in section 4, followed by conclusion and discussion in section 5.

2 Age Classification System

Fig. 2 shows the whole process of our system. Before feature extraction, we need preprocess the images. The initial image is cropped into two sizes to accommodate two feature extraction methods. To extract contour feature, the image should include the whole face, because the position of chin is very important. To extract texture feature, including the organs is enough, and the advantage is that we can avoid the impact of backgrounds.

Feature extraction methods are very important in pattern recognition. Contour feature is easy to modeling comparatively speaking, while there are many methods to describe texture feature, like Local Binary Patterns, Gabor Feature, Local Gabor Binary Mapping Pattern(LGBP) [5]. We tried several methods of them and find LBP is the most stable and efficient. Then several classifiers depend on different features are trained with SVM-POs, for the preparation of classifier combination.

At last, we combine the output of SVM-POs and get the final result. Almost all of the combination mechanisms belong to two categories of information integration techniques. One is to combine the features before classification, the other is to combine the results of classifiers. All the combination methods we use in our experiment belong to the latter categories, for example, choquet fuzzy integral. The probabilistic outputs are combined into a single composite score with trained fuzzy measure or hierarchical classifiers, and the class with highest probability will be output.

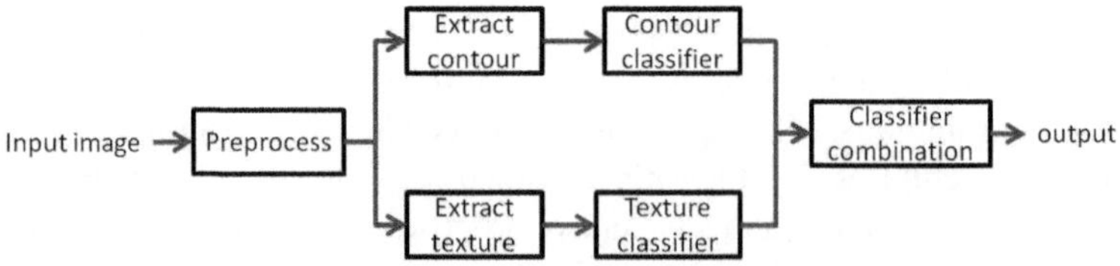

Fig. 2. The proposed age classification system by combining contour and texture classifiers

3 Feature Extraction

In this section, we will briefly introduce the feature extraction methods we use and discuss their characteristic.

3.1 Contour Feature

Kwon and Lobo did researches on age classification first. They consulted studies in cranio-facial research, art and theatrical makeup, plastic surgery and found with the growth of a people, the shape of head turns from circle to oval. So they put forward utilizing the proportion of distance between organs to decide whether a facial image belongs to child or adult [3]. We also use this information in our experiment, but we do not calculate the proportion, instead, we more accurately use 58 points to describe the contour of a face.

To detect the contour, we adopt Active Appearance Model (AAM) [6], a statistical model which derives from Active Shape Model (ASM) [7]. Before using AAM, we should normalize the face, otherwise the detection result will be imprecise. We first detect the position of two eyes [8], then rotate and scale the face to locate the eyes at the same position. After normalization, AAM can easily find the contour with 58 points $P_1, P_2, \ldots, P_{58}$, as show in Fig. 3. We don't use AAM to find the position of calvaria because it will be affected by hair seriously.

Then we just stretch the x and y coordinates and get a 116 dimension contour feature vector $\{P_{1x}, P_{1y}, \ldots, P_{58y}\}$, which is also the base for texture feature extraction.

3.2 Texture Feature

Texture feature has better performance than contour feature under many circumstances. Many researches have been done on this and LBP has been proved powerful on texture description. The LBP operator value can be calculated as Fig. 4.

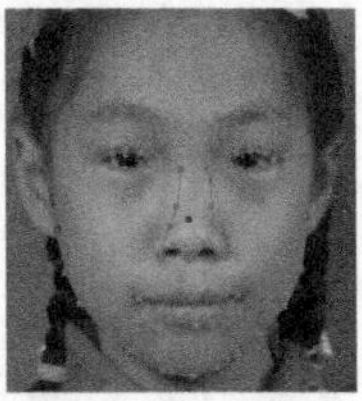

Fig. 3. AAM detection result

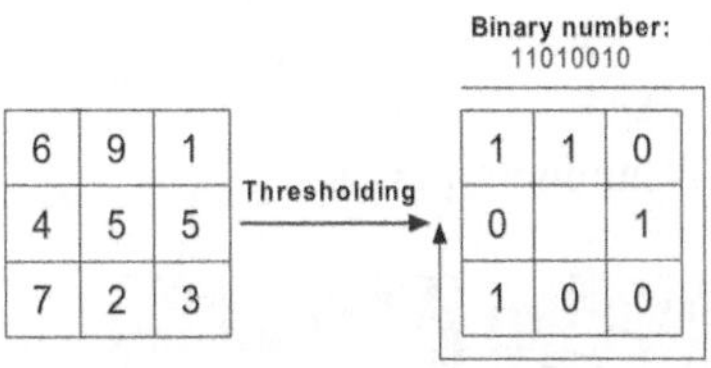

Fig. 4. Illustration of the LBP value computation

We apply LBP operator on every pixel and divide the image into m non-overlapping rectangular regions $\{R_0, R_1, \ldots, R_m\}$, the histogram of j-th region is:

$$H_j = \{h_{0,j}, h_{1,j}, \ldots, h_{59,j}\} \tag{1}$$

where 59 is the number of bins for uniform LBP operator. At last, we concatenate all H_j together and get the final LBP feature:

$$V = \{H_0, H_1, \ldots, H_{m-1}\} \tag{2}$$

There are two ways of dividing the image into regions. The traditional way is to cut the image into $n \times n$ regions equally (and get the LBP_n feature), as shown in Fig. 5(a). This way of dividing is easy, because we do not need to know the position of organ and it performs good too. However, it is not pure texture feature. As we can see the mouth of right image is upper than that of the left one. So the same region doesn't correspond to the same part of face, and the final LBP_n vector will also contains contour information. We can extract texture feature more explicitly with the contour information, and we call it Located Local Binary Patterns(LLBP).

Fig. 5(b) shows how we decompose the image in our experiment. Skin around eyes is the most important part of face in age classification [2], so we first locate the regions of eyes by points of canthi. Here we set a fixed height of eye regions to prevent from getting too narrow region caused by squinting. The process of mouse and nose regions is similar to that of eyes, and the remaining regions are divided averagely according to the number of regions. We can certainly divide the image in a better way, but our method has already surpassed all others in the experiment. Before classification, we should zoom the regions to have same sizes, otherwise, the obtained feature will still contain contour information.

Bin Xia proposed LGBP feature in [5]. They use gabor filters on image first [9], then extract LBP feature on transformed images. Before classification, feature dimension will be decreased on every region. We also implemented this method as a comparison.

4 Classifier Combination

Originally SVM only predict class labels, we can use strategies like Majority Voting Rules or Borda Count to integrate the outputs, but it's a rough estimation. In our experiment, the SVM output probability for every class instead of single class label, and the combination results are better. To get probabilistic output, our goal is to estimate

$$p_i = p(y = i|\text{x}), i = 1, \ldots, k \tag{3}$$

where k is number of classes.

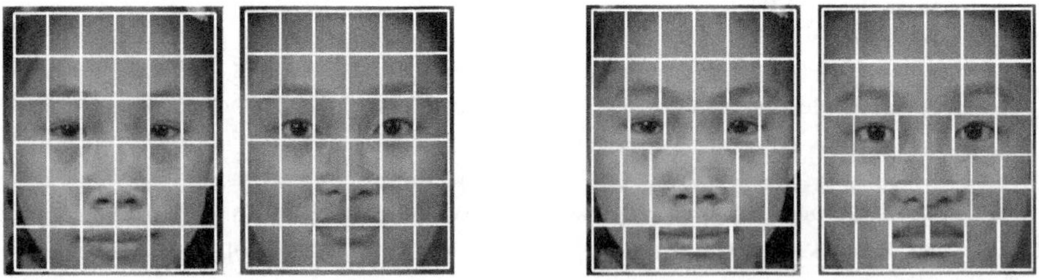

(a) (b)

Fig. 5. Two divide method

Since we use one-against-one strategy, we can get the probability of multi-class problem from pairwise class probabilities $r_{ij} \approx p(y = i|y = i \text{ or } j, \mathrm{x})$, which is estimated as Eq. 4 proposed in [10].

$$r_{ij} \approx \frac{1}{1 + e^{A\hat{f}+B}} \tag{4}$$

where A and B are estimated by minimizing the negative log-likelihood function using training data and decision values $\hat{f}$.

Then p_i can be obtained from all the r_{ij}'s solving the following optimization problem [11]:

$$\min_{\mathbf{P}} \frac{1}{2} \sum_{i=1}^{k} \sum_{j:j\neq i} (r_{ji}p_i - r_{ij}p_j)^2 \quad \text{subject to} \sum_{i=1}^{k} p_i = 1, p_i \geq 0, \forall i. \tag{5}$$

Noticing the equality

$$p(y = j|y = i \text{ or } j, \mathrm{x}) \cdot p(y = i|\mathrm{x}) = p(y = i|y = i \text{ or } j, \mathrm{x}) \cdot p(y = j|\mathrm{x}), \tag{6}$$

then the objective function can be reformulated as

$$\min_{\mathbf{P}} \frac{1}{2} \mathbf{P}^T Q \mathbf{P} \tag{7}$$

where

$$Q_{ij} = \begin{cases} \sum_{s,s\neq i} r_{si}^2 & if\ i = j, \\ -r_{ji}r_{ij} & if\ i \neq j. \end{cases} \tag{8}$$

Simple combination rules like sum or product rule were proved efficient in [12]. There are a bit more complicated methods that need training, such as weighted sum, hierarchical classifiers. Here we use another widely used combination strategy: fuzzy integral.

Fuzzy integrals are integrals relies on the concept of fuzzy measures. Let $X = \{x_1, x_2, \ldots, x_n\}$ be a finite set and let $P(X)$ indicates the power set of X. Then a fuzzy measure g over set X is defined as:

Definition 1. $g : P(x) \rightarrow [0, 1]$ *such that:*
(1) $g(\emptyset) = 0, g(X) = 1;$
(2) $A \subseteq B \Rightarrow g(A) \leq g(B)$

The choquet fuzzy integral we use can be based on any fuzzy measure[13]. Given an unknown sample T, the confidence of T belongs to class c_j can be calculated by $C_g(h_j(x_1), \ldots, h_j(x_n))$, denoted as $C_g^j(T)$, where $h_j(x_i)$ is the confidence of T belongs to class c_j given by classifier x_i. The calculation of $g(h_i)$ can be solved by quadratic programming [14].

5 Experiments

Our experimental data come from frontal faces of BCMI-Omron age database. We set 18 years old as the boundary of adults and children, because laws in most countries do

Table 1. Experimental data

Training data		Test data	
Age Group	Number	Age Group	Number
≤18	215	≤18	71
19∼23	221	19∼23	73
24∼50	219	24∼50	74
≥50	149	≥50	50
Total	804	Total	268

like this. So this kind of set has practical value. The other two boundaries are set as 23 and 50 years old to make the database has a good distribution. (See Tab. 1). One-fifth of the data are chosen randomly as testing data.

To have a comparison, we also asked three participants to classify the data. The participants did the test twice. At the first time, they directly classify the image according to their life experience. We find the two younger participants (22 years old both) didn't do well (with precision 67.16% and 70.52%), while the elder(50 years old) reached 76.12%, as expected [15]. So they were asked to do the test again. This time, they saw the training data before classification and reached 81.72% on average.

Tab. 2 shows the accuracy of different feature extraction methods. The classifiers are all SVM with RBF kernel and probability output. For Kwon's system, we set 18 years old as the boundary between children and adults. From the results, we can see the LLBP we proposed outperforms all other methods. LBP_6 and LBP_7 also performs good, because they contain both texture and contour information, as mentioned before. In addition, although contour feature get a low accuracy relatively speaking, it's still good than the easiest gray feature, so it's still useful for combination.

Then, we take a look at the combination results. Besides fuzzy integral, weighted sum, product rule and hierarchical classifiers are also chosen in our experiment. Tab. 3 shows 6 combinations of different classifiers. We first compare different combination methods, we can see from the results that fuzzy integral is the best among them, slightly better than weighted sum.

At last, we compare the combination results of different sets. From Tab. 2 and Tab. 3 we can see the integration of LBP_6 and LBP_7 doesn't make much progress, the reason is that these two features contain similar information, so they don't complement each others. To prove LLBP we propose is really better, we combine LLBP and LBP_6 with contour feature separately. As we expected, the raise of LLBP is greater than LBP_6 on fuzzy integral and weighted sum, but they both get decreased on hierarchical classifiers and product rule. This is because the performance of contour classifier is not very good, so it becomes a drag. We then combine LLBP with LBP_n. This time, the precision get increased a lot, because traditional LBP feature is nearly as strong as LLBP and it contains contour information at the same time. Although it seems that the performance of combination just get improved slightly, the best result which combines contour feature, LBP and LLBP together through fuzzy integral achieves 80.23%, only a bit lower than human's decision, so we think it's an encouraging result.

Table 2. Accuracy of different feature extraction method

Methods	Accuracy	Methods	Accuracy
Gray	60.82%	Kown's	74.63%
Contour	69.03%	LGBP	76.12%
LBP_6	77.24%	LLBP	**77.99%**
LBP_7	77.24%	Human	81.72%

Table 3. Classifier combination results

	Hierarchical	Product	Weighted Sum	Fuzzy Integral
LBP_6+LBP_7	77.61%	77.61%	77.99%	77.99%
Contour+LBP_6	76.49%	77.24%	77.61%	77.61%
Contour+LLBP	77.24%	77.61%	78.73%	78.73%
LBP_6+LLBP	78.73%	79.10%	79.48%	79.48%
LBP_7+LLBP	78.30%	79.10%	79.10%	**79.48%**
Contour+LBP_6+LLBP	79.10%	79.48%	79.48%	**80.23%**

6 Conclusions and Future Work

We improve the traditional LBP feature extraction method in this paper, pure texture feature is extracted by dividing the image more reasonably and it outperforms all the baselines. Moreover,we integrate contour feature with texture feature by fuzzy integral and the accuracy of age classification is increased. Here, we still divide the image into rectangles. In fact, the regions can be irregular shape and will fit the shape of face better. A further extension of our work is to utilize hair information, we plan to extract the color and hairstyle information to get better performance.

Acknowledgments

This work was partially supported by the National Natural Science Foundation of China (Grant No. 60773090 and Grant No. 90820018), the National Basic Research Program of China (Grant No. 2009CB320901), the National High-Tech Research Program of China (Grant No. 2008AA02Z315), and the Science and Technology Commission of Shanghai Municipality (Grant No. 09511502400).

References

1. Iga, R., Izumi, K., Hayashi, H., Fukano, G., Ohtani, T.: A gender and age estimation system from face images. In: SICE Annual Conference in Fukui, pp. 202–209 (2003)
2. Lanitis, A.: On the significance of different facial parts for automatic age estimation. In: 14th International Conference on Digital Signal Processing, vol. 2, pp. 1027–1030 (2002)
3. Kwon, Y., Lobo, N.: Age classification from facial images. Computer Vision and Image Understanding 74(1), 1–21 (1999)
4. Horng, W., Lee, C., Chen, C.: Classification of age groups based on facial features. Tamkang Journal of Science and Engineering 4(3), 183–192 (2001)

5. Xia, B., Sun, H., Lu, B.: Multi-view gender classification based on local gabor binary mapping pattern and support vector machiness. In: IJCNN 2008, pp. 3388–3395 (2008)
6. Edwards, G., Taylor, C., Wolfson, C.: Interpreting face images using active appearance models. In: Int. Conf. on Face and Gesture Recognition, pp. 300–305 (1998)
7. Cootes, T., Taylor, C., Cooper, D., Graham, J.: Active shape models–their training and application. Computer Vision and Image Understanding 61(1), 38–59 (1995)
8. Viola, P., Jones, M.: Rapid object detection using a boosted cascade of simple features. In: Accepeted Conference On Computer Vision And Pattern Recognition, vol. 1, pp. 511–518 (2001)
9. Lades, M., Vorbruggen, J., Buhmann, J., Lange, J., Malsburg, C., Wurtz, R., Konen, W.: Distortion invariant object recognition in the dynamic link architecture. IEEE Transactions on Computers, 300–311 (1993)
10. Lin, H., Lin, C., Weng, R.: A note on platts probabilistic outputs for support vector machines. Technical report, Department of Computer Science, National Taiwan University (2003)
11. Wu, T., Lin, C., Weng, R.: Probability estimates for multi-class classification by pairwise coupling. The Journal of Machine Learning Research 5, 975–1005 (2004)
12. Heerden, C., Barnard, E.: Combining multiple classifiers for age classification. In: 20th Annual Symposium of the Pattern Recognition Association of South Africa, pp. 59–64 (2009)
13. Murofushi, T., Sugeno, M.: An interpretation of fuzzy measures and the choquet integral as an integral with respect to a fuzzy measure. Fuzzy Sets and Systems 29(2), 201–227 (1989)
14. Grabisch, M.: Fuzzy integral for classification and feature extraction. In: Fuzzy Measures and Integrals:Theory and Applications, pp. 415–434 (2000)
15. Miyamoto, N., Jinnouchi, Y., Nagata, N., Inokuchi, S.: Subjective age estimation system using facial images- how old we feel compared to others. In: IEEE International Conference on Systems, Man and Cybernetics Conference Proceedings, vol. 964, pp. 3449–3453 (2005)

A Salient Region Detector for GPU Using a Cellular Automata Architecture

David Huw Jones, Adam Powell,
Christos-Savvas Bouganis, and Peter Y.K. Cheung

Imperial college London, Electrical and Electronic Engineering
Princes Gardens, London, SW7 2AZ

Abstract. The human visual cortex performs salient region detection, a process critical to the rapid understanding of a scene. This is performed on large arrays of locally interacting neurons that are slow to simulate sequentially. In this paper we describe and evaluate a novel, bio-inspired, cellular automata (CA) architecture for the determination of the salient regions within a scene. This parallel processing architecture is appropriate for implementation on a graphics processing unit (GPU). We compare the performance of this algorithm against that of CPU implemented salient region detectors. The CA algorithm is less subject to variation due to changing scale, viewpoint and illumination conditions. Also due to its GPU implementation, this algorithm is able to detect salient regions faster than the CPU implemented algorithms.

Keywords: Cellular automata, Graphic Processing Units, Evolution, Saliency, Low-level vision.

1 Introduction

The human vision system incorporates fast image processing algorithms that are a function of low-level, local interactions between large quantities of parallel processing neurons. These include orientation, frequency and color filters, edge, as well as motion and salient region detection. We have had little success imitating their function to a similar scale or speed on sequential systems[1]. New parallel processing platforms such as the graphic processing unit (GPU) have more comparable architectures to the biological visual cortex that may allow us to simulate its function faster and on larger scales. However designing large-scale, locally interacting parallel systems such that they perform some function or display certain complex properties is computationally difficult.

Saliency is the measure of object conspicuity within the scene. This is determined by the V4 region of the primary visual cortex and is used to direct the rapid movements of the eye (saccades). In this paper we will show that a large array of low-level processes can be effectively designed on, and for, a GPU to determine the salient regions of any given scene. We will then compare the speed and transformation invariance of this algorithm to that of three other

K.W. Wong et al. (Eds.): ICONIP 2010, Part II, LNCS 6444, pp. 501–508, 2010.

salient region detectors: the classic Itti and Koch algorithm[2] which uses a combination of color, intensity and orientation filters, feature maps and local centre-surround difference calculations; a directed, weighted graph-based visual saliency (GBVS)[3] approach which uses a Markov chain analysis to detect conspicious nodes; and a recently published algorithm that uses local entropy analysis to determine regions for attention based on information maximization (AIM)[4].

2 Low-Level Architecture for Image Processing

One class of locally interacting arrays of small processors that have been used to simulate biological computation is cellular automata. Cellular automata (CA) are dynamic systems in which space and time are discrete. CA consist of a number of identical cells in an array. Each cell can be in one of a number of states. The next state of each cell is determined at discrete time intervals according to the current state of the cell, the current state of the neighbouring cells and a next-state rule that is identical for each cell.

Von Neumann developed CA to study self-reproducing systems. Since then CA have been extensively used to study or mimic the capabilities of biological systems [5]. CA have also been used to mimic some of the low-level capabilities of biological vision systems. These include edge detection[6] and feature extraction[7].

We propose to use a CA architecture with a set of bio-inspired image-processing filters local to each cell. We will use an evolutionary algorithm to determine how each cell calculates its next state from the output of these filters.

2.1 Proposed CA Architecture

The CA will be a bounded rectangular array that has the same dimensions as the input image. The luminescence of each pixel of the input image is the initial state (at time $t = 0$) of each cell.

Saliency is an output of the V4 region of the primary visual cortex. Between this stage and the optic nerve are the V1 and V2 stages. These perform orientation, gradient and edge detection. We will use similar filters (applied locally to a 10x10 neighbourhood about each cell) as inputs to each cell:

1. Orientation filters. Calculate the minimum, ψ_{min}, maximum, ψ_{max}, and standard deviation, ρ_ψ, of the response of a bank of five Gabor filters defined by the convolution kernel:

$$k(i,j) = e^{-\dfrac{i^2+j^2}{20}} \cos\left(\frac{2\pi}{5}(i\cos\theta + j\sin\theta)\right) \qquad (1)$$
$$i,j \in [-5,5],\ \theta \in \{0,30,60,90,120\}$$

2. Gradient filters. The mean response of ten gradient responses, $\bigtriangledown(x, y, \theta)$ about each cell $c_{x,y}$.

$$\bigtriangledown(x,y) = \sqrt{(x'c_{x+x',y+y'})^2 + (y'c_{x+x',y+y'})^2} \tag{2}$$
$$x' = 5\cos(\theta) \;\; y' = 5\sin(\theta)$$
$$\theta \in \{0, 18, 36, 54, 72, 90, 108, 126, 144, 162\}$$

3. The standard deviation, ρ_c, and mean,$\bar{c}$, state values within the cell neighbourhood.

The scale and coefficients of each of these filters has been chosen for three reasons: to maximise the difference in response to unique and common input sources, to achieve an approximate conservation of sum pixel states across the automata, and to achieve fast processing times. The first iteration ($t = 1$) will be determined by a sum of five of these variables,λ_i (we exclude $\bar{c}$ to improve illumination invariance) weighted by various coefficients, k_i. We are going to use an evolutionary algorithm to choose values for k_i.

$$c_{t=1,x,y} = \sum_i k_i \lambda_i (c_{t=0,x,y}) \tag{3}$$

Further iterations are determined by the previous iteration, the six variables and a decay function α.

$$c_{t+1,x,y} = c_{t,x,y} + \alpha \sum_i k_i \lambda_i (c_{t,x,y}), \;\; \alpha = e^{-\frac{t}{2T}} \tag{4}$$

Where T is the total number of iterations to be performed by the CA. The decay function ensures the CA converges and the location of any output peaks is at the centre of its corresponding inputs. By treating each cell in the neighbourhood equally we ensure some degree of rotation invariance.

2.2 Implementation

As well as being an appropriate medium on which to mimic biological systems, CA are particularly suited for implementation on graphics processing units (GPU). The GPU architecture allows us to execute many thousands of parallel threads, but each thread must run the same code. This is comparable to a CA, in which each of many thousands of cells run the same program synchronously.

3 The Evolutionary Design of Cellular Automata for Image Processing

As saliency is such a subjective measure there may be one or more effective solutions, making it an interesting task for evolutionary algorithms. We need an evolutionary algorithm appropriate for exploring search spaces with many possible solutions. Here we describe the fitness function and mutation strategy for the coefficients, k_i, of equation (4).

3.1 Training Set and Fitness Function

To train our CA we create a set of test input grayscale images and use their pixel saturations to set the initial states of each cell within the CA. We then repeatedly iterate the CA, each cell using equations (3), (4) and evolved values for k_i to determine its next state. The final states $c_{t=T,x,y}$ of each cell within the CA form a map of the salient regions in the image.

The test images we use are chosen for their different scale of features and subject matter. For each image, I, in our training set we determine a saliency map from Itti's algorithm. The fitness of the evolved solution is determined as a sum-of-squared differences between the final state of the CA and the saliency map. Note that though this constrains the potential of our algorithm to detect salient regions to that of the performance of Itti's algorithm, we expect our algorithm to outperform it by other metrics, such as speed, translation and illumination invariance.

By loading the training images and their corresponding salient maps on the device, we can test each solution with minimal host-device or device-host data transfer; significantly speeding up the training stages of the evolution algorithm.

3.2 Mutation Algorithm

We use a variant of the HereBoy algorithm [8] for the mutation of this CA because its version of constrained simulated annealing is particularly suited to exploring search spaces with many possible solutions.

The HereBoy algorithm uses a population of two solutions: the best solution so far and a mutation of this solution. In most cases, if the mutated solution performs better than the current solution, the current is replaced with the mutated. Otherwise the mutated solution is discarded and another mutation of the best solution is evaluated.

The HereBoy algorithm first attempts to determine the general structure of the solution with high mutation rates, then tries to refine it with lower mutation rates. Thus the probability of mutating each coefficient k_i is function of predefined limits ($p_{m,min}, p_{m,max}$), the fitness of the solution and the expected maximum fitness, f_{max}:

$$p_m = \frac{f_{max} - f}{f_{max}} \times (p_{m,max} - p_{m,min}) + p_{m,min} \tag{5}$$

In order to ensure the evolutionary algorithm doesn't settle on local maxima of the search space, occasionally the algorithm will select the sub-optimal solution to mutate. The probability p_r of this occurring is determined according to the same formula for determining p_m. We determine the limits of (5) by trial and error.

3.3 Evolved Solution

This algorithm took approximately three days[1] to find the following maximum in the search space:

$$c_{t=1,x,y} = 0.95\rho_c + 0.1 \bigtriangledown + 0.75\psi_{min} - 1.0\psi_{max} - 0.9\rho_\psi \quad (6)$$

$$c_{t+1,x,y} = c_t + \alpha(0.95\rho_c + 0.77\bar{c} + 0.1 \bigtriangledown + 0.75\psi_{min} - 1.0\psi_{max} - 0.9\rho_\psi) \quad (7)$$

Figure 1 shows the salient regions detected by this algorithm over 30 iterations. Figure 2 compares the salient regions detected by this algorithm with those detected by other salient region detectors. The input images for both figures are from a set of images used to evaluate the performance of the AIM algorithm. None of these images are part of the evolutionary algorithm training set.

Fig. 1. Example output of iterations t=5,10,15,20

Fig. 2. Input images from [4], then salient regions detected by Itti, GBVS, AIM and CA algorithms

4 Performance of CA Salient Region Detectors

The salient regions detected by the CA are comparible in location and detail to those of the Itti and GBVS algorithms; the detail returned by the AIM algorithm is much greater than the others but at a cost to the speed of the algorithm. We now need to test the sensitivity of this algorithm to transformations of the input, that is, the salient regions of a scene should be the same regardless of the point of view, illumination conditions and quality of the image capture device.

[1] On an Intel Core 2, 2.8Ghz with an Nvidia C2050 GPU.

4.1 Performance Metric

We test our salient region detector using a standard sequence of images provided by Mikolajczyk[9]. This set includes various textured and structured images and their transforms under illumination, scale, viewpoint, JPEG compression and blur. This test was designed to compare detectors that return feature locations and descriptions, so we have adapted the test to compare region detectors instead.

Each set of images consists of the same scene subject to an increasingly large transformation. Where this affects the image geometry this transformation is described by a known affine transformation, $(\mathbf{A}, \overline{b})$. The error introduced by the transformation is calculated as the sum-of-squared differences between the salient regions detected in the original image and the transformed image.

1. The salient regions, $s(I)$ of both the original image I_0 and the transformed image I_t are determined.
2. The affine transformation, $\overline{x_t} = \mathbf{A}\overline{x} + \overline{b}$, of I_t is used to correct for the distortion of I_t, giving $s(I_t)'$.
3. The affine transformation of I will have affected the size and location of the scene present in I_t. The part of I that is visible in I_t is calculated by taking the inverse affine transform of the corners, c of the bounding box of I_0:

$$\begin{pmatrix} c_x^t \\ c_y^t \\ 1 \end{pmatrix} = \begin{pmatrix} \mathbf{A}^{-1} & -\mathbf{A}^{-1}\overline{b} \\ 0 & 1 \end{pmatrix} \begin{pmatrix} c_x \\ c_y \\ 1 \end{pmatrix} \tag{8}$$

4. The sum of squared differences,e_t between $s(I_0)$ and $s(I_t)'$ is calculated for every pixel in this common part.
5. To compensate for the reduced area,a_t, (and thus potential error) of I_t, the error introduced by the transformation,e is determined as $e = e_t a_0 / a_t$.

4.2 Results

Figure 3(a) shows the speed of this algorithm compared to that of the Itti, AIM and GBVS algorithms. Note that the CA algorithm speed is primarily executed on a GPU (Nvidia C2050) whereas the other algorithms are executed, using MATLAB code provided by the authors, on a 2.8Ghz CPU. Figures 3(b-f) compare the sensitivity of this algorithm to various input transformations. From these results we can see the CA salient region detector performs well under conditions of varying illumination, scale and viewpoint. However it does not perform as well under compression and blur variance. We believe this is due to the dependence of the algorithm on changes to local gradient and orientation and might be corrected by the use of various de-blurring convolution filters. Furthermore we believe the error due to viewpoint change could be improved by the use of circular neighbourhoods about each cell, instead of the rectangular neighbourhoods the algorithm currently uses. The speed of the CA algorithm decreases relative to image size faster than the Itti and GBVS algorithm. This is because as the CA gets larger, it also requires more iterations to distribute feature information across the automata.

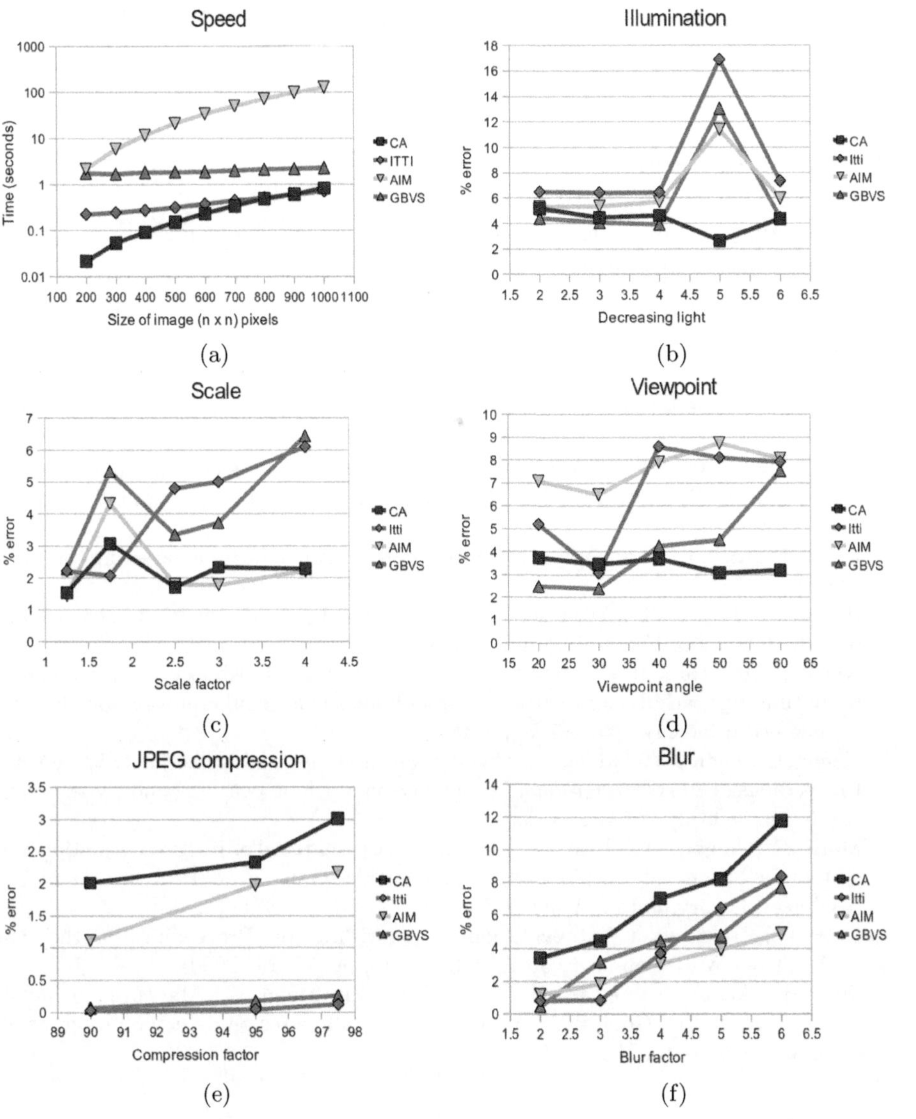

Fig. 3. (a) Speed vs Image size; (b-f) Error due to various input image transforms

5 Conclusions

The CA on a GPU architecture we propose is a biologically plausible emulation of the architecture of the visual cortex and runs considerably faster than CPU alternatives. With the assistance of an evolutionary design algorithm we have designed this CA to perform salient region detection. This salient region detector has proved more invariant to illumination, scale and viewpoint transforms than three other salient region detectors (Itti, AIM and GBVS). However this algorithm was more susceptible to variation under compression and blur transformations.

Acknowledgements

The authors acknowledge the support received from EPSRC on grants EP/C549481 and EP/E045472.

References

1. Markram, H.: The blue brain project. Nature Reviews Neuroscience 7, 153–160 (2006)
2. Itti, L., Koch, C., Niebur, E.: A model of saliency-based visual attention for rapid scene analysis. IEEE Transactions on Pattern Analysis and Machine Intelligence 20, 1254–1259 (1998)
3. Harel, J., Koch, C., Perona, P.: Graph-based visual saliency. In: Proceedings of NIPS (2006)
4. Bruce, N., Tsotsos, J.: Attention, and visual search: An information theoretic approach. Journal of Vision 9(3), 1–24 (2009)
5. Kansal, A.R., Torquato, S., Harsh, G.R., Chiocca, E.A., Deisboeck, T.S.: Simulated brain tumor growth dynamics using a three-dimensional cellular automaton. Journal of theoretical biology 203, 367–382 (2000)
6. Chang, C., Zhang, Y., Gdong, Y.: Cellular automata for edge detection of images. In: Proceedings of 2004 International Conference on Machine Learning and Cybernetics, vol. 6, pp. 3830–3834 (August 2004)
7. Morie, T., Nagata, M., Iwata, A.: Design of a pixel-parallel feature extraction vlsi system for biologically-inspired object recognition methods. In: Proc. Int. Symp. on Nonlinear Theory and its Applications, pp. 371–374 (October 2001)
8. Levi, D.: Hereboy: a fast evolutionary algorithm. In: Proceedings of the 2nd NASA/DoD Workshop on Evolvable Hardware, pp. 17–24 (2000)
9. Mikolajczyki, K., Tuytelaars, T., Schmid, C., Zisserman, A., Matas, J., Schaffalitzky, F., Kadir, T., Gool, L.V.: A comparison of affine region detectors. Internation Journal of Computer Vision 65, 43–72 (2005)

VG-RAM WNN Approach to Monocular Depth Perception

Hélio Perroni Filho and Alberto F. De Souza

Laboratório de Computação de Alto Desempenho, Universidade Federal do Espírito Santo
Av. Fernando Ferrari, 514, 29075-910, Vitória-ES, Brazil

Abstract. We have examined Virtual Generalizing Random Access Memory Weightless Neural Networks (VG-RAM WNN) as platform for depth map inference from static monocular images. For that, we have designed, implemented and compared the performance of VG-RAM WNN systems against that of depth estimation systems based on Markov Random Field (MRF) models. While not surpassing the performance of such systems, our results are consistent to theirs, and allow us to infer important features of the human visual cortex.

Keywords: Monocular depth perception, weightless neural networks.

1 Introduction

Humans perceive the world visually in three dimensions; however, the eye itself is blind to depth, capturing the environment around us in images as two-dimensional as a photograph. It is the brain that enables our sense of depth, inferring three-dimensional world models from the two-dimensional visual data it receives [1].

The human visual system is extremely good at estimating depth from static monocular images [2]. To do so, it uses monocular cues such as texture gradient variations, occlusions, known object sizes, haze, etc [2, 3]. Saxena, through his research on machine-learning algorithms based on Markov Random Field (MRF) models, showed that it is possible to reliably estimate depth maps based solely on static monocular information [3]; however, his results lack biological plausibility, since, as far as we know, there is no relationship between MRF and the workings of the visual cortex. In this paper we examine Virtual Generalizing Random Access Memory Weightless Neural Networks (VG-RAM WNN) as a platform for depth map inference from static monocular images.

VG-RAM WNN model the excitatory/inhibitory decoding performed by the dendritic trees of biological neurons [4]. As a machine-learning approach, they are more closely related to the biological domain than MRF. Previous success stories as computer vision tool (see, e.g. [5, 6]) also indicate its viability as a basis for production systems development, such as robot vision systems. Here, we introduce the VG-RAM WNN Hori**Z**ontal Slid**E**r Mul**T**ichannel **A**rchitecture (ZETA), designed for estimating depth from static monocular images. While not surpassing the performance of Saxena's systems [3] – an objective for further research – our results with ZETA are consistent to his and allow us to infer important features of the human visual cortex.

K.W. Wong et al. (Eds.): ICONIP 2010, Part II, LNCS 6444, pp. 509–516, 2010.

2 Visual Cues for Depth Perception

The visual system reconstructs a three-dimensional representation of the world around us from the two-dimensional images projected onto the retinas. This reconstruction is based on cues both *monocular* and *stereoscopic*. Monocular cues include some – such as motion parallax and defocus – that are better suited for depth estimation in continuous visual streams; others, such as texture differences, texture gradients, color distribution, and edges are more appropriate for depth estimation from static images. Stereoscopic cues are basically the binocular disparity caused by slight projection differences of the images on our two retinas (stereoscopic depth perception is out of the scope of this paper and is not discussed further).

A surface's texture *"is related to periodical luminosity fluctuations in the image, which let us interpret the surface as a homogenous structure"* [7]. The visual perception of a surface's texture also varies with observer distance; as the point of view recedes over a surface, its texture's features soften, eventually becoming undistinguishable. This is called a *texture gradient* [8]. For example, in Fig. 1(a), the street pavement's stones become ever smaller and less defined as the point of view retreats, thus forming a texture gradient that produces a particular sensation of depth.

Edges (i.e. the limits between differently colored or textured surfaces) and color information also provide important cues for depth estimation. Sudden depth variations often lie in the limits between scene objects, while surfaces filled with the same or very similar colors tend to belong to the same object and to possess constant depth.

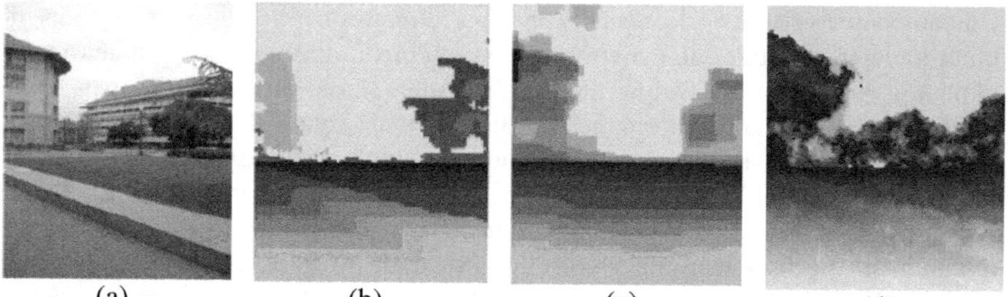

(a) (b) (c) (d)

Fig. 1. (a) Monocular image with natural depth information. (b) Ground truth depth. (c) Best performing MRF estimation. (d) Best performing VG-RAM WNN estimation.

3 Markov Random Field Approach for Monocular Depth Perception

The Saxena's [3] MRF-based depth estimators work by sectioning input images into rectangular patches and then calculating two sets of *features* (*absolute* and *relative*) for each patch. Those features attempt to capture three types of local visual cues: texture variations, texture gradients, and color. Also, in order to capture global features, the slicing is repeated at three different scales. Finally, summary features are calculated for each patch column. So, for each patch, a feature vector is constructed from its own absolute and relatives features, plus those of its immediate neighbors at all three scales and the summary features of the column it belongs to [3].

A patch's absolute features are given by

$$E_i = \{E(L_1, H_i), E(L_1, S_i), E(L_1, V_i), ..., E(L_9, V_i), E(P_1, V_i), ..., E(P_6, V_i)\} \tag{1}$$

where L_1, ..., L_9 are the nine Laws Filters [3], P_1, ..., P_6 are six oriented edge detectors spaced at 30° intervals [3], H_i, S_i and V_i are the Hue, Saturation and Value (brightness) image channels (HSV image format) of the i-th input image patch, and

$$E(F, I) = \left(\sum\nolimits_{(x,y)} |I * F|, \sum\nolimits_{(x,y)} (I * F)^2\right) \tag{2}$$

is the absolute and quadratic sum of the output of each filter over the i-th input image patch. Also, for each patch i taken from the image $I(x, y)$ at spatial scale s, a 10-column histogram is calculated for each filter output $|I*F|$, resulting in a 170-dimensional feature vector

$$y_{is} = \begin{Bmatrix} H_{10}(L_1 * H_{is}), H_{10}(L_1 * S_{is}), \\ H_{10}(L_1 * V_{is}), ..., H_{10}(L_9 * V_{is}), \\ H_{10}(P_1 * V_{is}), ..., H_{10}(P_1 * V_{is}) \end{Bmatrix} \tag{3}$$

where L_1, ..., L_9 are the nine Laws Filters, P_1, ..., P_6 are the six Prewitt Filters, H_i, S_i and V_i are the H, S and V channels of the i-th patch of the input image at the s-th spatial scale, and H_{10} is the filter output's 10-dimensional histogram. Relative characteristics between two patches i and j are calculated as the difference between their histograms, i.e. $y_{ijs} = y_{is} - y_{js}$.

The absolute and relative feature vectors are fed to statistical models trained to maximize the probability $P(d|X)$ (conditional probability of depth d, given the input feature vector X) using images and corresponding ground truth. Both Gaussian and Laplacian probability distributions were examined by Saxena according to the models below [3]:

$$P_G(d \mid X; \theta, \sigma) = \frac{1}{Z_G} \exp\left(-\sum_{i=1}^{M} \frac{(d_i(1) - x_i^T \theta_r)^2}{2\sigma_{1r}^2} - \sum_{s=1}^{3} \sum_{i=1}^{M} \sum_{j \in N_s(i)} \frac{(d_i(s) - d_j(s))^2}{2\sigma_{2rs}^2} \right) \tag{4}$$

$$P_L(d \mid X; \theta, \lambda) = \frac{1}{Z_L} \exp\left(-\sum_{i=1}^{M} \frac{|d_i(1) - x_i^T \theta_r|}{2\lambda_{1r}} - \sum_{s=1}^{3} \sum_{i=1}^{M} \sum_{j \in N_s(i)} \frac{|d_i(s) - d_j(s)|}{2\lambda_{2rs}} \right) \tag{5}$$

The Gaussian distribution (4) has parameters θ_r and σ^2_{1r} for patches on row r, and σ^2_{2rs} for patches on row r at scale s. For the Lapacian distribution (5), the parameters are θ_r and λ_{1r} for patches on row r, and λ_{2rs} for patches on row r at scale s. On both cases, M is the total number of patches. See [3] to learn how these parameters are adjusted to minimize (4) and (5) for a set of images and corresponding ground truth. To compute the depth map of an input image, its patches are feed to (4) or (5) and the depth d that minimizes (4) or (5) for each patch is taken.

4 VG-RAM WNN Approach for Monocular Depth Perception

VG-RAM WNN neurons work by associating, during the training phase, input bit-arrays (collected from the input by the neurons synapses) to outputs (in our case, depth values); the pairs {bit-arrays, depth} seen during training are stored into the neurons' RAM. During the test phase, when presented with a (potentially unseen) input, the neurons collect a bit-array from the input and respond with the output associated to the closest (according to the Hamming distance) learned pair. Despite its remarkable simplicity, VG-RAM WNN is a powerful machine learning tool [4, 5, 6, 9].

The ZETA VG-RAM WNN architecture (see Fig. 2) estimates, for images of dimension (n, m), depth maps of dimension (l, k), such that $l < n$ and $k < m$. It is composed of z neural layers, each containing k VG-RAM WNN neurons, and an input window of dimensions $(5\sigma, m)$ (we discuss σ below), which "slides along" input images horizontally. As the window advances in steps of n/l pixels, each neuron samples a *patch* of dimensions $(5\sigma, m/k)$ using its set of synapses (Fig. 2(b)), whose size is determined by the ZETA parameter w. Before being sampled, each patch is decomposed across four dimensions – the (i) Hue, (ii) Saturation and (iii) Value channels of the HSV image format, plus the (iv) output of a Sobel Edge Detection filter applied to the Value channel (Fig. 2(c)).

The section of the input window from which a neuron samples via its synapses is its *receptive field.* The receptive fields of neurons on the *same* layer are stacked contiguously as a column of height m, while the receptive fields of neurons of equivalent positions on *different* layers are coincident (Fig. 2 (b)). However, the latter sample their receptive fields through randomly-distributed synapses; therefore, even neurons looking at the same region will have different inputs. As a result, during testing, each image patch may be assigned as many as z distinct depth estimates (one for each neuron assigned to the patch).

At setup time, a neuron's synapses are randomly laid over its visual field according to a Gaussian distribution of mean $(5\sigma/2, m/2k)$ and standard deviation σ – a connection pattern frequently found in biologic neural networks [1, 6] – and retain that layout for the remainder of the session. Further, in a way analogous to that of biological *on center / off center* retinal neurons, synapses convert their numeric inputs to bit outputs by forming pairs called *minchinton cells* [10]. In the ZETA architecture, each synapse w_t forms a minchinton cell with the next synapse w_{t+1} (w_w forms a cell with w_1). Each minchinton cell returns the bit 1 if w_t's value is greater than w_{t+1}'s, and 0 otherwise [6].

ZETA's input window and neuron layout are inspired by the fact that depth map value distributions often start with larger values at the image's "top" (mostly due to far-off structures, such as the sky and distant buildings) and decrease towards the "bottom" (usually dominated by the scene's ground or close-by objects). On the other hand, the patch decomposition is inspired in the human visual system, which has neural circuits to capture a scene's luminance (HSV's Value channel) and colors (Hue and Saturation), as well as to detect edges [1].

During training, the network scans input images horizontally and each neuron associates the multi-dimensional view of the input patch currently centered on its receptive field to the corresponding ground truth depth value; then the focus moves on to the next column. The testing process is very similar – the network scans the test input horizontally, collecting neural depth estimations as it moves along – with the important difference that the Winner-Take-All algorithm is used to resolve a single depth estimate for each patch out of the possible z hypotheses (Fig. 2(d)).

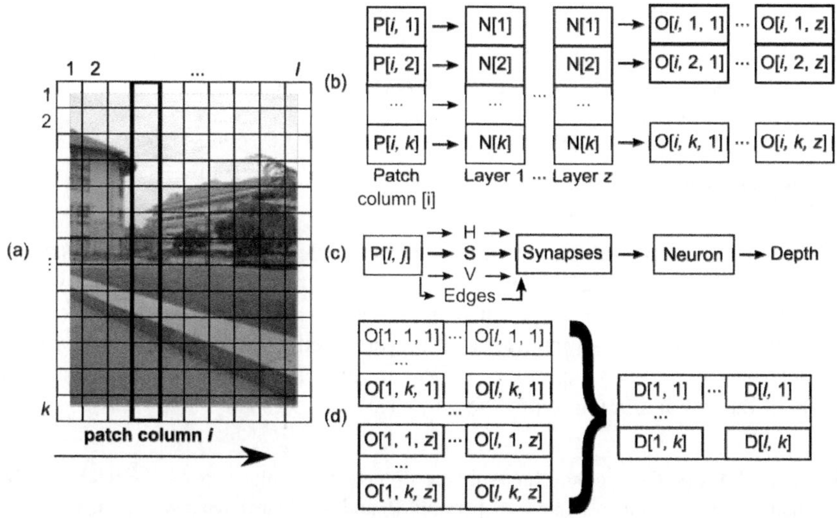

Fig. 2. Schematics of the ZETA VG-RAM WNN architecture. As the network's view window (patch column i) slides across the input image (a) of dimension (n, m), each neuron N[1]...N[k] of each Layer 1...z is fed with an input patch P[i, $1 \ldots k$] according to its position and host layer (b). Patches are decomposed into four dimensions (three HSV channels plus an edge map) which are sampled by the neuron's synapses (c). After all input patches have been processed, the Winner-Take-All algorithm is used to collapse the multiple depth estimates into a single depth map (d).

5 Methodology

In order to enable direct quantitative comparison with Saxena's results, we have adopted the same data base and metrics used in his work [3]. The data base contains 425 samples, each composed by one input image and accompanying ground truth depth map, obtained with a high resolution camera and laser range scan SICK LMS-291 mounted on a robot. Images have 1704x2272-pixel resolution and depth maps have 86x107-pixel size.

Saxena employs the *Mean Absolute Error* (MAE) metric to evaluate his systems' performance. In order to emphasize multiplicative rather than additive errors, all depth values are transformed to $\log_{10}$ scale prior to metric calculation. The resulting *logarithmic MAE* (logMAE) metric for a given estimated depth map S_k and ground truth depth map R_k, both of dimensions (m, n), can be defined as:

$$logMAE(S_k) = \frac{1}{mn}\sum_{i=1}^{m}\sum_{j=1}^{n} |\log_{10}(S_k[i,j]) - \log_{10}(R_k[i,j])| \quad (6)$$

As previously mentioned, most depth maps display a value distribution in which values at the image's "top" typically belong to far-off structures, while those at the "bottom" typically belong to the ground or close-by objects. In order to allow closer examination of the impact of this distribution on the overall estimation performance, we also calculate the *logarithmic MAE of the i-th row* ($logMAE_i$):

$$logMAE_i(S_k) = \frac{1}{n}\sum_{j=1}^{n} |\log_{10}(S_k[i,j]) - \log_{10}(R_k[i,j])| \qquad (7)$$

6 Experiments

To find ZETA's optimal parameters, we executed several validation sessions, in which 190 (44%) of the 425 samples were used for training and 94 (22%) were used to evaluate the network's performance. Once the optimal parameters were determined, all 284 validation cases (66%) were used to train the network, while the hold-out 141 cases (33%) were used to measure its performance in a final test session. Results of the validation phase are shown in Table 1.

Table 1. ZETA's performance in the validation phase. Cell values are the network's logMAE performance as a function of synapse dispersion (σ) and number of synapses per neuron (w).

σ	W			
	32	**64**	**128**	**256**
5	0.193	0.190	0.188	0.186
10	0.193	0.190	0.189	0.187
20	0.190	0.190	0.188	0.186
40	0.192	0.188	0.185	0.180

For a fixed layer depth z = 10 (the maximum value supported by the machine employed in the experiments), the synapse dispersion factor (σ) was varied within the set of values {5, 10, 20, 40} and the number of synapses per neuron within the set of values {32, 64, 128, 256}. As Table 1 shows, ZETA's performance improves as both the synapse count and field of view (σ) of each neuron increases. In fact, we were not able to find a point after which this trend would seem to weaken due to the limitations of the machine available for the experiments (an Intel machine with 2GB of RAM).

Once the best parameters supported by our test environment were determined, we proceeded with the final test round, whose results are compared with Saxena's in Fig. 3 (see also Fig. 1 for a visual comparison with the depth map estimates for one image of the hold-out set). Additionally to overall logMAE results, we also plot mean $logMAE_i$ values to further illustrate ZETA's performance; as Saxena has not published row-wise results for his MRF estimators, comparing row-wise ZETA's performance with that of MRF is not possible.

In the graph of Fig. 3, the y-axis is the $logMAE_i$, while the x-axis is the line of the depth map. As the graph shows, ZETA's performance is poorest around lines 30-60, which roughly corresponds to the "middle" of the images – not surprisingly, this is the region where the most complex visual structures are found. ZETA's performance improves near the "bottom" of the images (right side of the graph), which mostly depict the ground and close-by structures. It's also noticeable that, while much more reliable than Saxena's baseline MRF system, on average ZETA does not outperforms the best (Laplacian) MRF estimator.

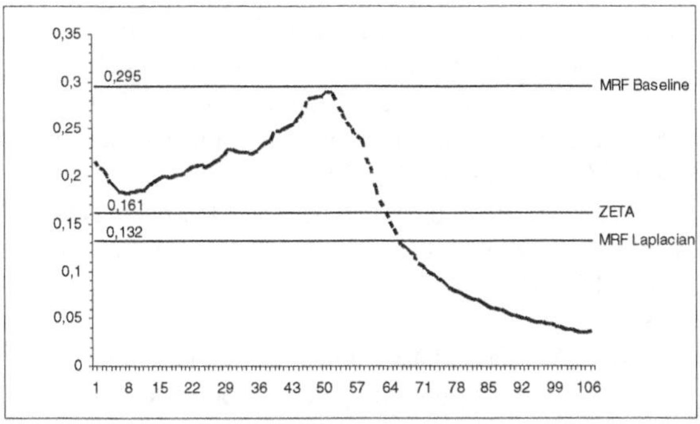

Fig. 3. Comparison between ZETA's and Saxena's results. The dotted line plots ZETA's $logMAE_i$ (y-axis) as a function of the depth map line (x-axis). Full lines in the graph represent the logMAE of ZETA and Saxena's MRF for the whole hold-out set.

7 Conclusions

In this work we developed the ZETA VG-RAM Weightless Neural Network (WNN) architecture, which demonstrates one way how WNN neurons can be wired to learn – and from then on consistently estimate – relationships between visual inputs (in the form of static monocular images) and depth values. This is a novel approach since, to the best of our knowledge, WNN's were not previously considered as a depth estimation tool; furthermore, it raises the question of whether – and if yes, where and how – biological neural circuits involved in depth perception employ architectural solutions similar to those adopted in our implementation of ZETA.

In order to assess ZETA's performance, its depth estimations were quantitatively compared to those of Saxena's MRF systems [3] in terms of deviation from a laser-produced ground-truth using the *logarithmic Mean Absolute Error* (logMAE) metric. While the VG-RAM WNN estimator did not outperform the best MRF estimator in average, their performances were similar; besides, the overall results are promising, and motivate further work on improving the VG-RAM WNN estimator.

Since the input decomposition across the dimensions of color and edges – processes known to occur in the human brain – was found to enhance our network's performance, we conjecture that adding other dimensions could further that trend. In particular, we believe that enabling the network to simultaneously "see" inputs across several scales, as well as enabling vertical image screening movements, could enhance our system's generalization significantly.

References

1. Kandel, E., Schwartz, J., Jessell, T.: Principles of Neural Science, 4th edn. McGraw-Hill, New York (2000)
2. Loomis, J.M.: Looking down is looking up. Nature 414, 155–156 (2001)
3. Saxena, A., Chung, S.H., Ng, A.Y.: 3-D Depth Reconstruction from a Single Still Image. International Journal of Computer Vision 76(1), 53–69 (2008)
4. Aleksander, I., De Gregorio, M., França, F.M.G., Lima, P.M.V., Morton, H.: A Brief Introduction to Weightless Neural Systems. In: 17th European Symposium on Artificial Neural Networks, pp. 299–305. d-side publications, Evere (2009)
5. Rohwer, R., Morciniec, M.: A Theoretical and Experimental Account of n-Tuple Classifier Performance. Neural Computation 8(3), 629–642 (1996)
6. De Souza, A.F., Badue, C., Pedroni, F., Dias, S.S., De Souza, H.O., De Souza, S.F.: Face Recognition with VG-RAM Weightless Neural Networks. In: Kůrková, V., Neruda, R., Koutník, J. (eds.) ICANN 2008, Part I. LNCS, vol. 5163, pp. 951–960. Springer, Heidelberg (2008)
7. Krylov, A.S., Kutovoi, A., Leow, W.K.: Texture parameterization with Hermite functions. In: 5th International Conference on Computer Graphics & Vision, University of Nizhny Novgorod, Russia (2002)
8. Krantz, J.H.: Texture Gradient, `http://psych.hanover.edu/krantz/art/texture.html`
9. Carneiro, R.V., Dias, S.S., Fardim Júnior, D., Oliveira, H., Garcez, A.A., De Souza, A.F.: Improving VG-RAM Neural Networks Performance Using Knowledge Correlation. In: King, I., Wang, J., Chan, L.-W., Wang, D. (eds.) ICONIP 2006. LNCS, vol. 4232, pp. 427–436. Springer, Heidelberg (2006)
10. Mitchell, R.J., Bishop, J.M., Box, S.K., Hawker, J.F.: Comparison of Some Methods for Processing Grey Level Data in Weightless Networks. In: Austin, J. (ed.) RAM-Based Neural Networks, pp. 61–70. World Scientific, Singapore (1998)

Semi-supervised Classification by Local Coordination

Gelan Yang[1], Xue Xu[2], Gang Yang[3], and Jianming Zhang[4]

[1] Hunan City University, Department of Computer Science, Yiyang, China
[2] University of Science and Technology of China, Department of Automation, Hefei, China
[3] Department of Power & Energy Systems, Ecole Supérieur d'Electricité (Supélec), Gif-sur-Yvette Cedex, Franc
[4] College of Computer and Communication Engineering
Changsha University of Science and Technology, Changsha, China
glyang@mail.ustc.edu.cn

Abstract. Graph-based methods for semi-supervised learning use graph to smooth the labels of the points. However, most of them are transductive thus can't give predictions for the unlabeled data outside the training set directly. In this paper, we propose an inductive graph-based algorithm that produces a classifier defined on the whole ambient space. A smooth nonlinear projection between the sample space and the label value space is achieved by local dimension reduction and coordination. The effectiveness of the proposed algorithm is demonstrated by the experiment.

Keywords: mixture of factor analyzers; local linear coordinate; semi-supervised classification; manifold learning.

1 Introduction

Traditionally, the knowledge over labeled data is agglomerated and applied to new data. However, in many real life machine learning tasks, such as web page classification and document indexing, the acquisition of labeled data is costly, while large amount of unlabeled samples can be obtained easily. It means that, the structural information that can be inferred from the labeled data may be very limited; however, by taking advantage of the geometrical knowledge of both labeled and unlabeled data, it is possible to solve the problem. Semi-supervised learning context that learns from both labeled and unlabeled samples has drawn great attentions over the past few years. The relatively comprehensive surveys are given by [1] and [2], respectively. An older version can be found in [3], and a special discussion on inductive methods is presented in [4].

The graph-based technique provides a useful approach for modeling the relationship between labeled and unlabeled data. In this modeling scenario, data is represented by the nodes of a graph, and the edges in the graph are labeled with the pairwise distances of the incident nodes. The graph-based methods are based on the assumption that the nearby points are likely to have the close label. Most of them are intrinsically transductive, which means that they give predictions only for the

K.W. Wong et al. (Eds.): ICONIP 2010, Part II, LNCS 6444, pp. 517–524, 2010.

unlabeled data in the training set, and can't be easily extended to the new test point. Several approximation methods have been proposed. Zhu et al. advised that the nearest neighbor label value could be used for the test example [5]. Similarly in [6], the affine transformation of the nearest neighbor was computed and applied to the test point. Chapelle et al [7] approximately represented the test points as a linear combination of training set in the feature space. These approximation methods are effective when the training set is large enough, but they can't present a smooth projection on the ambient space. Natural out of sample extensions are realized by manifold regularization [8], where the graph is merely used to regularize an inductive kernel. Finding a linear mapping that minimizes the cost function of the nonlinear eigenmap can also produce an inductive classifier [9]. The harmonic mixture models that naturally handle the induction as standard mixture models are proposed in [10].

Similar to harmonic mixture models, our method is also based on mixture models and graph technique. The main idea is to transform the sample space into a 1-dimensional space under the label restriction. Mixture of Factor Analyzers (MFA) [11] is used to perform local dimension reduction, and local linear coordination [12] helps to achieve the smooth nonlinear transformation. Experiments on several data sets show the effectiveness of our method.

2 Semi-supervised Local Coordination

2.1 MFA for Local Dimension Reduction

Many high-dimensional data in real-world applications can be modeled as data points lying close to a low-dimensional nonlinear manifold. The mixture of factor analyzers is used here to capture the complex structure embedded in the sample data.

In the factor analysis, a D dimensional observed data x is modeled by a corresponding d_z dimensional latent variable z. The generative model is represented as $x = \Lambda z + \mu$, where Λ denotes the factor loading matrix and μ accounts for the independent noise. The probability models for factors z and random variable μ are separately assumed to be $N(0,I)$, and $N(0,\Psi)$, where Ψ is set as a diagonal matrix. A natural extension of factor analysis on complex data is MFA, where the high-dimensional data x is modeled as

$$p(x) = \sum_{q=1}^{Q} \int P(x|z,\varpi_q)P(z|\varpi_q)P(\varpi_q)dz \tag{1}$$

Here, $\varpi_q, q = 1,\cdots,Q$ denotes a mixture of Q factor analyzers. $\pi_q = P(\varpi_q)$ is the average mixing proportion of the qth factor analyzer in the whole space. $z \in R^{d_z}$ is the latent variable or called local low-dimensional data satisfying $P(z|\varpi_q) = N(0,I)$. The intrinsic dimensionality d_z can be measured by the maximum likelihood estimator[13]. If the measure value is $\tilde{d}$, we set $d_z = \tilde{d} + 1$ to allow for some redundancy. Similar to FA, the observed data x and the low-dimensional data z are connected by $P(x|z,\varpi_q) = N(\mu_q + \Lambda_q z, \Psi)$, where μ_q and Λ_q are separately the mean value and the loading matrix. MFA's parameters π_q, Λ_q, μ_q, Ψ, $q=1,\cdots,Q$ can be obtained by EM algorithms[11]. As a result, the local counterpart of x in the qth factor analyzer can be represented as

$$z_q = \Lambda_q^T(\Psi + \Lambda_q\Lambda_q^T)^{-1}(x-\mu_q) \tag{2}$$

with probability

$$\begin{aligned} h_q &= E\left[\omega_q \middle| x\right] \\ &= \frac{P(x,\omega_q)}{\sum_{r=1}^{Q} P(x,\omega_r)} \\ &= \frac{\pi_q N\left(x-\mu_q, \Lambda_q\Lambda_r^T + \Psi\right)}{\sum_{r=1}^{Q} \pi_r N\left(x-\mu_r, \Lambda_r\Lambda_r^T + \Psi\right)} \end{aligned} \tag{3}$$

We take $\hat{z} = \begin{bmatrix}1 & z^T\end{bmatrix}^T$ as the augmented vector form of z. At last, the sampled high-dimensional data $X = \{x_1,\cdots,x_n\} \in R^D$ can be changed into Q set of low-dimensional data. Let $\hat{z}_{qi}$ and h_{qi} separately denote x_i's counterpart in the qth factor analyzer and the corresponding probability.

2.2 Aligning Local Models by Local Coordination

In the above part, the Q sets of low-dimensional data are obtained by MFA. Next the local coordinate will be aligned together to recover a global parameterization of the manifold. Roweis et al[14] added a regularizing term to the standard maximum likelihood objective function to encourage the agreement of the internal coordinates. Brand [15] introduced a cost function that measured the amount of disagreements between the linear models on the global coordinates. Local coordination is proposed in [11] where affine transformation is used to align the local models.

Local coordination is relatively easy to realize and is adopted here to align the Q sets of local coordinates. Let T_q denote the affine transformation to align the qth factor analyzer, the global coordinate can be represented as

$$y = \sum_{q=1}^{Q} h_q T_q \hat{z}_q \tag{4}$$

Let $Y = \{y_1,\cdots,y_n\} \in R^{d_y}$ denote the global coordinate, $T_q = \begin{bmatrix}T_{q,0} \cdots T_{q,d_z}\end{bmatrix}$, $\hat{z}_{qi} = \begin{bmatrix}\hat{z}_{qi,o} \cdots \hat{z}_{qi,d_z}\end{bmatrix}^T$ is the expand form. Then, the global coordinate is expanded as

$$y_i = \sum_{q=1}^{Q} h_{qi} T_q \hat{z}_{qi} = \sum_{q=1}^{Q}\sum_{m=0}^{d_z} h_{qi} T_{q,m} \hat{z}_{qi,m} \tag{5}$$

Here, q and m can be combined to form a new index as $j = j(q,m)$, Let $\tilde{z}_{ji} = h_{qi}\hat{z}_{qi,m}$, $\tilde{T}_j = T_{q,m}$. Then, the former representation is rewritten as

$$y_i = \sum_{q=1}^{Q}\sum_{m=0}^{d_z} h_{qi} T_{q,m} \hat{z}_{qi,m} = \sum_{j=1}^{Q\times(d_z+1)} \tilde{T}_j \tilde{z}_{ji} \tag{6}$$

Its matrix form is $Y = \tilde{T}\tilde{Z}$. Here $\tilde{T}$ is composed by the local affine transformation T_q .

2.3 Label Smoothness

Label smoothness is achieved by graph-based technique. Let Γ_i be a vector of indices of points in the $(k-1)$ neighbor of x_i , and $\overline{\Gamma}_i = \begin{bmatrix} i \\ \Gamma_i \end{bmatrix}$ is a vector including i and Γ_i . Let e be the vector of all $1's$, I_k be the identify matrix with rank k . Then , $J = (I - \frac{1}{k} ee^T)$ is the mean removal operator. $Y_{\Gamma_i} J \begin{bmatrix} 1 \\ -w_i \end{bmatrix} \to 0$ means that the nearby points are likely to have the same label. Here, $w_i = \frac{1}{\sum_{j\in\Gamma_i} K(x_i, x_j)} \begin{bmatrix} K(x_i, x_{\Gamma_i(1)}) \\ \vdots \\ K(x_i, x_{\Gamma_i(k-1)}) \end{bmatrix}$ is the property extracted from the high-dimensional space, and $K(x_i, x_j)$ is a function to measure the similarity of two points, usually it's defined by the Gaussian weight

$$K(x_i, x_j) = \begin{cases} \exp\left(-\left\| x_i - x_j \right\|^2 \Big/ 2\sigma^2 \right) & \cdots \quad j \in \Gamma_i \\ 0 & \cdots \quad j \notin \Gamma_i \end{cases} \tag{7}$$

Here, σ is a scale parameter set by $\sigma = \left\| x_i - x_{\Gamma_i(k-1)} \right\|$, and $x_{\Gamma_i(k-1)}$ is the $(k-1)^{th}$ nearest neighbor of x_i . Noting that $J\begin{bmatrix} 1 \\ -w_i \end{bmatrix} = \begin{bmatrix} 1 \\ -w_i \end{bmatrix}$, and the cost function of local approximation at y_i is expressed in Eq. (8).

$$err_{pi} = \left\| Y_{\Gamma_i} \begin{bmatrix} 1 \\ -w_i \end{bmatrix} \right\|^2 \tag{8}$$

The total cost function of all sample data is

$$err_p = \sum_{i=1}^{n} err_{pi} = \sum_{i=1}^{n} \left\| Y_{\Gamma_i} \begin{bmatrix} 1 \\ -w_i \end{bmatrix} \right\|^2 = \left\| YB_p \right\|_F^2 \tag{9}$$

Here, $\|\cdot\|_F$ denotes the Frobenius norm, B_p is a sparse matrix satisfying $B_p(\overline{\Gamma}_i \quad i) = \begin{bmatrix} 1 \\ -w_i \end{bmatrix}$. The cost function err_p is defined only on the geometry information. Afterwards, the label constraints should be considered. Let L be the collection of indices of labeled points, and s_i is the flag to identify the labeled points satisfying $s_i = \begin{cases} 1 & \cdots & i \in L \\ 0 & \cdots & i \notin L \end{cases}$, and $F = [f_1, \ \cdots \ , f_n]$ denote the given label value. Then, the label error of x_i can be defined as

$$err_{li} = s_i \| y_i - f_i \|^2 \tag{10}$$

The loss function $Err_p(Y)$ defined both on approximation error and label error can be written as

$$\begin{aligned} Err_p &= \sum_{i=1}^{n} \left((1-a_i)^2 err_{pi} + a_i^2 err_{li} \right) \\ &= \sum_{i=1}^{n} \left((1-a_i)^2 \left\| Y_{\overline{T}_i} \begin{bmatrix} 1 \\ -w_i \end{bmatrix} \right\|^2 + a_i^2 s_i \| y_i - f_i \|^2 \right) \\ &= \| YB_p(I_n - A) \|_F^2 + \| (Y - F)A \|_F^2 \\ &= \| \tilde{T}\tilde{Z}B_p(I_n - A) \|_F^2 + \| (\tilde{T}\tilde{Z} - F)A \|_F^2 \end{aligned} \tag{11}$$

Here, $a_i = (\frac{l}{n}(1-\beta) + \beta) s_i$ is the tradeoff parameter at y_i, l is the number of labeled points, β is the minimal weight coefficient set by the user, and $A = diag\{a_1, \ \cdots \ , a_n\}$ is a diagonal matrix. When minimizing (11), the optimal solution can be easily obtained, shown in Eq. (12).

$$\tilde{T}^* = \arg\min_{\tilde{T}} Err_p = FAA^T \tilde{Z}^T C_p^{-1} \tag{12}$$

Where $C_p = \tilde{Z}B_p\overline{AA}^T B_p^T \tilde{Z}_T + \tilde{Z}AA^T\tilde{Z}^T + \lambda I_{d_z}$, $\overline{A} = (I_n - A)$, and $\lambda \in R^+$ is a small const value. We take $\gamma = \frac{1}{n-l}\sum_{i \notin L} y_i$ as the decision threshold for classification.

3 Experiment Results

There are three different types of data sets used in our experiments. MNIST is a dataset of 60,000 handwritten digit images, and each 28 × 28 image is considered as a point in the 784-dimensional space. Every time, 2000 images of 2 classes are randomly selected for training, and rest of the images are used for testing. The final accuracy is averaged by all 45 combinations of the 10-class digits. The ADA and GINA datasets are both binary classification datasets. The ADA dataset originates from the marketing domain, and consists of 4,147 datapoints in a 48-dimensional space. The GINA dataset is a handwriting recognition dataset that has 3,153 datapoints described by 970 features.

In the experiments, SLC (Semi-supervised local coordinate) is compared with other three inductive methods: LPP, NPE [16] and LLTSA [17]. Here, LPP is performed in semi-supervised manner [18]. NPE and LLTSA are semi-supervised extended with prior information [19]. The final results on all data sets are separately listed in tables 1, 2 and 3, and the best performance is shown in bold. The result in table 1 shows that SLC outperforms other methods on MNIST in all cases. Table 2 demonstrates the superiority of SLC and LPP on ADA. In table 3, the classification accuracy reveals the benefits of both SLC and LLTSA. From these results, we can see that, SLC is a relatively robust classification algorithm, and achieves good performance on all the data sets.

Table 1. Semi-supervised classification on MNIST dataset

Label proportion	**Classification Accuracy**			
	LPP	***NPE***	***LLTSA***	***SLC***
0.1%	0.736	0.733	0.685	**0.969**
0.3%	0.849	0.865	0.731	**0.984**
0.5%	0.902	0.899	0.742	**0.984**
1%	0.917	0.945	0.830	**0.983**
3%	0.937	0.943	0.864	**0.985**
5%	0.959	0.961	0.913	**0.987**
10%	0.971	0.971	0.945	**0.990**
20%	0.981	0.979	0.965	**0.992**

Table 2. Semi-supervised classification on ADA dataset

Label proportion	**Classification Accuracy**			
	LPP	***NPE***	***LLTSA***	***SLC***
0.1%	**0.528**	0.525	0.381	0.518
0.3%	**0.543**	0.532	0.428	0.484
0.5%	**0.623**	0.599	0.477	0.620
1%	0.679	0.642	0.628	**0.720**
3%	0.663	0.590	0.648	**0.700**
5%	0.713	0.699	0.662	**0.714**
10%	**0.741**	0.738	0.719	0.723
20%	**0.755**	0.746	0.744	0.731

Table 3. Semi-supervised classification on gina dataset

Label proportion	Classification Accuracy			
	LPP	*NPE*	*LLTSA*	*SLC*
0.1%	0.532	0.526	**0.597**	0.486
0.3%	0.581	0.565	0.618	**0.647**
0.5%	0.584	0.577	**0.642**	0.610
1%	0.588	0.592	**0.645**	0.630
3%	0.646	0.650	0.669	**0.684**
5%	0.682	0.692	0.701	**0.729**
10%	0.710	0.725	0.713	**0.767**
20%	0.763	0.775	0.775	**0.799**

4 Conclusions

An inductive semi-supervised classification algorithm SLC is presented in this paper. It combines the label constraint and the graph-based constraint to establish a smooth nonlinear transformation for classification. Mixtures models are used here to handle the induction, and local approximation ensures the label smoothness on the graph. The experimental results demonstrate the effectiveness of the proposed approach.

References

1. Zhu, X.: Semi-supervised learning literature survey. Technical Report 1530, Computer Sciences, University of Wisconsin-Madison (2005)
2. Chapelle, O., Weston, J., Schölkopf, B.: Cluster kernels for semisupervised learning. In: Proceedings of Neural Information Proccessing Systems, vol. 15, pp. 873–880. MIT Press, Cambridge (2003)
3. Seeger, M.: Learning with labeled and unlabeled data. Technical report, Edinburge (2000)
4. Haffari, G.: A Survey on Inductive Semi-Supervised Learning. Technical report, School of Computing Science, Simon Fraser University (2005)
5. Zhu, X., Ghahramani, Z., Lafferty, J.: Semi-supervised learning using Gaussian fields and harmonic functions. In: Proceedings of the 20th International Conference on Machine Learning, Washington, DC, USA, pp. 912–919 (2003)
6. Teng, L., Li, H., Fu, X., Chen, W., Shen, I.F.: Dimension reduction of microarray data based on local tangent space alignment. In: Proceedings of the 4th IEEE International Conference on Cognitive Informatics, University of California, Irvine, USA, pp. 154–159 (2005)

7. Chapelle, J.W., Schölkopf, B.: Cluster kernels for semi-supervised learning. In: Advances in Neural Information Processing Systems, vol. 15, pp. 585–592. MIT Press, Cambridge (2003)
8. Belkin, M., Niyogi, P., Sindhwani, V.: Manifold regularization: A geometric framework for learning from examples. Technical Report TR-2004-06. University of Chicago (2004)
9. He, X., Yan, S., Hu, Y., Niyogi, P., Zhang, H.J.: Face recognition using laplacianfaces. IEEE Transactions on Pattern Analysis and Machine Intelligence 27(3), 328–340 (2005)
10. Zhu, X., Lafferty, J.: Harmonic mixtures: combining mixture models and graph-based methods for inductive and scalable semi-supervised learning. In: The 22nd International Conference on Machine Learning, Bonn, Germany, pp. 1052–1059 (2005)
11. Ghahramani, Z., Hinton, G.E.: The EM algorithm for mixtures of factor analyzers. Technical Report CRG-TR-96-1, Department of Computer Science, University of Toronto (1996)
12. Th, Y.W., Roweis, S.T.: Automatic alignment of hidden representations. In: Sixteenth Annual Conference on Neural Information Processing Systems, Vancouver, Canada, vol. 15, pp. 841–848. MIT Press, Cambridge (2002)
13. Levina, E., Bickel, P.J.: Maximum likelihood estimation of intrinsic dimension. In: Eighteenth Annual Conference on Neural Information Processing Systems, Vancouver, Canada, pp. 777–784 (2004)
14. Roweis, S., Saul, L., Hinton, G.E.: Global coordination of local linear models. In: Sixteenth Annual Conference on Neural Information Processing Systems, Vancouver, Canada, pp. 889–896 (2002)
15. Brand, M.: Charting a manifold. In: Sixteenth Annual Conference on Neural Information Processing Systems, Vancouver, Canada, pp. 985–992 (2002)
16. He, X., Cai, D., Yan, S., Zhang, H.J.: Neighborhood preserving embedding. In: Tenth IEEE International Conference on Computer Vision, Beijing, vol. 2, pp. 1208–1213 (2005)
17. Zhang, T., Yang, J., Zhao, D., Ge, X.: Linear local tangent space alignment and application to face recognition. Neurocomputing 70, 1547–1583 (2007)
18. He, X.: Incremental semi-supervised subspace learning for image retrieval. In: Proceedings of the ACM Conference on Multimedia, New York, pp. 10–16 (October 2004)
19. Yang, X., Fu, H., Zha, H., Barlow, J.L.: Semisupervised nonlinear dimensionality reduction. In: ICML 2006, Pittsburgh, PA, pp. 1065–1072 (2006)

RANSAC Based Ellipse Detection with Application to Catadioptric Camera Calibration

Fuqing Duan[1], Liang Wang[2], and Ping Guo[1]

[1] College of Information Science and Technology, Beijing Normal University, Beijing 100875, P.R. China
[2] College of Electronic Information and Control Engineering, Beijing University of Technology, Beijing 100080, P.R. China

Abstract. In this paper, a simple method for ellipse detection is proposed and applied in central catadioptric camera calibration. It consists of two phases. Firstly it locates ellipse center candidates using center symmetry of ellipses, and the detected edge points are grouped into several subsets according to the center candidates. Then all the ellipses are fitted by performing RANSAC for each subset. We also present an approach for calibrating a central catadioptric camera based on the bounding ellipse of the catadioptric image. Using the proposed ellipse detection method, we can easily detect the bounding ellipse. As a result, a simple self-calibration can be realized, which can be used in some applications where high accuracy of the calibration is not required. Experiments show the proposed method is effective.

Keywords: Ellipse detection, central catadioptric camera, camera calibration.

1 Introduction

Detection of ellipses or circles from an image of a scene is very useful in many vision applications such as object location, tracking, and camera calibration etc. Hutter and Brewer [10] fit an ellipse to the image of the vehicle wheel for vehicle pose determination. Kwolek [12] uses an ellipse model to approximate the image of the human head in head tracking. Ying et al [16, 17] use the images of spheres, which are ellipses, to calibrate central catadioptric cameras. For detecting structures in images, Hough transform and RANSAC are generally used robust techniques. However, an ellipse has five parameters which include location, shape, and orientation. So directly applying Hough transform to ellipse detection requires a five-dimensional accumulator. It is impractical because of the huge computation and storage. On the other hand, it is well known that RANSAC also needs a huge computation if the ratio of the inliers is low.

In papers[1,9],ellipse detection is decomposed into two stages. The first stage detects the ellipse center, and the second stage determines the other parameters. The method in [1] requires accurate calculation of the gradients and tangents of the edge pixels, which is sensitive to the image noise and time-consuming. The

K.W. Wong et al. (Eds.): ICONIP 2010, Part II, LNCS 6444, pp. 525–532, 2010.

method in [9] firstly determines two symmetry axes by the Hough transform following the horizontal and vertical scanning, whose cross is the ellipse center, and then uses the geometry symmetry to determine the other parameters. Xu et al.[15] develop the Randomized Hough Transform (RHT) which randomly selects five points in each iteration and use them to vote on the ellipse parameters. The accuracy and speed of this algorithm depend on the ratio of the inliers, which is similar to RANSAC. If the ratio is low, the number of the random sampling should be large, and then the computation is heavy.

Many computer vision applications expect a large field of view such as robot navigation, surveillance, teleconferencing and virtual reality etc. One effective way to enlarge the field of view is to combine mirrors with conventional cameras, which is called a catadioptric imaging system [2]. A catadioptric system with a unique viewpoint is called a central catadioptric system. A central catadioptric system [2] can be built by setting a parabolic mirror in front of an orthographic camera, or a hyperbolic, elliptical, planar mirror in front of a perspective camera, where the single viewpoint constraint can be fulfilled via a careful alignment of the mirror and the camera. Previous calibration methods [3,4,13,16,17] for central catadioptric cameras are mainly based on the projections of lines. Nearly all these approaches need conic fitting since a line in space is projected to a conic in a central catadioptric image, and the accuracy of the calibration highly depends on the accuracy of the conic fitting. The conic is called the line image. In general, only a small segment of the conic is visible in the catadioptric image due to the partial occlusion, which makes the conic estimation hard to accomplish. Wu et al [14] present a calibration method of no conic estimation, but it is mainly for para-catadioptric cameras. Moreover, there are few algorithms for self-calibration. Kang[11] proposes a nonlinear self-calibration method to calibrate a para-catadioptric system by tracking feature correspondences across multiple views. Since the omni-directional images have large distortion, finding feature correspondences is not a trivial task.

In this work, similar to several previous approaches, ellipse detection is decomposed into two phases. Firstly ellipse center candidates are detected using center symmetry of ellipses, and the detected edge points are grouped into several subsets according to the center candidates. Then all the ellipses are fitted by performing RANSAC for each subset. Using the proposed ellipse detection method, we can easily detect the bounding ellipse of the catadioptric image. As a result, a simple self-calibration method is presented, which can be used in some applications where high accuracy of the calibration is not required or serve as a good initial estimation. Section 2 shows the ellipse detection. Section 3 describes the proposed calibration method. Experimental results are reported in Section 4, and followed are some conclusions in Section 5.

2 Ellipse Detection

An ellipse is a conic, and can be generally represented as follows:

$$ax^2 + 2bxy + cy^2 + 2dx + 2ey + f = 0 \tag{1}$$

Given an image, the task of the ellipse detection is to decide whether the image contains an ellipse, and to determine the ellipse equation (1) if an ellipse is present. Similar to previous approaches [1,9], we decompose the ellipse detection into two phases. The ellipse center candidates are detected in the first phase, and other parameters are determined in the second phase.

It is well known that an ellipse is symmetrical on the ellipse center, and the middle point of any two symmetric points is the ellipse center. For an image containing ellipses, the edge map can be easily obtained using the Canny or other edge operator. Since each pair of symmetric points of one ellipse will cast a vote on its center, while each pair of dissymmetrical points casts a vote dispersedly, ellipse center candidates can be detected by the voting from the middle point of each two points. For each ellipse center candidate, there is an associated subset of symmetric points.

In the second phase, we use RANSAC to estimate the ellipse equation (1) for each ellipse candidate. RANSAC is an iterative method for robust fitting of a model from a set of observed data which contains many outliers. Assume that the ratio of the inliers is ρ, the minimal data set for determining the model includes k data points, and the probability that there is a good sampling among the M samplings is z, then [6]

$$\mathrm{M} = \frac{\log(1-z)}{\log(1-\rho^k)} \tag{2}$$

From the equation (2), we can see that the number M increases with the ratio ρ decreasing and with the data number k increasing. Generally speaking, the number k should be 5 for ellipse parameter estimation when using RANSAC directly, since an ellipse has 5 parameters. So the sampling number M will be very large and the computation is impractical if the ratio of inliers is very low. However, here the point number of the minimal data set for determining the ellipse is 3 because only three geometry parameters need to be determined. Since the ratio of inliers in each subset is very high for each ellipse center candidate, the sampling number will be very small in the propose method, and the parameters can be detected accurately and quickly by RANSAC.

Finally, we need to verify whether each detected ellipse candidate is a real ellipse since we use a threshold in detecting the center candidates. For each ellipse candidate, we can determine the following parameters from equation (1):

$$\mathrm{E} = ac - b^2, \ \ \mathrm{F} = a + c, \ \ \mathrm{H} = \begin{vmatrix} a & b & d \\ b & c & e \\ d & e & f \end{vmatrix} \tag{3}$$

An ellipse should subject to the condition [15]:

$$\mathrm{H} \neq 0, \mathrm{E} > 0, \mathrm{FH} \leq 0 \tag{4}$$

where, the ellipse is a circle if $\mathrm{E} = 0$.

3 Central Catadioptric Camera Calibration Based on the Boundary Ellipse

3.1 Central Catadioptric Camera

Geyer and Daniilidis [8] show that the central catadioptric imaging process is equivalent to the following two-step mapping by a sphere (see Fig. 1):

1) Under the viewing sphere coordinate system Oxyz, a 3D point $\mathbf{X} = [x,\ y,\ z]^T$ is projected to a point $\mathbf{X}^s$ on the unit sphere centered at the viewpoint $\mathbf{O}$ by $\mathbf{X}^s = [x/r,\ y/r,\ z/r]^T, r = ||\mathbf{X}||$.

2) The point $\mathbf{X}^s$ on the viewing sphere is projected to a point $\mathbf{m}$ on the image plane Π by a pinhole camera through the perspective center $\mathbf{O}^c$. The image plane is perpendicular to the line going through the viewpoints $\mathbf{O}$ and $\mathbf{O}^c$, and it is also called the catadioptric image plane.

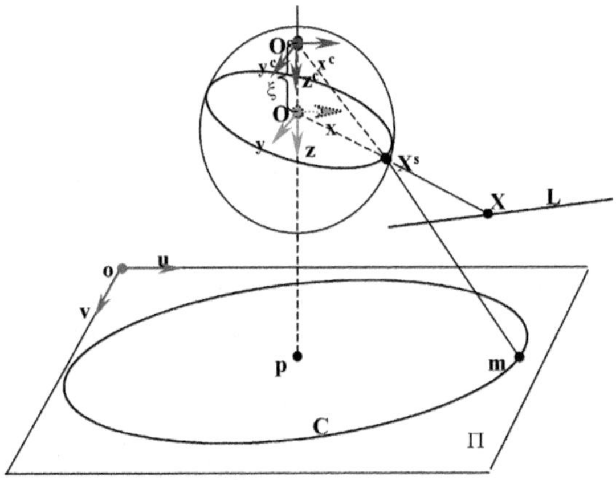

Fig. 1. Central catadioptric camera

In this camera system, the optical axes of the pinhole camera is the line $\mathbf{O}^c\mathbf{O}$, and thus its principal point is the intersection, $\mathbf{p} = [\mathrm{u}_0, \mathrm{v}_0, 1]^T$, of the line $\mathbf{O}^c\mathbf{O}$ with the image plane Π. The distance from point $\mathbf{O}$ to $\mathbf{O}^c$, $\xi = ||\mathbf{O} - \mathbf{O}^c||$ is called the mirror parameter, which determines the mirror used in the central catadioptric camera. The mirror is a paraboloid if $\xi = 1$, an ellipsoid or a hyperboloid if $0 < \xi < 1$, and a plane if $\xi = 0$. The details can be found in [8]. In this paper, we assume $0 < \xi \leq 1$, i.e. do not consider the case of plane mirror. Let the intrinsic matrix of the pinhole camera be

$$\mathbf{K} = \begin{bmatrix} \mathrm{rf\,s\,u_0} \\ \mathrm{0\ f\ v_0} \\ \mathrm{0\ 0\ 1} \end{bmatrix}. \tag{5}$$

Where f is the effective focal length; r is the aspect ratio; $\mathbf{p} = [\mathrm{u_0}, \mathrm{v_0}, 1]^T$is the principal point; s is the parameter describing the skew of the two image axes. Then, the catadioptric image of a space point $\mathbf{X}$ is

$$\mathbf{m} = \lambda \mathbf{K}[\mathbf{I},\ \xi \mathbf{e}] \begin{bmatrix} \mathbf{X}^s \\ 1 \end{bmatrix} = \lambda \mathbf{K}(\mathbf{X}^s + \xi \mathbf{e}), \tag{6}$$

with λ being a scalar, $\mathbf{I}$ the 3×3 identical matrix, and $\mathbf{e} = [0,\ 0,\ 1]^T$.

In the catadioptric camera calibration, there are totally six parameters $\{\mathrm{r}, \mathrm{f}, \mathrm{s}, \mathrm{u_0}, \mathrm{v_0}, \xi\}$ to be determined.

3.2 Calibration Based on the Boundary Ellipse

It's well known that the mirror boundary of the central catadioptric camera is a circle and the projection of the mirror boundary is an ellipse (see Fig.2). The optical axis is perpendicular to the plane containing the circle and goes through the center of the circle, so the center of the bounding ellipse is the principal point $[\mathrm{u_0}, \mathrm{v_0}, 1]^T$. In [7, 11], partial intrinsic parameters of the camera can be determined from the bounding ellipse of the catadioptric image. In fact, since both the eccentricity ε of the mirror and the field of view (FOV) are usually known, we can determine all intrinsic parameters from the bounding ellipse. Here the mirror parameter ξ can be obtained from the eccentricity ε of the mirror as [8]:

$$\xi = \frac{2\varepsilon}{1 + \varepsilon^2} \tag{7}$$

Let the distance from the optical center $\mathbf{O}^\mathrm{c}$ to the center $\mathbf{O}^\mathrm{b}$ of the boundary circle be h and the radius of the circle be q. As Fig.2 shows, q and h can be decided from FOV, i.e. $q = \sin\theta, h = \xi + \cos\theta,\ \theta = \frac{FOV}{2}$. Assume that the coordinate of a point on the mirror boundary under the perspective coordinate system $\mathrm{O^c x^c y^c z^c}$ is $\mathbf{X} = (x, y, h)^T$, and its image point is $\mathbf{m} = (u, v, 1)^T$. Then,

$$\mathbf{K}^{-1}\mathbf{m} = \frac{1}{h}\mathbf{X}, \mathbf{m}^\mathrm{T}\omega\mathbf{m} = \mathbf{m}^\mathrm{T}\mathbf{K}^{-\mathrm{T}}\mathbf{K}^{-1}\mathbf{m} = \frac{\mathbf{X}^\mathrm{T}\mathbf{X}}{h^2} = 1 + \frac{q^2}{h^2} \tag{8}$$

Where $\omega = \mathbf{K}^{-\mathrm{T}}\mathbf{K}^{-1}$ is the image of the absolute conic (IAC) of the pinhole camera. Then the bounding ellipse can be expressed as:

$$\mathbf{m}^\mathrm{T}\mathbf{C}\mathbf{m} = \mathbf{0}. \tag{9}$$

where,$\mathbf{C} \approx (\omega - \lambda \mathbf{E_3})$,$\omega_{\mathbf{11}} = \frac{\mathbf{1}}{\mathbf{rf}}$,$\lambda = 1 + \frac{q^2}{h^2}$, $\mathbf{E_3} = \begin{bmatrix} 0\,0\,0 \\ 0\,0\,0 \\ 0\,0\,1 \end{bmatrix}$. In addition, the catadioptric image of the 3D mirror point $\mathbf{X}_0 = (\mathrm{q}, 0, \mathrm{h})^\mathrm{T}$ is

$$\mathbf{m}_0 = \frac{1}{h}\mathbf{K}\mathbf{X}_0 = (\frac{\mathrm{rf}q}{h} + \mathrm{u_0}, \mathrm{v_0}, 1)^\mathrm{T} \tag{10}$$

It is one intersection of the bounding ellipse and the line $v = \mathrm{v}_0$ on the image plane. So we can determine the item ω_{11} of the IAC from the image point $\mathbf{m}_0$, and the IAC can be uniquely determined by the item ω_{11} and the bounding ellipse equation (9). Thus, all camera intrinsic parameters can be obtained by using Choleskey decomposition of the IAC.

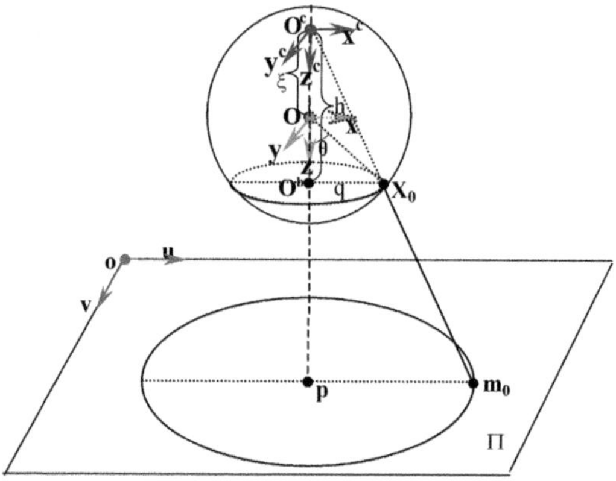

Fig. 2. Mirror boundary and its projection

4 Experimental Results

Fig.3a shows an image with a resolution of 2048×1536, which is taken by a catadioptric system consisting of a perspective camera with a hyperbolic mirror. The mirror is designed by the Center for Machine Perception, Czech Technical University, its FOV is 217.2 degree, and the eccentricity of the hyperbolic mirror is 1.302, corresponding to $\xi = 0.966$. Fig.3b shows the edge map. The detected ellipse center and the symmetrical point set are also shown in Fig.3b. The boundary ellipse detected by RANSAC is shown in Fig.3a, and Fig.3c shows the rectified image using the calibration result.

We can see from the figures that there are many outliers among the symmetrical point sets of the two images, and the ellipse detection using RANSAC is very accurate. From the rectified images, we can see that the rectified lines are roughly straight, which shows that the calibration approach is effective. Furthermore, we optimize the calibration results by using the constraint that a space line is projected onto a great circle on the view sphere. The used line image points are manually chosen. The rectified image using the optimizing results is shown in Fig.3d. By comparing the rectified lines, we can see that the optimization improves the calibration results and the calibration accuracy of the proposed algorithm is not high. However, it realizes a simple self-calibration and can be used in some applications where high accuracy of the calibration is not required or serve as a good initial estimation.

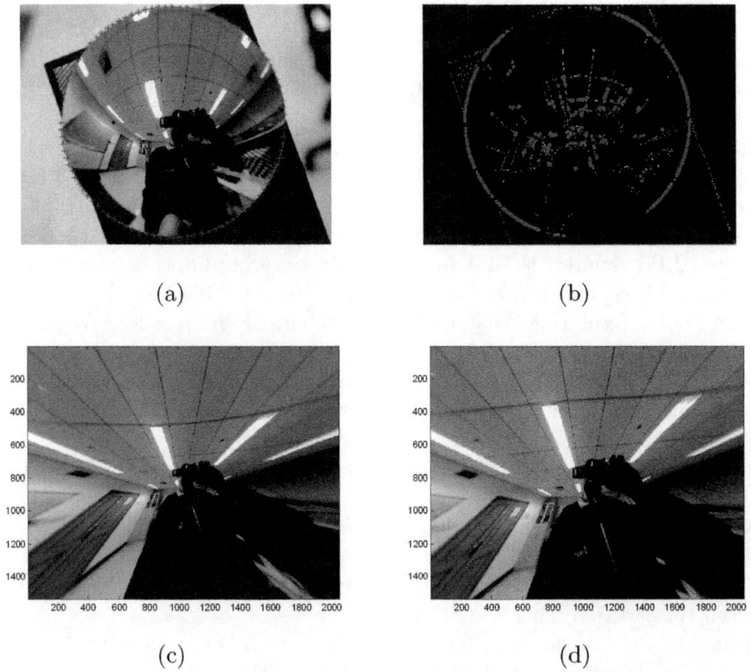

Fig. 3. The hypercatadioptric camera: (a) The image and the boundary ellipse. (b) The edge map and the symmetrical point set. (c) The rectified image using the proposed calibration. (d) the rectified image after optimization.

5 Conclusion

In this work, a simple method for ellipse detection is proposed and applied in catadioptric camera calibration. Similar to several previous approaches, the ellipse detection is decomposed into two phases. Firstly ellipse center candidates are detected using center symmetry of ellipses, and the detected edge points are grouped into several subsets according to the center candidates. Then all the ellipses are fitted by performing RANSAC for each subset. Using the proposed ellipse detection method, we can easily detect the bounding ellipse of the catadioptric image. As a result, a simple self-calibration method for central catadioptric cameras is presented, which can be used in some applications where high accuracy of the calibration is not required. Experiments demonstrate the efficiency of the proposed method. Of course, the calibration accuracy can be improved by an optimizing process automatically. This is our future work.

Acknowledgments. This work was supported by the National Natural Science Foundation of China (Grant No.60872127).

References

1. Aguado, A., Montiel, M.E., Nixon, M.S.: On using directional information for parameter space decomposition in ellipse detection. Pattern Recognition 29, 369–381 (1996)
2. Baker, S., Nayer, S.: A theory of single-viewpoint catadioptric image formation. IJCV 35, 175–196 (1999)
3. Barreto, J.P., Araujo, H.: Geometric properties of central catadioptric line images and their application in calibration. IEEE Trans. on PAMI 27, 1327–1333 (2005)
4. Barreto, J.P., Araujo, H.: Fitting conics to paracatadioptric projection of lines. Computer Vision and Image Understanding 101, 151–165 (2006)
5. Vossler, D.L.: Exploring Analytic Geometry with Mathematica. Academic Press, London (1999)
6. Fischler, M.A., Bolles, R.C.: Random Sample Consensus: A Paradigm for Model Fitting with Applications to Image Analysis and Automated Cartography. Comm. of the ACM 24, 381–395 (1981)
7. Geyer, C., Daniilidis, K.: Catadioptric camera calibration. In: Proc.7th ICCV, vol. I, pp. 398–404 (1999)
8. Geyer, C., Daniilidis, K.: Catadioptric projective geometry. IJCV 45(3), 223–243 (2001)
9. Ho, C.T., Chen, L.H.: A fast ellipse/circle detector using geometric symmetry. Pattern Recognition 28, 117–124 (1995)
10. Hutter, M., Brewer, N.: Matching 2-D Ellipses to 3-D Circles with Application to Vehicle Pose Identification. IVCNZ, 153–158 (2009)
11. Kang, S.: Catadioptric self calibration. CVPR I, 201–207 (2000)
12. Kwolek, B.: Stereo-vision based head tracking using color and ellipse fitting in a particle filter. In: Pajdla, T., Matas, J(G.) (eds.) ECCV 2004. LNCS, vol. 3024, pp. 192–204. Springer, Heidelberg (2004)
13. Wu, F.C., Duan, F.Q., Hu, Z.Y., Wu, Y.H.: A new linear algorithm for calibrating central catadioptric cameras. Pattern Recognition 41, 3166–3172 (2008)
14. Wu, Y., Li, Y., Hu, Z.: Easy calibration for para-catadioptric-like camera. In: IEEE International Conference on Intelligent Robots and Systems, pp. 5719–5724 (2006)
15. Xu, L., Oja, E.: Randomized Hough Transform(RHT): Basic Mechanisms, Algorithms and Complexities. Computer Vision, Graphics, and Image Processing: Image Understanding 57, 131–151 (1993)
16. Ying, X.H., Hu, Z.Y.: Catadioptric camera calibration using geometric invariants. IEEE Transactions on Pattern Analysis and Machine Intelligence 26(10), 1260–1271 (2004)
17. Ying, X.H., Zha, H.B.: Identical projective geometric properties of central catadioptric line images and sphere images with applications to calibration. IJCV 78, 89–105 (2008)

Speed Up Image Annotation Based on LVQ Technique with Affinity Propagation Algorithm

Song Lin[1,3], Yao Yao[1], and Ping Guo[1,2]

[1] School of Com. Sci. & Tech., Beijing Institute of Technology, Beijing 100081, China
[2] Lab. of Image Proc. & Patt. Recog., Beijing Normal University, Beijing 100875, China
[3] School of Com. & Info., Fujian Agriculture and Forestry University, Fujian 350002, China
pinege@sina.com, yaoyao19880812@gmail.com, pguo@ieee.org

Abstract. For a support vector machine (SVM) classifier applied to image annotation, if too many training samples are used, the training speed might be very slow and also bring the problem of declining the classification accuracy. Learning vector quantization (LVQ) technique provides a framework to select some representative vectors which can be used to train the classifier instead of using original training data. A novel method which combines affinity propagation algorithm based LVQ technique and SVM classifier is proposed to annotate images. Experimental results demonstrate that proposed method has a better speed performance than that of SVM without applying LVQ.

Keywords: automatic image annotation, learning vector quantization, affinity propagation, support vector machine, training speed.

1 Introduction

Automatically describing images with some semantic words is the goal of automatic image annotation (AIA). In recent years, various methods of AIA have been proposed [1][2][3], among them, the most common method was to regard each annotation word as a category label, thus, classifiers could be used to annotate images. Besides, in AIA, one image is often divided into some blocks, and each block is regarded as a training sample for classifiers instead of segmented regions [2][3][4], because there is still no general method that can achieve perfect segmentation results at present.

As a kind of efficient classifier, support vector machine (SVM) [5], which is especially dominant on small sample size data, has been widely used in AIA[3][4][6]. Among these studies, Shao *et al* [6] adopted SVM to implement image annotation by dividing the visual descriptors into different image categories. Chen *et al* [4] combined diversity density algorithm and SVM for image categorization, Cusano *et al* [3] randomly selected 1500 points from the training set to train SVM for AIA. In spite of the improvement of classification accuracy in those research works, the training set still has many samples, which does not satisfy the small sample size characteristics of SVM, and too many samples will slow the training process.

K.W. Wong et al. (Eds.): ICONIP 2010, Part II, LNCS 6444, pp. 533–540, 2010.

Therefore, if we select some representative samples rather than all samples (e.g. selecting hundreds of samples from tens of thousands of samples) for SVM training, it may not only speed up the training process observably, but also can accelerate the annotation process because of producing sparse support vectors.

Learning vector quantization (LVQ) [7] is a framework to acquire a small number of representative vectors from the training data. Chang *et al* [8] adopted LVQ technique to initialize RBF network for texture image classification. In [9] and [10], LVQ was used to reduce the feature vector dimension, while Jiang *et al* [11] utilized LVQ to refine training data. However, in those works, LVQ was realized in a traditional way, namely, Self-Organizing Feature Map (SOM) [7]. SOM has two obvious shortcomings: low speed and poor robustness. As another realization of LVQ, affinity propagation (AP) algorithm [12] could overcome those shortcomings. Frey and Dueck [12][13] verified that AP algorithm achieves competitive classification performance and robust results in image categorization. Yang *et al* [14] applied AP algorithm to obtain Gaussian mixture model parameters in AIA. Considering the excellent performance of AP and the characteristics of SVM, we proposed a combination method of AP-based LVQ and SVM to speed up the training and annotation process of AIA.

2 Background

Here we briefly review the algorithms which are referred in this work.

2.1 Learning Vector Quantization

The basic idea of LVQ is to map k-dimensional vectors in the vector space R^k into a finite set of vectors $\mathbf{Y}=\{y_i|i=1, 2, \ldots, N\}$. Each vector y_i is called a codeword. And the set of all the codeword is called a codebook. Thus, this idea can be realized by applying clustering algorithms to select codeword. The earliest realization of LVQ is achieved by using SOM algorithm [7].

SOM is a model of neural network, which consists of an input layer and a competitive layer, which is usually considered as output layer. It clusters the input vectors through a competitive learning process: First, it computes the distances between the input vector and all connection weights vectors. Then, the neuron closest to the input vector as the winning neuron is selected. Final result is achieved through adjusting the connection weight vectors iteratively.

2.2 Affinity Propagation Algorithm

AP algorithm [12] can be applied as a clustering algorithm. The similarity set between all pairs of data points is taken as input. Similarity $s(i,k)$ indicates how well the data point k is suited to be the exemplar for point i. $s(k,k)$ for each data point k is known as "preference", it is a parameter required to input by user. The larger values of $s(k,k)$, the more clusters will be generated.

Two kinds of messages are exchanged between data points. The responsibility $r(i,k)$ reflects the accumulated evidence for how well-suited point k is to serve as the exemplar for point i. The availability $a(i,k)$ reflects the accumulated evidence for how appropriate it would be for point i to choose point k as its exemplar. Messages are iteratively updated according to some rules until a set of exemplars emerges.

2.3 SVM

As a kind of efficient classifier, SVM [5] is known having good generalization capability, and it is especially suitable for the problem of small number amount of samples with high dimension setting.

SVM distinguishes different classes through looking for their boundaries, which are determined by some points near boundaries known as support vectors. So, when training sample set is too large, it will need a lot of time to train, and many support vectors are generated. Consequently, generalization performance and classification accuracy may be degraded, and the speed of the classification will also be slow.

3 Proposed Scheme

The proposed scheme is illustrated as Fig. 1, which contains training stage and annotation stage.

3.1 Training

Feature Extraction. Each image is divided into a set of blocks for feature extraction, and feature extracted from each block is represented as a vector.

LVQ-A (LVQ with AP algorithm). In this stage, a few representative feature points (i.e. codebook) are selected for each class from the full training sample set with AP algorithm, and these representative points compose a concise training set for SVM. Assuming there are C categories in the training set, M images for each class, K blocks of every image, and the number of representative points for each class is L ($L<<K$), thus, the original training set has $M \times C \times K$ points, it can be decreased to $L \times C$ points, size of training set is reduced $M \times K/L$ times.

Directly refining $M \times C \times K$ points to $L \times C$ points would take huge memory space, also greatly increase the computational complexity, therefore, this stage should be further divided into following two sub-stages:

1) LVQ applied to single-image. Empirically, for each training image, we select about 30 representative points by AP from K feature points. So, the "preference" parameter should be determined in advance for the approximately fixed 30 points. Yang et al [14] found there was an approximately linear relationship between logarithm of the preference and logarithm of the cluster number. Corresponding preference parameter can be determined by that relationship.

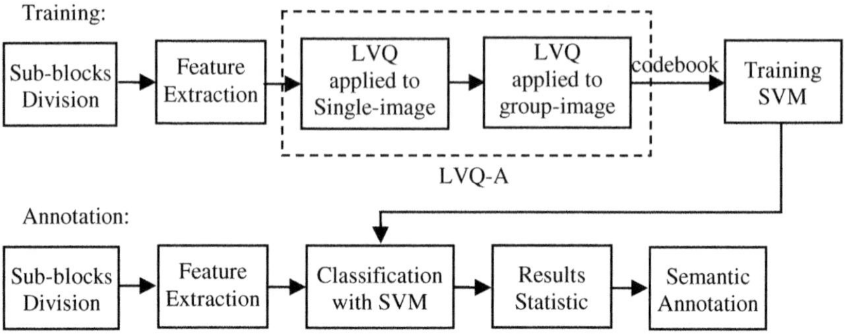

Fig. 1. Framework of the proposed scheme

2) LVQ applied to group-image. In the previous step, representative points have been selected for each image. Then, for selecting more general codebook for each class, the representative points chosen from the images of same class are gathered as the input of AP algorithm. After executing AP once more, codebooks for all categories can be obtained. According to result of experiments, we set 100 representative points for each class (i.e. let $L \approx 100$). Likewise, the preference parameter needs to be determined. At this point, the previous method is no longer applicable, because it is found that the number of clusters was sensitive to the preference parameter when $L \geq 50$, the preference value changed slightly could lead to significant changes in number of clusters. Via experiments, the empirical estimation of preference parameter can be obtained.

Training SVM. The refined representative point set is adopted as training samples to train SVM with **LIBSVM algorithm**.

3.2 Annotation

For each test image, it will take the same method to extract the features as in training phase. Then, with the trained SVM, each block is classified into corresponding class. And for each class, we calculate the number of blocks which have been classified into it, and annotate the test image with the class label including most of blocks.

4 Experimental Results

In this section, we first describe the experiment setup, then analyze the experimental results on the three aspects: preference parameters setting for AP, selecting the size of codebook and comparative evaluation of the proposed method.

4.1 Experimental Setup

Image dataset are collected from two parts: 1) a subset of Corel Database downloaded from [15] , including almost 4,000 images belonging to 4 categories; 2) some pictures from the Internet. We choose a subset of 500 images which covers 5 classes: "tiger", "elephant", "horse", "car" and "airplane". Each class contains 100 images, half of them is used for training and half of for testing.

In following experiments, all images were rescaled to the size of about 100,000 pixels. Then, each image was divided into a set of overlapping 32×32 blocks, with a sliding window moving by 16 pixels separately in the horizontal and vertical direction. Since all images contain roughly the same number pixels, they also contain approximately same number of blocks (about 350). Two types features are extracted for each block: Haralick feature [16] and Tamura feature [17]. Haralick feature is a kind of texture feature based on gray co-occurrence matrix, which leads to a 256-dimensional feature vector. Tamura feature describes texture properties from psychological perspective, which lead to a 6-dimensional feature vector. In this work, we combine these two types of features, and a 262-dimensional vector can be acquired for each block.

Notice that in description of the following three experiments, the values of training time does not contain the time on sub-blocks division and feature extraction, because it is the same for each compared algorithm.

In addition, for the experiment an available SVM algorithm in MATLAB named LIBSVM was downloaded from [18].

4.2 Experiment 1: Preference Parameter Setting for AP

To get the approximately fixed cluster number, we have to set the value of preference parameter for AP in advance. However, the method in [14] is no longer suitable for image-level clustering at which larger cluster number is desired. Especially, for data in different classes, setting the same preference parameter may result in a big difference in cluster number, and we found that the cluster number depends on the median of the input similarities. So, we assign

$$pref = fac \times medi \,, \tag{1}$$

where *pref* denotes the value preference parameter, *medi* represents the median of the input similarities, and *fac* is a positive number. Apolynomial is adopted to describe the relationship between *fac* and the number of cluster, defined as:

$$y = ax + b + cx^{-1} + dx^{-2}, \tag{2}$$

where x represents the number of clusters, y denotes *fac* value, a, b, c, d are the parameters need to be determined. In this experiment, we found that for the same cluster number, the *fac* values of tiger, elephant, car and airplane could be roughly equated, and the *fac* of horse was 1.1 times of others. Therefore, we can use one polynomial to get the desired number of clusters for every class. Here, a=0.001, b=-0.25, c=130, d=-300 are set respectively. The fitting curve is shown in Fig. 2.

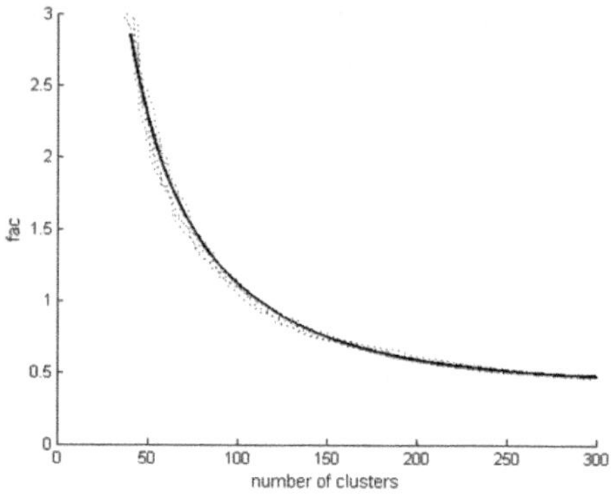

Fig. 2. *fac* vs. number of clusters

4.3 Experiment 2: Selection the Size of Codebook

To analyze the effect of the size of the codebook, we considered five codebook sizes of 50, 70, 100, 150 and 200 representative points for each class, respectively. The training time and classification accuracy of these sizes are listed in Table 1.

Table 1. The training time and accuracy for different codebook size

Codebook size	50	70	100	150	200
Training time(sec)	362	372	388	426	499
Accuracy (%)	73.5	76.5	79	77.8	79

As shown in Table 1, the codebook size of 100 had the highest classification accuracy of 79%, and the training time was only a litter longer than the smallest size of 50. Therefore, the codebook size of 100 was selected as a standard size in the proposed method which compared with other methods in the following experiment.

4.4 Experiment 3: Comparative Studies

In order to evaluate the speed performance of the proposed method, we compared it with LVQ-B (i.e. LVQ realized by SOM neural network) preprocessing plus SVM [11], and the SVM trained by using original sample set.

Owing to greatly reducing size of training sample set (e.g. 87462 points are reduced into 492, in this experiment, almost 180 times reduction) and high efficiency of the AP algorithm, proposed method (SVM combined with LVQ-A) win out obviously at training speed (shown as Fig.3.(a)). Fig. 3(b) illustrates LVQ based methods are also much faster than the SVM without LVQ on classification because of the reduction of support vector number. And proposed method performs more outstandingly with the more number of categories.

As for the accuracy of classification, besides to the three methods above, the Quadratic Discriminant Analysis (QDA) [19] with LVQ-A and QDA without LVQ are compared also. QDA is another classifier deferent from SVM, it requires large sample size with low dimensionality data. Here, the dimension of feature vectors was reduced to 30 by principal component analysis (PCA). In Table 2, we can see that there is no decline in accuracy when SVM is combined with LVQ, even the accuracy is improved in [11], while the relatively large accuracy loss of QDA is caused by a combination of LVQ. The results demonstrate that such a LVQ method is not applicable to those classifiers requiring large number of training samples, which in turn prove the feasibility of the LVQ method for SVM.

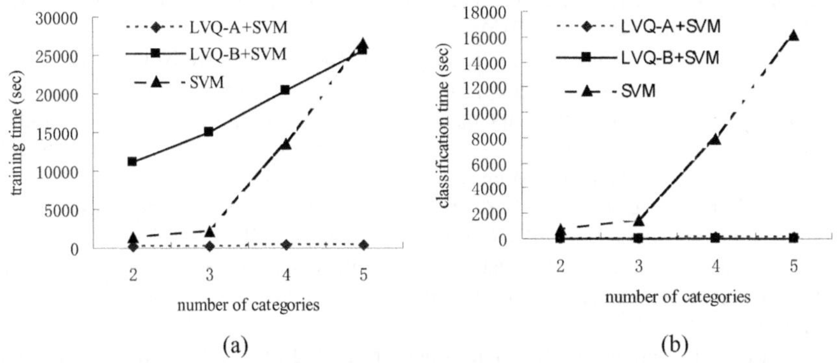

Fig. 3. The propose method is compared with LVQ-B +SVM and original SVM. (a) training time; (b) classification time.

Table 2. The accuracy comparison for various algorithms

Number of categories	2	3	4	5
LVQ-A+SVM	100%	98.5%	85.1%	79%
LVQ-B+SVM	99.7%	97.5%	82.6%	73.5%
SVM	100%	99.3%	86.9%	82%
LVQ-A+QDA	100%	99.2%	74.5%	68.5%
QDA	98.8%	100%	82.5%	76%

5 Conclusion

In this paper, a fast image annotation method, which incorporates AP-based LVQ technique and SVM classifier, is proposed. In this method, AP algorithm based LVQ technique is used to reduce the size of the training set, and a general method for AP is suggested to get desired cluster number. Experimental results confirm that the proposed annotation method has an excellent acceleration performance both on training and annotation phrase compared with the SVM on original training set and the SVM combined with SOM-based LVQ. And the advantage of the proposed method is more obvious for more samples and categories. Moreover, the comparison of accuracy also shows the feasibility of the proposed method.

Acknowledgments. The research work described in this paper was fully supported by the grants from the National Natural Science Foundation of China (Project No. 90820010, 60911130513). Prof. Ping Guo is the author to whom all correspondence should be addressed, his e-mail is pguo@ieee.org.

References

1. Perronnin, F.: Universal and Adapted Vocabularies for Generic Visual Categorization. IEEE Trans. PAMI 30(7), 1243–1256 (2008)
2. Carneiro, G., Chan, A.B., Moreno, P.J., Vasconcelos, N.: Supervised Learning of Semantic Classes for Image Annotation and Retrieval. IEEE Trans. PAMI 29(3), 394–410 (2007)
3. Cusano, C., Ciocca, G., Schettini, R.: Image annotation using SVM. In: Proceedings of SPIE, vol. 5304, pp. 330–338 (2004)
4. Chen, Y., Wang, J.Z.: Image categorization by learning and reasoning with regions. J. Machine Learning Research 5, 913–939 (2004)
5. Vapnik, V.N.: The Nature of Statistical Learning Theory. Springer, New York (1995)
6. Shao, W.B., Phung, S.L., Naghdy, G.: A multi-class image classification system using salient features and support vector machines. In: Proceedings of International Conference on Intelligent Sensors, Sensor Networks and Information Processing, pp. 431–436 (2007)
7. Kohenen, T.: Self-Organizing Maps. Springer, Berlin (2001)
8. Chang, C.Y., Wang, H.J., Fu, S.Y.: Texture Image Classification Using Modular Radial Basis Function Neural Networks. J. of Electronic Image 19(1), 1–11 (2010)
9. Viitaniemi, V., Laaksonen, J.: Evaluating the performance in automatic image annotation: Example case by adaptive fusion of global image features. Image Comm. 22(6), 557–568 (2007)
10. Vailaya, A., Figueiredo, M.A.T., Jain, A.K., Zhang, H.J.: Image Classification for Content-Based Indexing. IEEE Trans. Image Processing 10(1), 117–130 (2001)
11. Jiang, Z., He, J., Guo, P.: Feature Data Optimization with LVQ Technique in Semantic Image Annotation, accepted to ISDA 2010 (2010)
12. Frey, B.J., Dueck, D.: Clustering by Passing Messages between Data Points. Science 315, 972–976 (2007)
13. Dueck, D., Frey, B.J.: Non-metric Affinity Propagation for Unsupervised Image Categorization. In: Proceedings of the IEEE International Conference on Computer Vision, pp. 1–8. IEEE Press, Brazil (2007)
14. Yang, D., Guo, P.: Image Modeling with Combined Optimization Techniques for Image Semantic Annotation. Neural Computing and Applications (in press, 2010)
15. Stanford vision lab, `http://vision.stanford.edu/resources_links.html`
16. Haralick, R.M., Shanmugam, K., Dinstein, I.: Texture features for image classification. IEEE Trans. on Sys. Man and Cyb. 3(6), 610–621 (1973)
17. Tamura, H., Mori, S., Yamawaki, T.: Texture features corresponding to visual perception. IEEE Trans. on Sys., Man and Cyb. 8(6), 460–473 (1978)
18. LIBSVM-A, `http://www.csie.ntu.edu.tw/~cjlin/libsvm`
19. Guo, P., Jia, Y.D., Lyu, M.R.: A Study of Regularized Gaussian Classifier in High-Dimension Small Sample Set Case Based on MDL Principle with Application to Spectrum Recognition. Patt. Recog. 41(9), 2842–2854 (2008)

Dictionary of Features in a Biologically Inspired Approach to Image Classification

Sepehr Jalali*, Joo Hwee Lim, Sim Heng Ong, and Jo Yew Tham

Institute for Infocomm Research (I^{2}R), 1 Fusionopolis Way #21-01 Connexis (South Tower) Singapore 138632
National University of Singapore, 21 Lower Kent Ridge Road, Singapore 119077
{stusj,joohwee}@i2r.a-star.edu.sg,
eleongsh@nus.edu.sg, jytham@i2r.a-star.edu.sg
http://www.i2r.a-star.edu.sg, http://www.nus.edu.sg

Abstract. We introduce new methods for creation of a dictionary of features for a biologically inspired model of visual object classification that is shown to handle the recognition of several object categories. We provide a new method for creation of this features dictionary using non-supervised cortex like methods. Different clustering approaches were experimented and improved performance is achieved on image centers which results in real time classification of images by HMAX model.

Keywords: Dictionary of features, hierarchical structure, biologically inspired, clustering, HMAX.

1 Introduction

Object recognition in cortex is thought to be mediated by the ventral visual pathway running from primary visual cortex, V1, over extrastriate visual areas V2 and V4 to inferotemporal cortex, IT. Over the last decades, several physiological studies in non-human primates have established a core of basic facts about cortical mechanisms of recognition that seem to be widely accepted and that confirm and refine older data from neuropsychology. A brief summary of this consensus knowledge begins with the ground-breaking work of Hubel and Wiesel first in the cat [1] and then in the macaque [2]. Starting from simple cells in primary visual cortex, V1, with small receptive fields that respond preferably to oriented bars, neurons along the ventral stream show an increase in receptive field size as well as in the complexity of their preferred stimuli.

HMAX is a hierarchical computational model of object recognition in cortex proposed by Riesenhuber and Poggio [3]. The standard model simulates the feed-forward path of the visual cortex and has been used to classify animal vs. non-animal images and paper clip images first and is similar to Neocognitron [4] in using both simple and complex cells. This model is used to find a good tradeoff between invariance and selectivity. A dictionary of features is created

* Corresponding Author.

K.W. Wong et al. (Eds.): ICONIP 2010, Part II, LNCS 6444, pp. 541–548, 2010.

by randomly sampling over the images in the higher levels of this structure and is used for performing classification of images in the supervised level of the structure. In this paper, we investigate different non-random methods for sampling and compare its performance with existing models.

In this paper, we introduce the HMAX model in Section 1.1. In Section 2 we introduce our implementation of the model and provide the experimental results of our modifications to the dictionary of features created by the model and discuss the achievements of modification followed by conclusion and future works proposals in Section 3.

1.1 Related Work

The standard HMAX model consists of *S1*, *C1*, *S2* and *C2* layers, followed by a classifier such as Support Vector Machine. The structure begins with a gray scale input image which is fed into the S1 layer. In S1 layer a fixed size of Gabor filters is implemented on different scales of the images which provides the same invariance to scale for Gabor filters [5,6]. In this model, an image is fed into the structure and 10 different scales of the image are created as input to S1 layer. Gabor filters in 4 directions in their standard model, and 16 directions in their extended model are created based on Eq. 1 and convolved on the images.

$$G(x,y) = \exp\left(-\frac{(X^2+\gamma^2Y^2)}{2\sigma^2}\right)\cos\left(\frac{2\pi}{\lambda}X\right) \tag{1}$$

These outputs are sent to C1 layer, which performs a local max operation on both size and position of the filter responses. The response of a patch of pixels X to a particular S1 filter G is given by Eq. 2

$$R(x,y) = \left|\frac{\sum X_iG_i}{\sqrt{\sum X_i^2}}\right| \tag{2}$$

The output of this layer will be between 500-2000 different patches of size 4×4, 8×8, 12×12 and 16×16 depending on the size of the input image. In this level, a dictionary of features is randomly sampled from these patch windows. One or two samples are randomly sampled from each training image, and a feature's dictionary of size 4075 of prototypes is created. The response of a patch of C1 units X to a articular S2 feature/prototype P, of size n×n, is given by a Gaussian radial basis function in Eq. 3.

$$R(x,P) = \exp\left(-\frac{\| X - P \|^2}{2\sigma^2\alpha}\right) \tag{3}$$

In order to train the SVM, the distance of each sample from each training image with each entry on the dictionary is calculated and the max is taken in C2 layer. These features are sent to the SVM for training. For testing images the same hierarchical procedure is repeated and the performance of the system is calculated. They proposed a few modifications to improve the performance of the system such as creating sparse prototypes and take the max response from all directions for each window. The other proposed modification is to run a SVM normals method [7] to select the features with higher weights. SVM is run a few times, and each time features with lower weights are dropped. In the previous model proposed by Serre et.al [8] this hierarchical structure is repeated to create S3 and C3 layers, and in another implementation, S2b and C2b layers are also created to bypass S2 and C2 layers in creating the dictionary [9].

2 Implementation and Results

The use of random sampling to create the dictionary of features for the model, was a prospective investigation possibility which motivated us to compare the performance of a non random sampling method with random sampling and investigate the role of compressive sampling in this case. The image dataset used for these experiments was *Caltech101* [10], which includes 101 classes of objects plus a background category. Each class contains between 35 to 800 color images, in different sizes. Most categories have about 50 images. The size of each image is roughly 300 x 200 pixels. We used 30 randomly chosen images for training from each class and the rest of the images were used in the test phase. In [5, 11] random sampling is performed on each image, and one or two samples from each image are added to a dictionary to create a dictionary of size 4075. Samples are selected from each image in different positions and scales using a random generator function with a Gaussian distribution, based on the number of images per class and by taking a different number of samples from each image.

We performed different non-random sampling methods and compared their performance in an extensive set of experiments. Clustering can be considered the most important unsupervised learning problem; so, as every other problem of this kind, it deals with finding a structure in a collection of unlabeled data. We performed a K-means clustering after sampling more samples from each image in different approaches. Different number of samples and different number of clusters were tested in a series of approaches. For clustering we used a standard K-means algorithm. Whenever an empty cluster was created in the batch update phase, we created a new cluster consisting of the one point furthest from its centroid. Squared Euclidean distance was chosen as the distance measure, so that each centroid is the mean of the points in the corresponding cluster. We used the K-means function in Statistical Toolbox of Matlab(R) for clustering.

2.1 Clustering over Images from All Classes

In the first approach, we sampled between 5 to 20 random samples per image individually, to achieve a more dense sampling and added these samples to the dictionary of features, resulting in a very big features dictionary of size 15000 to 60000. We then performed a clustering over the whole dictionary and created a dictionary of sizes 1000 to 9000. We evaluated the performance of the system 8 times for each sample number and dictionary size and the average performance on each size was calculated. The results can be seen in Fig. 1.

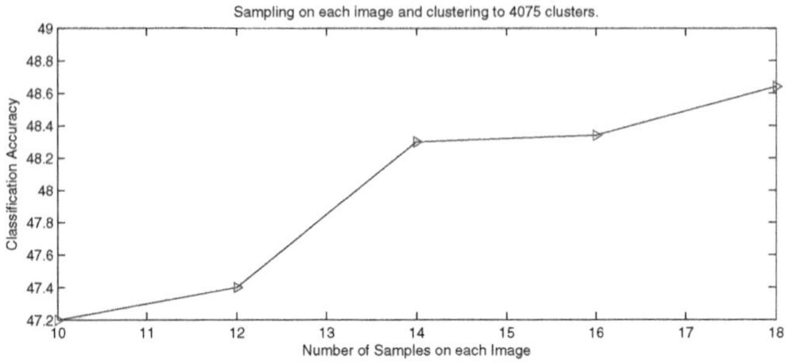

Fig. 1. Sampling over all images and performing clustering over the whole samples to create the dictionary of features

2.2 Performing Clustering on Images Individually

In another approach, we sampled all the possible positions of C1 features for each image non-randomly and added between 3 to 10 clusters per image to the dictionary and evaluated the performance of the system. Furthermore we performed sampling on more features from each image in another set of experiments, and performed another clustering on the whole dictionary again to reduce the number of features to 4075. As can be seen in Fig. 2, the classification accuracy achieved in this method, is around 52.69% in the best case.

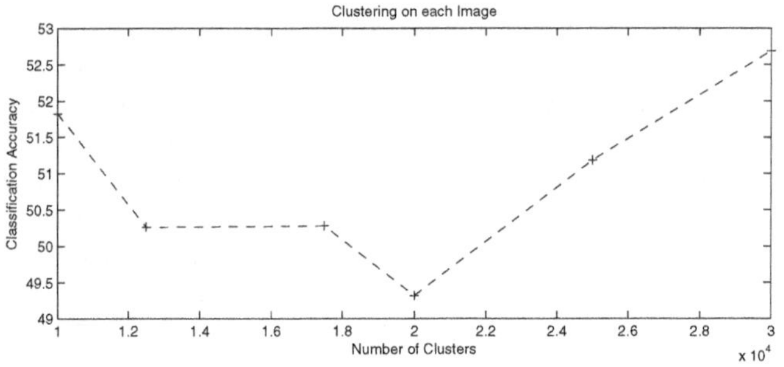

Fig. 2. Sampling over one single image and performing clustering in image level

2.3 Clustering over Images of Each Class and Investigating K-NN Algorithm Performance

In the third approach, we performed sampling on images of each class separately. Different number of samples were chosen for each image both randomly and non-randomly. In the case of non-random sampling, a regular scanning over all possible sizes and positions was conducted with a step size depending on the ratio of number of possible positions for sampling over number of samples to be sampled from each image. We sampled different numbers of samples from each

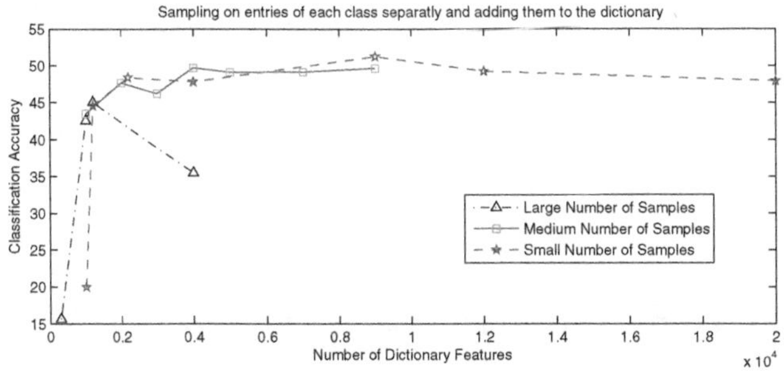

Fig. 3. Sampling over each class of images

image and created 3 different categories. Small number of samples standing for 100-300 samples in each image, 300-600 samples per image for medium number of sampling, and 600-2000 for large number of samples. For all the samplings the ratio of the number of clusters over each class to number of samples per image, was kept around 0.1 as it showed best performance. The results of these experiments can be seen in Fig. 3 In another experiment, we tried skipping the higher levels of the hierarchy to evaluate the performance of S2 dictionary in a K-NN method. We created a dictionary of features as described above (sampling over each class of images) and labeled each sample according to its class of images. In the test phase, we sampled from each image 10-90 random samples, and found the K-NN matches (K varying between 10 to 100) with the existing dictionary of features, and assigned the image to the class based on a majority voting of the minimum distance with dictionary prototypes. In the classification phase, k is a user-defined constant, and an unlabeled vector (a query or test point) is classified by assigning the label which is most frequent among the k training samples nearest to that query point. Different approaches were taken here such as sampling randomly, non-randomly, sampling more features from each image, and performing a clustering afterwards, but none of them resulted in a performance better than 10 percent, which is very low in comparison with the 52+ % performance we achieved with the previous method.

The high performance achieved here, emphasizes on the role of attention and selection of samples which have more information on each class. In another experiment in this level, we used only images of one class, and created a dictionary from those images, and used this dictionary to classify the whole dataset, and we achieved high performances of correct classification. However, the performance was slightly lower than sampling over all the classes, this suggests that these features can be used to create a universal dictionary of features for classification in this model.

2.4 Sampling over Center of Images

Since most of the images in Caltech101 are focused on the center, in another approach we tried sampling from the center of images which we assumed there

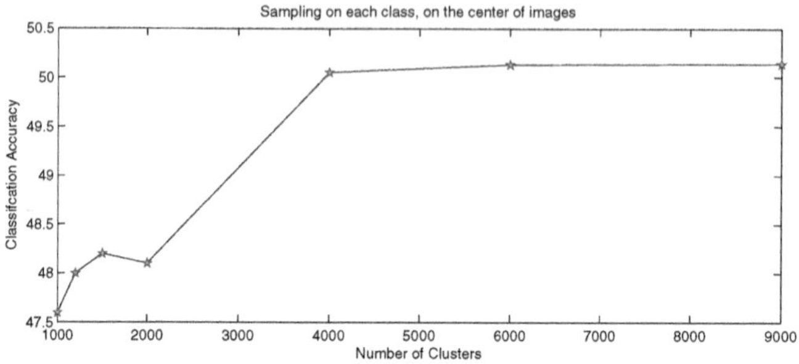

Fig. 4. Creating the dictionary of features from the center of images rather than the whole image

exist more meaningful information related to objects, and less information regarding background. We tried different sizes of center of image from a quarter to half, and tested it on two different modes. One is to create the dictionary of features from the center of images, and in the testing and testing phase, work with full size images. The results of this approach can be seen in Fig. 4. One explanation to the reason of why clustering did not improve the performance can be because the clusters created from images are both from background categories and object categories. Since in *Caltech101* dataset, most objects are in the center of the images, we tried clustering over only center of the images for creating the dictionary of features. In this case, we sampled from the center quarter of the image only, and created a dictionary of features from there. The results of this approach can be seen in Table. 5. This approach did not beat the performance of random sampling but we could achieve better performances with smaller sizes for dictionary of features in comparison with random sampling. All experiments were based on the GPU CUDA implementation code provided in [5, 11] for hierarchical structures. In all samplings, the relationship between

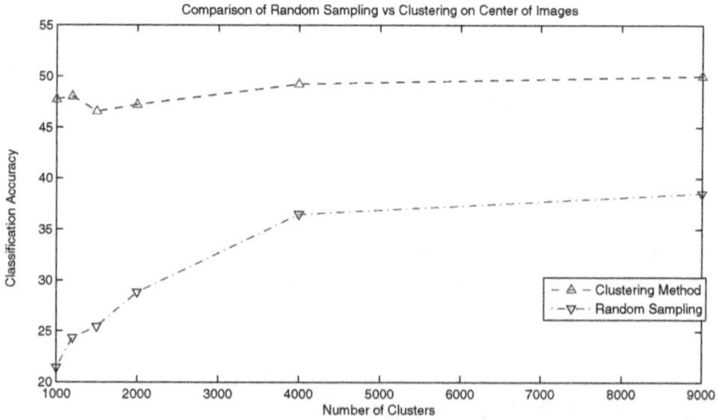

Fig. 5. Comparing the Clustering method with Random Sampling on center quarter of the images

number of samples and number of clusters is kept reasonable. For instance in the sampling of all the images, we sample 10 to 20 times more than number of clusters. If the number of clusters is too high, the clustering has not much effect, and if it is too low, the clustering performs an averaging on the features which results in blurring and losing important information. Different number of clusters were implemented and where the number of clusters were more than 10% of number of samples, the performance was much lower, which with very low ratio, we were getting good performances in the training set with over-fitting effect.

3 Discussion and Conclusions

As can be seen in Table 1, By sampling more features from each image and clustering in image level, we achieve performances slightly higher than previous random sampling method. During the previous experiments, we came to the point that with sampling only from the center of the images, a better performance is acquired when clustering, in comparison with randomly sampling from the images. The reason random sampling performs as good as clustering, remains a matter of debate despite some similarities it has with compressive sampling [12]. As shown in section 2.4, using clustering on the center quarter of the image, gives much better classification performance in comparison with random sampling. The lower performance of random sampling may be due to the non-sparseness of the sampling domain, which is a requirement for the compressive sampling. In compressive sampling, it is suggested that sampling randomly over a sparse representation of a signal, will achieve the same performance with dense sampling over the non-dense space, which results in less computational costs.

Table 1. Comparison between Random method and Clustering Method Performance (all numbers are the average of 8 random runs.)

Method	Performance
Random Sampling	52.35
Clustering on each Image	52.69
Clustering on each Class	51.22
Clustering on all Images	48.62
Clustering on Center of Images	50.04

On the other hand, the averaging and blurring effects of clustering can be the reason of equal performance with random sampling which are not necessarily the best performance. The clusters created from images, are both from background categories, and object categories, and clustering features of the center of the images, where most features are from objects rather than background, resulted in better performance in comparison with random sampling on the same dictionary size as can be seen in Fig. 5. This encourages us to find better interest points in images and perform this hierarchical structure on those interest points, rather than random sampling or clustering. One prospective way is to use wavelet transform to find points with higher information as interest points which are also used in compressive sampling methods. Another possible approach will be selecting Difference of Gaussian(DoG) function to generate interest points which are also used in SIFT method [13].

References

1. Hubel, D., Wiesel, T.: Receptive fields of single neurones in the cat's striate cortex. The Journal of Physiology 148(3), 574 (1959)
2. Hubel, D., Wiesel, T.: Receptive fields and functional architecture of monkey striate cortex. The Journal of Physiology 195(1), 215 (1968)
3. Riesenhuber, M., Poggio, T.: Hierarchical models of object recognition in cortex. Nature Neuroscience 2, 1019–1025 (1999)
4. Fukushima, K.: Neocognitron: A self-organizing neural network model for a mechanism of pattern recognition unaffected by shift in position. Biological Cybernetics 36(4), 193–202 (1980)
5. Mutch, J., Lowe, D.: Object class recognition and localization using sparse features with limited receptive fields. International Journal of Computer Vision 80(1), 45–57 (2008)
6. Mutch, J., Lowe, D.: Multiclass object recognition with sparse, localized features
7. Mladenić, D., Brank, J., Grobelnik, M., Milic-Frayling, N.: Feature selection using linear classifier weights: interaction with classification models. In: Proceedings of the 27th Annual International ACM SIGIR Conference on Research and Development in Information Retrieval, pp. 234–241. ACM, New York (2004)
8. Serre, T., Riesenhuber, M.: Realistic modeling of simple and complex cell tuning in the HMAX model, and implications for invariant object recognition in cortex. In: Massachusetts Institute of Technology Computer Science and Artificial Intelligence Laboratory (2004)
9. Serre, T., Oliva, A., Poggio, T.: A feedforward architecture accounts for rapid categorization. Proceedings of the National Academy of Sciences 104(15), 6424 (2007)
10. Fei-Fei, L., Fergus, R., Perona, P.: Learning generative visual models from few training examples: an incremental Bayesian approach tested on 101 object categories. In: IEEE Workshop on Generative-Model Based Vision, CVPR 2004 , vol. 2 (2004)
11. Mutch, J., Knoblich, U., Poggio, T.: CNS: a GPU-based framework for simulating cortically-organized networks. Technical Report MIT-CSAIL-TR-2010-013 / CBCL-286, Massachusetts Institute of Technology, Cambridge, MA (February 2010)
12. Candes, E., Wakin, M.: People hearing without listening: An introduction to compressive sampling. IEEE Signal Processing Magazine 25(2), 21–30 (2008)
13. Lowe, D.: Object recognition from local scale-invariant features. In: ICCV, p. 1150. IEEE Computer Society, Los Alamitos (1999)

A Highly Robust Approach Face Recognition Using Hausdorff-Trace Transformation

Werasak Kurutach, Rerkchai Fooprateepsiri, and Suronapee Phoomvuthisarn

Information Science and Technology, Mahanakorn University of Technology
140 Cheumsam rd., Nongchok, Bangkok, Thailand
{Werasak,Rerkchai,Suronapee}@mut.ac.th

Abstract. Face recognition research still face challenge in some specific domains such as pose, illumination and Expression. In this paper, we proposes a highly robust method for face recognition with variant illumination, scaling, rotation, blur, reflection and expression. Techniques introduced in this work are composed of two parts. The first one is the detection of facial features by using the concepts of Trace Transform and Fourier transform. Then, in the second part, the Hausdorff distance is employed to measure and determine of similarity between the models and tested images. Finally, our method is evaluated with experiments on the AR, ORL, Yale and XM2VTS face databases and compared with other related works (e.g. Eigen face and Hausdorff ARTMAP). The extensive experimental results show that the average of accuracy rate of face recognition with variant illumination, scaling, rotation, blur, reflection and difference emotions is higher than 88%.

1 Introduction

Biometric identification system makes use of either physiological characteristics (such as a fingerprint, iris pattern, or face) or behavior patterns (such as hand-writing, voice, or key-stroke pattern) to identify a person. Because of human inherent protectiveness of his/her eyes, some people are reluctant to use eye identification systems. Face recognition has the benefit of being a passive, non intrusive system to verify personal identity in a "natural" and friendly way. In general, there are two approaches to face recognition systems: 1) Brightness-based, which make use of the pixel brightness directly or features in low dimensionality manifolds without shape information, such as PCA-based approaches, and 2) Feature-based, which involve the use of geometric features such as positions of facial features. The proposed method combines elements from both approaches. We compare its results with several traditional methods in typical experiments and demonstrate the superiority of the face representations proposed. The well-known approaches used for face recognition is based on the use of eigenfaces [1,2], elastic matching [3,4], neural networks [5, 6], waveletfaces [7], fisherfaces [8, 9], hausdorff ARTMAP [10] and trace transform [11,12,13]. This paper presents a face feature extraction and recognition method that employs the texture representation derived from the Trace transform. The rest of this paper is as follows. An introduction to the trace transform, its properties and its

K.W. Wong et al. (Eds.): ICONIP 2010, Part II, LNCS 6444, pp. 549–556, 2010.

relationship to other well known transforms are given in Section 2. The method of extracting the binary string and similarity measure are described in Section 3. Section 4 presents the experimental results and then conclusions and some areas of further work are described in Section 5.

2 Feature Extraction

2.1 The Trace Transform

Let F denote an image. A method to represent the characteristics of image F decided by $l(\theta, \rho)$ onto the horizontal axis θ and the vertical axis ρ, is called the *trace transform* [14]. The trace-line l is decided using the distance from the origin to l is denoted by ρ, and the directional vector denoted by θ, as shown in Fig. 1(a). The trace-line l is represented by $l = \{(x, y) : \rho = x \cos\theta + y \sin\theta\}$, and a function used in the trace transform is represented as $g(F : \theta, \rho, l) = T(F : \theta, \rho, l)$. A matrix (or image) generated by the trace transform is called trace image as shown in Fig. 1 (b).

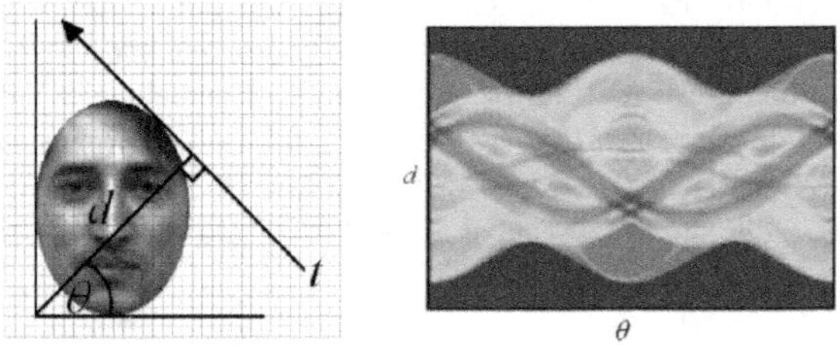

(a) (b)

Fig. 1. (a) Parameters of the trace transform. (b) A trace image visualized in 3D space.

The trace image generated by the trace transform method has the following characteristics. If the original image rotates, its trace image shifts along the horizontal axis θ. If the original image translates to a certain vector, its trace image undergoes changes as follows. For convenience they are stated in terms of a trace matrix. Columns remain unchanged and stay in their places, though may shift up or down. A shift vector specifies numbers a and b such that a column with coordinate θ_i shifts vertically to $a \cos(\theta_i - b)$.This is because of the characteristics, feature values extracted from an input image by trace transform are always invariant to transition and rotation.

2.2 Robust Identifier

The feature values for identifying face are calculated by the combination of values in a trace image decided by the trace transform using three functions called *trace function* T, *diametric function* P, and *circus function* Φ. The trace function T is used to produce a trace image using an input image; the diametric function P is used to produce a diametric matrix using the trace image; the circus function Φ is used to produce the final feature values using the diametric matrix. The procedural processing steps to extract the features are show in table below:

Table 1. The processing steps for extract the values of T, P and Φ

Step 1: *Trace function,* $T = T(F:\theta,\rho,l)$ (1)

- Trace transform is determined by the trace function T.
- Trace image is generated by the trace function. When rang of θ is $[0,2\pi]$, and the rang of ρ is $[\rho_{min}, \rho_{max}]$.

Step 2: *Diametric function,* $P = P(T(F:\theta,\rho,l))$ (2)

- The feature values are acquired by the diametric function P using the column values of the trace image.
- The diametric is generated by the diametric function P using the parameter ρ of diametric moving direction.

Step 3: *Circus function,* $\Phi = \Phi(P(T(F:\theta,\rho,l)))$ (3)

- The features are acquired by the circus function Φ using the diametric matrix and the parameter θ.

For invariance to shift and amplitude scaling can be achieved by taking the Fourier transform of Eq. (3)

$$\Pi(\Phi) = \mathcal{F}\left[\kappa P\left(T(F:\theta,\rho,l)\right)\right], \quad (4)$$

then to exploiting the linearity identify and translation property of the Fourier transform give

$$\Pi(\Phi) = \kappa\, e^{-j\theta\Phi}\mathcal{F}\left[P\left(T(F:\theta,\rho,l)\right)\right]. \quad (5)$$

Taking the magnitude of $\Pi(\Phi)$ gives

$$|\Pi(\Phi)| = \left|\kappa\mathcal{F}\left[P\left(T(F:\theta,\rho,l)\right)\right]\right|. \quad (6)$$

From Eq. (6) means that the original image and the modified image give equivalent descriptors except for the scaling factor κ.

A binary string is acquired by taking the sign of the difference between neighboring coefficients,

$$b_\omega = \begin{cases} 0 & if\ |\Pi(\omega)| - |\Pi(\omega+1)| < 0 \\ 1 & otherwise. \end{cases} \quad (7)$$

The image identifier is then made up of these values $B = \{b_1, b_2, \dots, b_n\}$ for $n \in N$. The results are further improved by using different diametrical functionals to extract multiple component identifiers and concatenating them to obtain complete identifier as shown in Fig. 2.

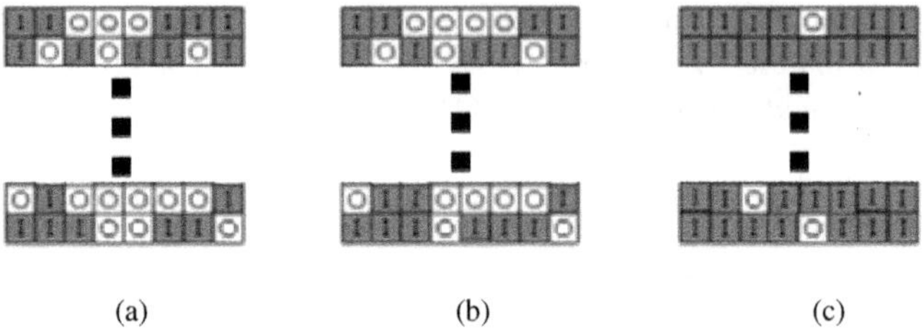

Fig. 2. The binary identifier for an image (a) and its rotated version (b). The difference between the identifiers is show in (c). The identifier is 1D but has been mapped to 2D for presentation purposes only.

3 Similarity Measure

3.1 The Classical Hausdorff Distance

Given two point sets *A* and *B*, the Hausdorff distance [15] between *A* and *B* is defined as

$$H(A,B) = \max\big(h(A,B), h(B,A)\big), \tag{8}$$

$$h(A,B) = \max_{a\in A} \min_{b\in B} \|a-b\|, \tag{9}$$

$$h(B,A) = \max_{b\in B} \min_{a\in A} \|a-b\|, \tag{10}$$

where $\|\cdot\|$ denotes some norm of points of *A* and *B*. This measure indicates the degree of similarity between two point sets. It can be calculated without an explicit pairing of points in their respective data sets. The conventional Hausdorff distance, however, is not robust to the presence of noise. Dubuisson et. al. [16] have studied 24 different variations of the Hausdorff distance in the presence of noise. A modified Hausdorff distance (MHD) using an average distance between the points of one set to the other set gives the best result. This measure is the most widely used in the task of object identification and defined as

$$\mathrm{h(B,A)} = \frac{1}{\mathrm{n}}\sum_{\mathrm{a}\in\mathrm{A}} \min_{\mathrm{b}\in\mathrm{B}} \|\mathrm{a}-\mathrm{b}\|, \tag{11}$$

with *h(B,A)* defined similarly. This modified Hausdorff distance is less sensitive to noise than the conventional one. It is possible, however, to end show the Hausdorff distance with even more attractive features as it is shown in the next section.

3.2 The Hausdorff- Shape Context

In this section, we propose an alternative way to find the minimum distance between point *a* and set *B* to overcome the above problem. Instead of finding the nearest distance, in our approach, the point descriptor, shape context, is used to find the best matching between point a and set B. We, therefore, call this shape similarity measure as "Hausdorff-Shape context".

$$h_{HSC}(A,B) = \sum_{a \in A} \omega(a,b') \min_{b \in B} C_{sc}(a, N(b)), \tag{12}$$

and

$$\omega(a,b') = \frac{\mathcal{D}(a,b')}{\sum_{a \in A} \mathcal{D}(a,b')} \text{ and } \sum \omega(a,b') = 1 \tag{13}$$

where $b' = \arg\min_{b \in B} C_{sc}(a, N(b))$ and $C_{sc}()$ is χ^2 test statistic. In the example shown in Fig. 3, the candidate point b' is the one marked by □ which is the correct corresponding point between point a and a point in set B.

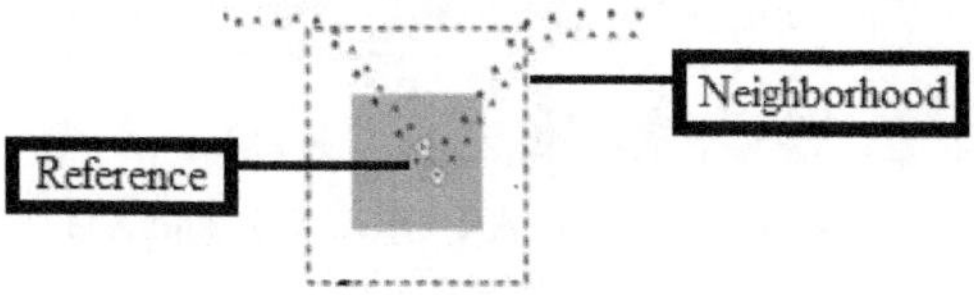

Fig. 3. The grey shade indicates the neighborhood area. The point marked by ○ is a sample point a of the first shape A. The points marked by ▲ and □ are the candidate matching points of the second shape B.

The cost of matching between two points a and b, $C_{sc}(a, N(b))$ is weighted by their distance, *(a, b')*. Therefore $\omega()$ is a normalized distance between points a and b' over the entire distance between sets A and B. Furthermore, the neighborhood *N()* is designed to reduce the computation time of the shape matching, since it finds the best point matching only in the neighborhood area. Thus faster performance improvement can be achieved. The $h_{HSC}(B,A)$ is defined in a similar way. The shape similarity measure in (12) with the maximum Hausdorff matching in (10) is defined to be a confidence level of matching:

$$\text{dist}_{HSC}(A,B) = 1 - H(A,B) \tag{14}$$

4 Experimental Results

In this section, we describe a face database we used and then present a face recognition result under variant illumination, scaling, rotation, blur and difference emotions (smiling, angry and screaming). Our proposed method was implemented on the AR[17], ORL[18], Yale[19] and XM2VTS[20] face databases. In the real world applications, the face recognition should be invariant to rotation, size variation and facial expression. In our proposed method, we only use single image from the face database for training. The face images for testing were generated by applying a random scaling, rotation, blurring and illumination factors to the face images, which was distributed within [1-50, 1+50]%, [0°, 360°], [1-15, 1+15]% and [1-15, 1+15]%.

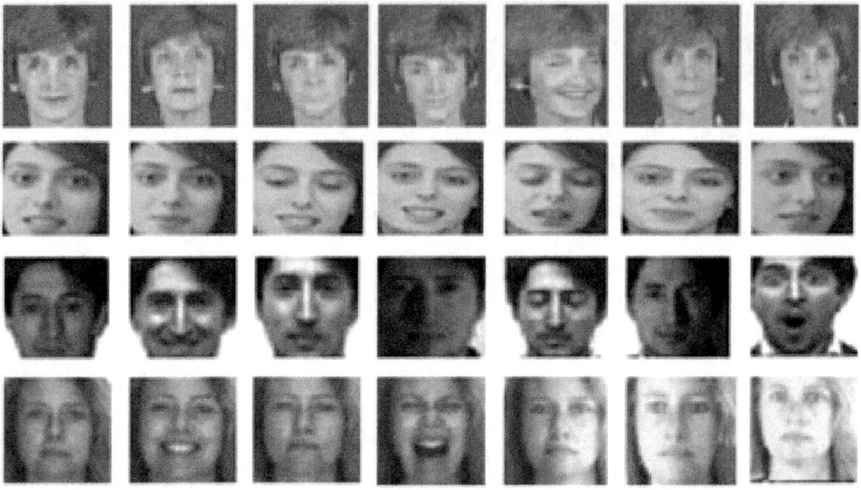

Fig. 4. Some example of facial images in AR, ORL, Yale and XM2VTS face databases

In summary, our proposed method is robust to rotation, size variation and facial expression. From the inspection of the table 2, it was found that our proposed method performed better than the Eigen face method in all cases. Such robustness comes from the use of Trace transform, Fourier transform, Circle Function and matching measure in section 3. It also comes from the fact that only the flagged line is used rather than entire face representation which helps us maximize the matching between reference and test images. Another advantage of our approach is that, when new subjects are added to the system we do not need to retrain on the whole face database, in fact only images of the new subject are used to find the new optimal parameter of the algorithm. This may not be the case for Eigen face and Hausdorff ARPMAP: when new subjects are added to the face database, these systems must be retrained over the whole face database, which is a barrier for real applications.

Table 2. Performance of our method

Condition	Success rate (%)		
	Eigen face	**H-ARTMAP**	**Our Method**
Normal case	≈90	≈94	≈96
Scaling ± 50%	≈67	≈58	≈84
Rotation ± 360°	≈54	≈52	≈96
Scaling + Rotation	≈46	≈49	≈82
Blur ±15%	≈70	≈87	≈81
Illumination	≈67	≈84	≈85
Reflection	≈90	≈92	≈96
Smiling	≈82	≈88	≈89
Angry	≈73	≈82	≈88
Screaming	≈32	≈31	≈71

5 Conclusions

This paper proposes a highly robust method for face recognition. Techniques introduced in this work are composed of two parts. The first one is the detection of facial features by using the concepts of Trace Transform, Fourier Transform and Circle Function. Then, in the second part, the notions of Hausdorff distance and Image identifier algorithm are employed to measure and to determine the similarity between the models and the tested images. Our approach is evaluated by experimenting with 5,362 face images in the AR, ORL, Yale and XM2VTS face databases. The experimental result has shown that the average accuracy rate is higher than 88%.

References

1. Martinez, A.M., Kak, A.C.: PCA versus LDA. IEEE Trans. Pattern Analysis and Machine Intelligence 23(2), 228–233 (2001)
2. Moghaddam, B.: Principal Manifolds and Probabilistic Subspaces for Visual Recognition. IEEE Trans. Pattern Analysis and Machine Intelligence 24(6), 780–788 (2002)
3. Belhumeur, P.N., Hespanha, J.P., Kriegman, D.J.: Eigenfaces vs. Fisherfaces: Recognition using Class Specific Linear Projection. IEEE Trans. Pattern Analysis and Machine Intelligence 19(7), 711–720 (1997)
4. Tefas, A., Kotropoulos, C., Pitas, I.: Using Support Vector Machines to Enhance the Performance of Elastic Graph Matching for Frontal Face Authentication. IEEE Trans. Pattern Analysis and Machine Intelligence 23(7), 735–746 (2001)
5. Lin, S.H., Kung, S.Y., Lin, L.J.: Face Recognition/Detection by Probabilistic Decision-Based Neural Network. IEEE Trans. Neural Networks 8(1), 114–132 (1997)
6. Lawrence, S., Giles, C.L., Tsoi, A.C., Back, A.D.: Face Recognition: A Convolutional Neural-Network Approach. IEEE Trans. Neural Networks 8(1), 98–113 (1997)
7. Chien, J.T., Wu, C.C.: DiscriminantWaveletfaces and Nearest Feature Classifiers for Face Recognition. IEEE Trans. Pattern Analysis and Machine Intelligence 24(12), 1644–1649 (2002)
8. Liu, C., Wechsler, H.: A Shape- and Texture-Based Enhanced Fisher Classifier for Face Recognition. IEEE Trans. Image Processing 10(4), 598–608 (2001)
9. Belhumeur, P.N., Hespanha, J.P., Kriegman, D.J.: Eigenfaces vs. Fisherfaces: Recognition using Class Specific Linear Projection. IEEE Trans. Pattern Analysis and Machine Intelligence 19(7), 711–720 (1997)
10. Thammano, A., Rungruang, C.: Hausdorff ARTMAP for Human Face Recognition. WSEAS Trans. Computers 3(3), 667–672 (2004)
11. Srisuk, S., Petruu, M., Petrou, M., Fooprateepsiri, R., Sunat, K., Kurutach, W., Chopaka, P.: A Combination of Shape and Texture Classifiers for a Face Verification System. In: Zhang, D., Jain, A.K. (eds.) ICBA 2004. LNCS, vol. 3072, pp. 44–51. Springer, Heidelberg (2004)
12. Fooprateepsiri, R., Kurutach, W.: A Fast and Accurate Face Authentication Method Using Hamming-Trace Transform Combination. IETE Technical Review 27(5), 365–370 (2010)
13. Fooprateepsiri, R., Kurutach, W.: Facial Recognition using Hausdorff -Shape -Radon Transform. International Journal of Digital Content Technology and its Applications 3(2), 67–74 (2009)

14. Kadyrov, A., Petrou, M.: The Trace Transform and Its Applications. IEEE Transactions on Pattern Analysis and Machine Intelligence 23(8), 811–828 (2001)
15. Huttenlocher, P., Klanderman, G., Rucklidge, W.: Comparing Images using the Hausdorff Distance. IEEE Trans. PAMI 15(9), 850–863 (1993)
16. Dubuisson, M., Jain, A.K.: A Modified Hausdorff Distance for Object Matching. In: Proc. ICPR, pp. 566–568 (1994)
17. AR Database, `http://cobweb.ecn.purdue.edu/~aleix/aleix_face_DB.html`
18. ORL Face Database, Retrieved from `http://www.uk.research.att.com/facedatabase.html`
19. Yale Face Database, Retrieved from `http://cvc.yale.edu/projects/yalefaces/yalefaces.html`
20. The XM2VTS database, `http://www.ee.surrey.ac.uk/CVSSP/xm2vtsdb/`
21. Fooprateepsiri, R., Kurutach, W., Duangphasuk, S.: A Hybrid Method for Facial Recognition Systems. In: The 2009 IEEE Symposium on Computational Intelligence for Multimedia Signal and Vision Processing, USA, pp. 53–60 (April 2009)
22. Fooprateepsiri, R., Kurutach, W.: An Image Identifier Based on Hausdorff Shape Trace Transform. In: Leung, C.S., Lee, M., Chan, J.H. (eds.) ICONIP 2009. LNCS, vol. 5863, pp. 788–797. Springer, Heidelberg (2009)

Blind Image Tamper Detection Based on Multimodal Fusion

Girija Chetty[1], Monica Singh[2], and Matthew White[2]

[1] Faculty of Information Sciences and Engineering,
University of Canberra, Australia
girija.chetty@canberra.edu.au
[2] Video Analytics Pty. Ltd. Melbourne, Australia

Abstract. In this paper, we propose a novel feature processing approach based on fusion of noise and quantization residue features for detecting tampering or forgery in video sequences. The evaluation of proposed residue features – the noise residue features and the quantization features, their transformation in optimal feature subspace based on fisher linear discriminant features and canonical correlation analysis features, and their subsequent fusion for emulated copy-move tamper scenarios shows a significant improvement in tamper detection accuracy.

Keywords: image tampering, digital forensics, feature selection, image fusion.

1 Introduction

Digital Image tampering or forgery has become major problem lately, due to ease of artificially synthesizing photographic fakes- for promoting a story by media channels and social networking websites. This is due to significant advances in computer graphics and animation technologies, and availability of low cost off-the-shelf digital image manipulation and cloning tools. With lack of proper regulatory frameworks and infrastructure for prosecution of such evolving cyber-crimes, there is an increasing dissatisfaction about increasing use of such tools for law enforcement, and a feeling of cynicism and mistrust among the civilian operating environments.

Another problem this has lead to, is a slow diffusion of otherwise extremely efficient image based surveillance and identity authentication technologies in real-world civilian operating scenarios. In this paper we propose a novel algorithmic framework for detecting image tampering and forgery based on extracting noise and quantization residue features, their transformation in cross-modal subspace and their multimodal fusion for intra-frame and inter-frame image pixel sub blocks in video sequences. The proposed algorithmic models allow detecting the tamper or forgery in low-bandwidth video (Internet streaming videos), using blind and passive tamper detection techniques and attempt to model the source signatures embedded in camera

K.W. Wong et al. (Eds.): ICONIP 2010, Part II, LNCS 6444, pp. 557–564, 2010.

pre-processing chain. By sliding segmentation of image frames, we extract intra-frame and inter-frame pixel sub-block residue features, transform them into optimal cross-modal subspace, and perform multimodal fusion to detect evolving image tampering attacks, such as JPEG double compression, re-sampling and retouching. The promising results presented here can result in the development of digital image forensic tools, which can help investigate and solve evolving cyber crimes.

2 Background

Digital image tamper detection can use either active tamper detection techniques or passive tamper detection techniques. A significant body of work, however is available on active tamper detection techniques, which involves embedding a digital watermark into the images when the images are captured. The problem with active tamper detection techniques is that not all camera manufacturers embed the watermarks, and in general, most of the customers have a dislike towards cameras which embed watermarks due to compromise in the image quality. So there is a need for passive and blind tamper detection techniques with no watermark available in the images.

Passive and blind image tamper detection is a relatively new area and recently some methods have been proposed in this area. Mainly these are of two categories [1, 2, 3, 4]. Fridrich [4] proposed a method based on hardware aspects, using the feature extracted from photos. This feature called sensor pattern noise is due to the hardware defects in cameras, and the tamper detection technique using this method resulted in an accuracy of 83% accuracy. Chang [5] proposed a method based on camera response function (CRF), resulting in detection accuracy of 87%, at a false acceptance rate (FAR) of 15.58%. Chen et al. [6] proposed an approach for image tamper detection based on a natural image model, effective in detecting the change of correlation between image pixels, achieving an accuracy of 82%. Gou et al [7] introduced a new set of higher order statistical features to determine if a digital image has been tampered, and reported an accuracy of 71.48%. Ng and Chang [8] proposed bi-coherence features for detecting image splicing. This method works by detecting the presence of abrupt discontinuities of the features and obtains an accuracy of 80%. Popescu and Farid [3] proposed different CFA (colour filter array) interpolation algorithms within an image, reporting an accuracy of 95.71% when using a 5x5 interpolation kernel for two different cameras. A more complex type of passive tamper detection technique, known as "copy-move tampering" was investigated by Bayram, Sencar, Dink and Memon [1,2] by using low cost digital media editing tools such as Cloning in Photoshop. This technique usually involves covering an unwanted scene in the image, by copying another scene from the same image, and pasting it onto the unwanted region. Further, the tamperer can use retouching tools, add noise, or compress the resulting image to make it look genuine and authentic. Finally, detecting tampers based on example-based texture synthesis scheme was proposed by Criminisi et al[9] that is based on filling in a region from sample textures. It is one of the state-of-the-art image impainting or tampering schemes. Gopi et al in [10] proposed a pattern recognition formulation and used auto regression coefficients and neural network classifier for tamper detection.

One of the objectives of the work reported here is development of robust and automatic tamper detection framework for low bandwidth Internet streamed videos where most of the fingerprints left by tamperer can be perturbed by heavy compression. However, by fusing multiple image tampering detectors, it could be possible to uncover the tampering in spite of the heavy compression, as different detectors use cues and artifacts at different stages of the image formation process. So if an image lacks certain cues, a complementary detector would be used for making a decision For example, a copy move forgery might have been created with two source images of similar quantization settings but very different cameras. In this case, the copy move forgery can be successfully detected by a different detector. We thus benefit from having several tamper detection modules at hand rather than only using the one type of detector. Another advantage of fusing several detector outputs to make a final decision is that, if one of the detector outputs noisy and erroneous scores, the other detectors could complement and enhance the reliability of the tamper decision. Therefore, the advantage of fusion is twofold: to handle images which were subjected to multiple, diverse types of tampering, and to boost the detection robustness and accuracy by making different modules work with each other. The challenge, however, lies in the synergistic fusion of diverse detectors as different detectors are based on different physical principles and segmentation structures.

We formulate the fusion problem in a Bayesian pattern recognition framework and use well known Gaussian Mixture Models for the task. The approach is based on detecting the tamper from the multiple image frames, by extracting noise and quantization residue features in intra-frame and inter-frame pixel sub blocks, transforming them into correlation subspace to extract the maximal correlation properties, and establish possible tampering of video. The approach extends the noise residue features reported by Hsu et al in [11] and is blind and passive, based on the hypothesis, that a typical tampering attacks such as double compression, re-sampling and retouching can inevitably disturb the correlation properties of the pixel sub-blocks within a frame (intra-frame) as well as between the frames (inter-frame) and can distinguish the fingerprints or signatures of genuine video from tampered video frames. The rest of the paper is organized as follows. Next Section describes the formulation of fusion problem. The details of the experimental results for the proposed fusion scheme is described in Section 4. The paper concludes in Section 5 with some conclusions and plan for further work.

3 Formulating the Fusion Problem

The processing pipeline once the images or video is captured consists of several stages. First, the camera sensor (CCD) captures the natural light passing through the optical system. Generally, in consumer digital cameras, every pixel is detected by a CCD detector, and then passed through different colour filters called Color Filter Array (CFA). Then, the missing pixels in each color planes are filled in by a CFA interpolation. Finally, operations such as demosaicing, enhancement and gamma correction are applied by the camera, and converted to a user-defined format, such as RAW, TIFF, and JPEG, and stored in the memory.

Since the knowledge about the source and exact processing (details of the camera) used is not available for application scenarios considered in this work (low-bandwidth Internet streaming video), and which may not be authentic and already tampered, we extract a set of residual features for pixel sub-blocks within the frame and between adjacent frames from the video sequences. These residual features try to model and extract the fingerprints for source level post processing within any camera, such as denoising, quantization, compression, contrast enhancement, white balancing, image sharpening etc. In this work, we use only two types of residual features: noise residue features and quantization residue features.

The noise and quantization residue features were first extracted from 32 x 32 pixel intra-frame and inter-frame pixel sub-blocks of the video sequences. A feature selection algorithm was used to select those features that exhibit maximum correlation. We used feature selection techniques based on two different techniques: Fisher linear discriminant analysis (FLD), and canonical correlation Analysis (CCA). The details of the two feature selection techniques is described in [12] and [13].

4 Experimental Results

The video sequence data base from Internet streamed movies was collected and partitioned into separate subsets based on different actions and genres. The data collection protocol used was similar to the one described in [14]. Figure 1 shows screenshots corresponding to different actions, along with emulation of copy move tampered scenes and the detection of tampered regions with the proposed approach.

Different sets of experiments were conducted to evaluate the performance of the proposed residue features in FLD (Fisher linear discriminant) and CCA(Canonical Correlation analysis) sub-space and their fusion in terms of tamper detection accuracy. The experiments involved a training phase and a test phase. In the training phase a Gaussian Mixture Model for each video sequence from data base was constructed [15]. In the test phase, copy-move tamper attack was emulated by artificially tampering the training data. The tampered processing involved copy cut pastes of small regions in the images and hard to view affine artefacts. Two different types of tampers were examined. An intra-frame tamper, where the tampering occurs in some of the pixel sub-blocks within the same frame, and inter-frame tamper, where pixel sub-blocks from adjacent frames were used. However, in this paper, we present and discuss results for the intra-frame tamper scenario only.

As can be seen from Table 1, which show the tamper detection results in terms of % accuracy, the performance of single mode noise residue and quantization residue features can be enhanced by using optimal feature selection subspace and their subsequent multimodal fusion. Figure 1 shows some sample results for intra-frame tamper scenario. We compared the performance of proposed feature selection techniques with features based on autoregressive coefficients and neural network classification proposed by Gopi et al in [10].

Fig. 5. Row 1: Screenshots from Internet streamed video sequences; Row 2: Copy-move tamper emulation for the scene; Row 3: Detection of tampered regions in the scene

Table 1. Evaluation of noise and quantization residue features for emulated copy-move tamper attack (% Accuracy); $\tilde{f}_{Intra-Inter}$ (noise residue features); $f_{Intra-Inter}$ (quantization residue features)

Internet streamed movie data subset	**% Accuracy**		
Different Residue features and their fusion	**CCA**	**FLD**	**ARC[10]**
f_{Intra} (Intra-frame noise residue features)	83.2	83.6	80.2
f_{Inter} (Inter-frame noise residue features)	83.8	83.4	83.1
$\tilde{f}_{Intra}$ (Intra-frame quant. residue features)	77.28	76.23	74.33
$\tilde{f}_{Inter}$ (Inter-frame quant. residue features)	72.65	71.44	69.45
$f_{Intra-Inter}$ (feature fusion- noise residue)	86.6	85.27	83.78
$\tilde{f}_{Intra-Inter}$ feature fusion- quant residue)	80.55	79.66	77.22
$f_{Intra-Inter} + \tilde{f}_{Intra-Inter}$ (hybrid fusion)	89.56	86.22	84.33

As can be seen in Table 1, the single mode noise residue features perform better than quantization residue features. For both noise residue and quantization residue features, the CCA and FLD features perform better than ARC features. CCA features result in better accuracy as they are based on canonical correlation analysis that can extract correlation properties better than features based on Fisher linear discriminant analysis. By fusing intra-frame and inter-frame pixel sub block features, we can see a better performance is achieved. This shows that better correlation information can be extracted when multiple frames are used for detecting tampers. In general quantization residue features perform worst compared to noise residue features. This can be expected as quantization artefacts for low-bandwidth video can significant damage tamper related correlation properties. However, by using a hybrid fusion of quantization and noise residue features, we can see a better performance is achieved.

In a pattern recognition framework, the classifier is also equally important in addition to a feature selection technique. Hence the next experiment involved examining the performance of GMM classifier with neural network (NN) classifier proposed in [10]. The results from this experiment are shown in Table 2. Since the experiments reported in Table 1 resulted in CCA features as the best performing features, we used CCA features for experimental results shown in Table 2.

Table 2. (% Accuracy) Performance for noise and quantization residue features and their fusion for GMM vs. NN classifier

% Accuracy	**GMM Classifier**	**NN Classifier [10]**
Different Residue features and their fusion	CCA features	CCA features
f_{Intra} (Intra-frame noise residue features)	83.2	81.4
f_{Inter} (Inter-frame noise residue features)	83.8	80.6
$\tilde{f}_{Intra}$ (Intra-frame quant. residue features)	77.28	75.77
$\tilde{f}_{Inter}$ (Inter-frame quant. residue features)	72.65	70.53
$f_{Intra-Inter}$ (feature fusion- noise residue)	86.6	83.22
$\tilde{f}_{Intra-Inter}$ feature fusion- quant residue)	80.55	77.23
$f_{Intra-Inter} + \tilde{f}_{Intra-Inter}$ (hybrid fusion)	89.56	83.45

The GMM classifier performs better than NN classifier from the results shown in Table 2 suggesting Bayesian framework allows better modelling of tampering. Further experiments are in progress to model other type of tampers specially those corresponding to optical properties of cameras, and interlacing and de-interlacing processing in cameras.

5 Conclusions

In this paper, we investigated a novel approach for video tamper detection in low-bandwidth Internet streamed videos using residue features from intra-frame and inter frame pixel sub-blocks, their transformation in optimal correlation subspace, and subsequent multimodal fusion. The evaluation of two different residue features, the noise and the quantization residue features for emulated copy-move tamper scenario show the potential of proposed blind and passive tamper detection approach for applications where the establishing the identity of the camera source is not available. The feature transformation of residue features in CCA and FLD subspace and their subsequent multimodal fusion of intra-frame and inter-frame features models the camera source signatures better, and allows blind and passive tamper detection. An accuracy of around 89.56 % was achieved for hybrid fusion of quantization and noise residue features extracted from intra-frame and inter-frame pixel sub blocks with a copy-move tamper emulation from low-bandwidth Internet streamed movie sequences.

References

[1] Bayram, S., Sencar, H.T., Memon, N.: An Efficient and Robust Method For Detecting Copy-Move Forgery. In: Proceedings IEEE International Conference on Acoustics, Speech and Signal Processing, ICASSP 2009, Taipei Taiwan (June 2009)

[2] Dirik, A.E., Memon, N.: Image Tamper Detection Based on Demosaicing Artifacts. In: Proceedings IEEE ICIP 2009, Cairo Egypt (November 2009)

[3] Popescu, A.C., Farid, H.: Exposing Digital Forgeries by Detecting Traces of Re-sampling. IEEE Transactions on signal processing 53(2) (February 2005)

[4] Fridrich, J., David, S., Jan, L.: Detection of Copy-Move Forgery in Digital Images, `http://www.ws.binghamton.edu/fridrich/Research/copymove.pdf`

[5] Hsu, Y.F., Chang, S.F.: Detecting Image Splicing Using Geometry Invariants and Camera Characteristics Consistency. In: ICME, Toronto, Canada (July 2006)

[6] Shi, Y.Q., Chen, C., Chen, W.: A natural image model approach to splicing detection. In: Proc. ACM Multimedia Security Workshop, Dallas, Texas, pp. 51–62 (September 2007)

[7] Gou, H., Swaminathan, A., Wu, M.: Noise Features for Image Tampering Detection and Steganalysis. In: Proc. of IEEE Int. Conf. On Image Processing (ICIP 2007), San Antonio, TX (September 2007)

[8] Ng, T.T., Chang, C.S.F., Lin, Y., Sun, Q.: Passive-blind Image Forensics. In: Zeng, W., Yu, H., Lin, C.-Y. (eds.) Multimedia Security Technologies for Digital Rights. Elsevier, Amsterdam (2006)

[9] Criminisi, A., Perez, P., Toyama, K.: Region filling and object removal by exemplar-based image inpainting. IEEE Trans. Image Process. 13(9), 1200–1212 (2004)

[10] Gopi, E.S., Lakshmanan, N., Gokul, T., KumaraGanesh, S., Shah, P.R.: Digital Image Forgery Detection using Artificial Neural Network and Auto Regressive Coefficients. In: Proceedings Canadian Conference on Electrical and Computer Engineering, Ottawa, Canada, May 7-10, pp. 194–197 (2006)

[11] Hsu, C., Hung, T., Lin, C., Hsu, C.: Video Forgery Detection Using Correlation of Noise Residues, `http://www.ee.nthu.edu.tw/~cwlin/pub/mmsp08forensics.pdf` (retrieved on 11/3/2010)

[12] Martinez, A.M., Kak, A.C.: PCA versus LDA. IEEE Transactions on Pattern Analysis and Machine Intelligence 23(2), 228–233 (2001), `http://www.ece.osu.edu/~aleix/pami01.pdf`, doi:10.1109/34.908974 (retrieved on 11/3/2010)
[13] Borga, M., Knutsson, H.: Finding Efficient Nonlinear Visual Operators using Canonical Correlation Analysis. In: Proc. of SSAB 2000, Halmstad, pp. 13–16 (2000)
[14] Laptev, I., Marszałek, M., Schmid, C., Rozenfeld, B.: Learning realistic human actions from movies. In: Proc. CVPR 2008, Anchorage, USA (2008)
[15] Chetty, G., Wagner, M.: Robust face-voice based speaker identity verification using multilevel fusion. Image and Vision Computing 26(8), 1249–1260 (2008)

Orientation Dependence of Surround Modulation in the Population Coding of Figure/Ground

Keiichi Kondo and Ko Sakai

Graduate School of Systems and Information Engineering, University of Tsukuba,
1-1-1 Tennodai, Tsukuba, Ibaraki, 305-8573 Japan
kondo@cvs.cs.tsukuba.ac.jp, sakai@cs.tsukuba.ac.jp

Abstract. Recent physiological studies have reported Border-Ownership (BO) selective cells that signal the direction of figure along a contour, which appear to be a basis for figure-ground segregation. Surround modulation has been proposed as an underlying neural mechanism of BO determination. The crucial question to the model is its orientation specificity: whether BO could be determined only from iso-orientation (with respect to the preferred orientation of the classical receptive field) that has been reported dominant in the modulation. We investigated computationally the dependence of surround modulation on the orientation characteristics during the determination of BO with natural images. The results showed that, even when modulation was limited to iso-orientation, population responses obtained through integration were unchanged while the responses of individual cells were varied, indicating a dominant role of iso-orientation suppression and an effectiveness of population coding in BO determination.

Keywords: Figure/Ground Segregation, Surround Modulation, Border-Ownership.

1 Introduction

Recent physiological studies [e.g. 1] have reported that a number of neurons in early-to-intermediate-level visual areas are selective to the direction of figural region along a contour. Such neurons are termed Border-Ownership (BO) selective neurons, as they signal which side of a border owns the contour. BO-selective cells are expected to play a crucial role in figure-ground segregation. Nishimura et al. [2] have proposed a computational model of BO-selective cells based on surround modulation apparent in early vision.

Surround modulation that has been reported elsewhere in physiology is suppressive and facilitatory responses of cells during the presentation of a particular stimulus around its Classical Receptive Field (CRF) while another stimulus is presented onto the CRF. The spatial characteristics of surround modulation in V1 include suppression dominant over facilitation, asymmetry with respect to the CRF, and diversity among individual cells [3]. A recent study has also reported orientation dependence in modulation: iso-orientation suppression and cross-orientation facilitation with respect to the preferred orientation of the CRF [4].

K.W. Wong et al. (Eds.): ICONIP 2010, Part II, LNCS 6444, pp. 565–573, 2010.

Watanabe et al. [5] considered that such characteristics of surround modulation were suitable for BO determination and hypothesized that the surround modulation underlay the selectivity of BO. These authors constructed a computational model of BO-selective cells applicable for natural images and evaluated the model response by the comparison with human perception for 100 natural images. The simulation results achieved 70% correct that was comparable with the human perception through a small window of the size same as the spatial extent of the model cells.

The present study investigates the mechanisms of surround modulation during the determination of BO. Specifically, we focused on the orientation characteristics in surround modulation. Physiological studies have reported iso-orientation suppression and cross-orientation facilitation with respect to the preferred orientation of the CRF. However, since suppression is dominant over facilitation, cross-orientation facilitation may be less effective in BO determination. Asymmetry of surround modulation that is the basis of BO computation could be realized by the asymmetry of the strength of suppression rather than orientation-dependent facilitation and suppression. However, orientation is not exchangeable in essence, in contrast to that the suppression and facilitation are exchangeable, as the modulation is linear. Furthermore, the necessity of cross-orientation facilitation is not derived from the analysis of Gestalt factors that have been known as phenomenological rules for the perception of the direction of figure. Therefore, it is crucial for the BO models based on surround modulation to provide evidence of the orientation independence in modulation. As the orientation component for input changes, it is inevitable that responses of individual cells change. If orientation independence is observed, it must be apparent in a behavior of population. Thus, we investigated the population behavior as well as individual responses of BO-selective model cells, by controlling orientation characteristics of surround facilitation, to examine the crucial characteristics of surround modulation in BO determination.

2 The Model of BO Selective Cells

The direction of BO is determined from three consecutive stages: (i) contrast detection, (ii) surround modulation, and (iii) BO determination. This section describes the essence of each step, with Fig. 1 illustrating the overview of the model.

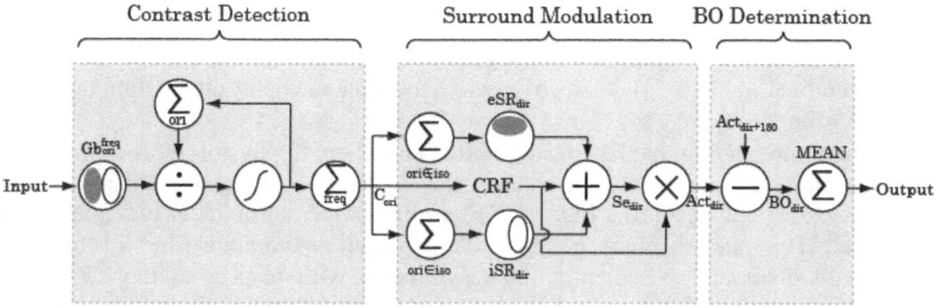

Fig. 1. An overview of the proposed model

2.1 Contrast Detection

The contrast detection stage consists of the following three steps: (i) detection of luminance contrast, (ii) contrast normalization and static nonlinearity, and (iii) integration of frequency channels. Detection of luminance contrast is achieved by Gabor filters that approximate the CRF of V1 neurons. To detect luminance contrasts of various orientations and frequencies in natural images, we use 48 types of Gabor filters with 8 preferred orientations in multiples of 45 deg and 6 frequencies in multiples of ${}^{-1/}\sqrt[5]{7/3}$ between 3 and 7 cpd. Note that input images comprise 320x480 or 480x320 pixel, and we set 35 pixel corresponding to 1 deg in visual angle. The luminance contrast detected by Gabor filters is half-wave rectified and passed through iterative, divisive normalization and static nonlinearity to reproduce contrast characteristics of V1 neurons [6]. We denote the output of this stage as:

$$S_{ori}^{freq}(x_0, y_0) \ , \tag{1}$$

where x_0 and y_0 indicate the center of a Gabor filter, and *ori* and *freq* denote its optimal orientation and frequency, respectively. Six frequency channels are integrated by weighted summation:

$$C_{ori}(x_0, y_0) = \sum_{freq} k_{freq} S_{ori}^{freq}(x_0, y_0) \ , \tag{2}$$

where k_{freq} is the weight according to human contrast sensitivity function [7].

2.2 Surround Modulation

Physiological studies on monkey V1 have reported that surround modulation is orientation selective: suppression is specific to the iso-orientation component with respect to the preferred orientation of the CRF, and facilitation is specific to the other components. Spatial extent of surround modulation is diversely localized and asymmetric with respect to the CRF. It has been reported that neurons in intermediate-level visual areas including V2 exhibit similar structure in surround modulation. The magnitude of suppression in the model is given by pooling iso-orientation contrasts detected by the previous stage, within one of pre-determined surround regions. To realize diversity of the surround regions, we use 40 distinct regions that are selected from a set of randomly generated regions. We use another 40 regions for facilitation from the same randomly generated set. A pair of suppressive and facilitatory regions is comprised in a model cell. Surround suppression and facilitation are given, respectively, by:

$$Inh_{dir}^{j} = \sum_{ori \in dir \pm 90^\circ} (C_{ori} * iSR_{dir}^{j})(x, y) \ , \tag{3}$$

$$Exc_{dir}^{j} = \sum_{ori \notin dir \pm 90^\circ} (C_{ori} * eSR_{dir}^{j})(x, y) \ , \tag{4}$$

where iSR^j_{dir} and eSR^j_{dir} are inhibitory and excitatory regions (filters), respectively, for the convolution (*) with the contrast detected by the previous stage. Each pair of surround regions (j) is rotated in multiples of 45 degrees to obtain other sets that are denoted by *dir*. There are a total of 320 (40 pairs of surround (j) x 8 directions (*dir*)) sets of regions. The combination of equations 3 and 4 will be denoted as *iso-cross* condition. In this study, we investigate the orientation specificity of surround modulation. Specifically, we examine the replacement of cross-orientation with iso-orientation for facilitation (*iso-iso* condition). An illustration of these combinations is shown in Fig. 2. The surround facilitation with iso-orientation contrast is given by:

$$Exc^j_{dir} = \sum_{ori \in dir \pm 90^\circ} (C_{ori} * eSR^j_{dir})(x, y) \ . \tag{5}$$

Note that the sets of surround regions are identical to those used in *iso-cross* condition.

The modulation of CRF response is given by a linear combination of the surround modulations (*Exc* and *Inh*) and the CRF response (*HM*). As the detection of contours has been known as extremely difficult in natural images, we use Human Marked Contours (HMC) available in Berkeley Segmentation Dataset [e.g. 8, 9] that were drawn by ten human participants. Note that HMC are used only for the CRF response. Gray-scale natural images are used for surround modulation. The modulated response ($HM + (Exc - Inh)$) is multiplied by the CRF response (*HM*) to realize the definition of CRF that the activity of a cell is zero if no stimulus is presented on the CRF. As the response is normalized and compressed, the response ranges between 0 and 1. Thus the activity of a cell (j) at a location (x_0, y_0) is given by:

$$Act^j_{dir}(x_0, y_0) = HM_{ori}(x_0, y_0)\{HM_{ori}(x_0, y_0) + (Exc^j_{dir}(x_0, y_0) - Inh^j_{dir}(x_0, y_0))\} \ , \tag{6}$$

where HM_{ori} is the iso-orientation contrast detected by the CRF from HMC. Surround modulation is achieved by subtracting inhibitory modulation from excitatory modulation (*Exc - Inh*). *ori* is the orientation orthogonal to *dir* so that BO is examined only in the directions orthogonal to the preferred orientation of the CRF, with 40 sets of surround regions. We integrate the responses of 320 (40 pairs of surrounds x 8 directions) cells at each location for evaluating the direction of BO that is coded by the 320 cells. The difference of responses between cells with identical surrounds (j) but the opposite BO preferences (*dir*, *dir* + 180) is:

$$BO^j_{dir}(x_0, y_0) = \begin{cases} Act^j_{dir}(x_0, y_0) - Act^j_{dir+180}(x_0, y_0) & \text{for } BO^j_{dir} > 0 \\ 0 & \text{for } BO^j_{dir} \le 0 \end{cases} \ . \tag{7}$$

$BO_{dir}{}^j$ represents the strength of BO in the direction (*dir*) determined by the j-the cell. The strength of BO for each direction is defined by an arithmetic average among the cells with distinct surrounds:

$$BO_{dir}(x_0, y_0) = \frac{1}{n} \sum_j BO^j_{dir}(x_0, y_0) \ , \tag{8}$$

where n (= 40) represents the number of surround regions for each BO direction. Finally, the direction of BO at a location, (x_0, y_0), is defined by:

$$BO(x_0, y_0) = vecsum(BO_{dir}(x_0, y_0)) \ , \tag{9}$$

where *vecsum* indicates a vector summation among 8 directions.

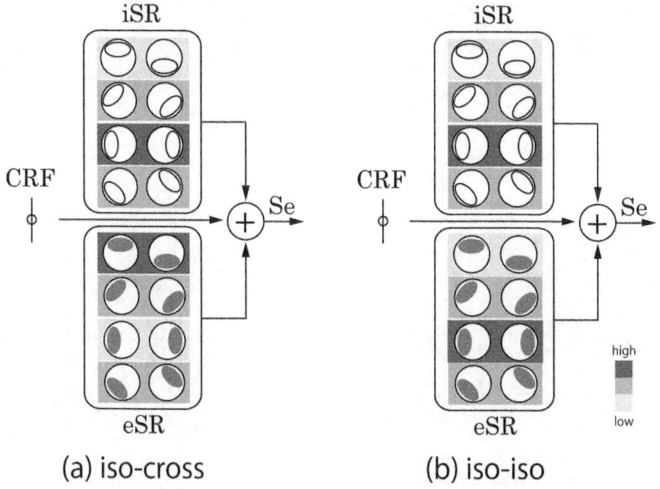

Fig. 2. Comparison between *iso-cross* and *iso-iso* conditions

3 Results

To investigate the orientation specificity of surround modulation in realistic BO determination, we carried out simulations of the model cells with a number of natural images by varying the orientation specificity of facilitation and compared the results with human perception available in Berkeley Segmentation Dataset [e.g. 8, 9]. Fig. 3 shows a few examples of the computed BO directions superimposed onto the original natural images. The overall consistencies for 100 natural images (517,718 locations) are given in Table 1. The results between the model and human perception for *iso-cross* (iso- and cross-orientations for suppression and facilitation, respectively) and *iso-iso* conditions were almost the same, with very similar variance (SD). The results indicate that the population response does not depend on orientation specificity in surround facilitation.

It is expected that, although the responses of individual cells differed significantly between *iso-cross* and *iso-iso* conditions, the population responses obtained through an integration of individual cells are almost identical, as illustrated in Fig. 4. To examine this hypothesis, we computed the difference between the responses of individual cells and that between the overall responses after integration. Under *iso-cross* and *iso-iso* conditions, we computed BO directions through vector summation of the responses of cells with the same set of surrounds from equation 7:

$$BO^j = vecsum(BO^j_{dir}) \ , \tag{10}$$

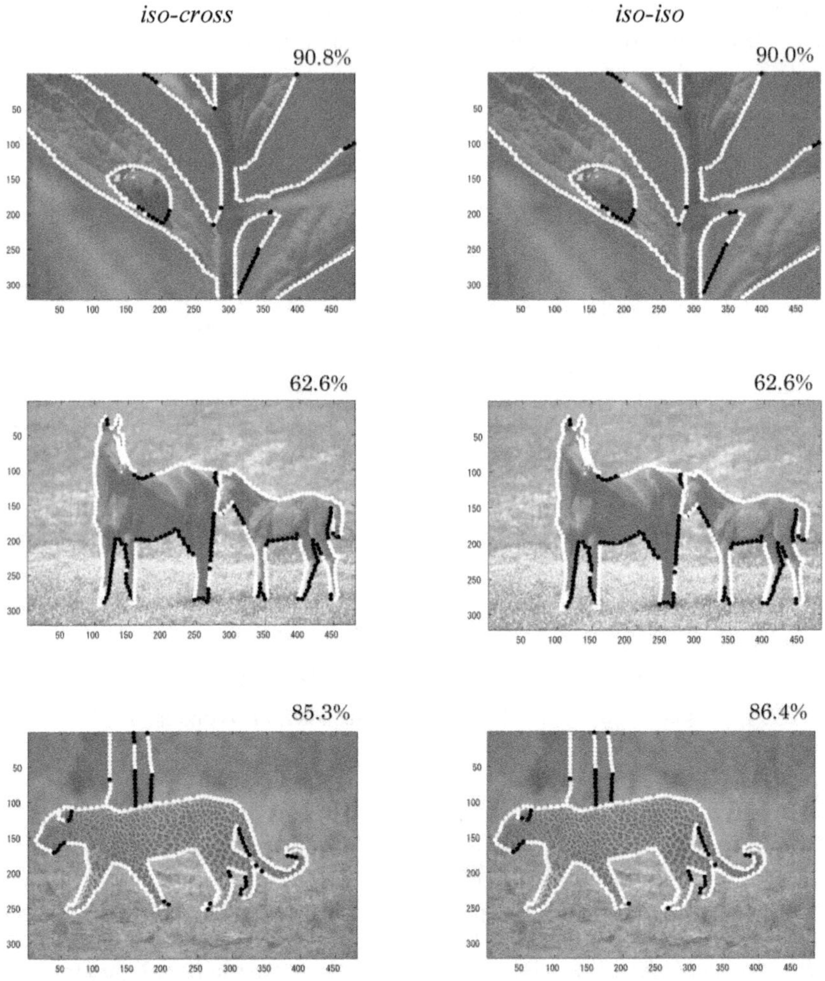

Fig. 3. Three examples of the comparison between human and model responses. White and black dots indicate points where the model response agreed with the human perception, and those without agreement, respectively, superimposed on the original images from Berkeley Segmentation Dataset [8, 9].

Table 1. Computed consistency and SD of *iso-cross* and *iso-iso* conditions

	iso-cross	*iso-iso*
Consistency	67.1%	67.1%
SD	14.2%	13.9%

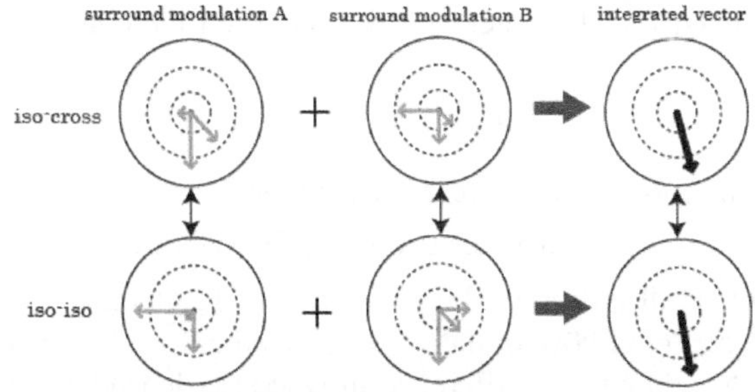

Fig. 4. An example of the comparison between integrated and pair-wised BO responses. Gray arrows indicate the three strongest responses among eight orientations for surround types A and B. The direction of the arrows shows the preferred orientation of the cell. Although the responses of individual cells are different between *iso-cross* and *iso-iso* conditions, the integrated responses (black arrows) show similar responses (black arrows).

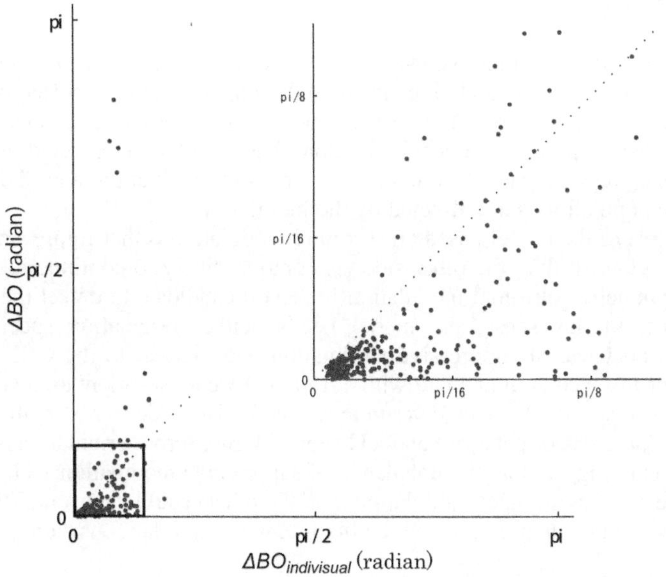

Fig. 5. A scatter plot of correlation between $\Delta BO_{indivisual}$ (horizontal axis) and ΔBO (vertical axis). The inset is the enlargement near the origin as indicated by a square

The difference in computed BO directions (in radian) between the conditions is defined by:

$$\left| BO^{j}_{iso\text{-}cross} - BO^{j}_{iso\text{-}iso} \right| , \tag{11}$$

thus, the mean difference among individual cells due to the conditions is given by:

$$\Delta BO_{indivisual} = \frac{1}{40} \sum_{j} \left| BO^{j}_{iso\text{-}cross} - BO^{j}_{iso\text{-}iso} \right| . \tag{12}$$

Similarly, the difference of the integrated BO between the conditions is given from equation 9:

$$\Delta BO = \left| BO_{iso\text{-}cross} - BO_{iso\text{-}iso} \right| . \tag{13}$$

We computed the correlation between the difference in BO directions by individual cells ($\Delta BO_{individual}$) and the difference in overall BO directions computed by the population (ΔBO), at all locations along contours in 100 natural images. The results are shown in Fig. 5. About 90% of data were located below the unity line, indicating that individual cells show more difference in response depending on the orientation compared with the population responses computed through integration. Together with the overall consistency between the model and human responses, we conclude that BO determination does not depend on the orientation in surround facilitation, specifically as a population.

4 Discussions

The present simulation study showed almost no difference in BO determination between iso- and cross-orientation in modulation, indicating ineffectiveness of surround facilitation. Our result is consistent with physiological studies that have reported stronger suppression over facilitation. The result is interpreted as that iso-orientation suppression cancels out cross-orientation facilitation and the resultant response as a population is not affected by the facilitation.

The principle of the model based on surround modulation is that figure side tends to include more contrast than the other side, as contours likely to continue to the figure side. The asymmetric surround modulation is a good candidate to detect the side with more contrast. In this sense, the model needs neither orientation specificity nor facilitation. Unbalance of suppressive modulation with respect to the CRF would be sufficient. For instance, a model cell with strong and weak iso-orientation suppression on each side, respectively, can determine correctly BO side for stimuli with two vertical lines (no cross-orientation) and a C-shaped figure (cross orientation dominant). These arguments suggest that the unbalance of suppressive modulation with respect to the CRF (one side has stronger modulation than the other) could determine BO, instead of the combination of suppression and facilitation, with a similar consistency.

Acknowledgement

We greatly thank C. Fowlkes, J. Malik, D. Martin, X. Ren and D. Tal for providing the Berkeley Segmentation Dataset. We also thank Y. Tsuji for the preliminary psychophysical experiment. This work was supported by grant-in-aid from MEXT of Japan (KAKENHI; Jouhou Bakuhatsu 19024011, 21013006; 22300090).

References

1. Zhou, H., Friedman, H.S., von der Heydt, R.: Coding of Border Ownership in Monkey Visual Cortex. J. Neuroscience 20, 6594–6611 (2000)
2. Sakai, K., Nishimura, H.: Surrounding Suppression and Facilitation in the Determination of Border Ownership. J. Cognit. Neurosci. 18, 562–579 (2006)

3. Jones, H.E., Grieve, K.L., Wang, W., Sillito, A.M.: Surround Suppression in V1. J. Neurophysiology 86(4), 2011–2028 (2001)
4. Jones, H.E., Wang, W., Sillito, A.M.: Spatial Organization and Magnitude of Orientation Contrast Interactions in Primate V1. J. Neurophysiology 88, 2796–2808 (2002)
5. Watanabe, S., Nishimura, H., Sakai, K.: Direction of Figure in Natural Images - An algorithm based on context dependency effect in visual cortex. In: Proc. 70th National Convention of IPSJ, vol. 5, pp. 255–256 (2008)
6. Wilson, H.R.: Nonlinear processes in visual pattern discrimination. Proc. Natl. Acad. Sci. 90, 9785–9790 (1993)
7. Sakai, K., Hirai, Y.: Neural grouping and geometric effect in the determination of apparent orientation. J. Opt. Soc. Am., A 19, 1049–1062 (2002)
8. Berkeley Segmentation Dataset, `http://www.eecs.berkeley.edu/Research/Project/CS/vision/grouping/segbench/`
9. Fowlkes, C.C., Martin, D.R., Malik, J.: Local Figure-Ground Cues are Valid for Natural Images. J. Vision 7(8):2, 1–9 (2007)

Increased Robustness against Background Noise: Pattern Recognition by a Neocognitron

Kunihiko Fukushima

Fuzzy Logic Systems Institute, Japan
fukushima@m.ieice.org
http://www4.ocn.ne.jp/~fuku_k/index-e.html

Abstract. The *neocognitron* is a hierarchical multi-layered neural network capable of robust visual pattern recognition. It has been demonstrated that recent versions of the neocognitron exhibit excellent performance for recognizing handwritten digits. When characters are written on a noisy background, however, recognition rate was not always satisfactory. This paper proposes several modifications, by which the neocognitrons can be much more robust against background noise.

1 Introduction

The *neocognitron* is a hierarchical multi-layered neural network capable of robust visual pattern recognition [1]. It acquires the ability to recognize patterns through learning.

It has been demonstrated that neocognitrons of recent versions exhibit excellent performance for recognizing handwritten digits [2,3]. Most of the experiments for these neocognitrons have been made using characters written on backgrounds containing little noise. When characters are written on a background contaminated with noise (e.g., the input pattern in Fig. 4 or patterns in the bottom of Fig. 5), however, the recognition rate of these neocognitrons is not always satisfactory.

This paper proposes several modifications, by which the neocognitrons can be much more robust against background noise. Main items of the modifications reside in the characteristics of *S-cells* and *C-cells* in the neocognitron, and we focus our discussion mainly on these improvements.

2 Architecture of the Network

The neocognitron is a hierarchical multi-layered network. It consists of layers of *S-cells*, which resemble simple cells in the primary visual cortex, and layers of *C-cells*, which resemble complex cells. These layers of S-cells and C-cells are arranged alternately in a hierarchical manner. In other words, a number of modules, each of which consists of an S-cell layer and a C-cell layer, are connected in a cascade in the network.

The new neocognitron proposed in this paper consists of four stages of S- and C-cell layers: $U_0 \rightarrow U_G \rightarrow U_{S1} \rightarrow U_{C1} \rightarrow U_{S2} \rightarrow U_{C2} \rightarrow U_{S3} \rightarrow U_{C3} \rightarrow$

K.W. Wong et al. (Eds.): ICONIP 2010, Part II, LNCS 6444, pp. 574–581, 2010.

$U_{S4} \to U_{C4}$. Here we use notation like U_{Sl}, for example, to indicate the layer of S-cells of the lth stage.

Each layer of the network is divided into a number of sub-layers, called *cell-planes*, depending on the difference in the features to which cells respond preferentially. Incidentally, a cell-plane is a group of cells that are arranged retinotopically and share the same set of input connections [1]. As a result, all cells in a cell-plane have receptive fields of an identical characteristic, but the locations of the receptive fields differ from cell to cell.

The stimulus pattern is presented to the input layer (photoreceptor layer) U_0. A layer of contrast-extracting cells (U_G) follows layer U_0. It consists of two cell-planes: one consisting of cells with concentric on-center receptive fields, and one consisting of cells with off-center receptive fields. The former cells extract positive contrast in brightness, whereas the latter extract negative contrast from the images presented to U_0. The output of U_G is sent to U_{S1}.

The S-cells of U_{S1} extract edge components of various orientations from the input image. To be more specific, layer U_{S1} has 16 cell-planes, each of which consists of edge-extracting cells of a particular preferred orientation.

The input connections of S-cells of higher stages are variable and are modified through learning. After having finished learning, S-cells come to work as feature-extracting cells. In higher stages, they extract more global features.

In each stage of the hierarchical network, the output of layer U_{Sl} is fed to layer U_{Cl}. C-cells, whose input connections are fixed, exhibit an approximate invariance to the position of the stimuli presented within their receptive fields. In other words, a blurred version of the response of U_{Sl} appears in U_{Cl}. The blurring operation is essential for endowing the neocognitron with an ability to recognize patterns robustly, with little effect from deformation, change in size, or shift in position of input patterns. The C-cells in the highest stage work as recognition cells, which indicate the result of the pattern recognition.

3 Feature-Extracting S-cells

3.1 Response of an S-cell

Input Signals to an S-cell: To show the essence of the process of feature extraction, we extract the circuit converging to a single S-cell and analyze its behavior. Fig. 1(a) shows the circuit. The S-cell receives excitatory signals directly from a group of C-cells, which are cells of the preceding layer. It also receives an inhibitory signal through a V-cell, which accompanies the S-cell. The V-cell receives fixed excitatory connections from the same group of C-cells as does the S-cell, and always responds with the average intensity of the output of the C-cells.

In the new neocognitron, the inhibitory signal from the V-cell works in a subtractive manner. (In old neocognitrons, the inhibitory signal worked in a shunting or divisional manner). This means that the S-cell is almost the same as the cells usually used in conventional artificial neural networks. What is different

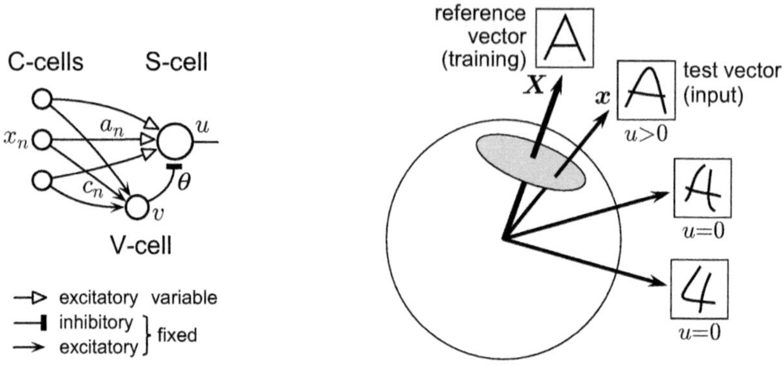

(a) Input connections converging to an S-cell.

(b) Response of an S-cell in a multi-dimensional feature space.

Fig. 1. Feature extraction by an S-cell

from conventional artificial neural networks is that the V-cell calculates the average, not by a linear summation, but by root-mean-square.

Let a_n be the strength of the excitatory variable connection to the S-cell from the nth C-cell, whose output is x_n. The output u of the S-cell is given by

$$u = \frac{1}{1-\theta} \cdot \varphi\left[\sum_n a_n x_n - \theta v\right], \tag{1}$$

where $\varphi[\ ]$ is a function defined by $\varphi[x] = \max(x, 0)$. The strength of the inhibitory connection is θ, which determines the threshold of the S-cell ($0 < \theta < 1$). The response of the V-cell is given by

$$v = \sqrt{\sum_n c_n x_n{}^2}, \tag{2}$$

where c_n is the strength of the fixed excitatory connection from the nth C-cell.

We now use vector notation $\boldsymbol{x} = (x_1, x_2, \cdots, x_n, \cdots)$ to represent the response of all C-cells, from which the S-cell receive excitatory signals. We define *weighted* inner product of arbitrary two vectors $\boldsymbol{x}$ and $\boldsymbol{y}$ by

$$(\boldsymbol{x}, \boldsymbol{y}) = \sum_n c_n x_n y_n, \tag{3}$$

where the strength of the input connections to the V-cell, c_n, is used as the weight for the inner product. We also define the norm of a vector, using the *weighted* inner product, by $\|\boldsymbol{x}\| = \sqrt{(\boldsymbol{x}, \boldsymbol{x})}$.

Renewing Input Connections: For the training of S-cells of layer U_{Sl}, the response of C-cells of the preceding layer U_{Cl-1} works as a training stimulus.

When a training pattern is presented to the input layer U_0 of the network, S-cells of U_{Sl} compete with each other, and several S-cells are selected as winners.

Although the methods of competition are slightly different from layer to layer, the process of renewing input connections of the winners are the same for all layers. We use *winner-kill-loser* rule to select winners for intermediate layers U_{S2} and U_{S3}, and a supervised competitive learning for the highest stags U_{S4}. Since they are almost the same as in the previous neocognitron [3], we discuss here only the process of renewing input connections of the winners.

Each S-cell usually becomes a winner several times during the training phase. Suppose an S-cell has become a winner at the tth time. We use vector $\boldsymbol{X}^{(t)}$ to represent the output of the C-cells presynaptic to this S-cell. Namely, $\boldsymbol{X}^{(t)}$ is the training vector for this S-cell at this moment. Connection a_n is renewed through an auxiliary variable a'_n, which increases in proportion to $X_n^{(t)}$. Namely, the amount of increase of a'_n is

$$\Delta a'_n = c_n X_n^{(t)} , \tag{4}$$

where c_n is the value of the fixed input connection to the inhibitory V-cell.

Let $\boldsymbol{X}$ be the sum of the training vectors that have made the S-cell a winner:

$$\boldsymbol{X} = \sum_t \boldsymbol{X}^{(t)} . \tag{5}$$

After having become winners for these training vectors, the strength of the auxiliary variable a'_n of this S-cell becomes

$$a'_n = \sum_t c_n X_n^{(t)} = c_n X_n . \tag{6}$$

The excitatory connection a_n is calculated from the value of a'_n by

$$a_n = a'_n / b , \tag{7}$$

where

$$b = \sqrt{\sum_n \frac{{a'_n}^2}{c_n}} = \|\boldsymbol{X}\| . \tag{8}$$

Response of an S-cell: Using weighted inner product defined by (3), we have

$$\sum_n a_n x_n = \frac{(\boldsymbol{X}, \boldsymbol{x})}{\|\boldsymbol{X}\|} . \tag{9}$$

from (6), (7) and (8). We also have $v = \|\boldsymbol{x}\|$ from (2).

Hence the response of the S-cell, which is given by (1), can be expressed as

$$u = \|\boldsymbol{x}\| \cdot \frac{\varphi[s - \theta]}{1 - \theta} , \tag{10}$$

where

$$s = \frac{(\boldsymbol{X}, \boldsymbol{x})}{\|\boldsymbol{X}\| \cdot \|\boldsymbol{x}\|} \,. \tag{11}$$

In the multi-dimensional feature space, s shows a kind of similarity between $\boldsymbol{x}$ and $\boldsymbol{X}$ (Fig. 1(b)). We call $\boldsymbol{X}$, which is the sum of the training vectors, the reference vector of the S-cell. Using a neurophysiological term, we can also express that $\boldsymbol{X}$ is the preferred feature of the S-cell.

The second term $\varphi[s - \theta]/(1 - \theta)$ in (10) takes a maximum value 1 if the stimulus vector $\boldsymbol{x}$ is identical to the reference vector $\boldsymbol{X}$, and becomes 0 when the similarity s is less than the threshold θ. In the multi-dimensional feature space, the area that satisfies $s < \theta$ becomes the tolerance area in feature extraction by the S-cell, and the threshold θ determines the size of the tolerance area. In other words, a non-zero response is elicited from the S-cell, if and only if the stimulus vector $\boldsymbol{x}$ is within a tolerance area around the reference vector $\boldsymbol{X}$.

S-cells of the Previous Neocognitron: In the previous neocognitron [3], S-cells have divisional inhibition, and the response of an S-cell is given by

$$u = \frac{\varphi[s - \theta]}{1 - \theta} \,. \tag{12}$$

Comparing (10) and (12), we can see that the response of an S-cell of the new neocognitron is equal to that of the previous neocognitron multiplied by $\|\boldsymbol{x}\|$.

3.2 Effect of Noise on S-cells

If there is no background noise, the characteristics of (12) is desirable for feature-extracting S-cells. The response of the S-cell is determined only by the similarity s between the input stimulus $\boldsymbol{x}$ and the training feature $\boldsymbol{X}$. It is not affected by the strength of the input stimulus $\boldsymbol{x}$. Hence S-cells can extract features robustly without being affected, say, by a gradual non-uniformity in thickness, darkness or contrast in an input pattern.

If an input character is written on a noisy background, like the pattern in Fig. 2, however, interference from the background character becomes serious when we use S-cells of (12). Features of the faint background character elicit large responses from some S-cells whose receptive fields cover only background features. This largely increases the recognition error of the neocognitron.

It is more desirable for S-cells to have characteristics like (10) under noisy environment, because the strength of extracted features come to be proportional to the intensity of input stimuli in the receptive fields, $\|\boldsymbol{x}\|$. The strength of irrelevant features from the interfering pattern remains low, in proportion to the weak intensity of the background noise in the input pattern.

4 C-cells: Blur by Root-Mean-Square Operation

A C-cell has fixed excitatory connections from a group of S-cells of the corresponding cell-planes of S-cells. Through these connections, each C-cell averages

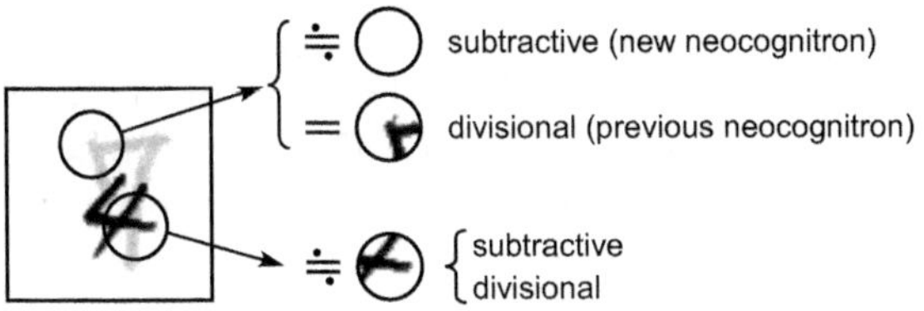

Fig. 2. Responses of S-cells with subtractive and divisional inhibition

the responses of S-cells whose receptive-field locations are slightly deviated. The averaging operation is important, not only for endowing neural networks with an ability to recognize deformed patterns robustly, but also for smoothing additive random noise contained in the responses of S-cells. This is another advantage of the averaging operation over the MAX-operation [4], which is very vulnerable to noise, because the output of a C-cell is determined by the response of a single maximum-output S-cell only.

Fig. 3 shows a block diagram of a C-cell. In the new neocognitron, a C-cell averages its input signals, not by a weighted linear summation, but by a root-mean-square. To reduce the computational cost, the spatial density of S-cells in a cell-plane is usually designed to be sparse, and a C-cell averages the responses of a small number of S-cells. Hence the output of the C-cell fluctuates with the shift in location of its preferred feature. The fluctuation can be made smaller by the root-mean-square than by a linear summation or a MAX operation.

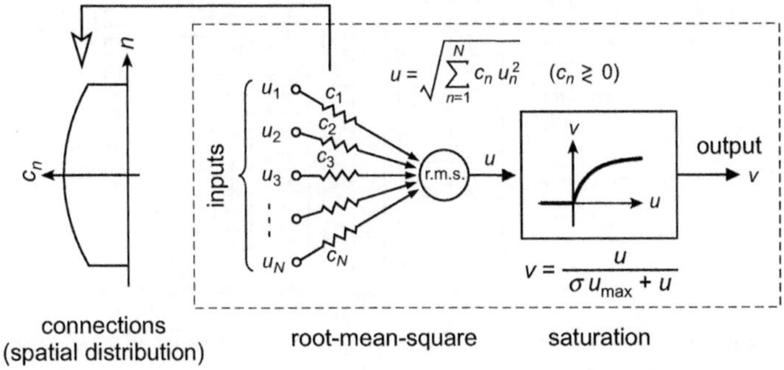

Fig. 3. A block diagram of a C-cell, which calculate the root-mean-square of its inputs

Since a role of a C-cell is to detect whether any of its presynaptic S-cells is active, it is better to have some saturation in the input-to-output characteristic.

In the previous neocognitron [3], in which C-cells calculate mean values by linear summation, the saturation is determined by a square root function, namely $v = \sqrt{u}$. If input patterns do not contain background noise, saturation by square-root works well. If input patterns contain some background noise, however, the square-root nonlinearity is not desirable. A small background noise is exaggerated by the square-root nonlinearity, because $dv/du = d\sqrt{u}/du \to \infty$ for $u \to 0$.

In the new neocognitron, the saturation is controlled adaptively by the maximum value of input signals to C-cells of the layer. In Fig. 3, u is the weighted root-mean-square of the input signals to a C-cell. Let u_{max} be the maximum value of u among all C-cells of the layer, and let σ be a positive constant. The output of a C-cell of the layer is given by

$$v = \frac{u}{\sigma u_{\mathrm{max}} + u} . \tag{13}$$

5 Computer Simulation

We tested the behavior of the new neocognitron by computer simulation. Fig. 4 shows a typical response of the network that has finished the learning. The responses of layers U_0, U_G and layers of C-cells of all stages are displayed in series from left to right. The rightmost layer, $\mathrm{U}_{\mathrm{C}4}$, shows the final result of recognition. The input character is written on a noisy background. In this example, character '4' in the foreground is recognized correctly, although there is a faint disturbing character '7' in the background.

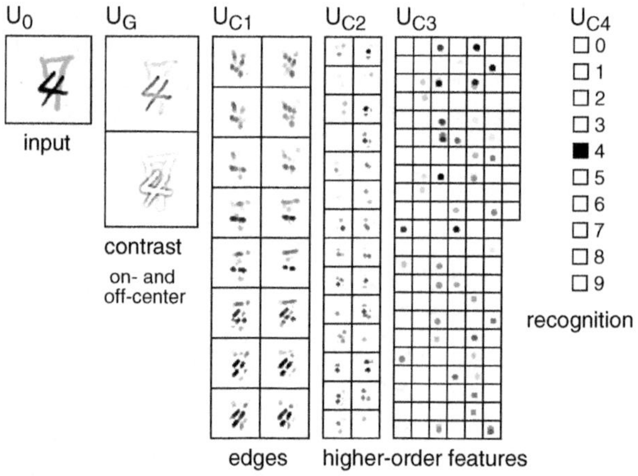

Fig. 4. An example of the response of the neocognitron. Character '4' in the foreground, which is disturbed by the faint background character '7', is recognized correctly.

We measured recognition error using handwritten digits (free writing) randomly sampled from the ETL1 database [5]. To be more specific, we measured the recognition error for a blind test set of 5000 patterns, where we used 3000 patterns (300 patterns for each digit) for the learning. We made this experiment twice for each condition, using different learning and test sets randomly sampled from the ETL1, and averaged the results of the two experiments. Fig. 5 shows how the recognition error changes with different levels of background noise.

We tested two different types of background noise. In one case, which is shown in Fig. 5(a), the background noise is a faint image of a different digit, which is

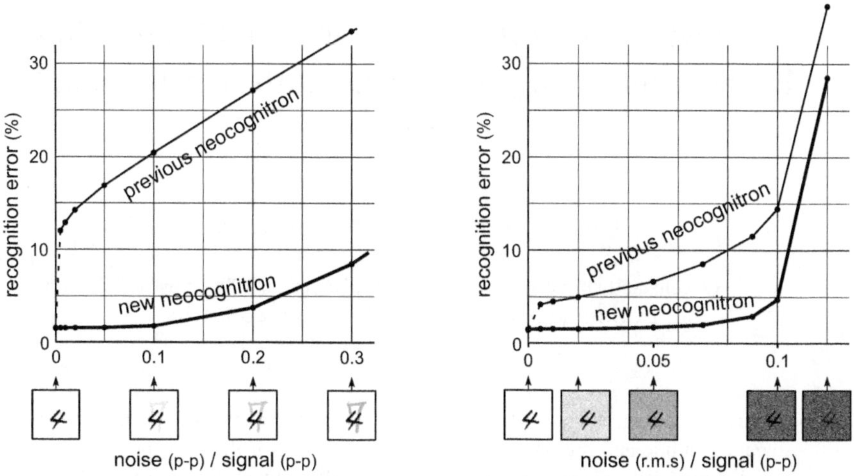

(a) The background noise is a faint image of a different digit.

(b) White noise is superimposed on the test digit.

Fig. 5. Recognition error of the new and previous neocognitrons under different levels of background noise

also sampled randomly from the ETL1 database. In the other case, which is shown in Fig. 5(b), a white noise is superimposed on the test digit.

Although the recognition error of the new neocognitron under noiseless condition (1.47%) is slightly higher than that of the previous one (1.40%), the increase in recognition error under noisy background is much smaller. Under noisy conditions, the new neocognitron exhibits much better recognition rate than the previous one.

Acknowledgements. The author thanks Prof. Isao Hayashi (Kansai University), Dr. Hayaru Shouno (University of Electro-Communications) and Dr. Masayuki Kikuchi (Tokyo University of Technology) for helpful comments. This work was partially supported from Kansai University by Strategic Project to Support the Formation of Research Bases at Private Universities: Matching Fund Subsidy from MEXT, 2008–2012.

References

1. Fukushima, K.: Neocognitron: a self-organizing neural network model for a mechanism of pattern recognition unaffected by shift in position. Biological Cybernetics 36, 193–202 (1980)
2. Fukushima, K.: Neocognitron for handwritten digit recognition. Neurocomputing 51, 161–180 (2003)
3. Fukushima, K.: Neocognitron trained with winner-kill-loser rule. Neural Networks 23, 926–938 (2010)
4. Riesenhuber, M., Poggio, T.: Hierarchical models of object recognition in cortex. Nature Neuroscience 2, 1019–1025 (1999)
5. ETL1 database, `http://www.is.aist.go.jp/etlcdb/#English`

Improving the Performance of Facial Expression Recognition Using Dynamic, Subtle and Regional Features

Ligang Zhang and Dian Tjondronegoro

Queensland University of Technology, Brisbane, 4000, Australia
ligzhang@gmail.com, dian@qut.edu.au

Abstract. Human facial expression is a complex process characterized of dynamic, subtle and regional emotional features. State-of-the-art approaches on facial expression recognition (FER) have not fully utilized this kind of features to improve the recognition performance. This paper proposes an approach to overcome this limitation using patch-based 'salient' Gabor features. A set of 3D patches are extracted to represent the subtle and regional features, and then inputted into patch matching operations for capturing the dynamic features. Experimental results show a significant performance improvement of the proposed approach due to the use of the dynamic features. Performance comparison with pervious work also confirms that the proposed approach achieves the highest CRR reported to date on the JAFFE database and a top-level performance on the Cohn-Kanade (CK) database.

Keywords: Facial expression recognition, Adaboost, support vector machine.

1 Introduction

Facial expression recognition (FER) is becoming an increasingly active research field in recent years due to its potential to be applied in many areas. FER supports many practical applications, such as human-computer interaction, patient and driver state detection. However, robust FER is still a challenging task as facial expression is a complex process, which is characteristic of different dynamic, subtle and regional facial changes (e.g. wrinkles) and is easy to be influenced by various environmental changes (e.g. illumination and occlusions). Therefore, an important step for FER is to accurately extract the useful dynamic, subtle and regional emotional features.

Current FER approaches can be classified into 4 categories based on the used feature types: motion-based, feature-based, model-based and appearance-based. Appearance-based approach can analyze facial images using multi-resolution information and has shown a significant advantage over other approaches in terms of capturing subtle features. The features used include Gabor feature [1], local binary patterns (LBP) [2], Haar [3], discrete Fourier transform (DFT) [4] etc. However, Current appearance-based approaches suffer from the drawback of using point-based Gabor or DFT features, which lack the ability to capture regional features. At the same time, the LBP and Haar are essentially based on statistics and can not capture

K.W. Wong et al. (Eds.): ICONIP 2010, Part II, LNCS 6444, pp. 582–589, 2010.

the subtle features with pixel accuracy. In addition, appearance-based approaches in static images have yet considered the dynamic information of feature position, scale and shape changes, which also represent useful information for FER.

This paper proposes an FER approach to improve the recognition performance based on patch-based 'salient' Gabor features. The novelty of our approach lies in the adoption of patch-based Gabor features and the definition of patch matching to solve point-based Gabor features' limitation in capturing the dynamic, subtle and regional features. The experimental results demonstrate big performance improvements as well as the state-of-the-art performances of the proposed approach on both the JAFFE and the Cohn-Kanade (CK) databases. The rest of the paper is organized as follows. Section 2 presents the proposed approach. Section 3 gives the experimental results. The conclusions are drawn in Section 4.

2 Proposed Approach

Fig. 1 illustrates the proposed approach, which is composed of the pre-processing, training and test stages. At the pre-processing stage, facial regions are manually cropped to imitate rough face detection and scaled to a resolution of 48*48 pixels. Then multi-resolution Gabor images are attained by convolving Gabor filters with these scaled facial regions. 2D Gabor filter with 8-scale (5:2:19 pixels) and 4-orientation (-45°, 90°, 45°, 0°) is used, and the other parameters are set based on [5]. During the training stage, a set of 3D patches are extracted from the Gabor images to represent the subtle and regional features. Patch matching operations are then performed to convert all extracted patches to distances for capturing the dynamic features of position, scale and shape changes. At the test stage, the distance features in a new image are attained by performing the same patch matching operations, and fed into support vector machine (SVM) to recognize 6 emotions: anger (AN), disgust (DI), fear (FE), happiness (HA), sadness (SA) and surprise (SU). The use of Gabor filter and SVM is due to their excellent performance reported in previous work [1]. Two SVMs with linear and radial basis function (RBF) kernels are used.

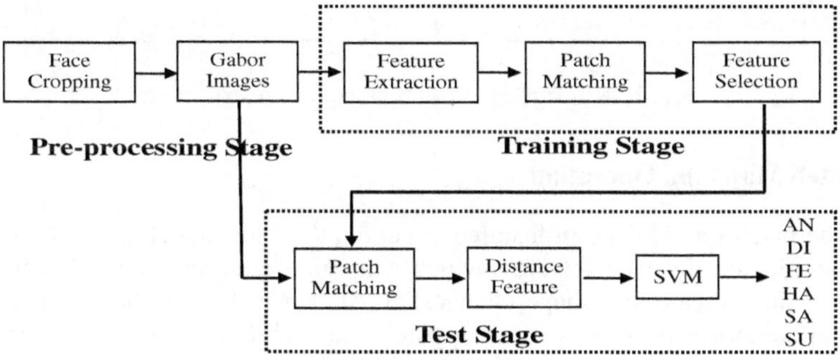

Fig. 1. Framework of the proposed approach

2.1 Patch-Based Feature Extraction

Feature extraction generates a set of discriminating 3D patches, which have an advantage over point-based Gabor features to represent the subtle and regional emotional features. As shown in Fig. 2, the algorithm can be described as follows: (1) all training images are classified into 10 sets. For each emotion, each Gabor scale, and each patch size, one Gabor image is randomly selected from all images of the given emotion E_k. (2) Given one patch with the size of $P_j*P_j*O_{num}$, move this patch across the row and column pixels of this Gabor image, a set of patches can be extracted (the line *a*). (3) Record the corresponding matching area and matching scale (the line *b* and *c*, details are explained in Section 2.2). (4) The final patch set can be constituted by concatenating the extracted patches of all emotions, all scales and all patch sizes.

To reduce the feature dimension and increase the processing speed, we only extract part of all patches by moving the patch P_a with a step (i.e. *Move_step* in Fig.2). The patch sizes (width*height*orientation) and the corresponding moving steps are set to be 2*2*4, 4*4*4, 6*6*4, 8*8*4, and to be 1, 2, 3, 4 pixels respectively. Given 48*48 images and 8-scale, 4-orientation Gabor filters, the final set contains 148,032 patches.

```
Input:Image set Si (i=1,…,10); patch size Pj (j=1,…,4);
      emotion index Ek (k=1,…,6); scale SCm (m=1,…,8);
      orientation On (n=1,…,4); orientation number Onum;
      image width W, height H (W=H=48).
Output:Extracted patches, matching area and scale.
For each Ek,each SCm,each Pj
    Choose one set Si randomly from 10 sets;
    Choose one image of emotion Ek randomly from Si;
    Move_step = Pj/2;
    For ih = 1 to (H-Pj+1) by Move_step
      For iw = 1 to (W-Pj+ 1) by Move_step
        Extract patches with sizes of Pj*Pj*Onum;        (a)
        Record the matching area (Lx,Ly,Rx,Ry)           (b)
        Lx = Max(ih-0.5*Pj,1);  Rx = Min(ih+1.5*Pj,H);
        Ly = Max(iw-0.5*Pj,1);  Ry = Min(iw+1.5*Pj,W).
        Record the matching scale SCm;                   (c)
      End
    End
End
Return patches, matching area and scale.
```

Fig. 2. Pseudo code of patch-based feature extraction

2.2 Patch Matching Operation

Given the patch set obtained in feature extraction, the patch matching operation tends to convert it into distance features, which can capture the dynamic information of feature position, scale and shape changes. As shown in Fig. 3, the patch matching operation comprises of 4 steps for each patch and each training image: (1) matching area and matching scale are defined to provide a bigger matching space (Fig. 3 (c)). Based on this space, the emotional feature, which varies its position, scale and shape in different images, still can be captured provided that it is located within the space. (2) The distances are obtained by matching this patch with all patches within its matching space in the training image. This step takes two patches as inputs and yields one distance value based on a distance metric (Fig. 3 (b,d,e)). (3) The minimum distance is chosen as the distance feature for this patch and this training image (the black block in Fig. 3 (f)). (4) The distance features of all patches are combined into the final set with 148,032 elements (Fig. 3 (f)).

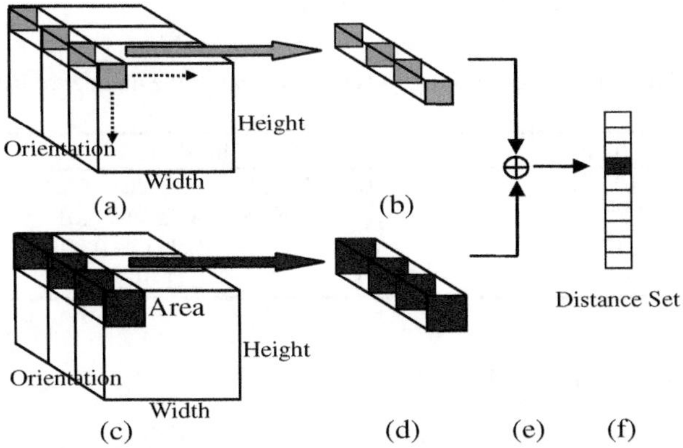

Fig. 3. Patch matching operation. One patch (b) is extracted from Gabor images (a); the corresponding matching area 'Area'(c); one distance (black block in (f)) is obtained by matching two patches (e).

The definition of matching area and matching scale plays a key role in capturing the dynamic features of position, scale and shape changes. The idea of them stems from the observation that position and scale of one feature do not have big changes in different facial images once these images are roughly located by a face detector. Thus, the invariance to position and scale changes can be accomplished by defining one such matching space for each feature. In this paper, given a patch P_a with size $P_j{*}P_j{*}O_{num}$, its matching area is set 2 times of P_a in width and height, but with the same orientation number and centre point. That is $Area = (2{*}P_j){*}(2{*}P_j){*}O_{num}$. While the matching scale is the same scale as patch P_a because most cropped facial regions belong to the same scale. There are 4 distance metrics used: dense L1 (DL1), dense L2 (DL2), sparse L1 (SL1) and sparse L2 (SL2). Sparse distance uses the maximum value of all orientations, while dense distance uses all values of all orientations.

2.3 Salient' Patch Selection

The feature extraction step produces a patch set that contains a big number of features and redundant information; therefore, this paper adopts the widely used and efficiency proved Adaboost for discriminative (called 'salient' here) patch selection. Since Adaboost was designed to solve two-class problems, in this research, the one-against-rest strategy is used to solve the six-emotion-class problem. The training process stops when the empirical error becomes below 0.0001 with the initial error of 1. This setting is inspired by the stopping condition in [1] that the generalization error becomes flat. Regarding the training set, the JAFFE database includes all selected images, whereas the CK database is only composed of the peak frames.

To explore the relationship between the CRR and the number of features, a group of error thresholds as listed in Table 1 are used to control the number of 'salient' patches. These thresholds are set based on our experimental observation that the empirical errors of Adaboost decrease with a factor of 10 and its numbers are evenly distributed between decimal intervals (e.g. 0.01 to 0.02). Accordingly, 38 numbers of features with 38 CRRs can be obtained by selecting patches with empirical errors bigger than the corresponding error thresholds.

Table 1. The 38 error thresholds used to control the number of patches

Index	Error thresholds
1st -10th	(10, 9, 8, 7, 6, 5, 4, 3, 2, 1) * 0.1
11th -19th	(9, 8, 7, 6, 5, 4, 3, 2, 1) * 0.01
20th -28th	(9, 8, 7, 6, 5, 4, 3, 2, 1) * 0.001
29th -38th	(9, 8, 7, 6, 5, 4, 3, 2, 1, 0) * 0.0001

3 Experimental Results

3.1 Databases

The JAFFE database [6] contains 213 gray images of 7 expressions posed by 10 Japanese females. Each object has 3 or 4 frontal face images for each expression and their faces are approximately located in the middle of the images. All images have been rated on 6 emotion adjectives by 60 subjects. The released portion of the Cohn-Kanade (CK) database [7] includes 2105 digitized image sequences from 182 subjects ranged in age from 18 to 30 years. The 6 basic expressions were based on descriptions of prototypic emotions. Image sequences are shown from neutral to target emotion.

In this paper, all the images of 6 expressions from JAFFE are used. For CK, 1,184 images that represent one of the 6 expressions are selected, 4 images for each expression of totally 92 subjects. The images are chosen from the last image (peak) of each sequence, then one every two images. The face regions of all selected images are cropped by taking the nose as a center point, and scaled to 48*48 pixels. The resulting regions contain features with different positions, scales and shapes.

3.2 Performance on the JAFFE Database

Fig. 4 shows the relationship between the CRR and the number of features on the JAFFE database. The CRR is the average performance of 10-set cross-validation. As can be seen, the proposed approach achieves the highest CRR of 93.48% using DL2 and linear SVM when the error threshold equals to 0.0001 and the number of features equals to 185. The overall performances of 4 distances grow up rapidly at the starting stage, however, the performances begin to level off when the number of features exceeds 150 for linear and 80 for RBF. For the performances of SVMs, linear performs better than RBF for all distances. Regarding the overall performances of distances, for both linear and RBF, the best performances are achieved by DL2, which is followed by SL2. On the other hand, SL1 and DL1 rank the last two.

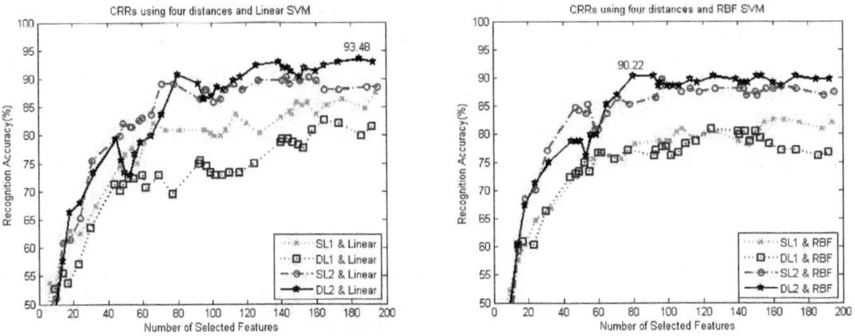

Fig. 4. Relationship between the CRR and the number of features on JAFFE

3.3 Performance on the CK Database

Fig. 5 illustrates the relationship between the CRR and the number of features on the CK database. Seen from this figure, the proposed approach obtains the highest CRR of 94.48% using DL2 and RBF SVM when the error threshold is 0 and the number of features is 180. This may imply that a performance improvement can be achieved once using a larger number of features. The relationship is similar to that of JAFFE in that the CRR grows up rapidly at starting stage and L2 distances outperform L1 distances for both linear and RBF. On the other hand, the CRR reaches a plateau with a speed quicker than that of JAFFE and DL1 performs better than SL1. Moreover, the performance difference between linear and RBF is smaller than that of JAFFE.

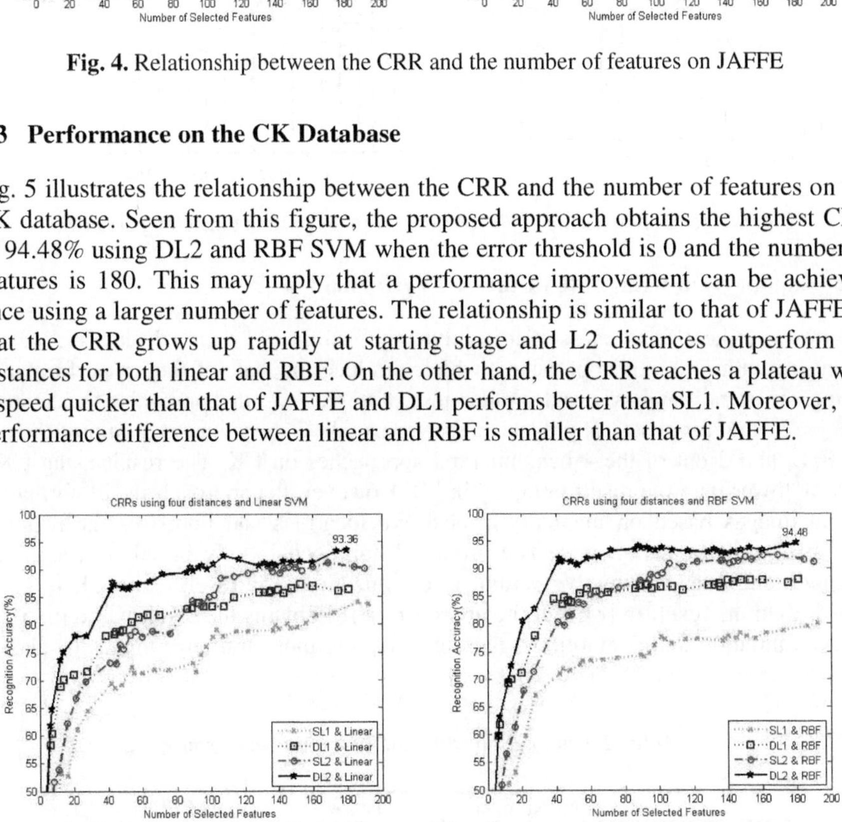

Fig. 5. Relationship between the CRR and the number of features on CK

3.4 Performance with and without Matching Area

To evaluate the performance improvement rising from the use of the dynamic features, we compare the performances obtained with and without matching area. Note that the features obtained without matching area are supposed to not include dynamic information of position, scale and shape changes. Fig. 6 shows the comparison results when the error threshold equals to 0. The results of JAFFE (left) and CK (right) are obtained using linear and RBF SVMs respectively. As can be seen, for JAFFE, the performances of 4 distance metrics are greatly boosted due to the use of matching area. There is a CRR increase of 11.41% using DL2. For CK, the CRRs of DL1 and DL2 are improved about 2.5% due to the use of matching area, while the CRRs of SL1 and SL2 does not benefit from using matching area. Considering the highest CRR of 4 metrics, we can see that dynamic features help to improve the performance.

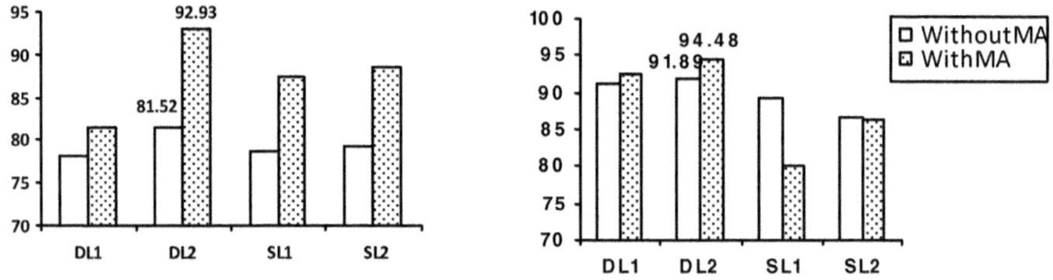

Fig. 6. Recognition accuracy (%) obtained with and without matching area (MA)

3.5 Comparison with State-of-the-Art Performance

To evaluate the effectiveness of using dynamic, subtle and regional features, we also compare with previous approaches, which produce the state-of-the-art performances using the same databases and similar test strategies to our approach. As shown in Table 2, the proposed approach outperforms all 7 benchmarked approaches on JAFFE, and 2 out of the 4 benchmarked approaches on CK. The result using CK is 0.62% lower than the result obtained in [2]. However, the approach in [2] normalizes facial images based on manually-labeled eye locations and improves the result by optimizing SVM parameters. The proposed approach is only based on rough face location and does not involve parameter optimization. The result using CK is 1.39% lower than the result in [8]. But the approach in [8] obtains the result based on 5-fold cross validation and 5 emotions, therefore, it uses more training images to classify less emotions compared to our approach.

Table 2. Comparison with state-of-the-art performance

	Emotion Number	Feature	JAFFE	CK
Proposed	6	patch-based Gabor	**93.48%**	**94.48%**
[9], 2005	6	fuzzy integral	83.2%	-
[10], 2006	6	KCCA	77.05%	-
[11], 2008	7	WMMC	65.77%	-
[12], 2009	7	SFRCS	85.92%	-
[13], 2005	7	Gabor + FSLP	91.0%	-
[8], 2008	7(JAFFE), 5(CK)	FEETS + PRNN	83.84%	95.87%
[2], 2009	7(JAFFE), 6(CK)	boosted-LBP	81.0%	95.1%
[1], 2006	7	Gabor	-	93.3%
[14], 2008	7	Gabor + Haar	-	93.1%

4 Conclusion

The paper proposes a novel FER approach to improve the recognition performance using dynamic, subtle and regional features, which are obtained based on patch-based Gabor features and patch matching operations. The experimental results demonstrate good performances of the proposed approach on both the JAFFE and CK databases, and show big performance improvements due to the use of dynamic features. In addition, the comparison with previous approaches confirms the state-of-the-art performance using the proposed dynamic, subtle and regional features. The future work includes adopting real face detectors and testing on seven facial expressions.

References

1. Littlewort, G., Bartlett, M.S., Fasel, I., Susskind, J., Movellan, J.: Dynamics of facial expression extracted automatically from video. Image and Vision Computing 24, 615–625 (2006)
2. Shan, C., Gong, S., McOwan, P.W.: Facial expression recognition based on Local Binary Patterns: A comprehensive study. Image and Vision Computing 27, 803–816 (2009)
3. Yang, P., Liu, Q., Metaxas, D.N.: Boosting encoded dynamic features for facial expression recognition. Pattern Recognition Letters 30, 132–139 (2009)
4. Xiang, T., Leung, M.K.H., Cho, S.Y.: Expression recognition using fuzzy spatio-temporal modeling. Pattern Recognition 41, 204–216 (2008)
5. Serre, T., Wolf, L., Bileschi, S., Riesenhuber, M., Poggio, T.: Robust Object Recognition with Cortex-Like Mechanisms. IEEE Transactions on Pattern Analysis and Machine Intelligence 29, 411–426 (2007)
6. Lyons, M., Akamatsu, S., Kamachi, M., Gyoba, J.: Coding facial expressions with Gabor wavelets. In: Proceedings of the Third IEEE International Conference on Automatic Face and Gesture Recognition, pp. 200–205 (1998)
7. Kanade, T., Cohn, J.F., Yingli, T.: Comprehensive database for facial expression analysis. In: Proceedings of the Fourth IEEE International Conference on Automatic Face and Gesture Recognition, pp. 46–53 (2000)
8. Wong, J.-J., Cho, S.-Y.: A face emotion tree structure representation with probabilistic recursive neural network modeling. Neural Computing & Applications (2008)
9. Yuwen, W., Hong, L., Hongbin, Z.: Modeling facial expression space for recognition. In: 2005 IEEE/RSJ International Conference on Intelligent Robots and Systems (IROS 2005), pp. 1968–1973 (2005)
10. Wenming, Z., Xiaoyan, Z., Cairong, Z., Li, Z.: Facial expression recognition using kernel canonical correlation analysis (KCCA). IEEE Transactions on Neural Networks 17, 233–238 (2006)
11. Zhengdong, C., Bin, S., Xiang, F., Yu-Jin, Z.: Automatic coefficient selection in Weighted Maximum Margin Criterion. In: 19th International Conference on Pattern Recognition, ICPR 2008, pp. 1–4 (2008)
12. Kyperountas, M., Tefas, A., Pitas, I.: Salient feature and reliable classifier selection for facial expression classification. Pattern Recognition 43, 972–986
13. Guo, G., Dyer, C.R.: Learning from examples in the small sample case: face expression recognition. IEEE Transactions on Systems, Man, and Cybernetics, Part B 35, 477–488 (2005)
14. Chen, H.Y., Huang, C.L., Fu, C.M.: Hybrid-boost learning for multi-pose face detection and facial expression recognition. Pattern Recognition 41, 1173–1185 (2008)

Identity Retrieval in Biometric Access Control Systems Using Multimedia Fusion

Girija Chetty[1], Renuka Biswas[2], and Julian Goodwin[2]

[1] Faculty of Information Sciences and Engineering,
University of Canberra, Australia
girija.chetty@canberra.edu.au
[2] Video Analytics Pty. Ltd. Melbourne

Abstract. In this paper, we propose a novel multimedia sensor fusion approach based on heterogeneous sensors for biometric access control applications. The proposed fusion technique uses multiple acoustic and visual sensors for extracting dominant biometric cues, and combines them with non-dominant cues. The performance evaluation of the proposed fusion protocol and a novel cascaded authentication approach using a 3D stereovision database shows a significant improvement in performance and robustness.

Keywords: Multimedia processing – Face and Speech signal analysis, Recognition and verification.

1 Introduction

Most of the currently deployed biometric access control systems for civilian applications are based on voice modality (also known as speaker recognition from telephone speech), and they are based on modeling a speaker based on unimodal information, i.e. either audio or acoustic features. Audio-based identification achieves high performance when the signal-to-noise ratio (SNR) is high. However, the performance degrades quickly as the test SNR decreases (referred to as a train/test mismatch), as shown in [1],[2], and [3]. Using heterogeneous sensors, such as set of video camera and microphone sensors for example, it is possible to capture the visual dynamics of the orafacial articulators during speech production, allowing inherent multimodality to be exploited. The information from two heterogeneous sensor sources (visual information in addition to voice information) can make the system robust against mismatch between training and test operating environments. It must be noted use of camera sensors along cannot result in much performance improvement, as the visual sensor data also could be highly susceptible to pose, illumination and expression variations. Several techniques have been proposed so far to address some of the above mentioned mismatch scenarios, [1], [2], and [3]. Many of the conventional techniques proposed so far were based on fusion of features extracted from dominant and discriminatory measurements from biometric sensor data corresponding to voice signals and 2D or 3D face images.

However, several studies from cognitive psychology [4] and psychophysical analysis of human visual speech [5] suggest that, dominant biometric cues or the

K.W. Wong et al. (Eds.): ICONIP 2010, Part II, LNCS 6444, pp. 590–597, 2010.

primary identifiers on its own cannot model the identity of a person in its entirety. The non-dominant cues from weaker measurements can also contain the identity specific information. The secondary identifiers have weaker discriminatory power and hence cannot be used on their own for making decisions on the identity of the subjects in question. Nevertheless, they could be used to supplement the decision taken by identifier modules with higher discriminatory power. Some such important secondary cues could be subject's demographic information such as gender, dialect, ethnicity, age or certain subtle individual nuances, such as facial expressions, lip gestures, eyebrow tweaks and raising and lowering of head during speaking or talking.

One of the most significant finding by Hani et al in [6]and Kroos et al in [7], suggest that a speaking face is a kinematic-acoustic system in motion, and the shape, texture and acoustic features during speech production are correlated in a complex way, and a single neuro-motor source controlling the vocal tract behavior is responsible for both the acoustic and the visible attributes of speech production. Hence, for a speaking face not only the facial motion and speech acoustics are inherently correlated multimedia signals, but contain contributions from non-dominant secondary information such as the head motion and fundamental frequency (F0) produced during speech production (as shown in Figure 1).

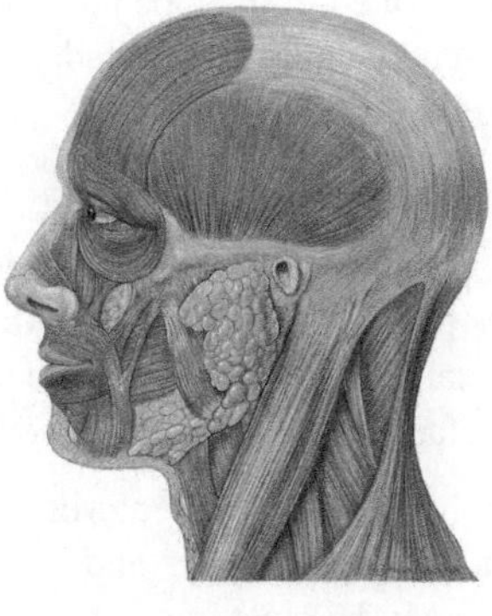

Fig. 1. Facial Muscles (source: http://en.wikipedia.org/wiki/Facial_muscles)

These findings from face-speech anatomy provide clues that facial movements during speech involve highly complex biomechanics with depth, motion and correlation variations. In this paper we propose a novel sensor fusion protocol and authentication technique to address some of the shortcomings of the current biometric access control systems for real world operating conditions.

2 Multimedia Sensor Fusion

The operation of proposed scheme, for an example access control scenario is described. For accessing a secure facility or building with a hallway and a door for example, the user would be verified first with the 2D face verification module (using the first video camera sensor) - unobtrusively without any user co-operation when approaching the door. If the face verification module accepts the user, the door will open automatically when the users approaches the door. If it fails, the second module – 3D face verification module (using the 2nd video camera sensor) springs into action. If this module fails, the voice verification module comes into effect and the system asks the user some questions and expects voice response from users to establish the identity.

The proposed authentication strategy provided by the cascaded authentication structure can provide several benefits. They can satisfy different performance measures: User acceptance measures and authentication accuracy measures-normally represented in terms of error rates, that is the false accept rates (FARs). false reject rates (FRRs) or Equal Error Rates (EERs). A system requiring a high level of security requires a lower FAR and the one requiring a high level of user acceptability requires as low FRR. A trade-off in general is needed to balance the two conflicting requirements.

3 Multimedia Feature Processing

In this Section, details of the multimedia processing of co-occurring audio and video signals from two cameras are described. It must be noted that the details of the complete processing modules is not discussed due to limited space in the paper. The detail treatment of the techniques used is described in [9], [10], [11], [12], [13] and [14].

2D and 3D Face Modules

These two face modules are the primary biometric identifier modules and use face images from single camera for the 2D facial feature extraction, and face images from both the cameras for the 3D facial feature extraction. We used two different methods for extracting feature vectors from face images for these modules. These methods are:

i) *ICA* (Independent Component Analysis) features - a description of face images by their projection on independent base images, and

ii) *AAM* (Active Appearance Model) features – a description of face images by an Active Appearance Model which describes the shape and grey value variations of the face images. The extracted feature vectors were classified with a Gaussian Mixture Model (GMM) classifier for making the decision as a genuine client or impostor. The dimensionality of ICA/AAM feature vectors is reduced to keep 95% of the variations intact. The details of extracting ICA/AAM feature vectors are briefly described next.

The positions of facial landmarks were manually labeled to provide normalized data to the feature extraction methods.

Independent Component Analysis (ICA)

The ICA feature extraction process used is describe in detail in [11].

Active Appearance Models (AAM)

The AAM process requires several facial landmarks (more than 100) along the several salient face regions such as lips, eyes, eyebrows and face contour. After adaptation of the model to a given face image, the resulting appearance parameters describing the shape and the grey value distribution of a given face are used as feature vector for GMM classifier. Further details on AAM are available in [14].

Audio Module

This module is also used as a primary biometric identifier and comprises extraction of MFCC features from speech signal of the speaking face video sequences. The details of MFCC features is given in [10] and [15].

Similar to the 2D and 3D face modules, each speaker based on MFCC acoustic feature vectors is represented by a GMM (Gaussian Mixture Model) model λ built for the client. We used three different types of feature vectors for building the speaker models. – *audio only* module, based on acoustic (MFFC) features vectors, and the AV module (based on fusion of MFFC + ICA+AAM features). The test speaker or the speaker utterance that is to be classified (the unknown pattern) is a sentence, represented by a sequence (O_P) of MFCC speech feature vectors, or ICA / AAM face feature vectors or AV feature vectors (fusion of ICA-MFCC-AAM) by,

$$O_P = \{o_1, o_2, \cdots o_t \cdots, o_{T_P}\}, \quad (1)$$

where o_t is the feature (speech/face observation (frame) at time t and T_P denotes the number of observation vectors in the sentence. We obtain N class-conditional joint probabilities

$$p(O_P|\lambda) = p(o_1, o_2, \cdots, o_t, \cdots o_{T_P}|\lambda) \quad (2)$$

that the observation sequence O_P was produced by the client speaker model λ.. $p(O_P|\lambda)$ is referred to as the likelihood that O_P was caused by λ. For GMM classifiers, the output scores are in log-likelihood form, denoted by $ll(O_P|\lambda)$.

Kinematic Features Module

This module is used as a secondary biometric module (because on its own, the secondary features cannot ascertain the identity reliably), and consists of extraction of kinematic or motion features from several regions of the face – lip region, eye region and complete frontal face region. The details of this module is presented in detail in [10].

The three types of visual features described above were concatenated to form a 24 dimensional visual motion feature vector. This is shown in Eqn. (3) where f_{DCT}, f_{GRD}, f_{CTR} represent the DCT, grid, and contour based motion features respectively from the lip region, and o_t refers to the observation feature vector for the frame at time t.

$$o_t = [o_t^{DCT}, o_t^{GRD}, o_t^{CTR}] \quad (3)$$

For entire sentence we have several visual frames and the secondary motion feature vector for the entire frame consists of several o_t vectors with f_{DCT}, f_{GRD}, *and* f_{CTR}. O_{S1}, the kinematic feature vector for the first secondary biometric module is denoted as:

$$o_{S1} = \{o_1, o_2, \cdots, o_t, \cdots, o_{T_v}\} \quad (4)$$

The rest of the secondary modules for eye region and entire frontal face region is obtained in a similar fashion by deriving the observation feature vector o_t from the explicit motion features - f_{DCT}, f_{GRD}, f_{CTR}, and can be denoted as:

$$o_{S2} = \{o_1, o_2, \cdots, o_t, \cdots, o_{T_v}\} \tag{5}$$
$$\vdots$$
$$o_{Sn} = \{o_1, o_2, \cdots, o_t, \cdots, o_{T_v}\}$$

$$n = 1....4$$

The audio-video sentence observation from a person's talking face is decomposed into its primary and secondary constituent parts.

As described before, the observations from primary and secondary biometric modules are processed by the independent GMM classifier modules to give individual sets of likelihoods, $ll(O_{P1}|\lambda)$, $ll(O_{P2}|\lambda)$, $ll(O_{P3}|\lambda)$, $ll(O_{S1}|\lambda)$, $ll(O_{S2}|\lambda$, $ll(O_{S3}|\lambda)$ and used for making a decision on the identity of the person being a genuine client or impostor. By setting the thresholds appropriately for every stage, different user and security requirements can be addressed in terms of FARs (false accept rates), FRRs (false reject rates) and EERs (Equal Error Rates). For high security requirements, FAR has to be maintained low, at the cost of losing user acceptance, and for high user acceptance, FRR has to be maintained low at the cost of losing security requirements. EER is a measure, where threshold is set such that FAR equals FRR. Most the systems designed with EERs less than 5% are acceptable for moderate security applications and systems with EERs more than 10% are useless and cannot be deployed. We now describe the details of the experiments that were carried out.

4 Experimental Results

We evaluated the performance of the proposed fusion protocol and cascaded authentication approach with thorough experimental investigation. First we report the performance of primary biometric modules –2D face, 3D face, audio-2D face, (MFCC-Eigenface) only, and one of the secondary biometric module (lip region - visual speech) only module results, followed by the performance of the fusion of the primary and one secondary biometric module and then of three modules (one primary and two secondary biometric modules).

A. Performance of Primary Biometric Module

The audio-only module performs a best EER of 2.4% was achieved at 48dB. At 21dB the EER dropped to worst possible EER of 50%.

B. Performance of Secondary Biometric only Module

To examine the inability of secondary biometric module to make reliable authentication decisions on its own, we show the performance of the secondary biometric module – lip region module performance here. In this set of single mode experiments, the effect of the GMM mixtures on the performance of the four lip features (visual speech feature types) - f_{DCT}, f_{GRD}, f_{CTR}, and feature fused (concatenated f_{DCT}-f_{GRD}-f_{CTR}) was tested initially. These tests were carried out using matched training and testing data sets, i.e., the original "clean" images. To examine whether the dynamic lip motion features, such as the f_{GRD} and f_{CTR}, features, would perform better with a larger number of GMM mixtures, we increased the number of mixtures from one until a performance trend became apparent. For each lip feature type, a trend in the EERs with respect to the number of mixtures can be seen. The number of mixtures that maximised the visual speech features performance for each of the four feature types, are given in Table 1.

Table 1. Number of gaussian mixtures that maximises the EER performance for each of the four types of secondary biometric (visual speech) features across ten levels of JPEG Q

Features	GMM mixtures	Clean	QF 50	QF 25	QF 18	QF 14	QF 10	QF 8	QF 6	QF 4	QF 3	QF 2
f_{DCT}	2	13.5	14.3	15.1	15.9	15.9	17.5	19.9	20.7	39.8	49.4	50.0
f_{GRD}	15	23.5	25.9	31.9	35.5	47.4	50.0	50.0	50.0	50.0	50.0	50.0
f_{CTR}	18	20.7	22.7	29.9	32.3	43.8	50.0	50.0	50.0	50.0	50.0	50.0
f_{DCT}-f_{GRD}-f_{CTR}	4	8	9.6	10.8	12.0	12.4	16.7	24.3	32.7	50.0	50.0	50.0

C. *Performance of Fusion of Secondary Biometric Module (2D Mouth-3D Face Features)*

In this Section the performance of fusion of two secondary biometric modules is examined to show that the secondary modules on their own or with fusion with each other yield a moderate improvement in authentication performance. The face gallery (training) set, comprising three images, was formed by arbitrarily extracting the first image frame from each of the first three training sentences from AVOZES module 6. These were used to form a face template for each of the N subjects.

For extracting 3D facial features, we have explored a fairly exhaustive set of features that extract discriminative information from 3D faces. The detailed description of each of the 3D facial features in a review in [17]. The fusion of 3D features, TEX-GABOR for texture module and CURV-PD for the shape module was used in score-level fusion, and the performance of the 2D Mouth - 3D face fusion module w.r.t. JPEG QF is given in Table 2.

Table 2. The mouth, face and face-mouth EERs for ten levels of JPEG QF

JPEG QF	50	25	18	14	10	8	6	4	3	2
Mouth [%]	14.1	14.9	15.7	15.7	17.3	19.8	20.6	39.5	50.0	50.0
Face [%]	1.2	1.2	0.4	0.4	1.2	1.2	2.0	8.1	14.1	25.0
Mouth-Face[%]	0.0	0.8	0.0	0.0	0.0	0.0	0.0	1.6	7.3	12.5

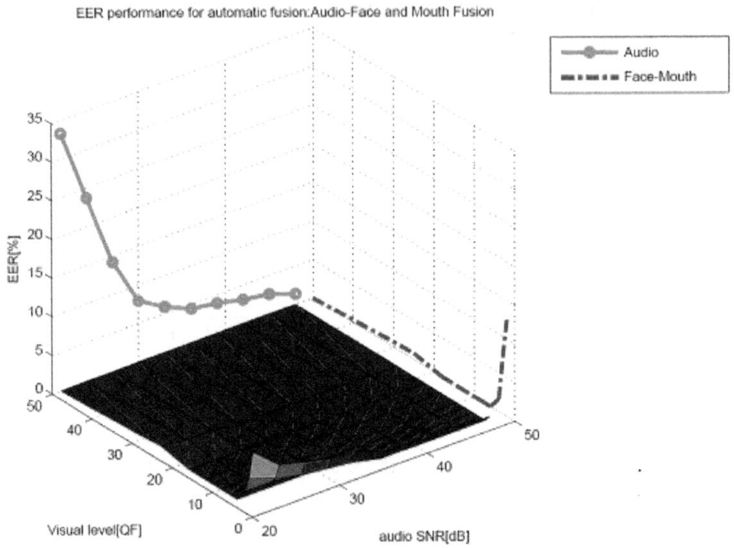

Fig. 2. Primary-Secondary Module Fusion Performance

D. Performance of Primary and Secondary Biometric Module Fusion

In this Section the performance of fusion of primary and two secondary biometric modules is examined to show that a significant robustness and enhancement in performance can be truly achieved by fusion of primary and several secondary biometric modules. For this set of experiments, we examined the performance of the fusion of primary biometric (Eigenface-MFCC), 2D mouth features (f_{DCT}-f_{GRD}-f_{CTR} and 3D face features (TEX-GABOR and CURV-PD) in a cascaded authentication strategy. The results for this set are shown in Figure 2.

5 Conclusions

In this paper, a novel sensor fusion technique for combining multiple heterogeneous sensors is proposed for biometric access control applications. The approach combines information from primary (dominant biometric cues) and several secondary biometric (non-dominant, subtle) modules, namely audio, face, visual speech, and 3D face information in an automatic unsupervised fusion, adapting to the local performance of each module, and taking into account the output-score based reliability estimates of each of the modules. The results as a whole are important for remote authentication applications, where bandwidth is limited and uncontrolled acoustic noise is probable, such as video telephony and online authentication systems.

References

[1] Potamianos, G.G., Neti, C., Gravier, G., Garg, A., Senior, A.W.: Recent Advances in the Automatic Recognition of Audiovisual Speech. Proceedings of the IEEE 91, 1306–1324 (2003)

[2] Chelubishi, C.C., Deravi, F., Mason, J.S.D.: A Review of Speech-Based Bimodal Recognition. IEEE Transactions on Multimedia 4, 23–35 (2002)

[3] Brunellil, R., Falavigna, D.: Person Identification Using Multiple Cues. IEEE Transactions on Pattern Analysis and Machine Intelligence 17, 955–966 (1995)

[4] Santi, A., Servos, P., Vatikiotis-Bateson, E., Kuratate, T., Munhall, K.: Perceiving biological motion: Dissociating talking from walking. Journal of Cognitive Neuroscience 15, 800–809 (2003)

[5] Callan, D., Jones, J.A., Munhall, K.G., Kroos, C., Callan, A., Vatikiotis-Bateson, E.: Neural processes underlying perceptual enhancement by visual speech gestures. Neuroreport 14, 2213–2218 (2003)

[6] Hani, C.Y., Kuratate, T., Vatikiotis-Bateson, E.: Linking facial animation, head motion, and speech acoustics. Journal of Phonetics 30(3), 555–568 (2002)

[7] Kroos, C., Kuratate, T., Vatikiotis-Bateson, E.: Video-based face motion measurement. Journal of Phonetics 30(3), 569–590 (2002)

[8] Goecke, R., Millar, J.B.: The Audio-Video Australian English Speech Data Corpus AVOZES. In: Proceedings of the 8th International Conference on Spoken Language Processing INTERSPEECH 2004 - ICSLP, vol. III, pp. 2525–2528 (2004)

[9] Chetty, G., Wagner, M.: Automated lip feature extraction for liveness verification in audio-video authentication. In: Proc. Image and Vision Computing, New Zealand, pp. 17–22 (2004)

[10] Chetty, G., Wagner, M.: Audio-Visual Speaker Identity Verification using Lip Motion Features. In: Proc. INTERSPEECH 2007 Conference (2007)

[11] Hyvarinen, A., Oja, E.: Independent Component Analysis: Algorithms and Applications. Neural Networks 13(4-5), 411–430 (2000)

[12] Dutagaci, H., Sankur, B., Yemez, Y.: 3D face recognition by projection-based features. In: Proc. SPIE Conf. on Electronic Imaging: Security, Steganography, and Watermarking of Multimedia (2006)

[13] Gokberk, B., Irfanoglu, M.O., Akarun, L.: 3D shape-based face representation and facial feature extraction for face recognition (in press). Image and Vision Computing (2006)

[14] Kahraman, F., Stegmann, M.B.: Towards Illumination-invariant Localization of Faces using Active Appearance Models. In: IEEE NORSIG 2006 (2006)

[15] Quatieri, T.F.: Discrete Time Speech Signal Processing. Signal Processing Series. Prentice Hall, Englewood Cliffs (2002)

[16] H.264/MPEG-4 AVC standard, (retreived on 15/5/2009) `http://www.itu.int/ITU-T/worksem/vica/docs/presentations/S3_P1_Sullivan.pdf`

[17] Bower, K.W., Chang, K., Flynn, P.: A survey of approaches and challenges in 3D and multimodal 3D + 2D face recognition. Computer Vision and Image Understanding 101(1), 1–15 (2006)

Improvement of Reuse of Classifiers in CBIR Using SVM Active Learning

Masaaki Tekawa and Motonobu Hattori

Interdisciplinary Graduate School of Medicine and Engineering,
University of Yamanashi, 4-3-11 Takeda, Kofu, Yamanashi, 400-8511 Japan
{g10mk024,m-hattori}@yamanashi.ac.jp

Abstract. In content-based image retrieval, relevance feedback is often adopted as the method of interactions to grasp user's query concept. However, since this method tasks the user, a small amount of relevance feedback is desirable. For this purpose, Nakajima *et al.* have proposed a method in which classifiers learned by using relevance feedback are reused. In this paper, we improve the criterion for reuse of classifiers so that retrieval becomes more accurate and quick. Experimental results show that our method performs much better than the conventional methods.

Keywords: CBIR, support vector machine, image retrieval, relevance feedback.

1 Introduction

Traditional image retrieval methods use indices which are made by manual annotation. However, this approach becomes difficult with an increase in the number of images by the development of digital technology. In addition, there is a concern that subjective annotation influences retrieval results. Content-based image retrieval (CBIR) is one of the systems which can solve these problems. In typical CBIR, the system expresses information of an image by a feature vector, and searches similar ones in an image database by comparing the degree of similarity between the feature vectors.

In CBIR, there is a need for a user to inform an image retrieval system of his or her desired output or query concept. In other words, an image retrieval system must grasp a user's intention for results of high quality. To address this requirement, relevance feedback (RF) can be used as a query refinement scheme to derive or learn a user's query concept. In RF, the system displays a few image instances and the user labels each image as "relevant" or "irrelevant". Based on the answers, another set of images are brought to the user for labeling. After some such interactions, the system may grasp the user's intention. The construction of such a query refinement scheme can be regarded as a machine learning task, and various machine learning techniques have been applied to CBIR. Among them, Tong *et al.* have proposed the use of a support vector machine active learning algorithm for RF [1]. The algorithm selects the most informative images to query

K.W. Wong et al. (Eds.): ICONIP 2010, Part II, LNCS 6444, pp. 598–605, 2010.

a user and quickly learns a boundary separating the images that satisfy the user's query concept from the rest of the dataset. Although a classifier once learned is discarded every time after retrieval is finished in Tong's method, Nakajima *et al.* have proposed a method to keep classifiers learned and to reuse them in future retrieval [2]. They have shown better performance in comparison with Tong's method.

In this paper, we aim at improving the criterion used in Nakajima's method so that an appropriate classifier can be selected when reusing classifiers and retrieval becomes more accurate and quick. Experimental results show effectiveness of the proposed method.

2 Relevance Feedback

In a typical CBIR system, an image is expressed by a feature vector in which each coordinate shows a visual content of the image. These include characteristics such as color, texture, shape, color layout and so on. When a query image is given, its features are extracted so that a group of similar images to the query image can be retrieved. A retrieval by low-level concepts which are similar to visual characteristics is achievable easily, and some systems such as QBIC [3] and SIMPLIcity [4] have been developed.

On the other hand, when a user wants to retrieve images by high-level concepts, for example, "beautiful flowers", it is necessary to derive them from low-level visual features. It is, however, usually quite difficult to do so because there are various semantic concepts in an image. In order to solve this problem, relevance feedback (RF) is often a critical component in CBIR. The typical process of RF in CBIR is as follows:

(1) A CBIR system provides an initial retrieval result.
(2) The user provides feedback on the above result to sort out relevant ones and irrelevant ones by high-level concepts.
(3) The user's feedback is learned by the system.
(4) New results are shown. Then go back to (2).

Steps (2)-(4) are repeated till the user is satisfied with results. In this way, the system can grasp high-level concepts of the user query in real-time by performing such interactions repeatedly. It is, however, most preferable to reduce the number of RF because this method tasks the user.

3 SVM Active Learning for CBIR

A support vector machine active learning (SVM_{Active}) is an approach that queries points so as to attempt to reduce the size of the version space as much as possible [1]. A superior classifier is generated by acquiring only informative data to be learned. Informative data for learning are labeled data which are distributed near the separating hyperplane. The system can decide the ideal decision boundary quickly by learning such data sequentially.

The algorithm of SVM_{Active} is summarized as follows:

(1) Given a relevant image and an irrelevant one as inquiry by a user (Query Images in Fig.1).
(2) Learn an SVM on the current labeled data.
(3) Show the top-r most relevant images to the user (Return Set in Fig.1). If the user satisfies with these images, terminate the retrieval. Otherwise, go to the next step.
(4) Ask the user to label l images closest to the SVM boundary (Label Set in Fig1). Return to (2).

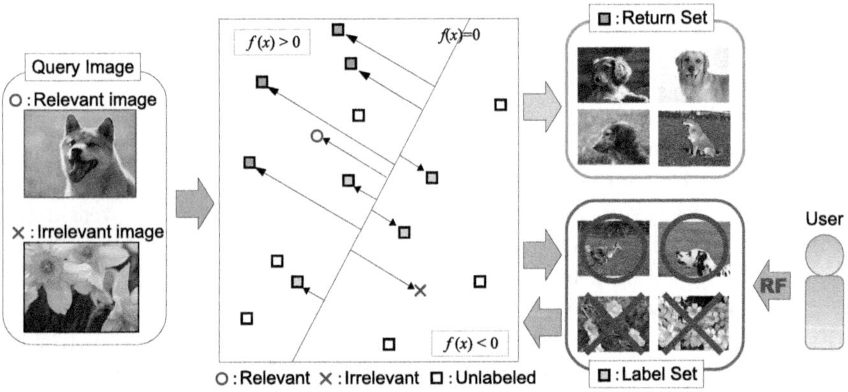

Fig. 1. SVM_{Active} for CBIR [1]

4 Reuse of Classifiers

Although a classifier learned in Tong's method is destroyed after a user satisfies retrieved images, Nakajima *et al.* have proposed saving classifiers once learned and reusing one of them if it is promising for a new retrieval [2].

The following is the algorithm of the Nakajima's method:

(1) Given a relevant image $\boldsymbol{x}_R$ and an irrelevant one $\boldsymbol{x}_{IR}$ as inquiry by a user.
(2) Choose a classifier which satisfies

$$f_k(\boldsymbol{x}_R) > 0 \quad \cap \quad f_k(\boldsymbol{x}_{IR}) < 0 \tag{1}$$

and has the greatest value of the following equation

$$|f_k(\boldsymbol{x}_R)| + |f_k(\boldsymbol{x}_{IR})| \tag{2}$$

where $f_k(\boldsymbol{x})$ shows the output of the kth classifier for input data $\boldsymbol{x}$. If there are no classifiers which satisfies Eq.(1), generate a new one.
(3) Learn an SVM on the current labeled data.

(4) Show the top-r most relevant images to the user. If the user satisfies with these images, terminate the retrieval and go to (6). Otherwise, go to the next step.
(5) Ask the user to label l images closest to the SVM boundary. Return to (3).
(6) Save the classifier with support vectors.

They have shown that their method could grasp users' intention more quickly and could retrieve images with higher precision rate than Tong's method [2].

5 Improvement of Criterion for Reuse of Classifiers

In general, the number of irrelevant images is far larger compared with that of relevant images for a certain query. Therefore, an irrelevant image of a query has a significant effect on a choice of a classifier in Nakajima's method. When the system reuses an inappropriate classifier, it may become an undesirable search result for a user. In addition, there is a concern about depression of the original performance of the classifier by learning inappropriate labeled data additionally.

Therefore, in order to choose a more appropriate classifier, we propose a novel method which uses a kind of relevance feedback in a choice of classifiers. In the proposed method, a classifier is chosen as follows:

Step 1. Calculate the following value for each classifier:

$$f_k(\boldsymbol{x}_R) \tag{3}$$

and choose classifiers which have the top-m largest values as candidates for reuse.

Step 2. Obtain l/m positive support vectors randomly in each candidate classifier. In total, l images are obtained.

Step 3. Ask the user to label l images.

Step 4. Reuse the classifier which has the highest classification accuracy to the labeled images if the number of the positive examples classified correctly by it is more than N. When there are plural classifiers which satisfy these requirements, reuse the classifier which has the largest value of the following formula:

$$\sum_{i=1} f_k(\boldsymbol{x}_{Ri}) \tag{4}$$

where each $\boldsymbol{x}_{Ri}$ shows an image labeled as relevant by the user out of l images.

As shown in the above procedure, we apply a kind of RF to decide which classifier to be reused. Moreover, the irrelevant image, $\boldsymbol{x}_{IR}$ given by the user as inquiry is not used to choose a classifier to be reused because it is much less reliable than the relevant image, $\boldsymbol{x}_R$. Thus, we may choose more appropriate classifier in comparison with Nakajima's method.

In the proposed method, the above procedure is followed by (3)-(6) of Nakajima's method described in **4**.

6 Experiments

For empirical evaluation of the proposed criterion, we used two image datasets. One has 2,000 images which are categorized into 20 groups such as flower, dog, sea, sky and so on. Each category contains 100 images of essentially the same semantic concept. The other has 5,000 images categorized into 25 groups, and each category contains 200 images. The source of both image sets is PHOTO BIBLE 20,000 published by Datacraft Co., Ltd.

For each image, two classes of major visual features are used:

(1) Color features. Two kinds of color features are utilized in the experiments. One is the 9-dimensional color moment. The other is the 120-dimensional color coherence vector [5].
(2) Texture features. We performed 3D wavelet transform (WT) for each image [6]. Thus, we extracted texture features by average and variance of horizontal, vertical and diagonal direction. The dimension of WT was 18.

We applied Gaussian kernel and soft margin to SVM learners [7]. The same parameters of SVMs were used for all experiments. To enable an objective measure of performance, we assumed that a query concept was an image category [1]. Each image was picked up as a positive example of inquiry, and a negative example of that was chosen randomly from other categories.

Parameters were set as follows: $r = 20$, $l = 20$, $m = 5$ and $N = 2$. Accuracy is computed by looking at the fraction of the $20(= r)$ returned result that belongs to the target image category. This is equivalent to computing the precision on the top-20 images.

In the proposed method, the relevance feedback used for choosing classifier was counted as one RF to make user's task the same with the conventional methods.

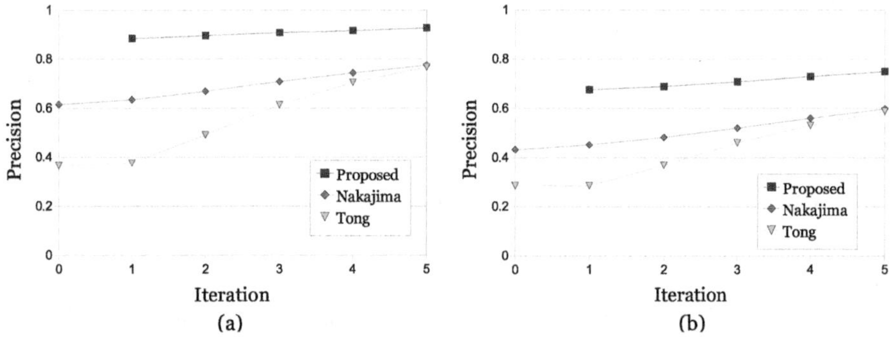

Fig. 2. (a) Average of precision for 2000 images. (b) Average of precision for 5000 images.

6.1 Precision

Figures 2 (a) and (b) show the average of precision for the two different datasets based on 10 trials. As shown in the figures, the proposed method performs much better for both datasets than the conventional methods. Since the initial precision of the proposed method is quite high, we can say that more appropriate classifiers were reused in the proposed method in comparison with the Nakajima's method.

6.2 Leaning Time

Figures 3 (a)-(b) show the average of learning time for each method measured by a PC with Core 2 Duo (3.0GHz) CPU and 2GB memory. Since the proposed and Nakajima's methods reuse classifiers and each reused classifier has support vectors as labeled images to be learned, the learning time of them is longer than Tong's method. However, it is still of practical use. Moreover, the learning time of the proposed method is almost comparable with that of the Nakajima's method.

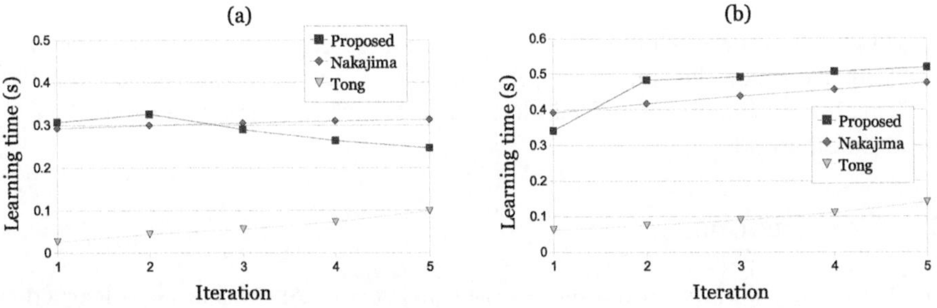

Fig. 3. (a) Average of learning time for 2000 images. (b) Average of learning time for 5000 images.

6.3 Searching Time

We made a comparison of searching time of classifiers between the proposed and Nakajima's method. For the proposed method, time to perform the first RF for choosing classifier was included in learning time as shown in Fig. 3, so it was omitted from searching time of classifiers. Table 1 shows the average based on 10 trials.

The result shows that searching time of both methods are almost the same and the average is less than 0.1 (sec) even for 5000 images dataset.

Table 1. Average and maximum of searching time (sec.)

	Nakajima's method		Proposed method	
Dataset	Average	Maximum	Average	Maximum
2000 images	0.038	0.071	0.034	0.071
5000 images	0.088	0.175	0.094	0.185

6.4 Reuse Rate

Here, we compared the number of classifiers generated and reuse rate. The reuse rate is defined as follows:

$$\text{reuse rate} = \frac{Q - N_C}{Q} \qquad (5)$$

where Q is the number of retrieval times in one trial that equals the number of images in each dataset, and N_C is the number of classifiers generated in one trial.

Table 2. The number of classifiers generated and reuse rate

	Nakajima's method		Proposed method	
Data set	classifiers	reuse rate	classifiers	reuse rate
2000 images	391.2	0.804	134.1	0.932
5000 images	1060.8	0.788	712.2	0.858

Table 2 shows the average based on 10 trials. This result shows the proposed method decreases the number of classifiers and improves the reuse rate. That is, learning efficiency for each individual classifier was much improved in the proposed method.

7 Conclusions

In this paper, we have proposed a novel method to reuse classifiers learned by SVM_{Active} for CBIR. In order to choose an appropriate classifier for reusing, we have adopted a kind of relevance feedback for the choice of classifiers. Moreover, we have proposed a criterion in which we don't evaluate the given irrelevant image, because irrelevant images are less reliable than relevant images. Experimental results show that the proposed method performs much better than the conventional methods. In addition, the proposed method reuses classifiers once generated effectively, and as a result the number of classifiers generated is much reduced in comparison with Nakajima's method.

References

1. Tong, S., Chang, E.: Support vector machine active learning for image retrieval. In: Proceedings of ACM International Multimedia Conference, pp. 107–118 (2001)
2. Nakajima, S., Hattori, M.: Reusing Classifiers in Support Vector Machine Active Learning for Content Based Image Retrieval. In: The 70th National Convention of IPSJ, vol. 1R-5, 1, pp. 477–478 (2008)
3. Niblack, W., Barber, R., Equitz, W., Flickner, M., Glasman, E., Petkovic, D., Yanker, P., Faloutsos, C., Taubin, G.: Querying images by content using color, texture, and shape. In: Proceedings of the SPIE Conference on Storage and Retrieval for Image and Video Databases, San Jose, CA, February 2-3, pp. 173–187 (1993)

4. Wang, J.Z., Li, J., Wiederhold, G.: SIMPLIcity: Semantics-Sensitive Integrated Matching for Picture LIbraries. IEEE Transactions on Pattern Analysis and Machine Intelligence 23(9) (September 2001)
5. Pass, G., Zabith, R.: Comparing images using joint histograms. Multimedia Systems 7, 234–240 (1999)
6. Eric, S.J., Tony, D.D., David, S.H.: Wavelets for computer graphics: A primer, part 1. IEEE Computer Graphics and Applications 15(3), 76–84 (1995)
7. Cristianini, N., Taylor, J.S.: An Introduction to Support Vector Machines. Cambridge University Press, Cambridge (2000)

Realizing Hand-Based Biometrics Based on Visible and Infrared Imagery

Goh Kah Ong Michael[1], Tee Connie[1], Teo Chuan Chin[1], Neo Han Foon[1], and Andrew Teoh Beng Jin[2]

[1] FIST, Multimedia University, Jalan Ayer Keroh Lama, 75450 Malacca, Malaysia
{michael.goh,tee.connie,ccteo,hfneo}@mmu.edu.my
[2] School of Electrical and Engineering, Yonsei University, Seoul Korea
bjteoh@yonsei.ac.kr

Abstract. This paper describes a hand-based biometric system by using visible and infrared imagery. We develop an acquisition device which could capture both color and infrared hand images. We modify an ordinary web camera to capture the hand vein that normally requires specialized infrared sensor. Our design is simple and low-cost, and we do not need additional installation of special apparatus. The device can capture the epidermal and subcutaneous features from the hand simultaneously. In specific, we acquire four independent, yet complementary features namely palm print, knuckle print, palm vein, and finger vein, from the hand for recognition. As a low-resolution sensor is deployed in this study, the images quality may be slightly poorer than those acquired using high resolution scanner or CCD camera. The line and ridge patterns on the hand may not appear clear. Therefore, we propose a pre-processing technique to enhance the contrast and sharpness of the images so that the dominant print and line features can be highlighted and become disguisable from the background. After that, we use a simple feature extractor called Directional Coding to obtain useful representation of the hand modalities. The hand features are fused using Support Vector Machine (SVM). The fusion of these features yields promising result for practical multi-modal biometrics system.

1 Introduction

With the advent of modern computing technology, there is increased reliance on biometrics to provide stronger personal authentication. Among the variety of biometric solutions in the market, hand-based system is the oldest, and perhaps the most successful form of biometric technology [1]. A number of features can be extracted from the human hand for recognition. The most prevalent identifiers used include hand geometry, fingerprint, palm print and hand vein. These hand properties are stable and reliable. Once a person has reached adulthood, the hand structure and configuration remain relatively stable throughout the person's life [2]. These hand features (except fingerprints) can be captured using common off-the-shelf imaging devices. This advantage has greatly facilitated the deployment of hand-based biometrics in large-scale applications [1]. Apart from that, the hand-scan technology is generally perceived as nonintrusive as compared to iris- or retina-scan systems [3]. The users do not need to

K.W. Wong et al. (Eds.): ICONIP 2010, Part II, LNCS 6444, pp. 606–615, 2010.

be cognizant of the way in which they interact with the system. Therefore, it will be less likely for the users to have visceral fear or discomfort when they use hand-based system.

This paper investigates the application of visible and infrared light technologies to capture the intrinsically different, yet complimentary, characteristics present on the hand. We design a hand device which could acquire a number of features, namely palm print, knuckle print, palm vein, and finger vein, from the hand simultaneously for recognition. Palm print refers to the smoothly flowing pattern formed by alternating creases (ridges) and troughs (valleys) on the palmar surface of the hand, while knuckle prints are characterized by the horizontal-like lines distributed on the palmar area of the knuckles (joints of the fingers). Several researches have been devoted to study palm print which include Eigenpalms [4], Fisherpalms [5], Fourier spectrum [6], Gabor phase [7]-[8], line features [9], and feature points [10]. Knuckle print is a relatively new biometric modality as compared to palm print. The works reported for knuckle print are eigenfinger [11], linear discriminant analysis [12], Radon and Haar wavelet [13], and also location and line features [14]. Palm print and knuckle prints can be imaged using visible light as they lie on the epidermal surface of the hand.

On the other hand, palm vein and finger vein refer to the vascular or blood vein patterns recorded from underneath the human skin. Due to biological composition of the human tissues, the vein pattern can be observed under infrared light. In the entire electromagnetic spectrum, infrared refers to a specific region with wavelength typically spanning from 0.75μm to 1000μm. This region can be further divided into four sub-bands, namely near infrared (NIR) in the range of 0.75μm to 2μm, middle infrared in the range of 2μm to 6μm, far infrared (FIR) in the range of 6μm to 14μm, and extreme infrared in the range of 14μm to 1000μm. In the literature, the NIR [15]-[18] and FIR [19]-[20] sources were used to capture the vein images.

2 Proposed Solution

In this research, we endeavor to develop an online contactless acquisition device which can capture both the visible and infrared hand images. In specific, we want to acquire the different hand modalities, namely palm print, knuckle print, palm vein and finger vein images from the hand simultaneously without incurring additional sensor cost or adding user complication. We design the acquisition device in such a way that the users do not need to touch or hold on to any peripheral for their hand images to be acquired. We believe such setting helps to address the hygiene and social issues faced by contact-based biometric applications [21]. When a hand is presented above the acquisition device, the regions of interest (ROI) of the different hand features will be tracked and extracted. ROIs contain the important information of the hand which can be used for recognition. The ROIs are pre-processed so that the print and vein textures become distinguishable from the background. After that, distinguishing features in the ROIs are extracted using a technique called *directional coding* [22]. The hand features are mainly made up of line-like texture. The directional coding technique encodes the discriminative information of the hand based on the orientation of the line primitives. The extracted hand features are then fused at score level to yield better recognition accuracy. The framework of our proposed system is shown in Fig. 1.

The contributions of this paper are two-fold. Firstly, we develop a hand scanner which is able to capture both color and infrared hand images. We modify an ordinary web camera to capture the hand vein that normally requires specialized infrared sensor. Our design is simple and low-cost, and we do not need additional installation of special apparatus. The detail of such design will be presented in the subsequent discussion. Secondly, we propose a robust method to obtain good contrast hand images. This method is particularly useful for vein images. Due to the optical property of human tissue, the infrared light cannot penetrate very deeply under the human skin. Therefore, only superficial vein pattern can be detected by the sensor. Some users have thick skin (especially female), making it harder for the sensor to capture clear vein images. In this research, we deploy a novel image processing technique to obtain vivid line and ridge pattern from the hand. The proposed method removes illumination error while keeping good contrast between the line structure and the surrounding hand tissue.

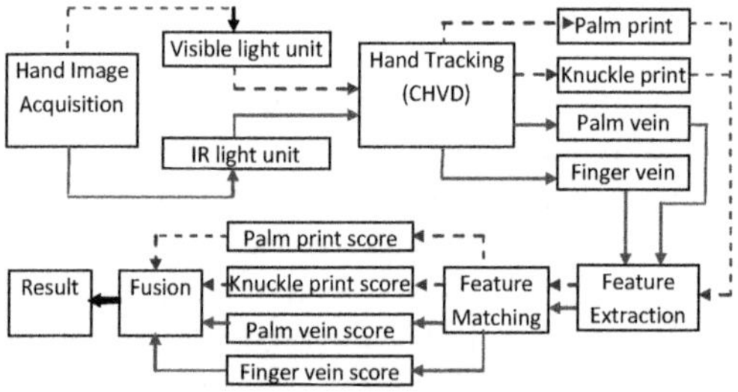

Fig. 1. Framework of the proposed system

3 Design and Implementation of Acquisition Device

The design and implementation of an efficient real-time hand acquisition device must contend with a number of challenges. Firstly, the acquisition device must be able to provide sufficient contrasted images so that the hand features are discernable and can be used for processing. The arrangement of the imaging sensor and design of the lighting units also have great impact on the quality of the images acquired. Therefore, the capturing device should be carefully calibrated to obtain high contrasted images. Secondly, a single acquisition device should be used to capture multiple image sources (e.g. visible and infrared images). It is not efficient and economical for a multimodal biometric system to install multiple capturing devices. Therefore, an acquisition device with low development cost is expected for a multimodal biometric system from the system application view.

In this research, we design a capturing device that aims to fulfill the requirements above. The hardware setup of the capturing device is shown in Fig. 2a and 2b. Two low-cost imaging units are mounted side by side on the device. The first imaging unit is used to capture visible light images while the second for obtaining infrared images. The warm-white light bulbs are placed around the imaging units to emit yellowish light source which helps to enhance the line patterns on the palm and fingers under visible light. To acquire IR image, we modify the ordinary webcam to be an IR-sensitive camera. The webcam used for infrared imaging is fitted with an infrared filter. The filter blocks the visible (non-IR) light and allows only the IR light to reach the sensor. A number of infrared LEDs are arranged on the board to serve as the infrared cold source to illuminate the vein pattern. We have experimented with different types of infrared LEDs and those emitting light in the range of 880nm to 920 nm provide relatively good contrast of the vein pattern. A diffuser paper is used to attenuate the IR source so that the radiation can be distributed more uniformly around the imaging unit.

During image acquisition, we request the user to position his/her hand above the sensor with the palm facing the sensor (Fig. 2c). The user has to slightly stretch his/her fingers apart. There is no guidance peripheral to constraint the user's hand. The user can place his/her hand naturally above the sensor. We do not restrict the user to place his/her hand at a particular position above the sensor nor limit them to pose his/her at a certain direction. Instead, we allow the user to move his/her hand while the images are being acquired. Besides, the user can also rotate his/her hand while the images are being taken. The optimal viewing region for the acquisition sensor is 20cm from the surface of the imaging unit. We allow a tolerable focus range of 20cm ± 4cm to permit more flexibility for the users to interact with the system.

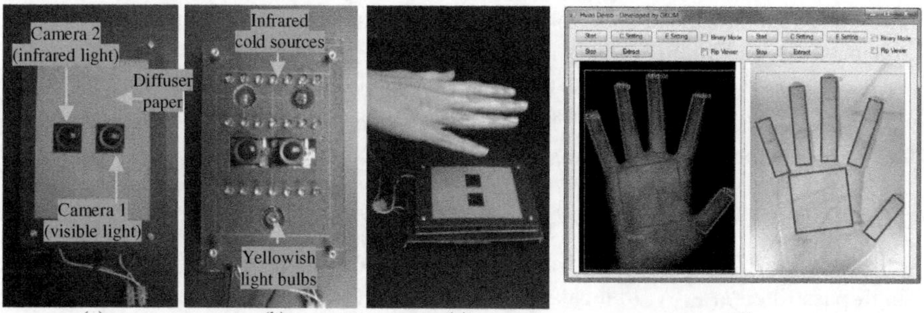

(a) (b) (c) (d)

Fig. 2. Hardware setup for the image acquisition device

In this study, a standard PC with Intel Core 2 Quad processor (2.4 GHz) and 3072 MB RAM is used. The program was developed using Visual Studio .NET 2008. The application depicted in Fig. 2d shows a live video of the hand image sequence recorded by the sensor. Both of the visible light and IR images of the hand can be captured simultaneously. The interface provides direct feedback to the user that he/she is placing his/her hand properly inside the working volume. After the hand is detected in the working volume, the ROIs of the palm and fingers will be captured and stored as bitmap format from the video sequence. The hand image was detected in real-time video sequence at 30 fps. The image resolution is 640 x 480 pixels, with color output type in 256 RGB (8 bits-per-channel). The delay interval between capturing the current and the next ROI was 2 seconds. We used the setup described above in an office environment to evaluate the performance of the proposed multimodal hand-based biometric system. We have recorded the hand images from 136 individuals. 64 of them are females, 42 of them are less than 30 years old. The users come from different ethnic groups such as Chinese, Malays, Indians, and Arabians. Most of them are students and lecturers from our university. Ten samples were captured for each user. The samples were acquired in two different occasions separated at a mean interval of two months.

4 Imagery Pre-processing and Feature Representation

4.1 Pre-processing

We adopt the hand tracking algorithm proposed in [21] and [22] to detect and locate the region of interest (ROI) of the palm and fingers. After obtaining the ROIs, we enhance the contrast and sharpness of the images so that the dominant hand features can be highlighted and become distinguishable from the background. Gamma correction is first applied to obtain better image contrast [23]. To bring out the detail of the line and ridge patterns, we have investigated a number of well-known image enhancement methods like Laplacian filters, Laplacian of Gaussian, and unsharp masking method. Although these techniques work well for sharpening the images, the noise elements tend to be over-enhanced. For this reason, we propose a local-ridge-enhancement (LRE) technique to obtain a sharp image without overly amplifying the noise. This method discovers which part of the image contains important lines and ridge patterns, and amplifies only these areas.

The proposed LRE method uses a "ridge detection mask" to find the palm vein structures in the image. LRE first applies a low-pass filter, $g(x, y)$, on the original image, $I(x, y)$, shown in Fig. 3a to obtain a blur version of the image, $M(x, y)$,

$$M(x, y) = g(x, y) * I(x, y) \tag{1}$$

In this research, Gaussian filter with σ=60 is used for this purpose. After that, we use a high-pass filter, $h(x, y)$, to locate the ridge edges from the blur image,

$$M'(x, y) = h(x, y) * M(x, y) \tag{2}$$

Note that since the trivial/weak ridge patterns have already been "distilled" in the blur image, only the edges of the principal/strong ridges show up in $M'(x, y)$. In this work, the Laplacian filter is used as the high-pass filter.

At this stage, $M'(x, y)$ exhibit the edges of the primary ridge structure (Fig. 3c). We binarize $M'(x, y)$ by using a threshold value, τ. Some morphological operators like opening and closing can be used to eliminate unwanted noise regions. The resultant image is the "mask" marking the location of the strong ridge pattern.

We "overlay" $M'(x, y)$ on the original image to amplify the ridge region,

$$I'(x, y) = \begin{cases} c \cdot I(x, y) & \text{if } M'(x,) = 1 \\ I(x, y) & \text{otherwise} \end{cases} \tag{3}$$

where $I'(x, y)$ is the enhanced image and c is the coefficient to determine the level of intensity used to highlight the ridge area. The lower the value of c, the more the ridge pattern will be amplified (the darker the area will be). In this work, the value of c is empirically set to 0.9. Fig. 3f shows the result of the enhanced image. We wish to point out that more variations can be added to determine different values for c in order to highlight the different ridge areas according to their strength levels. For example, gray-level slicing can be used to assign larger weight, c, to stronger ridge pattern, and vice versa. We do not perform this additional step due to the consideration for computation overhead (computation time is a critical factor for an online application).

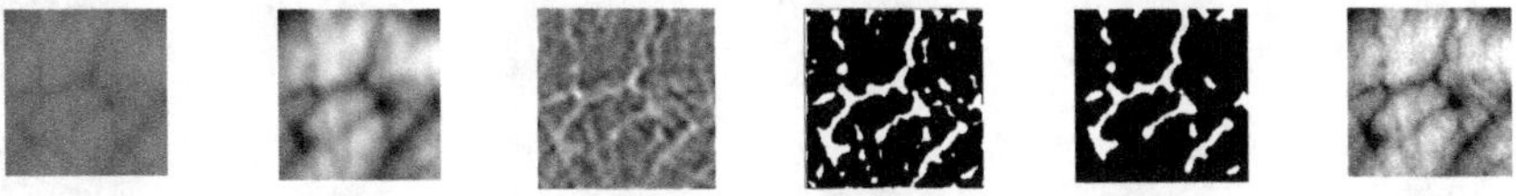

(a) (b) (c) (d) (e) (f)

Fig. 3. Processes involved in the proposed LRE method

4.2 Feature Extraction

We apply a scheme named Directional Coding method to extract the palm print and palm vein features. These hand features contain similar textures which are primarily made up of line primitives. For example, palm prints are made up of strong principal lines and some thin wrinkles, whilst knuckle prints comprise asymmetry creases and wrinkles. The patterning of the hand vein which contains vascular network also resembles line-like characteristic. Therefore, we can deploy a single method to extract the discriminative line information from the different hand features.

The proposed Directional Coding technique aims to encode the line pattern based on the proximal orientation of the lines. We first apply Wavelet Transform to decompose the palm print images into lower resolution representation. The Sobel operator is then used to detect the palm print edges in horizontal, vertical, +45°, and -45° orientations. After that, the output sample, $\Phi(x, y)$, is determined using the formula,

$$\Phi(x, y) = \delta\left(\arg\max_f (\omega_R(x, y))\right) \tag{4}$$

where $\omega_R(x, y)$ denotes the responses of the Sobel mask in the four directions (horizontal, vertical, +45°, and -45°), and $\delta \in \{1,2,3,4\}$ indicates the index used to code the orientation of $\omega_R(x, y)$. The index, δ, can be in any form, but we use decimal representation to characterize the four orientations for the sake of simplicity. The output, $\Phi(x, y)$, is then converted to the corresponding binary reflected Gray code. The bit string assignment enables more effective matching process as the computation only deals with plain binary bit string rather than real or floating point numbers. Besides, another benefit of converting bit string to Gray code representation is that Gray code exhibits less bit transition. This is a desired property since we require the biometric feature to have high similarity within the data (for the same subject). Thus, Gray code representation provides less bit difference and more similarity in the data pattern. Fig. 4b to 4e shows the gradient responses of the palm print in the four directions. Fig. 4f is the result of taking the maximum gradient values obtained from the four responses. This image depicts the strongest directional response of the palm print and it closely resembles the original palm print pattern shown in Fig. 4a. The example of directional coding applied on palm vein image is also illustrated in Fig. 4g to 4l.

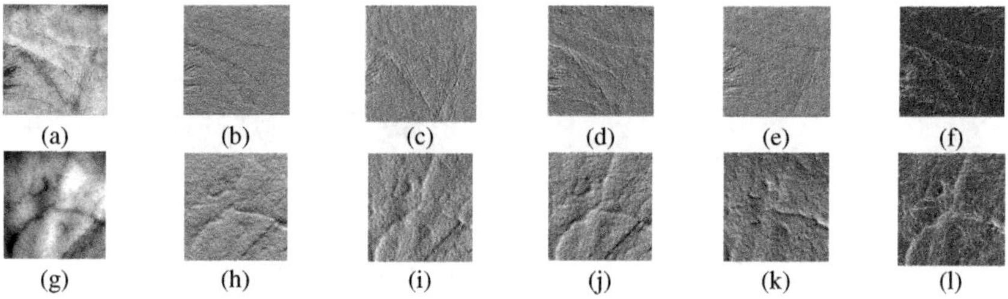

Fig. 4. Example of Directional Code applied on palm print and vein image

4.3 Fusion

In this paper, Support Vector Machine (SVM) is adopted as the fusion mechanism to consolidate the matching scores produced by the palm print and knuckle print modalities. Hamming distance is used to count the fraction of bits that differ between two code strings generated by Directional Coding. In this research, the Radial Basis Kernel (RBF) function is explored. RBF kernel is defined as [24]-[25],

$$K(x, x_i) = \exp\left(-\frac{(x - x_i)^2}{2\sigma^2}\right) \tag{5}$$

where $\sigma > 0$ is a constant that defines the kernel width.

5 Results and Discussion

5.1 Verification Performance Using Directional Code

An experiment was carried out to assess the effectiveness of the proposed Directional Coding technique applied on the different hand modalities. The performance of the proposed method is recorded in Table 1. The three common performance measures namely false acceptance rate (FAR), false rejection rate (FRR) and genuine acceptance rate (GAR) are used in this test. We also include the GAR without LRE for comparison. The Directional Coding technique yielded satisfactory results by giving equal error rate (EER) of 1.96%, 2.5%, 0.73%, and 1.87% for palm print, knuckle print, palm vein, and finger vein, respectively. The result showed that the proposed method was able to encode the discriminative information on the hand well. The pre-processing step had indeed helped to improve the overall performance by 5% over GAR without LRE.

Table 1. Applying Directional Coding on the different hand modalities

Modalities	FAR	FRR	GAR	GAR without LRE
Palm print	1.98	1.95	98.02	93.42
Knuckle print	2.32	2.68	97.32	94.31
Palm vein	0.51	0.95	99.05	94.89
Finger vein	1.77	1.98	98.02	93.23

5.2 Analysis of Biometric Features

Correlation analysis of individual experts is important to determine their discriminatory power, information complementary ability and data separability. A common way to identify the correlation which exists between the experts is to analyze the errors made by them. The fusion result can be very effective if the errors made by the classifiers are highly de-correlated (with higher independency). In other words, the higher the de-correlation between the errors made by the classifiers, the more effective the fusion will become. This is due to the reason that more new information will be introduced when the de-correlation between the errors increases [26].

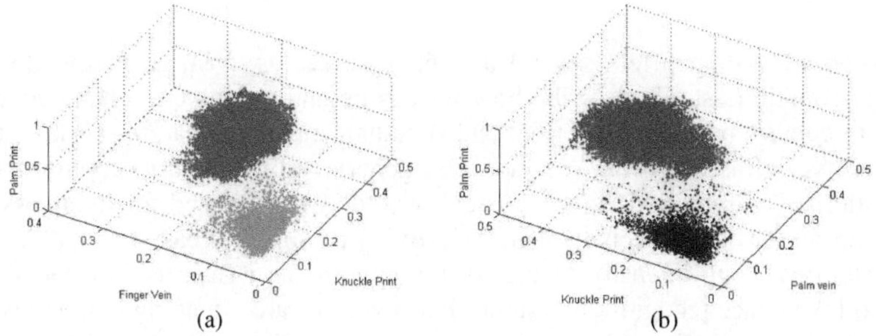

Fig. 5. Visual representation of the correlation of the different hand modalities

One way to visualize the correlation between two classifiers is to plot the distribution graph of the genuine and imposter populations. In the correlation observation shown in Fig. 5, the distribution of the genuine and imposter scores for the four hand features take the form of nearly independent clusters. This indicates that the correlation between the individual hand modalities is low. In other words, we found that the biometrics are independent and are suitable to be used for fusion.

5.3 Fusion of Biometric Features

The scores obtained from the palm print, knuckle print, palm vein, and finger vein experts were fused using SVM. We assessed SVM with Gaussian Radial Basis Function (RBF) kernels. The bandwidth of the σ parameter in the RBF kernel had been evaluated in the range of 0.01 to 1. We determined the best parameter in the test using the development set and applied it to our experiment set. EER of 0% was achieved when we fused the four modalities together. The receiver operating curves illustrated in Fig. 6 depicts the improvement gain when all the biometric traits are fused.

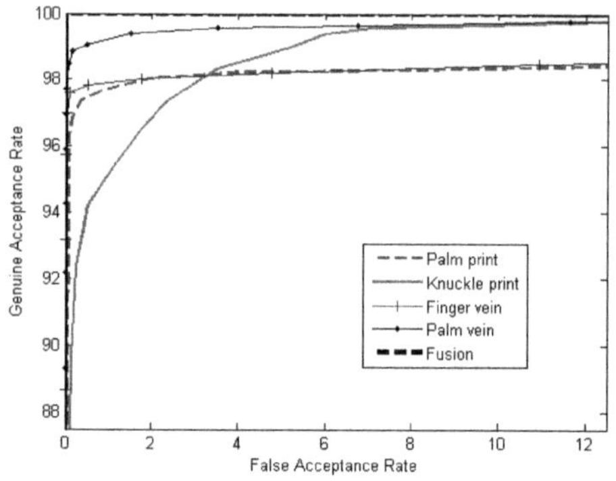

Fig. 6. Receiver operating curve for the fusion of different hand modalities

6 Conclusion

The proposed system offers several advantages like low-cost, accuracy, flexibility, and user-friendliness. We describe how we design and implement a hand acquisition device to capture both the epidermal and subcutaneous hand features without the use of expensive infrared sensor. We modify a generic web camera to capture the hand vein pattern. The modified "infrared" sensor could even be used for liveliness test in which the sensor only detects the hand of a living person. It is easy to spoof a biometrics system by using intensity images of the genuine user (e.g. using a facial photograph to fake a face recognition system). However, infrared imaging could only detect a live sample when a living hand with incessant blood flow is presented to the sensor. Apart from this, we also introduce the LRE method to obtain good contrast print and vein images. To obtain useful representation of the hand modalities, we apply a technique called directional coding. This method represents the biometric features in bit string format which enable speedy matching and convenient storage. Extensive experiments had been conducted to evaluate the performance of the system in terms of speed and accuracy. Our approach produced promising result to be implemented in a practical biometric application.

References

1. Hand-based biometrics. Biometric Technology Today 11(7), 9–11 (2003)
2. Yörük, E., Dutağaci, H., Sankur, B.: Hand biometrics. Image and Vision Computing 24(5), 483–497 (2006)
3. Jain, A.K., Ross, A., Prabhakar, S.: An Introduction to biometric recognition. IEEE Transactions on Circuits System and Video Technology 14(1), 4–20 (2004)

4. Lu, G., Zhang, D., Wang, K.: Palmprint recognition using eigenpalms features. Pattern Recognition Letters 24(9-10), 1473–1477 (2003)
5. Wu, X., Zhang, D., Wang, K.: Fisherpalms based palmprint recognition. Pattern Recognition Letter 24, 2829–2838 (2003)
6. Li, W., Zhang, D., Xu, Z.: Palmprint Identification by Fourier Transform. Int. J. Pattern Recognition Artif. Intell. 16(4), 417–432 (2003)
7. Kong, W.K., Zhang, D., Li, W.: Palmprint feature extraction using 2-D Gabor filters. Pattern Recognition Letters 36(10), 2339–2347 (2003)
8. Nanni, L., Lumini, A.: On selecting Gabor features for biometric authentication. International Journal of Computer Applications in Technology 35(1), 23–28 (2009)
9. Zhang, D., Shu, W.: Two novel characteristics in palmprint verification: datum point invariance and line feature matching. Pattern Recognition 32, 691–702 (1999)
10. Duta, N., Jain, A.K., Mardia, K.V.: Matching of palmprint. Pattern Recognition Letters 23, 477–485 (2002)
11. Ribaric, S., Fratric, I.: A biometric identification system based on eigenpalm and eigenfinger features. IEEE Trans Pattern and Machine Intelligence 27(11), 1698–1709 (2005)
12. Savic, T., Pavesic, N.: Personal recognition based on an image of the palmar surface of the hand. Pattern Recognition 40, 3152–3163 (2007)
13. Nanni, L., Lumini, A.: A multi-matcher system based on knuckle-based features. Neural Computing and Applications 18(1), 87–91 (2009)
14. Li, Q., Qiu, Z., Sun, D., Wu, J.: Personal Identification using knuckleprint. In: Sinobiometrics, Guangzho, pp. 680–689 (2004)
15. Cross, J., Smith, C.: Thermographic imaging of the subcutaneous vascular network of theback of the hand for biometric identification. In: Proceedings of IEEE 29th International Carnahan Conference on Security Technology, pp. 20–35 (1995)
16. Miura, N., Nagasaka, A., Miyatake, T.: Feature extraction of finger-vein patterns based on repeated line tracking and its application to personal identification. Machine Vision and Applications 15, 194–203 (2004)
17. Wang, J.G., Yau, W.Y., Suwandya, A., Sung, E.: Person recognition by fusing palmprint and palm vein images based on "Laplacianpalm" representation. Pattern Recognition 41, 1514–1527 (2008)
18. Toh, K.A., Eng, H.L., Choo, Y.S., Cha, Y.L., Yau, W.Y., Low, K.S.: Identity verification through palm vein and crease texture. In: International Conference on Biometrics (2005)
19. Wang, L., Leedhamb, G., Cho, D.S.Y.: Minutiae feature analysis for infrared hand vein pattern biometrics. Pattern Recognition 41, 920–929 (2008)
20. Lin, C.L., Fan, K.C.: Biometric verification using thermal images of palm-dorsa vein patterns. IEEE Transactions on Circuits and Systems For Video Technology 14(2), 199–213 (2004)
21. Michael, G., Connie, T., Andrew, T.: Touch-less palm print biometrics: Novel design and implementation. Image and Vision Computing 26(12), 1551–1560 (2008)
22. Michael, G., Connie, T., Andrew, T.: An Innovative Contactless Palm Print and Knuckle Print Recognition System. Pattern Recognition Letters (2010)
23. Gonzalez, R.C., Woods, R.E.: Digital Image Processing, 2nd edn. Prentice-Hall, Inc., New Jersey (2002)
24. Saunders, C.: Support Vector Machine User Manual. RHUL, Technical Report (1998)
25. Vapnik, V.: Statistical Learning Theory. Wiley-Interscience publication, Hoboken (1998)
26. Verlinde, P.: A Contribution to Multi-Modal Identity Verification Using Decision Fusion. PhD dissertation, Department of Signal and Image Processing, Telecom Paris, France (1999)

Visual Object Detection by Specifying the Scale and Rotation Transformations

Yasuomi D. Sato[1,2], Jenia Jitsev[2], and Christoph von der Malsburg[2]

[1]Department of Brain Science and Engineering, Graduate School of Life Science and Systems Engineering, Kyushu Institute of Technology,
2-4, Hibikino, Wakamatsu-ku, Kitakyushu, 808-0196, Japan
sato-y@brain.kyutech.ac.jp
[2]Frankfurt Institute for Advanced Studies, Johann Wolfgang Goethe University,
Ruth-Moufang-Str. 1, 60438 Frankfurt am Main, Germany
{sato,jitsev,malsburg}@fias.uni-frankfurt.de

Abstract. We here propose a simple but highly potential algorithm to detect a model object's position on an input image by determining the initially unknown transformational states of the model object, in particular, size and 2D-rotation. In this algorithm, a single feature is extracted around or at the center of the input image through 2D-Gabor wavelet transformation, in order to find not only the most likely relative size and rotation to the model object, but also the most appropriate positional region on the input image for detecting the correct relative transformational states. We also show the reliable function on the face images of different persons, or of different appearance in the same person.

Keywords: Visual Object Detection, Gabor Filter Decomposition, Transformation Specific Similarities, Feature Correspondence.

1 Introduction

In order to understand a whole visual object recognition process, we need to know fundamental mechanisms to detect a position and transformational states for size and 2D-rotation of a single object. This is because under general assumption about the natural view conditions, the transformational state of the object cannot be accessed in advance. Many of object recognition systems constructed so far were under a restricted assumption that the transformations of the object, in particular, the size and rotation, have already been known [1]. Little is seen for recognition demonstrations without this unnatural assumption. How to match the input image to the representation of the model reference is also still unclear to specify the models position on an input image. So, understanding of mechanisms for specifying the initially unknown transformational states of the size, rotation as well as position is an important and requisite technical step to a construct visual object recognition system.

In this work, we are importantly interested in understandings of a mechanism how an uncertain position of one single object is specified on an input image,

K.W. Wong et al. (Eds.): ICONIP 2010, Part II, LNCS 6444, pp. 616–624, 2010.

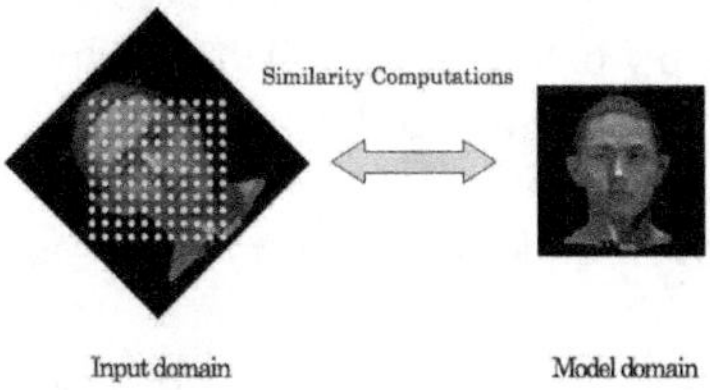

Fig. 1. A system for detecting correct relative size and rotation transformations between two images of the input (I) and model (M) of the same object, 11×11 rectangular grids are placed on the image I. On each grid point of the image I, the single Gabor feature can be extracted for computing the similarity to a single Gabor feature M. It can be dedicated to detect the most likely transformational states for each grid on the image I.

simultaneously finding the other unknown transformational states for scale and rotation. The information about such transformational states of the object is frequently discarded when visual object recognition is achieved, but we address that it is very valuable for the disambiguation of the visual object representation, with developments of the visual object recognition system without discarding such transformation information.

2 System Organizations

We assume that there exist two face images of the same or different person taken at a different scale and/or orientation. Two domains (called input (I) and model (M)) are set up for holding the images to be matched one to another (see, Fig.1).

For the I domain, Gabor features J^{I} are extracted from N different grid points, $\boldsymbol{x}_p^{\mathrm{I}} = (x_p^{\mathrm{I}}, y_p^{\mathrm{I}})$ with the square grid coordinate located at a center of the image I. For the M domain, a single Gabor feature J^{I} is extracted at a center of the image M. The Gabor feature components are defined as convolution of the image with a family of Gabor functions. The Gabor Function and its intrinsic parameters are referred to [2,3] as usual Since the parameter values referred previously are different from those in this work, we thus describe as follows: $k_{max} = 14.2$, $\sigma = 14.0$, the orientation parameter, $\theta = \pi k/8$ ($k \in \{0, 1, \ldots, 7\}$), and the spatial frequency parameter, (the scale factor $a_0 = 0.8$, $\mathrm{L} = \{\mathrm{I}, \mathrm{M}\}$, $l_{\mathrm{I}} \in \{0, 1, \ldots, 11\}$ and $l_{\mathrm{M}} \in \{2, \ldots, 9\}$), in order for the Gabor feature to be sampled discretely over the image. As mentioned here, the frequency range on the I domain is extended further to the fundamental frequency range of the M domain for accounting for possible scaling up or down.

We will explain a basic concept of how initially unknown transformational states of the face can be detected. Considering the image M as a reference model, the size transformation can be interpreted as the radial shift of a face toward a center or an edge of the image I, while the rotation transformation corresponds to the circular shift at the center. To make a distinction between scaling up and scaling down of the face, we introduce a notion, $s \in \{-2, -1, 0, +1, +2\}$ with $+$ or $-$ for scaling up or down operations respectively, where every index number indicate the scale of the image I relative to the image M with $[1/a_0]^s$ (for example, $s = 0$ means no change to the reference model). Similarly, the rotation transformation is also given as followings: $[r/8] \times \pi$ and $r \in 0, 1, \ldots, 7$.

3 Size and Rotation Transformation-Specific Similarities

Two images of the same or different face at different scale and/or orientation are matched in the system. For this, it would be necessary to find a proper similarity measure between both the images, specifying the most likely transformations applied to the I. We propose so-called transformation-specific similarities between two domains. A set of two vectors in each domain are constructed for computation in terms of all given transformation parameters. Each is a gathered average of the corresponding Gabor feature component. The similarity computations can be performed by aligning the vectors relative to each other and taking the scalar product for each transformation parameter. The similarities with the highest value determine the most likely scale and rotation transformations.

3.1 Decomposition of Scale and Rotation

In order to explain the similarity computation, we use one single Gabor feature extracted from a center point of the M image. The M Gabor feature can be compared to the other Gabor feature extracted from the image I, which is systematically located around the preferred center. Spatial frequency and orientation components are calculated, averaging over the respectively orientation and spatial frequency components:

$$X_k^{\mathrm{L}} = \frac{1}{|l_{\mathrm{L}}|} \sum_{l \in l_{\mathrm{L}}} J_{p,l,k}^{\mathrm{L}}, \tag{1}$$

$$Y_l^{\mathrm{L}} = \frac{1}{|k|} \sum_{k=0}^{n-1} J_{p,l,k}^{\mathrm{L}}. \tag{2}$$

where p is the position index for the Gabor feature extraction point. For L=M, the position the center of the image. For L=I, p takes somewhere within a rectangular coordinate with an arbitrary size around the center. The detail explanation and result obtained by using these equations will be given in the next section. We obtain two vectors in each domain, a scale group and a rotation group vector. For the I domain, rotation group and scale group vectors are $X^{\mathrm{I}} = (X_0^{\mathrm{I}}, \ldots, X_7^{\mathrm{I}})$ and $Y^{\mathrm{I}} = (Y_0^{\mathrm{I}}, \ldots, Y_{11}^{\mathrm{I}})$. For the M domain accordingly $X^{\mathrm{M}} = (X_0^{\mathrm{M}}, \ldots, X_7^{\mathrm{M}})$ and $Y^{\mathrm{M}} = (Y_2^{\mathrm{M}}, \ldots, Y_9^{\mathrm{M}})$. We normalized the respective group vector by an L^2-Norm.

Let us compute the scale-(S) and rotation-specific (R) similarities between two images. All possible transformation-specific alignments of the vectors are generated to apply the scalar product operation. The similarity specific for the scale transformation is given by:

$$S_s^{\mathrm{S}}(\boldsymbol{Y}^{\mathrm{I}}, \boldsymbol{Y}^{\mathrm{M}}) = \frac{\sum_{l \in l_{\mathrm{M}}} Y_{l+s}^{\mathrm{I}} Y_l^{\mathrm{M}}}{\sqrt{\sum_{l \in l_{\mathrm{M}}+s} \left(Y_l^{\mathrm{I}}\right)^2 \sum_{l \in l_{\mathrm{M}}} \left(Y_l^{\mathrm{M}}\right)^2}}.$$

Each shift alignment of the scale vectors corresponds to a scale parameter s, making possible the computation of the scale specific similarity for this particular parameter. In the similarity computation for the rotation transformation, the specific similarity for a given rotation degree r between the I and M domains is given by

$$S_r^{\mathrm{R}}(\boldsymbol{X}^{\mathrm{I}}, \boldsymbol{X}^{\mathrm{M}}) = \frac{\sum_{r' \in k} X_{r'-r+n}^{\mathrm{I}} X_{r'}^{\mathrm{M}}}{\sqrt{\sum_{r' \in k} (X_{r'}^{\mathrm{I}})^2 \sum_{r' \in k} (X_{r'}^{\mathrm{M}})^2}}, \quad (r' - r) \mod n.$$

This means, for example, that for the rotation $r = 1$, both the vectors are shifted relative to each other so that kth component of the rotation vector for the M matches against the $(k+1)$th component of the rotation vector for the I. Doing this for all given transformation parameter, we get all transformation specific similarities to start the detection process of the face position and to determine the initially unknown transformational states.

3.2 Similarity Computation and Transformation Detection

The group vectors established in Sec.3.2 give us representational basis on which the transformation specific similarities are computed. To make the size and rotation transformation detection more reliable, the computed similarities should reflect the actual degree of similarity between the I and the M given their respective specific transformation. So, the I image is scaled and/or rotated with an arbitrary size in the range of $[-2, 2]$ and an arbitrary rotation angle in the range of $[0, 7]$, in relation to the M object. A natural approach to estimate the computed similarities is to look on the similarity values by presenting transformed I object with the corresponding degrees within the continuous transformation regions. We then expect transformation-specific similarities taking the highest value for the corresponding transformation parameter. The similarity values fall down when the transformed images are changed gradually from the corresponding given parameter. The similarity values calculated in this way, tuning curves for each transformation specific similarity can be constructed.

The tuning curves are constructed from similarity values for each scale and rotation transformation s_p and r_p, centered on the value for the preferred transformation parameter. Taking a glance at the tuning curves, the prominent peaks marking the maximum similarities for the preferred transformation are obvious. This should give enough discrimination basis for the competitive evaluation of the likelihood for the rotation (Fig.2(a)) and the scale transformation (Fig.2(b)).

Let us return again to how the correct transformation in scale and rotation can be detected. We have to notice that the uncertainty about the transformations

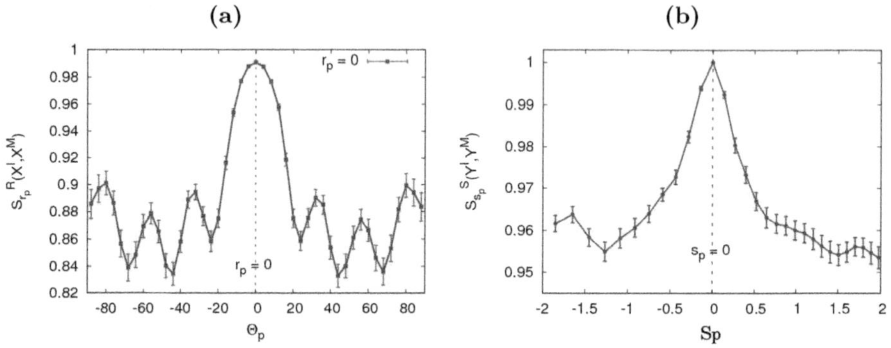

Fig. 2. Tuning curves based on specific similarities for (a) rotation and (b) size transformation. The similarity tuning curves are plotted as functions of deviation from preferred transformation (scale (s_p) and rotation (r_p). The preferred scale and rotation degrees are relative size and rotation to the degrees taking the maximum similarities. Because of using 100 different identity comparisons between centers of the image, an average over the similarities at each parameter and the error $\sigma/\sqrt{100}$ can be calculated where s is the standard deviation.

come to be broken by selecting transformation parameters r_0 and s_0, which take the maximum similarity value among all similarities computed from the group vectors of both the domains:

$$r_0 = \arg\max_r \{S_r^{\mathrm{R}}(\boldsymbol{X}^{\mathrm{I}}, \boldsymbol{X}^{\mathrm{M}})\}, \tag{3}$$

$$s_0 = \arg\max_s \{S_s^{\mathrm{S}}(\boldsymbol{Y}^{\mathrm{I}}, \boldsymbol{Y}^{\mathrm{M}})\}. \tag{4}$$

The obtained observation about the discriminative power of the transformation specific similarities supports the idea of their utility in the competitive evaluation of the transformation parameters. We have elucidated that the specific scale and rotation transformation similarities computed from group vectors have an intimate relation to the corresponding scales and rotations of the object on the image I, and constitute a suitable measure for estimating the transformation parameters between the two images.

4 Simulation Results

In the section, we examine a general ability for transformation detection in scale and rotation and find a good positional range of the correct transformation detection to specify a model reference on the image I. We use photographic images of human faces, rotated and scaled within the continuous range described above. Depending on the outcome, the system performance can be considered as correct, supposing the right transformation parameters, or incorrect, failing to do so. Each task involves presentation of an arbitrarily transformed face image of the I, while the M image (no scale and no rotation) of (1) the same face,

(2) the same face in different appearance or (3) the different face is put on the M domain. Employing this simple case, we can probe detection performance of the system under various conditions.

4.1 Correct Transformation Detection on Facial Images

100 images with faces of different persons were prepared for the experiment. Each image size is 256×256 [pixel]. Taking this image size on the I domain, the object sizes, $D^2 = 83 \times 83$, 104×104, 130×130 (standard scale), 163×163 and 204×204 [pixel] are set. The image can be additionally transformed with 8 different rotations. So, each of the 100 different objects has total 39 transformed instances of its reference image.

Reference face positions on the I domain cannot be specified in advance. We have to cope with such a positional uncertainty. A high probability correct against pixel positions of the I image is thus demanded. To probe for this quality, we set up a center point for feature extraction in the reference image M, while the Gabor feature on the image I is extracted from an arbitrary position chosen within 60×60 square grid coordinate set around the domain's center. Alongside with the absolute coordinates, the position can also be expressed as a distance from the center point computed relative to the size of the face on I. For each position, the relative distance is thus $d = D$, where d is the distance from the center measure in absolute pixel number. By measuring the rate of the correct transformation detection for each point inside the region, we are able to indicate correctness of transformation detection on pixel positions of the I image, showing the gradually slow decrease of system performance with the increasing distance from the original position.

In Fig. 3, for each object size the probabilities for the correct transformation detection are plotted as functions of a relative distance ($d = D$). The performance is above 95% level inside of the region as big as 0.147 of the object size, which is a considerable amount of shift bearing in mind rather heterogeneous

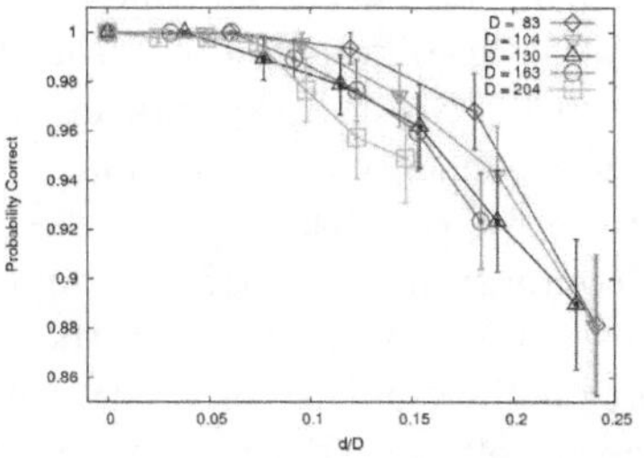

Fig. 3. Correct transformation detection on positions over the relative distance ($d = D$) from the reference location, independent of an object size in the I domain. The probability is calculated for each shifted position by probing transformed face images of 100 different persons for a detection of the correct transformation. During each single detection trial, reference image of the person is put on the M domain while an arbitrary transformed version of the same person is presented on the I.

structure of a face image. Shifting towards longer relative distances will surely disturb transformation detection still the system stays well above 85% transformation detection rate even for shifts larger than 20% of the object size.

So, taking a single feature as reference on an M domain does not require the exact knowledge of the corresponding feature's position on the I to detect correct transformation parameters. Moreover, there seem to be a well-defined similarity landscape which can be exploited by the system to locate the position of the reference feature under initial uncertainty by moving in the direction of the increasing similarity towards its maximum. The established correctness of the transformation detection against pixel positions constitutes a basis for a further development towards multiple reference feature approach, where each reference feature will be able to find its transformed counterpart on the input on the basis of transformation-specific similarities computed across multiple positions on the input domain.

4.2 Detection Performance under Different Conditions

In the next experiment, we are trying to test the transformation detection ability of our system under some different visual conditions on the face image comparisons: (A) Two same persons without any distinguishing appearance are respectively located on the center of the I or M images. (B) The two same persons look slightly different, for instance, one of them is smiling while the other is smiling. But they are positioned on the center of each image. The relative size, rotation and illumination between the two persons is approximately same. (C) One of the two same persons in the different has already been under different image conditions of the other. The conditions are different scaling, rotation, illumination and position of the object. (D) The different persons on the I and M images.

For this, 100 different pairs of the same persons are prepared from the FERET database [4]. There are hundreds of face image pairs. However in many of the pairs, at glance, one of the pair is already scaled, inclined the head, or shifted on the image, relative to the other. We should carefully realize even the case for different illumination. For this, we should have selected 100 different pairs with the same face size and the different looks from the FERET database in the experiments of (A) and (B). In the experiment (C), the 100 different face pairs without any consideration about transformation and translation should randomly be picked up from the database.

When simulating one comparison, we extract a single feature from the center point on each image. This is because as shown in the previous section, we have demonstrated 100.00% probability of the correct transformation in terms of the center-to-center comparison between the two images of the same person. This result is just as the one of (A) in Table 1. In the (B) experiment, we have obtained an acceptable value of the transformation detection rate, 93.00%, even though the two persons in the pair look slightly different.

If a collection of the pairs should be repaired to involve cases with different scaling, rotation and so on, the transformation detection performance gradually becomes decreased to 79.00% as shown in the (C) in Table 1. One of the reasons

Table 1. Detection performance for face image comparisons between the input and model domains. (A) Two same persons without any distinguishing appearance located on the center of each image. (B) The two same persons placed on the center of each image look slightly different. The relative size, rotation and illumination between the two persons are approximately same. (C) One of the two same persons has already been under different image conditions such as different scaling, rotation, illumination and position. (D) The different persons.

Item	Probability Correct
A	100.00%
B	93.62%
C	79.00%
D	5.42%

why is apparently due to a relative shift of the face in the pair must be out of an appropriate size of the higher robustness of the correct transformation detection. Thus, if we should take a good care of transformational or translational difference of the two images, even though they are slightly different in their look, this result will suggest that our algorithm used here may possess a preliminary capability for finding identity of a person.

Indeed, as shown in (D) of Table 1, we have obtained much less detection performance, which presents a potential to support the functionality of finding the personal identity. Of course, in this work, to use our proposed algorithm is just at a fundamental step toward practicable application. So there still ample discussion about the advanced utility of our algorithm to detect the correct transformation in scale and rotation.

5 Discussions and Conclusions

If attempting to shed the light on mechanisms underlying invariant visual recognition, the key point is to understand the general principles how invariance is achieved, establishing the relevant computational recognition systems [5]. In the current study, our main focus lies on the proof of a general principle for a transformation-tolerant feature processing. In the processing, the most likely relative transformation in scale and rotation can be detected under a natural condition of its initial uncertainty.

A method proposed here has brought the useful ability that object recognition can be achieved without loss of information about the initially uncertain transformational states. We have shown that this computational method has a supportive functionality for finding the correct transformation with a certain region of feature extraction positions on an input image. Results of face detection performance have given us an important step toward more practical application in the future. In particular, the maximum operation of the transformation-specific similarities may be a relatively powerful tool to construct an architecture for a translation-invariant particular face recognition.

This is because our position detection results on the object image allow us to propose a concept of dealing with translation invariance analog to the detection of the scale and rotation. The rough idea is to find the location corresponding to the reference feature by maximum operation of its similarities over all possible positions in the image, determining the area where the model feature is most likely to be found. We should not forget that making conclusions about the identity of the object is also the case. We intend to establish a gallery of different model objects in the model domain to be able to decide for a particular object identity as well. This would require a substantial step in the detection of transformation and recognition of the identity, achieving the coherent global decision about the image presented on the input.

Acknowledgments. This work was supported by the "Bernstein Focus: Neurotechnology through research grant 01GQ0840" funded by the German Federal Ministry of Education and Research (BMBF) and by the Hertie Foundation. Y.D.S was financially supported by the Grant-in-Aid for Young Scientist (B) No. 22700237.

References

1. Wolfrum, P., von der Malsburg, C.: What is the optimal architecture for visual information routing? Neural Computation 19(12), 3293–3309 (2007)
2. Wiskott, L., von der Malsburg, C.: Face Recognition by Dynamic Link Matching. In: Sirosh, J., Miikkulainen, R., Choe, Y. (eds.) Lateral Interactions in the Cortex: Structure and Function, vol. 11 (1996)
3. Wiskott, L., Fellous, J.-M., Krüger, N., von der Malsburg, C.: Face recognition by elastic bunch graph matching. IEEE Transactions on Pattern Analysis and Machine Intelligence 19(7), 775–779 (1997)
4. Okada, K., Steffens, J., Maurer, T., Hong, H., Elagin, E., Neven, H., von der Malsburg, C.: The Bochum/USC Face Recognition System and How it Fared in the FERET Phase III Test. In: Wechsler, H., Phillips, P.J., Bruce, V., Fogelman Soulifie, F., Huang, T.S. (eds.) Face Recognition: From Theory to Applications, pp. 186–205. Springer, Heidelberg (1998)
5. Hummel, J., Biederman, I.: Dynamic binding in a neural network for shape recognition. Psychological Review 99(3), 480–517 (1992)

Multi-view Gender Classification Using Hierarchical Classifiers Structure

Tian-Xiang Wu[1] and Bao-Liang Lu[1,2,*]

[1]Center for Brain-Like Computing and Machine Intelligence
Department of Computer Science and Engineering
[2]MOE-Microsoft Key Laboratory for Intelligent Computing and Intelligent Systems
Shanghai Jiao Tong University
800 Dong Chuan Rd., Shanghai 200240, China
tcwtxster@gmail.com, bllu@sjtu.edu.cn

Abstract. In this paper, we propose a hierarchical classifier structure for gender classification based on facial images by reducing the complexity of the original problem. In the proposed framework, we first train a classifier, which will properly divide the input images into several groups. For each group, we train a gender classifier, which is called expert. These experts can be any commonly used classifiers, such as Support Vector Machine (SVM) and neural network. The symmetrical characteristic of human face is utilized to further reduce the complexity. Moreover, we adopt soft assignment instead of hard one when dividing the input data, which can reduce the error introduced by the division. Experimental results demonstrate that our framework significantly improves the performance.

Keywords: Hierarchical classifiers, Gender classification, Multi-view facial images.

1 Introduction

Gender classification using facial images is widely used in human-computer interaction and the applications depending on it, such as demographics and visual surveillance. Most of the existing approaches do not consider, or design some features which are claimed to be robust to the pose variation of faces. They will fail in practical use facing unconstrained face poses, or say, multi-view faces.

To ease this multi-view problem, Toews and Arbel [1] proposed the idea of relative location information of the organs, which is used to infer the most likely position of the face. The result of gender classification was obtained by combining the results of organs. Takimoto et al. [2] extracted the features around the eyes and mouths which requires the positions of eyes and mouths to be exactly located in advance. In their work, local information is used to facilitate multi-view problems. Lian and Lu [3] aligned the facial images based on the position of eyes, and apply LBP [4] to feature extraction and SVM to gender classification directly.

In this paper, we propose a framework which decomposes the multi-view problem into several single-view subproblems and hence reduces the complexity. Under this

* Corresponding author.

K.W. Wong et al. (Eds.): ICONIP 2010, Part II, LNCS 6444, pp. 625–632, 2010.

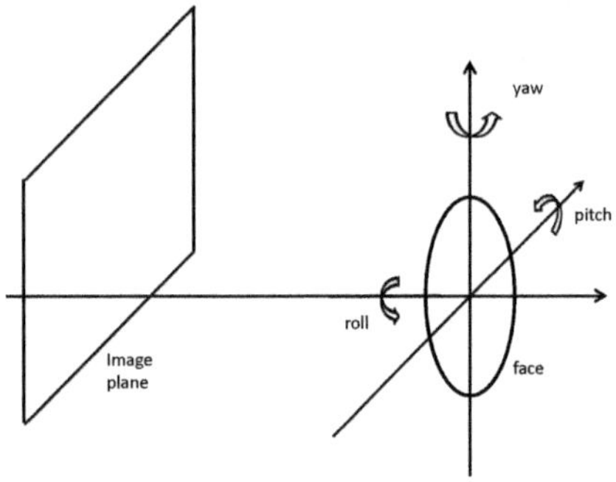

Fig. 1. The pose is decomposed into three rotations: pitch, roll and yaw

framework, any traditional feature extraction methods and classifiers (e.g. SVM, neural network) can be used. The pose is decomposed into three rotations: pitch, roll and yaw (See Fig. 1). To simplify the problem and without lose of generality, we only consider yaw rotation. The extension to the other two is trivial. The framework has two layers. In the first layer, we discretize the continuous angle space into K bins. A classifier whose output is in $\{1, \ldots, K\}$, is trained to predict which bin the input facial image falls in. We call it orientation classifier. One problem of the discretization process is the boundary effect. It is unreasonable to simply put a image on a boundary to either side. Therefore we adopt soft assignment, allowing the partitions overlap on the boundary. Moreover, to reduce the number of categories the orientation classifier should deal with, we make use of the symmetrical characteristic of human face, horizontally flipping the images whose faces toward right. By doing so the accuracy of the orientation classifier is increased. Then in the second layer, for each bin we train a classifier which specializes in gender classification of images from that bin. These classifiers are called gender classifiers.

The rest of the paper is organized as follows: Section 2 describes the main idea in the proposed hierarchical classifiers framework. Section 3 introduces some tricks to improve the accuracy. Section 4 shows the gender classification procedure using our framework. Experiment results are presented in Section 5. Some conclusions and future work are outlined in Section 6.

2 Hierarchical Classifiers

Traditional gender classification algorithms always work well on images with the same pose, since the alignment is easy. For multi-view facial images, the issue becomes much more complex. The feature space is much larger and it is difficulty to design orientation-invariant features. An efficient solution is to divide the feature space into several subspaces according to face orientations, which decomposes the multi-view problem into easier classification tasks on simpler subspaces.

In this paper we propose a two-layered classifier (See Fig. 2). The first layer includes a classifier, which extracts the feature vectors from the original image and classifies it into several categories according to its orientation. Then the task of gender classification

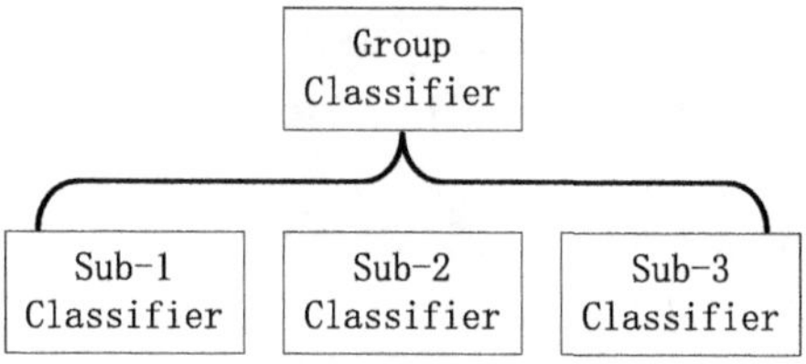

Fig. 2. Hierarchical classifiers structure

is passed to the next layer where we make use of experts of gender classification for certain orientations. It is obviously that the accuracy of the classifier in the first layer is important for the whole problem. Classification error in the first layer leads to error answers in the second layer. So this method is suitable when the initial classification problem has a high precision.

3 Angle Categories Selection

In this section, we show some technologies that can be used in angle category selection to improve the accuracy of gender classification.

Since the faces of human beings are bilaterally symmetric, the images in which the persons face right turn to be the ones facing left after a horizontal flip. If we get the information of the face direction in the images, the original space of input images should be reduced by a half. An easy classifier is trained for the direction classification in this paper to reduce the complexity.

The hierarchical classifiers structure is suitable for the problem whose initial classification problem has a high precision. Classification error of the initial separation leads to error answers in the individual classification. We found in the experiment that the images whose angles are near the dividing line of the two angle categories are easily to be misclassified. The error answer caused by the misclassification is due to the lack of information in the certain individual classifier. We get another trick which is to add the samples whose angles are near the dividing line into the training data of the two neighboring categories. The detail is refer to Section 5.2.

By using the combination of results generated in the gender classifiers, the risk in the first classification layer is apportioned.

$$result_i = \sum_{k=1}^{K} p_i^k \times result_i^k \tag{1}$$

$result_i$ is the possibility of the ith sample to be a male, and $result_i^k$ is the result from the sub classifiers. p_i^k means the possibility for the ith sample to belong to the category. Moreover, Weighted Relevance Aggregation Operator(WRAO) [5] helps hierarchical model with hierarchical fuzzy signatures to work better.

In our problem, the classification is the first layer is not so hard. Huang and Shao [6] use SVM to achieve perfect performance on face pose classification problem on the standard FERET data base. With tolerance on the dividing region of the neighboring

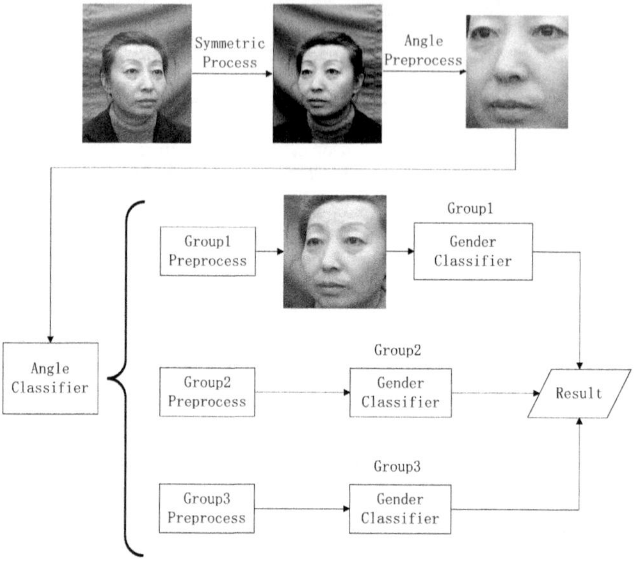

Fig. 3. The gender classification process using hierarchical classifiers structure

categories, the uncertainty in the first layer won't make trouble for the gender classification.

4 Gender Classification Procedure

In this section, we introduce the whole hierarchical classifier based gender classification system(See Fig. 3).

Alignment is important for gender classification based on facial images [7]. First, faces are fixed in the center of the result pictures. Facial components are in the certain places for feature extraction after alignment. To obtain the facial components, we adopt Active Shape Model (ASM) [8], a statistical model of the shape of the deformable object, to get the locations of eyes and mouth, and then cut the rectangle out of the facial image.

The bilaterally symmetrical characteristic of human face is make use of. The images facing right are turned left. Then, the images will be classified into some classes according to the angle of the face. Images in different angle classes are taken to their own gender classifiers. Now we have converted the original problem to gender classifications based on facial images of fixed angle, a well-studied problem with many good approaches.

Gender classification processes in the different categories are similar. The facial images are re-cutting, using the information of angles which is the label of the category, in order to put most of the human face into the picture, and align the organ positions precisely. Some information of hair is also taken into images for classification. Feature extraction is done in different categories and the suitable gender classification is prepared for the facial images.

5 Experiment

5.1 Data Set

To compare the performance, we select the gender classification problem based on multi-view facial images in the CAS-PEAL face database [9](See Table 1). We take all the images labeled "PM" from the "POSE" section. We put different people into the training and test table, i.e., we partition the set according to people instead of single images so that two different images of a person cannot stay in both training or test set. This would help avoid over-fitting, or the similarity of the two images will take unnecessary information to the test set. The total 7273 different-pose facial images are so organized into 11 groups, such that within each of them, the number of the training samples is 70% the number of female facial images, which are fewer than male images. We let the percentage be 50% if there are not many in that group.

Table 1. Description of training and test data based on facial images

Data Set	Description	Total	Male	Female	Training	Test
	PM-67	101	79	22	11*2	79
	PM-45	1039	595	444	306*2	427
	PM-30	938	516	422	295*2	348
	PM-22	101	79	22	11*2	79
	PM-15	938	516	422	295*2	348
CAS-PEAL	PM+00	1039	595	444	306*2	427
	PM+15	938	516	422	295*2	348
	PM+22	101	79	22	11*2	79
	PM+30	938	516	422	295*2	348
	PM+45	1039	595	444	306*2	427
	PM+67	101	79	22	11*2	79
	TOTAL	7273	4165	3108	4284	2989

5.2 Implementation

The images in the data set are of many different angles. We want to use this prior knowledge to separate them into some categories. However, as we have pointed out, too many categories will introduce complexity to the division problem. So some of the images in different angles must be put into one category. We use the symmetric property of human face to reduce the total amount of angles from 11 to 6. In this case, 3 categories are enough.

The division illustrated in Fig. 4(a) looks quite natural. Given such a division, none of the two corresponding classifiers would be able to solve the angle between the two regions. For we use the structure of hierarchical classifiers, classification error in the first layer probably leads to error answers in the second layer, especially near the division of the two neighboring regions. In this paper, we use the division in Fig. 4(b). Images of PM00 and PM15 are used for training the first classifier. Images of PM15, PM22 and PM30 are for the second classifier. And PM30, PM45 and PM67 are for the

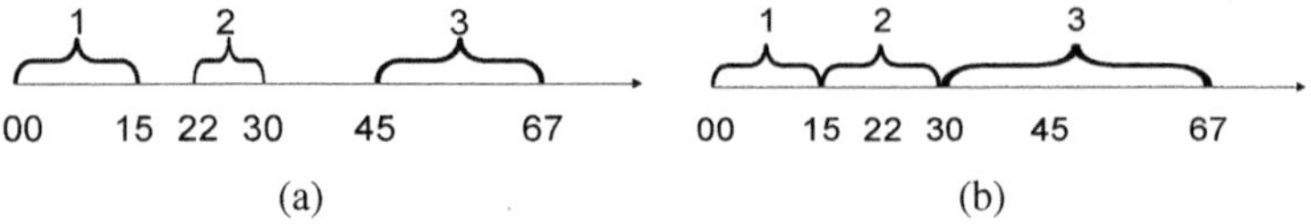

Fig. 4. Angle region division about gender classification: (a)naive way, 6 angles are divided equally into three groups; (b)the way to get all angles in and improve accuracy

Table 2. Gender classification with or without symmetric process

Method	Accuracy (RBF kernel)	Accuracy (linear kernel)
Symmetric accuracy	99.63%	99.60%
With symmetric process	93.34%	92.64%
Without symmetric process	92.31%	91.50%

third. The training data of the division of the two neighboring regions are trained in both corresponding classifiers. The images whose angles are near the division can be classified correctly in both corresponding classifiers. This will also help to reduce the risk of the accumulative error introduced by the classification error in the first layer.

For feature extraction, we use multi-resolution local Gabor binary pattern (MLGBP) to extract the features of each facial image. The MLGBP [10] feature, which is the input of the SVM classifiers, is derived by combining multi-resolution analysis, Gabor characteristic and uniform LBP histograms [11]. All experiments were performed on a Pentium fourfold CPU (2.83GHz) PC with 8GB RAM.

All the classifiers in this paper are Support Vector Machines, lib-svm v.2.86 in detail. RBF kernel and linear kernel are used for comparison.

5.3 Result

The effect of symmetry is shown in Tab. 2. We make use of the symmetry of human face to reduce the originally 11 angles to 6. The accuracy of the whole process including symmetric transformation and gender classification is better than that without the process. It means the symmetric transform provides the classification with less complexity and more accuracy but less harm.

The classifications in small fields show great advantage over the one in the large complex space(See Tab. 3). The result shows the additional classification step won't

Table 3. Gender classification in different angle categories

Angle Category	Accuracy (RBF)	Total SV	Accuracy (linear)	Total SV
PM00,PM15	97.51% (1095/1123)	416	97.51% (1095/1123)	346
PM15,PM22,PM30	98.26% (1523/1550)	423	97.94% (1518/1550)	352
PM30,PM45,PM67	97.78% (1670/1708)	586	97.48% (1665/1708)	348
Classify directly	92.31% (2759/2989)	1480	91.50% (2735/2989)	840

Table 4. Gender classification using hierarchical classifiers structure

Method	Accuracy (RBF)	Total SV	Accuracy (linear)	Total SV
Angle classify	98.43% (2942/2989)	1161	98.43% (2942/2989)	965
Classify directly	92.31% (2759/2989)	1480	91.50% (2735/2989)	840
Whole System	97.89% (2926/2989)	–	97.59% (2917/2989)	–

harm the accuracy but increase it(See Tab. 4). In the experiment, we find most of the testing data which are misclassified in the angle classification are the facial images laying near the dividing line and being classified into the neighboring category. The selected classifiers are still suitable for these images and prepare better classification. The angle categories selection helps to solve the main problem in hierarchical classifiers framework. So the accuracy of the gender classification with symmetric process and angle classification is close to the performance of the expert classifiers in their fields.

5.4 Complexity Analysis

As we know, the time complexity of a standard SVM QP solver is $O(M^3)$, where M denotes the number of training samples. In our hierarchical classifiers framework, we cut the training samples into K groups, where K is the number of classifiers in the second layer of the structure. In each group, the corresponding classifier only needs to deal with its own training samples, so they can be trained in parallel, meaning that the running time could be improved to $O((\frac{M}{K})^p)$. Even if we run the training in serial, it will only take $O(K(\frac{M}{K})^p)$. Thus time complexity is reduced in both situations.

During recognition, the time eater is to calculate the kernel of test and support vectors especially in high dimension space. So we suppose the time complexity of SVM is $O(v)$, where v is the number of support vectors. The statistic shows that the sub-problems need much less time than the original problem.

6 Conclusions and Future Work

We have proposed a novel framework for gender classification based on multi-view facial images, i.e., hierarchical classifiers. The most important advantage of our framework over traditional SVM is that prior knowledge is used to get the input image to the expert of that field who can be easily get trained and give an answer. Experimental results show the effectiveness of our framework. A future extension of our work is to use the combination of some classifiers of organs or other parts to make our classifier more robust and improve accuracy.

Acknowledgments

This work was partially supported by the National Natural Science Foundation of China (Grant No. 60773090 and Grant No. 90820018), the National Basic Research Program of China (Grant No. 2009CB320901), the National High-Tech Research Program of China (Grant No. 2008AA02Z315), and the Science and Technology Commission of Shanghai Municipality (Grant No. 09511502400).

References

1. Toews, M., Arbel, T.: Detection, localization, and sex classification of faces from arbitrary viewpoints and under occlusion. IEEE Transactions on Pattern Analysis and Machine Intelligence 31, 1567–1581 (2009)
2. Takimoto, H., Mitsukura, Y., Fukumi, M.: Robust Gender and Age Estimation under Varying Facial Pose. IEE J. Trans. 127(7), 1022–1029 (2007)
3. Lian, H., Lu, B.: Multi-view gender classification using local binary patterns and support vector machines. In: Proceedings of the Third International Symposium on Neural Networks, pp. 202–209 (2006)
4. Ahonen, T., Hadid, A., Pietikainen, M.: Face recognition with local binary patterns. In: Pajdla, T., Matas, J(G.) (eds.) ECCV 2004. LNCS, vol. 3021, pp. 469–481. Springer, Heidelberg (2004)
5. Mendis, B., Gedeon, T., et al.: Generalised weighted relevance aggregation operators for hierarchical fuzzy signatures. In: International Conference on Computational Intelligence for Modelling Control and Automation (2006)
6. Huang, J., Shao, X., Wechsler, H.: Face pose discrimination using support vector machines (SVM). In: Proc. of 14th International Conference on Pattern Recognition (ICPR 1998) (1998)
7. Makinen, E., Raisamo, R.: Evaluation of Gender Classification Methods with Automatically Detected and Aligned Faces. IEEE Trans. Pattern Analysis and Machine Intelligence 30(3), 541–547 (2008)
8. Cootes, T., Taylor, C., Cooper, D., Graham, J., et al.: Active shape models-their training and application. Computer vision and image understanding 61(1), 38–59 (1995)
9. Gao, W., Cao, B., Shan, S., et al.: The cas-peal large-scale chinese face database and baseline evaluations. Technical report of JDL, `http://www.jdl.ac.cn/~peal/peal_tr.pdf`
10. Lian, H., Lu, B.: Multi-view gender classification using multi-resolution local binary patterns and support vector machines. International Journal of Neural Systems 17(6), 479–487 (2007)
11. Xia, B., Sun, H., Lu, B.: Multi-view gender classification based on local Gabor binary mapping pattern and support vector machines. In: IEEE International Joint Conference on Neural Networks, pp. 3388–3395 (2008)

Partial Extraction of Edge Filters by Cumulant-Based ICA under Highly Overcomplete Model

Yoshitatsu Matsuda[1] and Kazunori Yamaguchi[2]

[1] Department of Integrated Information Technology,
Aoyama Gakuin University,
5-10-1 Fuchinobe, Chuo-ku, Sagamihara-shi, Kanagawa, 252-5258, Japan
matsuda@it.aoyama.ac.jp
http://www-haradalb.it.aoyama.ac.jp/~matsuda
[2] Department of General Systems Studies,
Graduate School of Arts and Sciences, The University of Tokyo,
3-8-1, Komaba, Meguro-ku, Tokyo, 153-8902, Japan
yamaguch@graco.c.u-tokyo.ac.jp

Abstract. It has been well known that ICA can extract edge filters from natural scenes. However, it has been also known that the existing cumulant-based ICA can not extract edge filters. It suggests that the simple ICA model is insufficient for explaining the properties of natural scenes. In this paper, we propose a highly overcomplete model for natural scenes. Besides, we show that the 4-th order covariance has a positive constant lower bound under this model. Then, a new cumulant-based ICA algorithm is proposed by utilizing this lower bound. Numerical experiments show that this cumulant-based algorithm can extract edge filters.

1 Introduction

Independent component analysis (ICA) is a well-known technique in blind source separation [1,2]. Besides, it has been known that ICA can extract edge filters from natural scenes [3]. Therefore, ICA is also expected to give a general framework for explaining both the characteristics of natural scenes and the early visual system in the brains [4]. However, the ICA-based framework is still controversial and has not been investigated thoroughly yet. In this paper, we pay attention to the fact that the existing cumulant-based ICA algorithms can *not* extract edge filters. Cumulants are the higher order statistics such as kurtosis, which are estimated through the average of some polynomial functions of given signals. The utilization of cumulants is one of the most basic approaches in ICA, and many cumulant-based ICA algorithms have been proposed [5,6,7]. However, it has been known empirically that cumulant-based ICA algorithms can *not* extract local edge filters from natural scenes, where the extracted components are globally noisy patterns. The edge extraction requires highly nonlinear functions such as tanh. The reason has not been clarified completely yet. One possibility

K.W. Wong et al. (Eds.): ICONIP 2010, Part II, LNCS 6444, pp. 633–640, 2010.

is that cumulant-based ICA is not robust to outliers [6,7]. Although such effect of outliers is reduced by giving a sufficiently large number of samples, the edge filters can not be extracted by increasing the number of natural scenes. It indicates that this hypothesis is inadequate. Another possibility is that the simple ICA model is insufficient to completely explain the characteristics of natural scenes and the early visual system. Some works have focused on the overcompleteness [8,9,10], which means that the number of sources may be larger than the dimension of the observed images. Though they suggest the significance of the overcompleteness, they also have to use highly nonlinear functions instead of cumulants.

In this paper, we assume a highly overcomplete ICA model and propose a new cumulant-based ICA algorithm which can extract edge filters from natural scenes. In this model, the number of sources is assumed to be much larger than the dimension of observed images. Because the estimation of all the independent components is difficult, we attempt to extract only a part of components accurately. Under this highly overcomplete model, we derive a lower bound of the 4-th order covariance. Then, a cumulant-based algorithm is derived by utilizing the lower bound. In consequence, the new algorithm is an improvement of the maximization of the 4-th order covariance with the Jacobi method. The significant point is the employment of a lower bound condition in each pair optimization of the Jacobi method. In this model, the reason why previous cumulant-based algorithms can not extract edge filters can be explained by an over-fitting problem.

This paper is organized as follows. In Section 2.1, a linear overcomplete ICA model is defined and the concept of the partial estimation is described. In Section 2.2, we derive a lower bound of the 4-th order covariance under this model. In Section 3, a new cumulant-based ICA algorithm with the Jacobi method is proposed by utilizing the lower bound. In Section 4, numerical experiments show that the proposed algorithm can extract edge filters from natural scenes. Lastly, this paper is concluded in Section 5.

2 Model

2.1 Linear Overcomplete ICA Model and Partial Estimation

The linear ICA model is given as

$$\boldsymbol{X} = \boldsymbol{A}\boldsymbol{S} \tag{1}$$

where $\boldsymbol{X} = (x_{ik})$, $\boldsymbol{A} = (a_{ij})$, and $\boldsymbol{S} = (s_{ik})$ correspond to known observed signals (original natural images in the edge extraction), the mixing matrix (edge filters), and independent source signals, respectively. $\boldsymbol{X}$ is an $N \times M$ matrix where N and M are the number of signals and that of samples, respectively. $\boldsymbol{S}$ is an $L \times M$ where L is the number of sources, and $\boldsymbol{A}$ is an $N \times L$ matrix. If $N = L$, the separating model is given as follows:

$$\boldsymbol{Y} = \boldsymbol{W}\boldsymbol{X} \tag{2}$$

where $\boldsymbol{Y} = (y_{ik})$ and $\boldsymbol{W}$ correspond to the separated signals and the $N \times N$ separating matrix. $\boldsymbol{W} = \boldsymbol{A}^{-1}$ can be estimated by maximizing a measure of the independence of $\boldsymbol{Y}$ (such as the sum of kurtoses of $\boldsymbol{Y}$ [6,7]). In the overcomplete model ($L > N$), however, such simple estimation can not be employed because $\boldsymbol{A}$ is not invertible. Here, the partial estimation is introduced. First, $\boldsymbol{A}$ is divided as follows:

$$\boldsymbol{A} = \left(\tilde{\boldsymbol{A}}\boldsymbol{B}\right) \tag{3}$$

where $\tilde{\boldsymbol{A}}$ is a square $N \times N$ matrix and $\boldsymbol{B} = (b_{im})$ is the complementary $N \times (L-N)$ matrix. $\tilde{\boldsymbol{A}}$ can consist of arbitrary N columns of $\boldsymbol{A}$ by any permutation. The partial estimation is defined as the estimation of $\tilde{\boldsymbol{A}}$ instead of $\boldsymbol{A}$. The optimal separating matrix is given as

$$\boldsymbol{W} = \boldsymbol{\Lambda}\tilde{\boldsymbol{A}}^{-1} \tag{4}$$

where $\boldsymbol{\Lambda} = (\lambda_i)$ is a diagonal matrix for scaling. Note that $\boldsymbol{Y} = \boldsymbol{W}\boldsymbol{X}$ is no longer the estimation of sources.

2.2 Lower Bound of 4-th Order Covariance in Partial Estimation

Here, a lower bound of the 4-th order covariance on $\boldsymbol{Y}$ is derived in the partial estimation $\boldsymbol{W} = \boldsymbol{\Lambda}\tilde{\boldsymbol{A}}^{-1}$. In this paper, the following 4-th order covariance is focused on:

$$\mathrm{cov4}\,(i,j) = \mathrm{ave}\left(y_{ik}^2 y_{jk}^2\right) - 1 \tag{5}$$

where $i \neq j$ and ave() is the average operator over samples k. Besides, the variance of each y_{ik} is normalized to 1 by choosing a suitable scaling matrix $\boldsymbol{\Lambda}$. The 4-th order covariance cov4$\,(i,j)$ is strongly related to kurtoses [11]. One advantage of cov4$\,(i,j)$ is that both its true optimum and its lower bound are supposed to be 0 in non-overcomplete ICA with super-Gaussian sources. By using the independence of $\boldsymbol{S}$ and letting $\boldsymbol{U} = (u_{il})$ be $\boldsymbol{W}\boldsymbol{A}$, cov4$\,(i,j)$ is transformed as follows (see [12]):

$$\mathrm{cov4}\,(i,j) = \sum_l u_{il}^2 u_{jl}^2 \gamma_l + 2\left(\sum_l u_{il}u_{jl}\right)^2 \tag{6}$$

where $\gamma_l = \mathrm{ave}\left(s_{lk}^4\right) - 3$ (the kurtosis of s_{lk}) and the variance of each s_{lk} is normalized to 1. If Eqs. (3) and (4) hold, each element u_{il} is given as

$$u_{il} = \begin{cases} \lambda_i \delta_{il} & l \leq N, \\ \sum_p w_{ip} b_{p(l-N)} & \text{otherwise} \end{cases} \tag{7}$$

where δ_{il} is the Kronecker delta. Then, Eq. (6) is rewritten as follows:

$$\begin{aligned} \mathrm{cov4}\,(i,j) = & \sum_{m=1}^{L-N} \left(\sum_p w_{ip} b_{pm}\right)^2 \left(\sum_p w_{jp} b_{pm}\right)^2 \gamma_{(N+m)} \\ & + 2\left(\sum_{m=1}^{L-N} \sum_p w_{ip} b_{pm} \sum_p w_{jp} b_{pm}\right)^2 . \end{aligned} \tag{8}$$

Besides, because the variance of each y_{ik} is normalized to 1, the following equation on u_{il} holds:

$$\sum_l u_{il}^2 = \lambda_i^2 + \sum_m \left(\sum_p w_{ip} b_{pm} \right)^2 = 1. \tag{9}$$

Now, the following assumptions are employed:

Assumption 1: w_{ip} and b_{pm} $(p = 1, \cdots, N)$ are assumed to be independent of each other. It approximately means that edge filters are assumed to be independent of each other. Under this assumption, the following approximation holds:

$$\sum_p w_{ip} b_{pm} \simeq \frac{\sum_p w_{ip} \sum_p b_{pm}}{N}. \tag{10}$$

Thus, Eq. (8) is approximated as follows:

$$\begin{aligned} \mathrm{cov4}\,(i,j) \simeq & \frac{\left(\sum_p w_{ip}\right)^2 \left(\sum_p w_{jp}\right)^2 \sum_m \left(\sum_p b_{pm}\right)^4 \gamma_{(N+m)}}{N^4} \\ & + \frac{2\left(\sum_p w_{ip}\right)^2 \left(\sum_p w_{jp}\right)^2 \left(\sum_m \left(\sum_p b_{pm}\right)^2\right)^2}{N^4}. \end{aligned} \tag{11}$$

Besides, the following equation is derived from Eq. (9):

$$\frac{\left(\sum_p w_{ip}\right)^2 \sum_m \left(\sum_p b_{pm}\right)^2}{N^2} = 1 - \lambda_i^2. \tag{12}$$

Assumption 2: b_{pm} and $\gamma_{(N+m)}$ $(m = 1, \cdots, L - N)$ are independent of each other. It means that the edge filters are independent of the kurtoses. Then, the following approximation holds:

$$\begin{aligned} \sum_m \left(\sum_p b_{pm}\right)^4 \gamma_{(N+m)} & \simeq \frac{\sum_m \left(\sum_p b_{pm}\right)^4 \sum_m \gamma_{(N+m)}}{L-N} \\ & \simeq \sum_m \left(\sum_p b_{pm}\right)^4 \bar{\gamma} \end{aligned} \tag{13}$$

where $\bar{\gamma}$ is the average of $\gamma_{(N+m)}$ over m.

Assumption 3: $\bar{\gamma}$ is positive. It means that almost all of the sources are super-Gaussian. Then, the following inequality on $\sum_m \left(\sum_p b_{pm}\right)^4 \bar{\gamma}$ holds:

$$\sum_m \left(\sum_p b_{pm}\right)^4 \bar{\gamma} \geq \frac{\left(\sum_m \left(\sum_p b_{pm}\right)^2\right)^2 \bar{\gamma}}{L-N}. \tag{14}$$

By Eqs. (12) and (14), a lower bound of Eq. (11) is given as

$$\mathrm{cov4}\,(i,j) \geq \left(1-\lambda_i^2\right)^2 \left(\frac{\bar{\gamma}}{L-N}+2\right). \tag{15}$$

Assumption 4: The model is highly overcomplete. It means that $L-N$ is sufficiently large. Then, the first term of Eq. (15) is negligible. Therefore, the following inequality holds:

$$\mathrm{cov4}\,(i,j) \geq 2\left(1-\lambda_i^2\right)\left(1-\lambda_j^2\right). \tag{16}$$

Assumption 5: The characteristics of all the sources are similar. It means that λ_i is approximately the same value λ irrespective of i. Then, Eq. (16) is approximated as

$$\mathrm{cov4}\,(i,j) \geq \alpha \tag{17}$$

where $\alpha = 2\left(1-\lambda^2\right)^2$.

Eq. (17) gives a common lower bound α of $\mathrm{cov4}\,(i,j)$ in the highly overcomplete model, where α is the only parameter to be set manually in the range of $[0, 2]$. Because the lower bound of $\mathrm{cov4}\,(i,j)$ is 0 in non-overcomplete cases, almost all of the previous cumulant-based ICA algorithms are supposed to achieve the optimal state where every $\mathrm{cov4}\,(i,j)$ is equal to 0. However, Eq. (17) suggests that such optimization may cause serious over-fitting problems under highly overcompleteness.

3 Method

Because the partial estimation does not use the orthogonal constraints, the non-orthogonal local pair optimization algorithm (which has been proposed in [12] by the authors) is employed as the basic algorithm. It is a variation of the Jacobi method, where $\mathrm{cov4}\,(i,j)$ was optimized with respect to each pair (i,j). By sweeping the simple optimization over all the pairs, the Jacobi method can optimize all of $\mathrm{cov4}\,(i,j)$'s. In consequence, the following algorithm with the lower bound α is proposed in this paper:

---

cumulant-based ICA algorithm using the lower bound α

1. (Initialization):
   (a) Center each row of $\boldsymbol{X}$ (set its mean to 0).
   (b) Let $\boldsymbol{W}$ be the $N \times N$ diagonal matrix which normalizes each row of $\boldsymbol{X}$ (set its variances to 1).
   (c) Let $\boldsymbol{Y}$ be $\boldsymbol{W}\boldsymbol{X}$.
2. (Sweeping all the pairs):
   For each pair (i,j) of the rows of $\boldsymbol{Y}$ with $\mathrm{cov4}\,(i,j) > \alpha$,
   (a) Whiten y_i and y_j by the Gram-Schmidt method and update $\boldsymbol{W}$ according to the whitening.

(b) Calculate the following 4-th order statistics on $\boldsymbol{Y}$: $\kappa_{iiii}^{Y} = \text{ave}\left(y_{ik}^{4}\right)$, $\kappa_{iiij}^{Y} = \text{ave}\left(y_{ik}^{3} y_{jk}\right)$, $\kappa_{iijj}^{Y} = \text{ave}\left(y_{ik}^{2} y_{jk}^{2}\right)$, $\kappa_{ijjj}^{Y} = \text{ave}\left(y_{ik}^{1} y_{jk}^{3}\right)$, and $\kappa_{jjjj}^{Y} = \text{ave}\left(y_{jk}^{4}\right)$.

(c) Calculate $\hat{\theta}_1$ and $\hat{\theta}_2$ by repeating the following equations alternatively until convergence:

$$\theta_1 = -\frac{\text{atan2}\left(-B\left(\theta_2\right), -A\left(\theta_2\right)\right)}{2} \text{ and } \theta_2 = -\frac{\text{atan2}\left(-B\left(\theta_1\right), -A\left(\theta_1\right)\right)}{2}. \tag{18}$$

where

$$A\left(\phi\right) = \kappa_{jjjj}^{Y}\left(\cos 2\phi - 1\right) - 2\kappa_{ijjj}^{Y} \sin 2\phi - 2\kappa_{iijj}^{Y} \cos 2\phi + 2\kappa_{iiij}^{Y} \sin 2\phi + \kappa_{iiii}^{Y}\left(\cos 2\phi + 1\right), \tag{19}$$

$$B\left(\phi\right) = \kappa_{ijjj}^{Y}\left(2 - 2\cos 2\phi\right) + 4\kappa_{iijj}^{Y} \sin 2\phi + \kappa_{iiij}^{Y}\left(2 + 2\cos 2\phi\right), \tag{20}$$

and $\text{atan2}\left(v, u\right)$ is the arctangent function of v/u with the range of $\left(-\pi, \pi\right]$.

(d) Transform linearly the i-th and j-th rows of $\boldsymbol{Y}$ and $\boldsymbol{W}$ by the 2×2 matrix $\left[\cos\theta_1, \sin\theta_1; \cos\theta_2, \sin\theta_2\right]$

3. Repeat the above Step 2 until the number of the iterations of sweepings reaches the maximum.

The linear transformation in each pair optimization minimizes κ_{iijj}^{Y} (see [12] for the details). It is crucial that the condition $\text{cov4}\left(i, j\right) > \alpha$ on each pair optimization is added in this paper. Though it does not guarantee that every $\text{cov4}\left(i, j\right) > \alpha$, it can simply avoid the over-fitting.

4 Experiments

Here, the proposed algorithm is applied to the edge extraction from natural scenes. 30000 samples of natural scenes of 12×12 pixels were used as $\boldsymbol{X}$. The lower bound α in the range of $[0, 2]$ was set to 0, 1, 1.5, and 2. If $\alpha = 0$, the method is approximately equivalent to the previous cumulant-based ICA methods with the non-overcomplete model. The maximum of iterations of sweepings was set to 5. For comparison, fast ICA algorithms with tanh and kurtosis [6] were applied to the same data. Fig. 1 shows the extracted components in the partial estimation of the mixing matrix $\boldsymbol{A}$. In Figs. 1-(a) ($\alpha = 0$) and 1-(b) ($\alpha = 1$), noisy patterns were dominant as in the previous cumulant-based ICA (Fig. 1-(f)). On the other hand, local edge filters could be extracted by $\alpha = 1.5$ in Fig. 1-(c) as in the highly nonlinear tanh-based ICA (Fig. 1-(e)). If $\alpha = 2$ (the maximum), no pair optimization was carried out (Fig. 1-(d)). Those results verify that the proposed cumulant-based ICA algorithm under the highly overcomplete assumption is useful for extracting edge filters if the lower bound α is given appropriately.

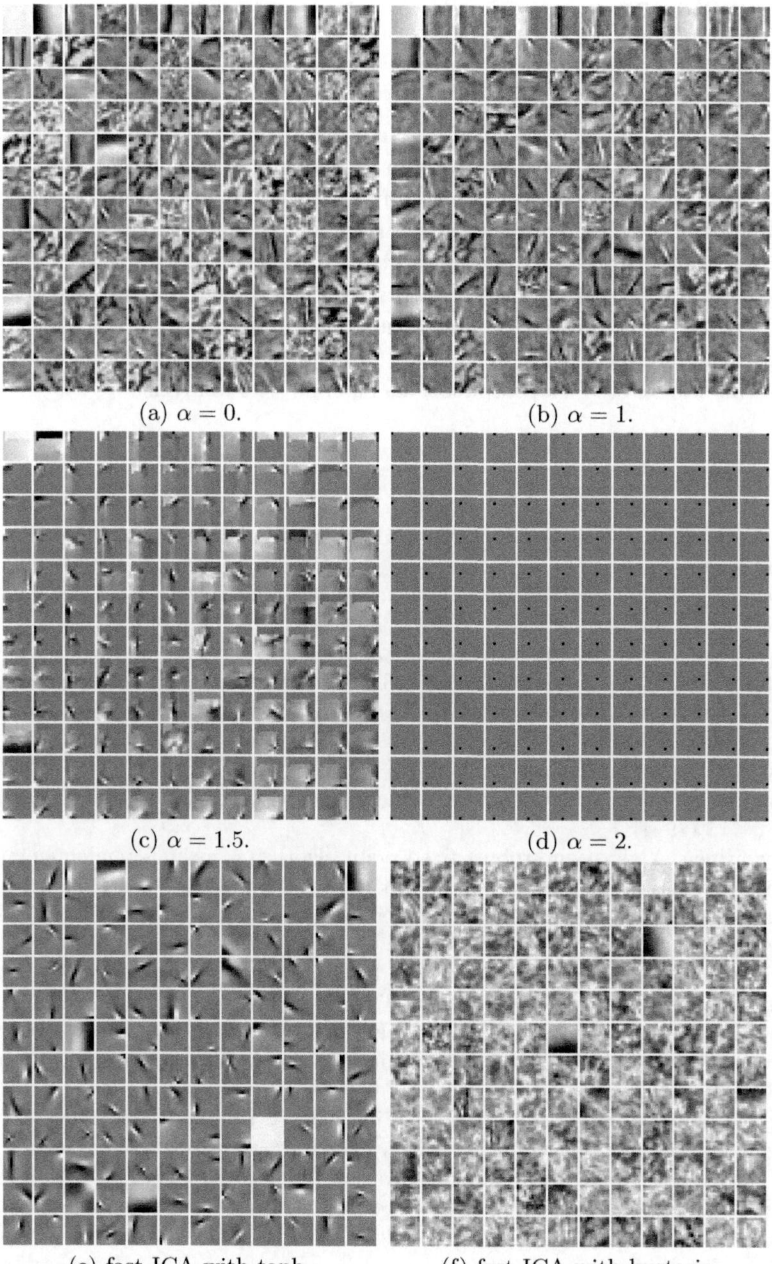

(a) $\alpha = 0$. (b) $\alpha = 1$. (c) $\alpha = 1.5$. (d) $\alpha = 2$. (e) fast ICA with tanh. (f) fast ICA with kurtosis.

Fig. 1. Extracted independent components from natural scenes: They show 144 extracted components from natural scenes of 12×12 pixels. (a) The proposed algorithm with the lower bound $\alpha = 0$. (b) $\alpha = 1$. (c) $\alpha = 1.5$. (d) $\alpha = 2$. (e) fast ICA with a highly nonlinear function tanh. (f) fast ICA with kurtosis.

5 Conclusion

In this paper, we proposed a highly overcomplete model for natural scenes and derived a lower bound of the 4-th order covariance under this model. Numerical experiments verified that the proposed cumulant-based ICA algorithm with the lower bound can extract edge filters in contrast to the existing ones. In order to determine the lower bound appropriately, we are now planning to consolidate the foundation of this model both theoretically and experimentally. Especially, the validity of the assumptions in Section 2.2 should be investigated further. One significant advantage of cumulants over highly nonlinear statistics is the facility in the optimization. So, we are also planning to develop an efficient ICA algorithm in image processing by this approach. Besides, we are planning to utilize this model for other applications such as text analysis.

References

1. Hyvärinen, A., Karhunen, J., Oja, E.: Independent Component Analysis. Wiley, Chichester (2001)
2. Cichocki, A., Amari, S.: Adaptive Blind Signal and Image Processing: Learning Algorithms and Applications. Wiley, Chichester (2002)
3. Bell, A.J., Sejnowski, T.J.: The "independent components" of natural scenes are edge filters. Vision Research 37(23), 3327–3338 (1997)
4. Hyvaerinen, A., Hurri, J., Hoyer, P.O.: Natural Image Statistics: A Probabilistic Approach to Early Computational Vision (Computational Imaging and Vision), 1st edn. Springer, Heidelberg (2009)
5. Comon, P.: Independent component analysis - a new concept? Signal Processing 36, 287–314 (1994)
6. Hyvärinen, A.: Fast and robust fixed-point algorithms for independent component analysis. IEEE Transactions on Neural Networks 10(3), 626–634 (1999)
7. Cardoso, J.F.: High-order contrasts for independent component analysis. Neural Computation 11(1), 157–192 (1999)
8. Olshausen, B.A., Field, D.J.: Sparse coding with an overcomplete basis set: a strategy employed by v1? Vision research 37(23), 3311–3325 (1997)
9. Hyvärinen, A., Hoyer, P.O., Inki, M.: Estimating overcomplete independent component bases for image windows. Journal of Mathematical Imaging and Vision 17, 139–152 (2002)
10. Teh, Y.W., Welling, M., Osindero, S., Hinton, G.E.: Energy-based models for sparse overcomplete representations. J. Mach. Learn. Res. 4, 1235–1260 (2003)
11. Matsuda, Y., Yamaguchi, K.: Semi-invariant function of jacobi algorithm in independent component analysis. In: Proceedings of IJCNN 2004, Budapest, Hungary, pp. 2147–2151. IEEE, Los Alamitos (2004)
12. Matsuda, Y., Yamaguchi, K.: A simple overcomplete ICA algorithm by non-orthogonal pair optimizations. In: Proceedings of IJCNN 2009, Atlanta, Georgia, USA, pp. 2027–2031. IEEE, Los Alamitos (2009)

Random Projection Tree and Multiview Embedding for Large-Scale Image Retrieval

Bo Xie[1], Yang Mu[1], Mingli Song[2], and Dacheng Tao[3]

[1] School of Computer Engineering, Nanyang Technological University
[2] College of Computer Science, Zhejiang University
[3] Centre for Quantum computation & Intelligent Systems, Faculty of Engineering and Information Technology, University of Technology, Sydney

Abstract. Image retrieval on large-scale datasets is challenging. Current indexing schemes, such as k-d tree, suffer from the "curse of dimensionality". In addition, there is no principled approach to integrate various features that measure multiple views of images, such as color histogram and edge directional histogram. We propose a novel retrieval system that tackles these two problems simultaneously. First, we use random projection trees to index data whose complexity only depends on the low intrinsic dimension of a dataset. Second, we apply a probabilistic multiview embedding algorithm to unify different features. Experiments on MSRA large-scale dataset demonstrate the efficiency and effectiveness of the proposed approach.

1 Introduction

Content-based image retrieval on large dataset has attracted a lot of attention recently with the development of the Internet. A lot of practical systems [1][2] and novel methods [3][4][5] have been developed which exploited different cues of the images and various indexing techniques. However, the size of the dataset and high dimensional image features present a great challenge for efficient image retrieval. Conventional indexing schemes, e.g. k-d tree [6], can effectively reduce the retrieval time by using a tree structure. Unfortunately, the complexity of k-d tree grows rapidly with the feature dimensions, making it ineffective for high dimensional features. On the same time, rich complementary information resides in images. For instance, the sunset scene is better recognized by color while bicycles may be easier classified by shape. Currently, there is no principled way to incorporate these different features, or multiple views for better image retrieval.

We propose a novel image retrieval system that can deal with the two problems, respectively. First, to efficiently index high dimensional data, random projection trees [7] are used in our system. Compared with k-d tree, random projection tree can automatically adapt to the data distribution, thus achieve a provably lower complexity. Second, to unify heterogeneous features, we present a multiview embedding algorithm based on a probabilistic framework. Compared with traditional manifold embelding methods [8][9], our algorithm can learn different weightings for different features adapted to their contribution to

K.W. Wong et al. (Eds.): ICONIP 2010, Part II, LNCS 6444, pp. 641–649, 2010.

Algorithm 1. Algorithm to Construct a Subtree

Procedure MakeTree(S)
if $|S| < MiniSize$ **then**
  return $(Leaf)$
else
  $Rule \leftarrow$ ChooseRule(S)
  $LeftTree \leftarrow$ MakeTree($\{x \in S : Rule(x) = \text{true}\}$)
  $RightTree \leftarrow$ MakeTree($\{x \in S : Rule(x) = \text{false}\}$)
  return $([Rule, LeftTree, RightTree])$
end if

Algorithm 2. Algorithm to Generate a Splitting Rule

Procedure ChooseRule(S) $\{c > 0$ is a parameter$\}$
if $\Delta^2(S) \leq c\Delta_A^2(S)$ **then**
  Choose a random unit direction p
  Sort projection values: $a(x) = p^T x, \forall x \in S$, generating a list $a_1 \leq a_2 \leq \cdots \leq a_n$
  Compute $\mu_{i1} = \frac{1}{i}\sum_{j=1}^{i} a_j$, $\mu_{i2} = \frac{1}{n-i}\sum_{j=i+1}^{n} a_j$ and

$$c_i = \sum_{j=1}^{i}(a_j - \mu_{i1})^2 + \sum_{j=i+1}^{n}(a_j - \mu_{i1})^2$$

  Find i that minimizes c_i and set $\theta = (a_i + a_{i+1})/2$
  $Rule(x) := p^T x \leq \theta$
else
  $Rule(x) := \|x - \text{mean}(S)\| \leq \text{median}\{\|z - \text{mean}(S)\| : z \in S\}$
end if
return $(Rule)$

the final embedding. In this way, useful information is promoted and noise is suppressed. We note that our approach is related to hierarchical fuzzy trees with fuzzy aggregation operators [10].

2 Random Projection Tree Indexing

In this section, we present random projection tree [7] for high dimensional feature indexing, whose complexity only depends on the low intrinsic dimension of data.

2.1 Random Projection Tree Construction

Given a set of sample points $A = \{x_i \in \mathbb{R}^d\}_{i=1}^n$, we first randomly choose a unit vector $p \in \mathbb{R}^d$ and project the set of data points to obtain $v_i = p^T x_i$. Then, we choose an appropriate threshold t that partitions the set into two sets $B = \{v_i \leq t\}$ and $C = \{v_i > t\}$. We go on to partition sets B and C until the desired level is reached. In this way, we have created a random projection tree from the sample set A.

Detailed description on how to make a tree and choose the splitting rule is summarized in Alg. 1 and Alg. 2.

After building a random projection tree, it can be used to find approximate nearest neighbors. For a given query image represented by a high dimensional feature vector $q \in \mathbb{R}^d$, we traverse down the tree to find the query's corresponding leaf node. All the data samples used to construct the tree that fall in the same leaf node are the query's nearest neighbors. Therefore, instead of comparing n data points in the dataset, we only need to compute L inner products with the stored random unit vectors and comparison operations, where L is the number of levels of the tree which only depends on the low intrinsic dimension of the data.

2.2 Analytical Properties

The random projection tree can be viewed as a vector quantization technique that represents the cells associated with the leaf nodes by their means. We evaluate the quality of approximation by the diameters and averaged diameters of the cells.

In detail, denote the set of data samples in a cell as S. Then the squared diameter of S and the averaged squared diameter are

$$\Delta^2(S) = \max_{x,y \in S} \|x - y\|^2, \tag{1}$$

$$\Delta_A^2(S) = \frac{1}{|S|^2} \sum_{x,y \in S} \|x - y\|^2 = \frac{2}{|S|} \sum_{x \in S} \|x - \text{mean}(S)\|^2, \tag{2}$$

respectively, where $|S|$ is the cardinality of the set.

Note that we use two different splitting strategies according to the relationship between $\Delta^2(S)$ and $\Delta_A^2(S)$ in Alg. 2. The first one projects data in a random direction and splits them so that the average squared interpoint distance is maximally reduced. We name it split by projection. The second strategy is designed for data with very different scales, for example, large amount of data samples center around the origin and others are spread out. This rule can effectively separate these two distinct clusters and we call it split by distance.

We go on to define a statistical concept of dimension. For set $S \subset \mathbb{R}^d$, denote the sorted eigenvalues of its covariance matrix by $\sigma_1^2 \geq \sigma_2^2 \geq \cdots \geq \sigma_d^2$. We say the local covariance dimension of set S is (r, ϵ) if $\sigma_1^2 + \cdots + \sigma_r^2 \geq (1-\epsilon)(\sigma_1^2 + \cdots + \sigma_d^2)$. That is, most of the variance is concentrated in a r-dimensional subspace.

With this concept, we now show the reduction rate of cell diameters in random projection trees.

Theorem 1. *There exist constants $0 < c_1, c_2, c_3 < 1$ with the following property. Suppose a random projection tree is built with set $S \subset \mathbb{R}^d$. Consider any cell C for which $S \cap C$ has local covariance dimension (r, ϵ), where $\epsilon < c_1$. Select a point $x \in S \cap C$ at random, and let C' be the cell that contains it at the next level down.*

If C is split by distance then

$$\mathbb{E}[\Delta(S \cap C')] \leq c_2 \Delta(S \cap C);$$

If C is split by projection then

$$\mathbb{E}[\Delta_A^2(S \cap C')] \leq (1 - \frac{c_3}{d})\Delta_A^2(S \cap C).$$

In both cases, the expectation is over the randomization in splitting C and the choice of $x \in S \cap C$.

The proof of this theorem is in [11]. From this theorem, we can see the expected average diameter of cells is halved in every $O(d)$ levels.

3 Multiview Embedding

The formulation of our algorithm is based on stochastic neighbor embedding [12,13]. The key idea is to construct a probability distribution based on the pairwise distances. In detail, denote the set of original data points $\{x_i \in \mathbb{R}^d\}_{i=1}^n$ and the normalized distance matrix $P \in \mathbb{R}_+^{n \times n}$, where $p_{ii} = 0$ and $\sum_{i,j} p_{ij} = 1$. The matrix P encodes pairwise distance relationship between samples and can be interpreted as a probability distribution. Similarly, we define the probability distribution Q in the output feature space, with each element

$$q_{ij} = \frac{\left(1 + \|y_i - y_j\|^2\right)^{-1}}{\sum_{k \neq l} \left(1 + \|y_k - y_l\|^2\right)^{-1}}, \tag{3}$$

where $y_i \in \mathbb{R}^r$ is the output data corresponding to x_i.

To incorporate data with multiple views, we assume the probability distribution on the original space is a linear combination of all the different views, i.e.

$$p_{ij} = \sum_{t=1}^{v} \alpha^t p_{ij}^t, \tag{4}$$

where α^t is the combination coefficient for view t and p_{ij}^t is the probability distribution under view t. The coefficient vector $\alpha = \left[\alpha^1, \cdots, \alpha^v\right]^T$ lies on a simplex in $\mathbb{R}^v$, denoted as $\alpha \in \Delta^v$. This is the same as $\alpha^t \geq 0, t = 1, \cdots, v$ and $\sum_{t=1}^{v} \alpha^t = 1$. Obviously, p_{ij} is a probability distribution since $\sum_{i \neq j} p_{ij} = \sum_t \alpha^t \sum_{i \neq j} p_{ij}^t = \sum_t \alpha^t = 1$.

Finally, to learn the data embedding, we minimize the Kullback Leibler (KL) divergence of the two probability distributions:

$$\min_{y_i, \alpha} f = \min_{y_i, \alpha} \sum_{i \neq j} p_{ij} \log \frac{p_{ij}}{q_{ij}}. \tag{5}$$

We adopt alternating optimization to solve this problem. We fix α and learn the output data points y_i for $i = 1, \ldots, n$. Then, we optimize over α with fixed y_i. These two steps are repeated until convergence or maximum number of iterations is reached.

Algorithm 3. Algorithm to Learn Weighting for Each View

initialize: $\gamma, \alpha_0, t > 0$
repeat
  Set $t_k := t$
  while $f\left(p_{t_k}(\alpha_{k-1})\right) > g_{t_k}\left(p_{t_k}(\alpha_{k-1}), \alpha_{k-1}\right)$ **do**
    Set $t_k := \gamma t_k$
  end while
  Set $t := t_k$ and $\alpha_k = p_{t_k}(\alpha_{k-1})$
until convergence

Optimization over y_i is solved by gradient descent. The gradient with respect to an output data point is

$$\frac{\partial f}{\partial y_i} = 4\sum_j (p_{ij} - q_{ij})(y_i - y_j)\left(1 + \|y_i - y_j\|^2\right)^{-1}. \tag{6}$$

Finding optimal α is a convex optimization problem and can be efficiently solved. Here, we use an extended gradient descent approach. In iteration k, we approximate the object function by its first order expansion plus a proximal term

$$g_{t_k}(\alpha, \alpha_{k-1}) = f(\alpha_{k-1}) + \langle \nabla f(\alpha_{k-1}), \alpha - \alpha_{k-1}\rangle + \frac{t_k}{2}\|\alpha - \alpha_{k-1}\|^2,\ s.t. \alpha \in \Delta^v, \tag{7}$$

where α_{k-1} is the iterate in iteration $k-1$ and t_k is the step size in iteration k.

We update the kth iterate to be

$$\alpha_k = p_{t_k}(\alpha_{k-1}). \tag{8}$$

where $p_{t_k}(\alpha_{k-1}) = \underset{\alpha \in \Delta^v}{\operatorname{argmin}} g_{t_k}(\alpha, \alpha_{k-1})$.

This is a simple quadratic objective function with linear constraint. It can be easily solved by many standard convex optimization toolkit.

In every iteration round, we make sure t_k satisfy the following inequality:

$$f(\alpha_k) \le g_{t_k}(\alpha_k, \alpha_{k-1}). \tag{9}$$

In implementation, we employ a varying step size strategy: we repeatedly increase the current t_k by a multiplicative factor $\gamma > 1$ until the condition in Eq. 9 is satisfied.

The algorithm is illustrated in Alg. 3. And it has been shown [14] that the convergence rate of this algorithm is $O(1/k^2)$.

4 Experiments

We have conducted experiments on the object category of the second edition MSRA MM (Microsoft Research Asia Multimedia) dataset [15], which was collected by using Microsoft Live Image Search. The object category contains

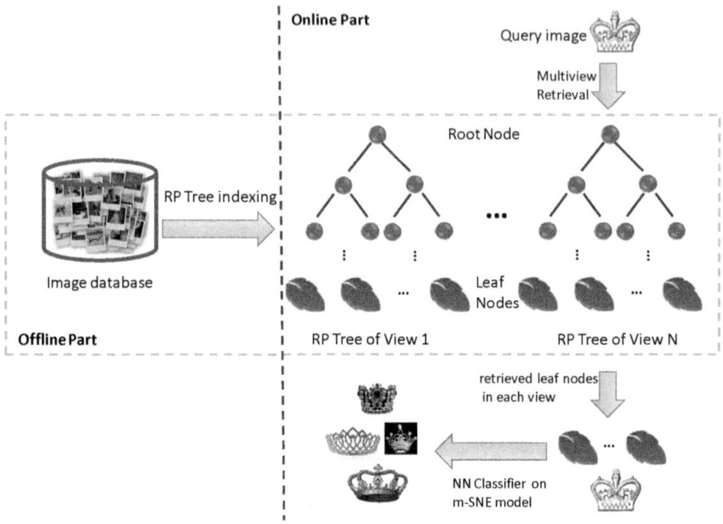

Fig. 1. System overview

257,060 images in total with 295 concepts. Each concept has around 800 images and these images were manually labeled with 2, 1 and 0 according to the relevance to the concept. In our experiments, we regarded relevant images, i.e., images with labels 1 or 2, as positive samples and irrelevant ones as negative samples.

We adopted seven low level features: 1) 225-D block-wise color moment; 2) 144-D color correlogram; 3) 75-D edge distribution histogram; 4) 7-D face feature; 5) 64-D HSV color histogram; 6) 256-D RGB color histogram and 7) 128-D wavelet texture feature. These features encode visual information from different perspectives, such as color, shape and texture. Also, we can see that most features have high dimensionality, which is difficult to build index with conventional methods.

For every view, we built a random projection tree of 12 levels to index the whole dataset, with each leaf node covering about one hundred images in the dataset. For retrieval, we randomly sampled 50 images per concept as query images. Every query image was associated with six to seven hundred retrieval candidates by pooling nearest neighbors retrieved by random projection trees from all views.

These retrieval candidates were diverse and provided complementary information since they were obtained from different views. To obtain a unified representation, we then performed multiview embedding to these retrieval candidates and the query image. Finally, we ranked the retrieval candidates by their Euclidean distances to the query image in the learned embedding. The whole process is illustrated in Fig. 1.

Precision results were calculated for top 10 to top 100 retrieval images. For comparison, we also evaluated other schemes: single view with best performance (BSV), mean performance of single views (MSV), concatenating all the features (CAT) and distributed approach of spectral embedding [16] (DSE). The proposed multiview learning method is denoted by m-SNE. In Fig. 2, we demonstrate the averaged precision curves for top 20 classes. It can be seen that our approach performed better than or comparable to other approaches.

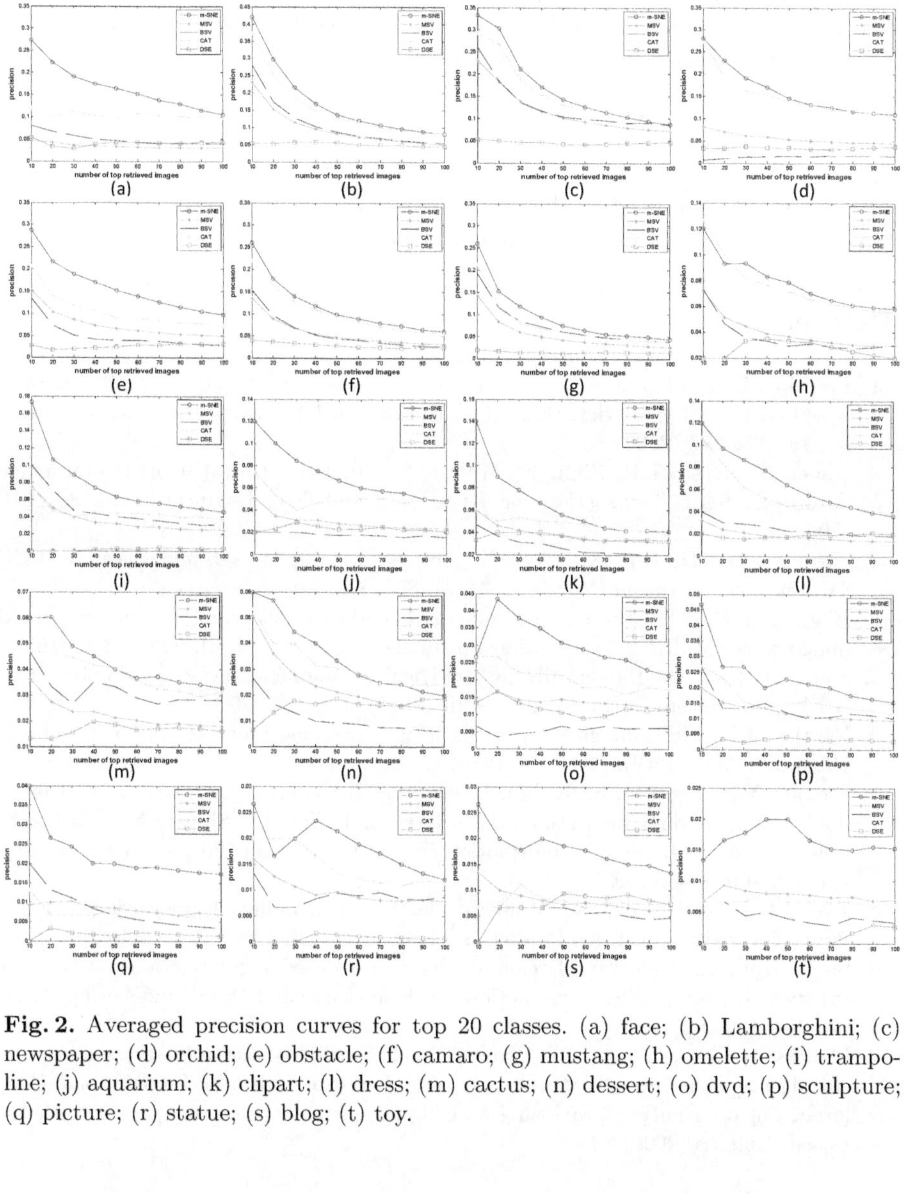

Fig. 2. Averaged precision curves for top 20 classes. (a) face; (b) Lamborghini; (c) newspaper; (d) orchid; (e) obstacle; (f) camaro; (g) mustang; (h) omelette; (i) trampoline; (j) aquarium; (k) clipart; (l) dress; (m) cactus; (n) dessert; (o) dvd; (p) sculpture; (q) picture; (r) statue; (s) blog; (t) toy.

5 Conclusion

We have introduced a novel system for large-scale image retrieval. We applied random projection trees to efficiently index data with high dimensions. This method can automatically adapt to the data distribution and thus its complexity only depends on the low intrinsic dimension of the data. To unify and exploit different features, we proposed a novel multiview embedding method based on a probabilistic framework. Compared with ad-hoc concatenation of features, our approach can learn different weightings for different views to promote important features and suppress noise. Experiments on MSRA dataset demonstrate the efficiency and effectiveness of our approach.

Acknowledgement

This work was supported by the Open Project Program of the State Key Lab of CAD&CG (Grant No. A1006), Zhejiang University.

References

1. Petrakis, E.G., Faloutsos, C., Lin, K.I.D.: Imagemap: An image indexing method based on spatial similarity. IEEE Transactions on Knowledge and Data Engineering 14, 979–987 (2002)
2. Natsev, A., Rastogi, R., Shim, K.: Walrus: A similarity retrieval algorithm for image databases. IEEE Transactions on Knowledge and Data Engineering 16, 301–316 (2004)
3. Tian, X., Tao, D., Hua, X.S., Wu, X.: Active reranking for web image search. IEEE Transactions on Image Processing 19(3), 805–820 (2010)
4. Bian, W., Tao, D.: Biased discriminant euclidean embedding for content-based image retrieval. IEEE Transactions on Image Processing 19(2), 545–554 (2010)
5. Song, D., Tao, D.: Biologically inspired feature manifold for scene classification. IEEE Transactions on Image Processing 19(1), 174–184 (2010)
6. Bentley, J.L.: Multidimensional binary search trees used for associative searching. Commun. ACM 18(9), 509–517 (1975)
7. Freund, Y., Dasgupta, S., Kabra, M., Verma, N.: Learning the structure of manifolds using random projections. In: Platt, J., Koller, D., Singer, Y., Roweis, S. (eds.) Advances in Neural Information Processing Systems 20, pp. 473–480. MIT Press, Cambridge (2008)
8. Zhou, T., Tao, D., Wu, X.: Manifold elastic net: a unified framework for sparse dimension reduction. Data Mining and Knowledge Discovery, 1–32 (2010)
9. Si, S., Tao, D., Geng, B.: Bregman divergence based regularization for transfer subspace learning. IEEE Transactions on Knowledge and Data Engineering 22(7), 929–942 (2010)
10. Mendis, B.S.U., Gedeon, T.D., Kóczy, L.T.: Flexibility and robustness of hierarchical fuzzy signature structures with perturbed input data. In: International Conference of Information Processing and Management of Uncertainty in Knowledge Based Systems (2006)

11. Dasgupta, S., Freund, Y.: Random projection trees and low dimensional manifolds. In: STOC 2008: Proceedings of the 40th annual ACM symposium on Theory of computing, pp. 537–546. ACM, New York (2008)
12. Hinton, G., Roweis, S.: Stochastic neighbor embedding. In: Advances in Neural Information Processing Systems 15, pp. 833–840. MIT Press, Cambridge
13. van der Maaten, L., Hinton, G.: Visualizing data using t-sne. Journal of Machine Learning Research 9, 2579–2605 (2008)
14. Nesterov, Y.: Introductory Lectures on Convex Optimization: A Basic Course (Applied Optimization), 1st edn. Springer, Netherlands
15. Li, H., Wang, M., Hua, X.S.: Msra-mm 2.0: A large-scale web multimedia dataset. In: IEEE International Conference on Data Mining Workshops, pp. 164–169 (2009)
16. Long, B., Yu, P.S., Zhang, Z.M.: A general model for multiple view unsupervised learning. In: Proceedings of the SIAM International Conference on Data Mining, Atlanta, Georgia, USA, pp. 822–833. SIAM, Philadelphia (2008)

Online Gesture Recognition for User Interface on Accelerometer Built-in Mobile Phones

BongWhan Choe, Jun-Ki Min, and Sung-Bae Cho

Department of Computer Science, Yonsei University
262 Seongsanno, Seodaemun-gu, Seoul 120-749, Korea
{bitbyte,loomlike}@sclab.yonsei.ac.kr, sbcho@cs.yonsei.ac.kr

Abstract. Recently, several smart phones are equipped with a 3D-accelerometer that can be used for gesture-based user interface (UI). In order to utilize the gesture UI for the real-time systems with various users, the diversity robust algorithm, yet having low training/recognition complexity, is required. Meantime, dynamic time warping (DTW) has shown good performance on the simple time-series pattern recognition problems. Since DTW is based on the template matching, its processing time and accuracy depend on the number of templates and their quality, respectively. In this paper, an optimized method for online gesture UI of mobile devices is proposed which is based on the DTW and modified k-means clustering algorithm. The templates, estimated by using the modified clustering algorithm, can preserve the time varying attribute while contain diversities of the given training patterns. The proposed method was validated on 20 types of gestures which are designed for the mobile contents browsing. The experimental results showed that the proposed method is suitable to the online mobile UI.

Keywords: Accelerometers, gesture recognition, dynamic time warping, k-means clustering.

1 Introduction

Gestures are natural and easy means as user interfaces (UI) because they have been employed in daily life. Recently developed mobile devices have built-in accelerometers of micro-electro-mechanical systems (MEMS) which allows the gesture inputs. Previous gesture UI was limited to pivoting the display or game controls based on simple motions such as a tilt and rotation [1].

In order to extend the gesture UI applications with complex motions, many pattern classification techniques like a hidden Markov models (HMMs) and dynamic time warping (DTW) have been studied. Especially, DTW has been highlighted on the mobile gesture recognition systems since it requires few training data and can be easily updated by modifying the matching templates [2]. However, the length of patterns and number of templates increase the processing time of the DTW. Let N be

K.W. Wong et al. (Eds.): ICONIP 2010, Part II, LNCS 6444, pp. 650–657, 2010.

the number of templates. The time complexity of the algorithm is represented as follows:

$$O(\sum_{i=1}^{N} m_i \cdot n) \tag{1}$$

where m_i is the length of the ith template and n is the length of the input sample. Therefore, it is difficult to apply DTW directly on the mobile devices because of the limitations of the resources.

In order to address the problem, this paper proposes an optimized method for the online gesture recognition based on the DTW and modified k-means clustering algorithm. It estimates N templates by using the clustering algorithm that reduces the number of templates, while it preserves the generality of the patterns. Moreover, the proposed method adaptively modifies the training templates by combining them with the user's input patterns.

2 Related Work

There were several studies on developing fast and accurate recognizer with more complex gestures. Pylvänäinen used continuous HMMs to recognize 10 gestures where he analyzed the correlation between accuracy and feature quantization [3]. Kela et al. quantized 3D-vectors acquired from a tri-axis accelerometer into 1D-symbols based on a k-means clustering algorithm and recognized them by using the HMMs [4]. Liu et al. proposed uWave that uses DTW as a matching algorithm, and validated it on the same gesture set with Kela's [2]. Wu et al. utilized SVMs with the frame-based features such as mean, energy, and correlation among the axes for classifying 12 gestures [5]. The previous methods described so far, however, depended on complicated algorithms that require high computational power, or did not consider the users' variations.

3 Gesture Recognition for Online User Interface

The proposed method is composed with three steps: pre-processing, local template estimation, and gesture matching. Fig. 1 shows the overall process of the proposed method. In the pre-processing phase, each gesture is segmented from the continuous input sequences based on the mean variations and the maximum values within a sliding window of 120ms that moves at a 60ms step. Here, we defined minimum length of the segment and assumed a movement shorter than the minimum as a noise. In pre-processing step, the quantization and smoothing are applied by averaging sequences within the sliding window. It reduces noises and the matching steps of the DTW algorithm. In order to minimize the affect of gravity, the initial acceleration (direction of the gravity) is subtracted from the input sequence. Remaining phases of the local template estimation and gesture matching are described in detail in the following subsections.

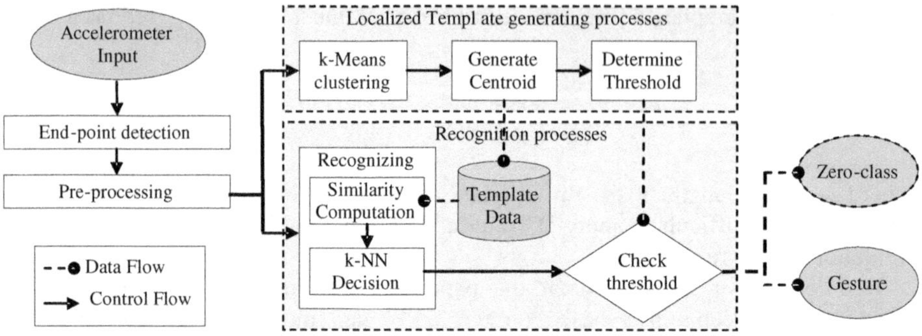

Fig. 1. The overview of the proposed method

3.1 Diversity Modeling with Local Templates

The accuracy of DTW based recognizer highly depends on the quality of its templates. Especially on the mobile environment, the templates should contain various patterns because of the dynamics of input gestures. If the hardware has enough computing resources, the accuracy of the recognizer could be increased by using huge amount of templates. Since the computational power of the mobile device, however, is limited, some representative templates have to be used which was chosen from the whole training set. Using *k*-means clustering algorithm is one of the simple and effective ways to model the local templates from the given set. Yet the acceleration data of varying length have to be converted into the fixed length patterns where some information would be lost.

In order to resolve this problem, we modified the distance measurement and averaging process of the *k*-means clustering algorithm. In other words, the proposed method calculates the distance between an input pattern and each centroid by using DTW algorithm, while it estimates the centroid based on the resized patterns of averaged length for each cluster. The modified clustering process is represented as follows:

1) Set randomly chosen samples as the clusters' initial centroids, $c_1, c_2, ..., c_k$.
2) For each sample, calculate distances to the centroids by using DTW and assign its cluster as the nearest one.
3) For each cluster, estimate the new centroid by:
   a) Calculate the average length of samples.
   b) Resize each sample to be the averaged length.
   c) Calculate the mean of the resized samples.
4) If centroids were changed, go back to step 2.

In step 3b, we resized samples based on a linear resampling algorithm.

3.2 Dynamic Time Warping-Based Gesture Recognition

DTW matches two time series patterns and gets the minimum distance in warped sequence from matching result [6-8]. In general, it is implemented by using a dynamic programming algorithm with two-dimensional matrix. Let $d(R_i, C_j)$ be the cost function of the two elements R_i and C_j where they are the ith and jth points of the input sequence R and the template sequence C, respectively. The distance between two sequences based on the 1st order warping method is calculated as follows:

$$D(i,j) = \min\begin{pmatrix} D(i,j-1) \\ D(i-1,j) \\ D(i-1,j-1) \end{pmatrix} + d(R_i, C_j) \quad (2)$$

where $D(0, 0) = 0$ and $D(i, 0) = D(0, j) = max$ (the maximum distance) [9]. In this paper, Euclidean distance was used as the cost function because of its simplicity and high accuracy.

The proposed method evaluates the similarities between an input sequence and the local templates using DTW algorithm. The input movement is then recognized as the most similar gesture based on k-nearest neighbor voting (k-NN) where $k \geq 1$. Since the unintended movement or outlier gestures can be sensed during the online gesture recognition, we defined certain threshold for each gesture. If the similarity is higher than the threshold, the gesture is accepted. Otherwise, it is rejected. Here the threshold is decided according to the average and standard deviation of the intra distances for each cluster.

4 Experimental Result

4.1 Gesture-Set Definition and System Implementation

We defined 20 gestures which can be intuitively used for browsing mobile contents. Fig. 2 shows the gestures and corresponding commands for a media player scenario. When a user wants to listen to music with the mobile device, he or she firstly unlocks the gesture interface function by shaking the device left-to-right (wakeup). After browsing play-lists by snapping gestures, the user select the song by shaking the device forward-to-backward (accept). While playing a song, the user can bounce up or down to control the sound volume. Finally, he or she locks the gesture interface function by shaking the device left-to-right in order to prevent unintended gesture inputs.

In order to validate the proposed method, the gestures were collected from four subjects of 25~29 years old over two days by using Samsung Omnia mobile phone (SCH-M490) with MS Windows Mobile 6.1 platform where the acceleration is sampled by 50Hz with +2g~-2g scale. On each day, the participants repeated each of the 20 gestures ten times (2,000 gestures in total) with an initial pose for browsing mobile contents. Fig. 3 shows a snapshot of the implemented application. It initializes the built-in accelerometer with the start button, and senses the acceleration while the user inputs the gesture. The proposed method automatically detects the start and end points of the gesture sequence, and identifies it. In order to confirm the recognition result, the application displays it at the top of the central box. Moreover, the user can register a new gesture or can modify existing templates manually with the setup menu.

Movement	Semantic label	Gesture class	UI scenario
 Snapping	Reciprocation	Snap left (NL) Snap right (NR) Snap forward (NF) Snap backward (NB)	Browse contents
 Bouncing	Reciprocation	Bounce up (BU) Bounce down (BD)	Volume up/down
 Rotating	Hold	Rotate landscape (RN) Rotate portrait (RP)	Display pivot
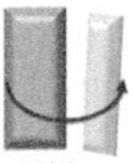 Tilting	Hold	Tilt left (LL) Tilt right (LR) Tilt forward (LF) Tilt backward (LB)	Fast forward /backward View tags
 Tapping	External impact	Tap left (TL) Tap front (TF) Tap right (TR) Tap back (TB) Tap top (TT) Tap bottom (TM)	Select/play corresponding contents
 Shaking	Shake	Shake left-right (SLR) Shake forward-backward (SFB)	Gesture UI lock/unlock

Fig. 2. Gesture set and scenarios

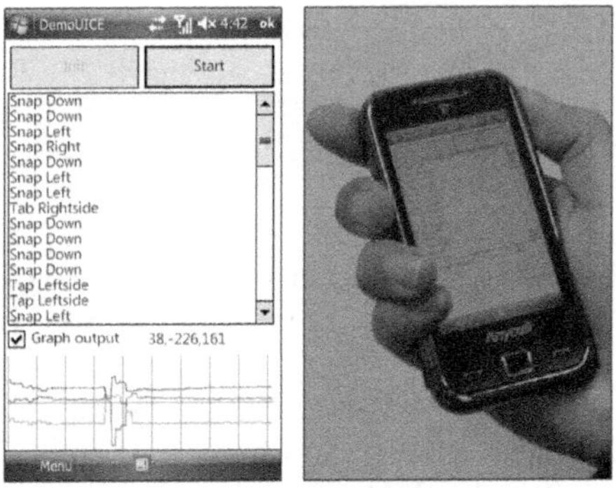

Fig. 3. The implemented gesture UI application

4.2 Gesture Recognition Accuracy Test

In order to evaluate the statistical accuracies of the proposed method, we implemented our algorithm on the PC platform of Core2 Duo 3GHz CPU as well. Here, we conducted experiments for the between-person variations and the day-to-day variations. Seven series of the experiments were considered according to the template generation approaches as follows: using whole learning data as templates (All), selecting three samples randomly for each gesture as its templates (Ran-3), generating three templates based on the k-means clustering with Euclidean distance measure (Euc-3) or with DTW distance measure (DTW-3), and using five templates for each gesture with the same fashion by Ran-5, Euc-5, and DTW-5. Here, all of them use DTW algorithm for the gesture matching stage. In case of Euc-3 and Euc-5, the training gesture sequences were resized into the same length before calculating the Euclidean distance. For the between-person variation test, 30 samples were available as the templates per class for the 'All', while 20 samples were used for the day-to-day variation test.

In order to calculate the recognition accuracy, we manually labelled the collected gestures which were automatically segmented as explained in Section 3. Accuracy and precision were calculated as follows:

$$\text{Accuracy} = \#\text{correctly recognized gestures} / \#\text{gestures} \quad (3)$$

$$\text{Precision} = \#\text{correctly recognized gestures} / \#\text{recognized gestures} \quad (4)$$

As shown in Fig. 4 and Fig. 5, DTW-5 achieved comparative accuracy against to All-case although the DTW-5 uses fewer templates than it. Moreover, the proposed method yielded better performance than the other approaches which use the same number of templates with the proposed method. Fig. 6 shows the processing time of the algorithms which represents the efficiency of the proposed method.

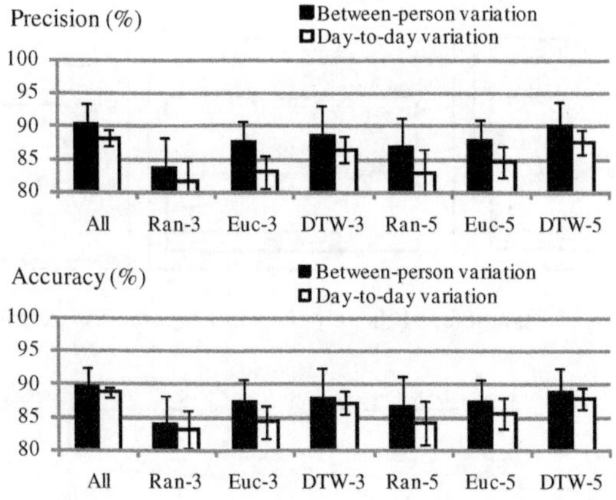

Fig. 4 & 5. Averaged precision and accuracy for each case

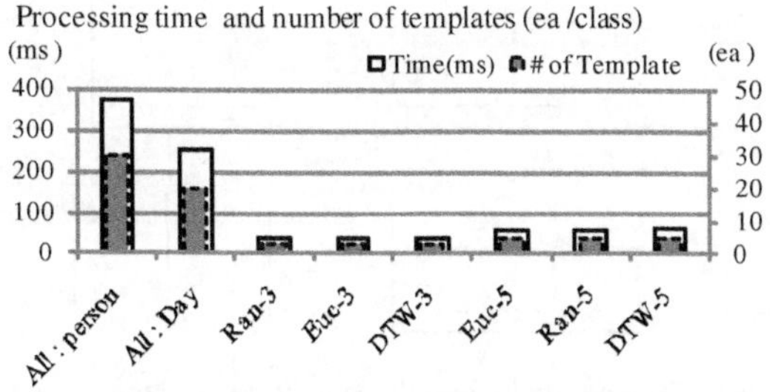

Fig. 6. Recognition time per sample

4.3 Template Analysis

We analysed the estimated templates and their matching results. Fig. 7 and Fig. 8 show the examples of the templates and input samples of the tapping-top gesture (the most difficult type to identify correctly among the given gesture set), respectively. Since the gestures are conducted under personal variations, some of them such as Fig. 8(d), (g), (i) and (j) are hard to recognize by using conventional algorithms like Euc-3 and Ran-3. On the other hand, as shown in error table of Fig. 8, the proposed method (DTW-3) recognizes the new input gestures correctly since it includes various patterns of templates as illustrated in Fig. 7.

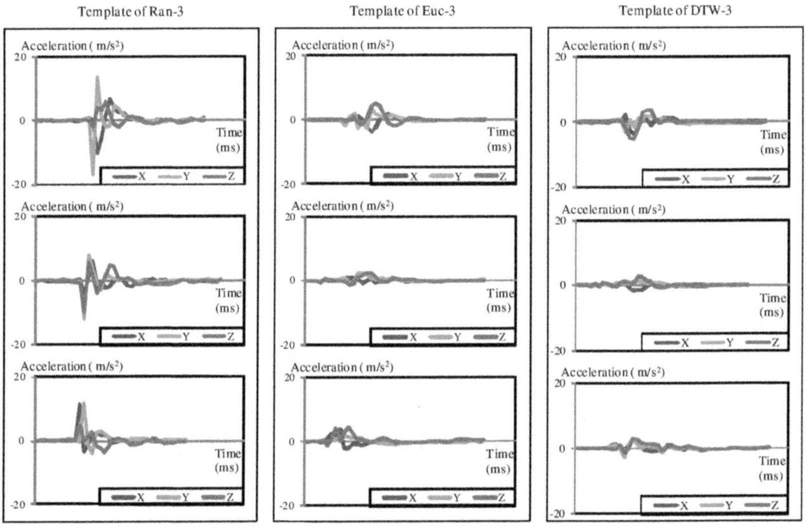

Fig. 7. Examples of the templates for tapping-top gesture

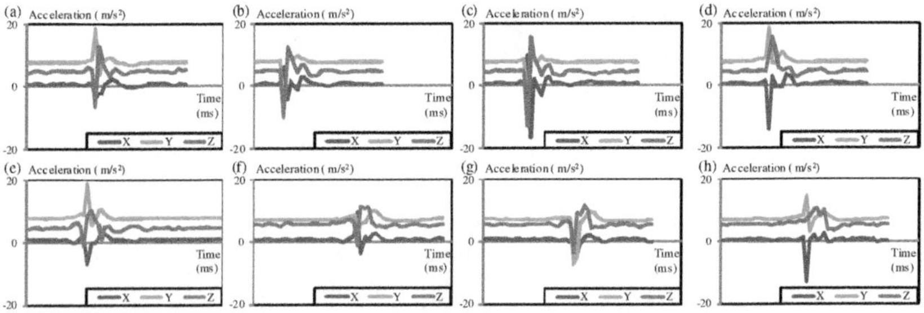

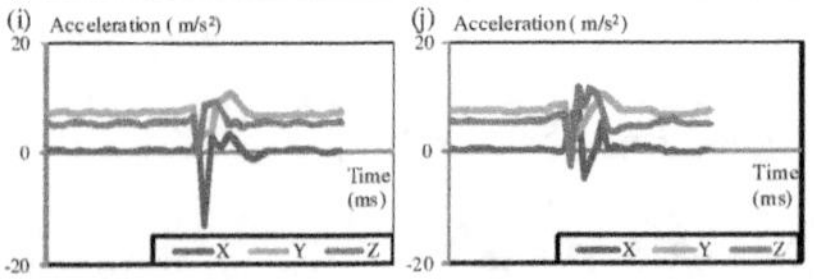

Error table of the result ; Ran-3 and Euc-3

Test Sample	Ran-3	Euc-3	Test Sample	Ran-3	Euc-3	Test Sample	Ran-3	Euc-3
a	X		b			c	X	
d	X	X	e			f		X
g	X	X	h	X		i	X	X
j	X	X						

Fig. 8. Examples of the test samples belong to the tapping-top gesture (a~j), and the corresponding error table for the templates represented in Fig. 7.

5 Conclusion

In the real world, the gestures have many dynamics such as variations and ambiguities. In order to address the problem, this paper proposed localized templates for gesture recognition algorithm. The localized templates are estimated by using modified k-means clustering algorithm where DTW is used for the distance measure which preserves the characteristics of the time series patterns. By using DTW as the gesture matching algorithm, the proposed system showed better performance than conventional algorithms in terms of its accuracy and processing cost. Since the performance of a clustering algorithm depends on the number of clusters, the localized template with a cluster validity method has to be investigated as our future work. In order to increase the accuracy, personalization based on template adaptation will be conducted as well. Finally, comparison experiments with other existing recognition approaches such as HMM and SVM have to be performed.

Acknowledgements. This research was supported by the Converging Research Center Program through the Converging Research Headquarter for Human, Cognition and Environment funded by the Ministry of Education, Science and Technology (2009-0093676). It was also supported by the Original Technology Research Program for Brain Science through the National Research Foundation of Korea (NRF) funded by the Ministry of Education, Science and Technology (2010-0018948).

References

1. Dachselt, R., Buchholz, R.: Natural throw and tilt interaction between mobile phones and distant displays. In: Proc. of the 27th Int. Conf. Extended Abstracts on Human Factors in Computing Systems, pp. 3253–3258 (2009)
2. Liu, J., Wang, Z., Zhong, L., Wickramasuriya, J., Vasudevan, V.: uWave: Accelerometer-based personalized gesture recognition and its applications. Pervasive and Mobile Computing 5(6), 657–675 (2009)
3. Pylvänäinen, T.: Accelerometer based gesture recognition using continuous HMMs. In: Marques, J.S., Pérez de la Blanca, N., Pina, P. (eds.) IbPRIA 2005. LNCS, vol. 3522, pp. 639–646. Springer, Heidelberg (2005)
4. Kela, J., Korpipaa, P., Mantyjarvi, J., Kallio, S., Savino, G., Jozzo, L., Marca, S.D.: Accelerometer-based gesture control for a design environment. Personal and Ubiquitous Computing 10, 285–299 (2006)
5. Wu, J., Pan, G., Zhang, D., Qi, G., Li, S.: Gesture recognition with a 3-D accelerometer. In: Zhang, D., Portmann, M., Tan, A.-H., Indulska, J. (eds.) UIC 2009. LNCS, vol. 5585, pp. 25–38. Springer, Heidelberg (2009)
6. Sakoe, H., Chiba, S.: Dynamic programming algorithm optimization for spoken word recognition. IEEE Transactions on Acoustics, Speech and Signal Processing 26(1), 43–49 (1978)
7. Salvador, S., Chan, P.: Fast DTW: Toward accurate dynamic time warping in linear time and space. Intelligent Data Analysis 11(5), 561–580 (2007)
8. Capitani, P., Ciaccia, P.: Warping the time on data streams. Data and Knowledge Engineering 62(3), 438–458 (2007)
9. Wobbrock, J.O., Wilson, A.D., Li, Y.: Gestures without libraries, toolkits or training: a $1 recognizer for user interface prototypes. In: Proc. of the 20th Annual ACM Symposium on User interface Software and Technology, vol. 20, pp. 159–168 (2007)

Constructing Sparse KFDA Using Pre-image Reconstruction

Qing Zhang[1] and Jianwu Li[2,*]

[1] Institute of Scientific and Technical Information of China,
Beijing 100038, China
[2] Beijing Key Lab of Intelligent Information Technology,
School of Computer, Beijing Institute of Technology,
Beijing 100081, China
zqbit@tom.com, ljw@bit.edu.cn

Abstract. Kernel Fisher Discriminant Analysis (KFDA) improves greatly the classification accuracy of FDA via using kernel trick. However, the final solution of KFDA is expressed as an expansion of all training examples, which seriously undermines the classification efficiency, especially in real-time applications. This paper proposes a novel framework to construct sparse KFDA using pre-image reconstruction. The proposed method (PR-KFDA) appends greedily the pre-image of the residual between the current approximate model and the original KFDA model in feature space with the local optimal Fisher coefficients to acquire sparse representation of KFDA solution. Experimental results show that PR-KFDA can reduce the solution of KFDA effectively while maintaining comparable test accuracy.

Keywords: KFDA, pre-image reconstruction, sparse approximation framework, kernel method.

1 Introduction

Fisher Discriminant Analysis (FDA) is a classical classification method, but only fits for solving linear classification tasks. To adapt to nonlinear cases, the kernel version of FDA (KFDA) is introduced by S. Mika et al. in 1999 [1]. The main idea in kernel-based methods, often called "kernel trick", is to map examples into a high dimensional feature space and then reformulate the problem into dot product form substituted by Mercer kernels [2].

However, similar to most kernel methods, all the training patterns are responsible for constructing the final expression of KFDA, which prevents KFDA from solving massive testing tasks and real-time stream classifying occasions [3], which need fast response. Previous attempts at addressing this issue pay much attention to the training phase using low rank matrices greedy approximation [4] [5] [6]. Also, in previous related work [7], C.J.C. Burges et al. proposes a Reduced-Set method to approximate the final kernel expansion of SVM and then B. Schölkopf et al. [8] extends [7] using

* Corresponding author.

K.W. Wong et al. (Eds.): ICONIP 2010, Part II, LNCS 6444, pp. 658–667, 2010.

fix-point iteration method for RBF kernel function instead of standard mathematical programming approaches. However, as mentioned in [9], these methods are easy to suffer numerical instability and local minima. In this paper, we propose a novel framework for constructing sparse KFDA using pre-image reconstruction, which incorporates multidimensional scaling (MDS) based method [10] and local fisher coefficient generating strategy into the basic procedure in [8] applied for SVM.

The rest of the paper is organized as follows: In section 2, KFDA is described briefly. The novel sparse KFDA method is proposed in section 3, followed by the experiments in section 4. The last section concludes this paper.

2 KFDA

Given a data set $X = \{x_1, \ldots, x_n\}$ containing n examples $x_i \in \mathbb{R}^d$, where n_1 examples belong to positive class denoted by $X_1 = \{x_1^1, \ldots, x_{n_1}^1\}$ and n_2 negative examples as $X_2 = \{x_1^2, \ldots, x_{n_2}^2\}$. The input data space $\mathbb{R}^d$ can be mapped into a high dimensional feature space F by a nonlinear mapping Φ. Then, KFDA constructs the Fisher criterion in F by maximizing

$$J(w) = \frac{w^T S_b^\Phi w}{w^T S_w^\Phi w}, \tag{1}$$

where $w \in span\{\Phi(x_1), \ldots, \Phi(x_n)\} \subset F$, i.e.,

$$w = \sum_{i=1}^{n} \alpha_i \Phi(x_i), \tag{2}$$

$S_b^\Phi = \left(m_1^\Phi - m_2^\Phi\right)\left(m_1^\Phi - m_2^\Phi\right)^T$ and $S_w^\Phi = \sum_{i=1,2} \sum_{x \in X_i} \left(\Phi(x) - m_i^\Phi\right)\left(\Phi(x) - m_i^\Phi\right)^T$ represent between-class scatter matrix and within-class scatter matrix respectively defined in F, where $m_i^\phi = \left(1/n_i\right) \sum_{x \in X_i} \Phi(x)$. The kernel trick is employed to compute the inner product $\langle \Phi(x), \Phi(y) \rangle$ in feature space by a kernel function $k(x, y)$ in original input space, e.g., RBF kernel function,

$$k(x, y) = \exp\left(\frac{-\|x - y\|^2}{2\sigma^2}\right). \tag{3}$$

Thus equation (1) can be written as

$$J(\alpha) = \frac{\alpha^T M \alpha}{\alpha^T N \alpha}, \tag{4}$$

where $M = (M_1 - M_2)(M_1 - M_2)^T$, $(M_i)_j = (1/n_i)\sum_{k=1}^{n_i} k(x_j, x_k^i)$, $\alpha = [\alpha_1, \ldots, \alpha_n]^T$, $N = \sum_{j=1,2} K_j (I - 1_{n_j}) K_j^T$, $(K_j)_{lm} = k(x_l, x_m^j)$, $l = 1, 2, \ldots, n$; $m = 1, 2, \ldots, n_j$, I is the identity matrix and 1_{n_j} the matrix with all entries $1/n_i$. There are several equivalent ways to address equation (4), e.g., solving the general-ized eigenproblem $M\alpha = \lambda N\alpha$ and then selecting the eigenvector α with the leading eigenvalue λ or computing $\alpha = N^{-1}(M_1 - M_2)$ [1] directly. Finally, the projection of a test pattern onto the KFDA model is computed by

$$\langle w, \Phi(x) \rangle = \sum_{i=1}^{n} \alpha_i k(x_i, x). \tag{5}$$

As the dimension of the feature space is usually much higher than the number of training samples [11], some form of regularization [12] is necessary by adding μI to N, where $0 \le \mu \le 1$.

3 A Novel Framework for Constructing Sparse KFDA

Now we suppose $\tilde{w}$ is our sparse solution of KFDA in F, which is expressed as

$$\tilde{w} = \sum_{j=1}^{k+1} \beta_j \Phi(\tilde{x}_j), \tag{6}$$

and compute the residual for $(k+1)$-th iteration

$$R_{k+1}^{\Phi} = w - \tilde{w} = \sum_{i=1}^{n} \alpha_i \Phi(x_i) - \sum_{j=1}^{k} \beta_j \Phi(\tilde{x}_j), \tag{7}$$

where the w is the original KFDA solution, as shown in (2) and $k \ll n$. Finding the $(k+1)$-th basis in (6) is equivalent to computing the pre-image of R_{k+1}^{Φ} in F via an inverse mapping $\widetilde{\Phi}^{-1}$ from feature space to input space,

$$\tilde{x}_{k+1} = \widetilde{\Phi}^{-1}\left(R_{k+1}^{\Phi}\right) = \widetilde{\Phi}^{-1}\left(\sum_{i=1}^{n} \alpha_i \Phi(x_i) - \sum_{j=1}^{k} \beta_j \Phi(\tilde{x}_j)\right), \tag{8}$$

then the revised (6) is

$$\tilde{w} = \sum_{j=1}^{k+1} \beta_j \Phi\left(\widetilde{\Phi}^{-1}\left(\sum_{i=1}^{n} \alpha_i \Phi(x_i) - \sum_{j=1}^{k} \beta_j \Phi(\tilde{x}_j)\right)\right). \tag{9}$$

The equation (9) above is obtained by a greedy iterative procedure, which is divided into two steps for each iteration in this framework: (1) "Basis-Generating-Step", for finding pre-images of residuals in F between the current approximate sparse model and original KFDA model. (2) "Coefficient-Generating-Step", for producing the current sparse model in the subspace spanned by approximated pre-images of the residuals in F.

In addition, this framework is capable of addressing any kernel-induced subspace vector approximation problems with any state-of-the-art techniques for pre-image computing and specified prior knowledge for the generation of effective coefficients.

3.1 Basis-Generating-Step

In this section, we pay much attention to how to find the inverse mapping $\tilde{\Phi}^{-1}$ in equation (9). To address this issue, recently, [10] [13] provide non-iterative algebra techniques using prior knowledge of training data for KPCA denoising problem [14], which avoid the numerical instability or local minima appeared in [8]. This paper mainly employs the MDS-based algebra approach [10], which directly finds locations of pre-images based on distance constraints in original space and feature space. If necessary, the fixed-point iteration [8] is further performed to refine the solution obtained by the MDS-based method.

According to [10], the distance constrained information by R_{k+1}^{Φ} and its i-th neighbor $\Phi(z_i)$ in F is computed firstly regarding our problem in $(k+1)$-th iteration as following:

$$\begin{aligned}
&\tilde{d}^2(R_{k+1}^{\Phi},\Phi(z_i))\\
&=\left\|R_{k+1}^{\Phi}-\Phi(z_i)\right\|^2\\
&=\left\langle\left(\sum_{i=1}^{n}\alpha_i\Phi(x_i)-\sum_{j=1}^{k}\beta_j\Phi(\tilde{x}_j)\right),\left(\sum_{i=1}^{n}\alpha_i\Phi(x_i)-\sum_{j=1}^{k}\beta_j\Phi(\tilde{x}_j)\right)\right\rangle\\
&-2\left\langle\left(\sum_{i=1}^{n}\alpha_i\Phi(x_i)-\sum_{j=1}^{k}\beta_j\Phi(\tilde{x}_j)\right),\Phi(z_i)\right\rangle+\left\langle\Phi(z_i),\Phi(z_i)\right\rangle\\
&=\sum_{i_1=1}^{n}\sum_{i_2=1}^{n}\alpha_{i_1}\alpha_{i_2}k(x_{i_1},x_{i_2})-2\sum_{j=1}^{k}\sum_{i=1}^{n}\beta_j\alpha_i k(\tilde{x}_j,x_i)\\
&+\sum_{j_1=1}^{k}\sum_{j_2=1}^{k}\beta_{j_1}\beta_{j_2}k(\tilde{x}_{j_1},\tilde{x}_{j_2})-2\sum_{i=1}^{n}\alpha_i k(x_i,z_i)\\
&+2\sum_{j=1}^{k}\beta_j k(\tilde{x}_j,z_i)+k(z_i,z_i)
\end{aligned} \quad . \tag{10}$$

Subsequently, we can obtain the distance relationship in input space by converting (9) in F to input space under the connection between two spaces for BRF kernel

$$d^2(R_{k+1}^{\Phi}, z_i) = -2\sigma^2 \ln\left(1 - \frac{1}{2}\tilde{d}^2(R_{k+1}^{\Phi}, z_i))\right). \tag{11}$$

Then according to the multidimensional scaling (MDS) principle, the approximate location of the pre-image can be computed by the known coordinates system in input space with the similar distance constrains information in feature space. More specifically, we construct the neighbor matrix $Z = [z_1, \ldots, z_n]$ in input space and then use centering matrix H to deal with it. Then by performing singular value decomposition (SVD) on ZH , firstly,

$$ZH = [E_1, E_2]\begin{bmatrix} \wedge_1 & 0 \\ 0 & 0 \end{bmatrix}\begin{bmatrix} V_1^T \\ V_2^T \end{bmatrix} = E_1 \wedge_1 V_1^T = E_1\Gamma, \tag{12}$$

the new coordinates system in input space is established based on the column vectors of orthogonal matrix E_1. Thus, we can compute the pre-image by satisfying the distance constrains relationship [15] in both spaces [16],

$$\left\|\widetilde{\Phi}^{-1}(R_{k+1}^{\Phi}) - z_i\right\|^2 \approx \left\|R_{k+1}^{\Phi} - z_i^{\phi}\right\|^2 = d^2(R_{k+1}^{\Phi}, z_i^{\phi}), \tag{13}$$

where $z_i^{\phi} = \Phi(z_i)$, $\widetilde{\Phi}^{-1}(R_{k+1}^{\Phi}) = E_1 c_{R^{\phi}} + \overline{z}$, $c_{R^{\phi}}$ is the coordinates in column space of E_1, $\overline{z}$ is the mean of the columns of Z. Then, the location of the pre-image can be derived (in detail, see [10]),

$$\tilde{x}_{k+1} = \widetilde{\Phi}^{-1}(R_{k+1}^{\Phi}) = \frac{1}{2} E_1 \wedge_1^{-1} v_1^T \left(d_0^2 - d^2\right)\overline{z} \tag{14}$$

where $d_0^2 = \left[\left\|c_1\right\|^2, \ldots, \left\|c_k\right\|^2\right]$, $d^2 = \left[d^2(R_{k+1}^{\Phi}, z_1), \ldots, d^2(R_{k+1}^{\Phi}, z_n)\right]$, and $\Gamma = \wedge_1 V_1^T = [c_1, \ldots, c_k]$.

If necessary, let the solution in (14) obtained by MDS-based approach as the starting point $\tilde{x}_{k+1}^1$, then carry out the fix-point iteration [8] to find the better solution by letting $\nabla_{\tilde{x}_{k+1}^*} \left\langle R_{k+1}^{\Phi}, \Phi(\tilde{x}_{k+1}^*)\right\rangle = 0$, and derive

$$\tilde{x}_{k+1}^{n+1} = \frac{\sum_{i=1}^{n} \alpha_i k(\left\|x_i - \tilde{x}_{k+1}^n\right\|^2) x_i - \sum_{j=1}^{k} \beta_j k(\left\|\tilde{x}_j - \tilde{x}_{k+1}^n\right\|^2) \tilde{x}_j}{\sum_{i=1}^{n} \alpha_i k(\left\|x_i - \tilde{x}_{k+1}^n\right\|^2) - \sum_{j=1}^{k} \beta_j k(\left\|\tilde{x}_j - \tilde{x}_{k+1}^n\right\|^2)}. \tag{15}$$

3.2 Coefficient-Generating-Step

After the $(k+1)$-th basis $\tilde{x}_{k+1}$ obtained in (14) or (15), the simple and convenient way [17] to generate coefficients in $(k+1)$-th iteration is only to update the new coefficient by minimizing $\left\| R_k^{\Phi} - \beta_{k+1}\Phi(\tilde{x}_{k+1}) \right\|^2$ and letting

$$\frac{\partial \left\| R_k^{\Phi} - \beta_{k+1}\Phi(\tilde{x}_{k+1}) \right\|^2}{\partial \beta_{k+1}} = 0, \tag{16}$$

then we can get

$$\beta_{k+1} = \frac{\left\langle R_k^{\Phi}, \Phi(\tilde{x}_{k+1}) \right\rangle}{\left\| \Phi(\tilde{x}_{k+1}) \right\|^2}. \tag{17}$$

In fact, for optimizing approximated KFDA model, minimizing the R_{k+2}^{Φ} in F for all previous coefficients will be more effective after a new basis is appended. The equation (9) can be formulated by a matrix form and we can obtain

$$\begin{bmatrix} \Phi_1(\tilde{x}_1)^T \\ \cdots \\ \Phi_{k+1}(\tilde{x}_{k+1})^T \end{bmatrix}^T \begin{bmatrix} \beta_1 \\ \cdots \\ \beta_{k+1} \end{bmatrix} = w. \tag{18}$$

The optimal coefficients $[\beta_1 \quad \ldots \quad \beta_{k+1}]^T$ can be acquired by the least square method, such that the new R_{k+2}^{Φ} is orthogonal to the current constructed sparse model,

$$\begin{bmatrix} \beta_1 \\ \cdots \\ \beta_{k+1} \end{bmatrix} = \begin{bmatrix} k(\tilde{x}_1, \tilde{x}_1) & \ldots & k(\tilde{x}_1, \tilde{x}_{k+1}) \\ \cdots & \cdots & \cdots \\ k(\tilde{x}_{k+1}, \tilde{x}_1) & \ldots & k(\tilde{x}_{k+1}, \tilde{x}_{k+1}) \end{bmatrix}^{-1} \begin{bmatrix} \sum_{i=1}^{n} \alpha_i k(x_i, \tilde{x}_1) \\ \cdots \\ \sum_{i=1}^{n} \alpha_i k(x_i, \tilde{x}_{k+1}) \end{bmatrix}. \tag{19}$$

However, due to the pre-image of a new residual is not an exact solution [9], the procedure using (18) does not work when two sparse bases are linearly dependent, which deprive the further basis-generating step of acquiring new pre-image of a smaller residual. Moreover, for our work, the ultimate goal is the approximately optimal projection direction. Thus, we consider projecting the weight vector $\tilde{w}$ into the approximate subspace spanned by current sparse bases and then compute the coefficients by maximizing Fisher criterion in (4) in original feature space spanned by all training data for each iteration. This approach clearly guarantees the most approximate angle between the current sparse model and KFDA model by the optimal Fisher coefficients locally and the further opportunity to search a smaller residual in next iteration heuristically, directed by the whole KFDA model information globally.

4 Experiments

4.1 Experimental Settings

In the following experiments, we compare the performance of PR-KFDA on 6 different datasets from UCI Repository [18] with KFDA and MPKFDA (T. Diethe et al., 2009) [5] as the sparse version of Regularized-KFDA by the idea of matching pursuit and low rank matrices approximation. The Gaussian RBF kernel is only used due to the convergence restriction [14] in fix-point method (15) and the procedure on each dataset is repeated 50 times over 50 different random splits of the dataset into "training" and "testing" parts as specified in Table 1.

In Table 1, there are seven columns including Data Set, Dim, N-OPT, N-Training, N-Testing, N-Neighbors, RBF- σ , which represent different binary classification problems, the rounds of fix-point iteration, the numbers of training and testing patterns, the neighbors for computing of pre-image in both spaces, the parameter σ for BRF kernel in (3), respectively.

The parameter σ for BRF kernel is selected by cross-validation, which is a little upper than the mean value to make the distance sensitive in both spaces for pre-image solving. Since there are no standard criteria for choosing optimal N-basis, the iteration of generating basis is stopped when the specified maximal iteration number or the given error threshold is reached. Here, we adopt the former way and set the same N-Basis listed in Table 2, which is the corresponding basis number of the optimal results for PR-KFDA within maximal iteration, to MPKFDA in order to compare the performance at the identical sparsity conditions. The regularize-tion variable μ in matrix N is chosen to be 0.001 and 0.0000001 for training and fisher coefficient generating phase, respectively.

Table 1. Experimental Settings for 6 UCI Benchmark datasets

Data Set	Dim	N-OPT	N-Training	N-Testing	N-Neighbors	RBF- σ
Heart	13	----	162	108	15	7.07
Breast-cancer	10	----	409	274	15	7.91
Australian	14	200	414	276	15	17.32
Diabetes	8	----	461	307	15	14.14
Ionosphere	34	200	211	140	13	8.66
Sonar	60	200	125	83	15	10.0

4.2 Experimental Results

The experimental results are listed in Table 2 and the performance on six datasets with incremental iterations procedure is given in Fig.1.a ~ f, respectively, whose *y*-axis shows the accuracy for classification problem and *x*-axis is the numbers of bases acquired in the process of constructing PR-KFDA.

Table 2. Experimental Results for PR-KFDA compared with MPKFDA and KFDA

Data Set	PR-KFDA		MPKFDA	KFDA	
	N-Basis	Accuracy	Accuracy	N-Basis	Accuracy
Heart	6	0.8139	0.8287	162	0.8267
Breast-cancer	1	0.9685	0.9426	409	0.9714
Australian	5	0.8207	0.8563	414	0.8551
Diabetes	6	0.7299	0.7455	461	0.7298
Ionosphere	5	0.9093	0.8484	211	0.9151
Sonar	10	0.7299	0.7595	125	0.7807

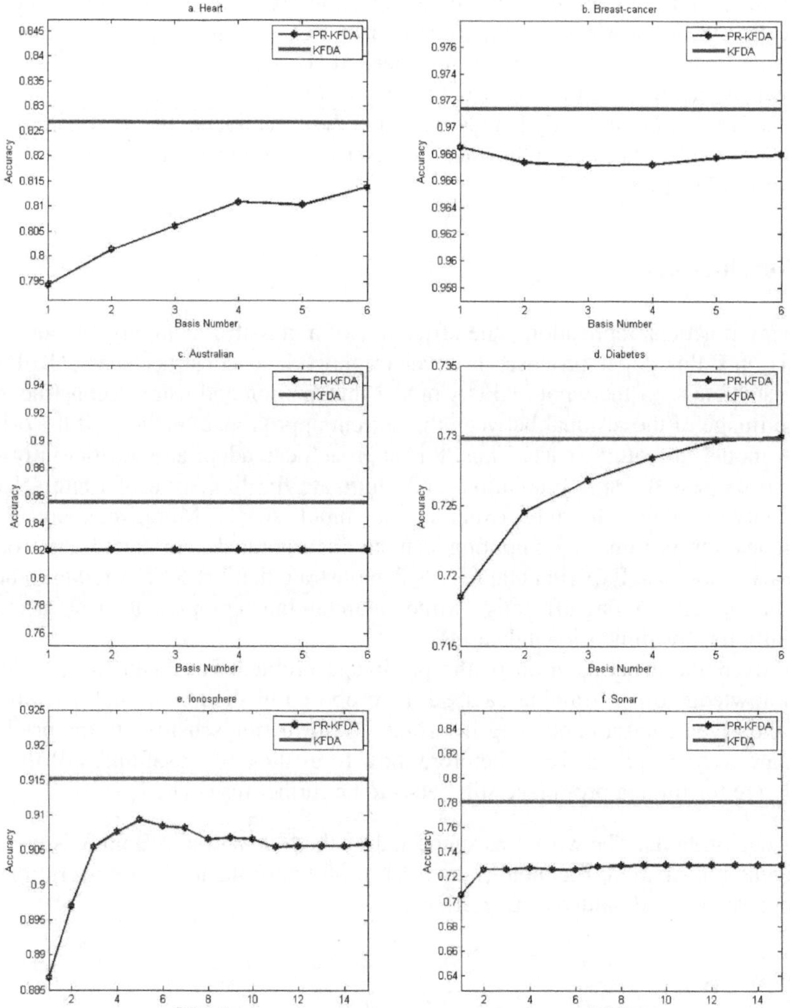

Fig. 1. The performance with incremental iterations procedure for constructing sparse KFDA on 6 UCI Benchmark datasets is given in Fig.1.a ~ f, respectively

4.3 Discussion

According to Table 2, the average performance of PR-KFDA is competitive to KFDA with considerable reduction of the original bases while maintaining comparable test accuracy, especially for low dimensional datasets. Moreover, PR-KFDA as the sparse version of Standard-KFDA also achieves the similar results of Regularized-KFDA in its sparse version at the identical sparsity conditions.

To further investigate the performance with incremental iteration procedure, we find that the proposed method has no homogeneous tendencies to the target KFDA accuracy line from Fig.1.a to Fig.1.f. This finding can be interpreted from the view of basis generating. Since the obtained pre-image is usually not an exact solution [10] [13] of the pattern in feature space, so the precision highly depends on the availability of effective prior knowledge of the training data, i.e., enough samples sensitive to the pre-image candidate in both spaces are needed. However, this prior knowledge is unknown in advance and changes with the transferred location of a new pre-image candidate in feature space dynamically. Thus, the effectiveness of this method may vary from diverse datasets or the same dataset in different iteration phrases, just as shown from Fig.1.a to Fig.1.f.

5 Conclusions

For many practical applications, the efficiency of a classifier is highly demanded [3]. To this end, this paper proposes a novel framework to construct sparse KFDA by taking advantage of the whole KFDA model information and using reconstruction of the pre-image of the residual between the current approximate model and the original KFDA model in feature space. Thus, this approach can adopt any vectors in original space to express the sparse solution, and eliminate the limitation of finite selection from only available training examples in input space. Moreover, any further techniques for pre-image computing can be incorporated into this framework as alternative modules. Experimental results demonstrate that PR-KFDA reduces the last solution of KFDA significantly while maintaining comparable test accuracy, especially for low dimensional datasets.

However, the exact solution of the pre-image problem is related to the effective prior knowledge of the training data, so the proposed method could not perform well when the given dataset containing insufficient information sensitive to the pre-image candidate in both spaces [16]. Therefore, how to evaluate the usefulness of the prior knowledge for finding pre-image still needs to be further researched.

Acknowledgments. The work was supported by the foundation of Beijing Key Lab of Intelligent Information Technology and We profoundly thank the reviwers for their thorough review and valuable suggestions.

References

1. Mika, S., Ratsch, G., Weston, J., Schölkopf, B., Mullers, K.R.: Fisher Discriminant Analysis with Kernels. In: IEEE Conference on Neural Networks for Signal Processing IX, pp. 41–48 (1999)

2. Vapnik, V.N.: Statistical Learning Theory. Wiley, New York (1998)
3. Dadgostar, F., Sarrafzadeh, A., Messom, C.H.: Multi-layered Hand and Face Tracking for Real-Time Gesture Recognition. In: Köppen, M., Kasabov, N., Coghill, G. (eds.) ICONIP 2008. LNCS, vol. 5506, pp. 587–594. Springer, Heidelberg (2009)
4. Mika, S., Smola, A., Schölkopf, B.: An Improved Training Algorithm for Kernel Fisher Discriminants. In: AISTATS, pp. 98–104 (2001)
5. Diethe, T., Hussain, Z., Hardoon, D.R., Shawe-Taylor, J.: Matching Pursuit Kernel Fisher Discriminant Analysis. In: International Conference on Artificial Intelligence and Statistics, pp. 121–128 (2009)
6. Xing, H.-J., Yang, Y.-J., Wang, Y., Hu, B.-G.: Sparse Kernel Fisher Discriminant Analysis. In: Wang, J., Liao, X.-F., Yi, Z. (eds.) ISNN 2005. LNCS, vol. 3496, pp. 824–830. Springer, Heidelberg (2005)
7. Burges, C.J.C.: Simplified Support Vector Decision Rules. In: 13th International Conference on Machine Learning, pp. 71–77 (1996)
8. Schölkopf, B., Knirsch, P., Smola, A., Burges, C.J.C.: Fast Approximation of Support Vector Kernel Expansions, and an Interpretation of Clustering as Approximation in Feature Spaces. In: DAGM Symposium Mustererkennung. LNCS, pp. 124–132. Springer, Heidelberg (1998)
9. Mika, S., Schölkopf, B., Smola, A., Müller, K.R., Scholz, M., Rätsc, G.: Kernel PCA and De-noising in Feature Spaces. In: Kearns, M.S., Solla, S.A., Cohn, D.A. (eds.) Advances in Neural Information Processing Systems, vol. 11, pp. 536–542. Morgan Kaufmann, San Mateo (1998)
10. Kwok, J.T., Tsang, I.W.: The Pre-Image Problem in Kernel Methods. IEEE Transactions on Neural Networks 15(6), 1517–1525 (2004)
11. Wang, L., Bo, L., Jiao, L.: Kernel Uncorrelated Discriminant Analysis for Radar Target Recognition. In: King, I., Wang, J., Chan, L.-W., Wang, D. (eds.) ICONIP 2006. LNCS, vol. 4233, pp. 404–411. Springer, Heidelberg (2006)
12. Girosi, F., Jones, M., Poggio, T.: Regularization Theory and Neural Network Architectures. Neural Computation 7, 219–269 (1995)
13. Honeine, P., Richard, C.: A Closed-form Solution for the Pre-image Problem in Kernel-based Machines. Journal of Signal Processing Systems, 1939–8018 (2010)
14. Teixeira, A.R., Tomé, A.M., Stadlthanner, K., Lang, E.W.: KPCA Denoising and the Pre-image Problem Revisited. Digital Signal Processing 18(4), 568–580 (2008)
15. Shin, H., Cho, S.: Invariance of Neighborhood Relation under Input Space to Feature Space Mapping. Pattern Recognition Letters 26(6), 707–718 (2005)
16. Scholkopf, B., Mika, S., Burges, C.J.C., Knirsch, P., Muller, K.-R., Ratsch, G., Smola, A.J.: Input Space Versus Feature Space in Kernel-based Methods. IEEE Transactions on Neural Networks 10(5), 1000–1017 (1999)
17. Vincent, P., Bengio, Y.: Kernel Matching Pursuit. Machine Learning 48, 165–187 (2002)
18. Blake, C.L., Merz, C.J.: UCI Repostitory of Machine Learning Databases. Department of Information and Computer Science. University of California, Irvine (1998), `http://www.ics.uci.edu/.mlearn/MLRepository.html`

Learning Basis Representations of Inverse Dynamics Models for Real-Time Adaptive Control

Yasuhito Horiguchi, Takamitsu Matsubara, and Masatsugu Kidode

Graduate School of Information Science,
Nara Institute of Science and Technology, Japan

Abstract. In this paper, we propose a novel approach for adaptive control of robotic manipulators. Our approach uses a representation of inverse dynamics models learned from a varied set of training data with multiple conditions obtained from a robot. Since the representation contains various inverse dynamics models for the multiple conditions, adjusting a linear coefficient vector of the representation efficiently provides real-time adaptive control for unknown conditions rather than solving a high-dimensional learning problem. Using this approach for adaptive control of a trajectory-tracking problem with an anthropomorphic manipulator in simulations demonstrated the feasibility of the approach.

Keywords: Learning Basis Representation, Inverse Dynamics, Adaptive Control.

1 Introduction

The dynamics model is crucial for precise control of fast movements on robotic manipulators [1]. However, the typical modeling of a robot as a rigid body dynamics system is not effective due to unknown friction and actuator dynamics. In recent years, researchers have applied statistical machine learning methods to this problem by dealing with it as a nonlinear regression problem [2,3,4]. This approach has solved the above problems; however, the machine learning methods require a certain amount of training data and have a large computational cost due to the high dimensionality associated with the number of joints on a robot. Also, since the training data for machine learning methods must be generated from actual movements of the robot in a real environment, it requires a long motion-execution time from the robot.

In the reality of our daily living environment, the dynamics of a robot can change due to actions such as holding an object and putting loads on the links that change inertial and kinematic parameters of the robot. We refer to such a change as *condition*, thus, the robot in a real environment meets multiple conditions.

The ultimate goal for such situations is to develop a learning method to rapidly track the changes in the dynamics (e.g., within a few seconds) and execute given tasks in the environment, that is, *real-time adaptive control*.

K.W. Wong et al. (Eds.): ICONIP 2010, Part II, LNCS 6444, pp. 668–675, 2010.

Some researchers attempted to use on-line tracking algorithms for changing robot dynamics(e.g., [2,5]); however, this approach still needs a certain amount of training data and a long motion-execution time due to high dimensionality.

While there are various changes in the robot dynamics in a real environment, there are still common factors because its mechanical structure and some of its characteristics (e.g., the length of most links) are invariant. If we could successfully extract such latent factors, it would simplify the adaptive control problem for the changes in the robot dynamics to be suitable for real-time adaptation even for unknown conditions.

In this paper, we propose a novel approach for adaptive control of robotic manipulators. Our approach uses a representation of inverse dynamics models learned from a varied set of training data with multiple conditions obtained from a robot. Since the representation contains various inverse dynamics models for the multiple conditions, adjusting a linear coefficient vector of the representation efficiently provides real-time adaptive control for unknown conditions rather than solving a high-dimensional learning problem. Our approach is inspired by a multi-task learning framework [6,7]. In that case, adaption to unknown conditions requires solution of a nonlinear optimization problem with a large computational cost, so it is an off-line process. Our approach focuses on real-time adaptation. Using this approach for adaptive control of a trajectory-tracking problem with an anthropomorphic manipulator in simulations demonstrated the feasibility of the approach.

Section 2 introduces the dynamics model for robot control, and previous work on machine learning adaptive control methods. Section 3 presents our proposed method for achieving real-time adaptive control. Section 4 presents the effectiveness of our method on a trajectory tracking problem with an anthropomorphic manipulator in simulations. Section 5 presents our conclusion for this paper.

2 Learning Dynamics for Adaptive Control

In this section, we briefly review dynamics model-based control methods and machine learning methods for dynamics models.

2.1 Computed Torque Control with Dynamics Model

A N-DoFs robotic manipulator attached to a base can be modeled by a rigid body dynamics system denoted as $\mathbf{u} = f(\mathbf{q}, \dot{\mathbf{q}}, \ddot{\mathbf{q}})$ where $\mathbf{q}, \dot{\mathbf{q}}, \ddot{\mathbf{q}} \in \mathbb{R}^N$ are joint angles, velocities, and accelerations, $\mathbf{u} \in \mathbb{R}^N$ denotes the input torque (e.g., [1]). In this paper, we call the model $f(\mathbf{q}, \dot{\mathbf{q}}, \ddot{\mathbf{q}})$ *Inverse Dynamics Model (IDM)*. To obtain torques $\mathbf{u}(t)$ needed at the joints to track a given trajectory $[\mathbf{q}(t), \dot{\mathbf{q}}(t), \ddot{\mathbf{q}}(t)]$ is often called *inverse dynamics problem.*

Computed torque control [1,3] is a popular solution for the problem with a rigid body dynamics system, which can be achieved by $\mathbf{u} = f(\mathbf{q}, \dot{\mathbf{q}}, \ddot{\mathbf{q}}_{ref})$ where $\ddot{\mathbf{q}}_{ref} = \ddot{\mathbf{q}}_d + \mathbf{K}_p\mathbf{e} + \mathbf{K}_v\dot{\mathbf{e}}$. $\ddot{\mathbf{q}}_d$, $\mathbf{K}_p$ and $\mathbf{K}_v$ are desired acceleration, position and velocity feedback gains respectively. $\mathbf{e}$ denotes the error between $\mathbf{q}_d$ and $\mathbf{q}$. The feedback term of the error $\mathbf{e}$ with a proper choice of both gains $\mathbf{K}_p$ and $\mathbf{K}_v$ is effective to compensate for modeling errors of the IDM. This approach requires all physical parameters of the robot to be known for f. Based on the structure of the rigid body dynamics, several schemes for efficient identification of the parameters have been developed (e.g., in [1]).

2.2 Learning IDMs from a Training Data Set

As denoted by [2,3,4], the rigid body dynamics modeling approach is not often effective for real robotic manipulators due to modeling errors in the physical parameters and difficulties for dealing with complex friction and actuator dynamics. For these difficulties, nonparametric regression techniques [2,4] can be a reasonable choice. Since the IDM is a function as $\mathbf{u}(t) = f(\mathbf{x}(t))$ where $\mathbf{x}(t) = [\mathbf{q}(t)^T, \dot{\mathbf{q}}(t)^T, \ddot{\mathbf{q}}(t)^T]^T \in \mathbb{R}^{3N}$, it can be estimated from a training data set $\mathcal{D} = \{\mathbf{X}, \mathbf{U}\}$ where $\mathbf{X} = [\mathbf{x}_1^T, \cdots, \mathbf{x}_d^T]^T$, $\mathbf{U} = [\mathbf{u}_1, \cdots, \mathbf{u}_d]^T$ and d is the number of data points. The effectiveness has been demonstrated by [2,4].

However, since such a training data set must be generated from actual movements of the robot in a real environment, thus, it requires a long motion-execution time from the robot. This problem is significant for real-time adaptive control. The ultimate goal of this study is to develop a learning method to rapidly track changes in the dynamics (e.g., within a few seconds) for daily use of the robot in a real environment. In the next section, we present our suggested approach focusing on this issue.

3 Learning Basis Representations of IDMs for Real-Time Adaptive Control

This section describes our proposed method for adaptive control that makes the adaptive control problem feasible in real time. In section 3.1, we first introduce a novel basis representation of IDMs. Section 3.2 describes a novel learning procedure for the basis representation from a varied set of training data. Section 3.3 shows a scheme for real-time adaptive control with the basis representation.

3.1 Basis Representation of IDMs

Our approach requires a varied set of training data $\bar{\mathcal{D}} = \{\mathcal{D}_1, \cdots, \mathcal{D}_M\}$ obtained from M conditions with a large variety. For a robot, the conditions can be artificially generated, for example by putting weights on links or having an object by the end-effector to generate a varied set of training data $\bar{\mathcal{D}}$. Then, by extracting a set of common factors $\{f_1^e, \cdots, f_J^e\}$ from the data set, based on the assumption that all dynamics models with multiple conditions share invariant characteristics, a basis representation of IDMs is formed as

$$\mathbf{u} = \hat{f}(\mathbf{x}; \mathbf{w}) = \mathbf{w}^T \boldsymbol{f}^e(\mathbf{x}) \tag{1}$$

where $\boldsymbol{f}^e = [f_1^e, \cdots, f_J^e]^T$ is referred to as the *Eigen Inverse Dynamics Models (EIDMs)*, $\mathbf{w} = [w_1, \cdots, w_J]^T$ is its linear coefficient vector and the function $\hat{f}$ ($\hat{f} : \mathbb{R}^{3N} \mapsto \mathbb{R}^N$) is referred to as *Parametric Inverse Dynamics Model (PIDM)*.

We assume that the PIDM, spanned by the EIDMs, contains various IDMs suitable for adaptive control in unknown conditions. If such an assumption is proper, adaptive control even for unknown conditions can be efficiently achieved by estimating the linear coefficient vector $\hat{\mathbf{w}}$ from a modest number of data set $\hat{\mathcal{D}}$ rather than directly solving a high-dimensional learning problem. Fig. 1 depicts the schematic diagram of the proposed method.

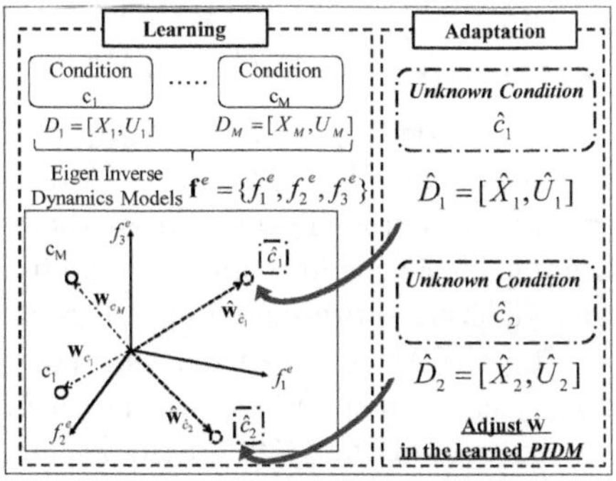

Fig. 1. The concept of the PIDM

3.2 Learning Procedure for PIDM

To obtain a compact representation of $\hat{f}$, the quantitative difference between two IDMs must be defined. By considering the IDM as a function where the input is $\mathbf{x}$ and the output is $\mathbf{u}$, the difference of two models f_1 and f_2 can be measured in the difference between the outputs corresponding to the same input as $||f_1(\mathbf{x}) - f_2(\mathbf{x})||$. With the definition of this measure, we find EIDMs from $\bar{\mathcal{D}}$. The learning procedure is composed of the following three steps:

(i) Data Alignment: We assume a varied set of training data $\bar{\mathcal{D}} = \{\mathcal{D}_1, \cdots, \mathcal{D}_M\}$ obtained from a robot with M conditions, where $\mathcal{D}_m = \{\mathbf{X}_m, \mathbf{U}_m\}$. As pre-processing for subsequent steps, the data alignment procedure generates the aligned torque matrix $\mathbf{U}^{all} \in \mathbb{R}^{M \times NC}$ from $\bar{\mathcal{D}}$, where N is the number of joints and C is the number of *contents*. A content is an input $\mathbf{x}$ commonly included among all conditions (in each $\mathcal{D}_m$ for all m). All the contents are represented by the content matrix $\mathbf{X}^c \in \mathbb{R}^{3N \times C}$, that is also generated by the alignment procedure. Thus, the (i, j) element of $\mathbf{U}^{all}$ is the torque $U(i, j)$ generated by the robot with i-th condition with ceil(j/N)-th content at mod(j/N)-th joint, that is, $U(i, j) = \mathbf{f}^i_{\mathrm{mod}(j/N)}(\mathbf{x}_{\mathrm{ceil}(j/N)})$. Note that, it is almost impossible to obtain such a data set directly because we can only apply $\mathbf{u}$ at the state $[\mathbf{q}, \dot{\mathbf{q}}]$ of the robot, that generates $\ddot{\mathbf{q}}$. It means that to obtain outputs with the same input for all conditions requires the exact IDMs for all conditions in advance. To avoid such a "chicken-and-egg" problem, we present an alternative data alignment procedure using the Mahalanobis distance in input space for approximately generating $\mathbf{U}^{all}$ from $\bar{\mathcal{D}}$. The algorithm list is shown in Fig. 2.

(ii) Extraction of EIDMs: The extraction of EIDMs $\mathbf{f}^e$ from $\mathbf{U}^{all}$ can be achieved by a Singular Value Decomposition (SVD) based matrix factorization and a nonlinear regression technique. The SVD for $\mathbf{U}^{all}$ leads to the following factorial representation as $\mathbf{U}^{\text{all}} = \mathbf{Y}\Sigma\mathbf{V}^T \approx \mathbf{W}\mathbf{F}_{basis}$. Define the linear coefficient matrix $\mathbf{W} = [\mathbf{w}^{1^T} \cdots \mathbf{w}^{M^T}]^T \in \mathbb{R}^{M\times J}$ to be the first $J(\leq M)$ rows of $\mathbf{Y}$, and the basis target matrix $\mathbf{F}_{\text{basis}} = [\mathbf{f}_{\text{basis}}^{1}{}^T \cdots \mathbf{f}_{\text{basis}}^{J}{}^T]^T \in \mathbb{R}^{J\times NC}$ to be the first J columns of $\Sigma\mathbf{V}^T$. The dimension J can be determined with the singular value spectrum. Typically $J \ll M$, if there would be certain similarity or correlation among IDMs for all conditions included in the training. This yields a compact and effective representation of the PIDM.

The EIDMs $\mathbf{f}^e$ are then learned with the content matrix $\mathbf{X}^c$ as inputs and $\mathbf{F}_{basis}$ as corresponding outputs using a nonlinear regression technique. With the success of the GPR in the learning IDMs [3,4], we utilize it for learning f_i^e as a smooth mapping from $\mathbf{x}$ to f_{basis}^j independently for all j.

(iii) Learning of PIDM: The PIDM $\hat{f}$ can be finally formed by the weighted linear combination of $\mathbf{f}^e$ as $\hat{\mathbf{u}} = \hat{f}(\mathbf{x};\mathbf{w}) = \mathbf{w}^T \boldsymbol{f}^e(\mathbf{x})$. By setting $\mathbf{w}$ as a row of $\mathbf{W}$, $\hat{f}$ approximately represents the IDM corresponding to a particular condition used in the training. Thus, the subspace contains a variety of the IDMs, and it may be suitable to achieve real-time adaptive control for the robot even in unknown conditions.

```
Input : D = {D_1, ··· , D_M}
Select the nominal condition number:
  b ∈ {1, ··· , M}
for n=1 to the number of data in D_b do
  for m=1 to M do
    for j=1 to the number of data in D_m do
      Compute Mahalanobis distance d(j)
      between x_n^b and x_j^m
    end for
    Find the index q that minimizes d(j):
      q ← arg min_j d(j)
    Insert d(q) in the minimal distance vector a:
      a(m) ← d(q)
    Insert u^m(q) in the torque candidate vector u_c:
      u_c(m) ← u^m(q)
  end for
  if max(a) < α then
    U^all ← [U^all, u_c]
    Add new row of X^c :
      X^c ← [X^c, x^b(n)]
  end if
end for
Output : U^all, X^c
```

Fig. 2. Algorithm for Data Alignment (1-DoF system). u^m indicates m-th conditions output. α is a threshold parameter.

Note that the learning procedure is inspired by the studies referred to as *style content separation* in several different contexts such as face recognition [8], the synthesis of human-like graphics [9] and learning stylistic movement primitives [10]. Our learning procedure can be interpreted as a modification of their methods to be particularly suitable for real-time adaptive control.

3.3 Real-Time Adaptive Control Method with PIDM

Achieving real-time adaptive control with the PIDM requires an adaptation process of the linear coefficient vector $\hat{\mathbf{w}}$ by a data set $\hat{\mathcal{D}}$ obtained from the robot with an unknown condition. Based on the linearity of $\mathbf{w}$ in $\hat{f}(\mathbf{x};\mathbf{w})$, such a process can be simply achieved by a least square method. A recursive least square algorithm is suitable for real-time adaptive control since it is a fully-recursive, computationally efficient method [11]. The linearity of $\mathbf{w}$ and its low dimensionality may cause a rapid convergence of the adaptation. The effectiveness of this approach is demonstrated in the next section through experiments.

Table 1. The load specifications on each link for learning and test

Joint Number	Condition Number																	
	Training															Test		
	c_1	c_2	c_3	c_4	c_5	c_6	c_7	c_8	c_9	c_{10}	c_{11}	c_{12}	c_{13}	c_{14}	c_{15}	c_{16}	c_{17}	c_{18}
1	0.0	2.0	0.0	0.0	0.0	0.0	2.0	0.0	0.0	2.0	1.0	2.0	1.0	0.0	0.0	0.5	0.0	0.0
2	0.0	0.0	2.0	0.0	0.0	0.0	0.0	2.0	2.0	2.0	1.0	2.0	0.0	1.0	0.0	0.5	0.0	2.25
3	0.0	0.0	0.0	1.0	0.0	0.5	0.5	0.5	0.0	1.0	0.5	0.0	0.0	0.0	0.0	0.25	1.25	1.25
4	0.0	0.0	0.0	0.0	1.0	0.5	0.0	0.0	0.5	1.0	0.5	0.0	0.0	0.0	0.5	0.25	1.25	1.5

4 Experiments

In this section, we investigate the effectiveness of our real-time adaptive control method through experiments with the model of a 4-DoFs manipulator (Barrett WAM in Fig. 3) in simulations. The simulator is built with the aid of the Robotics Toolbox [12] and its physical parameters are set from specifications supplied by the Barrett company. Section 4.1 describes data generation for learning the PIDM. Section 4.2 shows the learning result of PIDM from the generated data set and the feasibility of the real-time adaptive control with the PIDM for unknown conditions.

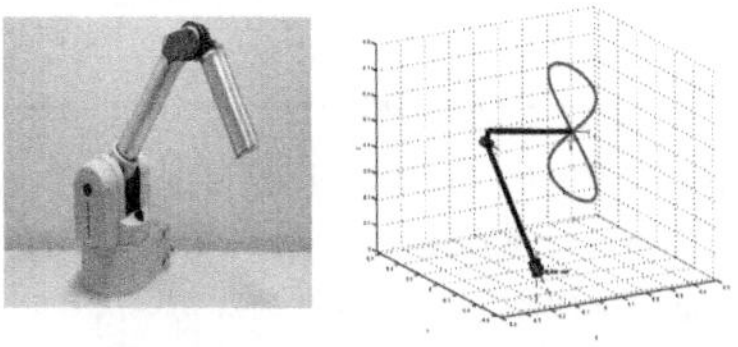

(a) (b)

Fig. 3. Anthropomorphic manipulator (4-DoFs Barrett WAM).(a) is the real robot and (b) is that of simulator with the figure-of-eight nominal trajectory.

4.1 Data Generation

In this experiment, fifteen different conditions (as training conditions) on the robot model for learning are produced by putting different weights of loads in the center of mass of each link, thus, we focus on changes in the inertial parameters of the robot in this experiment. Three conditions (as test conditions) are additionally set by the same manner for evaluations. Table 1 shows the details of loads for all conditions. The training conditions are denoted by $\{c_1, \cdots, c_{15}\}$, $\{c_{16}, c_{17}, c_{18}\}$ indicates the test conditions, respectively.

Figure 3 (b) shows a figure-of-eight nominal trajectory, which is placed at 0.15m along with the x axis, at 0.5m in z-axis. For each condition, data were captured at every 20ms under PD tracking control of the trajectory with seven different periods (evenly set from 7.25s to 8.75s per 0.25s) from randomly selected initial positions. 2800 data points were captured for each condition, that is, 42000 data points were prepared as a training data set $\bar{\mathcal{D}}$. The data alignment algorithm presented in Fig. 2 was applied for generating the aligned torque matrix $\mathbf{U}^{all}$ and $\mathbf{X}^c$.

4.2 Evaluation of the Real-Time Adaptive Control

The feasibility and validity of our real-time adaptive control method was evaluated. First, we applied our suggested procedure for learning PIDM $\tilde{f}$ from $\mathbf{U}^{all}$

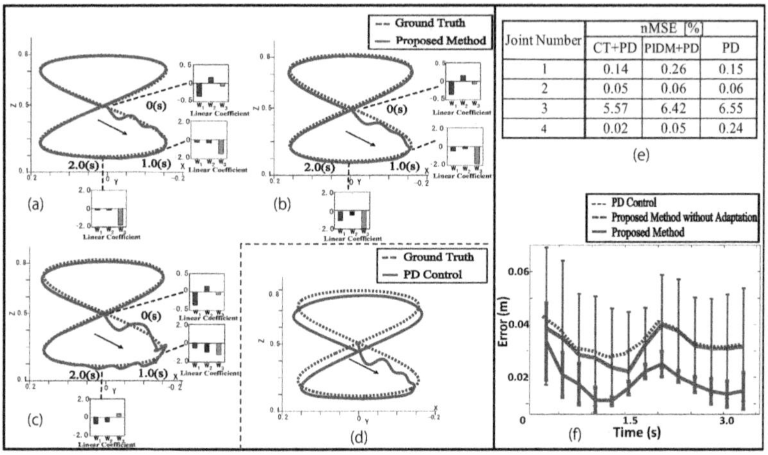

Joint Number	nMSE [%]		
	CT+PD	PIDM+PD	PD
1	0.14	0.26	0.15
2	0.05	0.06	0.06
3	5.57	6.42	6.55
4	0.02	0.05	0.24

Fig. 4. The result of the adaptive control by our proposed method with comparisons. (a)-(c) correspond to the result with the conditions c_{16}-c_{18}, respectively. (d) shows the tracking performance by a low-gain PD control for the condition c_{16}. (e) indicates the mean values of nMSE for all test conditions at each joint. (f) depicts the tracking errors through 10 trials with randomly selected initial values for $\hat{\mathbf{w}}$ in the condition c_{16}. Solid lines are the mean values and error-bars are the standard deviations for all methods.

and $\mathbf{X}^c$. As the result of SVD, the three dimensional bases explained more than 90% of $\mathbf{U}^{all}$, thus we set $J = 3$ and formed the PIDM $\hat{f}(\mathbf{x}; \mathbf{w})$.

Next, our real-time adaptive control method with the PIDM was evaluated for three test conditions. Adaptive control for trajectory tracking was performed by estimating the linear coefficient $\hat{\mathbf{w}}$ on-line from the incrementally observed training data set $\hat{\mathcal{D}}$. The performance was measured by both the tracking error and convergence time. To keep the position of the robot around the nominal trajectory during a transient phase of the adaptation, PD control with low gains was additionally applied with the adaptive control [1]. $\hat{\mathcal{D}}$ was captured at every 40ms without PD control. The adaptation procedure by an iterative least square method was achieved as $\hat{\mathbf{w}}(k+1) \leftarrow \hat{\mathbf{w}}(k) + g(\mathbf{x}(k), \mathbf{u}(k))$, where $g(\cdot)$ is a recursive update rule [11].

The tracking performance is plotted with the nominal trajectory for all test conditions in Fig. 4(a)-(c). For comparison, Figure 4(d) shows the tracking performance by the low-gain PD-control. The tracking performance measured in the normalized Mean Square Error (nMSE) is also plotted in Fig. 4(e) with the ground truth, i.e., the result of computed torque control with the exact IDMs, and the result of PD control. The time course of the tracking errors is depicted in Fig. 4(f). Our method rapidly adapted to test conditions and resulted in precise trajectory tracking for all cases. The tracking performance was very close to the ground truth (computed torque method with the exact IDM and PD control) and much better than just with the PD control. The linear coefficients quickly

[1] In this experiments, PD gains were commonly set as $\mathbf{k}_p = [45, 125, 30, 25]^T$ and $\mathbf{k}_v = [1.5, 3.0, 1.5, 0.375]^T$, where $\mathbf{k}_p$ is for position, $\mathbf{k}_v$ is for velocity, respectively.

adapted and the converged vector was significantly different for all conditions. Note that the adaptation mostly converged around 2.0s (with a data set corrected until a quarter of the figure-of-eight), while the original robot dynamics was a high dimensional system. These results demonstrate the feasibility of our real-time adaptive control method even for unknown conditions.

5 Conclusion

In this paper, we have proposed a novel approach for adaptive control of robotic manipulators. We have shown that our approach can achieve rapid adaptation (around 2.0s) for a robotic manipulator in unknown conditions in simulation, i.e., real-time adaptive control. Our suggested approach is applicable even in a real environment, thus, our future work includes its application to real robotic manipulators. An active selection of conditions for finding an effective subspace will be also addressed in near future.

References

1. Spong, M.W., Hutchinson, S., Vidyasagar, M.: Robot Dynamics and Control. John Wiley and Sons, New York (2006)
2. Vijayakumar, S., Schaal, S.: Locally weighted projection regression: An o(n) algorithm for incremental real time learning in high dimensional space. In: Proceedings of the Seventeenth International Conference on Machine Learning, pp. 1079–1086 (2000)
3. Nguyen-Tuong, D., Seeger, M., Peters, J.: Computed torque control with nonparametric regression models. In: American Control Conference (ACC), pp. 212–217 (2008)
4. Nguyen-tuong, D., Seeger, M., Peters, J.: Local gaussian process regression for real time online model learning and control. In: IEEE/RSJ International Conference on Intelligent Robots and Systems, pp. 365–372 (2008)
5. Nguyen-Tuong, D., Peters, J.: Incremental sparsification for real-time online model learning. In: Proceedings of Thirteenth International Conference on Artifical Intelligence and Statistics (AISTATS 2010), vol. 9, pp. 557–564 (2010)
6. Kemal Ciliz, M., Narendra, K.S.: Adaptive control of robotic manipulators using multiple models and switching. International Journal of Robotics Research 15(6), 592–610 (1996)
7. Ming, K., Chai, A., Williams, C.K.I., Klanke, S., Vijayakumar, S.: Multi-task gaussian process learning of robot inverse dynamics. In: NIPS, vol. 21, pp. 1–8 (2008)
8. Tenenbaum, J.B., Freeman, W.T.: Separating style and content with bilinear models. Neural Computation 12, 1247–1283 (2000)
9. Brand, M., Hertzmann, A.: Style machines. In: Proceedings of the 2000 SIGGRAPH, pp. 183–192 (2000)
10. Matsubara, T., Hyon, S.-H., Morimoto, J.: Learning stylistic dynamic movement primitives from multiple demonstrations. In: IEEE/RSJ International Conference on Intelligent Robots and Systems, IROS 2010 (accepted 2010)
11. Haykin, S.: Adaptive Filter Theory. Prentice Hall, Englewood Cliffs (2002)
12. Corke, P.I.: A robotics toolbox for matlab. IEEE Robotics and Automation Magazine 3(1), 24–32 (1996)

Feel Like an Insect: A Bio-Inspired Tactile Sensor System

Sven Hellbach[1,2], André Frank Krause[1], and Volker Dürr[1,2,⋆]

[1] University of Bielefeld, Faculty of Biology, Dept. Biological Cybernetics
[2] CITEC Center of Excellence Cognitive Interaction Technology
Universitätsstrasse 21–23, 33615 Bielefeld, Germany
http://www.uni-bielefeld.de/biologie/Kybernetik/
{sven.hellbach,andre_frank.krause,volker.duerr}@uni-bielefeld.de

Abstract. Insects use their antennae (feelers) as near range sensors for orientation, object localization and communication. This paper presents an approach for an active tactile sensor system. This includes a new type of hardware construction as well as a software implementation for interpreting the sensor readings. The discussed tactile sensor is able to detect an obstacle and its location in 3D space. Furthermore the material properties of the obstacles are classified by use of neural networks.

Keywords: Active Tactile Sensing, FFT, Material Classification, Object Localization, Acceleration Measurement.

1 Introduction

Insects are a widespread group of animals, inhabiting a wide range of ecosystems and hence being confronted with variate living conditions. One of the reasons why insects are able to adapt to such different living conditions is their ability for rapid and parallel object recognition and scene analysis. Researching the sensor systems of insects helps to understand the complexity of nature, since in animal near-range sensing, the active tactile sense is often of central importance. Many insects actively move their antennae (feelers) and use them for orientation, obstacle localisation, pattern recognition and even communication [1]; mammals like cats or rats use active whisker movements to detect and scan objects in the vicinity of their body. Here we use the antenna of the stick insect *Carausius morosus* [2] as the biological model for a bionic sensor for reasons summarised by Dürr and Krause [3]. This paper expands research efforts presented in [4] and some results are validated by application of a larger, hence more reliable data set. Furthermore, the algorithm is extended to provide even better results. While [4] can be seen as a proof of concept, this paper aims at a practical application.

Beyond the understanding of nature's principles, it offers a new type of sensor for mobile robot systems. In particular, in environments where other sensors are

⋆ This work was supported by the German Research Council (DFG) DU380/3 and by the German ministry of research and education (BMBF) 03/3766.

K.W. Wong et al. (Eds.): ICONIP 2010, Part II, LNCS 6444, pp. 676–683, 2010.

not able to provide reliable data, e. g. vision sensors in dusty or dark environments, a tactile sensor is able to support such sensors by providing additional data. To stick with the example, it is difficult for vision sensors to determine material properties. The sensor described within this paper is able to provide information about the material with additional spacial information.

This paper presents two different methods for processing the sensor readings from the acceleration sensor. The first one is based on previous work in [4], where it could be shown that the derived method works well for estimating the contact position, and that it is able to classify two different kinds of material. Both was shown on a small data set for which the start and end of the contact was given. the known method has been improved to be able to deal with a continuous data flow, as well as to classify an arbitrary set of materials.

Furthermore, a second method is presented, which reduces the pre-processing steps by withdrawing some limiting constrains and allowing a neural network to find the necessary information within the data. In that way, better results for distance estimation are gained, especially for contact positions closer to the antennal tip.

The next section presents a short overview of the field of tactile sensing. The sensor hardware is introduced in section 3, while the software part is discussed in section 4. Afterwards, experimental results are presented in section 5. Finally, the work is concluded in section 6.

2 Previous Work

While thinking about scene understanding, particularly the understanding of scene objects, the use of tactile sensors in the broadest sense plays an increasing role [5,6]. Insect-like tactile sensors have been pioneered by Kaneko and co-workers, who used either vibration signals [7] or bending forces [8], both measured at the base of a flexible beam, to determine contact distance. In contrast, we use a single acceleration sensor located at the tip of the probe [9]. Contact distance is determined using the peak frequency of the damped oscillations of the free end. Beyond the focus on single antenna-like sensors, their integration with vision has been researched, for example, in the AMouse project [10]. Instead of antennae rat inspired whiskers are used. Different to the approach in this paper the vibration is not measured at the tip of the sensor, but on its mounting point. In contrast to these works, our system is able to detect 3d contact position with a single antenna by measuring its vibration in two dimensions. The interpretation of the sensor readings is done in a bio-inspired way using neural networks.

3 Sensor Hardware

The biological counterpart is equipped with a large number of sensors. Handling such a large number of sensors is a challenging task. In particular, integrating all types of sensors into a single antenna-like device still is demanding. Here, we have decided to regard the antenna at a higher level of abstraction. One of the basic features of the biological archetype is the ability to detect the position of potential obstacles and to analyse their properties. A first step to be able to mimic these abilities is to use a two axis acceleration sensor (Analog Devices ADXL210E), which measures the vibration characteristics during object contact. The used sensor is mounted at the tip of a 33 cm poly-acrylic tube. The antenna is designed in a way that the entire probing rod can be exchanged easily [4].

The robotic feeler was based on major morphological characteristics of the stick insect antenna, such as two rotary joints that are slanted against the vertical plane. The scale is approx. 10:1 compared to the stick insect to match that of the Bielefeld insectoid walking robot TARRY (Fig. 1). The actuator platform consists of two orthogonal axes. Two 6V DC motors (Faulhaber 1331T 006SR) were used rather than servo motors to minimise vibrations of the probe due to discrete acceleration steps. The linkage of the hinges was designed to mimic the action range of the stick insect, amounting to 90° vertical range, centred 10° above the horizon, and to 80° horizontal range centred 40° to the side. Positioning accuracy is limited by slack in the motors and amounts to approx. 5 mm at the tip of a 40 cm probe (approx. 7°).

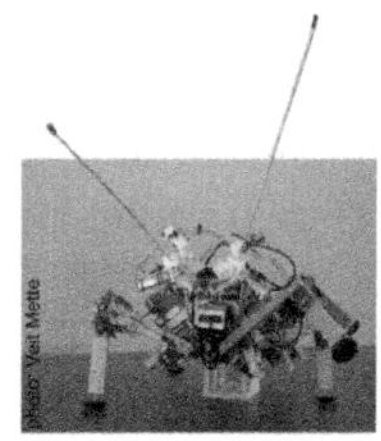

Fig. 1. The stick insect inspired walking robot TARRY

Both hinges are equipped with a position sensor (muRata SV01A). Hence the orientation of the sensor rod is known in two dimensions encoded in polar coordinates. Since the the hinge axes are orthogonal, no additional coordinate transformation is necessary. But, to describe the detected object's position in 3D space an additional coordinate is missing. It is gained by determining the contact position on the rod.

The control of the motion of the antenna as well as the sensor read out is implemented on an embedded system (ATMEL AT90CAN128). The raw sensor signal is available via RS232 for further processing.

4 Interpreting Sensor Readings

In the following two methods are described for the interpretion of the sensor signals. For both methods the principle idea stays the same. Depending on the position of contact, the free oscillating segment of the sensor tube differs. This results in different kinds of damped oscillations. Those characteristic properties of the damped oscillation are taken into account for estimating the position of contact. As an intuitive example, one can image a guitar string gripped at different position for playing different tone pitches.

In addition, the damped oscillation also carries information about the material involved. Back to the guitar string analogy this might be compared to the different tone caused by nylon or steel strings.

4.1 Method I: Constraint Based Input Dimension Reduction

The used acceleration sensor is able to measure the acceleration in two orthogonal dimensions. Hence, the data coming from the sensor is the projection of the actual oscillation onto both dimension vectors. This leads to different sensor readings, depending on the rotation of the antenna with respect to the axis defined by it. To align the rotated oscillation with one of the axes PCA[1] is applied. PCA computes a set of eigenvectors which are oriented with respect to the principle axes of the data distribution. The matrix of eigenvectors E can directly be used as an affine transform matrix applied on the data X: $X_{rotated} = E \cdot X$. The first dimension of the rotated data $X_{rotated}$ contains the part of the data with the largest variance. Only this part is used for further processing.

As a next step, it is necessary to know at which time a contact occurred. On a static system this is a trivial task, which can be solved with a simple threshold. However, it becomes more challenging while the active tactile sensor is in motion, since the motion induces an oscillation into the sensor rod as well. At the moment we stick to the threshold, keeping in mind that our further research will focus on that problem. For detecting the end of the oscillation the local maxima over time are considered. The end point is defined as the time, at which these maxima begin to drop below a dynamic threshold. The threshold is chosen to be 10% of the global maximum of the current damped oscillation. The window from the detected start to end point is taken into account for further processing after removing the mean.

As described above the basic idea is to take into account the frequency characteristics of the damped oscillation. Hence, the frequency spectrum of the time series within the window is computed using Discrete Fourier Transform.

Distance Estimation: Assuming that the fundamental oscillation and first harmonics are represented as significant peaks in the spectrum, two local maxima are determined. In doing so, the spectrum is divided into two intervals defined by the maximal and minimal occurring frequencies defined by the lengths of the rod. The intervals are chosen to be $(0\text{Hz}, 55\text{Hz}]$ and $(55\text{Hz}, 250\text{Hz}]$. For each of the intervals their global maximum is derived, assuming to represent the fundamental oscillation and first harmonics respectively. Both values are presented as input to a multi-layer perceptron. The network is a standard 2-layered network with a sigmoidal output function in the hidden layer and a linear function in the output layer using the Levenberg-Marquardt algorithm for training.

Material Classification: For material classification, the extracted frequencies used for distance estimation are not sufficient. Different material properties result in different decay characteristics of the damped oscillation. To extract these characteristics reliably for the fundamental oscillation and first harmonic, we remove disturbing frequencies first. For this, we apply two bandpass filters with the same limits as the already discussed intervals: $(0\text{Hz}, 55\text{Hz}]$ and $(55\text{Hz}, 250\text{Hz}]$. Both filtered spectra are transferred back into time domain. This results in two damped oscillation with different frequency and decay rate. For both, all local maxima are identified and an exponential decay function is fit:

$$f(t) = p_{0,j} + p_{1,j} \cdot e^{-\frac{t}{p_{2,j}} + p_{3,j}} \tag{1}$$

[1] Principle component analysis.

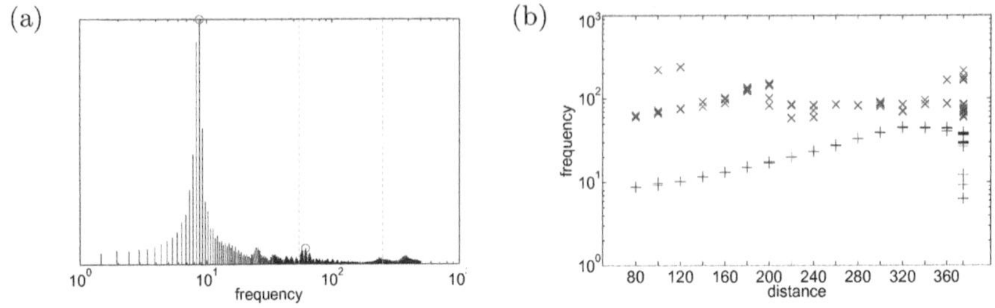

Fig. 2. (a) An example for a frequency spectrum of the damped oscillation caused by hitting an obstacle. Two dashed vertical lines indicate the intervals for the maximum frequency search. The circles show the position of the fundamental oscillation and the first harmonics. (b) The relation between the contact distance and the fundamental oscillation (+) and the first harmonics (×) is shown.

where $p_{0,j}$ to $p_{3,j}$ denote offset, amplitude, time constant and delay, respectively, computed for each of the filtered oscillations $j \in \{0, 1\}$. Those 8 parameters together with the extracted frequencies of the fundamental oscillation and the first harmonics are used as input for the neural classifier. As shown in [4], the parameters of the decay function not only depend on the material property but also on the position of contact. By also providing the two extracted frequencies, the neural network is able to derive the necessary information.

4.2 Method II: Let the Neural Network Do the Work

For the second method, the pre-processing doesn't go further as to calculate the frequency spectrum. So, as well as for method I, the data is aligned using PCA and the start and end point are derived. However, neither the search for the significant peaks nor the exponential fit are calculated.

Instead of using only a two dimensional input vector for distance estimation as it is done in method I, the network for method II gets a much higher dimensional input vector involving the entire spectrum. Experiments show that a sub-sampled spectrum is sufficient to learn the mapping, reducing the number of input dimensions. However, the network has to learn which part of the spectrum is important to solve the distance estimation task. It is clear that the mapping is more complex than for method I and thus a larger network is necessary. The network showing best results is a 3-layer network with 20 neurons for the first hidden layer and 5 for the second one.

Unlike method I, the input for material classification is the same as for the distance estimation task. The network as well is a multi-layer perceptron with 20, 30, and 30 neurons for the 3 hidden layers.

5 Results

The experiments show, that the active tactile sensor is able to discriminate different types of material as well as to derive the position of contact. Hence, six

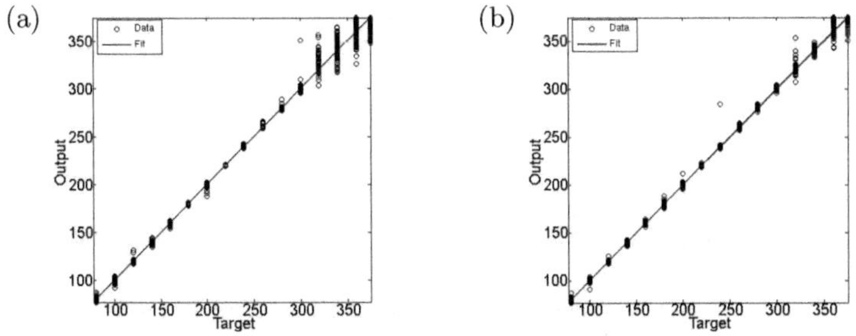

Fig. 3. Both plots show a regression plot, which indicates the relation between the network output distance and the desired target distance (both in mm). Each data sample is represented as a circle. An optimal solution would be located on the bisectrix $y = x$. Figure (a) shows the results for method I possessing larger errors for positions near the tip ($d > 300$) compared to the results of method II in Figure (b).

cylindrical objects with identical diameter were presented to the sensor, consisting of different materials, namely aluminium, wood, copper, brass, POM[2], and acrylic glass. These materials were chosen to represent a wide spectrum of materials with different damping characteristics. The selection includes such that are expected to be discriminated easily, e. g. aluminium and wood, as well as such that are much harder to distinguish, e.g. the two kinds of plastic. The impact occurred at 16 positions along the sensor tube, at 80mm to 360mm in steps of 20mm and at 375mm, as measured from the centre of rotation. Each impact was repeated 100 times to provide a large data set for network training.

In order to test the optimal performance of the of the active tactile sensor, the experiments in this paper are limited to a stationary case, i. e. all used data sets are recorded with a stationary antenna. To do so, the antenna was mounted on a working desk with the objects to be probed fixed in the desired distance. The contact at different distances always occurred with the same angle of impact. In the experiments presented here, the algorithm controlling the movement of the antenna ensures that the antenna stops and keeps applying a constant pressure to the probe, as soon as the probe has been hit. The application a constant pressure is necessary to avoid the rebounding of the antenna.

The curve in Figure 2 suggests that the first harmonics ($\times$) can not be used any more from a distance of 200 mm onwards. Even the fundamental oscillation ($+$) flattens for distant contacts. Since the network has to find a mapping from frequency values to distance (which is the inverse function to one being plotted). Performance is expected to deteriorate for contact distances beyond 300 mm. This can be confirmed in the regression plot in Figure 3. The network performance results in a root mean squared error (rmse) of 2.93 (about 0.7% of the antenna length). In contrast, when using the entire spectrum instead of the extracted frequencies performance improves to a rmse of 1.71. As a confirmation, whether this effect is not due to the changed network size, only the lower

[2] Polyoxymethylene.

(a)

Output Class \ Target Class	Aluminium	Wood	Copper	Brass	POM	Acrylic glass		
Aluminium	770	0	1	2	1	36	95,1%	4,9%
Wood	1	639	39	63	68	4	78,5%	21,5%
Copper	3	19	732	20	12	15	91,4%	8,6%
Brass	1	69	19	687	20	29	83,3%	16,7%
POM	1	73	7	28	699	0	86,5%	13,5%
Acrylic glass	24	0	2	0	0	716	96,5%	3,5%
	96,3%	79,9%	91,5%	85,9%	87,4%	89,5%	**88,4%**	
	3,8%	20,1%	8,5%	14,1%	12,6%	10,5%		**11,6%**

(b)

Output Class \ Target Class	Aluminium	Wood	Copper	Brass	POM	Acrylic glass		
Aluminium	1448	10	5	9	2	22	96,8%	3,2%
Wood	11	1414	61	80	45	6	87,4%	12,6%
Copper	15	56	1441	80	21	5	89,1%	10,9%
Brass	16	62	72	1384	32	11	87,8%	12,2%
POM	46	42	16	38	1496	4	91,1%	8,9%
Acrylic glass	64	16	5	9	4	1552	94,1%	5,9%
	90,5%	88,4%	90,1%	86,5%	93,5%	97,0%	**91,0%**	
	9,5%	11,6%	9,9%	13,5%	6,5%	3,0%		**9,0%**

Fig. 4. Confusion Matrix for material classification: The matrix summarizes the number of data samples assigned to a specific class by the network (output class) broken down into their target classes. The diagonal entries (light grey) contains number of the true positive classifications. The border columns and rows (dark grey) indicate the percentage of the correct and incorrect classified elements per class, while entries in the lower right right corner tells the overall performance. Figure (a) shows the results for method I and (b) for method II.

band of the spectrum serves as input. The results are comparable with the ones with extensive pre-processing. This can be explained by the fact that the upper part of the spectrum provides redundant information which makes the decision process more robust.

Furthermore, instead of using the entire spectrum a sub-sampled spectrum containing only each 10th frequency used as input. The classification results were not different from the ones using the entire spectrum, but the calculation time for training was 5 times faster (340 min vs. 79 min under MatLab on a Intel Core2Duo E8500).

In contrast to the experiments presented in [4] the material classification is performed with more than two materials and at different contact positions. The only restriction for the experiments on method I (Figure 4(a)) is to use only measurements up to 240 mm. This is done to avoid similar difficulties as for distance estimation and to focus on the limits of material classification. Experiments show that adding trials with larger distance leads to worse results and unstable convergence. In contrast, method II (Figure 4(b)) is able to handle distances larger than 240 mm and even to obtain better results than method I. However, when applying the same restrictions as for method I the results become even better (97.2% correct classified trials).

6 Conclusion

In this paper, a bio-inspired tactile sensor was presented. The system is able to detect the position of a possible obstacle and is furthermore able to classify its material properties. We were able to extend the method presented in [4]. Beyond this, a second method was introduced which leads to better results.

Experiments show that, if the contact position is close to the tip, both distance estimation and material estimation are less reliable. To cope with this limitation for a practical application, the search and detection strategy could be designed in an adequate way. This could be done, for example, by positioning the robot after the first contact in a way that for the second contact the expected contact position is below the critical distance.

However, before being able to include the antenna into a mobile robot, it is necessary to extend the pre-processing algorithm in a way that is able to handle self-induced noise by the motion of the robot.

Furthermore using only a tenth of the spectrum can be regarded as a first proof of concept. A deeper study on which part of the spectrum is sufficient, needs further investigations.

In this paper only simple multi-layer perceptrons were applied. Nevertheless, the data being processed is data with temporal characteristics, what suggests to apply recurrent neural network. Using recurrent networks would help to eliminate the start/stop-detection, which is done as the first pre-processing step. In that way distance estimation and material classification could run in an on-line system.

As a further perspective, it is desired to integrate multiple sensors onto a mobile platform. In doing so, a monocular vision-based system would benefit from the additional use of tactile sensors. The hypotheses gained from the vision system could be augmented with further information, like the detected material or the object's exact location in 3D Cartesian space. Additionally, the system is able to verify the visual object detection hypotheses.

References

1. Staudacher, E., Gebhardt, M., Dürr, V.: Antennal movements and mechanoreception: neurobiology of active tactile sensors. Adv. Insect Physiol. 32, 49–205 (2005)
2. Dürr, V., König, Y., Kittmann, R.: The antennal motor system of the stick insect *Carausius morosus*: anatomy and antennal movement pattern during walking. J. Comp. Physiol. A 187, 31–144 (2001)
3. Dürr, V., Krause, A.: The stick insect antenna as a biological paragon for an actively moved tactile probe for obstacle detection. In: Proc. of CLAWAR 2001, pp. 87–96 (2001)
4. Dürr, V., Krause, A.F., Neitzel, M., Lange, O., Reimann, B.: Bionic Tactile Sensor for Near-Range Search, Localisation and Material Classification. In: AMS, pp. 240–246 (2007)
5. Bebek, O., Cavusoglu, M.C.: Whisker sensor design for three dimensional position measurement in robotic assisted beating heart surgery. In: IEEE ICRA, Roma, Italy, pp. 225–231 (2007)
6. Kaneko, M., Kanayma, N., Tsuji, T.: Vision-based active sensor using a flexible beam. IEEE-ASME Trans Mechatronics I 6, 7–16 (2001)
7. Ueno, N., Svinin, M.M., Kaneko, M.: Dynamic contact sensing by flexible beam. IEEE-ASME Trans. Mechatronics 3, 254–264 (1998)
8. Kaneko, M., Kanayma, N., Tsuji, T.: Active antenna for contact sensing. IEEE Trans. Robot. Autom. 14, 278–291 (1998)
9. Lange, O., Reimann, B., Saenz, J., Dürr, V., Elkmann, N.: Insectoid obstacle detection based on an active tactile approach. In: Proc. of AMAM (2005)
10. Fend, M., Bovet, S., Hafner, V.: The artificial mouse - A robot with whiskers and vision. In: Proc. of the 35th ISR (2004)

Spectral Domain Noise Suppression in Dual-Sensor Hyperspectral Imagery Using Gaussian Processes

Arman Melkumyan and Richard J. Murphy

Australian Centre for Field Robotics, Rose Street Building J04
The University of Sydney, NSW, Australia, 2006
a.melkumyan@acfr.usyd.edu.au, richard.murphy@sydney.edu.au

Abstract. The use of hyperspectral data is limited, in part, by increased spectral noise, particularly at the extremes of the wavelength ranges sensed by scanners. We apply Gaussian Processes (GPs) as a preprocessing step prior to extracting mineralogical information from the image using automated feature extraction. GPs are a probabilistic machine learning technique that we use for suppressing noise in the spectral domain. The results demonstrate that this approach leads to large reductions in the amount of noise, leading to major improvements in our ability to automatically quantify the abundance of iron and clay minerals in hyperspectral data acquired from vertical mine faces.

Keywords: Hyperspectral, Gaussian Processes, Machine Learning, Feature Extraction, Absorption Feature, Iron minerals.

1 Introduction

Technological and methodological developments over the past 25 years have enabled remote identification, quantification and mapping of geological and biological materials on the Earth's surface using hyperspectral imagery. Hyperspectral imagery is most commonly acquired from airborne platforms. Continuing improvements in sensor technology have, however, enabled imagery to be acquired, cost-effectively, from field-based platforms for several applications including mapping of geology and mineraology [1]. Hyperspectral sensors typically measure reflected electromagnetic radiation in 10s to 100s of discrete, contiguous bands between 400 nm and 2500 nm.

Hyperspectral data enables absorption features, diagnostic of many biogeochemical materials, to be measured in a semi-continuous spectrum, enabling *identification* rather than mere *separation* of components [2] and [3]. Hyperspectral bands are narrow (< 6nm), and are often noisy in the spectral domain. Technological constraints mean that hyperspectral data are collected using different sensors to sample the Visible Near-InfraRed (VNIR; 400 – 1000 nm) and Short-Wave Infra-Red (SWIR; 1000 nm – 2500 nm) parts of the spectrum. Decreasing solar irradiance towards longer wavelengths, means that SWIR data are often acquired at a coarser spatial

K.W. Wong et al. (Eds.): ICONIP 2010, Part II, LNCS 6444, pp. 684–691, 2010.

resolution than are VNIR, allowing light to be collected over a greater area of ground. This means that data from the VNIR and SWIR sensors have to be spatially-registered after acquisition. The separate image cubes are then merged in the spectral domain so that each image pixel describes a complete spectral signature between 400 nm and 2500 nm.

Merging of these data presents problems for their subsequent analyses: 1) decreasing sensitivity of the sensors causes increasing noise towards the extremes of the VNIR spectrum (< 480 nm; > 960 nm); 2) the reflectance at long-wavelength terminus of the VNIR spectrum may not exactly match that of the short-wavelength terminus of the SWIR data, causing abrupt positive or negative changes in reflectance at this join; 3) the spectral join is located in the same part of the spectrum as a diagnostic absorption feature associated with iron minerals (Fig. 1), making the quantification of this feature difficult.

We propose a method based on Gaussian processes (GPs) [4] for suppressing noise in the spectrum at all wavelengths, with the primary objective of smoothing the spectrum across the wavelengths at, or close to, the junction of the data acquired by the two different sensors. This is a fully automated GP-based machine learning algorithm which uses the squared exponential covariance function [4] to learn the data provided by the sensors and suppress noise. Once spectral noise has been suppressed for each of the sensors, the algorithm compensates for the possible reflectance mismatch at the termini of VNIR and SWIR spectra, thus providing a single smoothed curve for all the wavelengths of interest. The smoothed spectral curve is then used to parameterize absorption features in the spectrum.

2 Materials and Methods

2.1 Hyperspectral Data

Hyperspectral imagery was acquired from a vertical mine face in an open-pit iron ore mine in Hamersley Province, Western Australia. Scanning VNIR and SWIR sensors (Specim, Finland) were mounted adjacently on a rotating stage. A reflectance panel (~ 99% Spectralon ®) was placed within the field of view of the sensors.

Data at each band, in each sensor, were corrected for dark current and converted to reflectance using pixel values over the calibration panel [5]. Data from the sensors were spatially-registered using multiple ground control points and merged into a single data-cube comprising 390 bands (400 – 2334 nm).

2.2 Gaussian Processes for Machine Learning

This section provides a brief introduction to GPs. Consider the supervised learning problem with a training set $D=(x_i, y_i)$, $i=1:N$, consisting of N input points $\mathbf{x}_i$ and the corresponding outputs y_i. The objective is to compute the predictive distribution $f(\mathbf{x}_*)$ at a new test point $\mathbf{x}_*$. The GP model uses a covariance function to place a multivariate Gaussian distribution over the space of function variables $f(\mathbf{x})$ mapping input to output spaces. This multivariate Gaussian distribution is then conditioned on the observed training dataset, resulting in a new predictive distribution for the points $\mathbf{x}_*$: $p(f_* \mid X_*, X, \mathbf{y}) = N(\boldsymbol{\mu}_*, \boldsymbol{\Sigma}_*)$ where $\boldsymbol{\mu}_*$ is the predicted mean and $\boldsymbol{\Sigma}_*$ is the predicted covariance.

During the learning stage GP model determines the hyper-parameters of the covariance function from the training dataset. In a Bayesian framework this can be performed by maximizing the log of the marginal likelihood (lml). The lml incorporates both data fit and complexity penalty to avoid possible overfitting of the dataset. This is a non-convex optimization task which can be performed using the gradient descent techniques with multiple starting points.

For further information on GPs and detailed mathematical derivations refer to [4].

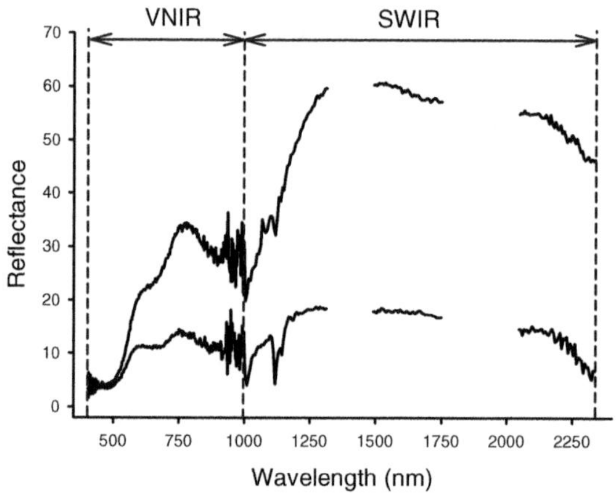

Fig. 1. Reflectance spectra of Goethite from individual image pixels. The spectral regions sensed by the VNIR and SWIR sensors are shown (*dotted vertical lines*).

2.3 Absorption Feature Extraction and Parameterization

To determine if Gaussian smoothing improved outcomes of spectral analysis, we compared the results from an Automated Feature Extraction (AFE) technique applied to the original and GP-smoothed image data. AFE automatically identifies absorption features and describes them in terms of a small number of parameters including, wavelength position, depth and width [6]. In the case of minerals, wavelength position is indicative of mineral type, depth is indicative of the mineral abundance and width is indicative of both type and abundance.

The basic concept of AFE is shown in Fig. 2 using a reflectance spectrum of goethite, acquired using a non-imaging field spectrometer (ASD, Boulder, Co.). In comparison to imaging spectrometers, field spectrometers produce relatively noise-free spectra (cf. Fig. 2a, Fig. 1). Several absorption associated with Fe^{3+} are evident

(Fig. 2a). The first stage of AFE is to remove the spectral continuum by dividing the spectrum by its upper convex hull (dotted line, Fig. 2a), at each wavelength. The resulting hull-quotients spectrum places all absorptions on the same plane of reference (Fig. 2b) [7]. The second stage identifies the wavelength position of the centre of an absorption feature and its shoulders (where the spectrum reaches unity); from these the other parameters are calculated. This is repeated for all absorptions in the spectrum.

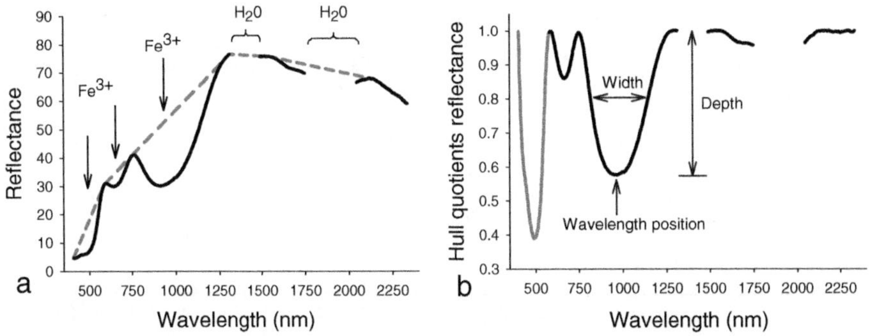

Fig. 2. Library spectrum of goethite: a) reflectance (*solid line*) showing 3 absorptions due to ferric iron (Fe^{3+}). The continuum is fitted over spectral maxima (*gray dashed line*). Data near 1400 nm and 1900 nm are not shown as they are affected by atmospheric water vapour. b) hull-quotients spectrum. The parameters (*wavelength position, depth and width*), derived from each absorption feature are indicated. The spectral region used to process the image data is indicated (*solid black line*).

Hyperspectral imagery from vertical mine faces can be used to determine their mineralogical composition and to separate ore from waste materials. Ore-bearing rocks have strong absorptions between 500 – 1300 nm. Some waste materials, mainly shale, can be distinguished by an absorption feature at 2208 nm caused by the clay mineral kaolinite. Parameterization of these absorption features using AFE enables ore to be separated from waste. Noise in image spectra strongly impacts all stages of AFE, making the determination of feature parameters inaccurate and imprecise.

3 Results

3.1 Effects on Image Spectra

Individual pixel spectra from areas of goethite and shale were extracted from the original and GP-smoothed images (Fig. 3). The original image spectra show large variations in reflectance caused by noise (Fig. 3, top panel). GP smoothing produces a seamless spectrum which is similar to the library spectrum acquired by the field spectrometer (cf. Fig. 3a, Fig. 2a). The hull-quotients spectra (lower panel) from GP-smoothed data are continuous and AFE now becomes straightforward.

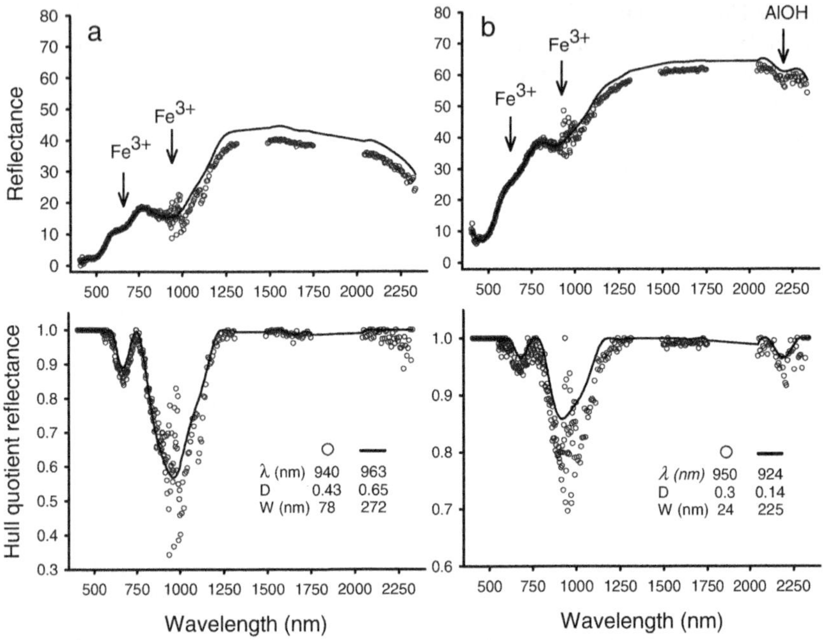

Fig. 3. Reflectance (top) and hull-quotients (bottom) spectra from individual image pixels; a) goethite, b) shale. Original (*circles*) and GP-smoothed (*solid line*) spectra show large differences. The parameters of the strongest absorption feature are shown in the bottom graphs: wavelength position (λ); depth (D) and width (W).

3.2 Effects on Parameter Images

GP smoothing of the image in spectral domain had no effect on the spatial domain. There were, however, major spatial improvements in the parameter images derived from GP-smoothed data. Mapping in the field indicated that the mine face was made up of distinct geological zones (Fig. 4).

An image of the depth parameter derived from the original image (Fig. 5a) shows greater depth, in zones 3 – 6, of the iron absorption at ~ 950-1000 nm. Some iron is present in the zones 1 & 2. There are no consistent changes in the depth of the feature among zones 3-6, indicating, incorrectly, that no single zone has more iron than another. The image of depth derived from the GP-smoothed image (Fig. 5b) showed an improved distinction between the ore and waste zones and zone 5 was correctly delimited from adjacent zones based on its iron content. Shales are distinguished by the presence of the ~2208 nm absorption due to kaolinite. The depth parameter of this feature, derived from the original image (Fig. 6a), showed increased amounts of kaolinite in zone 1 and, in particular, zone 2 but incorrectly quantified kaolinite in zones 3-6. In the less-noisy depth image, derived from GP-smoothed data (Fig. 6b), ore zones (3-6) are now accurately distinguished from the shales (1&2) and there is improved discrimination of linear variations in kaolinite in zone 2.

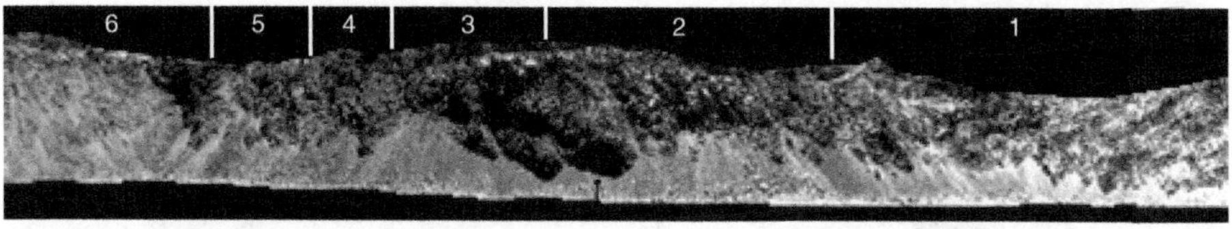

Fig. 4. Image of a mine face showing geological boundaries: 1) shale with moderate kaolinite; 2) manganiferous shale with abundant kaolinite; 3) mixed shale and goethite; 4) goethite; 5) martite-goethite; 6) martite and chert. Zones 1 & 2 are waste. Zones 3-6 are ore-bearing rocks, zone 5 being particularly abundant in iron.

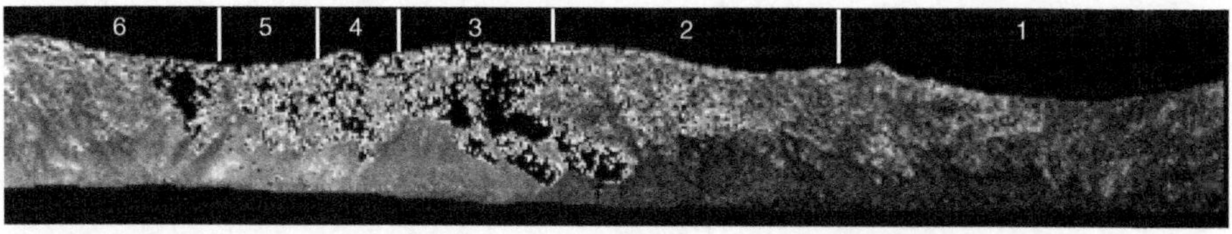

Fig. 5a. Depth parameter from the original image, for the deepest absorption feature. Pixel brightness in all images is proportional to the depth of the absorption feature.

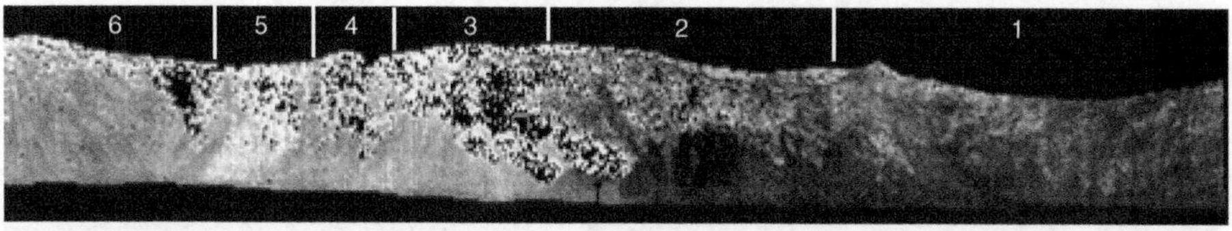

Fig. 5b. Depth parameter generated from the GP-smoothed image, for the deepest absorption feature. Zone 5, particularly iron-rich, is now distinguished.

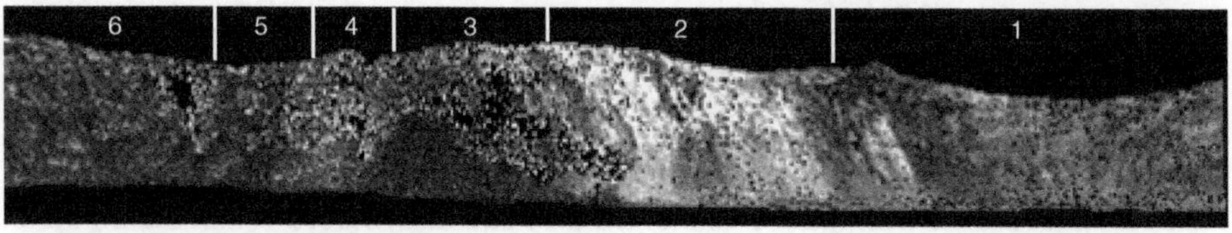

Fig. 6a. Depth parameter generated from original image, for the deepest absorption feature between 2000 nm and 2500 nm.

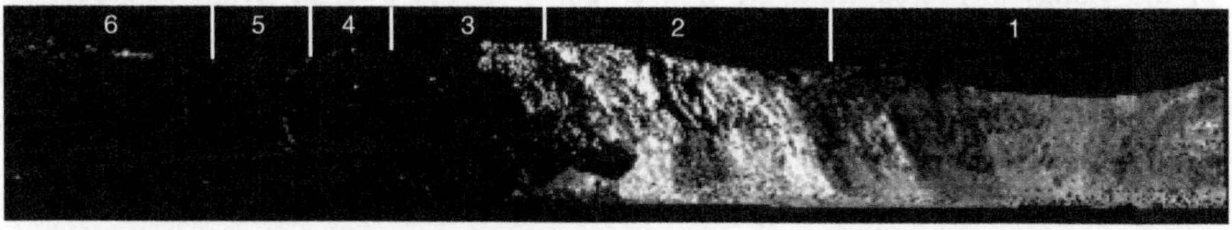

Fig. 6b. Depth parameter generated from the GP-smoothed image, for the deepest absorption feature in the spectrum between 2000 nm and 2500 nm.

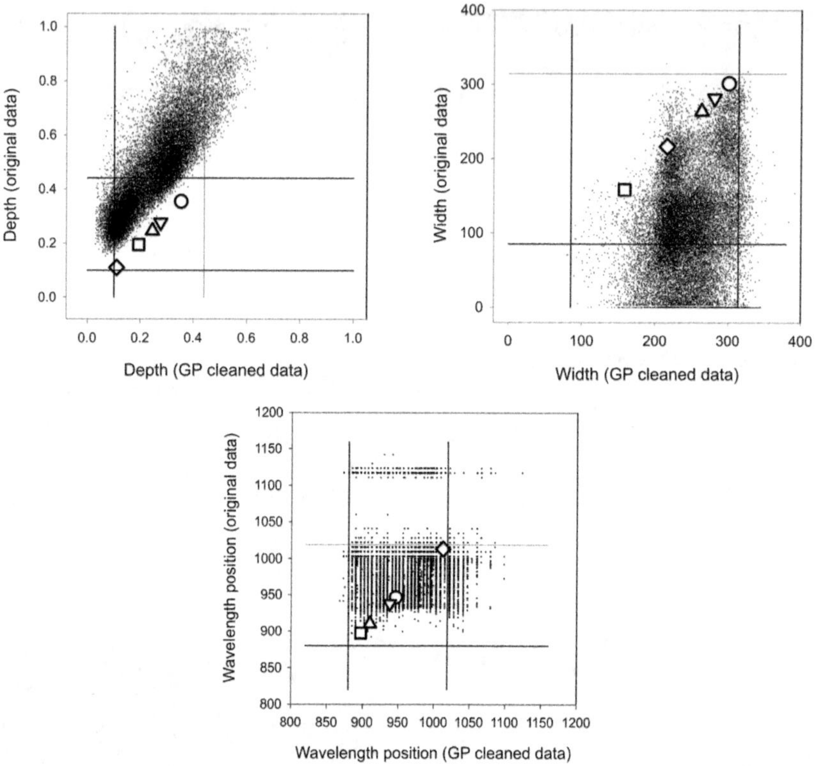

Fig. 7. Scatterplots of pixel values for each parameter, derived from the original and GP-smoothed data, superimposed with average parameter values for 5 rock types derived from the spectral library (large symbols). The minimum and maximum of values for each parameter derived from the library are indicated by straight lines.

3.3 Comparison of Parameters from the Image and Spectral Library

The images of wavelength position, depth and width were compared with the same parameters derived from a spectral library of 5 rock types found at the mine site (Fig. 7). The depth parameter from the GP-smoothed data had most values within the minimum and maximum values of the library spectra. Greater than 50% of pixels in the original image had values much greater than the maximal depth derived from the library spectra. The average depth of the absorption derived from library spectra, fell within the range of the values measured from the image. This was not true for depth derived from original image, where the depth parameter for one rock type - manganiferous shale (◊) - was below the limits defined by the pixel values for the original, but not the GP-smoothed data. Similarly, most pixel values for the width parameter derived from the GP-smoothed data fell within the limits of the library spectra but many from the original image fell below the minimal library limit. The majority of pixel values representing wavelength position derived from the GP-smoothed data, were within the limits derived from the library spectra, but were incorrectly partitioned into 2 discrete groups when derived from the original image. This is entirely due to spectral noise.

4 Conclusions

Noise in the spectral domain can have a deleterious impact on many techniques used to analyse hyperspectral imagery. The GP-smoothing method presented here greatly improved the discrimination and quantification of minerals in hyperspectral imagery of a vertical mine face. Used as a preprocessing step to AFE, the GP method enabled areas of abundant iron within the ore zones to be discriminated which were not distinguished in the original data. After application of GP-smoothing, absorptions indicative of kaolinite could be parameterized with high-specificity, enabling the separation of ore from waste materials and improving interpretation of the structure of the mine face. Further work is currently underway to improve results by incorporating spatial information.

Acknowledgments

This work has been supported by the Rio Tinto Centre for Mine Automation and the ARC Centre of Excellence programme, funded by the Australian Research Council (ARC) and the New South Wales State Government.

References

1. Herrala, R., Okkonen, J., Hyvärinen, T., Aikio, M., Lammasniemi, J.: Imaging Spectrometer for Process Industry Applications. In: European Symposium on Optics for Productivity in Manufacturing Optical Measurements and Sensors for the Process Industries, pp. 33–40. SPIE (1994)
2. Vane, G., Goetz, A.F.H.: Terrestrial Imaging Spectroscopy. Remote Sens. Environ. 24, 1–29 (1988)
3. Clark, R.N., Kind, T.V.V., Klejwa, M., Swayze, G.A.: High Resolution Reflectance Spectroscopy of Minerals. J. Geophys. Res. 95, 1265–12680 (1990)
4. Rasmussen, C.E., Williams, C.K.I.: Gaussian Processes for Machine Learning. MIT Press, Cambridge (2006)
5. Murphy, R.J., Underwood, A.J., Pinkerton, M.H.: Quantitative Imaging to Measure Photosynthetic Biomass on an Intertidal Rock Platform. Mar. Ecol. Prog. Ser. 312, 45–55 (2006)
6. Kruse, F.A.: Use of Imaging Spectrometer Data to Map Minerals Associated with Hydrothermally Altered Rocks in the Northern Grapevine Mountains, Nevada, and California. Remote Sens. Environ. 24, 31–51 (1988)
7. Clark, R.N., Roush, T.L.: Reflectance Spectroscopy: Quantitative Analysis Techniques for Remote Sensing Applications. J. Geophys. Res. 89, 6329–6340 (1984)

A High Order Neural Network to Solve Crossbar Switch Problem

Yuxin Ding[1], Li Dong[1], Ling Wang[1], and Guohua Wu[2]

[1] Department of Computer Sciences and Technology,
Harbin Institute of Technology, Shenzhen Graduate School, China
[2] Division of Mathematical Sciences, Nanyang Technological University
yxding@hitsz.edu.cn, dongli_hitsz@hotmail.com,
wangling_hitsz@hotmail.com, guohua@ntu.edu.sg

Abstract. High-order neural networks can be considered as an expansion of Hopfield neural networks, and have stronger approximation property and faster convergence rate. However, in practice high order network is seldom to be used to solve combinatorial optimization problem. In this paper crossbar switch problem, which is an NP-complete problem, is used as an example to demonstrate how to use high order discrete Hopfield neural network to solve engineering optimization problems. The construction method of energy function and the neural computing algorithm are presented. It is also discussed the method how to speed the convergence and escape from local minima. Experimental results show that high order network has a quick convergence speed, and outperforms the traditional discrete Hopfield network.

Keywords: Hopfield network, constraint satisfaction, crossbar switch problem.

1 Introduction

In 1985 Hopfield firstly used Hopfield neural network (HNN) to solve Traveling Salesman Problem which is a NP-complete combinatorial optimization problem [1]. Since then, Hopfield network has been widely applied to solve different combinatorial problems, such as map coloring [2], maximum cut problems [3], bipartite subgraph problems [4], crossbar switch problem [5-7]. Because high-order neural networks have stronger approximation property, faster convergence rate, greater storage capacity, and higher fault tolerance than lower-order neural networks, they have been intensively considered by researchers in recent years. In particular, there have been extensive results on the problem of the existence and stability of equilibrium points and periodic solutions of high-order Hopfield neural networks (HHNNs), for example, the references [8-12] all derive different sufficient conditions to guarantee the convergence of high-order neural network under different parameter settings. Due to the complexity of high-order network, the above researches mainly focus on the second-order continuous high-order Hopfield network. The reference [13] proves the stability of a special class of high-order discrete Hopfield neural network.

K.W. Wong et al. (Eds.): ICONIP 2010, Part II, LNCS 6444, pp. 692–699, 2010.

Compared with lower-order neural networks, high-order networks have their own advantages. However, in practice they are seldom used to solve combinatorial optimization problems. That is because constructing high order energy functions for optimization problems that satisfies the stability criteria of HHNNs is very difficult. So how to constructing high order energy functions for practical problems and whether using high-order HNN, instead of the first order HNN, to solve problems is valuable are two interesting questions. They are the research motivation of this paper. The rest of the paper includes the following parts. Section 2 introduces high-order discrete Hopfield neural network (HDHNN). In section 3, we propose the network construction method for the crossbar switch problem and compare its performance with the first order HNN and HHTN proposed by reference [5]. In this section we also discuss the strategy to escape from local minimum. The last section offers the conclusion of this paper.

2 High-Order Discrete Hopfield Network

Hopfield neural network has two models, continuous Hopfield neural network (CHNN) and discrete Hopfield neural network (DHNN). This paper focuses on DHNN. It is well known that DHNN operating in a serial mode will converge to a stable state which is corresponding to a local minimum of the Hopfield energy function if the connection weight matrix is symmetric and the diagonal elements of it are non-negative. However, DHNN can only handle optimization problem when the energy function can be expressed by a quadratic polynomial, if we want to deal with high-order problems, DHNN should be extend to high-order DHNN. Equation (1) is the energy function of high-order discrete Hopfield network [13]. Equation (2) is the state-evolving function of neurons [13].

$$E = -\frac{1}{n}\sum_{i_1}\sum_{i_2}\cdots\sum_{i_n} w_{i_1 i_2 \ldots i_n} x_{i_1} x_{i_2} \cdots x_{i_n} - \frac{1}{n-1}\sum_{i_1}\sum_{i_2}\cdots\sum_{i_{n-1}} w_{i_1 i_2 \ldots i_{n-1}} x_{i_1} x_{i_2} \cdots x_{i_{n-1}} - \ldots - \frac{1}{2}\sum_{i_1}\sum_{i_2} w_{i_1 i_2} x_{i_1} x_{i_2} - \sum_{i_1} I_{i_1} x_{i_1} \tag{1}$$

$$x_i(t+1) = f_h\left(-\frac{\partial E}{\partial x_i(t)}\right) \tag{2}$$

The connection weight of high-order DHNN is also symmetrical, that is, the value of $w_{i_1 i_2 \ldots i_n}$ is independent of the ordering of the index, for example, $w_{123} = w_{132} = w_{213} \ldots$. As the energy function is a very complicated high-order polynomial, even if the weights are symmetrical, the convergence can't be guaranteed. The reference [13] proves the stability of a special class of high-order DHNN operating in a serial mode when $x_i^k = x_i$ is hold. In general $f_h(y)$ is a binary function. If $y \le 0$, $f_h(y) = 0$ otherwise $f_h(y) = 1$.

3 Crossbar Switch Problem

3.1 Problem Description

Crossbar switch problem is an engineering problem in communication field; it is also an NP-complete problem. A $N \times N$ crossbar switch is a switch connecting a set of N inputs and N outputs where each input can be connected to any outputs as shown in fig.1. When there is a request from the input to output be satisfied the crosspoint switch will be closed. In each input line only one output line can be connected. Similarly, in each output line only one input line can be connected.

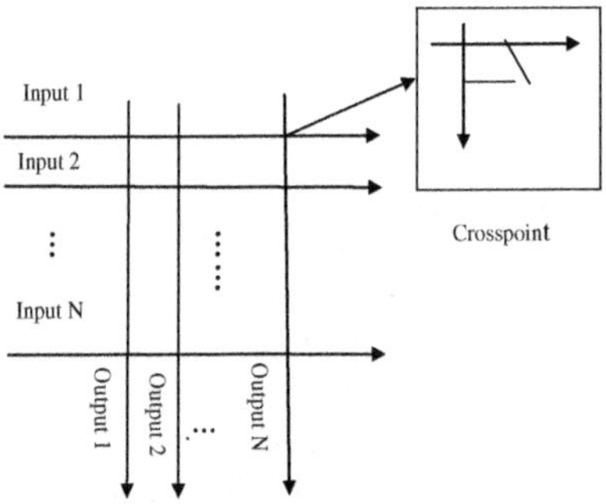

Fig. 1. $N \times N$ crossbar switch

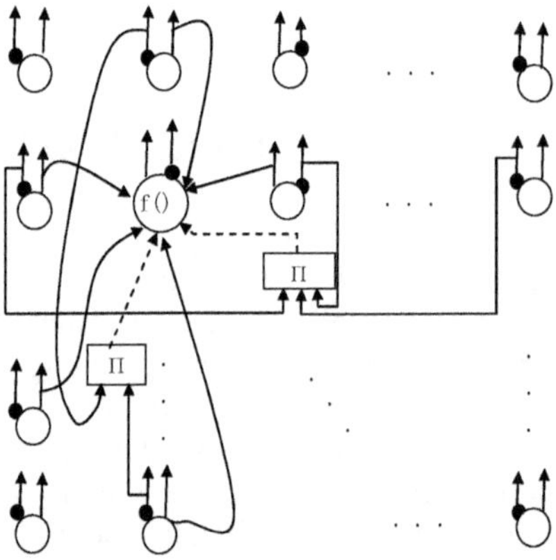

Fig. 2. Network topology graph

A $N \times N$ crossbar switch can be represented by a $N \times N$ binary request matrix R [5]. Rows and columns of the matrix R correspond to the inputs set and outputs set, respectively. There are two values 1 or 0 for each element in the matrix. $r_{ij} = 1$ means there is a request from i^{th} input line to the jth output line; $r_{ij} = 0$ expresses there is no request. The state of the switch can be represent by a $N \times N$ binary configuration matrix C, where $c_{ij} = 1$ indicates the request from i^{th} input line to the jth output line is satisfied, $c_{ij} = 0$ indicates that the request is discarded. For proper operation of the switch, there should be at most one request being satisfied in each row and each column. The throughput of the switch is optimal when the matrix C, which is a subset of the matrix R, contains at most a "1" in each row/column, and has a maximum overlap with R. this can be interpreted by the following example.

$$R = \begin{pmatrix} 1 & 0 & 0 & 0 \\ 1 & 1 & 1 & 0 \\ 0 & 0 & 0 & 0 \\ 1 & 0 & 0 & 1 \end{pmatrix} \qquad C_1 = \begin{pmatrix} 1 & 0 & 0 & 0 \\ 0 & 1 & 0 & 0 \\ 0 & 0 & 0 & 0 \\ 0 & 0 & 0 & 1 \end{pmatrix} \qquad C_2 = \begin{pmatrix} 1 & 0 & 0 & 0 \\ 0 & 0 & 1 & 0 \\ 0 & 0 & 0 & 0 \\ 0 & 0 & 0 & 1 \end{pmatrix}$$

R is an input request matrix for a 4×4 crossbar switch and its optimal configuration matrices will be C_1 or C_2. There are 6 requests, however only 4 requests are satisfied and the others are discarded.

Hopfield network and its variants [5]-[7] have been applied to solve this problem. These networks are all first order networks. The energy function used in these methods is described as Equation (3), where c_{ij} is the output of neuron i,j. In this section we use the crossbar switch problem as an example to discuss how to construct the high order energy function of this problem, how to construct its network topology, and how about its performance.

$$E=\frac{A}{2}\sum_{i=1}^{N}(\sum_{k=1}^{N}c_{ik}-1)^2+\frac{B}{2}\sum_{j=1}^{N}(\sum_{k=1}^{N}c_{kj}-1)^2 \tag{3}$$

3.2 Constructing Energy Function of Crossbar Switch Problem

As mentioned above, a $N\times N$ crossbar switch can be represented by a $N\times N$ binary request matrix. Let neuron $X_{i,j}$ corresponds the crosspoint switch at ith row and jth column($0\le i,j<N$). c_{ij} is the output of the neuron $X_{i,j}$, The constraint for ith row is represented as (4). In (4) the first item is zero when at least one request is satisfied in ith row; the second item is zero when at most one request is satisfied in ith row. Equation (4) equals zero, if and only if there is only one request be satisfied in ith row. In (4) the first item is a high order item. The constraint for all rows is represented as (5). Equation (5) equals zero, if and only if there is only one request be satisfied in each row. Equation (6) is the energy function for all columns. It equals zero if and only if there is only one request be satisfied in each column. The energy function for crossbar switch problem is the sum of E_1 and E_2, which is shown in (7). When (7) takes the minimum value 0, all c_{ij} are the solution of the problem.

$$\prod_{j=0}^{N-1}(1-c_{ij})+\sum_{j=0}^{N-1}c_{ij}\sum_{k=j+1}^{N-1}c_{ik}=0 \tag{4}$$

$$E_1=\sum_{i=0}^{N-1}(\prod_{j=0}^{N-1}(1-c_{ij}))+\sum_{i=0}^{N-1}\sum_{j=0}^{N-1}c_{ij}\sum_{k=j+1}^{N-1}c_{ik} \tag{5}$$

$$E_2=\sum_{j=0}^{N-1}(\prod_{i=0}^{N-1}(1-c_{ij}))+\sum_{j=0}^{N-1}\sum_{i=0}^{N-1}c_{ij}\sum_{k=i+1}^{N-1}c_{kj} \tag{6}$$

$$E=E_1+E_2 \tag{7}$$

We can expand (7) and simplify it by combing like terms, notice in the simplified polynomial the exponent of each variable c_{ij} is 1. For any item $Ac_{i_1j_1}c_{i_2j_2}...c_{i_nj_n}$, it can be transformed into $-\frac{1}{n}(-n)Ac_{i_1j_1}c_{i_2j_2}...c_{i_nj_n}$, n is the number of variables in the term.

So the energy function (7) has the same form as (1). c_{ij} satisfies $c_{ij}^k = c_{ij}$. This proves that (7) is a high order energy function of DHHNNs. $-nA$ is the value of one of high order weights among neurons X_{i_1,j_1} ,..., X_{i_n,j_n} .In this paper a high order weight $w_{i_1 j_1, i_2 j_2, \ldots i_n j_n}$ represents the weight from neurons X_{i_2,j_2} ,..., X_{i_n,j_n} to X_{i_1,j_1} . The DHHNN that is constructed according to (7) and works in a serial mode converges to a stable a point. However, in reality (7) is not required to be expanded, and can be directly used to construct the neural network.

3.3 Constructing High Order Hopfield Neural Network

In this paper the neurons are binary neurons; the state-evolving function of neurons is defined in (2). As discussed above if we do not expand *E*, there are two forms for any variable c_{ij} in *E*, c_{ij} and $1-c_{ij}$. So we expand the structure of neurons as following, each neuron has two outputs, one is the positive output c_{ij} , the other is the negative output $1-c_{ij}$. The input of a neuron c_{ij} is the sum of the following four parts:

1) The product of the negative outputs of neurons that are in the same row as c_{ij} ;

2) The negative of the sum of the positive outputs of neurons that are in the same row as c_{ij} ;

3) The product of the negative outputs of neurons that are in the same column as c_{ij} ;

4) The negative of the sum of the positive outputs of neurons that are in the same column as c_{ij} ;

The first and third parts represent high order weights. Take the neuron $X_{1,1}$ as an example. Its input is shown in fig.2. In fig.2 the neural network is represented as a $N \times N$ neuron matrix. Each neuron has two outputs; the black dot represents the negative output. Π is a multiplier. $X_{1,1}$ has two high order weights represented by the dot line, and their weight value is 1. From fig. 2 we can see that the high order network is not a fully connected network, the number of high-order connections is $2N^2$, the number of first order connections is $2(N-1)N^2$. The weight value of first order connections is -1. There is no the connection explosion problem that connections increase exponentially with the increase of neurons. The output of neurons is 0 or 1, so the output of the multiplier is 0 or 1, this shows the existence of high order weights do not affect the performance of the network. The neural computing algorithm working in serial mode is shown as following:

```
t=0
randomly initialize c_ij(0) to 1 or 0(i, j = 0,1,..N - 1)
3: For i = 0 to N - 1
For j = 0 to N - 1
```

```
{     d_ij(t) = -∂E / ∂c_ij(t)
      c_ij(t + 1) = f_h(d_ij(t) )
  let all c_mn(t + 1) = c_mn(t) ,  (m ≠ i or n ≠ j)
  t = t + 1;
  }
  if all c_ij(t) = c_ij(t - 1)
  return all c_ij(t)
  else  goto 3
```

In the above algorithm the threshold function $f_h(y)$ is defined in section 2. During the iteration process if all neurons' states are not changed and the energy is not equal to zero, this means the network traps into local minimum.

3.4 Strategy for Escaping from Local Minima

As gradient descent network Hopfield-type network is easy to fall into local minimum; two methods are usually adopted to escape from local minima: stochastic approach and deterministic approach. The deterministic approaches include the "divide and conquer" method [13], the "rock and roll" perturbation method [14], and neurons' competitive learning method [15]. Stochastic approaches include genetic algorithm [16], annealing theory [17], particle swarm optimization [18], and so on. Although theoretically stochastic approaches can reach the global optimum, practically it is very difficult to achieve. It not only takes long running time, but also very difficult to determine the termination conditions. All methods are suitable for the high-order gradient descent network. However, our target is to compare the difference in performance caused by the topologies of the high order network and the lower order network, so in this paper we do not use any global optimization strategies to escape from the local minimum discussed on above. We use the following method: select some variables $c_{i,j}$ randomly and reassign values to them, and then repeat the iteration process. This procedure is also called a disturbance. Every time only a small part of variables are selected and their values are changed to keep the energy value maintained at a relatively low level.

3.5 Simulation Results

In the experiments different $N \times N$ crossbar problems in which N ranges from 20 to 100 were simulated. For each $N \times N$ crossbar switch problem, we simulated it 100 times and in each simulation the request matrix was randomly initialized. Each simulation was terminated if a solution was found or the iteration step exceeded the maximum iteration step 1000. In this paper when all neurons are updated once, we call it an iteration step. For a simulation if a solution is found, the simulation is called a convergent simulation. In order to test the high order network's performance, we compare it with the traditional discrete Hopfield neural network (DHNN) [19]. The energy function of DHNN is defined as Equation (3). The parameters A and B were set to $A = 1, B = 1$. We also compared our results with Hysteretic Hopfield network with dynamic tunneling (HHTN) proposed by [5]. Their performance is evaluated by the following 2 criteria:

(1) Convergence rate: the ratio of the number of convergent simulations to the total number of simulations.

(2) Average iteration steps: the averaged iteration steps required for each simulation.

Table 1. Performance comparison for different size crossbar switch problems

N	Avg. Iteration steps			Converge Rate(%)		
	DHNN	HHTN	DHHNN	DHNN	HHTN	DHHNN
20	14	6	4	100	100	100
30	17	8	4	100	100	100
50	29	11	4	100	100	100
80	45	23	3	100	100	100
100	58	26	4	100	100	100

The result is shown in table 1. From the experimental results, for the $N \times N$ crossbar switch problems, the performance of the DHHNN is better than both the other two networks, especially for a larger N the improvement of performance is significant. It is because the high order network structure accelerates the convergence speed of the energy function. Furthermore, for DHHNN, the number of iteration steps is not increase greatly with the increase of N; it seems to be independent with N. However, the number of iteration steps of the other two networks is increased with the scale of the crossbar switch significantly.

4 Summary

In this paper $N \times N$ crossbar switch problem is used as an example to discuss how to construct high order networks to solve combinatorial optimization problems. In theory if we can find a high order energy function for any combinational optimization problem, which has the same form as (3), we can solve this problem by using a DHHNN. The experimental results show higher order network has a quicker convergence speed than the first order network and the number of its iteration steps is almost independent of the scale of the problem. It is valuable to construct high order network structures for applications. Our work is based on only one case study. Whether the result can be generalized to other problems is still need to be researched deeply.

Acknowledgments. This study is supported by the National high-tech research development plan (863) (No. 2007AA01Z194) and Key Laboratory of Network Oriented Intelligent Computation (Shenzhen).

References

1. Hopfield, J.J., Tank, D.W.: Neural computation of decisions in optimization problems. Biol. Cybern. 52(1), 141–152 (1985)
2. Galán-Marín, G., et al.: A Study into the Improvement of Binary Hopfield Networks for Map Coloring. In: Beliczynski, B., et al. (eds.) ICANNGA 2007. LNCS, vol. 4432, pp. 98–106. Springer, Heidelberg (2007)
3. Wang, J.: A Memetic Algorithm with Genetic Particle Swarm Optimization and Neural Network for Maximum Cut Problems. In: Li, K., Fei, M., Irwin, G.W., Ma, S. (eds.) LSMS 2007. LNCS, vol. 4688, pp. 297–306. Springer, Heidelberg (2007)
4. Wang, J., Tang, Z.: An improved optimal competitive Hopfield network for bipartite subgraph problems. Neurocomputing 61(5), 413–419 (2004)
5. Thangavel, P., Gladis, D.: Hopfield Hysteretic Hopfield network with dynamic tunneling for crossbar switch and N-queens problem. Neurocomputing 70, 2544–2551 (2007)
6. Xia, G., Tang, Z., Li, Y., Wang, J.: A binary Hopfield neural network with hysteresis for large crossbar packet-switches. Neurocomputing 67, 417–425 (2005)
7. Troudet, T.P., Walters, S.M.: Neural Network Architecture for Crossbar Switch Control. IEEE Trans. Circuits Syst. 38(1), 42–56 (1991)
8. W. Noguchi, C.-K. Pham.: A Proposal to Solve N-Queens Problems Using Maximum Neuron Model with A Modified Hill-Climbing Term. Neural Networks, IJCNN, 2679–2683 (2006)
9. Xu, B., Liu, X., Teoc, K.L.: Global exponential stability of impulsive high-order Hopfield type neural networks with delays. Computers and Mathematics with Applications 57(3), 1959–1967 (2009)
10. Yi, X., Shao, J., Yu, Y., et al.: Global exponential stability of impulsive high-order Hopfield type neural networks with delays. Journal of Computational and Applied Mathematics 219(3), 216–222 (2008)
11. Ou, C.: Anti-periodic solutions for high-order Hopfield neural networks. Computers and Mathematics with Applications 56(3), 1838–1844 (2008)
12. Gopalsamy, K.: Learning dynamics in second order networks. Nonlinear Analysis: Real World Applications 8(9), 688–698 (2007)
13. Cheung, K., Lee, T.: On the Convergence of Neural Network for higher Order Programming. IJCNN 2, 1507–1511 (1993)
14. Foo, Y.P.S., Szu, H.: Solving Large-Scale Optimization Problems by Divide-and-Conquer Neural Networks. IJCNN 1, 507–511 (1989)
15. Lo, J.T.-H.: A New Approach to Global Optimization and its Applications to Neural Networks. Neural Networks 2(5), 367–373 (1992)
16. Wang, L.P., Li, S., Tian, F.Y., Fu, X.J.: A Noisy Chaotic Neural Network for Solving Combinatorial Optimization Problems: Stochastic Chaotic Simulated Annealing. IEEE Trans. on Sys., Man, Cybern., Part B - Cybern. 34(5), 2119–2125 (2004)
17. Amartur, S.C., Piraino, D., Takefuji, Y.: Optimization Neural Networks for the Segmentation of Magnetic Resonance Images. IEEE Trans. on Medical Imaging 11(2), 215–220 (1992)
18. Salcedo-Sanz, S., Yao, X.: A hybrid Hopfield network-genetic algorithm approach for the terminal assignment problem. IEEE Trans on Systems, Man and Cybernetics 34(6), 2343–2353 (2004)
19. Li, Y.M., et al.: An Improvement to Ant Colony Optimization Heuristicp. In: Sun, F., Zhang, J., Tan, Y., Cao, J., Yu, W. (eds.) ISNN 2008, Part I. LNCS, vol. 5263, pp. 816–825. Springer, Heidelberg (2008)

Identification of Liquid State of Scrap in Electric Arc Furnace by the Use of Computational Intelligence Methods

Marcin Blachnik, Tadeusz Wieczorek, Krystian Mączka, and Grzegorz Kopeć

Silesian University of Technology, 8 Krasińskiego St., 40-019 Katowice, Poland
marcin.blachnik@polsl.pl
http://www.rm.polsl.pl

Abstract. A constant aspiration to optimize electric arc steelmaking process causes an increase of the use of advanced analytical methods for the process support. Optimization of the production processes lead to real benefits, which are, for example, lower costs of production. More often computational intelligence methods are used for this purpose. In this paper authors present three methods used for identification of liquid state of scrap in electric arc furnace using analysis of signals of the current of furnace electrodes.

Keywords: industrial application, process modeling, electric arc furnace, signal processing, noise estimation, classification.

1 Formulation of the Problem

Electric arc furnaces (EAFs) are widely used in steelmaking and in smelting of nonferrous metals. The electric-arc steelmaking process carried out in industrial circumstances is very complicated. The process and the final properties of steel depend on various, often difficult to precise factors. Hence, proper management of each melting process and optimal control of the process parameters are very important.

The liquefying of the charged materials is mainly performed by electric energy. Steel scrap, alloys and fluxes are charged into the EAF furnace with baskets or by means of charging barrows. After charging, the furnace is closed and melting process begins. The roof and the electrodes handling device move down, the roof is lowered and the furnace is switched on. The automation system predicts the required set points for the electric control, basing on either the defined production practice table for the steel grade or on a neural network controller [1].

This stage of the process is repeated by discharging additional baskets into the furnace and meltdown of these materials until enough liquid steel is available to reach the required tapping weight. Before tapping steel to the ladle, deslagging of the bath through the opened furnace door in the arc furnace is done. After deslagging, the liquid steel is tapped into the steel ladle and the EAF process starts again (fig. (1)). Due to stochastic behavior of the EAF load and intensity

K.W. Wong et al. (Eds.): ICONIP 2010, Part II, LNCS 6444, pp. 700–707, 2010.

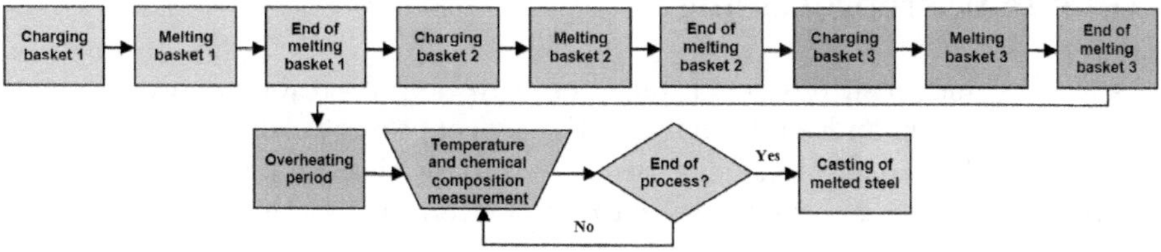

Fig. 1. Diagram of the melting process

of the disturbances, there has been an ongoing need for accurate and reliable predictors of the state of charge in the furnace (there is no possibility to control the process visually)[2]. The prediction problem is further complicated by the fact that the EAF characteristics are non-linear in most cases and fluctuation of the voltage amplitude is random and depends on the load of the furnace and its mode of operation. A number of non-linear influences and the temporary change in the process must be taken into account. Due to the variability and complexity of the EAF process, accurate optimization must be based on actual operating data, which is often noisy and requires significant preprocessing. The steelmaking process is continually changing, so is the charge of the furnace. Procedures and raw materials are changing too. Conventional control and optimization methods are not able to solve a task of controlling the EAF at the highest possible power with a low variance.

The electrical energy transferred to the melting steel should be optimally maintained during the entire melting process with the optimal distribution of radiated heat within the furnace (for protection of the furnace roof and walls). On the other hand, the scrap should be melted down in the shortest time possible. It is very difficult to identify the moment of time when the basket is melted down enough and discharging the next basket is possible. It is still uncertain to determine the melting process parameters. Nowadays, a believable observation of the process is almost impossible. This problem requires an adaptive data-based modeling system [4]. Computational intelligence methods (CI) can solve these difficulties and therefore, these methods are used for this kind of modeling problems [5,6,7].

The method used in the paper for identification of scrap liquid state was based on the observation that melted steel has a great impact on stabilizing electric arc. It is caused by the fact that when most of steel is melted down the arc burns calmly because the slag is more foamy and covers the arc. The covered arc burns in a more stable way. The increase of plasma gases also stabilizes the arc. Therefore, identification of scrap liquid state is equivalent to identification of the furnace operation state, which we called "calm arc".

In the paper, three methods of identification of the calm arc state basing on intelligent modeling are described. Section 2 presents our approach to preprocessing of signals, section 3 presents performed experiments and the obtained results and finally, section 4 presents discussion on the obtained results and conclusions.

2 Signal Preprocessing

As already has been described, an alternative approach for monitoring scrap melting process in an EAF is analysis of electrical properties of electrodes. There are three independent electrodes used for scrap melting, all powered by a single source. The figure (2) presents the current of electrodes. In this chart, six phases

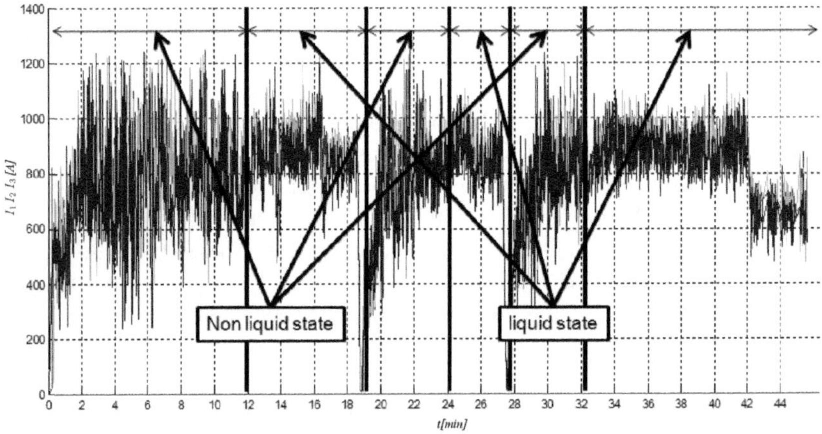

Fig. 2. Chart of the current of electrodes of the whole melting process, with marked process stages

of the melting process, including refining phase (described in the introduction), can be noticed. Each phase is marked by vertical lines separating the whole process into subprocesses.

The goal of analysis of the signals is to predict the liquid state of scrap and the beginning of the refining phase. Manual analysis of charts of the current of the electrodes performed by process engineers proved that the most important property, which distinguishes these individual phases is the level of noise of the electric arc. High amplitude of the noise is related to the non-liquid state and respectively low level of the noise is related to the liquid state. Assuming that the measured signals can be defined as a smooth function f of some input variables z_i and additive noise n and then, the resulting signal y can be defined as:

$$y = f(t, z_1, z_2, \ldots, z_k) + n \tag{1}$$

The next aim is to extract the noise and to use its amplitude as input feature/features for the classifier. Moreover, according to the previous paragraph, the amplitude of the noise should be a monotonically decreasing function so the obtained classifier could be a simple, threshold-based classifier (e.g. single node decision tree).

2.1 Estimation of Variance with Sliding Window

In our first and the simplest approach, we have assumed that the level of noise is measured by the local variance of the signal of each independent current of electrode. Due to the fact that analysis of variance required to be performed simultaneously, sliding window estimation was used. In this approach the last k samples (in our case the samples of the last 30s, because different sampling frequencies may have appeared) were used to estimate desired value - in this case it was the variance of the signal. Unfortunately, the sliding window estimation of the noise has some drawbacks that can lead to incorrect prediction. According to the definition (1): if the function f is not constant ($f \neq const.$) inside the given window $t = [t - 30s, t]$ the obtained variance is overestimated. This situation appears especially in the beginning of the melting process and after each break when high-voltage taps of the transformer are frequently changed, but also in other cases when the current is linearly changed. Sudden increases of variance value is another disadvantage of this method which can be observed during sudden changes of signal values. In our application such situation appears during the refining phase when the tap of transformer is changed radically reducing power of electrodes. Both situations are presented in the figure (3).

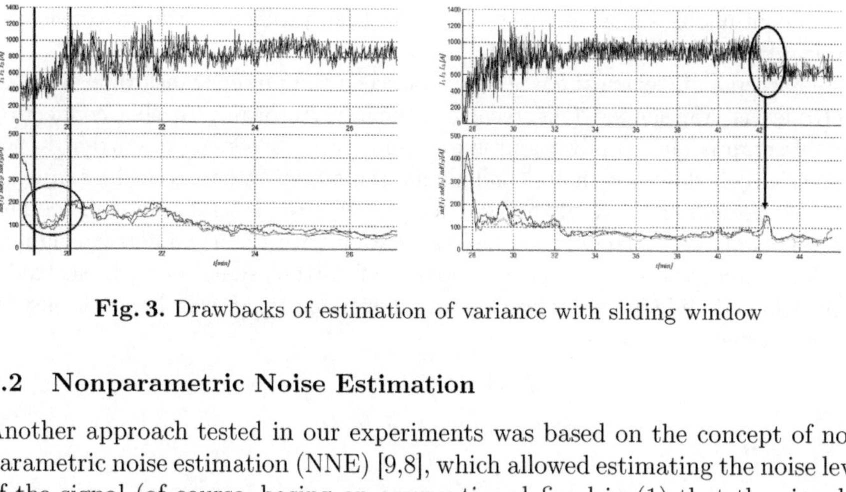

Fig. 3. Drawbacks of estimation of variance with sliding window

2.2 Nonparametric Noise Estimation

Another approach tested in our experiments was based on the concept of non-parametric noise estimation (NNE) [9,8], which allowed estimating the noise level of the signal (of course, basing on assumption defined in (1) that the signal is defined by a smooth function f of input variables z_i and additive noise n (1)). The goal of NNE is to estimate the variance of that noise $var(n)$. It is usually used to estimate the level of noise which disturbs the function f by calculating the desired error rate (mean square error) of function f estimator

$$MSE_{true}\left(M\left(z_i\right)\right) = MSE_{estimated}\left(M\left(z_i\right)\right) - NNE(z_i) \tag{2}$$

Where $MSE_{estimated}\left(M\left(z_i\right)\right)$ is the MSE of our model M trained by the use of features z_i, $NNE(z_i)$ is the variance of the noise $(var(n))$ and $MSE_{estimated}\left(M\left(z_i\right)\right)$ is the true MSE that should approach to zero without apprehension of overfitting.

The example of such simple noise estimator is so-called Delta-test. Delta-test is defined as:

$$var(n) \approx \gamma = \frac{1}{2M}\sum_{i=1}^{M}\left(y_i - y_{N[i,1]}\right)^2, \text{ with } var(\gamma) \rightarrow 0 \text{ for } M \rightarrow \inf \tag{3}$$

where $y_{N[i,1]}$ is the first nearest neighbor to sample y_i. Due to the assumption that the NNE is additivity of the noise it could be directly applied to resolve problems of estimation of the level of the noise in signals of the current of electrodes. In this approach, we have also applied the sliding window in such a way that the level of the noise was estimated in the given window.

The properties of NNE allow reducing or even removing one of the problems that appeared during estimating the variance described in the section 2.1. As NNE is insensitive to the function f, calculated noise level is not overestimated and more accurately reflects in the true value of noise. The only existing problem is the sensitivity to the step-like signal changes. None of already defined methods can deal with it.

2.3 Analysis of Symmetry between Signals

None of the presented methods of noise estimation is able to face the problem of non-continuous signal changes. To solve that problem we have considered physical properties of the modeled object.

The scheme of electrical connections between the power transformer and the electrodes is star-shaped (the receiver scheme). According to this connection type, if scrap is melted down and is in liquid state, dispersing of current in the receiver circuit does not change and should remain constant. According to what has been presented above, if the symmetry is disturbed, scrap is not in a liquid state. This allows replacing the noise estimation with the asymmetrical signals. Instead of analysis of the noise component of current signal of each electrode, we considered differences between current signals, what was less complex in computations:

$$\begin{aligned} I_a &= abs\left(I_1 - I_2\right) \\ I_b &= abs\left(I_1 - I_3\right) \\ I_c &= abs\left(I_2 - I_3\right) \end{aligned} \tag{4}$$

where I_a, I_b and I_c are new signals, and the I_1, I_2 and I_3 are original currents of these electrodes. The desired property of the new signals is insensitivity to non-continuous signals that were described in the section (2.1). The plot of new signal is presented in the figure (4).

3 Experiments and Results

3.1 Dataset Description

All methods described in (2) were used to construct datasets for training and validating the final classifier. The datasets were created by concatenation of 24

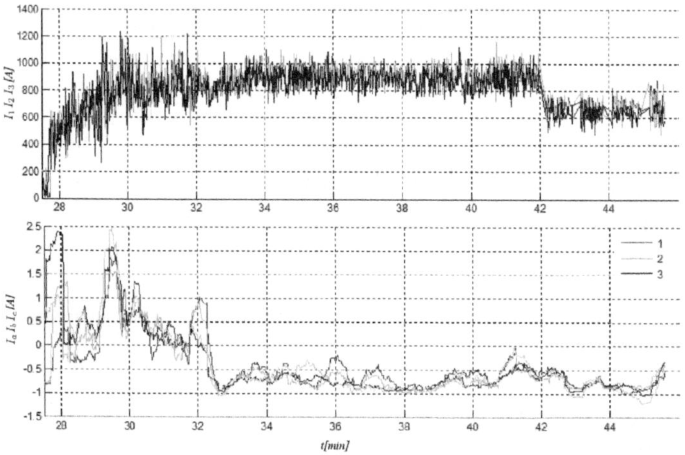

Fig. 4. Original current signals and asymmetrical signals

independent melting processes, each consisting of three phases described in the introduction and processed according to the schemes presented above. The total number of samples in the datasets used for training the classifiers was equal to 5501 and number of features was dependent on the preprocessing scenario.

Each melting process was performed independently with a different scenario. The first scenario was designed in such a way that an average noise signal was estimated for each electrode and then, the classifier had just one single input variable $\overline{n} = avr\left(n_1, n_2, n_3\right)$, where $n_{1\ldots3}$ were estimated noise of current signal of each electrode $I_1 \ldots I_3$ for given input signal. The second tested procedure was designed in such a way that estimation of noise was independent for each electrode. In that case the classifier had three input variables, each of them represented the noise level related to each electrode. In the next two scenarios, noise estimation was replaced by asymmetry analysis. As the input features for the classifier average of I_a, I_b, I_c and the asymmetry signals without averaging were used. The last tested scheme - called Meta Process - included combination of all variables from the previous steps.

3.2 Test Procedure

All the methods of noise level estimation described in this paper: estimation of variance with sliding window, the NNE approach and asymmetry analysis were tested with different classifiers with the same scenarios for all datasets. This allows determining the quality of the combination of noise extraction method and a given classifier. The testing scheme consisted of 5 steps:

1. determining the window size
2. estimation of noise level for each sample (see next subsection)
3. labeling each sample as *liquid* or *non-liquid*
4. determining classifier properties

5. estimating the classifier quality using 5 fold cross validation procedure
6. selecting the best parameters set (window size, classifier properties - C, sigma)

3.3 Results

In our experiments Gaussian and linear SVM and CART decision tree were used. The results collected in the table (1) represent only the best results obtained for a given classifier.

Table 1. Comparison of the results obtained for different preprocessing procedures and for different classifiers

	Preprocessing	Classifiers		
		Gaussian SVM	Linear SVM	C4.5
1	Sliding window variance (SWV)	96.8	95.7	97.2
2	Average of (SWV) signals	91.3	89.6	91.3
3	NNE preprocessing	96.8	96.7	96.7
4	Avarage of NNE signals	90.6	90.5	90.2
5	Asymmetry analysis	95.8	95.2	96.2
6	Average of asymmetry signals	96.0	96.0	95.8
7	Concatenation of 1,3,5	98.4	97.0	98.4

4 Discussion of the Obtained Results and Conclusions

Online determination of the state of scrap during the melting process is very important and can bring measurable benefits, both economical and ecological, by reducing energy consumption and speeding up the process. Moreover, it is very crucial that the approach proposed in this paper does not incur any extra expense or special equipment and uses only the information which is already stored in databases.

In this paper we have compared three different approaches of preprocessing of measured signals to create datasets used for training the classifier. Two of those methods were based on statistical analysis and one was based on physical properties of the electric scheme of the heating system. In our experiments all three methods were tested with two different scenarios. In the first one, estimated values were the average ones, what simplified the classification model. Such scenario was verified to check if that simplification could be applied without loss in accuracy. The obtained results presented in the table (1) pointed out that such simplification is not correct and leads to significant loss in accuracy (oscillating in the region 6%). In our experiments two statistical approaches obtained better results then the one based on asymmetry analysis. Surprisingly, the results obtained by estimation of variance with sliding window were better then the ones obtained by Delta-test. The best results of all classifiers were obtained by concatenation of all three preprocessing steps. In the Meta-Process scenario, each method was able to bring their own benefits which could downgrade any possible drawbacks of other methods. The results of all tested methods show that C4.5 decision tree and Gaussian SVM were better than the linear model. Moreover, C4.5 decision tree was usually slightly better then Gaussian SVM. This situation can be explained by the fact that the input features were independent signals (monotonically decreasing) and each of that signals could be thresholded.

Methodology proposed in this paper does not cover all possible statistical tests. We are planing to extend our research by other, more advanced NNE approaches. We are going to experiment with the Gamma test [9] that would bring more detailed information about noise. We also intend to process other measured and stored in the database signals, e.g. the voltage of electrodes.

Acknowledgements

The project was partially sponsored by the grants No N N508 486638 and N N516 442138 from the Polish Ministry of Education and Science (MNiSW).

References

1. Wieczorek, T.: Intelligent control of the electric-arc steelmaking process using artificial neural networks. Computer Methods in Material Science 6(1), 9–14 (2006)
2. Wieczorek, T., Pilarczyk, M.: Classification of steel scrap in the EAF process using image analysis methods. Archives of Metallurgy and Materials 53(2), 613–618 (2008)
3. Millman, M.S., Nyssen, P., Mathy, C., Tolazzi, D., Londero, L., Candusso, C., Baumert, J.C., Brimmeyer, M., Gualtieri, D., Rigoni, D.: Direct observation of the melting process in an EAF with a closed slag door. Archives of Metallurgy and Materials 53(2), 463–468 (2008)
4. Kendall, M., Thys, M., Horrex, A., Verhoeven, J.P.: A window into the electric arc furnace, a continuous temperature sensor measuring the complete furnace cycle. Archives of Metallurgy and Materials 53(2), 451–454 (2008)
5. Wieczorek, T., Blachnik, M., Mączka, K.: Building model for time reduction of steel scrap meltdown in the electric arc furnace. In: Rutkowski, L., Tadeusiewicz, R., Zadeh, L.A., Zurada, J.M. (eds.) ICAISC 2008. LNCS (LNAI), vol. 5097, pp. 1149–1159. Springer, Heidelberg (2008)
6. Wieczorek, T., Mączka, K.: Modeling of the AC-EAF process using computational intelligence methods. Electrotechnical Review 11, 184–188 (2008)
7. Blachnik, M., Mązka, K., Wieczorek, T.: A model for temperature prediction of melted steel in the electric arc furnace (EAF). LNCS, vol. 4839, pp. 371–378. Springer, Heidelberg (2010)
8. Jones, A.J.: New tools in non-linear modelling and prediction. Computational Management Science 1, 109–149 (2004)
9. Liitiäinen, E., Corona, F., Lendasse, A.: Nearest Neighbor Distributions and Noise Variance Estimation. In: ESANN, Belgium (2007)
10. Schölkopf, B., Burges, C., Smola, A.: Advances in Kernel Methods: Support Vector Machines. MIT Press, Cambridge (1998)
11. Quinlan, J.R.: C4.5: Programs for machine learning. Morgan Kaufman, San Francisco (1993)

Simulating Wheat Yield in New South Wales of Australia Using Interpolation and Neural Networks

William W. Guo, Lily D. Li, and Greg Whymark

School of Information and Communication Technology
Central Queensland University
North Rockhampton QLD 4702, Australia
{w.guo,l.li,g.whymark}@cqu.edu.au

Abstract. Accurate modeling of wheat production in advance provides wheat growers, traders, and governmental agencies with a great advantage in planning the distribution of wheat production. The conventional approach in dealing with such prediction is based on time series analysis through statistical or intelligent means. These time-series based methods are not concerned about the factors that cause the sequence of the events. In this paper, we treat the historical wheat data in New South Wales over 130 years as non-temporal collection of mappings between wheat yield and both wheat plantation area and rainfall through data expansion by 2D interpolation. Neural networks are then used to define a dynamic system using these mappings to achieve modeling wheat yield with respect to both the plantation area and rainfall. No similar study has been reported in the world in this field. Our results demonstrate that a four-layer multilayer perceptron model is capable of producing accurate modeling for wheat yield.

Keywords: Neural networks, multilayer perceptron, interpolation, wheat yield, plantation area, rainfall, New South Wales.

1 Introduction

New South Wales (NSW) is an important wheat growing state in Australia. A dynamic simulating system that is able to accurately predict wheat yield in advance can provide a great advantage for wheat growers, traders, and governmental agencies in planning for distribution of wheat yield. The commonly used approaches to model such temporal events are based on time series analysis by statistics and/or intelligent means, such as the autoregressive moving average model (ARMA) [1], generalized linear autoregression (GLAR) [2], artificial neural networks (ANN) [3][4], and their combinations [5][6]. Time series analysis treats the most recent sequence of events more important than the earlier ones in modeling, and thus focuses on the appearances of consecutive events, from which forecasting is drawn. However, time series analysis is concerned little about the factors that cause the sequences of events in the analysis. For example, wheat yield is an annual time series, but the wheat yield in a year is more likely to be affected by many factors, such as plantation area, rainfall, quality of seeds, temperature variation, fertilization, and level of disease occurrence. Some of these factors, if not all, should be considered in wheat yield simulation.

K.W. Wong et al. (Eds.): ICONIP 2010, Part II, LNCS 6444, pp. 708–715, 2010.

To incorporate other factors that affect wheat yield in simulation, we treat the historical data of annual wheat yield as an unknown function determined by the annual wheat plantation area and rainfall in wheat growing stage (mainly in autumn and winter), two factors that have a broad impact to the wheat yield in a large region. To model such unknown and nonlinear function, we employ multilayer perceptron (MLP) neural networks to approximate the wheat yield in respect to plantation area and rainfall, irrespective to the time factor.

Neural networks were used to map the nonlinear relation between wheat yield and plantation area without considering rainfall data [7]. However, there has been little information published in studies similar to what this research is brought in modeling wheat yield by incorporating both plantation area and rainfall together through neural networks. The purposes of this study are firstly to analyze the nonlinearity of this problem through statistical data analysis; secondly to approximate the nonlinear relation between wheat yield and both plantation area and rainfall using MLPs, through which their usability in wheat forecasting can be assessed; discussion and conclusion can then be made based on the outcomes for quantitative simulation of wheat yield.

2 Data Pre-processing and Statistical Analysis

The annual wheat plantation area in hectares and yield in tonnes in New South Wales (NSW) from 1861 to 2007 are available in the report of Australian Bureau of Statistics [8]. Among the 135 data entries, the 132 datasets from 1876 to 2007 are consecutive annual results and hence are selected for our study. The wheat yield over the years varies from tens of thousands to over eight million tones with correspondence to variations in plantation area ranging from tens of thousands to over four million hectares. For the convenience of neural network training and modelling, both the wheat plantation area and yield are normalized to the range between 0 and 1 using 5 million hectares and 10 million tonnes as the normalizing factors respectively. After the neural network simulation, all results can be converted back to their original scales. Since we use relative absolute mean error (RAME) to assess the accuracy of wheat yield forecasting, backward conversion is actually not necessary in our discussion.

Over these 132 years, in general, the wheat yield in NSW kept increasing, except two downward periods in the 1950s and the late 1980s to the early 1990s (Fig. 1). These variations correlate to the fluctuations in plantation area over the years. This shows that the wheat plantation area indeed largely affects the wheat yield. Correlation analysis reveals a quadratic correlation between them with a coefficient of 0.8967, and it is very useful in understanding the general relation between wheat yield and plantation area. However, it is too coarse to be used for quantitative prediction.

In southern Australia, including NSW, wheat sowing occurs in April after autumn rain. If the soil is moist, it will sprout in 5-7 days and takes 5-7 months to mature. This spans the late autumn (April-May), entire winter (June-August), and early spring (September-October). While growing, wheat requires 200-380 mm of rain, particularly in the early to middle growing stages. As a result, we also select the total rainfall of autumn and winter as another contributing factor to wheat yield in this study. Since the Australian Bureau of Meteorology (ABM) [9] keeps rainfall statistical data only from 1900 onwards, there are 108 rainfall datasets that can be mapped to the wheat yield records. Therefore, we have 108 entries for simulation with MLP.

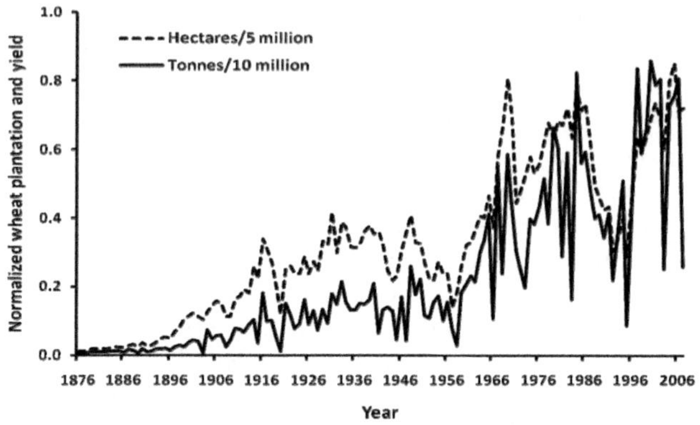

Fig. 1. Sequence of annual wheat plantation and yield in NSW (1876-2007)

The accuracy of nonlinear function approximation through MLPs requires a large base of training datasets that ideally covers the whole space of potential occurrences. If we choose both wheat plantation area and the rainfall in autumn and winter as the input to MLP to approximate wheat yield, the 108 entries from 1900 to 2007 are insufficient for training a reliable MLP. The 2D interpolation is used to expand the training dataset.

3 Expansion of Training Data Using 2D Interpolations

Since one wheat yield value is associated with both a plantation area value and a rainfall value, we can explore the usefulness of some 2D interpolation methods for data expansion through creating a 3D surface constrained by the 108 known mappings. A number of interpolation algorithms are considered for achieving this goal.

Bilinear interpolation approximates the value $V(x, y)$ at a given point $P(x, y)$ based on the known values at the four corner points of a unit square [10][11]. In the coordinate system illustrated in Fig. 2, in which the four corner points are located at $P(0, 0)$, $P(0, 1)$, $P(1, 0)$, and $P(1, 1)$, this interpolation is then expressed as

$$V(x,y) \approx \begin{pmatrix} 1-x & x \end{pmatrix} \begin{pmatrix} V(0,0) & V(0,1) \\ V(1,0) & V(1,1) \end{pmatrix} \begin{pmatrix} 1-y \\ y \end{pmatrix}. \quad (1)$$

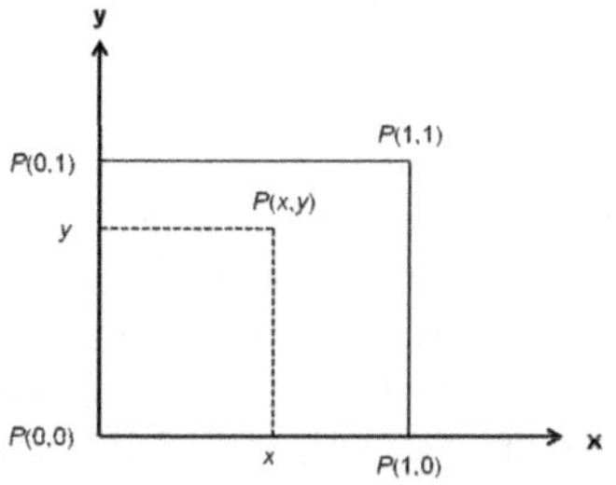

Fig. 2. Illustration of coordinate system for 2D interpolation

More sophisticated interpolation algorithms use nonlinear functions to approximate the unknown values based on some known values [10]. Polynomial interpolations are examples of these algorithms. The 2D Lagrange polynomial interpolation [12], based on $(n + 1)(m + 1)$ known points, is given by

$$V(x, y) \approx \sum_{i=0}^{n} \sum_{j=0}^{m} L_{ij}(x, y) V(x_{i,} y_j), \tag{2}$$

where its kernel function is defined as follows:

$$L_{ij}(x, y) = L_i(x) L_j(y) \qquad 0 \le i \le n,\ 0 \le j \le m, \tag{3}$$

$$L_i(x) = \prod_{s=0, s\neq i}^{n} \frac{(x - x_s)}{(x_i - x_s)} \quad \text{and} \quad L_j(y) = \prod_{s=0, s\neq j}^{n} \frac{(y - x_s)}{(y_i - x_s)}, \tag{4}$$

whereas

$$L_{ij}(x_r, y_s) = \begin{cases} 1 & i = r,\ j = s \\ 0 & otherwise \end{cases}. \tag{5}$$

Assuming the known values being at the four corners shown in Fig. 2, third-order polynomials can be used to interpolate a surface within the square by

$$V(x, y) \approx \sum_{i=0}^{3} \sum_{j=0}^{3} a_{ij} x^i y^j, \tag{6}$$

where a_{ij} denotes the 16 coefficients that must be determined by the four equations yielded by the original values at the four corners, eight equations yielded from derivatives in the x-direction and y-direction, and four equations resulted from the cross derivative. This approach is called the bicubic interpolation [10][13].

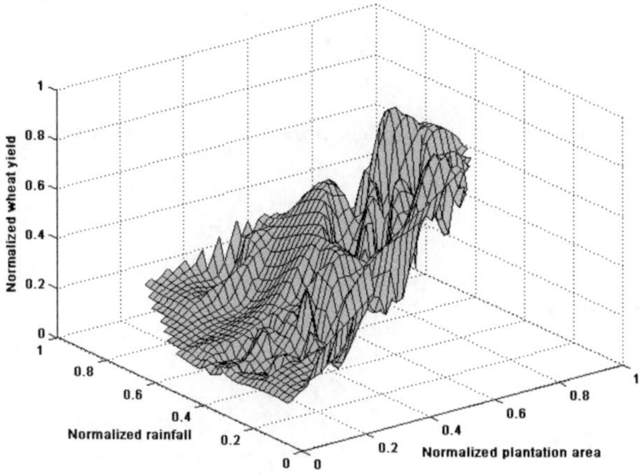

Fig. 3. The 3D surface created using the bicubic interpolation for this study

The 2D Lagrange polynomial requires $(n + 1)(m + 1)$ known points to support the interpolating process and may lead to the significant loss of data in the marginal regions, which is a disadvantage if it is applied to this problem with limited data available. As a result, the bicubic interpolation is chosen to expand the training data.

The 3D surface approximated using bicubic interpolation constrained by the 108 known datasets is shown in Fig. 3. Note that the rainfall values are also normalized to the range of 0 to 1 by 500 mm. By excluding the mappings either close to the edges of this surface or overlapped with the 108 known datasets, there are 1050 datasets extracted from this surface for MLP training. The original 108 datasets will be used to test the trained MLP for wheat yield simulation.

4 Simulating Wheat Yield Using Neural Networks

To choose an appropriate neural system for this simulation, three-layer and four-layer multilayer perceptrons (MLPs) are selected for training and testing. We selected these MLPs because many successful applications using these MLPs have been reported in various fields [14-16].

For the three-layer MLP (Fig. 4a), the first layer has two linear neurons for taking plantation area and rainfall as the input respectively. The two values are then fed to the hidden layer with a number of *tansig* neurons. The output layer has one linear neuron for wheat yield. The output of this three-layer MLP can be written as

$$O = \sum_{j=1}^{n} B_j \tanh\left(\sum_{i=1}^{2} A_{ij} x_i\right), \tag{7}$$

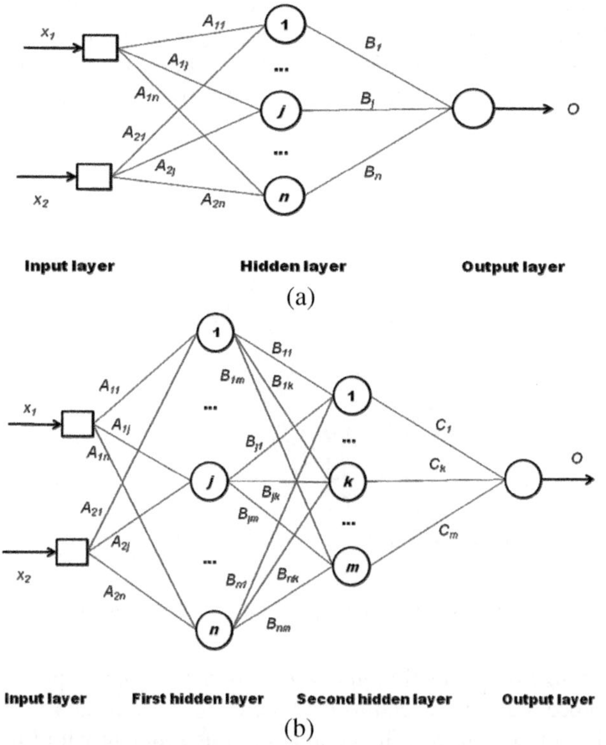

Fig. 4. Three-layer MLP (a) and four-layer MLP (b) used in this study

where x_i is either the plantation area or rainfall; A_{ij} is the weight from the input to the jth neuron in the hidden layer; B_j is the weight from the jth neuron in the hidden layer to the linear neuron as the output.

The four-layer MLP has two hidden layers with different numbers and types of neurons (Fig. 4b). The first hidden layer is constructed using *logsig* neurons whereas the second hidden layer consists of *tansig* neurons. Both the input and output layers are the same as the three layer MLP. The output of this four-layer MLP is defined as

$$O = \sum_{k=1}^{m} C_k \tanh(\sum_{j=1}^{n} B_{jk} (\log(\sum_{i=1}^{2} A_{ij} x_i)), \quad (8)$$

where B_{jk} is the weight from the jth neuron in the first hidden layer to the kth neuron in the second hidden layer; C_k is the weight from the kth neuron in the second hidden layer to the linear neuron as the output.

Many performance functions can be used to control the process of neural network training. The mean square error (MSE) defined below is chosen in this study:

$$MSE = \frac{1}{N} \sum_{t=1}^{N} (y_o(t) - y_s(t))^2, \quad (9)$$

where y_o and y_s are the original and simulated values, respectively.

In this study, the Levenberg-Marquardt algorithm [17] is chosen to train the selected MLPs because this algorithm has been reported to be the fastest method for training moderate-sized feedforward neural networks [18][19]. Our MLP models are built using the neural network tools in MATLAB® [20].

We choose three-layer and four-layer MLPs for simulating the wheat yield through plantation area and rainfall. The training is constrained by the 1050 expanded datasets and controlled by MSE. The 108 original datasets are then used to test the trained MLPs. Our experiments show that many MLPs can return consistent and satisfactory outcomes, with the four-layer MLP outperforming the three-layer MLP as illustrated in the cases shown in Fig. 5.

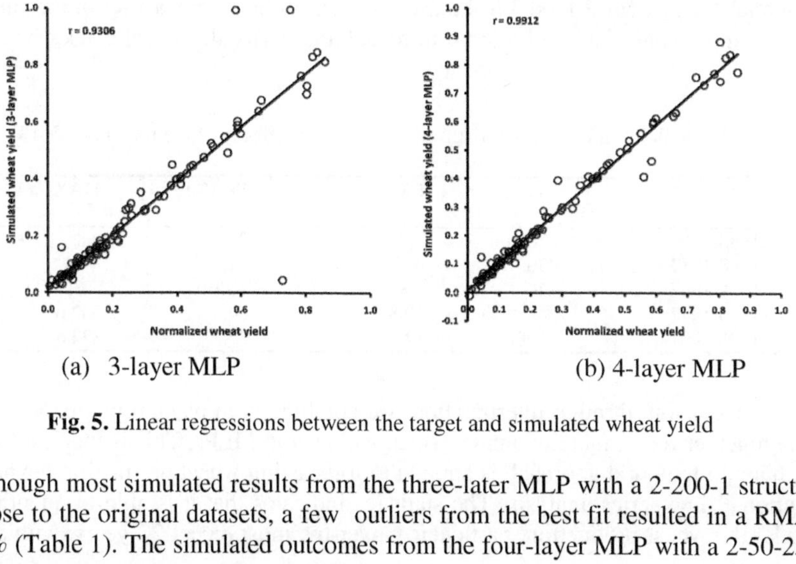

(a) 3-layer MLP (b) 4-layer MLP

Fig. 5. Linear regressions between the target and simulated wheat yield

Although most simulated results from the three-later MLP with a 2-200-1 structure are close to the original datasets, a few outliers from the best fit resulted in a RMAE of 20% (Table 1). The simulated outcomes from the four-layer MLP with a 2-50-25-1 structure correlate more closely with the original data across the whole range, with a lower RMAE of <13% (Table 1).

Table 1. Test results of MLP models

MLP	Structure	Training data	Testing data	Correlation	RMAE (%)	SD (%)	MAX (%)
3-layer	2-200-1	1050	108	0.9306	20.0	52.1	400.0
4-layer	2-50-25-1	1050	108	0.9912	12.5	37.3	332.6

5 Discussion and Conclusion

Although the three-layer MLP has more neurons in the hidden layer than the total number of neurons in both hidden layers in the four-layer MLP, its relatively simple structure may not adequately establish an approximating mechanism for approaching accurate mappings between wheat yield and both plantation area and rainfall. On the other hand, the four-layer MLP is likely to approach the solution in a different manner. In this four-layer structure, the outcomes from the first hidden layer are fed to the second hidden layer for further processing. The training in the second hidden layer using a different transfer function from that in the first hidden layer is able to distinguish the subtle variations passed from the first hidden layer. These subtle changes are further enlarged in the second hidden layer. As a result, the four-layer MLP returns a better solution in approximating wheat yield than the three-layer MLP does.

Another interesting observation on the outcomes is that on average the simulation accuracy of our MLPs improves temporally forwards. That is the simulation is more accurate for the wheat yields of recent years than for that of the earlier years. This is clearly shown in Table 2, in which for the four-layer MLP, the RMAE of the most recent 20 years from 1988 to 2007 is 3.7% with a maximum error within 10%, much less than all other periods. This is indeed a very encouraging trend indicating a great potential for this MLP model to be used for wheat yield forecasting in the future, although more study will be required to reveal and verify the actual causes behind this trend.

Table 2. Temporal comparison of simulation accuracy of the four-layer MLP

Period	Year	RMAE (%)	SD (%)	MAX (%)
1988-2007	20	3.7	3.3	9.8
1968-1987	20	8.4	11.5	37.8
1948-1967	20	7.8	10.1	39.1
1900-1947	48	19.8	54.5	332.6
1900-2007	108	12.5	37.3	332.6

In conclusion, through interpolation, we combine both plantation area and rainfall data together for simulating annual wheat yield using MLPs. This is the first time that such a non-temporal approach is applied to forecasting wheat production beyond conventional time series analysis. The statistics indicates that it is able to quantitatively model the wheat production, particularly over the most recent 20 years, with an error less than 4% on average. This MLP model can be further improved by taking into account other influential factors, such as temperature, soil nutrients, and diseases. Inclusion of new wheat data will also improve the simulation. These are the topics of our next project.

References

1. Box, G.E.P., Jenkins, G.M.: Time series analysis: forecasting and control. Holden Day, San Francisco (1970)
2. Shephard, N.: Generalized linear autoregression. Economics working paper 8. Nuffield College, Oxford (1995)
3. Liang, X., Zhang, H., Li, X.: A simple method of forecasting option prices based on neural networks. In: Chien, B.-C., Hong, T.-P., Chen, S.-M., Ali, M. (eds.) IEA/AIE 2009. LNCS, vol. 5579, pp. 586–593. Springer, Heidelberg (2009)
4. Zhang, G.P., Kline, D.M.: Quarterly time-series forecasting with neural networks. IEEE Transactions on Neural Networks 18, 1800–1814 (2007)
5. Yu, L., Wang, S., Lai, K.K.: A novel nonlinear ensemble forecasting model incorporating GLAR and ANN for foreign exchange rates. Computers and Operations Research 32, 2523–2541 (2005)
6. Taskaya-Temizel, T., Casey, M.C.: A comparative study of autoregressive neural network hybrids. Neural Networks 18, 781–789 (2005)
7. Guo, W.W.: Incorporating statistical and neural network approaches for student course satisfaction analysis and prediction. Expert Systems with Applications 37, 3358–3365 (2010)
8. Australian Bureau of Statistics (ABS): Agricultural commodities - Historical data. Canberra (2006)
9. Australian Bureau of Meteorology (ABM), `http://www.bom.gov.au/climate`
10. Hammerlin, G., Hoffmann, K.: Numerical mathematics. Springer, New York (1991)
11. Castleman, K.R.: Digital image processing. Prentice-Hall, Englewood Cliffs (1996)
12. Guo, W.W.: A novel application of neural networks for instant iron-ore grade estimation. Expert Systems with Applications 37, 8729–8735 (2010)
13. Keys, R.G.: Cubic convolution interpolation for digital image processing. IEEE Transactions on Acoustics, Speech, and Signal Processing 29, 1153–1160 (1981)
14. Hornik, K., Stinchcomb, M., White, H.: Multilayer feed forward networks are universal approximators. Neural Networks 2, 359–366 (1989)
15. Guo, W.W., Li, M.M., Whymark, G., Li, Z.X.: Mutual complement between statistical and neural network approaches for rock magnetism data analysis. Expert Systems with Applications 36, 9678–9682 (2009)
16. Samarasimghe, S., Kulasiri, D., Rajanayake, C., Chandraratne, M.: Three neural network case studies in biology and natural resource management. In: Proceedings of the 9th International Conference on Neural Information Processing, Singapore, pp. 2279–2283 (2002)
17. Marquardt, D.: An algorithm for least-squares estimation of nonlinear parameters. SIAM Journal on Applied Mathematics 11, 431–441 (1963)
18. Hagan, M.T., Menhaj, M.: Training feedforward networks with the Marquardt algorithm. IEEE Transactions on Neural Networks 5, 989–993 (1994)
19. Hagan, M.T., Demuth, H.B., Beale, M.H.: Neural network design. PWS Publishing, Boston (1996)
20. Demuth, H., Beale, M., Hagan, M.: Neural network toolbox 5. The MathWorks, Natick (2007)

Investment Appraisal under Uncertainty – A Fuzzy Real Options Approach

Shu-Hsien Liao and Shiu-Hwei Ho

Graduate Institute of Management Sciences, Tamkang University
No.151, Ying-chuan Road Tamsui, Taipei County 25137, Taiwan
Michael@mail.tku.edu.tw, succ04.dba@msa.hinet.net

Abstract. The main purpose of this paper is to propose a fuzzy approach for investment project valuation in uncertain environments from the aspect of real options. The traditional approaches to project valuation are based on discounted cash flows (DCF) analysis which provides measures like net present value (NPV) and internal rate of return (IRR). However, DCF-based approaches exhibit two major pitfalls. One is that DCF parameters such as cash flows cannot be estimated precisely in the uncertain decision making environments. The other one is that the values of managerial flexibilities in investment projects cannot be exactly revealed through DCF analysis. Both of them would entail improper results on strategic investment projects valuation. Therefore, this paper proposes a fuzzy binomial approach that can be used in project valuation under uncertainty. The proposed approach also reveals the value of flexibilities embedded in the project. Furthermore, this paper provides a method to compute the mean value of a project's fuzzy expanded NPV that represents the entire value of project. Finally, we use the approach to practically evaluate a project.

Keywords: Project valuation, Real options, Fuzzy numbers, Flexibility, Uncertainty.

1 Introduction

DCF-based approaches to project valuation implicitly assume that a project will be undertaken immediately and operated continuously until the end of its expected useful life, even though the future is uncertain. By treating projects as independent investment opportunities, decisions are made to accept projects with positive computed NPVs. Traditional NPV techniques only focus on current predictable cash flows and ignore future managerial flexibilities, therefore, may undervalue the projects and mislead the decision makers.

Since DCF-based approaches ignore the upside potentials of added value that could be brought to projects through managerial flexibilities and innovations, they usually underestimate the upside value of projects [1, 2]. In particular, as market conditions change in the future, investment project may include flexibilities by which project value can be raised. Such flexibilities are called real options or strategic options. The real options approach to projects valuation seeks to correct the deficiencies of the

K.W. Wong et al. (Eds.): ICONIP 2010, Part II, LNCS 6444, pp. 716–726, 2010.

traditional valuation methods through recognizing that managerial flexibilities can bring significant values to projects.

In DCF, parameters such as cash flows and discount rates are difficult to estimate [3]. These parameters are essentially estimated under uncertainty. With respect to uncertainty, probability is one way to depict whereas possibility is another. Fuzzy set theory provides a basis for the theory of possibility. By modeling the stock price in each state as a fuzzy number, Muzzioli and Torricelli [4] obtained a possibility distribution of the risk-neutral probability in a multi-period binomial model, then computed the option price with a weighted expected value interval, and thus determined a "most likely" option value within the interval. Muzzioli and Reynaerts [5] also addressed that the key input of the multi-period binomial model is the volatility of the underlying asset, but it is an unobservable parameter. Providing a precise volatility estimate is difficult; therefore, they used a possibility distribution to model volatility uncertainty and to price an American option in a multi-period binomial model. Carlsson and Fuller [3] mentioned that the imprecision in judging or estimating future cash flows is not stochastic in nature, and that the use of the probability theory leads to a misleading level of precision. Their study introduced a real option rule in a fuzzy setting in which the present values of expected cash flows and expected costs are estimated by trapezoidal fuzzy numbers. Carlsson et al. [6] also developed a methodology for valuing options on R&D projects, in which future cash flows were estimated by trapezoidal fuzzy numbers.

In addition to the binomial model, the Black-Scholes model [7] is another way to evaluate the option's value. Wu [8] applied the fuzzy set theory to the Black-Scholes formula. Lee et al. [9] adopted the fuzzy decision theory and Bayes' rule as a basis for measuring fuzziness in the practice of option analysis. The Black-Scholes models are used to evaluate simple real option scenarios such as delay decisions, research and development, licenses, patents, growth opportunities, and abandonment scenarios [10]. Despite its theoretical appeal, however, the practical use of real option valuation techniques in industry has been limited by the complexity of these techniques, the resulting lack of intuition associated with the solution process, or the restrictive assumptions required for obtaining analytical solutions. On the other hand, Cox et al. [11] developed a binomial discrete-time option valuation technique that has gained similar popularity to evaluate real options due to its intuitive nature, ease of implementation, and wide applicability to variety of option attributes. In addition, analytical models such as the Black-Scholes formula focus on a single option and cannot deal with multi-option situations.

2 The Valuation Approach

In considering option value, the traditional NPV can be expanded as: expanded NPV = static NPV + value of option from active management [1]. The expanded NPV is also called strategic NPV. Static NPV is the NPV obtained using the traditional discount method; it is also called passive NPV. In this study, a fuzzy binomial valuation approach is proposed to evaluate investment projects that are embedded with real options. The value of the project is represented by its expanded NPV, which can be evaluated by the valuation approach. However, the parameters are estimated by fuzzy numbers when the expanded NPV is estimated; thus, the expanded NPV is called fuzzy expanded NPV (*FENPV*) in this study.

The proposed valuation approach is based on Cox et al. [11]. Assuming there is a call option with the present value of underlying asset S_0 and exercising price K, the value of the underlying asset has P_u probability to rise to uS_0 or P_d probability to drop to dS_0 in the next period. The factors u and d represent the jumping up and down factors of the underlying asset's present value, respectively. The option will be exercised at period t = 1 if the underlying value is higher than K, and forgone if the underlying value is lower than K. The dynamics of the option value is shown in Fig. 1.

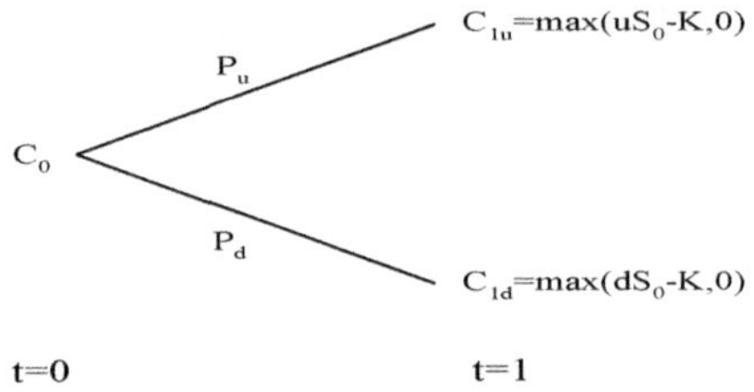

Fig. 1. The dynamics of option value

If the option is sold at price C_0, then the pricing approach is generally based on the assumption of replicating portfolio and can thus be determined by the following expression

$$C_0 = \frac{1}{(1+r)}\left[P_u C_{1u} + P_d C_{1d}\right] \tag{1}$$

in which r is risk-free interest rate, and P_u and P_d are risk-neutral probabilities, which are determined by the following formulas.

$$P_u = \frac{(1+r)-d}{(u-d)} \tag{2}$$

$$P_d = \frac{u-(1+r)}{(u-d)} = 1 - P_u \tag{3}$$

Therefore, the price or present value of the call option is the discounted result of the option values C_{1u} and C_{1d} with risk-neutral probabilities. Also, under the assumption of no arbitrage opportunities, the condition $0<d<1<(1+r)<u$ must be satisfied. Furthermore, the expected return of the underlying asset should be zero based on the no-arbitrage assumption:

$$P_u\left(\frac{uS_0}{1+r} - S_0\right) + P_d\left(\frac{dS_0}{1+r} - S_0\right) = 0 \tag{4}$$

That is

$$\frac{uP_u}{1+r} + \frac{dP_d}{1+r} = 1 \tag{5}$$

Thus, we have the following risk-neutral probabilities equations:

$$\begin{cases} P_u + P_d = 1 \\ \dfrac{uP_u}{1+r} + \dfrac{dP_d}{1+r} = 1 \end{cases} \tag{6}$$

From (1), (2) and (3), we know that the main factors affecting the call option value are jumping factors u and d; it is not easy, however, to estimate their values in a precise manner due to the uncertainty of the underlying volatility.

The cash flow models applied to many financial decision making problems often involve some degree of uncertainty. In the case of deficient data, most decision makers tend to rely on experts' knowledge of financial information when carrying out their financial modeling activities. The nature of this knowledge often tends to be vague rather than random. Hence, this study considers possibilistic uncertainty rather than probabilistic uncertainty and employs fuzzy numbers instead of statistics to estimate the parameters. For lightening computation efforts, we utilize the triangular fuzzy numbers $\tilde{u}=[u_1,u_2,u_3]$ and $\tilde{d}=[d_1,d_2,d_3]$ to represent the jumping factors of the underlying asset. Therefore, the risk-neutral probabilities equations can be rewritten as

$$\begin{cases}\tilde{P}_u \oplus \tilde{P}_d = \tilde{1} \\ \dfrac{\tilde{u}\otimes\tilde{P}_u}{1+\mathrm{r}} \oplus \dfrac{\tilde{d}\otimes\tilde{P}_d}{1+\mathrm{r}} = \tilde{1}\end{cases} \tag{7}$$

where $\tilde{P}_u=[P_{u1},P_{u2},P_{u3}]$ and $\tilde{P}_d=[P_{d1},P_{d2},P_{d3}]$. Thus, we have

$$\begin{cases}[P_{u1},P_{u2},P_{u3}]\oplus[P_{d1},P_{d2},P_{d3}]=[1,1,1] \\ \dfrac{[u_1,u_2,u_3]\otimes[P_{u1},P_{u2},P_{u3}]_u}{1+\mathrm{r}} \oplus \dfrac{[d_1,d_2,d_3]\otimes[P_{d1},P_{d2},P_{d3}]}{1+\mathrm{r}}=[1,1,1]\end{cases} \tag{8}$$

which are

$$\begin{cases}P_{ui}+P_{di}=1 \\ \dfrac{u_i\times P_{ui}}{1+\mathrm{r}}+\dfrac{d_i\times P_{di}}{1+\mathrm{r}}=1\end{cases} \quad \text{for } i=1,2,3 \tag{9}$$

It can be solved by considering the following relationship.

$$P_{ui}=\frac{(1+\mathrm{r})-d_i}{u_i - d_i} \tag{10}$$

$$P_{di}=\frac{u_i-(1+\mathrm{r})}{u_i - d_i} \tag{11}$$

Since the risk-free interest rate r and the exercising price K are usually known, they are crisp values, whereas, the option values C_{1u} and C_{1d} become fuzzy numbers as a result of the jumping factors being fuzzified. That is, $\tilde{C}_{1u}=\max(\tilde{u}\mathrm{S}_0-\mathrm{K},0)$ and $\tilde{C}_{1d}=\max(\tilde{d}\mathrm{S}_0-\mathrm{K},0)$. The ranking of two triangular fuzzy numbers $\tilde{A}=[a_1,a_2,a_3]$ and $\tilde{B}=[b_1,b_2,b_3]$ can be derived from $\max(\tilde{A},\tilde{B})=[\max(a_1,b_1),\max(a_2,b_2),\max(a_3,b_3)]$. Thus, the pricing formula for the fuzzy call option is

$$\tilde{C}_0=\frac{1}{1+\mathrm{r}}[\tilde{P}_d\otimes\tilde{C}_{1d}\oplus\tilde{P}_u\otimes\tilde{C}_{1u}] \tag{12}$$

In practical application, the present value of the underlying asset is determined by the NPV of the investment project; the exercising price is the additional outlay to exercise the embedded option.

Managerial flexibility to adopt future actions introduces an asymmetry or skewness in the probability distribution of the project NPV [2]. In the absence of such managerial flexibility, the probability distribution of project NPV would be considerably symmetric. However, in the existence of managerial flexibility such as the exercising of options, enhanced upside potential is introduced and the resulting actual distribution is skewed to the right.

In essence, identical results are obtained in the case of possibilistic distribution which is adopted by this study to characterize the NPV of an investment project. In other words, the characteristic of right-skewed distribution also appears in the *FENPV* of an investment project when the parameters (such as cash flows) are characterized with fuzzy numbers. Although many studies have proposed a variety of methods to compute the mean value [12, 13] and median value [14] of fuzzy numbers, these works did not consider the right-skewed characteristic present in the *FENPV*. Therefore, this study proposes a new method to compute the mean value of the *FENPV* based on its right-skewed characteristic. This mean value can be used to represent the *FENPV* with a crisp value. Moreover, different *FENPV*s can be compared according to their mean values.

Let $\tilde{C} = [c_1(\alpha), c_3(\alpha)]$ be a fuzzy number and $\lambda \in [0,1]$. Then, the mean value of $\tilde{C}$ is defined as

$$E(\tilde{C}) = \int_0^1 [(1-\lambda)c_1(\alpha) + \lambda c_3(\alpha)]d\alpha \tag{13}$$

The weighted index λ is called the pessimistic-optimistic index in [15], but the index is determined by a subjective decision in [15]. However, this study considers that the index can be determined objectively. Fig. 2 illustrates a case in which the *FENPV* is represented by a right-skewed triangular fuzzy number. The right-skewed characteristic of *FENPV*—meaning that the more skew to the right, the more optimistic the payoff of the project—provides a clue to determining λ with $\lambda = A_R / (A_L + A_R)$, where A_L and A_R are the left-part area and right-part area of the *FENPV*, respectively. Thus, when λ is determined objectively and substituted into (13), the mean value of the *FENPV* can be computed as follows

$$E(FENPV) = \frac{(1-\lambda)c_1 + c_2 + \lambda c_3}{2} \tag{14}$$

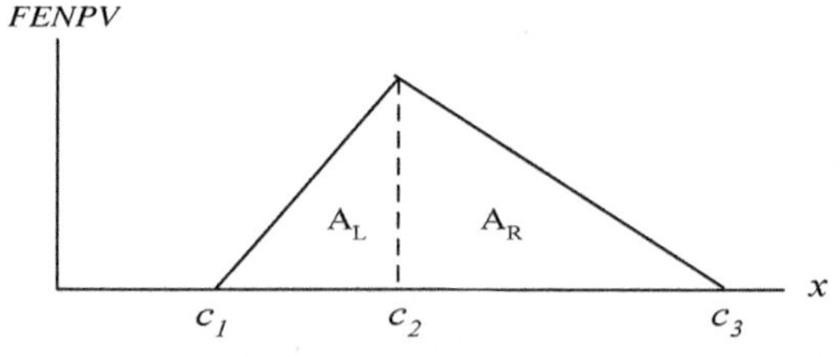

Fig. 2. A *FENPV* with right-skewed distribution

3 Illustrative Examples

An enterprise must continually develop new products and introduce them into the market to create profit. Therefore, evaluating projects of new product development is a crucial task that should be an ongoing effort of an enterprise. In this case, a local biotechnology company in Taiwan proposes a new product development project that needs evaluation. The project must go through two stages before the new product can be introduced into the market. Stage one of the project will require two years and an investment of $I_1 = 40$ (million NT$) toward product development. When this is done, the project will proceed to the second stage, which will require one year and an outlay of $I_2 = 80$ (million NT$) to acquire the equipment and raw material for mass production. Experts estimate that the project will create cash inflows with a present value of 100 (million NT$). If we use the biannual risk-free interest rate r = 3% as the discounting rate and frame six months as one period, the NPV of the project can be calculated as follows:

$$\mathrm{NPV} = 100 - 40 - \frac{80}{(1+0.03)^4} = -11.08\ (\text{million}) \tag{15}$$

This negative NPV suggests that the project should be rejected.

The above results are obtained under the assumption that cash inflows can be generated with certainty. However, this assumption is unrealistic. In fact, the cash inflows will vary with fluctuations in market conditions, such as the market demand of the new product. According to experts' estimation, the new product may have a rate of 20% × (1 ± 5%) fluctuation per year with regard to its market demand. Since the volatility is estimated under uncertainty, a triangular fuzzy number is employed to characterize the possibilistic uncertainty of the volatility. Based on the estimation, the triangular fuzzy number $\tilde{\rho} = [(1-0.05)\times 0.2, 0.2, (1+0.05)\times 0.2] = [0.19, 0.2, 0.21]$ is used to express the fuzzy volatility. From the fuzzy volatility $\tilde{\rho}$, the fuzzy jumping factors $\tilde{u}$ and $\tilde{d}$ can be determined as $\tilde{u} = \exp(\tilde{\rho} \otimes \sqrt{\tau})$ and $\tilde{d} = 1/\tilde{u}$, where τ is the chosen time interval expressed in the same unit as $\tilde{\rho}$ and exp denotes the exponential function. In this case, the value of τ is 0.5 because there are six months (0.5 year) in each period. As a result, we have $\tilde{u} = [1.1438, 1.1519, 1.1601]$ and $\tilde{d} = [0.8620, 0.8681, 0.8743]$. The fuzzy risk-neutral probabilities are $\tilde{P}_u = [0.5448, 0.5704, 0.5962]$ and $\tilde{P}_d = [0.4038, 0.4296, 0.4552]$, respectively. With the above conditions, a binomial tree of the project's cash inflows can be established, as shown in Fig. 3. (For simplicity, the numbers in the binomial tree are represented to two digits after the decimal point.)

Nevertheless, the project may have some decision flexibilities when the project is undertaken. For instance, when the market conditions are unfavorable, the project can be deferred one period to undertake or the project can abandon its second stage investment to prevent losses from mass production. Therefore, the project with deferring option and abandoning option will be evaluated in the following subsections, respectively. Moreover, the project with a sequential multiple options which is combined with deferring option and abandoning option will also be evaluated.

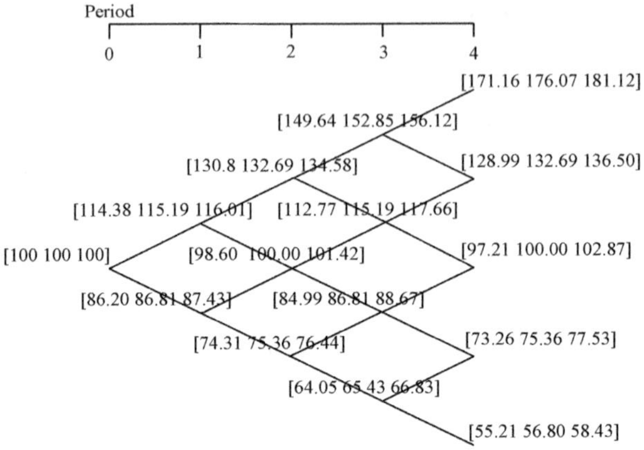

Fig. 3. Binomial tree of the project's cash inflows

3.1 Option to Defer

First of all, considering the situation that decision maker defers one period to undertake the first stage investment and commits to undertake the second stage investment. In this case, the project's total outlay that discounted to period one is calculated as follows:

$$I\_{defer} = 40 \times (1 + 0.03) + \frac{80}{(1 + 0.03)^3} = 114.41 \tag{16}$$

The decision tree is shown in Fig. 4, where V=100, $\tilde{I}\_{defer} = [114.41, 114.41, 114.41]$ and $\tilde{0} = [0, 0, 0]$. The root value in Fig. 4 is the *FENPV* of the project with deferring option and can be calculated as follows:

$$FENPV = [\tilde{P}_u \otimes \tilde{C}_{1u} \oplus \tilde{P}_d \otimes \tilde{C}_{1d}]/(1+0.03) = [0, 0.43, 0.92] \tag{17}$$

The mean value of the *FENPV* is 0.46 (million), and the value of the option to defer the first stage investment is 0.46 – (−11.08) = 11.54 (million NT$).

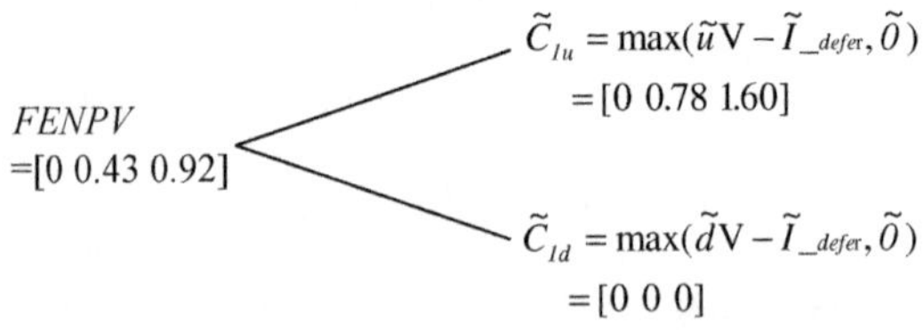

Fig. 4. The decision tree of the project with the option to defer

3.2 Option to Abandon

Furthermore, when the decision maker only possesses the option to abandon the second stage investment, this implies that the decision maker has already completed the first stage investment without deferring.

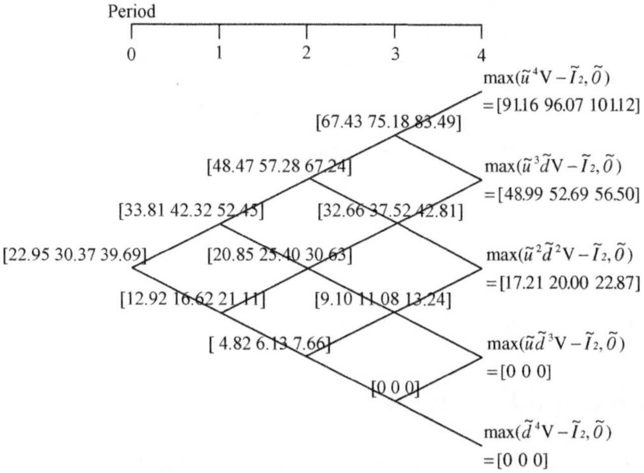

Fig. 5. The decision tree of the project with the option to abandon

The decision tree is shown in Fig. 5, in which $\tilde{I}_2=[80,80,80]$. From the root value in Fig. 5, we can conclude that the *FENPV* of the project with option to abandon the second stage investment is *FENPV* = [22.95, 30.37, 39.69] – $\tilde{I}_I$= [−17.05, −9.64, −0.31], where $\tilde{I}_I=[40,40,40]$. In this case, the mean value of the *FENPV* is −8.68 (million), and thus, the value of the option to abandon the second stage investment is −8.68 − (−11.08) = 2.4 (million).

3.3 Sequential Multiple Options

Finally, when the project involves these two options but with different expiration days, these two options form a sequential multiple options. The decision tree of the sequential multiple options is shown in Fig. 6.

In the sequential multiple options, decision makers have the options not only to abandon the second stage investment but also to defer the first stage investment. Therefore, the decision in period one is $\max(\tilde{C}_{Iu}-\tilde{I}_I,\tilde{0})$ and $\max(\tilde{C}_{Id}-\tilde{I}_I,\tilde{0})$, where $\tilde{C}_{Iu}$ and $\tilde{C}_{Id}$ are the project values in the up and down cases, respectively, during period one. Based on the values at period two, we can find that $\tilde{C}_{Iu}=[33.81,42.32,52.45]$ and $\tilde{C}_{Id}=[12.92,16.62,21.11]$. The *FENPV* of the project with sequential multiple options is *FENPV* = [0, 1.28, 7.21], its mean value is 3.60 (million), and the value of the sequential multiple options is 3.60 − (−11.08) = 14.68 (million).

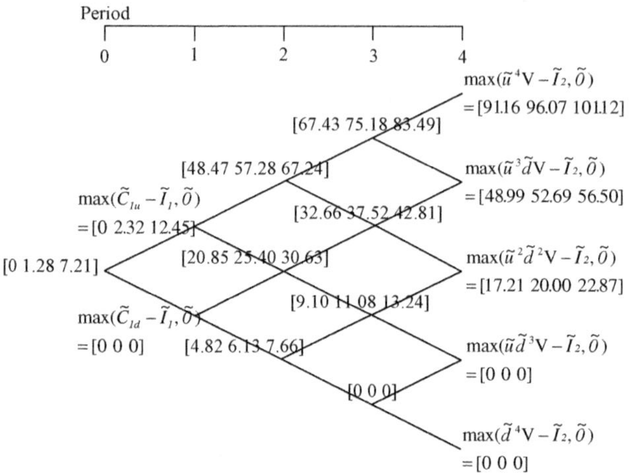

Fig. 6. The decision tree of the project with sequential multiple options

4 Discussion and Conclusion

In Table 1, we summarize the evaluation results of the new product development project that embedded with three different real options, respectively.

Table 1. A summary of the results (in million NT$)

Type of option	*FENPV* of the project	Mean value of the *FENPV*	Option value
Option to defer	[0, 0.43, 0.92]	0.46	11.54
Option to abandon	[-17.05, -9.635, -0.31]	-8.68	2.4
Multiple options	[0, 1.28, 7.21]	3.60	14.68

From the evaluation results, we can observe that if the project does not have any decision flexibility, the project's NPV is −11.08 (million NT$) and the project should therefore be rejected. However, when the project is embedded with some decision flexibilities, the decisions will be different. Confronting uncertain market conditions, the decision flexibilities, such as deferring investment in the first stage or abandoning investment in the second stage, have specific values. In this paper, we have verified the values of these flexibilities from the aspect of fuzzy real options.

When the project involves the option to defer investment in the first stage, the mean value of the project's *FENPV* is 0.46 (million NT$). The overall value of the project is positive, thus, the project become acceptable. Moreover, the value of the option to defer is 11.54 (million NT$). The option value stems from the flexibility that decision maker can defer investment in the first stage to avoid the downward losses at project initiation.

Moreover, when the project includes the option to abandon the second stage investment, the mean value of the project's *FENPV* is −8.68 (million NT$). Although this mean value is negative, it is still greater than the original NPV=−11.08 (million NT$). This reveals that the second stage option can still prevent losses when the market conditions are downward and can retain the upside potential of profit when the market conditions are upward. Therefore, this option to abandon the second stage investment has a value of 2.4 (million NT$)—lower than the value of option to defer. The reason is that the first stage investment has been completed without deferring, no matter what the market conditions are. Thus, even though the market conditions are downward at the initiation of the project, the decision maker will only be able to prevent losses at the second stage. Due to the smaller extent of hedging, the second stage option has a lower option value than the first stage option.

Lastly, when both options form a sequential multiple options, the mean value of the project's *FENPV* is 3.60 (million NT$), which represents the total value of the project. Since this value is positive, the project is acceptable. The value of the sequential multiple options is 14.68 (million NT$). This option value is higher than the value of a single option. This result shows that the multiple options provide greater value than a single option because multiple options provide more flexibility. However, the value of multiple options does not equate directly to the addition of the values of both options. The value cannot be raised linearly because of the nonlinear operations in the valuation model and the trade-off between both options in the hedging process.

In an uncertain economic decision making environment, information such as cash flows, interest rate, cost of capital, and so forth possess some vagueness but not randomness [16]. Consequently, this study has proposed the fuzzy binomial valuation approach to evaluate investment projects with embedded real options in uncertain decision making environments.

References

1. Trigeorgis, L.: Real options and interactions with financial flexibility. Financ. Manag. 22, 202–224 (1993)
2. Yeo, K.T., Qiu, F.: The value of managerial flexibility-a real option approach to investment evaluation. Int. J. Proj. Manag. 21, 243–250 (2003)
3. Carlsson, C., Fuller, R.: A fuzzy approach to real option valuation. Fuzzy Sets Syst. 139, 297–312 (2003)
4. Muzzioli, S., Torricelli, C.: A multiperiod binomial model for pricing options in a vague world. J. Econ. Dyn. Control 28, 861–887 (2004)
5. Muzzioli, S., Reynaerts, H.: American option pricing with imprecise risk-neutral probabilities. Int. J. Approx. Reason 49, 140–147 (2008)
6. Carlsson, C., Fuller, R., Heikkila, M., Majlender, P.: A fuzzy approach to R&D project portfolio selection. Int. J. Approx. Reason 44, 93–105 (2007)
7. Black, F., Scholes, M.: The pricing of options and corporate liabilities. J. Polit. Econ. 81, 637–659 (1973)
8. Wu, H.C.: Pricing European options based on the fuzzy pattern of Black-Scholes formula. Comput. Oper. Res. 31, 1069–1081 (2004)
9. Lee, C.F., Tzeng, G.H., Wang, S.Y.: A new application of fuzzy set theory to the Black-Scholes option pricing model. Expert Syst. Appl. 29, 330–342 (2005)
10. Miller, L., Bertus, M.: License valuation in the aerospace industry: A real options approach. Rev. Financ. Econ. 14, 225–239 (2005)

11. Cox, J., Ross, S., Rubinstein, M.: Option pricing: A simplified approach. J. Financ. Econ. 7, 229–263 (1979)
12. Carlsson, C., Fuller, R.: On possibilistic mean value and variance of fuzzy numbers. Fuzzy Sets Syst. 122, 315–326 (2001)
13. Fuller, R., Majlender, P.: On weighted possibilistic mean and variance of fuzzy numbers. Fuzzy Sets Syst. 136, 363–374 (2003)
14. Bodjanova, S.: Median value and median interval of a fuzzy number. Inf. Sci. 172, 73–89 (2005)
15. Yoshida, Y., Yasuda, M., Nakagami, J., Kurano, M.: A new evaluation of mean value for fuzzy numbers and its application to American put option under uncertainty. Fuzzy Sets Syst. 157, 2614–2626 (2006)
16. Kahraman, C., Ruan, D., Tolga, E.: Capital budgeting techniques using discounted fuzzy versus probabilistic cash flows. Inf. Sci. 142, 57–76 (2002)

Developing a Robust Prediction Interval Based Criterion for Neural Network Model Selection

Abbas Khosravi, Saeid Nahavandi, and Doug Creighton

Centre for Intelligent Systems Research (CISR)
Deakin University, Geelong, Australia
{akhos,saeid.nahavandi,doug.creighton}@deakin.edu.au

Abstract. This paper studies how an optimal Neural Network (NN) can be selected that is later used for constructing the highest quality delta-based Prediction Intervals (PIs). It is argued that traditional assessment criteria, including RMSE, MAPE, BIC, and AIC, are not the most appropriate tools for selecting NNs from a PI-based perspective. A new NN model selection criterion is proposed using the specific features of the delta method. Using two synthetic and real case studies, it is demonstrated that this criterion outperforms all traditional model selection criteria in terms of picking the most appropriate NN. NNs selected using this criterion generate high quality PIs evaluated by their length and coverage probability.

Keywords: Neural network, prediction interval, model selection.

1 Introduction

The learning capacity of NNs intimately hinges on the number of hidden layers and the number of neurons per layer. A NN model with many hidden units may have the capability to memorize the input patterns, however its generalization error often becomes large. Also, if the number of hidden layers and their neurons is very low, the training error will be large, which may result in even worse generalization error (over-fitting and under-fitting problems) [1]. While training NNs is well documented and practically straightforward in literature, there is no concrete rule of thumb for finding the optimal configuration of NNs. Generally, modelers use one of the following techniques for developing NN models using available data: heuristic constraint, trial and error, constructive, pruning, resampling, and evolutionary techniques [2] [3]. None of these techniques can guarantee satisfactory results in all cases. As there is no simple clear-cut technique for finding the optimal number of layers and neurons per layer, they have regarded NN design as more of an art than a science [4].

A common problem related to the point prediction (regardless of predictor type) is that predictions convey no information about different kinds of uncertainty affecting the prediction process [5]. Therefore, their application as a decision aiding tool for operation planning practically becomes unreliable. Researchers have tried to incorporate uncertainties into NN predictions with the aim of improving the reliability of NN point predictions. Calculating confidence intervals for point prediction of NNs or constructing PIs are alternative solutions proposed in literature. The delta [6],

K.W. Wong et al. (Eds.): ICONIP 2010, Part II, LNCS 6444, pp. 727–734, 2010.

bootstrap [7], Bayesian [1], and mean-variance estimation methods [8] have been proposed in the last two decades for PI construction. Applications of NN-based PIs are still rare compared to the NN application for point predictions. However, implementation of the delta techniques for PI construction has proliferated in recent years in different fields, such as load forecasting [9] [10], manufacturing [11], and material handling systems [12].

Unfortunately, literature is not rich in the field of NN-based PI construction and requires more attention from both academia and industry. One basic problem is to how select NNs that will yield the highest quality PIs. Selection criteria for NN point predictors are often concentrated on error terms. One should make note of the fact that for PIs, there are other factors that are not covered by traditional model selection criteria, such as BIC, AIC, and RMSE. Furthermore, no discussion has been made on the quality of PIs for training and test samples. This study aims at assessing performance of some traditional model selection criteria and coming up with a more efficient criterion that outperforms its traditional rivals. The proposed method will be tailored based on the specific features of delta technique for constructing PIs.

The rest of this paper is as follows. Section 2 briefly introduces the delta method and its extension for constructing PIs. The new criterion for selecting NN models is described in Section 3. Simulation results are demonstrated in Section 4. Section 5 concludes the paper with some remarks.

2 Theory and Background

In this paper, constructing PIs for outputs of NNs is done through implementation of the delta method [6]. The motivations for using this technique are its lower computational requirements as well as its mathematical foundation. Considering NNs as nonlinear regression models, one may represent them as below,

$$y = f(x,\theta^*) + \varepsilon \tag{1}$$

where y is the output of NN, $x_{m\times 1}$ represent the inputs, and $\theta^*_{p\times 1}$ are the true values of NN parameters. The term ε is error associated with the modeling function and its misspecification. $\hat{\theta}$, an estimate of θ^*, is obtained through minimization of the sum of squared error (SSE) cost function. Point prediction for the i-th sample is obtained using $\hat{\theta}$,

$$\hat{y}_i = f(x_i,\hat{\theta}) \tag{2}$$

Taylor expansion of (2) around the true values of model parameters (θ^*) can be expressed as,

$$\hat{y}_i \approx f(x_i,\theta^*) + f_0^T(\theta - \theta^*) \qquad (i = 1,\ldots,n) \tag{3}$$

f_0^T in (3) is a matrix containing derivatives of $\hat{y}_i$ with respect to the network parameters. With the assumption that ε in (1) is independently and normally distributed $N(0,\sigma^2)$, the $100(1-\alpha)\%$ PI for $\hat{y}_i$ is,

$$\hat{y}_i \pm t_{n-p,\frac{\alpha}{2}} s\sqrt{1+f_0^T(F^TF)^{-1}f_0} \tag{4}$$

where s and F are the standard deviation estimate and the Jacobian matrix of NN outputs with respect to its parameters, respectively. Also $t_{n-p,\frac{\alpha}{2}}$ is the inverse of the Student t cumulative distribution function with $n-p$ degrees of freedom.

De Veaux et al. [13] developed PIs for the case that NNs are trained using weight decay cost function,

$$\hat{y}_i \pm t_{n-p,\frac{\alpha}{2}} s\sqrt{1+f_0^T(F^TF+\lambda I)^{-1}F^TF(F^TF+\lambda I)^{-1}f_0} \tag{5}$$

where λ is a constantan determined by modeler. The procedure for calculation of s in (5) is different with the one used in (4). Detailed discussion on these issues can be followed in [13].

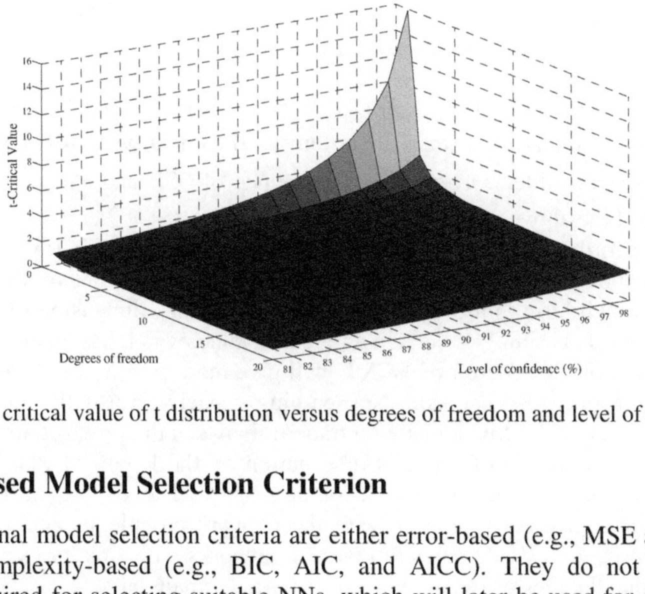

Fig. 1. The critical value of t distribution versus degrees of freedom and level of confidence

3 Proposed Model Selection Criterion

The traditional model selection criteria are either error-based (e.g., MSE and AMPE) or error-complexity-based (e.g., BIC, AIC, and AICC). They do not address all aspects required for selecting suitable NNs, which will later be used for constructing PIs. One can count the followings affecting the quality of PIs constructed using (5):

a) $t_{n-p,\frac{\alpha}{2}}$: Figure 1 represents evolution of t distribution critical values versus level of confidence and degrees of freedom. For a fixed α, the value of t quickly rises as complexity of the network increases. Therefore, it is reasonable to avoid using complex networks with many parameters.

b) s: This term is one of the main factors with a high impact on the lengths of PIs. Usually, the smaller the prediction errors, the narrower PIs. All aforementioned model selection criteria use a measure of this term for evaluating NNs' performance.

c) $(F^T F + \lambda I)^{-1} F^T F (F^T F + \lambda I)^{-1}$: This long term totally depends on the training samples. As the Jacobian matrix of NN is evaluated for the training data, one should take care of its entries' magnitudes. Large entries of this matrix for the training samples later are translated to wide PIs for the test samples. None of the previous model selection criteria considers magnitude of this term and its impact on the lengths of PIs.

d) f_0^T : This term relates to the derivatives of NN outputs with respect to the network parameters evaluated for out of sample data. Similarity between in and out of sample data results in small magnitude for entries of this $1 \times p$ vector. Generally there is no guarantee that the training and test data will have similar patterns. Due to this, there is not much control on this vector. However, this term can be effectively kept small through minimizing the size of NN parameters by using a weight decay training algorithm.

With regard to the above discussion, any new criterion for selecting NN models must include the first three items. The proposed model selection criterion is called Error-Complexity-Magnitude-based Criterion (ECMC) covering all these aspects,

$$ECMC = SSE \times \ln(\ln(K)) \tag{6}$$

where K is the norm of $(F^T F + \lambda I)^{-1} F^T F (F^T F + \lambda I)^{-1}$, and SSE is the sum of squared errors. $\ln(\cdot)$ is the natural logarithm used to deal with SEE and K , which differ over several orders of magnitude. Presence of SSE and $\ln(\cdot)$ terms in (6) is a direct result of the above discussions.

It is noteworthy to mention that the Jacobian matrix implicitly carriers some information about network complexity. From one side, as the network complexity is increased through hiring more hidden layers and neurons, dimension of the Jacobian matrix goes up. This simply results in a bigger K value, which has a negative effect on the narrowness of PIs. Therefore, ECMC will give more priority to the networks with smaller size. From the other side, NN learning capacity is usually improved as its dimension increases. ECMC looks for a tradeoff between these two contrasting issues.

ECMC also addresses effects of t-distribution on the length of PIs. This is done through managing the size of the Jacobian matrix. The degree of freedom in t distribution has an inverse relation with the network complexity and dimension of Jacobian matrix. As ECMC votes in favor of networks with smaller size, it is almost guaranteed that degrees of freedom of t-distribution remain large.

In practice, several NNs with different structures will be trained using training data and then ranked based on ECMC. These networks then will be used for constructing PIs for test samples. After projecting the test samples to the trained networks, PI Coverage Probability (PICP) will be first calculated,

$$PICP = \frac{1}{n} \sum_{i=1}^{n} c_i \tag{7}$$

where,

$$c_i = \begin{cases} 1 & t_i \in [LPI_i, UPI_i] \\ 0 & t_i \notin [LPI_i, UPI_i] \end{cases} \tag{8}$$

LPI_i and UPI_i are lower and upper bounds of the PI constructed for i-th sample. t_i is the i-th target value of the test samples. Delta-based PIs are computed at $100\,(1-\alpha)\%$ confidence level. Therefore, any NN model that its PICP is less than $100\,(1-\alpha-\gamma)\%$ will be discarded. γ can be a small multiple of α (e.g., $\frac{1}{2}\alpha$). For those networks that satisfy this condition, Normalized Mean PI Length (NMPIL) is then calculated:

$$NMPIL = \frac{1}{n}\sum_{i=1}^{n}\frac{UPI_i - LPI_i}{t_{\max} - t_{\min}} \tag{9}$$

where $t_{\min}$ and $t_{\max}$ represent minimum and maximum values of targets. NMPIL is an indication of the narrowness of PIs. The narrower the PIs are, the more useful they are. The remained networks are then ranked based on NMPIL. If the best network based on ECMC yields the smallest NMPIL, we conclude that ECMC has picked the best NN model for constructing PIs. This algorithm is summarized in Figure 2.

Step	
1	Begin
2	Train the NN model using training data.
3	Rank trained NNs based on ECMC.
4	Calculate PICP for all NNs.
5	If PICP $<(100-\alpha-\gamma)\%$, then discard the NN.
6	Calculate NMPIL for the remaining NNs.
7	Rank NNs based on NMPIL.
8	If $Rank^1_{ECMC} = Rank^1_{NMPI}$, then 'Good Selection'.
9	If $Rank^1_{ECMC} \neq Rank^1_{NMPI}$, then 'Bad Selection'.
10	End

Fig. 2. Algorithm for evaluating performance of ECMC

4 Simulation Results and Discussion

Two synthetic and real case studies are implemented to assess the usefulness of ECMC. In both case studies, different structures are examined for developing NN models. For a two layer NN with n_1 and n_2 neurons in the 1st and 2nd hidden layers, the representation will be In-n_1-n_2-Out. In the experiment, we first fix the number of neurons in the second layer and then vary the number of neurons in the first layer. The procedure described in Figure 2 is then implemented on these networks for assessing practicality of the proposed model selection criterion. PIs for each sample are then constructed using (5). The procedure of training and comparing NNs is repeated 100 times to avoid any subjective judgment due to the misspecification of NNs in the training stage. For the purpose of comparison, RMSE, MAPE, BIC, AIC, and AICC criteria for NN model assessment and selection are exploited in a similar manner. PIs are constructed with 90% confidence ($\alpha = 0.1$). γ is considered to be $\frac{\alpha}{2}$ for evaluating PICP. For both cases, data samples were separated into two training (75%) and test (25%) subsets. Training and test samples are scaled to have zero mean and unit variance.

4.1 Synthetic Case Study

A two-dimensional test problem, named Rastrigin's function, is used as the synthetic experiment in this paper: $f(x)=\sum_{i=1}^{2}[x_i^2-10\cos(2\pi x_i)+10]+\varepsilon$, where $-5.12\leq x_i\leq 5.12$. ε is additive noise with normal distribution $N(0,5)$. Four hundred samples are randomly generated for further analysis. A two layer NN is considered for modeling this function. The number of neurons in the first (n_1) and second (n_2) layer are varied in the range of one to ten. A summary of results is presented in Table 1. The last column includes the average results of the good NN model selection for each criterion. For instance, BIC selects the optimal model (best model selected based on the training data yields the narrowest PIs for the test data) only in 15% of all cases (totally 10000 cases).

The results clearly show that ECMC is more effective than BIC, AICC, RMSE, and MAPE. With the exception of one case, there is always a big difference between performance of ECMC and these criteria. The only serious rival of ECMC is AIC, which shows a good performance amongst traditional criteria. In four out of 10 cases, it does better than ECMC, in four cases it shows performance lower than ECMC, and in two cases, its performance is equal to ECMC performance. AIC performance is less than the proposed ECMC criterion in total average of performance.

Table 1. Model selection results for case study one

Model Selection Criterion	Number of neurons in the 2nd layer										Average
	1	2	3	4	5	6	7	8	9	10	
BIC	14	12	14	19	23	15	19	11	10	10	15
AIC	52	39	40	38	50	39	52	44	45	52	45
AICC	40	30	32	26	40	24	28	25	15	16	28
RMSE	78	50	37	34	25	30	38	38	48	52	43
MAPE	1	10	12	12	6	8	11	11	9	16	10
ECMC	82	53	40	39	26	32	41	39	51	52	46

4.2 Real Case Study

The data in the 2nd case study comes from a real world baggage handling system. Time required for processing 70%, 80%, and 90% of bags are considered as targets. These three variables are hereafter referred to as B70, B80, and B90 respectively.

Table 2, Table 3, and Table 4, respectively, report the average model selection results for B70, B80, and B90. Comparing the results for these targets and different NN structures reveals that in all cases, the proposed model selection criterion (ECMC) outperforms the traditional model selection criteria, especially BIC and AIC. The summary of all results and difference between results of ECMC and other model selection criteria are shown in Table 5. It is observed that in more than 66% of cases, models selected based on ECMC show the best performance for test sample as well.

Apart from ECMC, RMSE is the best model selection criteria among others. This is mainly due to the presence of SSE in (5). The result difference between RMSE and ECMC is 6.83%, demonstrating the superiority of ECMC over RMSE.

Table 2. Model selection results for B70

Model Selection Criterion	Number of neurons in the 2nd layer						Average
	3	4	5	6	7	8	
BIC	33	27	20	29	20	30	27
AIC	74	72	56	62	50	55	62
AICC	58	49	35	41	25	35	41
RMSE	86	88	80	79	68	54	76
MAPE	43	52	49	51	50	44	48
ECMC	86	90	81	87	69	58	79

Table 3. Model selection results for B80

Model Selection Criterion	Number of neurons in the 2nd layer						Average
	3	4	5	6	7	8	
BIC	29	23	25	23	20	10	22
AIC	70	55	52	50	44	32	51
AICC	55	35	32	28	27	19	33
RMSE	84	70	66	54	36	30	57
MAPE	36	46	38	36	31	28	36
ECMC	85	73	74	62	60	41	66

Table 4. Model selection results for B90

Model Selection Criterion	Number of neurons in the 2nd layer						Average
	3	4	5	6	7	8	
BIC	20	18	19	20	19	20	19
AIC	58	56	41	36	38	33	44
AICC	44	34	30	24	25	23	30
RMSE	66	66	45	32	35	30	46
MAPE	27	33	30	19	15	18	24
ECMC	65	68	53	48	53	39	54

Table 5. Total average of model selection results

Model Selection Criterion	Total Average	Difference with ECMC
BIC	22.50	43.72
AIC	51.89	14.33
AICC	34.39	31.83
RMSE	59.39	6.83
MAPE	35.89	30.33
ECMC	66.22	-

5 Conclusion

This study proposes a novel model selection criterion for neural networks used for constructing prediction intervals. In a comparative study, usefulness of different traditional model selection criteria was examined. An error-complexity-magnitude-based criterion was developed for selecting the optimal structure of neural networks. This criterion is based on elements used for constructing prediction intervals using the delta method. For the purpose of evaluation and comparison, two synthetic and real case studies were implemented. The obtained results showed that models selected using the proposed model selection criterion yield the highest quality prediction intervals for the majority of the cases.

Acknowledgments. This research was fully supported by the Centre for Intelligent Systems Research (CISR) at Deakin University.

References

[1] Bishop, C.M.: Neural Networks for Pattern Recognition. Oxford University Press, Oxford (1995)
[2] Lehtokangas, M.: Modelling with constructive backpropagation. Neural Networks 12, 707–716 (1999)
[3] Garcia-Pedrajas, N., Hervas-Martinez, C., Munoz-Perez, J.: COVNET: a cooperative coevolutionary model for evolving artificial neural networks. IEEE Transactions on Neural Networks 14, 575–596 (2003)
[4] Zhang, G., Eddy Patuwo, B., Hu, M.Y.: Forecasting with artificial neural networks: The state of the art. International Journal of Forecasting 14, 35–62 (1998)
[5] Lowe, D., Zapart, C.: Point-Wise Confidence Interval Estimation by Neural Networks: A Comparative Study based on Automotive Engine Calibration. Neural Computing & Applications 8, 77–85 (1999)
[6] Hwang, J.T.G., Ding, A.A.: Prediction Intervals for Artificial Neural Networks. Journal of the American Statistical Association 92, 748–757 (1997)
[7] Heskes, T.: Practical confidence and prediction interval. Advances in neural information processing systems 9, 176–182 (1997)
[8] Nix, D.A., Weigend, A.S.: Estimating the mean and variance of the target probability distribution. In: IEEE International Conference on Neural Networks, pp. 55–60 (1994)
[9] Khosravi, A., Nahavandi, S., Creighton, D.: Construction of Optimal Prediction Intervals for Load Forecasting Problems. IEEE Transactions on Power Systems 25, 1496–1503 (2010)
[10] Chiu, C.-C., Kao, L.-J., Cook, D.F.: Combining a neural network with a rule-based expert system approach for short-term power load forecasting in Taiwan. Expert Systems with Applications 13, 299–305 (1997)
[11] Ho, S.L., Xie, M., Tang, L.C., Xu, K., Goh, T.N.: Neural network modeling with confidence bounds: a case study on the solder paste deposition process. IEEE Transactions on Electronics Packaging Manufacturing 24, 323–332 (2001)
[12] Khosravi, A., Nahavandi, S., Creighton, D.: A prediction interval-based approach to determine optimal structures of neural network metamodels. Expert Systems with Applications 37, 2377–2387 (2010)
[13] Veaux, R.D.d., Schumi, J., Jason, S., Ungar, L.H.: Prediction Intervals for Neural Networks via Nonlinear Regression. Technometrics 40, 273–282 (1998)

Author Index

GPSR Compliance

The European Union's (EU) General Product Safety Regulation (GPSR) is a set of rules that requires consumer products to be safe and our obligations to ensure this.

If you have any concerns about our products, you can contact us on ProductSafety@springernature.com

In case Publisher is established outside the EU, the EU authorized representative is:

Springer Nature Customer Service Center GmbH
Europaplatz 3
69115 Heidelberg, Germany

Batch number: 10844611

Printed by Printforce, the Netherlands